# WEIGHT CONVERSIONS

There are 28.35 grams in one ounce.

## COMMON WEIGHT CONVERSIONS

| ounces | grams |
| --- | --- |
| 1 | 28 |
| 2 | 57 |
| 3 | 85 |
| 4 | 113 |
| 5 | 142 |
| 6 | 170 |
| 7 | 198 |
| 8 (½ pound) | 227 |
| 9 | 255 |
| 10 | 284 |
| 11 | 312 |
| 12 | 340 |
| 13 | 369 |
| 14 | 397 |
| 15 | 425 |
| 16 (1 pound) | 454 |
| 24 (1½ pounds) | 680 |
| 32 (2 pounds) | 907 |
| 35.3 (1 kilogram) | 1000 |
| 40 (2½ pounds) | 1124 |
| 48 (3 pounds) | 1361 |
| 64 (4 pounds) | 1814 |
| 80 (5 pounds) | 2268 |

# TEMPERATURE CONVERSIONS

To convert from Fahrenheit to Celsius: subtract 32, divide by 9, and multiply by 5.

To convert from Celsius to Fahrenheit: divide by 5, multiply by 9, and add 32.

## COMMON TEMPERATURE CONVERSIONS

| °F | °C |
| --- | --- |
| 32 (freezing point of water) | 0 |
| 110 | 43.3 |
| 120 (rare red meat) | 48.9 |
| 130 (medium-rare red meat) | 54.4 |
| 140 (medium red meat) | 60 |
| 145 (extra-moist poultry breast) | 62.8 |
| 150 (medium-well red meat) | 65.6 |
| 155 (medium-well standard poultry breast) | 68.3 |
| 160 (well-done meat) | 71.1 |
| 190 (subsimmering water) | 87.8 |
| 200 (simmering water) | 93.3 |
| 212 (boiling water) | 100 |
| 275 | 135 |
| 300 | 148.9 |
| 325 | 162.8 |
| 350 | 176.7 |
| 375 | 190.6 |
| 400 | 204.4 |
| 425 | 218.3 |
| 450 | 232.2 |
| 475 | 246.1 |
| 500 | 260 |
| 525 | 273.9 |
| 550 (max oven temperature) | 287.8 |

THE
FOOD
LAB

Eggs boiled for 30-second intervals from 0 to 12 minutes.

# THE
# FOOD
# LAB

## BETTER HOME COOKING
## THROUGH SCIENCE

### J. KENJI LÓPEZ-ALT
photographs by the author

W. W. Norton & Company
NEW YORK | LONDON

For information about permission to reproduce selections from this book,
write to Permissions, W. W. Norton & Company, Inc.,
500 Fifth Avenue, New York, NY 10110

For information about special discounts for bulk purchases, please contact
W. W. Norton Special Sales at specialsales@wwnorton.com or 800-233-4830

Manufacturing through Asia Pacific Offset
Book design by Level, Calistoga, CA
Production managers: Julia Druskin and Joe Lops

Library of Congress Cataloging-in-Publication Data

López-Alt, J. Kenji.
The food lab : better home cooking through science / J. Kenji López-Alt;
photographs by the author. — First edition.
pages cm
Includes index.
ISBN 978-0-393-08108-4 (hardcover)
1. Food—Experiments. 2. Food—Analysis.
3. Cooking—Technique. 4. Cooking—Research.  I. Title.
TX651.L526 2015
664'.07—dc23

2015016358

W. W. Norton & Company, Inc.
500 Fifth Avenue, New York, N.Y. 10110
www.wwnorton.com

W. W. Norton & Company Ltd.
15 Carlisle Street, London W1D 3BS

15 16 17 18 19 20

*To Adri, who loves me despite the burgers;*

*To Ed, Vicky, and the whole Serious Eats team, for helping me do my thing;*

*To my father, the scientist;*

*To my grandfather, the nutty professor;*

*To the one sister I like better than the other;*

*To my mother, who would have preferred a doctor;*

*To the other sister as well;*

*To Dumpling, Hambone, and Yuba, the best taste-testers a man could ever hope for;*

*And to my grandmother, who would have preferred a Tostitos jar.*

# CONTENTS

# CONVERSIONS

## COMMON INGREDIENTS BY VOLUME AND MASS*

| INGREDIENT | TYPE | AMOUNT | WEIGHT |
|---|---|---|---|
| **Water-Based Liquids** (including water, wine, milk, buttermilk, yogurt, etc.) | | 1 cup = 16 tablespoons | 8 ounces (227 grams) |
| **Eggs** | Jumbo<br>Extra Large<br>Large<br>Medium<br>Small<br>Peewee | | 2.5 ounces (71 grams)<br>2.25 ounces (64 grams)<br>2 ounces (57 grams)<br>1.75 ounces (50 grams)<br>1.5 ounces (43 grams)<br>1.25 ounces (35 grams) |
| **Flour** | All-purpose<br>Cake/pastry<br>Bread | 1 cup | 5 ounces (142 grams)<br>4.5 ounces (128 grams)<br>5.5 ounces (156 grams) |
| **Sugar** | Granulated<br>Brown (light or dark)<br>Confectioners' | 1 cup | 6.5 ounces (184 grams)<br>7 ounces (198 grams)<br>4.5 ounces (128 grams) |
| **Salt** | Table<br>Diamond Crystal kosher<br>Morton's kosher | 1 teaspoon | 0.25 ounce (7 grams)<br>0.125 ounce (3.5 grams)<br>0.175 ounce (5 grams) |
| **Instant Yeast** | | 1 teaspoon | 0.125 ounce (3.5 grams) |
| **Butter** | | 1 tablespoon = ⅛ stick | 0.5 ounce (14 grams) |

*Note: In standard U.S. recipes, liquids are measured in fluid ounces (volume), while dry ingredients are measured in regular ounces (weight).

## VOLUME EQUIVALENCIES

3 teaspoons = 1 tablespoon

2 tablespoons = 1 fluid ounce

16 tablespoons = 1 cup (8 fluid ounces)

2 cups = 1 pint (16 fluid ounces)

4 cups = 1 quart (32 fluid ounces)

1 quart = 0.95 liters

4 quarts = 1 gallon

# WEIGHT CONVERSIONS

There are 28.35 grams in one ounce.

## COMMON WEIGHT CONVERSIONS

| ounces | grams |
|---|---|
| 1 | 28 |
| 2 | 57 |
| 3 | 85 |
| 4 | 113 |
| 5 | 142 |
| 6 | 170 |
| 7 | 198 |
| 8 (½ pound) | 227 |
| 9 | 255 |
| 10 | 284 |
| 11 | 312 |
| 12 | 340 |
| 13 | 369 |
| 14 | 397 |
| 15 | 425 |
| 16 (1 pound) | 454 |
| 24 (1½ pounds) | 680 |
| 32 (2 pounds) | 907 |
| 35.3 (1 kilogram) | 1000 |
| 40 (2½ pounds) | 1124 |
| 48 (3 pounds) | 1361 |
| 64 (4 pounds) | 1814 |
| 80 (5 pounds) | 2268 |

# TEMPERATURE CONVERSIONS

To convert from Fahrenheit to Celsius: subtract 32, divide by 9, and multiply by 5.

To convert from Celsius to Fahrenheit: divide by 5, multiply by 9, and add 32.

## COMMON TEMPERATURE CONVERSIONS

| °F | °C |
|---|---|
| 32 (freezing point of water) | 0 |
| 110 | 43.3 |
| 120 (rare red meat) | 48.9 |
| 130 (medium-rare red meat) | 54.4 |
| 140 (medium red meat) | 60 |
| 145 (extra-moist poultry breast) | 62.8 |
| 150 (medium-well red meat) | 65.6 |
| 155 (medium-well standard poultry breast) | 68.3 |
| 160 (well-done meat) | 71.1 |
| 190 (subsimmering water) | 87.8 |
| 200 (simmering water) | 93.3 |
| 212 (boiling water) | 100 |
| 275 | 135 |
| 300 | 148.9 |
| 325 | 162.8 |
| 350 | 176.7 |
| 375 | 190.6 |
| 400 | 204.4 |
| 425 | 218.3 |
| 450 | 232.2 |
| 475 | 246.1 |
| 500 | 260 |
| 525 | 273.9 |
| 550 (max oven temperature) | 287.8 |

# PREFACE

**Longtime fans of J. Kenji López-Alt can celebrate.**

For years we've loved (and cooked from) his practical columns about kitchen science on the excellent seriouseats.com website. With this book—precise and serious, witty and relaxed—Kenji joins the glittering constellation of men and women who have, over the past thirty years, brought the ancient human art of feeding ourselves into the scientific age. What goes on within a cube of ice or a stew-pot has followed the three laws of thermodynamics, among many others, for the past forty thousand years—or however long you believe it's been since our species' first act of cooking—but we just never really knew it. Kenji stands on the shoulders of giants, of Achatz, Adria, Arnold, Blumenthal, Kurti, McGee, Myrhvold, Roca, and This—all of whom have brought the realm of pure thought into the scullery, where it materialized into something delectable. Kenji does it in his own way. He has a degree from MIT and eleven years in restaurant kitchens—in my mind, the two minimum qualifications for a man who would aim to make a better hamburger or, to my surprise, boil a better pan of water. Kenji's recipes produce simple, delicious specimens of home cooking. They are not difficult to carry out, but they can be extremely precise, while the thought behind them may be complex, and his testing obsessive. But Kenji's book is not about recipes. And I'll bet you can't read even ten pages of it without becoming a better cook.

—Jeffrey Steingarten

I am a nerd,
and I'm
proud of it.

# My grandfather was an organic chemist, my father was a microbiologist, and I was a little nerdling.

**I was never meant to be a cook.** Just ask my mom, she'll tell you. Doctor? Sure. Lawyer? Yep—I can argue with the best of 'em. Scientist? Definitely. In fourth grade, we were given an assignment: write a book about ourselves in the future. I distinctly remember my future life according to my ten-year-old self. I'd be married at twenty-four. I'd have my first kid at twenty-six. I'd get my PhD by twenty-nine (how I'd manage to get my PhD while trying to raise a kid was a question I never asked myself). By thirty, I'd discover a cure for cancer, winning a Nobel prize. Having made my mark on the world, I'd spend the next forty years fulfilling my duties as the President of LEGOLAND before finally retiring and leaving the world a better place at the age of eighty-seven.

Lofty dreams indeed, but things seemed to be going on track all through high school. I did well in math and sciences (and particularly poorly in English, for the record), spending my summers playing music (chamber music camp, *not* band camp, thank you very much!) or working in biology laboratories. Did I ever show an inclination to cook? Not really. I took an after-school cooking class in third grade, where I learned to make simple syrup and stone soup. My dad trained me in the art of making open-faced tuna melts on Saturdays. He also taught me a valuable lesson in how not to cut a block of frozen beef straight from the freezer into steaks—a memorable afternoon that included the line, "Kenji, go get me the hammer," and concluded with shards of knife all over the kitchen floor and beef still as blocky as it ever was.

My specialties all through high school were half-assed guacamole and perfectly heated frozen chicken potpies. The one time I did exert myself in the kitchen, I produced a batch of what I thought were some pretty awesome almond tuiles coated in chocolate and filled with raspberry preserves. Being the incurable romantic that I am, I'd slaved over them for my high school girlfriend for Valentine's Day, see? Turns out she wasn't as into romantic nerds as I thought she was. I got dumped on Valentine's Day, her dad ate the tuiles, and my fledgling cooking career was put on hiatus.

The time for me to move upward and onward with my college education finally came, and I entered the Massachusetts Institute of Technology—that temple of science where nerdfolk congregate en masse to talk hertz and bytes and the average student wears only two-thirds of a pair of shoes during winter (I brought down the average).

For a while I fit right in, finally at home amongst my fellow geeks, reveling in the fascinating subculture and learning more than I'd ever learned before (mostly about such scientific puzzles as precisely how many whiskey-and-Cokes it takes before the next morning's hangover will prevent you from attending an 11 a.m lecture). But slowly a grim reality dawned on me: I loved biology and science, but I *hated* working in biology labs. It was the slowness of it all, the months and months of experimenting that would finally reveal results that showed you were wrong all along—and could you please repeat those tests? I got restless. I got annoyed, and I did what all heroes should do in a time of crisis: I ran away.

That's right.

That summer, I made the conscious decision not to take another job in a biology lab. Here I was, in the prime of my youth, pissing it away playing with pipettes and DNA sequencers. I set out with the goal of taking as nonacademic a job as I possibly could. Waitering seemed like a good gig. Meet cute girls, eat good food, hang out with cooks, party every night because I don't have to show up to work until 3 p.m. Basically, a repressed college kid's dream. As it happened, the first restaurant I walked into—an abysmal Mongolian grill-style joint in Harvard Square—wasn't in need of waiters, but it *was* desperately in need of cooks.✳

Without hesitating, I signed on. And that was the beginning of the end for me. Like a head-injury patient who suddenly develops a brand-new personality, something snapped the moment my hand touched a knife in a professional kitchen. I was no longer in control of my own destiny. Since that day, since the first time I wore that silly baseball hat and the T-shirt that identified me in no uncertain terms as a Knight of the Round Grill (seriously), I was a cook. It didn't matter to me that I knew nothing about cooking and that my job mostly consisted of flipping asparagus spears with my double-fisted spatulas. I knew right then that I'd discovered what I was going to do with the rest of my life.

I was ravenous. I tore through every cookbook I could lay my hands on. Going to the beach? Forget the Frisbee—I'm bringing Pépin. Friends heading to a movie? I'll be in the kitchen with my dog-eared Chinese cookbook. I worked in restaurants as much as my class schedule would allow, making up for lack of experience with brute force and sheer willpower. Unfortunately, what with trying to attain a degree and a decided lack of a cooking mentor (the closest I had was our fraternity chef, who was better at snorting coke off the piano than *tourné*-ing a potato), cooking for me was filled with an endless series of unanswered questions.

Why do I have to cook pasta in a huge volume of water? Why does it take so much longer to bake a potato than to boil it? How come my pancakes always suck? And what's really in baking powder anyway? I made a pact with myself then and there that as soon as I was finished with college, I'd never again do anything that I didn't enjoy doing. I'd spend my life trying to answer these questions that so fascinated me. The fact that cooks make very little money and work crazy schedules—I might not see my friends and family on holidays ever again—didn't deter me. I'd found my passion, and even if it made me a pauper, I'd be damned if I didn't follow it.

My mother did not take the news well.

Sure, I finished up my schooling (eventually graduating with a degree in architecture) while working part-time in restaurants, and I learned quite a bit about good science along the way (never once did I lose my interest in science itself—just in the practice of biology), and after graduating, I started working for some of the greatest chefs in Boston, but to my mom, a cook was a cook was a cook. Carefully plating a perfectly sautéed fillet of striped bass with a caviar beurre blanc and cute little tournéed radishes was no different from flipping burgers to her. (Ironically, she was sort of right—these days I find flipping burgers more fascinating than fancy restaurant food.)

*At least*, I thought to myself, *working in these great restaurants, I'll finally have the answers I've been seeking.*

Not so fast.

---

✳To be fair, "cook" is a pretty generous term. Spatula-wielding monkey is more like it.

First day on the line, I was given a lesson in the traditional double-fry technique for French fries: a dip in low-temperature oil for a few minutes, followed by a second fry in high-temperature oil. My first question was one that I thought was obvious to any free thinker: if the purpose of the first fry is merely to cook the potatoes all the way through, as many people had told me, shouldn't it be possible instead to boil the potatoes first until cooked through, followed by a single fry?

The chef de cuisine's response: "Em . . ., it might be possible, but you just don't do it. Don't ask so many questions, I don't have time to answer them all." Right as his answer was, it was hardly the pinnacle of scholarly sentiment or scientific inquiry that I had hoped for. Truth be told, as a professional cook, and with the hours that came with the territory, I had even *less* time to pursue the answers to the cooking questions I had, which were now beginning to mount like order tickets on a busy Saturday night.

So, after eight years of working in restaurants, I decided to shift tracks: perhaps recipe development and publishing were where the answers lay. It was only after I made this shift that my curiosity finally began to be sated. As a test cook and editor at *Cook's Illustrated* magazine, I not only had the opportunity to start answering my questions, I was also paid to do it! Here was a job that finally combined the top three of my four greatest loves: tasting great food, the scientific pursuit of knowledge, and the physical act of cooking (my wife would be the fourth), and it was a truly liberating experience. I discovered that in many cases—even in the best restaurants in the world—the methods that traditional cooking knowledge teaches us are not only outdated but occasionally flat-out wrong.

Then I moved back to New York City with my wife and discovered a job even better than the one I had at *Cook's Illustrated*. As chief creative officer at Serious Eats (www.seriouseats.com) and the author of its popular "The Food Lab" column, I was finally 100-percent free to do exactly what I wanted to do, explore the questions I wanted to explore, test the things I wanted to test, and cook the food I wanted to cook. And the best part? Doing it for a community of food lovers every bit as passionate and thoughtful about what they put in their mouths as I am.

Sure, I earn my keep in a number of ways, and testing and writing recipes is only a small part of it. I push commas, I stick words together, I blab about pizza this or hamburger that online, I fake my way through the occasional annoying business meeting or schmooze-fest foodie event; heck, it turns out that I even write a book now and then. But in the end, I'm a cook, and that's all I really ever wanted to be.

# What's in this book?

**About twenty years ago,** celebrated food scientist, author, and personal hero Harold McGee made a simple statement: contrary to popular belief, searing meat before roasting it does not "lock in the juices."* Now, saying this to a cook was like telling a physicist that rocks fall upward or an Italian that pizza was invented in Iceland. Ever since the mid-nineteenth century, when German food scientist Justus von Liebig had first put forth the theory that searing meat at very high temperatures essentially cauterizes its surface and creates a moistureproof barrier, it had been accepted as culinary fact. And for the next century and a half, this great discovery was embraced by world-famous chefs (including Auguste Escoffier, the father of French cuisine) and passed on from mentor to apprentice and from cookbook writer to home cook.

You'd think that with all that working against him, McGee must have used the world's most powerful computer, or at the very least a scanning electron microscope, to prove his assertion, right? Nope. His proof was as simple as looking at a piece of meat. He noticed that when you sear a steak on one side, then flip it over and cook it on the second side, juices from the interior of the steak are squeezed out of the top—the very side that was supposedly now impermeable to moisture loss!

It was an observation that anyone who's ever cooked a steak could have made, and one that has since led restaurants to completely revise their cooking methods. Indeed, many high-end restaurants these days cook their steaks first, sealed in plastic,

in low-temperature water baths, searing them only at the end in order to add flavor. The result is steaks that are juicier, moister, and more tender than anything the world was eating before von Liebig's erroneous assertion was finally disproved.

The question is, if debunking von Liebig's theory was such a simple task, why did it take nearly a hundred and fifty years to do it? The answer lies in the fact that cooking has always been considered a craft, not a science. Restaurant cooks act as apprentices, learning, but not questioning, their chefs' techniques. Home cooks follow the notes and recipes of their mothers and grandmothers or cookbooks—perhaps tweaking them here and there to suit modern tastes, but never challenging the fundamentals.

It's only in recent times that cooks have finally begun to break out of this shell. Restaurants that revel in using the science of cookery to come up with new techniques that result in pleasing and often surprising outcomes are not just proliferating but are consistently ranked as the best in the world (Chicago's Alinea or Spain's now-closed El Bulli, for example). It's an indication that as a population, we're finally beginning to see cooking for what it truly is: a scientific engineering problem in which the inputs are raw ingredients and technique and the outputs are deliciously edible results.

Now, don't get me wrong. I'm not out to try and prove to you that foams are the way of the future or that your eggs need to be cooked in a steam-injected, pressure-controlled oven to come out right. I'm not here to push some sort of newfangled, fancified, plated-with-tweezers, deconstructed/reconstructed cuisine. Quite the opposite, in fact.

My job is simple: to prove to you that even the

---

\* McGee was not actually the first to debunk this theory, but it was the first time anybody took real notice.

simplest of foods—hamburgers, mashed potatoes, roasted Brussels sprouts, chicken soup, even a g&#amn salad—are every bit as fascinating, interesting, storied, and delicious as what the chefs wearing the fanciest pants these days are concocting. I mean, have you ever stopped to marvel about exactly what goes on inside a hamburger when you cook it? The simultaneous complexity and simplicity of a patty formed from the chopped muscle mass of selected parts of a remarkably intricate animal, seasoned with salt and pepper, seared on a hot piece of metal, and then slipped into a soft toasted bun? You haven't? Well, let me give you a quick rundown to show you what I'm talking about.

## ON HAMBURGERS

Hamburgers start as patties of beef . . ., no, let me back up a bit. Burgers actually start as ground beef that's then formed into . . ., no, sorry, even further back. Hamburgers start with whole cuts of beef that are then ground into . . . Wait a minute, let's get all *Inception* on this and go one level deeper: hamburgers start with cows—animals that live exceedingly complicated lives, that can differ not only in breed and feed, but also in terms of exercise, terrain they're exposed to, how and when they're slaughtered, and whether they live on grass their whole lives or are supplemented with grain. From these animals come many cuts of meat that vary in flavor according to fat content, their function during the animal's life, and its specific diet. Blending specially selected cuts will lead to ground beef with the optimal flavor and fat profile.

From there, it's just a simple matter of grinding, forming patties, and cooking, right?

Not so fast. How you grind your beef can have a profound impact on the texture of the finished burgers. Think all ground beef is created equal?

Think again. And what about salting? Do you salt the meat and blend it in, or do you salt the outside of the patties? How do you form those patties? Pressing the beef into a ball and flattening it works, but is that really the best way? And what causes those burgers to puff up into softball-shaped spherical blobs when you cook them anyway? Once you start opening your mind to the wonders of the kitchen, once you start asking what's *really* going on inside your food while you cook it, you'll find that the questions keep coming and coming, and that the answers will become more and more fascinating.

Not only does answering questions about burgers help you to cook your burgers better, but it also reveals applications to all sorts of other situations. We start big fat burgers off on the cooler side of the grill and finish 'em with a sear in order to get a nice, perfectly even medium-rare color throughout, along with a strong, crusty sear. Guess what the best way to cook a big fat steak is? You got it: the exact same method applies, because the proteins and fat in a steak are similar to those in a hamburger.

Still not too sure what I mean? Don't worry, we'll answer all of these questions and more in due time.

## THE FIRST STEP TO WINNING IS LEARNING HOW NOT TO FAIL

Have you ever made the same recipe a half dozen times with great results, only to find that on the seventh time, it completely fails? The meat loaf comes out tough, perhaps, or the pizza dough just doesn't rise. Oftentimes it's difficult to point to exactly what went wrong. If you're a tinkerer in the kitchen, you like to modify recipes a bit here and there to suit your own taste or mood. That's all well and good, and luckily, the first six times, your modifications didn't affect the outcome of the recipe. What changed on that seventh time? Could it be the extra salt you

added? Perhaps the temperature of the room? Or maybe it's that you ran out of olive oil and used canola instead. Perhaps your stand mixer was on the fritz, so you blended everything by hand.

The point is, there are many ways you can stray from a written recipe, but only some of those forays will cause the recipe to fail. Being able to identify exactly which parts of a recipe are essential to the quality of the finished product and which parts are just decoration is a practical skill that will open up your opportunities in the kitchen as never before. Once you understand the basic science of *how* and *why* a recipe works, you suddenly find that you've freed yourself from the shackles of recipes. You can modify as you see fit, fully confident that the outcome will be a success.

Take a recipe for Italian sausage, for instance. The recipe in this book has you combine pork shoulder with salt and some aromatics, let the meat rest overnight, and then grind it and knead it the next day. Now, you've tasted my Italian sausage, and fair enough, you think it's got too much fennel. OK, so you use less fennel and more marjoram the next time instead. Because you've read the sausage chapter and understand that the keys to a great-textured sausage are the interaction between the salt and meat and the method by which the ground meat is mixed, you're confident that changing the spicing will still allow you to produce a great-tasting link. At the same time, you know that salt is what dissolves muscle proteins and allows them to cross-link, giving your sausage that snappy, juicy texture, so you can't cut back on the salt the same way you can with the fennel. Likewise, you know that you can make your sausage out of turkey or lamb if you'd like, but you can't change the fat content if you want it to remain juicy.

Fact: Cooking by rote—even when your mentors are some of the greatest chefs in the world—is paralyzing. Only by understanding the underlying principles involved in cookery can you free yourself from both recipes and blindly accepted conventional wisdom.

Starting to get an idea of what I'm talking about? Freedom. That's what.

## WHY THIS BOOK?

In many ways, the blog format is ideal for the type of work I do. I get to write about things in a pretty informal way and in return, my readers tell me what they think, ask insightful questions, and let me know what they'd like to see me tackle next. It's communal, and I owe my success as a blogger as much to my readers and my ever-supportive, always fun, incredible coworkers and fellow bloggers as I do to myself.

That said, there are limits to what the blogging platform can support. It's great for short articles, it does pictures OK, but good charts? Good graphs? Good, easy-to-understand layout? Long-form content? Forget about it. That's where this book comes in. It represents the culmination of not just a decade and a half of cooking and studying the science of everyday foods, but of years of learning how to apply this science in ways that can help home cooks cook everyday food in better, tastier ways.

What you *won't* find in this book are fancy-pants recipes calling for exotic ingredients or difficult techniques or chemicals or even much special equipment beyond, say, a food processor or a beer cooler. You also won't find any desserts. They just aren't my thing, and rather than fake a few of 'em, I figured I'd just own up to the fact that they just don't interest me in the way savory food does. (Remember that whole thing about not doing anything that I don't love doing?)

What you *will* find here is a thorough examination of classic recipes. You'll find out why your fried chicken skin gets crisp, what's going on inside a potato as you mash it, how baking powder helps your pancakes rise. Not only that, but you'll discover that in many (most?) cases, the most traditional methods of cooking are in fact *not* the ideal way to reach the desired end results—and you'll find plenty of recipes and instructions that tell you how to get better results. (Did you know that you can parcook pasta in room-temperature tap water? Or that the key to perfect French fries is vinegar?)

You'll probably find that I talk about my wife and my dogs a bit too much, and that I'm an insane fan of both the Beatles and the pun, that lowest form of wit. I may rightfully be accused of making *abstruse* references to any or all of the following topics: *The Simpsons*. Cartoons and movies from the 1980s. *Star Wars*. British comedians. *The Big Lebowski*. *MacGyver*. To these crimes, I plead guilty, but I will not repent.

Occasionally you will come across an experiment designed for you to carry out yourself at home. All of these experiments are party-friendly, and most of them are kid-friendly too, so make sure you've got company around if you're going to attempt them!

Some of you may use this book solely for the recipes, and there's nothing wrong with that. I'll still like you. I've done my best to write them as clearly and concisely as possible, and I guarantee each and every one of them will work as advertised (provided you follow the instructions). If they don't work for you, I want to hear about it! Others may read through the entire book without ever cooking a single thing from it. I might even like you guys more than I like the recipe-only guys, for it's what's going on behind the scenes, or under that well-browned crust, that really interests me.

If you're the armchair-cook type, you're in luck. This book was written to work from front to back. Recipes in later chapters build on basic scientific principles discussed in earlier chapters. On the other hand, if you like skipping around—say, potato salad doesn't interest you but roast beef sure does—well, you won't have much trouble either. I've done my best to make each lesson self-contained, cross-referencing earlier chapters when necessary.

One thing I want to make clear here: *This book is nowhere near comprehensive.* Why would I put myself down like that? Well, it's because the whole point of science is that it's a never-ending quest for knowledge. No matter how much we know about the world around us, the world inside a block of cheese, or the world contained in an eggshell, the amount that we don't know will always be much greater than what we do. The moment we think we know all the answers is the moment we stop learning, and I truly hope that time never comes for me. In the words of Socrates Johnson: "All we know is that we know nothing."

If there are three rules that I think would make the world a better place if everyone followed them, it'd be these: challenge everything all the time, taste everything at least once, and relax, it's only pizza.

## SO WHY TRUST ME?

When I chime in on online message boards, when I write blog posts that make some pretty bold claims (like, say, that frying in hotter oil actually makes food absorb *more* grease, not less—see page 861), I often get the same questions shot back at me: Says who? Why should I trust you? I've been cooking my food [X] way since before you were born, who are you to say that there's a better way?

Well, there are a number of answers I *could* give to this question: It's my job to study food, test it, and answer questions about it. I have a degree from one of the top engineering schools in the country. I spent a good eight years cooking behind the stoves of some of the best restaurants in the country. I've edited recipes and articles in food magazines and on websites for almost a decade. These are all pretty good reasons to put your faith in what I say, but the truth of the matter is this: *you shouldn't trust me.*

You see, "Just trust me" was the way of the old cooks. The MO of the master-apprentice relationship. Do what I say and do it now, because I say so. And that's exactly the mentality we're trying to fight here. I *want* you to be skeptical. Science is *built* on skepticism. Galileo didn't come to the conclusion that the Earth revolves around the sun, not the other way around, by blindly accepting what everyone else was telling him. He challenged conventional wisdom, came up with new hypotheses to describe the world around him, tested those hypotheses, and then and only then did he ask people to believe in the madness that he was spouting from behind that awesome beard of his. He did, of course, die under house arrest after being tried by the Roman Inquisition for all of his troubles. (Let's hope that doesn't happen to any of you budding kitchen scientists.) And that was for something as trivial as describing the shape of the solar system. Meanwhile, we're here tackling the big issues. Pancakes and meat loaf deserve *at least as much* scrutiny!

The point is this: if at any time while reading this book you come across something I've written that just doesn't seem right, something that seems as if it hasn't been sufficiently tested, something that isn't rigorously explained, then I fully expect you to call me out on it. Test it for yourself. Make your own hypotheses and design your own experiments. Heck, just e-mail me and tell me where you think I went wrong. I'll appreciate it. Honestly.

The first rule of science is that while we can always get closer to the truth, there is *never* a final answer. There are new discoveries made and experiments performed every day that can turn conventional wisdom on its head. If five years from now somebody hasn't discovered that at least one fact in this book is glaringly wrong, it means that people aren't thinking critically enough.

But some of you might be wondering now, what exactly *is* science? It's a really good question, and a topic that's often misunderstood. Let's talk about it a bit.

# The keys to good kitchen science

**Science is not about big words.** It's not about lab coats and safety goggles, and it's *definitely* not about trying to make yourself sound fancy. Science is not an end in and of itself, but a path. It's a method to help you discover the underlying order of the world around you and to use those discoveries to help you predict how things will behave in the future. The scientific method is based on making observations, keeping track of those observations, coming up with hypotheses to explain those observations, and then performing tests designed to disprove those hypotheses. If, despite your hardest, most sincere efforts, you can't manage to disprove the hypotheses, then you can say with a pretty good deal of certainty that your hypotheses are true. *That* is what science is, and it can be as simple as observing that of the first three beers you had, the coldest one was the tastiest, and therefore it's probably a good idea to chill down the fourth before you crack it open, or as complex as determining the gene that decides whether your kid's eyes are going to be blue or brown.

Most of us practice science every single day, often without even knowing it. For instance, when I first got married, I noticed that there seemed to be a direct correlation between some of my wife's bad moods and my propensity to leave the toilet seat up. (Observation.) I then thought to myself, *perhaps if I put the toilet seat down more often, my wife's mood and therefore my own happiness would improve.* (Hypothesis.) I tried putting down the toilet seat a few times and waiting to see how my wife reacted. (Testing.) Noticing an improvement in her moods, I started putting the seat down almost every time, only occasionally leaving it up to test the continued validity of my hypothesis. Some folks would call this just being a good roommate/husband. I call it science.

Believe it or not, the kitchen is perhaps the easiest place for a regular person to practice science every day. You've certainly performed your own scientific experiments in the past. Here's an example: You buy a new toaster with a darkness knob that goes from one to eleven (just in case you want it one shade darker than ten) and then notice your toast is coming out too dark on level six, so you turn it down to level four. Now your toast is too light. Working from these two observations, you hypothesize that perhaps five is the right level for your bread. Lo and behold, your next slice of toast and every slice of toast after that come out just right.

Now, that may not be the most groundbreaking observation in the history of science, and it *is* admittedly very limited in its application (I mean, you can't even guarantee that the next toaster you own will have the same scale), but it's science nevertheless, and in that sense, it's no different from what professional scientists do every day.

Scientists know that bias can be a powerful force in experiments. Oftentimes scientists only see what they want to see and find the answers they want to find, even if they don't realize it. Have you ever heard the story of Clever Hans the counting horse? Early in the twentieth century, Hans made quite a name for himself by apparently being intelligent enough to understand German, do arithmetic, recognize the days of the week, differentiate between musical tones, and even read and spell. His trainer would ask him questions and Hans would respond by tapping his hoof. For instance, when asked, "What is eight plus twelve?" he would stamp his hoof twenty times. The horse was a sensation, touring around Germany and amazing crowds with his incredible abilities.

After an intensive study carried out by the German board of education, the observers came to a shocking conclusion: the whole thing was a scam. Turns out that Hans wasn't able to do any math at all. What he *was* quite good at was interpreting the facial expressions and attitudes of his trainer. As he slowly stamped his foot against the ground, he'd observe the tension in the trainer's face; when he reached the correct number, the trainer would relax, Hans would know that he was finished, and he'd stop tapping. It's a skill to be admired, for sure. Heck, most of my marital problems would be solved if I could tell when my wife was tense versus relaxed. But could the horse do math? Nope.

Yet here's the thing: *the trainer didn't even realize what he was doing.* He thought he had an amazingly intelligent horse. In fact, the horse was so good at reading faces that even when a complete stranger was asking him the questions, he would answer equally well. How did the members of the board finally prove that Hans's so-called abilities were fake? They designed a series of scientific experiments. The simplest involved blindfolding the horse or having the trainer ask his questions out of Hans's line of sight. As expected, suddenly his amazing intellect disappeared. The most interesting tests of all involved having the trainer ask the horse questions the trainer himself did not know the answers to. Guess what? If the trainer didn't know the answer, neither did Hans.

THE POINT here, of course, is that designing a successful experiment—whether it involves a mathematical horse or takes place in your own kitchen—is about eliminating the bias of the experimenter (in this case, you). This isn't always easy to do, but it's almost always possible.

Let me tell you about an experiment I carried out a little while back to illustrate the seven key steps to a good tasting in the kitchen: eliminate bias, introduce a control, isolate variables, stay organized, address palate fatigue, taste, and analyze.

## IS NEW YORK PIZZA REALLY ALL ABOUT THE WATER?

Just as jocks like to stick together and nerds travel in packs, obsessives bordering on the psychotic (like me) seek out the acquaintance of others like themselves, in a manner that some say borders on the, well, the obsessive. The first time I heard that the mineral content of water might have an effect on the properties of bread dough was about ten years ago, when I read Jeffrey Steingarten's gloriously

obsessive piece about Roman breads in the chapter titled "Flat Out" in his book *It Must Have Been Something I Ate:*

> *In the shower, the shampoo refuses to lather. This means that Roman water is high in minerals, which can be good for the color and texture of bread, but slows fermentation and tightens the dough. I reach for my scuba diver's underwater writing slate, as seen on* Baywatch, *indispensable for recording those flashes of insight that so often strike one in the bath. We must test the water of Rome.*

Unfortunately, despite the heroic efforts he went through to bring true *pizza bianca* and *pane Genzano* to the home cook, the water issue was never satisfactorily resolved.

Well, eight years later, I decided to try to resolve it for myself, along with the help of another obsessive: Mathieu Palombino, chef-owner of New York City's Motorino, who kindly volunteered to aid me in my little experiment. The idea is simple: the minerals dissolved in water (mostly magnesium and calcium) can help proteins in the flour bond together more tightly, forming a stronger gluten structure, the network of interconnected proteins that gives dough its strength and elasticity. So, the higher the mineral content of water (measured in parts per million, or ppm), the stronger and chewier the dough. In theory, it makes sense, and it is easily provable in a laboratory. The more interesting question to me was, are the effects of the minerals in the water (referred to as Total Dissolved Solids, or TDS) significant enough to be detected by a normal eater in a real-world situation?

To answer the question, I charged Mathieu with

making Neapolitan pizzas using waters with different TDS contents and brought in a panel of pizza experts to taste the finished pies. The problem is that the real world is, well, real, and as such, very difficult to control. In any scientific endeavor, there are a number of key principles that must be adhered to if you want to ensure that your results are accurate and repeatable—the hallmark of any sound experiment.

## Key to a Good Tasting #1: Eliminate Bias

Despite our best efforts, we have yet to invent a device that empirically measures precisely how delicious pizza crust is, so our best option is to resort to the crude analyses of our mouths. Humans are notoriously bad at separating emotional responses to foods and food brands from their actual eating qualities, and the only way to eliminate this bias is with a double-blind tasting—a tasting in which neither the tasters nor the people preparing and serving the food know which sample is which.

To do that, I first gathered my waters—five different varieties of bottled waters with TDS contents ranging from less than 10 ppm (the maximum allowable for "purified" waters) all the way up to 370 ppm (mineral water on the high end of the TDS scale), along with tap water. I chose the specific brands because they were available at my local grocery.

- **Aquafina:** less than 10 ppm
- **Dasani:** approximately 40 ppm
- **Tap water:** approximately 60 ppm
- **Rochetta:** 177 ppm
- **San Benedetto:** 252 ppm
- **Evian:** 370 ppm

I transferred all the waters to clean bottles marked simply with numbers, making sure to note which water was in each bottle. That way, when I dropped the bottles off at Motorino—which I did without actually seeing Mathieu, lest I unknowingly reveal any information, like Hans and his trainer—Mathieu would have no idea what he was working with.

Normally I systematically ignore the advice of my better half, a PhD student in cryptography (that's the study of encryption, not the study of tombs), but this time, I shut up and listened to her for a change. She suggested that in order to further decrease bias and, more importantly, to allow me to participate in the tasting as well, she—an unbiased third party—should rearrange the numbered caps on all the bottles, taking note of which caps were moved to which bottles.

The result: three levels of encryption involving three different keys, none of which were useful on their own. Neither myself, my wife, my tasters, nor our talented chef would know which pizza was made with which water until after the tasting, when my wife would reveal how the bottle caps were switched, Mathieu would reveal which bottle of water was used to make each dough, and I would reveal which number corresponded to which brand.

## Key to a Good Tasting #2: Introduce a Control

The concept of a control is a simple one, but one that is often overlooked. The idea is that you need to include at least one sample in your tasting for which the answer is already known. That way, you can be sure that the experiment went according to plan and that your other results are reliable. In the case of a double-blind experiment like this one, it means doubling up on at least one of your samples. If the results for both are the same, then you have a pretty strong case that the experiment went according to plan.

In this case, I doubled up on both tap water and Evian, making a total of eight water samples. If the testing procedures were sound, and our palates were as fine-tuned as I believed them to be, the crusts made with the same water should be ranked very close to each other in the tasting.

### Key to a Good Tasting #3: Know What You Are Asking (Isolate Variables)

In Douglas Adams's *Hitchhiker's Guide to the Galaxy*, a team of scientists builds a supercomputer that is finally able to answer the Big Question: the Answer to Life, the Universe, and Everything. The ultimate irony is that when they're finally given the answer—forty-two—they realize that they never really knew exactly what they were asking in the first place.

Now, these were very poor scientists. Rule Numero Uno when it comes to science is this: *Be very precise about what question you are trying to answer.* The more limited the scope of your question, the easier it will be to design an experiment to answer it. There are approximately a gajillion variables when it comes to pizza, each one of them interesting in its own way. But here, I was interested in only one: how does the water's mineral content affect the dough? What that meant was that in order to isolate that single variable, I'd have to ensure that every single other variable out of those gajillion remained exactly the same from sample to sample. Easier said than done.

In the real world—particularly with cooking—there are an insane amount of variables to try and control for. Perhaps that log in the wood-burning oven is gonna burn slightly hotter for pizza #2 than for pizza #1, raising the temperature by a couple of degrees. Or maybe Mathieu will have to wait for a server with a load of dishes to pass by before inserting pizza #5 into the oven, adding a few seconds to its trip. This is an inevitable, unavoidable reality when it comes to science. What we can hope, however, is that these tiny changes in method from sample to sample will make a negligible difference compared to the variable we are actually testing for. We can also do our best to make sure every sample is treated the same.

I asked Mathieu to weigh the ingredients for each batch of dough precisely and to ensure that each batch was kneaded for the same length of time and allowed to ferment at the same temperature. While normally the *pizzaioli* at the restaurant take turns shaping, saucing, and baking the pizzas, this time Mathieu himself made each one from start to finish, ensuring that the method used was as consistent as possible.

On top of those measures, I also decided to present each sample in two forms: as a completed Margherita pizza and as a simple disk of dough baked on its own, to eliminate any variability that differences in topping distribution might add.

### Key to a Good Tasting #4: Stay Organized

Who better to taste pizzas than New York's foremost pizza cognoscenti, Ed Levine and Adam Kuban? In addition, Alaina Browne of the Serious Eats team joined us, along with my wife (as a reward for her good advice), and—through a miraculously fateful act of good networking—Jeffrey Steingarten himself, the very man who had unknowingly started me down the path of pizza (and, indeed, of food writing, period). Before arriving at the restaurant, I drew up tasting sheets for my panel to fill out. Each pizza was to be evaluated in four categories, and each category was rated on a scale from one to ten:

- **Dough Toughness:** Is it tender like cake or as chewy as leather?
- **Dough Crispness:** Does it crackle, or is it flaccid?
- **Oven Spring:** Did it form large, airy bubbles, or is it compact and dense?
- **Overall Quality:** How do you like it?

The first five tasters (including myself) arrived promptly at 4 p.m., with chef Mathieu waiting for us. Jeffrey, however, was nowhere to be seen, though he had warned me that he might be a bit late due to an important meeting. Ed phoned up his assistant. Evidently, his important meeting was in his bed with Sky King, his dog, but not to worry—his jacket was on, and he was hard at work on his shoelaces.

Meanwhile, Mathieu informed us that he'd dropped half of the first sample on the floor, meaning that for that batch, we'd only be tasting one pie, not two. Not to worry. Even Tesla must have dropped a few coils in his day, right? My perfectly orchestrated plans were beginning to slip, but a glass of Brachetto and a small plate of fingerling potatoes tossed with anchovies and olives helped me pull my senses back into focus.

## Key to a Good Tasting #5: Watch Out for Palate Fatigue

As soon as Mr. Steingarten arrived, fresh from his nap, I turned to Mathieu and gave him the go-ahead. Within three minutes, the first leopard-spotted, tender-crisp beauty was on the table. Eight pizzas among six people is a lot to consume in a single sitting, even for epic eaters like our humble tasting panel, and there was no way the pizzas in the latter half of the tasting were going to get a fair shake. Ideally, we'd have each taster taste samples in a different order. That way, one would be starting with number one, another with number six, another with number three, and so on, hopefully evening out the playing field. But given that the pizzas had to be baked one at a time, this was simply not possible. So we did the next best thing: one pie out of each batch of two was eaten straight out of the oven and the second one was saved until all eight were on the table. That way, we could go back and retaste to ensure that our original thoughts were sound, and we could taste all eight side by side.

Sparkling water and wine were provided (the latter after much deliberation) in order to rinse our mouths between bites.

## Key to a Good Tasting #6: Taste

Tasting is different from eating. I often get asked, "How can you review a restaurant fairly or how can you say one product is better than another? Doesn't it depend on how hungry you are at the time?" And, indeed, mood and appetite can have a powerful effect on how much you enjoy eating a particular food at any one time. But the goal of analytical tasting is to assess qualities of the food *beyond* your gut reaction of whether it's good or bad. With pizza, for example, I start by taking a slice from each pie with an across-pie-average of charring, bubbles, sauce, and cheese. I then bite just the tip, noting the pressure of the crust on my lower teeth to gauge its degree of crispness. As I pull the slice away from my mouth, applying just a bare soupçon of torque, I judge the effort it takes for the dough to tear. (In this case, Pizza #5 was clearly tougher than the rest, I thought triumphantly—one must never talk to his fellow tasters during a blind tasting lest your opinion influence those of others—*so it must certainly be one of those high-mineral samples.*)

After carefully working my way up the side of each slice, I evaluate the *cornicione* (the raised rim of the pizza). It's hard to find fault with any of them, but does #3 look just a shade paler than the rest, indicating a lower-mineral-content water? Could be. But if so, why isn't it also more tender? See, these are observations you can make relatively empirically—that is, free from bias. Sure, eating the same pizza when starving versus when stuffed will elicit two different reactions, but by reducing each one to basic elements that are more easily quantified—crispness, chewiness, degree of charring—you can get a more accurate picture of the pizza overall, divorcing it from the mental bias of your current state of mind.

With the pizzas tasted, we thanked Mathieu for his incredible pizza (the best in the city, for my money), and bravely made our separate ways through the night, several degrees more content and several pounds heavier.

## Key to a Good Tasting #7: Analyze

Once all the tasting has been done and all the data collected, it's your job to analyze it in order to make the most reasonable assessment you can about what factors are affecting your variable. In the case of the pizzas, that meant charting the data and listing it in order from lowest mineral content (which should

presumably deliver more more tender, softer, less-sprung, paler dough) to highest. If everything went according to theory, the lines for crispness, toughness, and airiness should all show distinct trends, going up or down as the mineral content of the water increases.

As it turns out, no such trends existed. True, the two batches made with Evian—the highest mineral content of all the water we tried—delivered the crispest crusts, but overall, there was not enough of a trend in the data to make a definitive statement. And you know what? Sometimes—more often than not—an experiment, no matter how closely controlled, does not produce the results you were looking for. Which is not to say that we got no results at all. In fact, taking a second look at the data, I saw that the rankings for crispness follow in step pretty closely with the rankings for overall enjoyment—implying that our enjoyment of pizza is directly related to how crisp it is. We all want a crisp crust, not a soggy one.

Earth-shattering, I know.

On the other hand, we can now pretty definitively say that the small differences that arise naturally in the course of making a good pizza by hand far outweigh any difference the mineral content of the water could make. That is to say, great New York pizza is most certainly *not* dependent on using New York City tap water, which is good news for everyone else in the world.

# What is cooking?

**I know you're eager to jump right in** and start cooking, but first answer this question: What is cooking?

If you're my wife, your answer will be, "It's that thing you do when that crazy look comes into your eyes." A great chef might tell you that cooking is life. My mom would probably say that it's a chore, while my wife's aunt would tell you that cooking is culture, family, tradition, and love. And, yes, cooking is all of those things, but here's a more technical way to think about it: Cooking is about transferring energy. It's about applying heat to change the structure of molecules. It's about encouraging chemical reactions to alter flavors and textures. It's about making delicious things happen with science. And before we can even begin to understand what happens when we grill a hamburger, or even what equipment we might want to stock our kitchen with, we have to get one *very* important concept into our heads first, as it'll affect everything we do in the kitchen, starting with which pots and pans we use. It's this: **Heat and temperature are not the same thing.**

At its most basic, cooking is the transfer of energy from a heat source to your food. That energy causes physical changes in the shape of proteins, fats, and carbohydrates, as well as hastens the rate at which chemical reactions take place. What's interesting is that most of the time, these physical and chemical changes are permanent. Once a protein's shape has been changed by adding energy to it, you can't change it back by subsequently removing that energy. In other words, you can't uncook a steak.

The distinction between heat and temperature can be one of the most confusing things in the kitchen, but grasping the concept is essential to helping you become a more rational cook. Through experience, we know that temperature is an odd measure. I mean, pretty much all of us have walked around comfortably in shorts in 60-degree weather but have felt the ridiculous chill of jumping into a 60-degree lake, right? Why does one but not the other make us cold, even though the temperature is the same? Let me try to explain.

**Heat** is energy. Third-grade physics tells us that everything from the air around us to the metal on the sides of an oven is composed of molecules: teeny-tiny things that are rapidly vibrating or, in the case of liquids and gases, rapidly bouncing around in a random manner. The more energy is added to a particular system of molecules, the more rapidly they vibrate or bounce, and the more quickly they transfer this movement to anything they are touching—whether it's the vibrating molecules in a metal pan transferring energy to a juicy rib-eye steak sizzling away or the bouncing molecules of air inside an oven transferring energy to the crusty loaf of bread that's baking.

Heat can be transferred from one system to another, usually from the more energetic (hotter) system to the less energetic (cooler). So when you place a steak in a hot pan to cook it, what you are really doing is transferring energy from the pan-burner system to the steak system. Some of this added energy goes to raising the temperature of the steak, but much of it gets used for other reactions: it takes energy to make moisture evaporate, the chemical reactions that take place that cause browning require energy, and so on.

**Temperature** is a system of measurement that allows us to quantify how much energy is in a specific system. The temperature of the system is dependent not only on the total amount of energy in that body, but also on a couple of other characteristics: density and specific heat capacity.

**Density** is a measure of how many molecules of stuff there are in a given amount of space. The denser a medium, the more energy it will contain at a given temperature. As a rule, metals are denser than liquids,* which, in turn, are denser than air. So metals at, say, 60°F will contain more energy than liquids at 60°F, which will contain more energy than air at 60°F.

**Specific heat capacity** is the amount of energy it takes to raise a given amount of a material to a certain temperature. For instance, it takes exactly one calorie of energy (yes, calories are energy!) to raise one gram of water by one degree Celsius. Because the specific heat capacity of water is higher than that of say, iron or of air, the same amount of energy will raise the temperature of a gram of iron by almost ten times as much and a gram of air four times as much. The higher the specific heat capacity of a given material, the more energy it takes to raise the temperature of that material by the same number of degrees.

Conversely, this means that given the same mass and temperature, water will contain about 10 times as much energy as iron and four times as much as

air. But even so, remember that air is far less dense than water, which means that the amount of heat energy contained in a given volume of air at a given temperature will be only a small fraction of the amount of energy contained in the same volume of water at the same temperature. *That's* the reason why you'll get a bad burn by sticking your hand into a pot of 212°F boiling water but you can stick your arm into a 212°F oven without a second thought (see "Experiment: Temperature Versus Energy in Action," page 33).

Confused? Let's try an analogy.

Imagine the object being heated as a chicken coop housing a dozen potentially unruly chickens. The temperature of this system can be gauged by watching how fast each individual chicken is running. On a normal day, the chickens might be casually walking around, pecking, scratching, pooping, and generally doing whatever chickens do. Now let's add a bit of energy to the equation by mixing a couple cans of Red Bull in with their feed. Properly pepped up, the chickens begin to run around twice as fast. Since each individual chicken is running around at a faster pace, the temperature of the system has gone up, as has the total amount of energy in it.

Now let's say we have another coop of the same size but with double the number of chickens, thereby giving it double the density. Since there are twice as many chickens, it will take double the amount of Red Bull to get them all running at an accelerated pace. However, even though the final temperature will be the same (each individual chicken is running at the same final rate as the first ones), the total amount of energy within the second coop is double that of the first. So, energy and temperature are not the same thing.

---

*Alright, Smarty-Pants. Yes, at high enough temperatures, metals will melt into very dense liquids, and yes, Even Smartier-Pants, mercury is a very dense metal that is liquid even at room temperature. Got that out of your system? OK, let's move on.

Now what if we set up a third coop, this time with a dozen turkeys instead of chickens? Turkeys are much larger than chickens, and it would take twice as much Red Bull to get one to run around at the same speed as a chicken. So the specific heat capacity of the turkey coop is twice as great as the specific heat capacity of the first chicken coop. What this means is that given a dozen chickens running around at a certain speed and a dozen turkeys running around at the same speed, the turkeys will have twice as much energy in them as the chickens.

To sum up:

- At a given temperature, denser materials generally contain more energy, and so heavier pans will cook food faster. (Conversely, it takes more energy to raise denser materials to a certain temperature.)
- At a given temperature, materials with a higher specific heat capacity will contain more energy. (Conversely, the higher the specific heat capacity of a material, the more energy it takes to bring it to a certain temperature.)

In this book, most recipes call for cooking foods to specific temperatures. That's because for most food, the temperature it's raised to is the primary factor determining its final structure and texture. Some key temperatures that show up again and again include:

- **32°F (0°C):** The freezing point of water (or the melting point of ice).
- **130°F (52°C):** Medium-rare steak. Also the temperature at which most bacteria begin to die, though it can take upward of 2 hours to safely sterilize food at this temperature.

- **150°F (64°C):** Medium-well steak. Egg yolks begin to harden, egg whites are opaque but still jelly-like. Fish proteins will tighten to the point that white albumin will be forced out, giving fish like salmon an unappealing layer of congealed proteins. After about 3 minutes at this temperature, bacteria experience a 7 log reduction—which means that only 1 bacteria will remain for every million that were initially there).
- **160° to 180°F (71° to 82°C):** Well-done steak. Egg proteins fully coagulate (this is the temperature to which most custard- or egg-based batters are cooked to set them fully). Bacteria experience a 7 log reduction within 1 second.
- **212°F (100°C):** The boiling point of water (or the condensation point of steam).
- **300°F (153°C) and above:** The temperature at which the Maillard browning reactions—the reactions that produce deep brown, delicious crusts on steaks or loaves of bread—begin to occur at a very rapid pace. The hotter the temperature, the faster these reactions take place. Since these ranges are well above the boiling point of water, the crusts will be crisp and dehydrated.

## SOURCES OF ENERGY AND HEAT TRANSFER

Now that we know exactly what energy is, there's a second layer of information to consider: the means by which that energy gets transferred to your food.

**Conduction** is the direct transfer of energy from one solid body to another. It is what happens when you burn your hand by grabbing a hot pan (hint: don't do that). Vibrating molecules from one surface will strike the relatively still molecules on another surface, thereby transferring their energy. This is by far the most efficient method of heat transfer. Here are some examples of heat transfer through conduction:

- Searing a steak
- Crisping the bottom of a pizza
- Cooking scrambled eggs
- Making grill marks on a burger
- Sautéing onions

Steaks searing—heat through conduction.

**Convection** is the transfer of one solid body to another through the intermediary of a fluid—that is, a liquid or a gas. This is a moderately efficient method of heat transfer, though in cooking its efficiency depends greatly on the way the fluid flows around the food. The motion of the fluid is referred to as **convection patterns**.

As a general rule, the faster air travels over a given surface, the more energy it can transfer. Still air will rapidly give up its energy, but with moving air, the energy supply is constantly being replenished by new air being cycled over a substance such as food. Convection ovens, for instance, have fans that are designed to keep the air inside moving around at a good clip to promote faster, more even cooking. Similarly, agitating the oil when deep

frying can lead to foods that crisp and brown more efficiently.

Here are some examples of heat transfer through convection:

- Steaming asparagus stalks
- Boiling dumplings in stock
- Deep-frying onion rings
- Barbecuing a pork shoulder
- The top of a pizza baking in an oven

Dumplings boiling—heat through convection.

**Radiation** is transfer of energy through space via electromagnetic waves. Don't worry, that's not as scary as it sounds. It doesn't require any medium to transfer it. It is the heat you're feeling when you sit close to a fire or hold your hand above a preheated pan. The sun's energy travels to the earth through the vacuum of space. Without radiation, our planet (and indeed, the universe) would be in a lot of trouble!

An important fact to remember about radiant energy is that it decays (that is, gets weaker) by the inverse square law—the energy that reaches an object from a radiant energy source is proportional to the inverse of the square of its distance. For example, try holding your hand 1 foot away from a fire, then move it 2 feet away. Even though you've only doubled the distance, the fire will feel only about one-quarter as hot.

Here are some examples of radiant heat transfer:

- Roasting a pig on a spit next to hot coals
- Toasting garlic bread under the broiler
- Getting a tan from the sun
- Broiling some marinated salmon

The top of a pizza cooks via radiation.

Most of the time, in cooking, all three methods of heat transfer are used to varying extents. Take a burger on the grill, for example. The grill grate heats the patty directly where it is in contact with it through conduction, rapidly browning it at those spots. The rest of underside of the patty is cooked via radiation from the coals underneath. Place a piece of cheese on the burger and pop the lid down for a bit, and convection currents will form, carrying the hot air from directly above the coals up and over the top of the burger, melting the cheese.

Grilled burgers cook through all three forms of energy transfer.

You might notice that these three types of heat transfer heat only onto the surface of foods. In order for food to cook through to the center, the outer layer must transfer its heat to the next layer, and so on, until the very center of the food begins to warm up. Because of that, the outside of most cooked foods will almost always be more well done than the center (there are tricks to minimizing the gradient, which we'll get to in time).

**Microwaves** are the only other standard method of energy transfer we commonly use in the kitchen, and they have the unique ability to penetrate through the exterior of food when heating it. Just like light or heat, microwaves are a form of electromagnetic radiation. When microwaves are aimed at an object with magnetically charged particles (like, say, the water in a piece of food), those particles rapidly flip back and forth, creating friction, which, in turn, creates heat. Microwaves can pass through most solid objects to a depth of at least a few centimeters or so. This is why microwaves are a particularly fast way to heat up foods—you don't need to wait for the relatively slow transfer of energy from the exterior to the center.

Phew! Enough with the science lesson already, right? Bear with me. Things are about to get a lot more fun!

# EXPERIMENT:
## Temperature Versus Energy in Action

**The difference between the definition of *temperature***
**and the definition of *energy* is subtle but extraordinarily important.**
**This experiment will demonstrate how understanding**
**the difference can help shape your cooking.**

## Materials

- **1 properly calibrated oven**
- **1 able-bodied subject with external sensory apparatus in full working order**
- **One 3-quart saucier or saucepan filled with water**
- **1 accurate instant-read thermometer**

## Procedure

Turn your oven on to 200°F and let it preheat. Now open the oven door, stick your hand inside, and keep your hand in the oven until it gets too hot to withstand. A tough guy like you could probably leave it in there for at least 15 seconds, right? 30 seconds? Indefinitely?

Now place a pan of cold water on the stovetop and stick your hand in it. Turn the burner to medium-high heat and let the water start to heat up. Stir it around with your hand as it heats, but be careful not to touch the bottom of the pan (the bottom of the pan will heat much faster than the water). Keep your hand in there until it becomes too hot to withstand, remove your hand, and take the temperature.

## Results

Most people can hold their hand in a 200°F oven for at least 30 seconds or so before it becomes uncomfortably hot. But let it go much above 135°F, and a pan of water is painful to touch. Water at 180°F is hot enough to scald you, and 212°F (boiling) water will blister and scar you if you submerge your hand in it. Why is this?

Water is much denser than air—there are many times more molecules in a cup of water than there are in a cup of air. So, despite the fact that the water is at a lower temperature than the air in the oven, the hot water contains far more energy than the hot air and consequently heats up your hand much more rapidly. In fact, boiling water has more energy than the air in an oven at a normal roasting temperature, say 350° to 400°F. In practice, this means that boiled foods cook faster than foods that are baked or roasted. Similarly, foods baked in a moist environment cook faster than those in a dry environment.

# Essential kitchen gear

**There's a lot of nonsense** to wade through out there when trying to stock your kitchen. Do you really need that $300 knife? How often are you gonna pull out that salad spinner? And exactly which one of those things they sell on TV is really going to take the place of every other piece of equipment in my kitchen? (Hint: none of them.) The problem with most of the people telling you to buy things is that they're usually the ones selling it. Who can you trust? Well, this chapter is a no-nonsense, no bullshit guide to point you toward what you *really* need in the kitchen and what is just noise.

Any Eddie Izzard fans in the audience? According to him, "The National Rifle Association says that guns don't kill people—people kill people. But I think the gun helps." Funny joke. But what's it got to do with cooking?

I remember a time back when I first started working in restaurants. One of my jobs was to reduce a couple quarts of heavy cream down to a couple of pints. I'd get a big heavy pot with an aluminum base, pour the cream into it, and cook it over the lowest possible heat so that it'd reduce without even bubbling. I'd do this every morning, and it would take a couple hours, but no big deal; I had plenty of other tasks to keep me occupied—peeling potatoes, peeling salsify, peeling carrots (ah, the life of a green cook). Then one morning, the pot I usually reduced the cream in was being used. Rather than wait for it to free up, I just grabbed one of the thinner stockpots off the shelf, poured in my cream, and heated it as usual.

What I ended up with was a pot full of greasy, broken cream, with a broken ego to match (the ego has since been repaired, but the cream was a lost cause). Tipping it out into the sink revealed a ½-inch-thick crust of brown crud on the bottom. The problem? My pot was too thin and its conductivity was too low. Rather than distributing the heat evenly over the bottom of the entire pot, the heat was concentrated in the areas directly above the flames. Those areas got overheated, causing the proteins in the cream directly above them to coagulate, stick to each other (and to the pot), and eventually burn. Without the emulsifying effect of the proteins, the fat in the cream separated out into a distinct, greasy yellow layer. Ick.

Obviously, it was the pot's fault, right? Well, not exactly. You often hear the expression "A bad cook blames his tools," and it's true: bad food is rarely bad because the pot was too thin or the blender was broken. But I think this is often misinterpreted. Nobody is saying that a good cook should be able to cook any dish regardless of the quality of their equipment. Reducing cream without a heavy, adequately conductive pot or an adequately low flame is nearly impossible, no matter how good a cook you are. Thin pots don't burn cream—people burn cream. But I think the thin pot helps.

In reality, bad food is often bad because the cook chose to try and cook something that he didn't have the proper tools for. This is, of course, just a more complicated way of saying, "Don't be stupid." And that's good advice for all walks of life, whether or not they involve homogenized emulsions of butterfat, water, phospholipids, and milk proteins

All of this is just a roundabout way of saying that the physical hardware you stock your kitchen with is just as important as the ingredients you choose or the techniques you use when you cook. Good equipment is the third side of the Triforce of cooking: good ingredients + good equipment + good technique = good food.

## POTS AND PANS

Now that we know all about heat transfer, let's talk about the tools we use to transfer heat from a heat source (your burner or your oven) to your food. I'm talking pots and pans. There's a stunning array of sizes and types available intended for a variety of uses, some of them highly specialized (think long, narrow fish poachers or tall, skinny asparagus pots), and others much more versatile. Unless you're the type who poaches fish and boils asparagus for every single meal, the latter are the type you should go for.

### Materials

When it comes to the performance of the given material in a pan, there are really two things that matter: its ability to distribute heat evenly across its entire surface (its conductivity) and its ability to retain heat and transfer it efficiently to food (its specific heat capacity and density).

Here are some common metals, along with their properties:

**Stainless steel** is very easy to maintain—as its name implies, it won't rust or pit, no matter how much you mistreat it. But it also is an extremely poor heat conductor. What this means is that heat will not travel rapidly through it. Stainless steel pans tend to develop distinct hot and cold spots that match the heat pattern of your burners. This can lead to uneven cooking, resulting in, for example, an omelet that's burned in some spots and still raw in others.

How do you gauge the heat distribution performance of a pan? The easiest way is to spread a thin layer of sugar evenly over the bottom, then heat it over a burner. The pattern in which the sugar melts will indicate the pan's hot and cold spots. A great pan will melt sugar very evenly.

**Aluminum** is a far better conductor of heat—one of the best, in fact. It's also a very inexpensive material. Why aren't all pans made of aluminum, you might ask? Well, there are two problems. It's not very dense, which means that despite its high-heat capacity, you'd need a pan that's a ridiculously unwieldy thickness for it to retain a reasonable amount of heat. Furthermore, it discolors and pits if exposed to acidic ingredients: wine, lemon juice, tomatoes, etc.

**Anodized aluminum** has been treated to give it a ceramic-like finish that is reasonably nonstick, as well as resistant to acid. This is the ideal metal for cooking foods that don't require an extraordinarily high level of heat. You wouldn't want to sear a steak in an anodized aluminum pan, but nothing is better for cooking an omelet.

**Copper** is even more conductive than aluminum. It's also quite dense, with a great heat capacity. But copper pans are very expensive. I'd love to have a great set of copper pots. I'd also love to have a lifetime supply of Stilton and a yacht with an onboard petting zoo. It ain't gonna happen. If you can afford a set of copper pots, you are a much richer person than I. For the rest of us, let's move on.

**Laminated,** or **tri-ply, pans** offer the best of both worlds. Generally, they are constructed with a layer of aluminum sandwiched between two layers of stainless steel. They have the high density of a stainless steel pan, with the great conductivity of aluminum, making them the pan of choice for most home cooks (including me!).

Time was that **nonstick pans** were pretty tough to recommend. Coatings that flake off or give off noxious fumes when heated too much are not something you want to be cooking with. These days, nonstick coatings are more durable and far safer. You'll want to own at least one good nonstick pan for egg cookery.

The subject of **cast-iron cookware** is so divisive that I feel the need to go into a little bit more detail about it. Being a proud owner of both a puppy named Hambone and some really nice cast-iron cookware, I've found that they are remarkably similar in many respects. They both require a little work, a little patience, and a whole lot of loyalty. The main difference is that in return for my investment, my cast-iron pans give me golden brown fried chicken, sizzling bacon, corn bread, apple pies, well-charred hash, perfectly seared steaks, bubbly pizzas, and crisp dumplings. Hambone, on the other hand, gives me mostly licks, chews, and a whole lot of poop. You do the math.

As far as retaining heat goes, nothing beats a good cast-iron pan. Its specific heat capacity is lower than of aluminum, but because it is so dense, for the same thickness of pan, you get about twice the heat retention capability. This is important: the pan doesn't cool down when you add food to it. While the temperature in a thin aluminum pan may drop by as much as 300 degrees when you add a half-pound rib-eye steak to it, a cast-iron pan will stay close to its original temperature, delivering a thicker, crisper, more evenly browned crust. Similarly, you can get away with using a little less oil when frying chicken, since the heat retained by the cast iron will rapidly reheat the oil when the chicken you add cools it down.

The fact that cast iron is oven-safe means that you can braise and bake in it just as well as you can fry or sear. Corn bread comes out with a beautiful golden brown crust, and pies, even with moist fillings, come out wonderfully crisp on the bottom. Its heat retention abilities mean that even when your oven's temperature fluctuates (as most thermostat-driven ovens do), the pan's heat will stay fairly constant.

And talk about durability! Cast-iron cookware is one of the few items in your kitchen that actually gets better as it gets older. Some of the very best pans have been passed down through multiple generations, their well-used surfaces worn as smooth and nonstick as a Teflon-coated pan—without the toxic chemicals. And because cast-iron pans are cast from a mold as a single piece of metal, there are no welded joints or even rivets to wear out.

There are, of course, a few downsides to cast iron:

- **Until a good layer of seasoning has built up, food will stick to it.** This goes for even the "preseasoned" skillets on the market now, which have a mediocre level of seasoning at best. With everyday use, a cast-iron skillet will be perfectly seasoned (I define this as nonstick enough to cook eggs in) within a few weeks. With less frequent use, you can expect the process to take a couple of months. It's a long haul, but think of how proud you'll be (just like housebreaking a puppy) when that first egg slides magically off the bottom.
- **It heats unevenly.** Contrary to popular belief, iron is a poor conductor of heat, which means that the heat doesn't travel far from its source. Trying to use a 12-inch cast-iron skillet on a 3-inch burner ring is an exercise in futility: the edges of the pan will never get hot. To heat a cast-iron pan effectively, you need a burner equal in size to the pan, and plenty of time for even heat distribution. Alternatively, a cast-iron pan can be preheated in a hot oven before transferring it to the stove. (Don't forget to use a kitchen towel or pot holder!)
- **It can rust.** While a good layer of seasoning will

prevent this, carelessness (like scrubbing the pan or not allowing it to dry thoroughly before storing it) can lead to rust spots.

- **You can't cook overly acidic foods in it.** Acidic foods will pick up flavor and color from the iron, turning them dingy and metallic-tasting. This means that until a very good layer of seasoning has developed, even quick wine-based pan sauces are out of the question, as are acidic recipes like tomato sauce.
- **It's heavy.** There's no getting around this one. The density of the material is what makes cast iron so good at retaining a large amount of heat energy in a small amount of volume. Innovations like helper handles help, but smaller cooks will probably struggle with tasks like flipping food or pouring a sauce from a cast-iron pan.
- **It requires special cleaning.** Because the cooking qualities of cast-iron cookware are dependent on how well seasoned it is, care must be taken while cleaning to prevent accidentally removing the layer of seasoning, or you're gonna have to start from scratch.

All that said, there's really not much to it when it comes to seasoning, maintaining, and storing your cast-iron cookware.

## HOW TO SEASON AND MAINTAIN CAST-IRON COOKWARE

### Initial Seasoning

When you first get a cast-iron pan, it will have either a bullet-gray dull finish (if it's an unseasoned pan) or a slick-looking black surface (a preseasoned pan). Unless you bought a seventy-five-year-old pan at a garage sale, it will also have a pebbly-looking surface, like this:

Modern cast iron is bumpy like that because it is not polished the way old cast iron was and retains some texture from the mold. I've compared my shiny, totally smooth 1930s Griswold pan (acquired at a flea market) to my ten-year-old Lodge skillet (which I bought new and seasoned myself) and found slight advantages with the old pan, but the new one does just fine.

So the key is all in seasoning it properly. How does it work?

Well, if you look at the surface of a cast-iron pan under a microscope, you'll see all kinds of tiny pores, cracks, and irregularities. When you cook food in the pan, it can seep into these cracks, causing it to stick. Not only that, but proteins can actually form chemical bonds with the metal as they come into contact with it. Ever have a piece of fish tear in half as you try to turn it because it seems like it's actually bonded with the pan? That's because *it has.*

To prevent either of these things from happening, you need to fill in the little pores, as well as create a protective layer in the bottom of the pan to prevent proteins from coming into contact with it. Enter fat.

When fat is heated in the presence of metal and oxygen, it polymerizes. Or, to put it more simply, it forms a solid, plastic-like substance that coats the pan. The more times oil is heated in a pan, the thicker this coating gets, and the better the nonstick properties of the pan.

Here's how to build up the initial layer of seasoning in your pan:

- Scrub the pan by pouring ½ cup kosher salt into it and rubbing it with a paper towel. This will scour out any dust and impurities that have collected in it. Then wash it thoroughly with hot, soapy water and dry thoroughly. If your oven has a self-cleaning cycle, one trip through with the pan left inside will demolish even the toughest cooked-on crud and give you a bare pan to start with.
- Oil your pan by rubbing every surface—including the handle and the bottom—with a paper towel soaked in a highly unsaturated fat like corn, vegetable, or canola oil. Unsaturated fats are more reactive than saturated fats (like shortening, lard, or other animal-based fats), and thus polymerize better. It's an old myth that bacon fat or lard makes the best seasoning agent, probably borne of the fact that those fats were very cheap back in cast-iron's heyday.
- Heat your pan in a 450°F oven for 30 minutes (it will smoke), or until its surface is distinctly blacker than when you started. An oven will heat the pan more evenly than a burner will, leading to a better initial layer of seasoning.
- Repeat the oiling and heating steps three to four times, until the pan is nearly pitch-black. Pull it out of the oven and place it on the stovetop to cool. Your pan is now seasoned and ready to go.

Until you've got a good layer of seasoning built up, avoid using too much soap or cooking acidic sauces, as both can make the seasoning process take longer.

## Maintenance

Many people are irrationally afraid of caring for cast iron. The truth is, once you've got a good layer of seasoning, cast iron is pretty tough. You can't scratch it with metal utensils. You can't destroy it by using soap (modern dish soaps are very gentle on everything except for grease). To maintain the seasoning and build on it, just remember a few key points:

- **Use the pan often.** A good layer of polymers should build up slowly in a succession of very thin layers. This means using your pan as much as possible—particularly for oil-based cooking such as frying or searing. Avoid making liquid-based dishes in the pan until it has acquired a reasonably good layer of seasoning.
- **Clean the pan immediately after use.** Removing food debris is much easier with a hot pan than one that has cooled. If you clean your cast-iron skillet while it is still hot, chances are all you'll need is a tiny bit of soap and a sponge.
- **Avoid tough abrasives.** These include metal scouring pads and cleaners like Comet or Bar Keepers Friend. The scrubby side of a sponge should be plenty for most tasks. I'm particularly wary about this at dinner parties, when a well-intentioned guest may decide to chip in after the meal and get a little too generous with the elbow grease, scrubbing out some of my seasoning.
- **Dry the pan thoroughly and oil it before storing.** After rinsing the pan, set it on a burner and heat it until it dries and just starts to smoke, then rub the entire inside surface with a paper towel dipped lightly in oil. Take it off the heat and let it cool to room temperature. The oil will form a protective barrier, preventing it from coming into contact with moisture until its next use.

A good rub-down with oil prevents rusting.

## Worst-Case Scenarios

There are basically only two really bad things that can happen to your cast-iron cookware, scaling and rust—and neither of them is that bad.

Scaling happens when you heat the pan too often without adding extra oil to it. Rather than coming off in microscopic bits, as the seasoning normally will, the layer of polymers sloughs off in large flakes. For the pan to reach this state, I stored it in the oven for a month's worth of heating cycles, without ever oiling the surface. It's easy to avoid the problem by oiling the pan after each use and not overheating it (if you're storing it in the oven, don't leave it there during the cleaning cycle, for instance), but once it happens, there's no turning back—you'll have to reseason it from the start.

Rust can appear on a cast-iron pan that is not seasoned well enough and is left to air-dry. Unless the entire pan has rusted (in which case, you'll have to reseason the whole thing), a rust spot is not much to worry about. Rinse out the pan, heat it until it dries and smokes, and rub it with oil. After

a few uses, the rusted spot should be perfectly seasoned again.

## Which Pan Should I Buy?

If you're lucky enough to come across a reasonably priced cast-iron pan (under $50 or so) from the early twentieth century at a yard sale or flea market, scoop it up immediately. You can also occasionally find good deals on eBay and sites like it.

I personally find it ridiculous to pay the $150-plus that some sellers are asking for old cast iron when a new cast-iron pan, like the 10¼-inch Seasoned Cast-Iron Skillet from Lodge costs a mere $16.98 and will give you an equally lustrous nonstick surface with just a bit of time and care.

## THE CORE: THE EIGHT POTS AND PANS EVERY KITCHEN NEEDS

I'm a hoarder by nature. I get pleasure out of acquiring as vast an array of pots and pans as possible, always telling myself that I'm going to use them regularly, that they really aren't a waste of money. But, in all honesty, the only real use I get out of ninety percent of my pans is a purely aesthetic one. They're like a necktie for my pot rack—and I *never* wear a tie.

The majority of the time, I find myself reaching for the same eight pans. I can't think of a single dish that can't be made using one of these, or a combination, and they're all you'll need to cook the recipes in this book. Here they are, the cornerstones of any well-stocked kitchen.

## 1. A 12-Inch Tri-Ply (Laminated) Straight-Sided Lidded Sauté Pan

A large skillet is the true workhorse of the kitchen. It's perfect for rapidly browning large quantities of vegetables or meat. Pan-roasting a whole chicken?

This is the pan of choice. Need to brown a pork tenderloin or a 3-rib beef roast? No problem. It's also excellent for braising and for reducing sauces. It has a tight-fitting lid and is oven-safe, which means you can brown your short ribs, add the liquid, cover, and braise in the oven, then reduce the sauce on the stovetop and serve all out the same pan.

Why is tri-ply construction important? Stainless steel is heavy and can retain a lot of heat, but it's slow conductor. Aluminum is lightweight (and retains less heat per unit volume), but transfers heat really fast. Combine the two in a single pan by sandwiching the aluminum in the center, and you've got a skillet that can retain heat for maximum browning and will distribute that heat evenly over its entire surface, eliminating hot and cold spots.

All-Clad is the benchmark for great tri-ply cookware, but it can be prohibitively expensive. In side-by-side tests, I've found that Tramontina-brand All-Clad knockoffs perform almost as well for every task, at about a third of the price. The choice is a no-brainer.

## 2. A 10-Inch Cast-Iron Skillet

Nothing beats cast iron for searing a steak or a nice skin-on, bone-in chicken breast. I actually keep a collection of cast-iron skillets in all sizes so that I can do everything from frying a single egg and serving it directly from its tiny skillet to baking pies, but the one I use most is my 10-inch pan. It's just the right size to sear a couple of steaks for me and the wife (I'll sear in batches or use two pans and two burners if I've got more people to cook for, to maximize heat transfer to the steaks), it's just the right size for corn bread, it's a beautiful serving vessel. The possibilities are really limitless.

If you don't have a well-seasoned cast-iron pan passed down by thoughtful grandparents, Lodge

brand is the easiest to find. If you're hunting at antique stores and flea markets, Griswold and Wagner are the best.

## 3. A 10-Inch Anodized Aluminum or Tri-Ply Nonstick Skillet

People will tell you that a well-seasoned cast-iron skillet that's properly seasoned will be as slick as a true nonstick skillet. Heck, I've probably said the same thing myself. Maybe even right here in this book. Well, here's the sad news: that ain't really true. Even the very best cast-iron skillets will never be as slick as a nonstick skillet. Any materials-science engineer can tell you that. Not only that, but unlike a cast-iron skillet, a nonstick pan can be light enough to maneuver easily when, say, rolling an omelet or flipping a couple of sunny-side-ups.

And that's why a medium nonstick skillet is a *must* in your arsenal. It's the best vessel for all kinds of egg cookery, from perfect golden omelets to fluffy scrambles to crisp-edged fried eggs. Brunches would be a much messier, more hectic, and altogether less pleasurable affair in my apartment without one.

The only downside to nonstick? You can't heat it past 500°F or so, as the coating will begin to vaporize, sending toxic fumes into the room. Newer materials are far safer, but even with them, you're at a disadvantage: it's tough to form a good, meaty crust on food cooked on a nonstick surface, and you're limited by the types of utensils you can use. Metal will scratch off the coating. Stick with wood, nylon, or silicone utensils made specifically for working with nonstick pans.

Here's the thing with nonstick: unlike other pans, these aren't going to last you your whole life, which means that spending a boatload of money on one is not a wise move. You want a midrange pan: something with enough heft that it retains heat fairly well, but not one that you'll be so scared of scratching that it ends up sitting in the corner of kitchen cabinet. I currently have a Cuisinart stainless steel nonstick skillet, but I'm not heavily committed to it. You should never become committed to a nonstick pan.

### 4. A 2½- to 3-Quart Saucier

The difference between a saucepan and a saucier is subtle but important. Saucepans have straight sides; sauciers are designed to keep their contents easily whiskable and stirrable, so they have gently sloped sides. This is a major advantage when cooking. It means that you don't have to try to shove a round spoon or whisk into a square corner.

I use a saucier for small batches of soup or stew, for cooking short pasta shapes (you don't need a big pot for this—see page 674), for reheating leftovers, for making cheese sauces or sausage gravy, for simmering tomato sauce or sweating a few vegetables, and even for one-chicken-sized batches of stock.

As with a nonstick skillet, any brand will do as long as it's thick, heavy, oven-safe, and, preferably, tri-ply. I use the Farberware Millennium Clad Stainless Steel Saucier. It has a great lip for pouring and a nice deep shape. I've been in a deep relationship with it for about eight years, with not a single complaint from either party. That's more than I can say about any other relationship I've been in.

### 5. A 12- to 14-Inch Carbon Steel Wok

You're forgiven for not owning a wok if you grew up with a Western kitchen. But I'm here to try and convince you that *everybody*, not just those who like to stir-fry, can benefit from a good large wok. There's no better vessel for deep-frying, steaming,

or smoking. For more info on buying and caring for a wok, see page 42.

### 6. A 6- to 8-Quart Enameled Cast-Iron Dutch Oven

My enameled Dutch oven is the first pot I owned that made me think to myself, *Wow, you've really got something special here.* It's a blue oval Le Creuset number, and it's still alive and kicking today, working at least as well as it did the day my mom bought it for me fifteen years ago. A good enameled Dutch oven will stick around for life. Because of its weight and heft, it's the ideal vessel for slow braises, in or out of the oven. See, all that heavy material takes a long time to heat up or cool down. This means that even if your oven is cycling on and off with its temperature making sine waves that stretch a good 25 degrees hotter and cooler than the number on the dial, the interior of your pot will show barely any fluctuations at all. This is a good thing for dependability and predictability in recipes.

Le Creuset sets the standard for quality when it comes to enameled cast iron, but it's also insanely pricey. If you buy one, you'll cherish it forever, and only partly because you've spent so much money on it (they're kinda like kids in that way). Lodge makes a perfectly serviceable version for about a third of the price, but buyer beware: I've seen a couple chip and crack in my day.

### 7. A 3- to 4-Gallon Stockpot

The big daddy of pots: this is the guy you pull out when you want to make pasta for twenty, when you've got a half dozen lobsters to boil, or when you've got several carcasses' worth of chicken bones sitting in the freezer just waiting to be turned into awesome stock. Until you own a big stockpot, you will never realize how much you needed one. The good news here is that when it comes to stockpots, the absolute cheapest will do. You'll never be doing anything in here aside from boiling or simmering vast amounts of liquid, so all you need is something that will hold water and stay level. You shouldn't have to spend more than $40 or so on one.

## 8. Something to Roast In

Decent roasting pans are expensive; there's no two ways about it. Just like with skillets, the best roasting pans are made with layered metals—stainless steel sandwiched with an aluminum core. When choosing a roasting pan, I look for one that I can use directly on a burner on the stovetop as well as in the oven, something with comfortable handles, and something that is thick enough that it won't warp under the heat of the oven or the weight of a turkey. My Calphalon pan is large and sturdy, and it has a nice U-shaped rack for holding large roasts. It's about $140, and I use it about twice a year, when I cook big roasts on holidays.

Want to know the honest truth? I could easily live without it. What I couldn't live without is a heavy-duty aluminum rimmed baking sheet with a wire cooling rack set on it. It's lighter and cheaper, stores right in the oven, and has the added advantage that it's shallow, making it much easier for hot air to circulate around the food that's cooking. It's what I use for roasting the other 363 days of the year. Mine has seen countless roast chickens, and it is warped and bent beyond belief, but it still does its job just as well as it ever did. I bought it for about $10 at a cooking supply store, along with a rack that cost another $5 or $6. (You can get these pans online as well—they're called half sheet pans. Nordic Ware makes a fine one for about $15.)

## HOW TO BUY AND CARE FOR A WOK

A good wok is one of the most versatile pans in the kitchen. There are those who argue that on a Western stove, with its flat, relatively low-output burners, a regular nonstick skillet is a superior vessel for stir-frying; they may have even showed you some fancy charts proving that a skillet gets to a higher temperature and maintains its heat better. This is utter and complete nonsense. All the charts in the world won't tell you as much as your mouth, and the fact is, stir-fries do taste better when made in a wok, because a good stir-fry is not simply about the temperature the metal reaches. It's about correct tossing and aerosolization of fats and juices as they leap up beyond the edges of the wok and are touched by the flame of the burner. It's about the ability to rapidly heat and cool a piece of food as you flip it over and over through the different heat zones created by the pan as (much as flipping a burger frequently will improve its cooking—see page 557). It's about wok *hei*, the slightly smoky, charred, metallic flavor that only comes from a seasoned cast-iron or carbon steel pan heated to ripping-hot temperatures.

I digress. Obviously, woks are the best choice for stir-frying, but they're also the ideal vessel for deep-frying, steaming, and indoor smoking. My wok is by far the most commonly used pan in my kitchen.

As with most things, however, not all woks are created equal. They come in a dizzying array of sizes, shapes, metals, and handle arrangements. Fortunately for us, the best woks also happen to be on the inexpensive end of the scale. Here are some things to consider when purchasing one.

### Materials

- Stainless steel woks are a waste of money. Not only are they extremely heavy and difficult to maneuver, they also take a long time to heat up and cool down—a fatal flaw for something that requires rapid, on-the-fly heat adjustments like a stir-fry. And food—particularly protein—has a tendency to stick to steel.

- Cast iron is a better choice, though it still takes a relatively long time to heat up and cool down. It offers a better nonstick surface. The main problem with cast iron is that if it's thin, it is extremely fragile—I've seen cast-iron woks crack in half when set down too hard. And when made thick enough to be durable, they are extremely cumbersome to lift, which is essential for proper flipping during a stir-fry.
- Carbon steel is your best bet. It heats quickly and evenly, is highly responsive to burner input, is both durable and inexpensive, and, if properly cared for, will end up with a practically nonstick surface. Look for carbon steel woks that are at least 14-gauge (about 2 mm thick). They should not give when you press on the sides.

## Manufacture

Woks are made in three ways:

- **Traditional hand-hammered woks** (like the ones they used to sell in those infomercials in the 1980s) are an excellent choice. The slight indentations left by the hammering pattern allow you to push cooked food to the sides of the pan while adding ingredients to the center without them slipping back. And hand-hammered woks are inexpensive. The only problem is that it can be difficult (impossible?) to find one with a flat bottom and a handle (more on that later).
- **Stamped woks** are made by cutting out a circular piece of thin carbon steel and pressing it by

machine into a mold. They are extremely cheap, but they are completely smooth, making it difficult to stir-fry properly. And they are, without fail, made from low-gauge steel and prone to developing hot and cold spots, as well as seeming flimsy.

- **Spun woks** are produced on a lathe, giving them a distinct pattern of concentric circles. This pattern offers the same advantages as a hand-hammered wok, allowing you to easily keep food in place against the sides of the pan. Spun woks can be found in heavy gauges, with flat bottoms, and with flip-friendly handles. They are inexpensive.

## Shape and Handles

Traditional woks have a deep bowl shape, designed to fit into a circular opening directly over the hearth. Unless you have a custom wok insert in your range (and if you do, you probably aren't reading this), you want to avoid round-bottomed woks. They won't work, period, on an electric range and are tough to use on a gas range even with one of those wok rings. On the other hand, woks with bottoms that are too flat defeat the purpose of the pan, making it tough to flip food properly and to move it in and out of the high-heat zone.

Your best bet is a wok with a 4- to 5-inch-wide flattened area at the bottom and gently sloping sides that flare out to between 12 and 14 inches. This will give you plenty of high-heat space for searing meats and vegetables at the bottom, with ample volume and room to maneuver when flipping. As for handles, you have two choices: Cantonese-style woks have two small handles on either side, while northern-style woks have one long handle and usually a smaller helper-handle on the opposite side.

This is the type of wok you want. The long handle facilitates flipping and stir-frying, while the short handle makes it easy to lift.

Finally, avoid nonstick woks like the plague. Most nonstick coatings cannot handle the high heat necessary for a proper stir-fry. They start vaporizing, releasing noxious fumes, long before they reach the requisite temperature. They make browning difficult, and it's impossible to get food to stick in place against the sides of the wok when you want to clear a surface to cook in the middle.

### Care and Maintenance

Just like a good cast-iron pan, a carbon steel wok's performance will improve the more you use it. Most come with a protective film of oil to prevent them from rusting or tarnishing in the store. It's important to remove this layer before using it the first time. Scrub the wok out with hot soapy water, dry it carefully, and place it over a burner at the highest heat possible until it starts to smoke. Carefully rotate the pan so that every area of it—including the edges—is exposed to this super-high heat. Then rub it down with oil, using a paper towel held in a pair of tongs, and you're ready to go. After use, avoid scrubbing the wok unless absolutely necessary. Usually a rinse and a rubdown with a soft sponge is all that's necessary. Purists may tell you not to use soap. But I do, and my wok is still well seasoned and completely nonstick. After rinsing it, dry the wok with a kitchen towel or paper towels and rub some vegetable oil into the surface to give it a vaporproof coating that will prevent it from rusting.

With repeated use, the oil you heat in your wok breaks down into polymers that fill the microscopic pores in the metal's surface, rendering the material completely nonstick. As you break in your wok,

the material will gradually change from silver to brownish and, finally, to a deep black. This is what you are looking for.

With proper care, your wok will not only last a lifetime but also actually improve with age.

### BASIC WOK SKILLS

Stir-frying is the quintessential wok technique; however, we're not really gonna spend any time on that here, as there isn't a single stir-fry recipe in this book (maybe you can write to my publisher and convince them you'd like to see a *Food Lab: Chinese Classics* some time in the future). But it's also the best tool for deep-frying, steaming, and smoking food indoors. Here's how to do it all:

- **Deep-frying** in a wok is vastly superior to doing it in a Dutch oven. The wide sides means that there's less mess—any splattering oil hits the sides and falls back down to the center. The shape also makes it much easier to maneuver food, leading to crisper, more evenly cooked results. Boil-overs become a thing of the past, once again due to the wide, sloping shape, which allows for plenty of bubble expansion before the oil threatens to spill over the edges. Finally, it's much easier to filter out bits of debris and detritus from the slanted sides of a wok than from the sharp corners of a Dutch oven.

- **Steaming** in a wok is also much easier than in another vessel. You can use a standard steamer insert for a large pot. Simply rest it directly on the bottom of the wok over simmering water, and use the dome-shaped lid to cover the pan. The advantage, of course, is that in a wide wok, you have far more surface area for steaming. This advantage can be stretched even further if you get yourself a couple of bamboo steamers.

Bamboo steamers are designed to fit directly into a wok and are stackable, meaning that you can have two or three tiers of food all steaming in the same wok at the same time. Try doing *that* in a Dutch oven!

- **Smoking** is also easy in a wok. All you've got to do is line the bottom with a piece of foil that extends over the edges by at least three-quarters the total width of the wok, then place your smoking medium (wood chips, tea leaves, sugar, rice, spices, whatever) directly on the bottom and set your food on a rack or a steamer on top of it. Place the wok over high heat until the smoking material on the bottom begins to smolder, then fold over the edges of the aluminum foil and crimp them to make a pouch, trapping the smoke inside.

## KNIVES, SCISSORS, CUTTING BOARDS, AND OTHER ESSENTIAL CUTTING TOOLS

If married life has taught me anything, it's that you're never always right, even when you are. Case in point: choosing the best kitchen equipment. When I first started dating my wife, the only knife she owned was a tiny plastic-handled, unbalanced, dull knife from IKEA that looked like it'd be more at home sitting next to an Easy Bake oven. Indeed, I spent a good chunk of 2007 trying to surreptitiously coax her into switching to the incredibly sexy, hand-hammered Japanese Damascus steel santoku knife that I'd bought specifically to impress any future wives with my good taste.

She ended up choosing the IKEA knife every time, claiming that the large size and precisely hand-engraved maker's signature on the hilt of the santoku blade intimidated her (don't worry, she was still suitably impressed by my raw masculine energy

whenever I wielded it). I've since gotten her to upgrade to a fairly nonintimidating Wüsthof 5-inch granton-edged santoku, but the point remains the same: once you narrow your choices to those within a certain quality level, the best knife for you is the one that you are most comfortable using. Anyone who tells you different is selling something. Probably knives.

When buying a knife, there are three main characteristics to consider: material, shape, and ergonomics.

The material a knife is made out of determines several factors, including how sharp it can be, how long it retains its edge, how easy it is to resharpen once it's dull, and how it reacts with acidic foods. In general, you've got three options: carbon steel, ceramic, or stainless steel.

- **Carbon steel** is a softer metal that is easy to sharpen, and it can be ground down into an extraordinarily sharp edge. Its disadvantages are that it dulls relatively quickly, requiring you to resharpen it every few weeks or so to maintain a good cutting edge; it can rust if not cared for properly; and it will discolor if it comes into prolonged contact with acidic fruits or vegetables. You have to carefully clean, dry, and oil it after each use to preserve its luster. Carbon steel is the material of choice for knife geeks who take great pleasure in the process of sharpening a blade down to a cut-through-anything-without-even-noticing edge. Just like a dog, it requires plenty of hard work to keep it well disciplined and healthy, but it'll reward you with a lifetime of loyal, faithful service. And unlike a dog, your knife will never pee on the carpet. That's a good thing.
- **Ceramic** blades are generally a poor choice. It's true that they can be ground to a razor-sharp edge, and

that that edge doesn't dull even with prolonged use, but their major drawback is a doozy: they chip easily and are unrepairable. A metal knife is flexible, which means that on a microscopic level, the sharp edge of its blade is constantly bending and deforming according to the varying pressures being applied along its length. Because of its crystalline structure, a ceramic blade, on the other hand, is extraordinarily brittle. Even the slightest shearing motion with the blade can cause it to chip or crack along the edge, relegating it to the "completely useless but I'll keep it anyway because I'm still holding out hope" drawer. These knives are also very light, which, for some people (like me), can be a deterrent.

- **Stainless steel** used to be the material for suckers: hard, pretty, and easy to maintain but completely unable to form a suitably sharp edge. These days, as materials science continues to advance, stainless steel knives are becoming more and more attractive, since they combine the easy-sharpening characteristics of carbon steel with the easy cleanup and rust- and tarnish-free nature of stainless steel. I still love my carbon steel knives, but to be honest, I've got more stainless knives in the kitchen now.

## Knives and Cutting Boards

When you are purchasing new knives, the price range and variance in quality level can be truly staggering. I mean, you can hit the local megastore and find a twenty-four-piece set priced to give you knives at a couple bucks a shot, or you can spend hundreds or even thousands of dollars on a single knife. What gives?

**Here's the truth**: Once you get to a certain level of quality, knives are largely a matter of personal taste. Do you need to spend $300 to get a decent knife? Absolutely not. Are you likely to find a good knife under $35 or so? Probably not. But no matter what knife you choose, these are the qualities to keep an eye out for:

- **A full tang.** The tang is the extension of the blade into the handle. In a good knife, the tang should extend all the way to the end of the handle. This provides maximum durability and balance.
- **A forged, not stamped, blade.** Forged blades are made by pouring metal into a mold, pounding it, trimming it, sharpening it, and polishing it by hand. This creates a very strong, very versatile blade from edge to heel. A stamped blade is cut out of a single sheet of metal and sharpened on one edge. Stamped blades usually bear a telltale sign of parallel stripes (caused by the rollers used to flatten the metal) when you reflect light off it into your eyes. Stamped blades are generally unbalanced and flimsy. The lower-end knives of most major manufacturers are stamped.
- **A balanced handle and a comfortable grip.** When you hold a knife, it should feel balanced in your hand, neither heavy nor light on the blade end. It should also fit effortlessly in your grip. Remember—a knife should be an extension of your hand. As such, it should feel completely natural.

## THE CORE: THE 6½ KNIVES THAT EVERY KITCHEN NEEDS

Collecting knives is fun, but if I had to pick, there are 6½ knives (I figure that between them, a peeler and a steel account for 1½ knives) that I'd never want to be without. Here they are.

# THE ANATOMY OF A KNIFE

Aknife consists of two main parts: the blade and the handle. In a well-made knife, the metal that the blade is made of will extend all the way through the handle. Different parts of this single piece of metal serve different purposes. Here are the main features of most knives:

- **The Cutting Edge** is the sharpened, honed edge of the blade. It should be razor sharp—a well-sharpened knife will literally be able to take the hairs off your arm (don't try it). Chef's knife blades come in varying degrees of curvature, designed for various tasks, such as slicing or rock-chopping.
- **The Back, or Spine**, is the long side opposite the sharp blade. This is where you hold your non–knife hand when rocking the knife back and forth for rapid mincing. It can also be used as a makeshift bench scraper for moving pieces of food around on your cutting board (you should never do this with the cutting edge—it'll dull it).
- **The Tip** is the sharp point at the end of the blade. It's used primarily for precision work.
- **The Heel** is at the bottom of the blade. In many Western-style knives, the metal thickens significantly at the heel. This is to make it easier to grip the knife using the blade grip (see page 49).
- **The Bolster** is the part of the blade that meets the handle. It is thick and heavy, providing a good balancing point for the blade and the handle. In a well-balanced knife, the center of mass should be somewhere near the bolster, so that you can rock the knife back and forth with minimal effort.

*continues*

- **The Tang** is the extension of the blade that runs through the handle. It provides balance as well as sturdiness. A knife with a full tang (that is, metal that extends to the butt of the handle) is unlikely to ever lose its handle.
- **The Handle** is where your whole hand rests if using the handle grip, or where your three smaller fingers rest if using the blade grip (which I recommend). Handles can be made of wood, polycarbonate, metal, or various exotic materials. I like the feel and grip of a real wood handle, but there is no right or wrong here.
- **The Butt** is the fattened section at the very bottom of the handle.

## THE TWO GRIPS

The first step to perfect knife skills is learning how to hold a knife. There are two basic grips: the handle grip and the blade grip.

- **The Handle Grip:** With the handle grip, your hand is completely behind the heel of the knife, with all your fingers tucked behind the bolster. It is generally used by beginning cooks or cooks with exceptionally small hands. It's comfortable, but it offers only limited control when doing precision knife work.

- **The Blade Grip**: The blade grip is the preferred grip for more experienced cooks. Your thumb and forefinger should rest in front of the bolster, directly on the blade. This grip is a little intimidating, but it offers much better control and balance. It may be difficult and/or uncomfortable with cheaper stamped knives that don't have a bolster.

When I first started cooking, I used the handle grip. I mean, it just made sense to me. It's called the handle for a reason, right? But when I moved into professional kitchens and was immediately mocked for my amateur grip (professional kitchens are relentlessly macho), I made the switch and saw an immediate and dramatic improvement in my knife skills. I'm not one to judge someone based on how they hold their knife (or if I did judge, I'd do it silently), but here's the deal: If you've only ever used the handle grip, give the blade grip a try—you may find your cuts improving dramatically. In return, I promise to judge you only slightly if you go back to using the handle grip.

What about your non-knife hand? In general, there are two positions you'll find that hand in. The most common is known as "the claw," and when people cut themselves with a knife, it's mostly likely because they weren't using the claw. Use this grip when dicing and slicing. Protect the fingertips of your non-knife hand by curling them inward, using your knuckles to guide your knife. When cutting food, always place it

*continues*

in a stable position, preferably with a cut surface flat against the cutting board. Then guide the knife blade against the food with your claw hand.

For mincing, a different approach is required. Place the tip of your knife on the cutting board and hold it in place with your free hand. Rock the blade up and down to reduce herbs (or anything else) to a fine mince.

## EAST VERSUS WEST:
## Which Knife Style Is Superior?

The difference between Japanese- and Western-style knives used to be night and day. Western knives had gently sloped, curved blades that came to a point, with a relatively thick spine compared to their length, and could rock on a cutting board. Japanese-style knives had flat blades made for slicing and chopping, not rocking, with a slender profile and a relatively light weight.

These days, the divide is not so clear. Western knife makers now offer santoku-style knives, referring to

the staple knife of the Japanese home cook. Even knives with a Western shape have been slimmed down and made lighter in response to a growing market of folks used to the easier-to-maneuver Japanese-style blades. Japanese knife makers, on the other hand, have begun applying their skills to *gyutou*, Western-style knives produced with Japanese forging techniques to make what can often be the best of both worlds.

So which style is the best? There is no right answer to that. I first learned to cook with the Western-style knives that everyone was using at the time, so my early knife collection mostly consisted of heavy German knives like Wüsthofs and Henckels. But as I began experimenting with Japanese-style knives, I found that I much preferred the precision they offered, and that their inability to perform rocking tasks like mincing was worth the trade-off for me. These days, I use a mix of Western- and Japanese-style knives.

A Western-style chef's knife has a curved, tapered blade.

A Japanese-style santoku knife has a straighter cutting edge and a blockier tip.

The key difference in how the two types of knives are used is that with a Western-style knife, rocking, planting the tip of the knife on the cutting board and lifting only the heel end as you feed food underneath, is a very common motion. With a Japanese-style knife, this is impossible—the shape of the knife doesn't allow for rocking. Slicing and chopping are the more common movements, and mincing herbs becomes a matter of repeated slicing rather than rocking.

The only way to tell which knives you prefer is to go into a store and try them out.

## 1. An 8- or 10-Inch Chef's Knife or a 6- to 8-Inch Santoku Knife

This is my knife. There are many others like it, but this one is mine.

Your chef's knife should be an extension of your hand and so should feel completely natural. When I'm feeling down and I need a bit of physical support, I don't ask my wife to hold my hand. I don't rub my dog on his belly. Nope, I go to my knife and just hold it. We've spent a lot of time together. I know her every curve (I just now realized that my knife is female) and exactly how she fits into my hand and likes to be held, and in return, she is supportive, loyal, and wicked sharp.

The chef's knife is the one you're going to be using for 95 percent of your cutting tasks, so you'd better make damned sure that you're comfortable with it, and here's the key: forget every review you've ever read. Once you get past a certain quality level, no single knife is better than another. That said, there are certain characteristics you can look for, depending on your cooking style, size, and comfort level in the kitchen. Here are a few of my basic recommendations, but let me repeat: only you can decide which knife is best for you. Go to a store, try some out, and mull it over for a day or two. You and your chef's knife are going to have a long, beautiful, and mutually beneficial relationship. Choose wisely.

### Western-Style Chef's Knives

- **For the average cook: The 8- or 10-inch Wüsthof Classic Cook's Knife (about $140).** This was the first decent knife I owned, and I still have it to this day.
  - **Pros:** It's got a thick spine with plenty of heft, which helps it do a lot of the slicing work for you. It has a curved blade that allows you to rock back and forth for rapid mincing. And there's plenty of space under the handle for your knuckles when chopping.
  - **Cons:** Some cooks may find it too heavy, and small-handed cooks may find the handle uncomfortably large.
- **For the small-handed cook: The Global G-2 8-inch Chef's Knife (about $120).** Stylish and functional.
  - **Pros:** It's forged from a single piece of metal, meaning that it's basically indestructible. It has an extremely sharp, precise blade and a well-balanced handle (it's filled with sand) to help it stay balanced even while in motion.
  - **Cons:** There's no bolster or heel, so using the blade grip for a long time on this one may irritate your forefinger where it rubs against the spine. And there's not too much space under the handle when the blade is against your cutting board, so you might end up rapping your knuckles a few times. The all-metal handle can get slippery if it gets messy (though nobody should be cooking in a messy kitchen anyway!). Ideal for vegetarians who want precise veg work and don't deal with messy meats.
- **The best buy option: The 8-inch Victorinox Fibrox Chef's Knife (about $25).** This is a favorite among beginning cooks who aren't yet sure they want to lay down over $100 for a chef's knife.
  - **Pros:** It's very sharp right out of the box, and it's very light, which some users may like. Grippable handle, and plenty of knuckle space.

- **Cons:** It has a stamped blade, with no real weight or heft, and it's difficult to resharpen. Cheap feel and construction—this isn't a knife designed to last a lifetime. Poor balance can also instill bad habits.

### Japanese-Style Chef's Knives

- **For the average cook: The 7-inch Misono UX10 Santoku (about $180).** This is my personal favorite. It's not the first knife I ever felt attachment to, but it's the first one I ever fell in love with. Would that we never be apart.
  - **Pros:** It's perfectly balanced, with a very comfortable bolster that makes the blade grip a dream. The blade is Swedish steel, which is extremely sharpenable and will hold an edge for a long, long time. Although it is designed for slicing and chopping, the blade has a strong-enough curve that you can even do some Western-style rocking with it, giving you the best of both worlds. Strong, sturdy construction, and plenty of heft—a real beauty to behold.
  - **Cons:** Just one: price. It's not a cheap knife, but considering that it will last you a lifetime, $180 seems fair.
- **For the small-handed cook: The 7-inch Wüsthof Classic Hollow Ground Santoku (about $100).** I used this knife extensively in restaurants, where precision vegetable slicing was required—so much so that it lost a good centimeter of its width with repeated sharpenings. I grew quite fond of it in the process.
  - **Pros:** Like all top-of-the-line Wüsthof products, it's impeccably constructed. It has a much more slender blade than Western-style Wüsthofs, so it's easier to make small, precise cuts and more comfortable for some cooks. The hollow-ground granton edge (with dimples along both sides of the blade) means that foods like potato slices won't stick to it.
  - **Cons:** It's not big enough for most really heavy-duty tasks—say, splitting a butternut squash or hacking through a chicken. Luckily, your cleaver will take care of that (see page 55).
- **The best buy option: The MAC Superior 6½-inch Santoku (about $75).** A favorite among pros and home cooks alike.
  - **Pros:** A very sharp blade, comfortable handle, and easy maneuverability.
  - **Cons:** The blade is tough to sharpen, and at 6½ inches, it's too small for many kitchen tasks. It's got neither the heft of the Misono nor the granton edge and solid feel of the Wüsthof, but it's a great knife by most standards.

## 2. A 3- to 4-Inch Sheep's Foot Paring Knife

For many years, I used a classic curved 3-inch paring knife from Wüsthof, and at first glance, the shape of the classic paring knife seems to make sense. A big curved chef's knife is for cutting, hacking, and chopping large things, so to cut, hack, and chop small things, you'd want to use a small version of a chef's knife, right? Thing is, there's a fundamental difference between how you use a paring knife and how you use a chef's knife—so why would you want them both to be the same shape? The real problem with the common paring knife is the curvature of the blade. With a chef's knife, this curve is designed to allow you to rock the knife for mincing.

But for a paring knife, it makes no sense: nobody is rocking a paring knife.

The key to a good paring knife is precision, and that means having a superthin blade and the ability to make cuts with minimal hand motion (the more you have to move your hand, the more uneven the cut becomes). A flat sheep's foot–shaped knife is ideal for this task. With a sheep's foot knife, it's possible to make contact with the cutting board with nearly the entire length of the blade while the tip is firmly inserted into the food: the straightness of the cut is defined by the straightness of the blade. Quicker, more precise, and less chance for user error are all pluses in my book.

The same reasoning applies even more strongly if you are using the knife to peel small things, like little potatoes or grapes. When using a curved paring knife, the curve of the blade and the curve of the object you are peeling run in opposite directions, so almost none of the food actually comes in contact with the blade, requiring you to dig deeper and remove more flesh than is necessary. Those of you who are used to using santoku knives in place of chef's knives will immediately recognize these advantages.

The 3-inch Kudamono Hollow-Edge Paring Knife from Henckels ($50) is one of the cheapest decent knives of this kind you can get, with the added advantage of having a hollow-ground granton edge. You can get the 3-inch Sheep's Foot Paring Knife from Wüsthof for the same price. It lacks the granton edge, but it is slightly heavier, sturdier, and feels better in the hand. If you want what I consider to be the ultimate paring knife, tack on another $5 to get yourself the same Wüsthof but with a granton edge. That's the one my knife kit packs.

### 3. A 10- to 12-Inch Serrated Bread Knife

I'm far less picky about bread knives than I am about chef's knives. For one thing, I don't use them often. For cutting soft breads like burger buns or sandwich bread, my chef's knife is more gentle than a bread knife. In fact, just about the only thing I use my bread knife for is cutting crusty bread, like a baguette or a rustic Italian loaf. If you never eat these, you have no need for a bread knife. That's why I don't feel the need to make sure my bread knife fits my hand like a glove. And since serrated blades are difficult, if not impossible, to sharpen at home, a bread knife won't last you as long as your chef's knife will.

You'll find bread knives with pointed teeth, scalloped teeth, and microserrations. I find that the best knives have wide sharp teeth, a forged (not stamped) blade for better sharpness and weight, and a good length. My first bread knife was the Zwilling J. A. Henckels Twin Pro S 8-inch Bread Knife (about $85), and it served me well for about a decade. My current bread knife is the F. Dick Forged 8-inch Bread Knife (about $65). It works just as well as the Henckels. If you're on a tighter budget, you could do worse than the Victorinox Fibrox bread knife (around $25).

### 4. A 6-Inch Boning Knife

Sure, you don't *think* you're going to be doing a lot of boning in your kitchen. . . . Wait, that came out wrong. Let's start over: you may not be removing the bones from many chickens or pig's legs right now, but I hope I'll be able to convince you that those are both goods skills to have under your belt. It not only saves you money (lots of it), but it also increases the deliciousness you are able to produce in your kitchen (we'll get to why later on).

A boning knife should be thin and moderately flexible, with a very sharp tip. The idea is that you want to be able to get that knife in between all the meat and the bones, working your way in, out, and around structures that aren't necessarily straight. A thin, flexible blade aids in this process. A good boning knife should also be made with a foot—an extra bit of metal jutting out of the heel—which you can use to scrape meat and connective tissue off the bones to clean them. I've yet to find a more capable boning knife than the Wüsthof Classic 6-inch Flexible Boning Knife (about $85).

## 5. A Good Heavy Cleaver

First things first: avoid expensive Japanese or German cleavers, period. If they sell it at Williams-Sonoma, you don't want it. A cleaver is meant to be for only the toughest of the tough jobs, and it will get beat up. It doesn't require the razor-sharp edge-maintaining abilities of expensive German or Japanese steel, so there's no sense in paying a higher price for one when cheaper models are just as serviceable.

My favorite is a heavy-duty 2-pound, full-tang, 8-inch-bladed behemoth of a cleaver that I got for $15 at a restaurant supply store in Boston's Chinatown. I use it nearly daily for taking apart chickens, hacking through animal bones, mincing beef or pork for hand-chopped burgers or dumplings, cleaving hearty vegetables, and trying to look really badass in the mirror (it's not so good at that particular function). If you live near a restaurant supply store, check it out for similar deals. As with all knives, you're looking for solid construction and a full tang. A cleaver should be plenty heavy as well.

Alternatively, you can get the more mass-market 7-inch wood-handled cleaver from Dexter-Russell (about $40). It's a tad more expensive—you're paying for the label—but it does exactly what it's supposed to do: hack the shit out of things.

## 6. A Y-Shaped Vegetable Peeler

A regular vegetable peeler has a blade aligned with the handle, requiring you to hold both vegetable and peeler at an awkward angle, limiting your precision. With a Y-peeler, you hold the peeler as if you're picking up an iPod, giving you far greater accuracy. The result is prettier vegetables, faster prep (once you get used to using it), and less waste. The Kuhn Rikon Original Swiss Peeler ($10.95 for 3) comes in assorted colors, has a built-in potato-eye remover, and is cheap, sturdy, and very sharp. I bought a set of a half a dozen in 2002 and still have four of them in perfect working order. (For the record, the other two were lost, not broken or worn out.)

## 7. A 10-Inch Honing Steel

Honing steels (sometimes incorrectly referred to as sharpening steels) are the long, heavy, textured metal rods that butchers and serial killers run their knives over before going at their meat.

Many people confuse honing with sharpening, but there is a distinct difference. When you sharpen a knife, you're actively removing material from the blade, creating a brand new razor-sharp beveled edge. When you hone a knife, all you're doing is making sure that edge is straight. The thing about metal is, it's malleable. That means that with regular kitchen use, that thin sharpened edge can get microscopic dents in it that throw the blade out of alignment. Even if the blade is still sharp, it can feel dull because the sharp edge has been pushed off to the side. That's where a honing steel comes in.

When used properly, a steel will realign the edge of the blade so that the sharpened bit is all facing in the right direction. You should steel your knife with every cooking session to ensure that you're getting the best edge possible.

When purchasing a steel, look for a heavy model at least 10 inches long. I use the Wüsthof 10-inch steel, which costs about $20. Just like a good knife, a high-quality steel will last a lifetime. The ridges may wear out over time, but don't worry—it's still doing its job.

Diamond steels are gaining more popularity these days. These are honing steels that have fine diamond powder embedded in them. This allows them to shave off a microscopic amount of edge material every time you run your knife across one of them. In this sense, they truly are sharpening steels. The advantage of using them is that you'll be able to slightly increase the time between true stone sharpenings. High-quality models tend to run a little more than twice as much as regular honing steels.

## CUTTING BOARDS

A good cutting board is as important as good knives. The ideal cutting board is large enough to give you ample space to work on (at least 1 foot by 2 feet, preferably much larger); heavy enough that it doesn't slip, slide, or break under the pressure of a heavy slam from a cleaver; and made of a material that is soft enough that it won't dull your blade.

Of the types of boards on the market, plastic (polyethylene) and wood are the only ones you should consider. A glass cutting board is like death to your blade: slow, painful, agonizing death as, stroke after stroke, the perfect edge that you worked

so hard to achieve is relentlessly worn away. A few years back, if you'd asked a health expert which type to use, they would have said plastic, not wood. Plastic is inert and inhospitable to bacteria, they'd say, whereas wood can house dangerous bacteria and transfer them to your food.

Turns out those health experts were wrong. A number of recent airtight studies have shown that wood is actually *less* likely to be a means of transferring bacteria, due to its natural antimicrobial properties. A wooden cutting board can be a death trap for bacteria. So long as you give it a scrub and a thorough drying after each use (which, of course, you should do with plastic boards as well), it's a perfectly safe material.

As for its actual function as a cutting surface, wood also takes home the gold, with some modern plastic boards coming in a close second. Wood is very soft, meaning that your knife can make great contact with every stroke, but it also has some self-healing properties—stroke marks will close up and fade away (though with repeated use, your board will become thinner and thinner).

I'm lucky enough to have a few large, heavy, butcher-block–style boards, which I received as a gift from an old chef of mine, that exactly fit my prep area. The best commercial models I've seen are the ones made by Ironwood Gourmet. They have a 20-by-14-inch version for about $50 that'll last you at least half a lifetime. Don't have the dough to spend? A plastic one is not ideal, but it will do just fine. The OXO Good Grips 15-by-21-inch version is a quarter of the cost and a great value.

With a wooden board, you'll want a small bottle of mineral oil to rub into the surface with a soft cloth or paper towel after each use to prevent staining and enhance its life.

# FEELING SHARP

There is nothing more frustrating than a dull knife. Not only does it make prep work a chore and your finished product less attractive, it's also downright dangerous. A dull blade requires more pressure to cut into a food, and it can easily slip off a tough onion skin, for example, and into your finger. Ouch. Most home cooks should sharpen their knives at least twice a year, much more frequently if they use their knives every day. There are three ways to go about it.

**Method 1: Use an Electric Sharpener.** A good-quality electric sharpener is an option, but I strongly discourage their use. First off, they remove a tremendous amount of material from your edge. Sharpen your knife a dozen times, and you'll have lost a good ½ centimeter of width, throwing it off balance and rendering any blade with a bolster (i.e., most high-quality forged blades) useless. Second, even the best models provide only an adequate edge. If you don't mind replacing your knives every few years and are happy with the edge an electric sharpener gives you, this is an option. But there are much better choices.

**Method 2: Take It to a Professional.** Provided you have a good knife sharpener nearby and are willing to pay to have the service performed, this is a good option. But if you sharpen your blades a dozen or so times a year, as I do, this can get quite expensive. And all but the best pros use a grinding stone, which will take away much more material than is necessary from your blade, reducing its lifespan. Want to forge a stronger relationship with your blade? Choose the next option.

**Method 3: Use a Sharpening Stone.** The best method by far. Not only will it give you the best edge, but it will also remove the least amount of material. Additionally—and I'm not kidding about the importance of this one—the act of sharpening your own knife will help you create a much stronger bond with your blade, and a knife that is treated respectfully will behave much better. You won't believe the difference a sharp knife can makes in your cooking.

Stones are designed to either be lubricated with oil or with water. I prefer water stones.

## Shopping and Maintenance

When buying a water stone, look for a large one, at least 2½ inches wide and 8 inches long and an inch thick. Stones come in various grit sizes, ranging from around 100 up to 10,000+. The lower the number, the coarser the grit, and the more material it will take off your knife. The higher the grit, the sharper the edge you will get, but the more strokes it will take to get you there.

I recommend keeping two stones in your kit: one with a medium grit (around 800 or so) to perform major

*continues*

sharpening jobs and one with a fine grit (at least 2,000) to tune the edge to a razor-sharp finish. For real pros, a stone with an ultrafine grit (8,000 and above) will leave a mirror-like finish on your blade, but most cooks won't notice the difference in terms of cutting ability. If you only have the budget or space for a single stone, I'd recommend one with a grit between 1,000 and 1,200. Two-sided stones are also available (coarse and fine grit), but these are usually of inferior quality. You will also need a stone fixer to repair any unevenness in the surface of your sharpening stones. I've yet to go farther down the rabbit hole to purchase a stone-fixer fixer. Both stones and fixers are available through Amazon.com.

Carefully dry your stone after each use, and store it wrapped in a kitchen towel in a dry, grease-free environment. Oil can soak into the porous material, ruining its sharpening ability (and your chances of ever slicing your onions thin enough for that soup). And, again, remember to hone your knife on a steel every time you use it. While this process won't actually take any material off the blade (see page 55), it will help keep the blade aligned, making slicing and dicing much easier.

## Step-by-Step: How to Sharpen a Knife

**Step 1: Work in Batches.** Although it's worth the effort, knife sharpening can take a bit of effort and time. If you're going to be setting up a station to sharpen a knife, think ahead and sharpen every knife that may need sharpening to get the whole process done in one session instead of several.

**Step 2: Soak Your Stone(s).** When working with water stones, it's essential to submerge them in water for at least 45 minutes before using. If the porous stone is not fully saturated, it will dry out during sharpening, causing the knife blade to catch and giving your edge nicks and dings. If you have two, soak both your stones, as well as your stone fixer.

**Step 3: Set Up Your Station.** Place your stone on a towel laid on a cutting board. Keep a container of water nearby to keep your stone moistened during the sharpening process. The stone should be oriented with a short end parallel to the edge of the counter.

**Step 4a: Begin the First Stroke.** Hold your knife with the blade edge pointing away from you. Place the heel of your knife on the far edge of the stone and, holding the blade gently but firmly with both hands at a 15- to 20-degree angle and using even pressure, slowly drag the knife over the stone toward you down its length while simultaneously moving the knife so that the contact point moves toward the tip of the blade.

**Step 4b: Maintain the Angle.** Be careful to maintain the 15- to 20-degree angle as you pull the knife across the stone. Pressure should be firm but gentle, and the blade should glide smoothly across the stone.

**Step 5: Repeat.** Each stroke should finish with the tip of the knife touching the bottom edge of the stone. Lift the knife, reset the heel at the top edge of the stone, and repeat.

**Step 6: Look for Silty Water.** As you repeat the process, a thin film of silty-looking water should collect on top of the stone and on the blade. This abrasive liquid will gradually take material off the edge of your knife, sharpening it.

**Step 7: Check for Burr.** As you continue to repeat strokes on the first side, a tiny burr will eventually form on the other side of the blade. To check for it, place the blade on your thumb and pull it backward. If burr has formed, it should catch slightly on your thumb (with really-fine-grit stones, though say, 2,000 or above, you won't feel it). It may take up to 30 or 40 strokes before a burr forms, and that is the indication that you should switch and start sharpening the other side.

**Step 8: Start Sharpening the Second Side.** Turn the knife over so the edge is pointing toward you. Place the heel of the blade near the base of the stone, again maintaining a 15- to 20-degree angle, then gently push the blade away from you while simultaneously dragging it across the stone toward the tip.

**Step 9: Repeat.** Your stroke should end with the tip of the blade against the top edge of the stone, still maintaining a 15- to 20-degree angle. Moisten your stone between strokes if it begins to dry out. Repeat for as many strokes as it took you to form the burr on the first side. Flip the knife back over and repeat steps 4 through 8, using fewer and fewer strokes per side, until you are down to one. (The blade will not form a burr during this stage.)

**Step 10: Fix the Stone.** After repeated use, your stone will begin to develop grooves in it, which can lessen its sharpening power. To fix it, use a low-grit stone fixer. Place the fixer flat against the stone and push it back and forth to grind down the stone and create a new flat surface.

**Step 11: Clean Up.** You should have a dedicated towel for this purpose, as the grit from the stone will never come out. After thoroughly drying the stone (allow to dry on a rack for at least a day), store it wrapped in its towel.

**Step 12: Hone and Test Your Blade.** After sharpening, hone your blade on a honing steel in order to get the edge in alignment, then test it for sharpness. Some people recommend trying to slice a piece of paper in half by holding it up and cutting through it. I find that even a relatively dull knife will pass that test but fail at other kitchen tasks. The best test is to simply use the knife to prep a vegetable. Do you notice any resistance, or does it fly through that onion? Can you slice a ripe tomato thin enough to read through it? Yes? Then you're done!

# MY KNIFE KIT

**W**ant to get a real idea of the essentials for good cooking? This is my knife kit, the tools I take with me whenever I venture out into a foreign kitchen. These are the things I want to make sure I have on hand all the time.

*Top row*: honing steel; *middle row*: small offset spatula, Y-shaped vegetable peeler; *bottom row, from left*: Western-style chef's knife, serrated bread knife, santoku knife, sheep's hoof paring knife and wine key, boning knife, flexible fish spatula, wooden spoon, rubber spatula, microplane grater.

## ESSENTIAL SMALL ELECTRIC TOOLS

There's no shortage of fun gadgets for the kitchen, but there are only a few that you absolutely need. Here's your basic starter kit, in descending order of importance. Notice that the three most important items on this list are tools used for measuring. That is not an accident.

### 1. Instant-Read Thermometer

This is it, folks: the one thing that more than any other purchase you can make will really revolutionize your cooking (especially if you often cook, or have ever been afraid of cooking, proteins). A good instant-read thermometer is the only way to ensure that your roasts, steaks, chops, and burgers come out that perfect medium-rare every time. Forget about poking meat with your finger, relying on inaccurate timing guides, or the nick-and-peek method. Buy a high-quality digital instant-read thermometer, and never serve a piece of over- or undercooked meat again.

The Splash-Proof Super-Fast Thermapen by ThermoWorks has a hefty price tag ($86), but it's money well spent. It's head and shoulders above the competition, with a stunning range of –58° to 572°F (–50° to 300°C), one-tenth of a degree precision, unparalleled accuracy, and a read time of under three seconds. Because of its wide range, you won't need separate meat, candy, and deep-fry thermometers—a single tool does all three tasks, and how.

Aside from my knives, it's my favorite piece of kit. For the best inexpensive model, which is slower and more difficult to use but still perfectly serviceable, check out the CDN Pro Accurate Quick-Read Thermometer ($16.95).

### 2. Digital Kitchen Scale

If you're on the fence about whether or not you need a kitchen scale, jump to page 73, "Weight Versus Volume," and read that section. Got it? See why you want a digital scale? Once I got one, I've used it almost every single day. A good digital scale will make inaccuracies and inconsistencies a thing of the past. And if you're the obsessive type, a scale can also help you figure out how much moisture your chicken lost during roasting, or exactly how far you've reduced that stock. Hooray!

Things to look for in a good scale: at least 1-gram (⅛-ounce) accuracy; a capacity of at least 7 pounds; a tare (zero) function; measurements in both metric and imperial units; a large, easy-to-read display; and a fold-flat design for storage.

The OXO Good Grips Food Scale with Pull-out Display ($45.95) has got all of that, plus a neat pull-out display that allows you to read measurements with ease, even when weighing large, bulky items that would otherwise obscure the screen. The only problem? Annoying fractions in the display instead of decimal places. Who the heck wants to measure ⅜ ounce? The Aquatronic Kitchen Scale by Salter ($49.95) lacks the pull-out-display feature, but it uses easy-to-read decimals, which makes both math and looking cool in front of Europeans much easier.

If you don't mind fractions or intend to go all metric, then stick with the OXO (that's what I use). Otherwise, the Salter Aquatronic wins.

### 3. Digital Timer/Stopwatch

Did you know that in restaurant kitchens, croutons are the number-one item most burnt by line cooks?*

---

*I just made this fact up.

I can't tell you the number of times I've popped a tray of sliced bread in the oven for crostini only to pull it out thirty minutes later after it sets off the smoke alarm.

At least I *used* to.

These days, I keep a Polder 3 in 1 Timer, Clock, and Stopwatch ($13.95) around my neck at all times. It's got an easy-to-read display, an unobtrusive size, intuitive buttons, a loud alarm, a magnet for sticking it to the fridge, and a nylon lanyard for keeping it around your neck, so there's no way you can forget about your roasting peppers, even if you leave the kitchen. With both count-up and count-down functions, what more could you want in a kitchen timer?

### 4. Immersion Blender

*Really?* some of you might be saying. *You'd really say that your immersion blender is more important than your food processor or mixer?* Well, if you rate importance by frequency of use, then absolutely. I use my immersion blender so frequently that I have it mounted on a holster on the wall right next to my stove and cutting board, ready at a moment's notice to emulsify a sauce, whip up a batch of mayonnaise, roughly puree some whole canned tomatoes directly in the pot, blend a cheese sauce, puree soup, whip cream in no time flat . . . you get the picture. It's a versatile tool, and you don't need a silly infomercial from the 1980s to tell you that.

Want a pitcherful of margaritas? The regular blender's your friend. Need to make two quarts of pesto? OK, pull out the food processor. But for smaller, everyday blending tasks, an immersion blender is the tool for the job. Ever get annoyed at those ropy pieces of egg white you come across when breading food? Blend the eggs for a few seconds, and they'll be perfectly uniform and smooth.

You like froth on your hot chocolate? Heat it up in the pot and buzz it to create a luxurious foam. Lumps in your béchamel? All gone. How about if you want to make just a few ounces of perfectly smooth cauliflower puree or a half-cup of mayonnaise? Yep, you can do that with an immersion blender too.

The Braun PowerMax, which is only about $30, has been performing admirably at least three times a week in my kitchen for the past eleven years now. It's the most reliable sidekick I know. Unfortunately, it's not widely available these days, as I found out when stocking the kitchen at the Serious Eats World Headquarters. So there we use the KitchenAid Immersion Blender (about $50), which works just as well. You can get it as part of a package that includes a whisk attachment and a mini–food processor, but believe me, those are dust collectors and you don't need 'em.

### 5. Food Processor

At a bare minimum, a good food processor should be able to:

- **Finely chop dry ingredients** like nuts and bread crumbs. In order to do this, a processor must have an easy-to-use pulsing action and a motor that stops and starts on a turn.
- **Roughly puree vegetables** for things like marinades, dips, and rustic soups (for full-on smoothness, use a regular blender). Bowl shape, power, and blade design all affect how well a processor can accomplish this. It should also not leak.
- **Grind meat.** Short of a dedicated meat grinder or an attachment for a stand mixer, the food processor is the best way to grind fresh meat. Meat can be tough to chop, so a very sharp blade and powerful motor are necessary.

- **Easily form emulsions** when making sauces like mayonnaise or a light vinaigrette. Bowl design can affect the way the blade makes contact with liquids.
- **Knead bread dough** quickly and efficiently. This is the most-heavy-duty kitchen task of all, and the processor's effectiveness relies mostly on the power of the motor.

I also like to have at least an 11- to 12-cup-capacity processor, which makes grinding meat and making dough much easier. Some models come with a mini-prep bowl that can be inserted into the main bowl for small tasks. These are cute but essentially useless. Whatever the tiny bowl can do, I can do with a knife. That may take slightly longer, but if you consider the time it takes to wash the blade, bowl insert, and lid, it's no contest.

There's also no use for a processor that's going to get gummed up or jammed every time it hits a hard nut or sticky dough. Particularly prone to failure are models with a side-mounted motor that drives the blade via a belt. Failing at even the easiest of tasks, those processors aren't worth the box they come in. Instead, look for models with a solid-state motor attached directly to the blade shaft, with no intermediary belt or chain. These take up a little extra space in terms of height, because the motor must be placed underneath the processor bowl, but that's an easy trade-off.

The two best processors that fit all this criteria at a reasonable price are the KitchenAid 12-cup Food Processor ($199.95) and the Cuisinart Prep 11 Plus 11-cup Food Processor (about $165). And at such similar price points (admittedly much more expensive than many useless models), it all comes down to bowl design, and in this category, the Cuisinart wins: it's got a larger feed tube, as well as straight sides that

ensure that all your food falls back down into the blade. For some reason, the KitchenAid has sloping sides. Ingredients can ride up the sides more easily and may not be chopped or emulsified properly.

## 6. Stand Mixer, with Meat Grinder Attachment

A good stand mixer is a true workhorse for anyone who bakes more than occasionally. When selecting one, there are a few criteria that I look for:

- **It should have a dough hook attachment** and a motor powerful enough to mix at least 2 pounds of bread dough without straining, shaking, or burning out.
- **It should have a whisk attachment** to whip cream and to whip egg whites quickly and efficiently into frothy meringues and foams.
- **It should have a paddle attachment** to cream butter and sugar effortlessly, as well as make short work of mashed potatoes and sausage mixtures.
- **It should feature planetary motion**, meaning the whisk attachment spins around its axis in one direction and orbits around the work bowl in the opposite direction, to maximize contact and mixing power.
- **It should have a port** for attachments such as a meat grinder or pasta maker.

Once again, just as with food processors, the epic battle for kitchen superiority (at least for the home consumer) comes down to KitchenAid and Cuisinart. Despite the fact that many manufacturers boast their motor wattages in their advertising (for instance, Cuisinart does a side-by-side comparison of their 800-watt SM-55 mixer versus the 325 watts of the KitchenAid Artisan), these numbers mean very little. Within a given manufacturer's product lineup, it is an indicator of how powerful the motor will be, but the wattage is actually the power *consumed* by the mixer, not the power *produced* by the motor. It's a marketing gimmick, pure and simple. Given a choice between two motors that perform equally well (say the 325-watt motor of the KitchenAid Pro 500 versus the 800-watt motor of the Cuisinart SM-55), it's actually better to pick the one with lower wattage and save on electricity.

Both the KitchenAid and the Cuisinart have a meat grinder attachment available, an absolute must in my kitchen. It saves money and produces incalculably better results for burgers, sausages, meatballs, and meat loaves. Here, the Cuisinart's all-metal Large Meat Grinder Attachment ($128.95) has an advantage over KitchenAid's plastic-and-metal Food Grinder Attachment ($49.95). But for the price of the Cuisinart attachment, you could buy a full-on dedicated meat grinder. The KitchenAid grinder has served me fine for years.

While either brand will do you well, the Kitchen-Aid Pro 500 ($299.95) gets my vote for wedding-registry priority numero uno. It's ideal for both heavy-duty bakers who make bread at least a couple times a week and want a real powerhouse and for those who will be mostly mixing batters, whipping cream, or even grinding meat.

### 7. Powerful Blender

There are a ton of decent blenders on the market—far more than good stand mixers or food processors. Then again, there are also a ton of poor blenders out there. You want a blender that's powerful enough to puree soup to a completely smooth, velvety texture, with enough vortex action to thoroughly mix a thick blue cheese dressing or crush a pitcherful of ice for frozen drinks. You also want a blender with simple, easy-to-understand controls; the ability to

pulse; and the capacity to slowly and evenly build up from a slow speed to a fast one, in order to prevent the lid from blowing off when you blend hot foods too fast. (Raise your hand if you've done this. Yep, thought so.)

The cream of the crop when it comes to blenders, the one that'll turn your shoes into soup or scare the pants off the fat kid in *The Goonies*, the one that'll turn all your cheffy foodie friends spinach-green with envy, is one from the Vitamix Pro Series. This is what every professional kitchen I've ever worked in has used, and with good reason. It's crazy powerful, has a very large capacity, and is built like a rock. They clock in at around $450 and up, placing them firmly out of the reach of most home cooks. Nearly as good and way cooler looking is the BlendTec, which, for around $400, will turn everything from a carrot to a full-sized Alpine ski into dust. (Don't believe me? Just Google it. Seriously, it's a great video.)

For a blender that won't break the bank, I'd go with the KitchenAid Vortex 5-Speed Blender (around $150). It has an easy to clean wide polycarbonate pitcher and a blade that creates a big enough vortex that I can blend a full batch of cheese sauce for Cheesy Broccoli Casserole (page 420) in one go without it gumming up.

### 8. Rice Cooker

There's no easier, more foolproof way to cook rice and other grains than in a rice cooker. Sure, you can cook rice in a pot, carefully monitoring the flame, hoping that you've added the right amount of water and that your rice isn't burning on the bottom, and taking it off the heat at just the right moment, but if you're anything like me, you've burnt one too many batches to fuss with that method any more. With a rice cooker, you just add your rice and water, shut the lid, flip the switch, and go, with the added advantage that it'll keep the cooked rice (or other grain) hot for hours.

Even the cheapest rice cooker will do—I had a $25 model I picked up in Chinatown that lasted me all through college and a good five years afterward. When I got married, I upgraded to a fancypants model with a fuzzy logic* processor and a nifty latching top that keeps the moisture level inside at the exact right level. I love my rice cooker almost as much as I love my instant-read thermometer, which is just a hair more than I love my wife (just kidding, honey).

## ESSENTIAL KITCHEN HAND TOOLS AND GADGETS

A well-stocked vacation home probably has two to three drawers worth of tools, of which only half are even identifiable and perhaps three or four are ever used. The following is a list of tools that you'll use *all the time*. Get 'em.

### 1. Utensil Holder

First things first: if your tools are at the back of a drawer, you probably won't use them. And if you don't use them, you probably won't cook as often. And if you don't cook, what's the point of living, really? A utensil holder with a capacity of at least 2 quarts helps keeps your tools handy right where you need them. If style is what you're after, Le Creuset makes handsome ceramic models in a variety of colors for around $25. If, on the other hand, pure functionality is your goal, any old small bucket will do. I use a $5 metal version from IKEA.

---

*Fuzzy logic is an obscure branch of logic used in control theory, artificial intelligence, and rice cookers—the only thing that these three have in common.

### 2. Bench Scraper

A bench scraper is one of those tools whose advantages aren't obvious until you start using it regularly. I keep one on my cutting board whenever I'm doing prep work. It quickly transfers chopped mirepoix to my saucepan or carrot peels to the trash. I use it to divide dough when making pizzas, or ground beef when making burgers. For cleanup, a bench scraper makes short work of dough scraps that have dried onto the work surface, and it efficiently picks up tiny bits of chopped herbs and other debris. (By the way, you should *never* use the blade of your knife to pick this stuff up off your board. It's dangerous, and it will rapidly dull the edge of your knife.) A bench scraper also makes removing stickers from glass bottles or labels from plastic containers a snap.

With its comfortable handle, sturdy construction, convenient built-in 6-inch ruler, and an edge sharp enough to rough-chop vegetables, the OXO Good Grips Pastry Scraper ($8.99) is the first choice for home kitchens. In my knife kit, however, I keep a lightweight plastic C. R. Manufacturing scraper (50 cents), which performs most of those functions at a fraction of the cost, in a much more compact package.

### 3. Saltcellar and Pepper Mill

Why would anyone need a saltcellar? Underseasoning food is the most common culinary blunder. Ask me why your food tastes blander than you'd like it to, and 90 percent of the time, all it needs is a little pinch of salt. Having a container of a salt in a prominent spot by your prep station or stove serves as a constant reminder to season, taste, season, and taste again until you get it exactly right. I guarantee that if you don't already have one, putting a saltcellar on your counter will make you a better cook. Any wide-mouthed covered container with an easy-open lid will do, but a dedicated saltcellar does it with style. Mine is a wooden job with a flip-top lid to prevent dust, water, or oil from getting in.

And pepper? If you've been using preground pepper, do yourself a favor and buy an inexpensive jar of pepper with a built-in mill. Then taste the fresh-ground stuff side by side with the preground. Which would you rather be putting on your food? If that doesn't convince you to go out and buy yourself a pepper mill, I can only assume that you are dead from the tongue up.

You'll want to invest in a mill that has a solid metal grinding mechanism. Cheap ones are usually made of plastic and will stop grinding after a year or less of regular use. Although $35 to $60 might seem like a big chunk of change, a real pepper mill will improve practically every savory food item you cook. Peugeot is the Rolls-Royce of pepper mills. Perfectly crafted, luxuriously styled, and awesomely efficient, these mills look good and grind like a dream. They also run upward of $55. More affordable and equally good if totally utilitarian is the Unicorn Magnum Pepper Mill ($36.90). It has a tough nickel-plated grinding mechanism, an easy-to-load design, and a quick grind-size adjustment screw.

## 4. Prep Bowls of All Sizes

Here's a mantra for aspiring chefs: *An orderly kitchen is a good kitchen.*

Isn't it annoying trying to chop carrots on your cutting board when that little pile of parsley in the corner is getting in your way? Or what about frantically trying to scoop up the chopped ginger to get it into that stir-fry-in-progress before your bok choy wilts? I use several prep bowls with a small capacity (we're talking 1-cup or less) pretty much every time I cook to keep chopped aromatics, measured

spices, grated cheese, whatever, off my board, within easy reach, and organized. This is what fancy cooks call their *mise en place*. In the cabinet directly above my cutting board, I have a couple dozen 25-cent ceramic condiment and cereal bowls from IKEA for this very purpose. (If you want to go fancy, you can get sets of Pyrex clear glass prep bowls.)

Large mixing bowls are equally valuable. While the all-glass ones look nice up on the shelf, they're a total pain in the butt to work with. I remember many days at *Cook's Illustrated* magazine when we'd have to search through stacks and stacks of glass bowls while working on a photo shoot to find the one or two that weren't chipped on their edges. Where do these glass chips end up? On the floor? In the food? In my own kitchen, I'd rather not find out. Plastic bowls seem like a reasonable solution until you realize that plastic absorbs both stains and odors from oily and other foods. Pour a batch of olive-oil-and-butter-based marinara sauce (page 695) into a white plastic bowl, and you'll find that you're now the proud owner of an orange plastic bowl.

Instead, I use inexpensive stainless steel bowls that I picked up from a restaurant supply store (if you don't have a good one near you, try the ABC Valueline brand from amazon.com). I have about half a dozen in sizes ranging from a couple of quarts up to 5 quarts. They're lightweight and easy to handle, shatterproof, stainproof, break-proof, odorproof, and microwavable.* Add to that their shallow design, which makes whisking and tossing a snap, and you've done made yourself a new best friend.

---

*Yep—it's totally safe to use metal bowls in a modern microwave. Just don't throw any foil, fork or other sharp objects in there: pointed objects can cause electrical arcing.

### 5. Wooden Spoons

Short of being born a woman in Italy and waiting for your daughter to have a child, nothing makes you feel more like an Italian grandmother than slowly and deliberately stirring a lazily simmering pot of *ragù* with a wooden spoon. Blood runs deep between a good spoon and his cook. I nearly cried the day I cracked the handle on the spoon that had lasted me through nine years and thirteen different kitchens—a flat-headed beechwood model that I think I stole from my mother's hidden secondary utensil drawer. It was so well used that the handle had conformed to the shape of my hand, and the head had been worn into an angle that perfectly fit the corners of my Dutch oven.

Whether stirring sauces, tasting soups, or gently whacking cheeky spouses who disturb you in the kitchen, a wooden spoon is the tool you'll want 90 percent of the time when you're cooking on the stovetop. I have half a dozen of various shapes and sizes that I use almost every time I cook. But if I had to pick a single spoon to perform every task, I'd choose one with a cupped section for tasting and a head that comes to a point, rather than being completely round, making it easier to get into the corners of pots and pans.

Whether you want a spoon with a completely flat section on the head or a more triangular profile is totally up to you. Like my favorite Beatles album, my favorite wooden spoon tends to waffle back and forth among the different spoons in my set.

### 6. Slotted Flexible Metal Spatula

Flexible enough to flip tender pieces of delicate fish without breaking them yet sturdy enough to get every last bit of a smashed burger off the bottom of your pan, a slotted metal fish spatula is an absolute essential in your tool kit. It's ideal for blotting

excess grease off cooked steaks and chops. Just pick up the meat from the skillet and put it on a paper towel, still on the spatula, then transfer to the serving plate—the wide slots allow the grease to drain off easily. The spatula is lightweight and maneuverable enough to flip fragile eggplant slices in a skillet of oil, but it will also handle whole grilled pork chops with ease. Its slight flexibility lends it agility and control, unlike stiffer spatulas (which have their place in the kitchen—we'll get to that).

And here's some good news: most of the expensive models are far too stiff to do the job well. I keep a $25 Lamsonsharp model in my kit, and the even cheaper Peltex (around $15) is the standard in most restaurant kitchens.

### 7. Tongs

A sturdy pair of tongs is like a heatproof extension of your fingers. Robust construction, slip-proof grips (ever try to grab a pair of stainless steel tongs with greasy fingers?), a spring-loaded class-3 lever design, and scalloped edges perfect for grabbing everything from tender stalks of spring asparagus to the biggest bone-in pork roast are the qualities to look for in a good set of tongs. The OXO Good Grips 9-inch Stainless Steel Locking Tongs ($11.95) set the bar for quality.

### 8. Microplane Zester Grater

When you're talking fine-toothed graters, pretty much only one brand comes to mind: the Micro-

---

*Remember levers from fourth grade? Class 3 is when the fulcrum is at one end (that'd be the hinge in the tongs), the load is at the other (yep, that's the food), and the effort is applied in the center (where you grip the tongs). It's a much better design than scissor-like class-1 lever tongs, which have very limited gripping power and don't open wide enough.

plane Zester Grater ($14.95). It is more than just a useful gadget—it's the only one to get.

My favorite thing to do with a zester is to go to town with it on an orange and watch as the little mountain of zest effortlessly grows on my cutting board. Wait—my favorite thing to do is grate delicate wisps of Parmigiano-Reggiano over my Bolognese. No, I take that back. My favorite thing is to grate fresh nutmeg on top of my gin flip. Or is it to sprinkle chocolate shavings over my soufflé? Oh, but I do dearly love the lovely little mound of ginger that smells oh so lovely as it falls off the zester into my bowl. No, I've got it, and this time I'm sure: it's being able to throw out my confounded single-tasking garlic press and using my Microplane to grate garlic into tiny, even mince.

So many things to grate, so little time!

## 9. Whisks

They're essential for mixing quick-bread batters or emulsifying hollandaise. Use one in a large pot of soup to incorporate seasoning much more quickly than a wooden spoon can. And a whisk is the best tool to whip cream or foam egg whites into frothy meringue. Models with stiff wires require much more movement and hard work from your wrist. The OXO Good Grips 9-inch Whisk ($8.95) has thin, flexible wires, which make whipping vinaigrettes into shape an effortlessly enjoyable endeavor.

## 10. Salad Spinner

Yes, it will get your greens dry, and we all know that dry greens are better at holding dressing (right?), but the salad spinner is actually one of the truly great multitaskers in the kitchen. I fill mine with water and pick herb leaves directly into the bowl. Once they're picked, I swish them around,

lift them up in the basket, dump the sandy water, and spin dry.

You can wash delicate items like berries and then dry them in a salad spinner lined with a few layers of paper towels to extend their shelf life by a few days. Or take chopped tomatoes for a spin for easy seeding (the seeds slip through the basket while the flesh stays put). Washed mushrooms, sliced peppers, broccoli florets—anything you could think of stir-frying or sautéing—will cook better after a thorough drying in the spinner. Use the power of centrifugal force to whip away excess marinade from shrimp, chicken, or kebab meat. And if you've got a sturdy one with small slots, like the OXO's Good Grips Salad Spinner (about $30), there's no need to own a colander—just drain beans, pasta, and vegetables in the spinner basket.

## 11. Stiff Spatula

My Due Buoi Wide Spatula (about $35) is exceedingly sexy, in that mostly platonic inanimate metallic object kind of way. It's got a business end that's 5 inches long, a generous girth of 3.9 inches at the front, and a hefty weight of 7.76 ounces. It's a size that can't be beat—just large enough to smash a ball of beef into a 4-inch patty or flip a couple portions' worth of browning home fries, without being so large that it doesn't fit into a small skillet. I've picked up a whole pizza off a hot stone with this thing. I'd like to see your wimpy plastic spatula do that!

The blade and tang are formed out of a single piece of cast stainless steel, which clocks in at a thickness of 0.04 inch (1 mm, or approximately 18 gauge). This is important: it allows you to lift a whole turkey or rib roast with reckless abandon. If you flip the spatula over, its keen and sturdy front edge substitutes handily for a paint scraper, allowing you to ensure that every last bit of flavorful, crisp crust stays firmly attached to your burger or steak, instead of remaining in the pan. The handle is made from tough, durable polycarbonate and features a full tang, for optimal strength and balance. This baby's gonna last a lifetime.

**And there's a musical bonus:** When struck daintily against the cutting board, the spatula vibrates at precisely 587.33 hertz (really!), with an outstanding overtone series. Even Stradivarius would be proud to apply his famous varnish to it. It's the very thing during that all-too-common situation when I desperately need to tune the fourth string of my guitar while applying cheese to my burgers.

You'd be hard-pressed to find a better stiff spatula.

## 12. Japanese-Style Mandoline

Sure, you can train for years and spend hours a day sharpening and honing your knives to get to the point where you can whip out fennel wisps so thin you can read through them or slice through your prep work at a hundred onions per hour. And I'll be the first one to tell you that you're really, *really* cool. But for the rest of us, a mandoline makes quick work of repetitive slicing and julienning tasks. At one point in my life, I owned a fancy-pants $150 French model. But you know what? It was heavy, bulky, and a pain in the butt to clean. And, with its straight blade, it didn't really do a great job. The Benriner Mandoline Plus ($49.95), on the other hand, features a sharp angled blade that cuts much more efficiently than those awkward straight blades or clumsy V-shaped cutters. Walk into the kitchen of any four-star restaurant in the city, and I guarantee you'll find at least a couple Bennies (as they

are affectionately called by line cooks) occupying a prominent place.

**Random trivia:** "Benriner" means "Oh, how handy!" in Japanese (despite the fact that the Janglish on the box front proclaims "Dry cut radishes also OK.")

## 13. Spider

A spider/skimmer accomplishes almost everything a slotted spoon does, and better, at a fraction of the cost. It excels at fishing dumplings, vegetables, or ravioli out of a pot of boiling water. And its wire construction and relatively open mesh creates less turbulence in the liquid than a standard slotted spoon, making it much easier to fish out food.

As for the task it was designed for—dunking and stirring foods for deep-frying—the only thing that even comes close in terms of agility and control is a long pair of chopsticks, and even Mr. Miyagi would have trouble picking up peas from a pot of boiling water with a pair of chopsticks. Wire-mesh spiders with bamboo handles are available at most Chinese grocers and restaurant supply stores for a few bucks a pop, but if you want something that'll last a long time, go with an all-metal spider like the Typhoon Professional Cook's Wire Skimmer, available for about $10 online.

## 14. Small Offset Spatula

Though these diminutive 4½-inch-long spatulas are intended for applying frosting to small pastries like cupcakes, you'll find that they have a slew of other uses in both the sweet and savory kitchen. Ever find yourself trying to unstick a fragile piece of food from a skillet with a spatula three times too big? The thin, flexible blade of a small offset spatula can slip under food items that even a fish spatula is too thick for. Pan full of slender breakfast sausages to flip one at a time? This is your tool. It's also indispensable for plating and presentation. A lightweight feel, comfortable handle, and ultrathin blade make the Ateco Small Offset Spatula (about $2) the industry standard, offering precision, control, and finesse. More control means less mess and better-tasting food. Oh, and it's good for cupcakes as well, if that's your bag.

### 15. Fine-Mesh Strainer

A full-size colander is great if you've got a full pot of pasta to drain, but it rarely gets used otherwise (and even then, I just use the basket of my salad spinner). For smaller everyday tasks like draining a can of tomatoes or beans, or ensuring that your crepe batter is perfectly smooth, a small hand strainer is what you need. I keep one hanging on a hook alongside my pots and pans for easy access. Inferior models consist of just a round mesh basket attached to a handle, but the 8-inch Stainless Steel Strainer from OXO ($24.95) also has a loop of metal sticking on the opposite side of the basket. This allows you to set the strainer over a bowl for no-handed operation. It may seem a little pricey for a simple strainer, but its heavy-duty construction means it will last and last.

### 16. Chopsticks

I admit it: this one is a little controversial. Either you grew up using chopsticks and wouldn't be caught dead near a pot of simmering water or a wokful of hot oil without them or you didn't—and, if so, you will probably wonder, "Do I *really* need them?"

But precise tips and a gentle touch will treat small, delicate pieces of fried or grilled food (say, a tempura of squash blossoms or slender stalks of asparagus on the grill) far more gently than a relatively clumsy pair of tongs, which are better suited to large items like fried chicken or a rack of ribs. I use chopsticks for picking up bits of food from a stir-fry in progress to taste for doneness. They are also ideal for picking out a few slippery noodles from a pot of boiling water to make sure that they are perfectly al dente before draining.

While regular chopsticks will do in many circumstances, high-heat applications require extra-long sticks made specifically for cooking. If you are lucky enough to have an East Asian kitchen supply store nearby, you can pick these up for a couple bucks a pair. Otherwise, you can find acceptable models online, like the Extra-Long Chopsticks from Hong Kong Imports Ltd. ($2).

### 17. Wine Key

Regular corkscrews and $100 rabbit-shaped models will get your cork out, and fast. But with a little practice, a waiter's wine key will open wine bottles (and beers) just as fast, and make you look infinitely cooler. The key is to use it as a lever. If you are pulling on it hard, you're doing it wrong! I keep a few in my cutlery drawer (like pens and razors, they tend to wander off into the world on their own from time to time), as well as one in my knife kit.

### 18. Citrus Juicer

Every professional kitchen has its own hazing rituals, and as a young chef-in-training, I endured a period of time—a good eight months or so—when my first duty every single morning was to ream twenty-four limes, twenty-four lemons, and a dozen oranges for fresh juice to use on the line during service. And the only tool I was allowed to use to do the job (lest I risk being called a wimp—believe me, a wimp is the last thing you want to be in the macho world of pro-

fessional kitchens) was a wood lemon reamer from Scandicrafts, Inc. ($4). It was two weeks before I could complete the task from start to finish without taking a break to nurse my painfully swollen hands, and I went through four of the reamers in the course of those eight months, slowly wearing them down until the grooved edges on the business end were as smooth and soft as river stones.

This is not to say that it's a bad product—I'd strongly recommend it for the occasional juicer—but if you go through a lot of citrus juice (some people believe that lemon juice is as important as salt, just ask the Greeks!), there are a number of other options on the market. I use the Two-in-One Juicer from Amco ($19.95). You place the citrus cut side down in the perforated cup-shaped holder, then squeeze the handles together to extract the juice. It's fast, efficient, and much easier on the hands than a conventional reamer. The only issue is that it sometimes leaves a bit of juice behind, forcing you to manually squeeze the empty citrus shells for maximum extraction. And though it comes in small (green), medium (yellow), and large (orange) sizes, intended for limes, lemons, and oranges, the yellow one works fine for both lemons and limes, making it the one to get.

## 19. Cake Tester

I know many chefs and cooks who keep a cake tester tucked into the pen pocket of their whites and none who use them to test cakes. Not that you *can't* test a cake's doneness with them, it's just why would you, when there are so many more interesting assisted-poking tasks at which it excels? Essentially a heavy-gauge wire with a handle, it's about as simple as a tool can get. The idea is that you poke it into the center of a cake and pull it out. If it comes out clean, the cake is done. So, it's sort of like a glorified toothpick, but the fact that it's long and made of metal means that it's useful for all kinds of other things.

The most obvious is testing the doneness of vegetables. Have you ever been told to stick a paring knife into a boiling potato to check if it's tender all the way through? The problem is that even the thinnest of paring knives makes a large stab wound in the potato, releasing starch and vastly increasing the chances that it'll break apart, particularly if you've bucked up for those tiny, tasty fingerlings. A cake tester neatly takes care of that problem. Want to know if those simmering carrots are tender enough to puree? How about if those baby radishes are cooked through? With a cake tester, you can find out without leaving behind any incriminating evidence. My favorite way to cook beets is in a tightly sealed foil pouch—a method that absolutely prevents you from poking them with a paring knife. A knife makes a hole in the foil too large to recover from. Not so a cake tester.

I use my cake tester instead of a fork to decide whether or not my braising brisket or short ribs are "fork tender." If the cake tester slides in and out with ease, the meat is ready. Lots of fish have membranes between layers of flesh that only soften at around 135°F or so (a perfect medium-rare). Stick your cake tester into that poaching salmon fillet, and it if meets resistance (i.e., if it feels like punching through pieces of paper), it's undercooked. Barbecuing a pork shoulder low and slow? You can check if it's done without losing any juices through the grill grates. Finally, if you ever (god forbid!) find yourself without your trusty thermometer by your side, a cake tester is the next best thing. Stick it into the center of your meat and leave it there for about 5 seconds, then pull it out and hold it

under your lower lip (an area particularly sensitive to heat). You'll know instantly whether your steak is cold, warm, or hot in the center. As accurate as a thermometer? No. Good in a pinch? You bet.

You can go all out and pay the $5 for a cake tester from OXO, which has a grippy black handle, but you may risk being made fun of for being too fancy-pants. The cake tester from Fox Run ($1.29) is the cheapest I've found online.

## 20. Lots of Squeeze Bottles

I'm guessing a good 80 percent of you have read Anthony Bourdain's proclamation of love to his squeeze bottles in *Kitchen Confidential*:

> *The indispensable object in most chefs' shtick is the simple plastic squeeze bottle, . . . essentially the same objects you see at hot-dog stands, loaded with mustard and ketchup. Mask a bottom of a plate with, say, an emulsified butter sauce, then run a couple of concentric rings of darker sauce—demi-glace, or roast pepper puree—around the plate, and . . . drag a tooth-pick through the rings or lines.*

Sure, it's a good tool to have if outdated, over-wrought plating is your thing. But there are better reasons to own a squeeze bottle than aesthetics. Namely, they'll make you a better cook and a better eater.

Before squeeze bottles made their appearance in my kitchen, I'd eat salads perhaps once or twice a month, and only when I was hosting a dinner party. The hassle of making a fresh batch of vinaigrette just for myself and my wife was simply too much (forget about using bottled dressing). These days, I keep a couple of different vinaigrettes ready to go in 12-ounce squeeze bottles in the fridge. Stick your finger over the top, give it a good shake, squirt it onto your greens in a mixing bowl, and boom: *lunch is served.* (In order to make sure that chunky items like shallots or crushed nuts won't get caught in the tip, sometimes you've got to snip off the tip of the bottle with a paring knife or a good pair of kitchen shears.)

As far as condiments go, squeeze bottles are another lifesaver. Sure, you can fill 'em with the standards: mustard, ketchup, and mayo, and, of course, you save money by buying those things in bulk instead of in individual squeezy containers. They are also great for saving money on all kinds of sauces and oils: I buy olive oil, sesame oil, soy sauce, hoisin sauce, oyster sauce, tonkatsu sauce, and Chinkiang vinegar (to name a few) in big cans. Then I just store the cans out of the way under the sink or in the closet and refill my squeeze bottles as needed. It'll make the inside of your refrigerator look all cool, organized, and cheffy as well.

Want to throw a fancy cocktail party? Squeeze bottles are your friend. Fill a big one with simple syrup, smaller ones with fresh-squeezed citrus juices or flavored syrups. You'll be cleaner, neater, and more efficient, cutting the time it takes to make each cocktail by a not-insignificant degree, and your guests will marvel at how pro you look.

As far as buying them goes, no need to get fancy. I picked up a couple dozen at a Chinese restaurant supply store. Amazon sells them for a few bucks apiece. Buy a half dozen and see if they don't change your life for the better.

And, yeah, like Tony says, you can use them to make your plates all frou-frou if you desire.

# WEIGHT VERSUS VOLUME

You may notice that in most cases, the baking recipes in this book are given with weight measurements—ounces and pounds—as opposed to volume measurements—cups and spoons. Why is that?

- **First and foremost, accuracy.** Volume measurements are simply not accurate. To prove this, I asked ten friends of mine to measure out a cup of flour from a bowl. Each person used the same measuring cup dipped into the same bowl of flour. I even had each of them use the same method: dipping the cup into the bowl, then leveling off the excess with a knife. Then I weighed each batch. The difference was astounding: the cups of flour ranged from as low as 4 ounces to as high as 6 ounces, depending on the force used while scooping. That means that one person in this group would end up using a *full* 50 percent *more flour* than the person with the lightest cup for the same recipe. If, on the other hand, I'd asked each person to weigh out 5 ounces of flour (my standard conversion for a cup of all-purpose flour), there'd be no problem at all: each person would weigh out exactly the same amount, regardless of how they handled a cup measure.

  With its fluffy texture and easy-to-aerate nature, flour is probably the most extreme example of this lack of correlation between volume and weight, but it extends to other ingredients as well.

- **Easy cleanup.** Think of it this way: to make a pizza crust recipe that calls for a couple cups of flour, a half teaspoon of salt, a tablespoon of olive oil, a cup of water, and a teaspoon of yeast using volume measures requires you to dirty at the very least a mixing bowl, a dry cup measure, a liquid cup measure, a half-teaspoon measure, a tablespoon measure, and a full teaspoon measure. Total items to wash: six. Not too nice.

  Here's how I make my pizza crust: place the bowl on the scale and weigh the required ingredients one at a time directly into the bowl. Total items to wash: one. Get it?

- **Measuring sticky items is easy.** Ever try and measure out, say, 2 teaspoons of honey? It's not easy. Sure, getting the honey into the teaspoon measure is simple enough—it's getting it out that's the problem. You end up either getting about half of it out and eyeballing an extra squirt or so directly from the bottle or, if you're anything like I used to be, you end up with most of the honey stuck to your finger as you try to scoop it out in desperation. With a kitchen scale and volume measurements, fretting about sticky ingredients is a thing of the past.

Here's the thing: We all grew up in this country using volume measurements instead of weight. We also unfortunately grew up without a firm grasp of the metric system, a system of measurement vastly superior to our ridiculous feet, inches, cups, and gallons. It might be harder to convince people to switch to metrics, but using weights instead of volume should be a no-brainer for anyone. Trust me. Buy yourself a good kitchen scale.

# { The basic pantry }

**T**he pantry is the backbone of your kitchen. Many beginning cooks are intimidated by recipes because of the sheer number of ingredients that need to be purchased the first time they cook something. But pancakes are a convenient food precisely because they are made from ingredients you pretty much always have on hand. Imagine having to buy flour, butter, eggs, buttermilk, baking powder, sugar, oil, and vanilla extract every single time you wanted to make pancakes!

I like to keep a well-stocked kitchen, and, as such, my pantry is a large one. I recently completely emptied my kitchen shelves and refrigerator and reorganized them, in the process cataloguing every pantry item I had into a single document that lives online, where I can access it at any time to see exactly what I have to work with. (What? Doesn't everybody do that?) I came up with 357 different food items, including 8 types of salt and 63 different spices (yikes!).

There's no need for you to keep a pantry that large, but *every* kitchen should be stocked with some basics. Here you'll find some tips on how to best use your refrigerator, as well as a list of ingredients that'll help you get through most of the recipes in this book with only the need to purchase perishable ingredients fresh. I divided it into refrigerated goods, baking supplies, grains, canned goods, spices, and what I call wet pantry items.

## REFRIGERATED GOODS

Like cell phones and clean underwear, a refrigerator is one of those things that you never really consider the importance of until it stops doing its job

(like mine did last week).\* Organizing your fridge for maximum efficiency—in terms of food shelf life, food safety, and easy access to the things you reach for most—should be a top priority. It'll make all of your cooking projects go faster and more easily, and having more fun in the kitchen inevitably leads to more cooking. That's a good thing in my book.

A fridge is basically just a big cold box with a few shelves in it, right? Well, that's true, but where you store food in the fridge can have quite an impact on its shelf life. Most refrigerators have cold and hot spots, with temperatures that range from 33° to 38°F or so. In general, the back of the bottom shelf, where cooler, heavier air falls to, and the back of the top shelf, closest to the fan and condenser, are the coldest spots, while the middle of the door is the warmest. How you organize your food in the fridge should be based on how cold it needs to be kept.

First, some basic tips on getting the most out of your fridge space on a daily basis:

- **Get a fridge thermometer.** There are a number of things that can cause your fridge to break down or lose power: electrical shorts or surges, clogged ventilation, etc. So it's possible that even with your temperature dial adjusted to the correct position, your fridge might be far warmer than it should be. A simple dial thermometer helps you monitor things to ensure that you're never caught in the dark.
- **Transfer food to smaller containers.** I keep a stack of half-pint, pint, and quart plastic deli contain-

---

\*The fridge, not the underwear.

ers to store almost all food once it's come out of the original packaging. Air is the enemy of most foods and can increase their rate of spoilage. By transferring them to smaller containers, you not only minimize air contact, but you also help keep your fridge organized and easy to navigate.

- **Label everything.** As soon as you transfer food into a smaller storage container, label the container, using permanent marker on masking tape with the date of storage, as well as what's inside. As much as I promote good science, there are some things that simply aren't worth experimenting with: creating life inside your refrigerator is one of them.

- **Prevent drippage.** To avoid messes and dangerous cross-contamination, always store raw meat—no matter how well wrapped—on a plate or a tray to catch any drips.

- **Keep fish extra cold.** It's best to use fresh fish immediately, but if you must store it, wrap it in plastic and sandwich it between two ice packs on a tray to ensure that it stays at 32°F or colder until ready to use. (Don't worry—because of dissolved solids in its cell structure, it won't freeze until well below 32°F.)

## Where to Store Food in the Refrigerator

There are three overriding factors to consider when deciding what to store where in the fridge.

- **Food safety** is of utmost importance. Fridges keep food fresh for longer, but that doesn't mean that harmful bacteria can't multiply to dangerous levels given enough time. To minimize risk, here's a rule of thumb: the more likely the possibility a food could make you sick and the higher the final temperature you intend to cook it to, the lower in the refrigerator it should be stored,

both to keep it cooler and to prevent cross-contamination. For instance, don't store raw chicken above leftovers from the night before. Juices from the bird can drip down unnoticed, contaminating your food.

- **Temperature** varies throughout your refrigerator, with, as mentioned earlier, either the very back of the bottom shelf or the back of the top shelf, near the vent, being the coldest spot, depending on the model. For maximum storage life, your refrigerator should be set to hold a minimum temperature of 34°F in these spots. No part of your refrigerator should rise above 39°F.

- **Humidity** plays a role in the freshness of vegetables. The crisper drawers in the bottom of your refrigerator are designed to prevent fresh cold air from circulating into them. Vegetables naturally emit a bit of energy as they go about their normal energy cycles, heating up the space in the drawer, thus enabling it to retain more moisture. Moist air can help prevent vegetables from shriveling or drying out. Most crisper drawers have a slider that controls the ventilation so that you can adjust the moisture level inside the drawer. The key is to maximize it, up to just below the point that moisture would start beading up on the vegetables' surfaces.

To give you an idea of good refrigerator storage organization, allow me to take you on a little tour of my fridge. Here's what you'll usually find there:

## The Main Compartment

### The Top Shelf
- **Ready-to-eat prepared foods.** Roasted red peppers, jarred tomatoes, a can of white asparagus, sundried tomatoes.

- **Ready-to-eat condiments that I don't use too often.** A variety of Chinese bean and chile pastes, curry paste, a half can of coconut milk, cans or jars of tahini, harissa, tomato paste, chipotles in adobo, olive tapenade, anchovies.
- **Pickled products.** Dill spears and chips, bread-and-butter pickles, ramps, jalapeños, capers, olives.
- **Fridge-friendly fruits** like apples, oranges, berries, melons, and grapes.

## The Middle Shelf

- **Leftovers in sealed containers.** Leftover mac and cheese, a few pieces of roasted chicken, my dog's food, braised asparagus, pizza sauce, salsa.
- **Cheese (in its original packaging or wrapped in parchment and stored in a sealed baggie).** A half hunk of goat's-milk Gouda, crumbled Cotija, homemade American cheese slices, sharp cheddar, a big hunk of Parmesan, Gorgonzola.
- **Eggs in their carton.** If it takes you more than a couple of weeks to go through a carton of eggs, store them on the back of this shelf, where it's a little cooler to maximize shelf life. Otherwise, you can keep them in the door (despite what anyone tells you). They'll keep for at least a few weeks, even in this relatively warmer environment.
- **Cold cuts and sandwich bread.** Martin's potato rolls, Arnold multigrain bread. Sliced sandwich bread will keep fine in the fridge. However, lean breads like baguettes or Italian-style breads should be stored at room temperature or in the freezer—the refrigerator will promote staling.

## The Bottom Shelf

- **Raw meat and poultry, wrapped carefully and on a plate.** Ground beef, skirt steak, fresh pork belly, Italian sausage.

- **Raw fish, in its wrapper and placed on a tray.** I buy my fish the day it's going to be consumed, and you should too—but see the tip on page 378 if you must store it overnight.
- **Milk and other dairy products.** Heavy cream, sour cream, cottage cheese, cream cheese, homemade crème fraîche, buttermilk.

## The Vegetable Crisper

- **Vegetables, stored in breathable plastic bags or plastic bags with the tops left slightly open.** Broccoli, celery, carrots, cucumbers, scallions, asparagus, radishes, turnips.
- **Herbs.** Parsley, cilantro, chives, thyme, rosemary, basil (in the summer). I wash and pick my herbs as soon as they get home, then store them rolled up in damp paper towels in plastic zipper-lock bags.

## The Fridge Door

The fridge door is the best place to store frequently used items and those that don't require the coldest temperature.

## The Top Shelf

- **Eggs**—if you go through a carton within a few weeks.
- **Butter and frequently used cheeses.** Cabot 83 unsalted butter, inexpensive Danish blue (love it on toast), Brie and other soft cheeses. Butter stays slightly softer in the fridge door, which makes it easier to spread on toast. If you eat a lot of cheese, you might want to store it here as well, so that it's not quite as cold when you grab it.

## The Middle Shelf

- **Condiments in their original packaging or in squeeze bottles if homemade.** Ketchup, chili sauce, several types of mustard, homemade mayo, Japanese barbecue sauce.

- **Premixed vinaigrettes in squeeze bottles.** Simple red wine vinaigrette, soy-balsamic vinaigrette.

### The Bottom Shelf

- **Drinks.** Whole milk, freshly squeezed pineapple juice, pitchers of chilled tap water, the occasional Cheerwine or Mexican Coke. Milk should go on a shelf in the main fridge compartment if you don't use much, but for daily drinkers, the door is a fine place for it, as it is for juices, sodas, etc.

## The Freezer

Everyone, of course, keeps frozen meats and vegetables in the freezer, but it's also an excellent place to store any heat- or light-sensitive items that might go rancid. In my freezer, aside from meat and veg, you'll find nuts (which can be toasted or crushed straight out of the freezer); cured meats like salt pork, bacon, and guanciale; dried bay leaves (I buy them in bulk); chicken stock frozen in 1-cup portions; bread crumbs; extra butter; yeast; sausage casings; whole-grain flours (they contain fats that can turn rancid at room temp); and fresh pasta, among other things.

Here are some tips for better freezer storage:

- **Keep your vents clear.** Make sure you don't stack food against the air vents, or you'll strain the freezer, greatly reducing its efficiency and efficacy.
- **Transfer meat from its original packaging.** To prevent freezer burn as well as to freeze the meat as quickly as possible (the faster it freezes, the less damage it will incur in the process), transfer it to flat airtight packaging. Best of all is to use a vacuum-sealer like a FoodSaver, which will completely eliminate the possibility of freezer burn. Next best is to wrap the meat tightly in foil, followed by several layers of plastic wrap (plastic wrap on its own will be air-permeable), or to use a freezer bag designed for long-term storage.

- **Freeze flat.** Wide, flat shapes freeze faster and can be stacked more efficiently than bulky packages. Freeze meats in a single layer in vacuum-sealed packages or freezer bags. Not only will this help you organize your freezer space, it'll also greatly cut down on defrosting time.

- **Label everything!** All packages should have the contents and date written on them. Nobody likes to play the frozen-mystery guessing game.

- **Defrost safely.** The best way to safely defrost meat is on a plate or a rimmed baking sheet in the refrigerator. Be aware that it'll probably take longer than you think: allow at least overnight for thin items like steaks, burgers, chicken breasts, and the like; up to 2 days for beef and pork roasts or whole chickens; and up to 3 or even 4 days for large turkeys. In emergencies, thinner foods can be rapidly defrosted by placing them in a bowl of cold water under a slowly running tap or, better yet, placed on an aluminum tray or pan, which will very quickly transmit energy from the room to the food. Steaks will defrost about 50 percent faster on an aluminum tray than on a wooden or plastic cutting board. Turn them over every half hour or so as they thaw. Do not try to defrost large items rapidly—the risk of dangerous bacteria growing on the exterior before the interior defrosts is too great.

## ESSENTIAL PANTRY INGREDIENTS

### Cold Pantry

Here are the refrigerated items I have on hand at all times:

- Bacon, slab (will last several weeks in the fridge, can be frozen for longer storage)
- Butter, unsalted (will last several weeks in the fridge; I keep a few extra pounds in the freezer, where it will keep indefinitely)
- Buttermilk
- Cheese, Parmigiano-Reggiano
- Eggs, large
- Ketchup
- Maple Syrup, Grade A dark amber
- Mayonnaise
- Milk, whole or 2% (or, if you must, skim)
- Mustard, Dijon
- Mustard, brown

## Baking Pantry

Some people are bakers, some are not. I wasn't born a baker, but I've discovered that after orga-nizing my baking pantry, making bread and pastry has become far more pleasurable for me. I used to store my flours and such in their original bags in a cabinet. To bake something, I'd have to pull every-thing out, try and measure out of a paper bag with a narrow opening, and finally end up folding the bag back down, forcing it to release a puff of flour that'd get all over my clothes and kitchen. Baking was a chore.

Then I decided to invest in a couple of large, sealable, wide-mouthed plastic tubs to store basic baking pantry items such as flours and sugar. This allows me to quickly and easily scoop up as much flour as I need without making a mess. These days, I make many more pizzas than I used to.

All of the items in the chart on page 79 should be stored in a cool, dry place, first transferred to a sealed container if appropriate.

# WHOLE WHEAT VERSUS REFINED WHITE FLOUR

A kernel of wheat is a pretty complicated thing, but as far as cooking is concerned, it can be divided into three basic parts: the endosperm, the hull, and the germ. Whole wheat flour is exactly what it sounds like—the entire grain from the wheat plant, ground up. Refined white flour contains only the starchy, proteinaceous sections from the endosperm, with all of the hull and germ removed. Why would anyone want to do that? It's all about gluten formation. We'll be talking quite a bit about gluten in this book, but for now, all you need to know is that gluten is the stretchy matrix of proteins that gives doughs their flexibility. It's formed when the proteins *gliadin* and *glutenin*, found in the endosperm, are mixed together in the presence of water.

White flour is excellent at developing gluten, delivering breads that are fluffy, chewy, and well risen. Whole wheat breads, on the other hand, tend to be dense and relatively dry. This is because ground-up sections of the hull and germ act sort of like tiny razor blades, snipping through the developing gluten and preventing individual strands from growing too long. You can substitute whole wheat flours in recipes if you'd like, but don't expect to get the same light, well-risen breads you'd achieve with white flour.

| ITEM | HOW LONG WILL IT KEEP? |
|---|---|
| **Baking Powder** | 6 months to a year, depending on humidity; to test for activity, place a teaspoon in a bowl and add a teaspoon of water: it should bubble and fizz vigorously. |
| **Baking Soda** | 8 months to a year |
| **Cornstarch** | Indefinitely |
| **Dutch-Process Cocoa** | 1 to 2 years |
| **Flour, all-purpose** | Transferred to a sealed container up to a year |
| **Flour, bread** | Transferred to a sealed container up to a year |
| **Gelatin, powdered** | Indefinitely |
| **Sugar, brown** | In an airtight plastic bag, 3 to 4 months optimally—after that, it may harden; hard brown sugar can be restored by briefly microwaving. |
| **Sugar, granulated** | Transferred to a sealed container, indefinitely |
| **Vanilla Extract** | 1 to 2 years |
| **Yeast, instant (rapid-rise)** | If possible, purchase in bulk and transfer to a sealed container; individual packets are harder to use and far more expensive. Keeps indefinitely in the freezer; if stored at room temperature or in the fridge, it will need to be proofed occasionally: add 2 tablespoons warm water and 1 teaspoon sugar to ½ teaspoon yeast and let sit for 10 minutes—it should produce foam. If not, replace. |

## Grains and Legumes

Grains and legumes should be stored in a cool, dry place. Beans will keep from 6 months to a year, while regular pasta and white rice will last indefinitely. Whole wheat pasta and brown rice will go rancid after extended storage (usually 6 to 8 months): smell them before using. If there is any hint of a fishy aroma, discard.

- Beans, dried black
- Beans, dried cannellini
- Beans, dried kidney
- Pasta, lasagna
- Pasta, short and holey (like elbows or penne)
- Pasta, long (like linguine or spaghetti)
- Rice, white or brown

## Canned Goods

Canned goods will last almost indefinitely, but it's better not to expose them to severe temperature fluctuations.

- Anchovies, oil-packed: after the container has been opened, anchovies can be stored in a sealed container under a layer of olive oil in the fridge for up to a month; for longer storage, roll up individual fillets, transfer to a zipper-lock freezer bag, and store in the freezer. I use Ortiz or Agostino Recca brand.

- Chipotle chiles, packed in adobo sauce
- Evaporated milk
- Tomato paste: I buy my tomato paste in tubes as opposed to cans so that I can use only what is needed for a recipe, without having to find a way to store the excess.
- Tomatoes, whole canned. I use Cento brand.

## Spices and Salts

Do you have a can of paprika or oregano in your kitchen that's been around since *He-Man* and *Mac-Gyver* were still on television? Do yourself a favor: throw it out. Spices lose their flavor over time, even when stored in sealed containers out of direct sunlight (as they should be). Whole spices may keep for up to a year or so without significant flavor loss, but preground spices will become noticeably less flavorful in a matter of months.

For the best flavor, you have two options. The first is to buy your spices whole and in small batches, replacing them every 6 months to a year or so. The alternative is to buy whole spices in bulk, keeping small amounts of them in jars in your spice rack and storing the remainder in vacuum-sealed pouches (like those for a FoodSaver-type vacuum-sealer) in a cool, dark place or, preferably, in the freezer. Salt will last forever, so long as it's kept dry.

- Bay leaves, whole (store in the freezer)
- Black peppercorns
- Chili powder
- Cinnamon, ground
- Coriander seeds
- Cumin seeds
- Fennel seeds
- Nutmeg, whole
- Paprika
- Red pepper, crushed

- Oregano, dried
- Sage, dried
- Salt, kosher
- Salt, Maldon

## Oils, Vinegars, and Other Liquids

Oils are the most sensitive wet pantry item in your kitchen. Stored badly, they can go rancid within a span of weeks. The enemies of oil are heat and light, which means that the way most people store them—in clear bottles close to the stove—is just about the worst thing you can do. I store my cooking oil and everyday extra-virgin olive oil in dark green wine bottles that I've washed and dried, fitted with inexpensive pour spouts for the purpose. They stay on my counter, far away from the window and the stove. The oils last for about a month in those containers before I refill them.

I keep expensive extra-virgin olive oils in their original containers in a dark cabinet, where they will last for about 2 months. Remember, there's no point in having great olive oil if you don't use it before it starts to lose flavor or go rancid. I've learned this the hard way. Olive oil's for eating, no matter how expensive it is. Eat it.

- Honey, clover
- Marmite, Vegemite, or Maggi Seasoning
- Molasses, regular
- Oil, canola (for sautéing)
- Oil, extra-virgin olive (for flavoring)
- Oil, peanut (for deep-frying)
- Soy sauce (if you won't use up a bottle within 2 months, store it in the refrigerator). I use Kikkoman brand.
- Vinegar, cider
- Vinegar, balsamic (supermarket)
- Vinegar, distilled white
- Vinegar, white wine

# WHICH SALT SHOULD I USE?

These days you see more types of salt on supermarket shelves than there are tools under Inspector Gadget's trench coat. But really, there's only one that you absolutely need in your kitchen: kosher salt. I use Diamond Crystal brand because I like the size of its grains. For the record, kosher salt is not called kosher because it's OK to eat under Jewish dietary law—all salt is kosher in that sense. Kosher salt should really be called *koshering* salt, because its large grains efficiently draw blood out from flesh during the koshering process (which, by the way, makes it an extremely efficient salt for dry-brining—more on that later).

Why use kosher salt over regular table salt? One word: sprinkling. Table salt is fine if you use it out of a saltshaker, but you get a much better idea of how much salt you're actually putting into or on your food if you add the salt with your fingers, and kosher salt is simply easier to pick up and apply that way. To apply an even layer of salt to your food, pick up a pinch of kosher salt, then hold your hand *high* above the food before sprinkling it. Because of turbulence in the air, your salt will rain down upon your food in a pattern that shows a normal (bell curve) distribution from where you drop it. The higher you drop it from, the more even the distribution.

All of the recipes in this book were tested with Diamond Crystal kosher salt. If you *must* use table salt, you should use only two-thirds as much as is called for, as table salt packs more tightly into a measuring spoon (most of the time it's called for in amounts too small to effectively measure with a scale). In most savory recipes, you'll be able to taste the salt level as you cook, adjusting it to suit your own palate. Whenever appropriate (for baking projects, brines, etc.), I've given salt measurements in weight.

And what about all the fancy "designer" salts? The pink or black ones? The grayish sea salt from Guérande in France that comes in big, moist clusters or the white pyramid-shaped Maldon sea salt from England? I have a bad habit of collecting them, partly because they're pretty and I like the way they look on my food, but mostly just to compete with my wife's shoe collection. (One new salt per pair of shoes seems to keep her shopping habit at bay.)

But what are they good for? These are all *finishing salts*, salts that are meant to be applied just before serving or even at the table. Despite claims to the contrary, you'll find that flavorwise, there is almost no difference between these salts and regular or kosher salt. Dissolve the same weights of the stuff into glasses of water, and they all become essentially identical. It's their shape that makes them interesting—the crunch and intense burst of, well, saltiness that they provide. Think you won't notice the difference? Go out and get yourself a box of Maldon sea salt (the finishing salt that I use most often), a box of kosher salt, and a box of regular table salt, then place three identical slices of ripe tomato on a plate (or if you prefer, three identical slices of steak). Sprinkle a bit of table salt on the first and eat it. Next, sprinkle some kosher salt on the second and eat it. Notice the difference? See how much more easily you can sprinkle the salt evenly across the surface of the food? Finally, sprinkle a few shards of Maldon salt on the last and eat it. Notice the crackle of salt crystals under your teeth and the accompanying burst of flavor? That's why I keep kosher salt next to my stove and cutting board and a large-crystal sea salt on my dining-room table.

EGGS, DAIRY, AND

THE
SCIENCE
*of*
BREAKFAST

1

Bacon and eggs: two perfect foods

# EGGS, DAIRY, AND THE
# SCIENCE *of* BREAKFAST

## RECIPES IN THIS CHAPTER

IS THERE ANY
**FOOD**
SO PERFECT, SO COMPLETE, SO
PROFOUNDLY SIMPLE YET
STAGGERINGLY COMPLEX AS THE
EGG?

I t's easily the most versatile and useful ingredient in the pantry. Just think of what you can do with eggs: You can eat them fried, scrambled, soft-boiled, hard-boiled, poached, baked, or turned into an omelet. They make the breading stick to your chicken parm. Their proteins can be set into a dense matrix that thickens custards or whipped into an airy foam that leavens batters. They can bring together your meat loaf without weighing it down, or act as culinary ambassadors, helping turn oil and water into a stable, creamy mayonnaise. All this, and they come in their own convenient, easy-to-measure, easy-to-store packaging to boot. They practically sell themselves.

Eggs are truly a marvel, and it's no wonder that their culinary uses are so varied. Just think: given fertilization and enough time, an entire living, breathing creature can be formed from the contents of an eggshell. The start of life, the start of many recipes. I can't think of a better subject with which to start this book.

# The FOOD LAB's Complete Guide
## TO BUYING AND STORING EGGS

When I say eggs, I'm pretty much always referring to chicken eggs, by far the most prevalent type of avian egg in the world. But are all chicken eggs created equal? Do some taste better than others? What factors affect how they work in recipes, and how can I make sure to get the best out of them? Here are the answers to all those questions and more.

## Identification

**What exactly is an egg?**

An egg is a vessel for the developing embryo of an animal that reproduces through sexual reproduction. In the culinary sense, we're usually referring to eggs from avian animals that are expelled from the body, like chicken eggs.

**What's inside the egg that makes it so culinarily useful?**

There are two basic parts to an egg: yolk and white.

**The yolk** is the nutritive source for the developing embryo, and it accounts for about 75 percent of the calories in an egg. Yolks may appear rich and fatty, but, in fact, they are essentially sacks of water that contain dissolved proteins, along with larger masses of protein and fat linked together with *lecithin*, an emulsifying molecule that allows fat and water molecules get along together harmoniously. We'll get back to that in a moment.

**The white** is also mostly water, along with a few proteins—the most important being *ovalbumin*, *ovomucin*, and *ovotransferrin*, which give it its unique capacity to both set when cooked and be whipped into stiff, shaving cream–like peaks.

Because the proteins in eggs are already dissolved and spread out in a liquid, it is very easy to incorporate them into other foods—much more than, say, meat proteins, which are relatively firmly set in place in relation to one another. (Have you ever tried whipping a steak? I have. It doesn't work.) Additionally, the fact that eggs contain such a wide variety of proteins, each of which behaves in a slightly different way when heat or mechanical action is applied, means that as a cook, you have great control over the final texture of your finished dish. Eggs cooked to 140°F, for example, will be soft and custard-like, while those cooked to 180°F will be bouncy and firm.

## Labeling: Size and Quality

**Eggs come in a few different sizes at the supermarket. Which ones should I be reaching for?**

Any carton of eggs that displays the United States Department of Agriculture (USDA) shield on it was packed according to USDA weight standards, which define six different classes, as shown in the chart below.

| WEIGHT CLASS | MINIMUM WEIGHT PER EGG |
|---|---|
| Jumbo | 2.5 ounces |
| Extra large | 2.25 ounces |
| Large | 2 ounces |
| Medium | 1.75 ounces |
| Small | 1.5 ounces |
| Peewee | 1.25 ounces |

In reality, you're unlikely to see small or peewee eggs at the supermarket—chickens these days are bred to produce eggs medium-size and up. Large eggs are the standard in most recipes, including the ones in this book. I do like to have jumbo eggs on hand in my fridge, though, for those post-night-out mornings when I can really use that extra half ounce of fried egg to fill me up. You're also more likely to find one of the coveted double yolks in a larger egg.

**What about those letter grades on the side of the carton? Are Grade A eggs better than Grade B?**

Like sizing, grading of eggs is a voluntary action that most manufacturers choose to comply with in order to get the USDA stamp of approval on their boxes. USDA grading experts examine sample eggs from each batch to determine the grade based on the quality of the whites, yolks, and shells. Eggs with the firmest whites, tallest-standing yolks, and cleanest shells will get an AA stamp, while eggs with watery whites, flat yolks, and stained shells receive a B. Grade A lies in the middle and is what most retail stores carry for consumers. As far as cooking quality goes, a firm white and yolk are important for things like poached eggs and fried eggs where a nice, tight appearance is desired, but in most cooking or baking application, any grade'll do—it's a cosmetic difference alone.

## Egg Freshness

**You mentioned that lower-graded eggs have watery whites and so will tend to spread out more than higher-graded ones. But doesn't freshness play a role in this too?**

Indeed it does. Very fresh eggs have tighter yolks and whites that will hold their shape much better during poaching or frying, as well as yolks that will remain better centered when boiled. Because of the way their proteins break down, eggs become looser and looser as they age. There's another important change too: as eggs age, they become more and more alkaline. This is particularly important in meringue-based dishes, as the pH of egg whites can greatly affect their foaming power. Egg whites foam best in slightly acidic environments, which means that old eggs will produce looser, wetter foams. To counteract this, a pinch of acidic cream of tartar will help your meringues stay stiff and weep-free.

Loose whites in older eggs.

I've heard that older eggs are better for boiling because they are easier to peel. Is this true? Is there any culinary advantage to using older eggs?

I believed this for the longest time—until I actually tested it with a few cartons of eggs from different sources, comparing them with some eggs I got from my neighbor in Brooklyn's backyard that were less than a week old. Guess what? Whether the eggs were a week old or two and a half months old, they were just as likely to have shells that stuck to them when peeling. On top of that, with older eggs, the yolks become uncentered, gravitating toward the egg wall, making for unattractive slices.

No matter how you plan on cooking them, fresh eggs are better than old ones.

**Is there a trick to getting the shell off a hard-boiled egg without mutilating the white?**
I've tried every method known to man, ticking them off one at a time. Shocking the eggs in ice water? It makes no difference. Poking a hole in the shells before cooking them? Nope, sorry. Steaming or pressure-cooking them? Nuh-uh. Adding vinegar to the water? All that does is dissolve the outermost layer of shell.

Eggs lowered into boiling water or hot steam have the best chance of peeling easily.

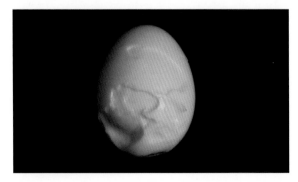

Eggs slowly heated in cold water will stick to their shells.

In fact, I discovered that the only thing that really seems to make a difference is the initial cooking phase. Drop the eggs into hot water, and they'll peel pretty easily (though even this doesn't work 100 percent of the time). Heat them up slowly, starting with cold water, and the egg proteins will end up fused to the inside of the shell.

As far as the actual peeling process, the easiest way is to peel the still-hot eggs under cool running water, starting from the fat end, where the air pocket is located. When the eggs are hot, the connection between the membrane and egg white is weaker, making it easier to remove the shell. The cool water not only helps gently dislodge stubborn bits of shell, it also prevents your fingers from getting burned. I put a fine-mesh strainer or colander in the sink to catch the shells, for easy cleanup.

Peeling under running water helps loosen the peel.

### How do I know how old an egg is?

You can try checking the carton label, which will give you a rough idea. On pretty much every package of eggs, you'll see a sell-by date, as well as a pack date (also known as the Julian date), the date on which the eggs were inspected, cleaned, and placed in the carton. The pack date is the three-digit number immediately above the sell-by date, starting with 001 for January 1 and ending at 365 for December 31. Legally, the sell-by date can be no more than 45 days after the pack date, but when properly refrigerated, eggs will remain wholesome for well beyond this 45-day period—60 to 70 days is reasonable.

While it's possible that the eggs you're buying were laid within a few days of their pack date, manufacturers have up to 30 days to clean and pack eggs, which means that, in theory, if you buy a carton of eggs on its expiration date, it may already be 75 days old! Clearly, checking the expiration date is not the most reliable way to guarantee fresh eggs. You're much better off checking the pack date.

### What if I buy eggs without a pack date or I've transferred the eggs to the egg compartment in my refrigerator door and no longer know the date?

First off, everyone tells you that if you want to maximize shelf life, you should get those eggs out of the fridge door and into the coldest part of your fridge. True. But what they fail to tell you is that even on a shelf in the door, eggs will last for several weeks beyond their pack date. So unless you eat or cook with eggs only on very rare occasions, go ahead and keep them in the door. You'll use 'em up long before they go bad.

That said, there's a quick and easy test to gauge the freshness of an egg: just drop it into a bowl of water. Eggshells are porous: they can lose about 4 microliters of water a day to evaporation while simultaneously taking air into the space between the shell and the inner membrane near the fat end. In very fresh eggs, the air space is tiny and the egg will sink to the bottom of the bowl and lie on its side. As eggs get older, the air space will grow, so old eggs will sink and then stand on their points as the air in the larger end tries to rise. If you've got an egg that floats, it's probably past its prime and should be discarded.

Old eggs stand up when submerged in water.

**My local farmers' market sells unrefrigerated eggs, and I've seen some supermarkets in Europe where the eggs just sit out on shelves. Are they crazy, or is it me?**
Most likely it's you. When eggs are first laid, they are covered in a thin wax-like coating called the cuticle. This cuticle is the egg's first barrier against bacterial infection and excessive moisture loss. In the United States, USDA-stamped eggs are all washed prior to packaging, a step that removes the cuticle. It may mean that our packaged eggs are cleaner to begin with, but it does mean that they have less protection against future bacterial infection as they sit in the supermarket—refrigeration is necessary to help

prevent this. But many eggs sold at farmers' markets or in European supermarkets have not been washed prior to packing. The cuticle remains intact, so refrigeration is unnecessary, but the eggs tend to have a shorter shelf life than refrigerated eggs.

**What about the "pasteurized eggs" I'm seeing on the market these days?**
Pasteurized eggs are a relatively new product. They are sterilized by submerging the eggs in a water bath at around 130°F, a temperature that, given enough time, is hot enough to kill any harmful bacteria on or inside the egg but cool enough that the egg won't cook. Pasteurized eggs are useful for people who like to eat their eggs runny or in raw preparations like mayonnaise but don't want to run the (very minimal) risk of getting sick from them. For most cooking purposes, pasteurized eggs will work fine, though you'll notice that the whites are runnier (making them difficult to poach or fry), and that they take about twice as long to whip into peaks. The yolks work just as well as those from regular eggs in mayonnaise or Caesar salad dressing.

**Is it true that brown eggs are healthier than white?**
Absolutely not. The color of the eggshell has to do with the breed of chicken, and it is largely controlled by market demands. In most of New England, brown eggs are the norm, while the majority of the rest of the country prefers white eggs. They are completely interchangeable.

## Egg Labeling
**I miss the old days, when I could walk into the supermarket and pick up a carton of eggs without feeling like I was making an important life decision. These days, there are dozens of varieties to choose from. What do all the labels mean?**

It is confusing, and it largely has to do with growing consumer awareness about the conditions in which egg-laying chickens are kept. Most spend their lives as little more than egg-producing machines, housed in batteries of individual cages, unable to spread their wings or even move, with little or no access to a space where a chicken could perform its natural behaviors. The label on the carton can be an indication of better welfare for the birds.

- **Natural** indicates that the eggs are minimally processed, but since all eggs are sold minimally processed, the label effectively means nothing. Similarly, the term Farm-Fresh carries with it no guarantees, because presumably nobody is selling rotten eggs that don't come from a farm.
- **Free-Range**, **Free-Roaming**, and **Cage-Free** eggs come from chickens that are not kept in battery cages, but instead in large open barns or warehouses. That is a major improvement in quality of life for the chickens, allowing them to engage in natural behaviors like pecking, dust-bathing, and spreading their wings. Free-Range and Free-Roaming chickens generally also have access to outdoor areas, but the labeling laws have no requirements as far as the size or quality of the area goes, nor for how long the chickens must be allowed out. Fact of the matter is, most of these chickens never set foot outside the barn. These labels are not audited—you're going on the word of the producer alone.
- **Certified Organic** eggs come from chickens kept in open barns or warehouses with an unspecified degree of outdoor access (again, for all intents and purposes, probably none). They must be fed an organic, all-vegetarian diet free of animal by-products, antibiotics, and pesticides, and farms are checked for compliance by the USDA.

- **Certified Humane** eggs have been verified by third-party auditors, and this label requires stricter controls on stocking densities, giving the chickens more space and the ability to engage in natural behaviors like nesting and perching. Producers are not allowed to engage in forced molting, the practice of inducing hens into a laying cycle by starving them (this practice is allowed for all other types of eggs).
- **Omega-3–Enriched** eggs come from chickens that have been fed supplements made from flaxseed or fish oil to increase the levels of omega-3 fatty acid—an essential fatty acid touted with several health benefits—in their yolks. While some people claim eggs high in omega-3s have a "fishy" aroma, in blind tastings, I've found no significant differences in the way these eggs taste.

If animal welfare is a concern, you are making a good step in the right direction by purchasing only Certified Organic or Certified Humane eggs. If you've got a local farmers' market where you can actually talk to the farmer producing the eggs you're purchasing, you're making an even better decision. Of course, the very best thing you can do is to build your own coop (or, better yet, convince your neighbor to do so) and keep a couple chickens. It won't save you much money in the long run, unless you keep a large flock and eat a *lot* of eggs, but you'll have the freshest-possible eggs and probably make plenty of friends in the process.

**That's all well and good for the chickens, but do Certified Organic or local eggs taste better, like the guys at the farmers' market would like you to think?**
That's a good question, and one that I've wondered about often. It seems natural that a happier, healthier chicken roaming around a backyard

poking, scratching, eating bugs and worms, clucking, and doing all the charming and funny things chickens do should produce tastier eggs, right? I mean, I *know* that some of the best-tasting eggs I've ever eaten have come fresh out of the coops or backyards of friends who keep their own flocks. The yolks were richer, the whites tighter and more flavorful, and it was just an all-around better experience. Or was it? What if all their greatness was simply in my head?

To test this, I organized a blind tasting in which I had tasters taste regular supermarket eggs, plain organic eggs, organic eggs with varying levels of omega-3, and eggs fresh from 100-percent free-roaming, pasture-raised chickens. All of the eggs were served scrambled. The results? Indeed the pastured eggs and omega-3–enriched eggs fared better than the standard supermarket eggs. But I

also noticed another correlation: the color of the eggs varied quite a bit, with the pastured eggs on the more intensely orange end of the spectrum. And the more omega-3s the eggs contained, the deeper orange the yolk. The plain organic eggs and standard factory eggs were the palest of the lot. This difference in pigmentation can be attributed to the varying diets of the chickens. Pastured hens eat bugs and flowers, both of which contribute color to yolks. Chickens bred for eggs with high omega-3 acids are fed with a diet enriched with flaxseeds and sea kelp, which contribute color. Chickens that lay these more expensive eggs are also sometimes fed pigmented supplements, like marigold leaves, that make their yolks nice and bright. Could it be that the flavor differences tasters were reporting had more to do with their reaction to the color than to the actual flavor of the egg?

These eggs were dyed green in order to figure out just how big a role color plays in our perception of flavor. Hint: it's a lot.

In order to eliminate color as a variable, I cooked up the same kinds of eggs, this time dying them green with some food coloring. When I re-administered the tasting with green eggs, *there was absolutely no correlation between flavor and provenance.* People liked the regular supermarket eggs just as much as the eggs that had come straight from the pasture.

Want to see the same effect for yourself? Take a look at these two (identical save for some Photoshop color tinkering) pans of eggs and tell me which one you'd rather eat:

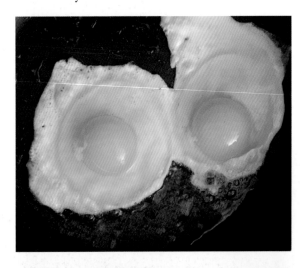

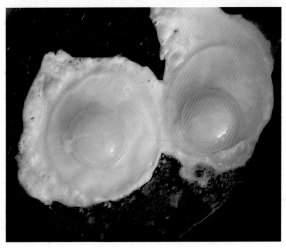

The old saying that you eat with your eyes? *It's true.*

**So you're telling me that it doesn't matter at all where I get my eggs from?**

No, I'm not saying that at all. Our minds are extraordinarily powerful, and our taste preferences have as much to do with our mental biases and upbringings as they do with real measurable physical characteristics in the food. You've probably noticed it yourself. Doesn't an ice-cold beer taste better when you're drinking it with friends on an outdoor patio on a warm summer evening than on those lonely nights when you're drinking solo? Doesn't the atmosphere and service in a restaurant affect the flavor of the food in your mind? Do you really think that your mom's apple pie is better than anyone else's? Chances are, the reason you like it so much is because it's your mom making it. The combination of physical appearance, weather, company, atmosphere, and even your mood can affect the flavor of food.

I like to think of it this way: I'm going to continue eating the freshest eggs I can find produced by the most humanely raised chickens because I care a bit about the chickens' well-being. The fact that my mind tricks me into thinking these eggs actually taste better is just icing on the cake. You mean I get to do the right thing *and* my eggs will taste better? Yes, please! One more advantage: eggs bought directly from the farmer at a farmers' market are generally much fresher (I've managed to buy eggs that had been laid the day they were sold), making them better to cook with and much easier to poach or fry.

# BOILED EGGS

**Boiled eggs are about the simplest recipe in any cook's repertoire, right?**

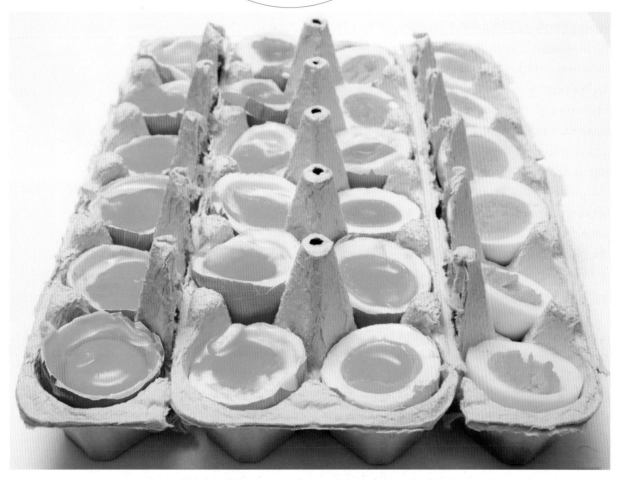

Eggs boiled for 30-second intervals from 0 to 12 minutes.

But how often do you get truly *perfect* boiled eggs? Hard-boiled eggs should have fully set, but not rubbery, whites surrounding yolks that are cooked through but still bright yellow and creamy, with no hint of chalkiness or crumbling, and certainly none of that dreaded sulfurous green tinge at the yolk-white interface that overcooked eggs acquire. Soft-boiled eggs, on the other hand, should have fully set whites with liquid yolks that ooze out like a soft custard, bathing your toast in their golden flow and enriching your crisp bacon. There's more to boiling an egg than meets the eye. Much more.

Nearly every basic cookbook offers a different technique for how it should be done: start the egg in cold water, or gently lower it into boiling water; add vinegar to the water to lower its pH, or add baking soda to the water to raise it; cover the pot, or don't cover it; use old eggs, or use new eggs; and on and on. But very few offer evidence as to why any one of these techniques should work any better than another. Apparently, boiling eggs is not . . . ahem . . . an *eggs*act science. Let's try and change that.

## What Is Boiling?

First things first: what exactly is boiling? The technical definition is that it is what occurs when the vapor pressure of a liquid is greater than or equal to the atmospheric pressure that surrounds it. Let's go back to the chicken coop analogy we used on page 29. Your pot of water is a coop full of chickens. The chickens tend to like each other and happily stick together inside the coop. Now, let's say we start adding energy to the mix by switching their water supply out with coffee. With the added energy, the chickens begin to become hyperactive—one or two of them might even be so energetic as to be able to jump the fence and escape. Add enough energy to the mix, and eventually the chickens will become so hyperactive that they'll tear down the fence and begin escaping very rapidly indeed.

Boiling water is the same thing. The water molecules are trapped in the pot and kept in place by their own fence—the pressure of the air in the atmosphere pushing down on them. Add energy to the pot in the form of heat, and water molecules begin to start leaping off the surface of the water. This is called *evaporation*. Eventually the pressure produced by the water molecules trying to escape becomes equal to or greater than the pressure of the atmosphere pushing down on it. The fence breaks, the floodgates open, and water molecules rapidly jump from a liquid state to gas, bubbling up violently. This conversion of liquid water to water vapor (steam) is what you see when you look at a pot of boiling water; with pure water at sea level, this occurs at 212°F (100°C).

Here's a quick rundown of what happens when you bring a pot of water to a boil:

- **Quivering: At between 130° and 170°F,** tiny bubbles of water vapor begin forming at nucleation

Quivering water below 170°F.

A subsimmer just below 195°F.

sites (more on those later) along the bottom and sides of the pot. They won't be large enough to actually jump and rise to the surface of the water, but their formation will cause the top surface to quiver a bit.
- **Subsimmer: At between 170° and 195°F,** the bubbles from the sides and bottom of the pot begin to rise to the surface. Usually you'll see a couple of streams of tiny, champagne-like bubbles rising from the bottom of the pot. For the most part, however, the liquid is still relatively still.
- **Simmer: At between 195° and 212°F,** bubbles break the surface of the water regularly, and from all points—not just a few individual streams, as in a subsimmer.

- **Full boil: At 212°F**, bubbles of water vapor escape extremely rapidly. This is the hottest that water can get at sea level without the aid of a pressure cooker.

## HEAT AND EGGS

Though some people claim that adding salt, vinegar, or baking soda to the water when you boil an egg can affect its final texture, in my testing, I found that the only factors that matter when boiling an egg in its shell are **time** and **temperature**.

To find out exactly how fast an egg cooks in boiling water, I cooked a dozen and a half eggs, removing them from the pot at 30-second intervals before splitting them open.

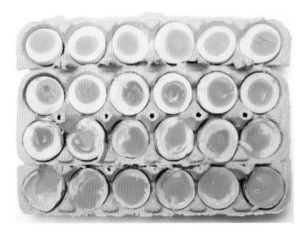

Now, there are a few things you'll notice immediately. First and most obvious is that the longer you keep an egg in boiling water, the hotter it gets. But here's something more important, which may seem trivial at first, yet, as we'll see, is instrumental in a perfectly cooked egg: foods in a hot environment cook from the outside in, and the bigger the temperature differential between the food and the environment, the more uneven the cooking will be.

What this means is that if you lower an egg into boiling water, it's possible to achieve an end result with a white that's tough, rubbery, and overcooked while the yolk is still just barely cooked in the center, like this one:

A bad egg.

So, what's the ideal temperature to cook an egg to? Here's what happens to an egg white as it heats up:

- **From 130° to 140°F:** As the white gets hot, its proteins, which resemble coiled balls of yarn, slowly start to uncoil.
- **At 140°F:** One of these uncoiled proteins, *ovotransferrin*, begins to bond with itself, creating a semisolid matrix that turns the egg white milky and jelly-like.
- **At 155°F:** The ovotransferrin has formed an opaque solid, though it is still quite soft and moist.
- **At 180°F:** The main protein in the egg white, *ovalbumin*, will cross-link and solidify, giving you a totally firm but still tender white.
- **Beyond 180°F:** The hotter you get the egg, the more tightly the egg proteins bond, and the firmer, drier, and more rubbery the egg white becomes. Eventually, hydrogen sulfide, or that "rotten-egg" aroma, begins to develop. Congratulations: your egg is overcooked.

# ALTITUDE AND BOILING

**B**ecause of gravity, the higher you go, the fewer air molecules there are in a given space—so the air is less dense. Lower density means lower atmospheric pressure, and lower atmospheric pressure means that water molecules in a pot need less energy to escape into the air. In Bogotá, Colombia, where my wife is from, for example, you're a good 8,000 feet above sea level and water boils at a temperature about 14 to 15 degrees lower than it does at sea level.

The graph below charts the boiling temperatures of water as you go into higher altitudes. The altitude effect can wreak havoc on recipes. Beans don't cook right. Pasta never softens. Stews take longer to braise. Pancakes can overrise and deflate. Go high enough, and you won't even be able to cook vegetables, which need to be heated to at least 183°F to break down.

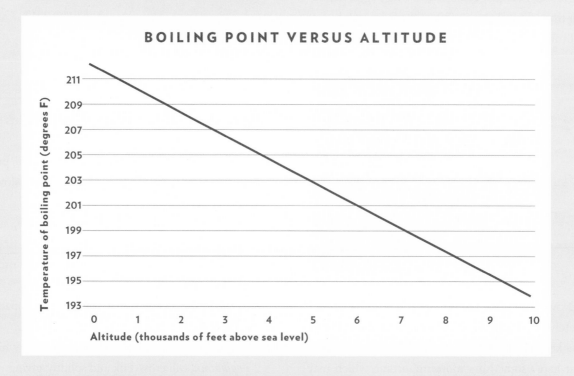

**BOILING POINT VERSUS ALTITUDE**

For some of these problems, most notably involving stews, dried beans, and root vegetables, a pressure cooker can be a lifesaver. It works by creating a vaportight seal around the food. As the water inside it heats up and converts to steam, the pressure inside the pot increases, because steam takes up more space than water. This increased pressure keeps the water from boiling, allowing you to bring it to a much higher temperature than you would in the open air. Most pressure cookers allow you to cook at temperatures between 240° and 250°F, no matter what the altitude. This is why pressure cookers are so popular throughout the Andes—no self-respecting Colombian home is without one.

# WATER-BOILING MYTHS

Myths about boiling water abound. Here are four of the most common:

- **Cold water comes to a boil faster than hot water.** False. It's absolutely untrue, but there is a good reason to use cold water instead of hot for cooking: hot water will contain more dissolved minerals from your pipes, which can give your food an off flavor.
- **Water that's been frozen or previously boiled will come to a boil faster.** False, though there is a little scientific reasoning behind this one. Boiling or freezing water removes dissolved gases (mostly oxygen), which can slightly affect the boiling temperature—so slightly, in fact, that neither my timer nor my thermometer could detect any difference.
- **Salt raises the boiling point of water.** True . . . sort of. Dissolved solids like salt and sugar will in fact increase the boiling point of water, causing it to come to a boil more slowly, but the effect is minimal (the amounts normally used in cooking effect less than a 1-degree change). For it to make any significant difference, you'd need to add it in really vast quantities. So for the most part, you can ignore this one.
- **A watched pot never boils.** Definitely true. Avert your eyes.

## On Salt and Nucleation

So, if salt doesn't lower the boiling point of water, how come throwing a handful of salt into a simmering pot will cause a sudden eruption of bubbles? It's because of tiny things called nucleation sites, which are, essentially, the birthplace of bubbles. In order for bubbles of steam to form, there has to be some sort of irregularity within the volume of water—microscopic scratches on the inside surface of the pot will do, as will tiny bits of dust or the pores of a wooden spoon. A handful of salt rapidly introduces thousands of nucleation sites, making it very easy for bubbles to form and escape. The same principle is used to "seed" clouds. Releasing dusty particles from an airplane causes millions of nucleation sites to be created in the moist atmosphere so water vapor droplets can coalesce and form clouds.

An egg yolk follows a different set of temperatures:

- **At 145°F:** The yolk proteins begin to denature, thickening the liquid yolk.
- **At 158°F:** The egg yolk is firm, able to hold its shape and to be cut with a fork or knife. Its appearance is still dark and translucent, with an almost fudge-like texture.
- **Between 158° and 170°F:** The yolk becomes firmer and firmer until eventually it suddenly shifts from translucent and fudge-like to pale yellow and crumbly as tiny spherical chambers invisible to the naked eye separate from each other.
- **Above 170°F:** The yolk becomes increasingly crumbly as the temperature goes up. The sulfur in the white rapidly reacts with the iron in the yolk, creating ferrous sulfide, tingeing the outside of the yolk an unattractive green.

Boiling eggs is all about balancing the differences between the way the whites and the yolks cook.

## SOFT-BOILED EGGS

For me, the ideal soft-boiled egg has a white that's completely opaque, but not to the point of rubberiness (somewhere in the range of 155° to 180°F), and a yolk that's pretty much 100-percent liquid (no hotter than 158°F). In this way, with each spoonful, you get tender bites of soft, velvety-smooth white bathed in a sauce of glorious, bright golden, rich, flavorful yolk.

So, remembering that foods cook from the outside in and that the hotter your cooking environment, the greater the temperature gradient that forms in your egg, you realize that for soft-boiled eggs, you want to start with *cold* eggs and submerge them in *hot* water, so that the whites cook and set while the yolks remain liquidy. I tried plunging the eggs directly into boiling water to cook until the whites were just set, but I ran into a problem: the outermost layers of the whites end up slightly overcooking. A much better way to do it is to bring a pot of water to a boil, shut off the heat, drop the eggs into it, cover the pot to help it retain some heat, and then start the timer. Since the water in the pot gets cooler as it sits, the eggs stand much less of a chance of overcooking and turning rubbery.

The other important thing to consider is the ratio of water to eggs—add too many eggs, and they'll cool the water down so much that they won't cook properly. So, 3 quarts is enough water to cook up to 6 eggs. Any more than that, and you'll want to cook in batches, or in a larger pot.

# FOOLPROOF SOFT-BOILED EGGS

**NOTE:** Depending on how hot your kitchen is and your cookware's heat-retention abilities, cooking times may vary slightly. It's a good idea to do a practice run with a single egg and adjust the time as necessary. Cooking times should be increased if you live at a moderately high altitude. At very high altitudes, you should maintain a boil for the first few minutes of cooking.

1 quart water for every 2 eggs

1 to 12 large eggs

Choose a lidded saucepan (or pot) small enough that the eggs will be fully submerged when you add them to the water. Bring the water to a boil over high heat. Add the eggs, cover the pan, and remove from the heat. Cook the eggs according to the times given in the chart, then remove with a slotted spoon and serve immediately.

| COOK TIME | DESCRIPTION | BEST USES |
|---|---|---|
| 1 to 3 minutes | Outer white set just enough to allow egg to retain its shape when carefully peeled | I use 1- or 2-minute eggs when I'm tossing the eggs with salad or pasta where the uncooked egg will emulsify with other ingredients; they're not pleasant to eat on their own. |
| 4 minutes | White is opaque nearly all the way through but retains a bit of translucency next to the yolk; yolk is barely warm and completely raw | Serve as a topping to vegetables or grains; place on top of blanched asparagus or green beans or in a bowl of noodle soup. |
| 5 minutes | White is opaque but still quivering and barely set toward the yolk; yolk is warm but completely raw | Breakfast |
| 6 minutes | White is opaque, firm all the way through; yolk is warm and starting to firm up at the edges | Breakfast |
| 7 minutes | White is fully cooked and as hard as that of a hard-boiled egg; yolk is golden and liquid in the center but beginning to set around the edges | Breakfast |

## HARD-BOILED EGGS

Hard-boiled eggs are a little more complicated. The goal is to have both the white and yolk at the point where they are opaque but not rubbery. The truly obsessive kitchen nerd's way to do it is to maintain the water at precisely 170°F so that the yolk comes out perfectly cooked and the white is still tender. And this method works. It's also a major pain in the butt. Luckily, there's an easier way.

We already know that if we drop the eggs directly into boiling water, the exterior heats up much faster than the interior, so that by the time the very center of the yolk reaches 170°F, the white and outer layers of yolk are hopelessly overcooked. You might be inclined to put the eggs in cold water and bring them to a boil gradually. This method works, but there's a problem: it causes the eggs to fuse to the shells.

So, cooking them gently gives you even results, but cooking them fast makes it easier to remove the shell. What I needed was a technique that bridged both of these. What if I were to start the eggs in a precise volume of boiling water and let them cook just long enough so that the whites set but remain separated from the shells, then lower the temperature of the water rapidly by adding a few ice cubes and finish cooking them?

It took a few dozen tries to get the exact timing and ice measurements down correctly, but guess what? *It works.* By using a fast start and a slow-and-steady cook to the end, you consistently get eggs that are both perfectly cooked through and through *and* easy to peel.

# FOOLPROOF HARD-BOILED EGGS

**NOTE:** Depending on how hot your kitchen is and your cookware's heat-retention abilities, cooking times may vary slightly. It's a good idea to do a practice run with a single egg and adjust the times as necessary. If you live at a high altitude, a thermometer is essential.

2 quarts water

1 to 6 large eggs

12 ice cubes

Pour the water into a lidded 3-quart saucepan and bring to a boil over high heat. Carefully lower the egg(s) into the water and cook for 30 seconds. Add the ice cubes and allow the water to return to a boil, then reduce to a subsimmer, about 190°F. Cook for 11 minutes. Drain the egg(s) and peel under cool running water.

## POACHED EGGS

What would eggs Benedict be without perfect soft-poached eggs, their snowy whites napped in a robe of gloriously thick, buttery real hollandaise and liquid, golden yolks ready to ooze out over your ham and into the nooks and crannies of your buttered and toasted English muffin?

Well, we already know all about soft-boiled eggs, and a poached egg is essentially a soft-boiled egg cooked in the nude (that's the egg, not the cook).

This does, of course, introduce all kinds of headaches. How does one keep the white from spreading in the pot of water? How do you prevent the yolk from breaking? How do you get the whole darn thing to keep its shape? Some recommend wrapping eggs in plastic wrap before lowering them into the water, to help them retain their shape. The eggs end up with ugly, creased surfaces. Other high-tech methods require hours-long baths in perfectly temperature-controlled water to partially solid-

ify the eggs, to help them keep their shape while poaching. I'm not prepared to wake up an extra hour early just to poach my eggs.

But take a look at your poached egg problems, then ask yourself a couple of questions. When do your poached eggs work? That answer is easy: when the eggs are fresh. The older an egg gets, the weaker the membrane that surrounds the white becomes, and the more likely it is that your egg will spread when it hits the water. This leads us to our first rule of poached eggs: use very fresh eggs.

What exacerbates the possibility of the white falling apart? Agitation. The more the egg shakes and shimmies, the more likely it is it'll separate. It's standard practice to poach eggs in simmering water, but we already know that eggs will coagulate even at a subsimmer, so there's no reason to keep the water anywhere near a boil. Bring it up to a boil, then turn the heat off when you add the eggs.

But what if you've got fresh eggs and a subsimmering pot of water, and your eggs still separate and get cloudy? The fact of the matter is that even if you are taking your egg straight from the chicken and into the pan, you're going to get some spreadage. With a supermarket egg that may have been laid up to 60 days earlier, that's an even bigger risk.

How do you get rid of the egg white that's already separated? The solution is an ingenious method that I first saw demonstrated by Heston Blumenthal of The Fat Duck, a restaurant in England. If you crack an egg and transfer it to a fine-mesh strainer, all of the loose bits of white will drain through while the tight white and yolk entrapped by their membrane will stay completely intact. You can then simply lower the strainer into the water (the hot water immediately surrounds the egg and starts the cooking process), and gently slide the egg out into the pan. The result is a perfectly shaped poached egg every time, with no "floaters."

Transfer eggs to individual small bowls.

Gently pour into a fine mesh strainer.

Let loose whites drip through.

Carefully lower into the subsimmering water.

Roll the egg out, and keep it moving gently with a wooden spoon.

## COMMON POACHED EGG QUESTIONS

I've read that adding vinegar to the water will help my eggs keep their shape better. Is this true?

Yes, . . . sort of. Eggs set up when their proteins denature and coagulate. Egg proteins can be denatured by heat, but they can also be denatured by acid. Adding vinegar to your water indeed causes them to set faster, but the effect is not quick. Rather than causing them to set faster in the short term, which is what you care about, vinegar just causes them to overset as they cook, becoming dry and tough. What's more, vinegar can make your eggs taste, well, vinegary.

Does salting make my eggs cook any better?

Nope. But there's a good reason to salt your water: *it makes your eggs tastier.* Just like pasta or potatoes, eggs absorb salt from the water as they cook, leading to a more evenly seasoned finished product.

*continues*

**Why are poached eggs so freaking delicious?**
This is a question that modern science has yet to answer and may well never get around to. Some scientists remark that the lack of progress on this particular front is due to the fact that other scientists don't spend the time to make and enjoy a good breakfast.

**Should I agitate my eggs as they cook, or swirl the water as I add them, as some books suggest?**
The strainer-to-pan technique completely eliminates the need to swirl the water before you add the eggs, a trick designed to help the eggs keep a nice, even torpedo shape. What you *do* want to do is to make sure the eggs move around after they've started to set up. If you cook your eggs with no motion at all, they will end up resembling fried eggs in shape, with flat bottoms and a pronounced dome around the yolks. You also run the risk of overcooking the bottoms and toughening them, as they are in direct contact with the hot bottom of the pan. By moving them around in the water and gently flipping them, you get more even cooking and a more even shape. I use a wooden spoon to flip them with the water currents, rather than trying to pick them up with the spoon.

**Diners have, what, fifty seats in them? How the heck can I serve more than a few eggs at a time?**
Diners are staffed by superhuman cooking machines known as short-order cooks, who have spent years practicing how to poach eggs perfectly. You want to get that good? One solution: practice.

OK, there's another way to get there, but don't tell anyone, promise? *Just cook the darn things in advance.* Poached eggs can be taken out of the pan right after cooking and transferred to cold water to chill. They'll stay there in a state of suspended animation for as long as you'd like. (Or as long as they don't begin to rot.) You can store them for a few hours or even a few nights in the fridge. Then, 15 minutes before you're ready to serve, just plop them into a bowl of hot water to warm up. Poached eggs by their very nature are never very hot—their yolks would solidify if they were. So 140°F, the temperature of hot water straight out of my tap, is just about the perfect temperature for reheating poached eggs.

# PERFECT POACHED EGGS

3 quarts water

2 tablespoons kosher salt

Large eggs (as many as desired)

1. Combine the water and salt in a large saucepan and bring to a boil over high heat, then reduce the heat to the lowest setting.

2. Carefully break the eggs into individual small bowls or cups. Carefully tip one egg into a fine-mesh strainer set over a bowl and allow the excess white to drain, swirling the strainer gently. You should be left with the yolk surrounded by tight egg white. Gently lower the strainer into the water, then tilt the egg out into the water. Repeat with the remaining eggs.

3. Allow the eggs to cook, swirling the water occasionally to keep them moving lazily around the pan and gently turning them, until the whites are fully set but the yolks are still runny, about 4 minutes.

4. To serve immediately, pick up the eggs one at a time with a perforated spoon and transfer to a paper-towel-lined plate to drain briefly. Serve.

5. Or, to save the eggs for later, pick up the eggs one at a time with a perforated spoon and transfer to a bowl of cold water to chill, then store submerged in the water in the refrigerator for up to 3 days. To reheat, transfer to a bowl of hot water and allow to stand until warm, about 15 minutes.

## HOLLANDAISE SAUCE

For many aspiring French chefs, great hollandaise is the bane of their existence. Far removed from the gloppy, greasy stuff you get at the typical diner, or worse, the powdered "just add milk" cafeteria version, a true hollandaise is creamy and rich, impossibly smooth, and perfectly well balanced with the flavors of eggs, butter, and a touch of lemon juice. It should flow slowly off a spoon so that it naps a poached egg in a thick robe. Never runny, and certainly never curdled, hollandaise has a delicate texture that's really tough to get right. At least, it used to be. I've figured out a way to make it perfectly every single time—even with no experience.

Hollandaise sauce, just like mayonnaise, is an egg-stabilized emulsion of fat in a water-based liquid (see "Obsessive-Emulsive," page 776). It's traditionally made by cooking egg yolks with a little water, whisking them constantly, until they've just begun to set, then slowly drizzling in melted clarified butter (see "Clarified Butter," page 109) and seasoning the sauce with lemon juice. With vigorous whisking, the butterfat gets broken up into microscopic droplets that are surrounded by the water from the lemon juice and the egg yolks. Both the acid in the lemon juice and the protein *lecithin* in the egg yolks prevent these fat droplets from coalescing and breaking down into a greasy pool. The result is a thick, creamy, delicious sauce.

Mayonnaise is relatively simple: it's made from a liquid fat (oil), and it's made and kept at room or fridge temperatures (for a foolproof recipe, see page 807). Hollandaise is more complicated. Butterfat begins to solidify below 95°F, so if you let your hollandaise get too cool, the solid chunks of fat will break the emulsion, turning it grainy. Reheat it, and it will separate into a greasy liquid (that's why leftover hollandaise can't be stored). On the other hand, if you let it get too hot, the egg proteins will begin to coagulate. You'll end up with a lumpy, curdled sauce with the texture of soft scrambled eggs. So the keys to a perfect hollandaise are two: careful construction of an emulsion by slowly incorporating butterfat into the liquid, and temperature control.

Once you realize this, the solution to foolproofing hollandaise becomes quite simple. Most classic recipes require you to heat both the butter and egg yolks before trying to combine the two. But what if you were to just heat one of them, so that when it is combined with the other, the final tempera-

A blender and hot butter make short work of hollandaise.

proteins and foam

pure butterfat

water and proteins

## CLARIFIED BUTTER

Solid butter may *look* like a single, homogeneous substance, but melt it in a pan, and it quickly becomes apparent that it's made up of a few different things.

- **Butterfat** makes up around 80 percent of the weight of butter (up to 84 percent for some high-end "European-style" butters, or as low as 65 percent for some fresh-churned, farm stand–style butters). Because there are many different fats that make up butterfat, each one of which softens and melts at a specific temperature range, butter goes through many textural changes as you heat it, slowly softening and becoming more and more malleable until finally, at around 95°F, all of the fats are liquefied.

- **Water** makes up another 15 percent (down to 11 percent for high-end butters, up to 30 percent for fresh-churned butters). In the cool environs of the fridge, the water and fat in a stick of butter commingle without any problem. But apply some energy to the situation by heating it in a skillet, and eventually the water converts to steam, forming small bubbles of vapor and causing your butter to foam. Once the foaming has subsided, you know that all of the water has made its escape and

*continues*

your butter has begun to climb above 212°F. Since water is denser than fat, when butter is melted in a large pot, this layer of water (and a few dissolved proteins) will sink to the bottom, where it will begin to bubble if heated long enough.

- **Milk proteins**, mainly casein, make up the remaining 5 percent (or so) of the butter. These proteins are the milky white scum that floats to the top of your butter as you melt it, and it's these proteins that will begin to brown and eventually burn and smoke as you heat butter in a hot skillet.

Because of its water and protein content, plain butter is not the ideal medium for searing food—it simply can't get hot enough without burning. For this reason, many chefs make *clarified butter*, butter from which the water and protein have been removed. It's the primary cooking fat in India, where it's known as ghee. You make it by melting butter, carefully skimming the white milk proteins off the top, and then pouring off the golden liquid fat, discarding the watery layer of proteins on the bottom. Once clarified, butter can be heated to much higher temperatures without fear of burning.

As I mentioned earlier, clarified butter is used to make a classic hollandaise, the thinking being that the water constituent in whole butter will dilute the sauce. A much easier way to avoid this water in your finished sauce is to just pour the melted butter slowly out of the pan until all that's left is the watery layer at the bottom, which you can then discard.

ture ends up in the correct range? I figured that if I heated my butter to a high-enough temperature, I should be able to slowly incorporate it into a mixture of raw egg yolks and lemon juice, gradually raising its temperature, so that by the time all the butter is incorporated, the yolks are cooked exactly how they need to be. Because the acidity of lemon juice can minimize curdling, there's a little bit of leeway as far as temperature is concerned: anywhere in the 160° to 180°F range for the finished sauce will work.

To test the theory, I heated up a couple of sticks of butter on the stovetop (the microwave also does just fine) to 200°F, then slowly drizzled the butter into my egg yolks and lemon juice, which I had run-

ning in the blender (adding a bit of water to the yolk mixture helps prevent it from sticking to the walls of the jar). A quick dash of salt and cayenne pepper, and there it was: perfect hollandaise without the headache. To make it even more foolproof, I tried it again using an immersion blender and its jar. I put the egg yolks, lemon, and water in the jar, stuck the wand of the hand blender down in there, poured in all the melted butter, and turned on the blender. As the vortex drew butter down into the whirling blades, a thick, stable emulsion formed like magic, until all the butter was incorporated and my sauce was thick, rich, and light, the way a great hollandaise should be.

# FOOLPROOF HOLLANDAISE SAUCE

**NOTE:** Cooled hollandaise can be very carefully reheated over the lowest-possible heat while whisking constantly. Hollandaise can't be refrigerated and then reheated.

## MAKES ABOUT 1 CUP

3 large egg yolks

1 tablespoon lemon juice (from 1 lemon)

1 tablespoon hot water

½ pound (2 sticks) unsalted butter, cut into rough tablespoon-sized chunks

Pinch of cayenne pepper

Kosher salt

### TO MAKE HOLLANDAISE WITH AN IMMERSION BLENDER

1. Add the egg yolks, lemon juice, and hot water to the blender cup (or a cup that will just barely hold the head of your blender).

2. Melt the butter in a small saucepan over medium-low heat and continue to heat until the butter just begins to bubble and registers 180° to 190°F on an instant-read thermometer. Transfer it to a liquid measuring cup, leaving the thin layer of whitish liquid behind (discard it).

3. Insert the head of blender into the bottom of the cup and run the blender. Slowly pour in the hot butter. You should see the sauce begin to form at the bottom of the cup. As the sauce forms, slowly pull the head of the blender up to incorporate more melted butter, until all of the butter is incorporated and the sauce has the consistency of heavy cream. Season with the cayenne and salt to taste. Transfer to a serving bowl or small saucepan, cover, and keep in a warm spot (not directly over heat!) until ready to serve.

### TO MAKE HOLLANDAISE IN A STANDARD BLENDER OR FOOD PROCESSOR

1. Add the egg yolks, lemon juice, and hot water to the blender or food processor and blend on medium speed until smooth, about 10 seconds.

2. Melt the butter in a small saucepan over medium-low heat and continue to heat until the butter just begins to bubble and registers 180° to 190°F on an instant-read thermometer.

3. With the blender running on medium speed, slowly drizzle in the butter over the course of 1 minute, stopping to scrape down the sides as necessary and leaving the thin layer of whitish liquid in the bottom of the pan (discard it). The sauce should be smooth, with the consistency of heavy cream. Season with the cayenne and salt to taste. Transfer to a serving bowl or small saucepan, cover, and keep in a warm spot (not directly over heat!) until ready to serve.

# EGGS BENEDICT

**SERVES 2 TO 4**

2 tablespoons kosher salt

2 teaspoons vegetable oil

4 slices Canadian bacon or
  thick-cut ham

4 large eggs

2 English muffins, split, toasted,
  and buttered

1 recipe Foolproof Hollandaise
  (page 111), kept warm

Dash of cayenne pepper (optional)

Minced fresh parsley or chives
  (optional)

1. Combine 3 quarts water and the salt in a large saucepan and bring to a
   boil over high heat.

2. While the water is heating, heat the vegetable oil in a 12-inch stainless steel or cast-iron skillet over medium heat until shimmering. Add
   the Canadian bacon (or ham) and cook, turning once, until browned on
   both sides, about 5 minutes. Transfer to a large plate and tent with foil
   to keep warm.

3. Carefully break the eggs into individual small bowls or cups. Turn off
   the heat under the boiling water. Carefully tip one egg into a fine-mesh
   strainer set over a bowl and allow the excess white to drain. You should
   be left with the yolk surrounded by tight egg white. Gently lower the
   strainer into the water, then tilt the egg out into the water. Repeat with
   the remaining eggs.

4. Allow the eggs to cook, swirling the water occasionally to keep them moving lazily around the pan and gently turning them, until the whites are fully set but the yolks are still runny, about 4 minutes. Remove the eggs with a slotted spoon and transfer to a paper-towel-lined plate to drain.

5. Top each English muffin half with a slice of Canadian bacon, followed by a poached egg. Spoon some hollandaise sauce over the eggs, sprinkle with the cayenne pepper and herbs, if using, and serve immediately, passing the extra hollandaise in a warm bowl on the side.

## EGGS FLORENTINE

*Don't dine on swine? Not to worry. Eggs and hollandaise go just as well with good sautéed spinach (asparagus would also be great here).*

**SERVES 2 TO 4**

Kosher salt

2 teaspoons vegetable oil

1 medium clove garlic, finely
  minced

1 bunch (about 4 ounces) spinach,
  trimmed, washed, and dried

Freshly ground black pepper

4 large eggs

2 English muffins, split, toasted,
  and buttered

1 recipe Foolproof Hollandaise
  (page 111), kept warm

Dash of cayenne pepper (optional)

Minced fresh parsley or chives
  (optional)

1. Combine 3 quarts water and 2 tablespoons salt in a large saucepan and bring to a boil over high heat.

2. Heat the vegetable oil in a 12-inch stainless steel or cast-iron skillet over medium-high heat until shimmering. Add garlic and cook, stirring constantly, until fragrant, about 30 seconds. Add spinach along with 2 tablespoons water, and cook, stirring occasionally, until the spinach is wilted and the water is mostly evaporated. Season to taste with salt and pepper. Transfer to a plate and set aside.

3. Cook the eggs as directed in steps 3 and 4 of the eggs benedict recipe, above.

4. Top each English muffin half with one-quarter of the spinach, followed by a poached egg. Spoon some of the hollandaise sauce over the eggs, sprinkle with the cayenne pepper and herbs, if using, and serve immediately, passing the extra hollandaise in a warm bowl on the side.

# FRIED EGGS

**If you're like me, mastering fried eggs was your very first culinary accomplishment.**

Or, I should say, fried eggs were my very first attempt at accomplishing something in the kitchen, because, truth be told, the eggs were never the same twice. This is not necessarily a bad thing. In fact, if I'd been taking notes on exactly how and why my eggs were never the same way twice, I'd call that science. I'd also most likely have developed a good technique years earlier.

Remember that egg whites start setting at around 155°F, while egg yolks start firming up as low as 145°F (see "Heat and Eggs," page 97). This gives us the rather tricky problem of trying to cook two different components of one food in the same cooking medium at completely different rates. It's not an easy trick to pull off, but it can be done.

We already know one thing for sure: if you want your eggs to be picture-perfect, the yolks standing tall, ready to be pierced by a fork so that their golden treasure cascades slowly across a plane of tight, clean whites, their edges showing just a bare hint of crispness, you have to start with the freshest-possible eggs. Draining the eggs in a fine-mesh strainer just like you did for poached eggs (see page 103) aids in achieving this result, though personally, I kinda like the bubbly, thin whites that spread around the pan and become extra crisp as they cook. Picture-perfect fried eggs are for advertisements.

A tight, tall yolk isn't just about looking good. With fresh eggs, the yolks are kept elevated above the hot surface of the pan, allowing them to set a little bit more slowly than the whites. This is key

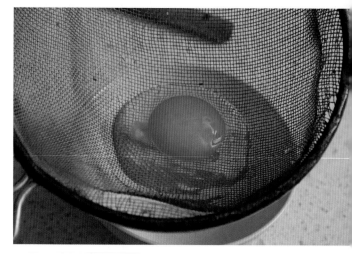

Straining before frying gets you a picture-perfect fried egg.

if you want your whites opaque before your yolks get hard.

What else can you do to keep the yolks from overcooking? Well, here's another thing we know: egg yolks contain far more fat than egg whites. Luckily for us here, fat is an excellent insulator—that is, it transfers energy less efficiently than water does (that's why whales are covered in blubber). We can use this fact to our advantage by adjusting the pan temperature. I cooked eggs in three different pans at three different heat settings just until the whites were set.

- **Over low heat,** the yolks came out completely firm and chalky at the bottom, with just a very thin layer of barely liquid yolk at the top. The whites were extremely rubbery and dry in all but the thickest part just around the yolk. With gentle heat and an extended cooking time, the differences in conductivity between the egg yolks and the egg whites don't have as strong an effect on their cooking rates—they basically cook in the same time. Additionally, the egg whites were pure white, with no crisping or browning at all on the bottom surface. Some folks like their egg whites this way. I think those people secretly just want poached eggs.
- **Over medium heat,** there was still a good amount of liquid yolk at the top of the egg, while the bottom half of the yolk became quite firm. The whites took on a touch of browning (even more if I used butter instead of oil—the milk proteins in the butter brown and stick to the eggs). This is a good compromise for folks who like some liquid yolk but don't want their whites to show any transparency at all.
- **Over high heat,** you can get whites that are completely set with yolks that are still almost completely liquid, but you run into another problem: the bottom of the eggs burns long before the rest of the egg is ready to eat.

For the simplest fried eggs, moderate heat is the way to go. Whether you're using butter or oil makes little difference in the cooking, as long as you make sure that the milk proteins in your butter don't burn before you slip the eggs into the pan; it's best to add them just after the foaming subsides. (This is an indication that the water in the butter has completely evaporated and the pan is somewhere in the 250°F range). Butter will give you richer flavor and deeper browning, while oil will give you cleaner egg flavor and slightly crisper bottoms—it's all down to personal preference.

For a long time, I was happy with my medium-heat eggs, sacrificing a bit of liquid yolk for the extra crispness in the whites, but then I saw a technique in Spain that made me rethink the way I fried eggs. There it's common to fry eggs not in a thin layer of fat, but in a shallow pool of it. Cooks would fill the pan with a half inch or so of olive oil and heat it up to deep-frying temperatures, then tilt the pan so the fat collected on one side, drop in the eggs, and baste them with the hot fat as they cooked. The eggs cooked quite rapidly from all sides, the whites quickly setting and transforming into a delicately lacy, lightly frizzled shell around the still-liquid yolk. What if I were to adapt part of this technique to my fried eggs at home?

By heating up a few tablespoons of oil in a skillet (I use nonstick or a good cast-iron pan), I can get a similar effect by then adding my eggs and using a spoon to baste the whites with the hot oil. The whites puff and crisp, setting rapidly, while leaving the yolk just barely heated through. It's my new favorite way to eat fried eggs, especially when I go for it and use the fancy-pants olive oil.

# EXTRA-CRISPY SUNNY-SIDE-UP EGGS

**SERVES 1**

2 large eggs

3 tablespoons olive oil (extra-virgin, if you prefer)

Kosher salt and freshly ground black pepper

1. Break one egg into a small cup, then transfer to a fine-mesh strainer set over a bowl and swirl gently until any excess white passes through. Return the egg to the cup. Repeat with the second egg.

2. Heat the olive oil in a medium nonstick or cast-iron skillet over medium heat until it registers 300°F on an instant-read thermometer. Carefully slip the eggs into the oil. Immediately tilt the skillet so that the oil pools on one side and use a spoon to spoon the hot oil over the egg whites, trying to avoid the yolks as much as possible. Continue doing this until the egg whites are completely set and crisp on the bottom, about 1 minute. With a spatula, transfer the eggs to a paper-towel-lined plate and season with salt and pepper. Serve immediately.

# SCRAMBLED EGGS, TWO WAYS

**There's a big divide in the world of scrambled eggs . . .**

. . . between those who like them rich, dense, and creamy (that's me), and those who like them light, relatively dry, and fluffy (that's my wife). This is the kind of stuff that can really tear a home apart, so in the interest of maintaining marital bliss, I decided that it was only right that I figured out how to make both types of scrambled eggs so that we could both enjoy our breakfast.

Just like with boiled eggs, cooking scrambled eggs is all about controlling the coagulation of egg pro-

---

\* To be perfectly honest, my wife likes her eggs cooked well past dry and fluffy. She'll cook them over high heat, poking them constantly with a wooden spoon, until they are reduced to dry, brown nuggets resembling, more than anything, mouse droppings. She calls them popcorn eggs. I call them gross.

teins, the difference being that with scrambled eggs, not only are the proteins in the whites and yolks mixed together, but you also have the opportunity to mix in additional ingredients—as well as to control the way the eggs come together by moving them as they cook. For my testing, I decided to begin with just plain eggs in order to gauge the effects of stirring and other mechanical actions. The only additive I used was a bit of butter in the pan to prevent them from sticking.

A few things became clear immediately. The difference between creamy scrambled eggs and fluffy ones has mostly to do with the amount of air they contain at the end. As beaten eggs are heated in a skillet, their proteins begin to set. At the same time, the moisture within them begins to evaporate, causing pockets of steam and air to build up within the eggs. Vigorous stirring or shaking will cause these pockets of steam and air to rupture, making the eggs denser. So, for the fluffiest scrambled eggs, your goal is to minimize the movement of the eggs in the pan, gently folding and turning them just enough to get them to cook evenly into large, golden, tender curds. For creamy eggs, constant stirring is preferable, to remove excess air and get the egg proteins to set up closely with one another, resulting in a dense, almost custard-like scramble.

Heat had a great effect on final texture as well. When cooked over very low heat, even gently folded eggs won't get too fluffy. This is because there's not enough energy in the pan to cause water vapor to form or to cause air bubbles to expand vigorously.

So, for fluffy scrambled eggs, you need to use relatively high heat (though if you let the pan get too hot, you risk overcooking—or, worse, browning—your eggs), while for creamy eggs, cooking over low heat gives you much more control over their texture.

## Additives

What about common additions to eggs—water, milk, and the like? There are basically two things they can do. First, they add some water to the mix, which makes for fluffier eggs (more water = more vaporization). Dairy ingredients also add fat, which can impede egg proteins from linking with each other, creating a more tender curd. This chart sums it all up:

| ADDITION | EFFECT ON TEXTURE AND FLAVOR | HOW IT WORKS |
|---|---|---|
| **Nothing** | Eggs cook fastest but are tougher | |
| **Water** | Increased fluffiness, diluted flavor | Extra water means more vaporization occurs, creating larger bubbles in the eggs and lightening them. |
| **Milk** | Increased fluffiness and tenderness | Milk is mostly water, which helps increase fluffiness, while the extra proteins and fats prevent the egg proteins from bonding too tightly, making them more tender. |
| **Cream** | Not as fluffy, but rich, with an almost cheesy flavor and texture | The high fat content of cream greatly reduces the bonding power of egg proteins. |
| **Cold butter** | Ultracreamy and dense | Cold cubes of butter not only add fat, for tenderness, but also help regulate temperature, cooling the eggs and letting them set more slowly; this leads to denser, creamier results. |

With all this data, my fluffy scrambled eggs were coming out great. I just had to make sure to whisk in some milk along with the eggs, to use relatively high heat, to keep the stirring and folding mini-

mal, and to make sure to get 'em out of the hot pan before they were completely cooked. Even once out of the pan, moisture will continue to evaporate from the eggs and the proteins will continue to set tighter and tighter. Removing the eggs from the pan when slightly undercooked ensures that they arrive at the table perfectly cooked.

My creamy eggs, on the other hand, were giving me more problems. They were coming out fine when I started them with cubes of cold butter, used low heat, and stirred constantly to break up curds and release air and vapor, but they were still not quite as rich and creamy as I'd like. Salting them well before cooking (see "Salting Eggs," below) and letting them rest helped, but if there's one thing I learned working in French restaurants, it's that when all else fails, add more fat. My solution was to add extra egg yolks to the mix, as well as to finish the dish with a touch of heavy cream. The cream serves two functions: it adds richness and smooths out the eggs' texture, and when added at the end of cooking, it also cools them down, preventing them from setting up too hard in the skillet. And how's this for gilding the lily?—use crème fraîche (see page 123) in place of heavy cream. The resulting eggs are the ultimate in luxury: rich, tender, almost custard-like in texture. Eggs-ceptional! (Sorry.)

## SALTING EGGS

Here's the scenario: You've just beaten a few eggs with a pinch of salt, getting ready to scramble them, when suddenly the dog gets stuck in the toilet, your mother-in-law calls, and the UPS guy rings the doorbell to deliver your brand-new digital thermometer. Thirty minutes later, you get back to those eggs and realize they've completely changed color. Once bright yellow and opaque, they're now dark orange and translucent. What's going on? And, more important, will it affect the way they cook?

Plain                    Salted

*continues*

Salt affects eggs by weakening the magnetic attraction that yolk proteins have for one another (yes, egg proteins *do* find each other attractive). Egg yolks are comprised of millions of tiny balloons filled with water, protein, and fat. These balloons are too small to see with the naked eye, but they are large enough to prevent light from passing through them. Salt breaks these spheres up into even tinier pieces, allowing light to pass through, so the salted eggs turned translucent. What does this mean for the way they cook? To find out, I cooked three batches of eggs side by side, noting their finished texture.

| SALTING TIMING | RESULTS |
|---|---|
| 15 minutes prior to cooking | The least watery and the most tender, with moist, soft curds |
| Just before cooking | Moderately tender and not watery |
| Toward the end of cooking | Toughest of the three, with a tendency to weep liquid onto the plate |

Turns out that salt can have quite a drastic effect on how eggs cook. When eggs cook and coagulate, the proteins in the yolks pull tighter and tighter together as they get hotter. When they get too tight, they begin to squeeze liquid out from the curds, resulting in eggs that weep in a most embarrassing manner. Adding salt to the eggs well before cooking can prevent the proteins from bonding too tightly by reducing their attraction to one another, resulting in a more tender curd and less likelihood of unattractive weeping. Adding salt immediately before cooking helps, but to get the full effect, the salt must have time to dissolve and become evenly distributed through the mixture. This takes about 15 minutes—just enough time for you to get your bacon cooked!

Eggs salted after cooking weep.

Eggs salted at least 15 minutes in advance retain their moisture.

# LIGHT AND FLUFFY SCRAMBLED EGGS

**SERVES 4**

8 large eggs

¾ teaspoon kosher salt

3 tablespoons whole milk

2 tablespoons unsalted butter

1. Combine the eggs, salt, and milk in a medium bowl and whisk until homogeneous and frothy, about 1 minute. Allow to rest at room temperature for at least 15 minutes. The eggs should darken in color significantly.

2. Melt the butter in a 10-inch nonstick skillet over medium-high heat, swirling the pan as it melts to coat evenly. Rewhisk the eggs until they are foamy, then transfer to the skillet and cook, slowly scraping the bottom and sides of the pan with a silicone spatula as the eggs solidify. Then continue to cook, scraping and folding constantly, until the eggs have formed solid, moist curds and no liquid egg remains, about 2 minutes (the eggs should still appear slightly underdone). Immediately transfer to a plate and serve.

# CREAMY SCRAMBLED EGGS

**SERVES 4**

6 large eggs

2 large egg yolks

¾ teaspoon kosher salt

2 tablespoon unsalted butter, cut
  into ¼-inch cubes and chilled

2 tablespoons heavy cream or
  crème fraîche (see page 123)

1. Combine the eggs, egg yolks, and salt in a medium bowl and whisk until homogeneous and frothy, about 1 minute. Allow to rest at room temperature for at least 15 minutes. The eggs should darken in color significantly.

2. Add the chilled butter to the eggs, then transfer the mixture to a 10-inch nonstick skillet, place over medium-low heat, and cook, stirring constantly, until the butter completely melts and the eggs begin to set. As the eggs become firmer, stir more rapidly to break up the large curds, and continue to cook until no liquid egg remains.

3. Remove the pan from heat, add the heavy cream, and, stir constantly for 15 seconds; the eggs should be completely tender with a custard-like texture that just barely holds a shape when you pile them up. Transfer to a plate and serve immediately.

## HOMEMADE CRÈME FRAÎCHE

Crème fraîche is made by allowing heavy cream to spoil in a controlled way. Bacteria introduced into the cream convert some of its sugar (mainly the complex carbohydrate lactose) into simpler sugars and acidic by-products. This lowers the pH of the cream, causing some of its proteins to coagulate, making it thicker. Good crème fraîche has a rich, creamy texture, stiff enough to form loose peaks, and a tangy, slightly cheese-like flavor. Store-bought crème fraîche is great, but it can be difficult to track down and pricey. When I found out that you can simply mix buttermilk (which has live bacterial culture) into heavy cream and let it thicken overnight to create a true crème fraîche at home, my mind was blown. I like to share my mind-blowing experiences, so here you go. Lucky you!

I played around with the ratios of cream to buttermilk quite a bit and in the end found that it doesn't really matter all that much. Add more buttermilk, and you'll need less time for it to thicken, but it'll be less creamy. Add less, and it takes longer, but tastes better. One tablespoon per cup (that's a 1:16 ratio) was about the perfect balance for me.

It gets superrich and creamy at right about the 12-hour mark. You can halt the process earlier by refrigerating it to stop the bacterial action—this is useful if you want a thinner Mexican-style *crema agria* for drizzling over your nachos or guacamole. For those of you worried about cream spoiling at room temp, that's the idea: it's the good bacteria from the buttermilk multiplying in there that prevent the dangerous bacteria from taking over.

And let the mind-blowing begin. *Commence countdown.*

# EASY HOMEMADE CRÈME FRAÎCHE

### MAKES 2 CUPS

**2 cups heavy cream**

**2 tablespoons buttermilk**

Combine the heavy cream and buttermilk in a glass jar or bowl. Cover and allow to rest at room temperature until thickened to the desired texture, 6 to 12 hours. Store in the refrigerator in a sealed container for up to 2 weeks.

# OMELETS

**Just as with scrambled eggs, there are two major types of omelet:**

. . . the hearty, big-as-your-face, stuffed-to-the-brim, fluffy, folded-in-half, light golden brown diner-style omelet and its refined French cousin, the moist, tender, pale-yellow variety, gently rolled like the world's most delicious cigar. And, just as with scrambled eggs, the method by which the eggs are heated and stirred is the primary factor that determines what you end up with.

For fluffy, diner-style omelets, the key is to start the eggs in hot butter and move them as little as possible during cooking. Rather than shaking the pan and breaking up the large curds, the best course of action is a move called the lift-and-tilt: use a silicone spatula to lift up the edges of the omelet and push them toward the center of the pan while tilting it, to allow the raw egg to run underneath. Repeating this technique means nearly all of the eggs can be set with minimal stirring. You'll still end up with a slick of raw egg across the top surface, which is easy to take care of: remove the skillet from the heat, add whatever toppings you like (ham and cheese are my favorite), cover it with a lid, and let the residual heat from the eggs gently cook the top through, then fold it in half and serve.

A tender, fancy-pants omelet—the kind I remember watching Jacques Pépin make look so easy—can

be made by cooking the eggs fast, but it's one of the most difficult techniques in cooking (really). Luckily, though, it doesn't actually require fast cooking, and once you've learned the cold-butter-cube trick you picked up with the creamy scrambled eggs to help regulate the cooking temperature (see page 122), you can actually make a tender French-style omelet in much the same manner, with slow cooking and constant stirring. The only difficult part about this style of omelet is the rolling. The trick is to make sure that one side of the egg disk is thicker than the other by rapping the pan sharply against the stove as you finish cooking so that the eggs collect on the end opposite the handle. Then let the bottom set slightly, and the omelet can be rolled up, starting at the thinner edge, before being turned out onto a plate.

A tender French omelet.

# DINER-STYLE HAM AND CHEESE OMELET

It's important to cook fillings without cheese before you add them to the eggs, or they will not heat up enough while the omelet cooks. Then tossing the cheese with the cooked filling will help get it started melting, so that it's nice and gooey by the time the omelet is done, without the need to overcook your eggs.

**MAKES 1 LARGE
OMELET, SERVING 2**

5 large eggs

¾ teaspoon kosher salt

¼ teaspoon freshly ground black
  pepper

2 tablespoons unsalted butter

4 ounces ham steak, diced

2 ounces cheddar cheese, grated

1. Combine the eggs, salt, and pepper in a medium bowl and whisk until homogeneous and frothy, about 1 minute. Allow to rest at room temperature for at least 15 minutes. The eggs should darken in color significantly.

2. Meanwhile, melt 1 tablespoon of the butter in a 10-inch nonstick skillet over medium heat and cook until lightly browned. Add the ham and cook, stirring frequently, until it has begun to brown on the edges, about 3 minutes. Transfer the ham to a small bowl, add the cheese, and toss to combine. Wipe out the skillet with a paper towel and return it to medium heat.

3. Add the remaining tablespoon of butter to the pan and cook until lightly browned. Rewhisk the eggs until foamy, then add to the skillet and cook, using a silicone spatula to push the edges in toward the center as they set and tilting the pan to spread the uncooked egg underneath. Continue pushing in the edges of the eggs and tilting the skillet, working all around the pan, until the omelet is almost set, about 45 seconds. Sprinkle the ham and cheese over half of the omelet, remove from the heat, cover, and let the omelet sit until it reaches the desired consistency, about 1 minute.

4. Using the silicone spatula, loosen the edges of the omelet from the skillet and shake the skillet to ensure that it's not stuck. Carefully fold the omelet in half, then slide it onto a serving plate and serve immediately.

### DINER-STYLE MUSHROOM, PEPPER, AND ONION OMELET

1. Omit the ham. Melt 1 tablespoon of the butter in a 10-inch nonstick skillet over medium-high heat and cook until lightly browned. Add ½ cup sliced mushrooms, season with salt and pepper, and cook, stirring and tossing frequently, until they've released liquid, the liquid has evaporated, and the mushrooms have started to sizzle again, about 3 minutes. Add ½ cup diced bell pepper and ½ cup diced onion, season with salt and pepper, and cook, stirring and tossing frequently, until the vegetables are softened and lightly browned, about 5 minutes longer. Transfer to a small bowl, add the cheese, and toss to combine.

2. Wipe out the skillet with a paper towel, return it to medium heat, and cook the omelet as directed. When the omelet is almost set, sprinkle the vegetables over half of it and proceed as directed.

### DINER-STYLE ASPARAGUS, SHALLOT, AND GOAT CHEESE OMELET

1. Omit the ham and cheddar cheese. Melt 1 tablespoon of the butter in a 10-inch nonstick skillet over medium-high heat and cook until lightly browned. Add 8 stalks asparagus, bottoms trimmed and cut into 1-inch segments, season with salt and pepper, and cook, stirring and tossing frequently, until tender and beginning to

brown, about 5 minutes. Add 1 large shallot, thinly sliced (about ½ cup) and cook until softened, about 3 minutes. Transfer the vegetables to a small bowl.

2. Wipe out the skillet with a paper towel, return it to medium heat, and cook the omelet as directed. When the omelet is almost set, scatter the asparagus, shallots, and 2 to 3 ounces fresh goat cheese, crumbled, over half of it and proceed as directed.

# KNIFE SKILLS:
## How to Cut a Bell Pepper

**There are two camps when it comes to cutting peppers:
those who cut skin side up and those who cut skin side down.**

I used to be in the former, finding that having the skin up was the only way I could break through it with my knife. The flesh would act as support while the knife cut through the skin. Then I realized that with a really sharp knife, cutting through the skin when the pepper faces down is not a problem, and cutting that way prevents you from compressing the flesh.

To cut a bell pepper, start by splitting it lengthwise in half through the stem with a sharp knife. Pull out the stem and seedy core and discard. Pull out the white ribs with your fingertips. For pepper strips, place each pepper half skin side down on the cutting board and slice lengthwise into strips of even width. For dice, hold a few strips at a time and slice across them.

1

2

3

4

# TENDER FANCY-PANTS OMELET

**SERVES 1**

3 large eggs

1 large egg yolk

½ teaspoon kosher salt

¼ teaspoon freshly ground black pepper

1 tablespoon whole milk

1 tablespoon chopped mixed fresh herbs, such as parsley, tarragon, and chives (optional)

1½ tablespoons unsalted butter, cut into ¼-inch cubes and chilled

1. Combine the eggs, egg yolk, salt, pepper, milk, and herbs, if using, in a medium bowl and whisk until homogeneous and frothy, about 1 minute. Allow to rest at room temperature for at least 15 minutes. The eggs should darken in color significantly.

2. Add two-thirds of the butter to the eggs. Melt the remaining butter in an 8-inch nonstick skillet over low heat. Add the egg mixture to the skillet and cook, stirring slowly and constantly with a silicone spatula, scraping the eggs off the bottom and sides of the skillet, until the eggs have begun to set, about 2 minutes. Then continue cooking, stirring and scraping, until the eggs are just firm enough to hold their shape when you draw the spatula through them. Shake the pan to distribute the eggs evenly over the bottom and then, holding the handle so that the pan rests at a slight angle, rap the pan against the stove so that the eggs are thicker on one side of the pan than the other. Remove from the heat, cover, and let the eggs set to the desired consistency, about 1 minute.

3. Remove the lid and, using the spatula, carefully roll the omelet, starting from the thicker side, then tuck the ends under. Carefully turn the omelet out onto a plate (it helps to hold the plate in one hand and the pan in the other), readjust the shape, and serve immediately.

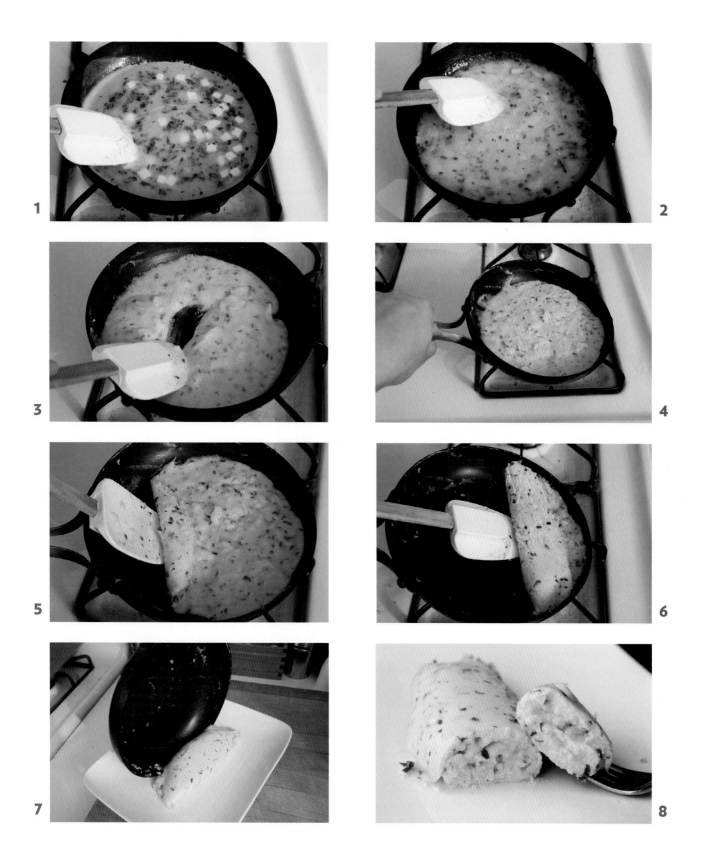

# OMELET FILLINGS

Omelets cook fast—far too fast for fillings to do much beyond warm up a bit. So the key to great filled omelets is to cook your fillings beforehand and have them warm and ready to go. Parcooking the filling while the salted eggs rest is a great way to do it. Your imagination is the only limit to what you can shove into an omelet, but here's a list of ingredients to get you started.

| INGREDIENT | HOW TO PREPARE |
| --- | --- |
| Young cheeses of all kinds (I like cheddar, Jack, blue, feta, Gruyère, Brie, and goat cheeses) | Grate or crumble. If using in conjunction with other cooked ingredients, toss with them in a small bowl after parcooking them; the residual heat will help start the melting process. |
| Hard grating cheeses like Parmigiano-Reggiano, Cotija, and Pecorino Romano | Grate on a Microplane and add to the raw eggs. |
| Cured meats like sausage, ham, and bacon | Cut into ½-inch pieces or nuggets and parcook in butter (let bacon cook in its own fat) until crisp on the edges and well browned. |
| Firm vegetables like onions, shallots, bell peppers, and hot peppers | Dice and soften in butter. |
| Tomatoes | Dice, salt, and drain. |
| Tender leafy vegetables like spinach and arugula | Sauté in butter, with a bit of minced garlic if desired. |
| Tender squashes like zucchini and summer squash | Sauté in butter. |
| Asparagus | Cut into ¼-inch slices on the bias and sauté in butter. |
| Scallions | Thinly slice whites and sauté in butter; thinly slice greens and incorporate into the filling or reserve for garnish. |
| Mushrooms | Slice thin and sauté in butter until the moisture has evaporated and the mushrooms are well browned. |
| Herbs | Add directly to the raw eggs. |

Some of my favorite omelet combinations are spinach with feta, asparagus and shallots with Gruyère, and onions, peppers, and ham with cheddar cheese. All classics, all delicious.

# BACON

**If there's one sure way to guarantee a relapse in an on-the-fence vegetarian, it's to dangle a strip of crisply fried bacon in front of him.**

I sometimes think that the only thing keeping my marriage harmonious is the unduly large number of make-up points I get every time I bring my wife bacon in bed. These days, I'm pretty good at cooking it, if I do say so myself, but this was not always the case. My bacon used to have a severe case of bipolar disorder: crispy and burnt in some spots, flaccid, rubbery, and undercooked in others.

Achieving perfectly crisp, evenly cooked bacon is all about patience. You see, bacon is made up of two distinct elements—the fat (which is actually a mixture of fat and connective tissue) and the lean—

## WET VERSUS DRY CURES

By now, you must have seen that high-falutin' bacon that seems to be invading every farmers' market and supermarket in the country (not to mention online sources). Is it worth its premium price? As far as flavor is concerned, that's simply a matter of personal preference. But there's a far more compelling reason to pick the fancy stuff over the standard supermarket brands, and it's in the cure.

All bacon is cured—that is, treated with salt in order to alter the structure of its proteins and preserve it. Traditionally, the cure was a *dry cure*: salt (often with other seasonings) was rubbed onto slabs of pork belly. Over the course of a few weeks, the salt worked its way into the belly at a leisurely pace, while the meat slowly lost moisture. The result was a dense hunk of deeply flavored belly with relatively little residual moisture. Many high-end bacons are now produced using this time-consuming method.

Most supermarket bacons, on the other hand, are cured with a *wet cure*: a saltwater solution is injected into the meat in many spots. With this technique, the salt can penetrate the meat much faster. What once took weeks is accomplished in a matter of days. Of course, with the injection of added water and insufficient time to dry, this shortcut bacon is far wetter than dry-cured bacon, with two results. First, it means that you're paying more for it than meets the eye. That 1-pound package contains at least an ounce or two of added water weight. Second—and more important—it won't cook the same way.

Try it: fry a piece of regular supermarket bacon side by side with a strip of high-end dry-cured bacon. The supermarket bacon will shrink and curl considerably more than the dry-cured bacon as its moisture evaporates. It'll also spit and sputter far more, due to the excess water droplets it's expelling as it cooks. So, putting flavor aside, if you're constantly enraged by bacon grease sputtering onto your stovetop or that darned strip that *just won't lie flat*, you may want to consider switching to a bacon produced the traditional way, with a dry cure.

and each cooks differently. Fat tends to shrink quickly when heated, but after the initial shrinking stage, it takes quite a bit of time to finish cooking as the connective tissue that remains is slowly broken down (undercooked connective tissue is what causes rubbery bacon). The lean, on the other hand, shrinks less than the fat, and because of this differential, your bacon twists and buckles (just like the bimetal strip inside a thermostat). This twisting in turn exacerbates the situation, because not only are your fat and lean shrinking at different rates, but entire sections of the strip are now cooking at different rates, depending on whether or not they are in direct contact with the pan.

By cooking bacon over low heat, the shrinkage differential can be minimized, keeping your bacon flatter and allowing it to cook more evenly. A large heavy skillet with even heat distribution is essential.

Want to cook bacon for a crowd? Do it in the oven. An oven heats much more evenly than a skillet does, delivering perfectly crisp bacon by the trayful.

# CRISPY FRIED BACON

**SERVES 2 TO 4**

8 slices bacon, cut crosswise in half

1. Place the bacon in an even layer in a 12-inch cast-iron or heavy-bottomed nonstick skillet and cook over medium heat until sizzling, about 4 minutes. Reduce the heat to medium-low and continue to cook until the fat is rendered and the bacon is crisp on both sides, flipping and rearranging the slices as necessary, about 12 minutes total.
2. Drain the bacon on a paper-towel-lined plate and serve.

# CRISPY OVEN-FRIED BACON
## FOR A CROWD

**SERVES 6 TO 10**

24 slices bacon (about 1 pound)

1. Adjust the oven racks to the lower- and upper-middle positions and preheat the oven to 425°F. Arrange the bacon slices in a single layer on two rimmed baking sheets. Roast the bacon until it's crisp and brown, 18 to 20 minutes, rotating the pans back to front and top to bottom halfway through cooking.
2. Drain the bacon on a paper-towel-lined plate and serve.

# CRISPY POTATO CAKE (AKA RÖSTI)

**Crispy and golden brown on the outside, creamy and tender in the middle,**

. . . with some good garlicky mayo (aioli, if you will) for dipping, the key to really great rösti is balancing the amount of starch that the finished potato cake retains. Too little starch, and it falls apart. Too much, and it comes out sticky. Potato cells contain their own starch, so this really becomes a question of how to cut the potatoes in order to release just the right amount. You can grate them on a box grater or in the food processor, but if you do, you'll end up rupturing potato cells, releasing a ton of liquid and starch from inside them. Then you're forced to squeeze the shreds dry, and your rösti will come out starchy and sticky, even with relatively low-starch potatoes like Yukon Golds.

Much better, though slightly more difficult, is to cut them on a mandoline. If you have one with extra teeth or blades (which you should!), it'll cut the potatoes directly into 1⁄16-inch shreds for you. If you

don't, it's easy enough to slice the potatoes into thin planks and then use a knife to get the matchsticks you need. A sharp mandoline (and a sharp knife) = fewer ruptured cells = less sticky starch release = better texture and more potato flavor in each bite. Some sources recommend rinsing the cut potatoes to rid them completely of their starch, then adding a measured amount of pure potato starch to them before cooking, but I find the results unsatisfactory. Rinsed potatoes don't soften properly when cooked, and you end up with rösti with al dente bits of crunchy potato inside.

The other key to great rösti is to parcook the potatoes before frying them. Why? Well, anybody who's worked the French fry station at a restaurant knows that potatoes begin to oxidize as soon as you cut them. Over the course of fifteen minutes or so, a cut potato will go from being pale white to reddish brown, and eventually to black. You don't want your potatoes to be black. Storing sliced potatoes in water will prevent this from happening (or at least slow it down), but it also rinses away lots of starch. Too little starch is just as bad as too much starch in rösti, so I avoid rinsing or submerging my taters in water at any point. Parcooking the potatoes accomplishes the goal of preventing them from browning and also leads to a better texture in the finished product—you don't have to worry about raw potato in the center of the potato cake. This is one case where the microwave is actually the best tool for the job. It allows you to cook the potatoes rapidly without either adding moisture or losing an excessive amount.

For a gussied-up version, I sauté onions and mushrooms until a deep golden brown and flavor them with a bit of thyme to form a central layer in my potato cake. You could use whatever sautéed vegetables you want.

After you've got your stuffing (if any), cooking the rösti is a simple matter of moderate heat and a good thick pan that will cook it gently and evenly. I use a well-seasoned cast-iron pan, but you can use a good nonstick pan if you don't have a cast-iron one. Crisping the potatoes properly takes a while, which gives you plenty of time to brew your coffee or squeeze your mangoes, or prepare whatever else it is your spouse likes served with their brunch.

# BASIC CRISPY POTATO CAKE
## (AKA RÖSTI)

**SERVES 2 OR 3**

3 medium russet (baking) potatoes (about 1 pound), rinsed and cut into 1/16-inch matchsticks or grated on the large holes of a box grater

¼ cup olive oil

Kosher salt and freshly ground black pepper

1. Spread the potatoes on a large microwave-safe plate and microwave on high until hot all the way through and softened but still slightly crunchy, about 5 minutes.

2. Heat 2 tablespoons of the oil in skillet over medium heat until shimmering. Add the potatoes and press into the bottom of the pan with a rubber spatula. Season with salt and pepper. Cook, swirling and shaking the pan occasionally, until the potatoes are deep golden brown and crisp on the first side, about 7 minutes. Carefully slide the rösti onto a large plate. Set another plate on top of it, upside down, grip the edges, and invert the whole thing so the rösti is now cooked side up.

3. Heat the remaining 2 tablespoons oil in the skillet and slide the rösti back in. Season with salt and pepper. Continue cooking, swirling and shaking the pan occasionally, until the rösti is deep golden brown and crisp on the second side, about 7 minutes longer. Slide onto a cutting board and serve immediately, with aioli or mayonnaise, or ketchup.

# CRISPY POTATO, ONION, AND MUSHROOM CAKE
## (AKA RÖSTI)

**SERVES 2 OR 3**

3 medium russet (baking) potatoes (about 1 pound), rinsed and cut into 1/16-inch matchsticks or grated on the large holes of a box grater

5 tablespoons olive oil

1 medium onion, finely sliced (about 1 cup)

4 ounces button mushrooms, finely sliced

2 medium garlic cloves, minced or grated on a Microplane (about 2 teaspoons)

1 teaspoon fresh thyme leaves

Kosher salt and freshly ground black pepper

1. Spread the potatoes on a large microwave-safe plate and microwave on high until hot all the way through and softened but still slightly crunchy, about 5 minutes.

2. Meanwhile, heat 1 tablespoon of the oil in a 10-inch cast-iron or heavy-bottomed nonstick skillet over medium-high heat until shimmering. Add the onions and mushrooms and cook, stirring and tossing occasionally, until softened and starting to brown, about 8 minutes. Add the garlic and thyme and cook, stirring frequently, until fragrant, about 30 seconds. Season to taste with salt and pepper. Transfer to a small bowl and wipe out the skillet.

3. Heat 2 tablespoons oil in the skillet over medium heat until shimmering. Add half of the potatoes and press into the bottom of the pan with a rubber spatula. Season with salt and pepper. Spread the onion/mushroom mixture evenly over the potatoes and top with the remaining potatoes. Press down into an even disk, using the spatula. Season with salt and pepper. Cook, swirling and shaking the pan occasionally, until the potatoes are deep golden brown and crisp on the first side, about 7 minutes. Carefully slide the rösti onto a large plate. Set another plate on top of it, upside down, grip the edges, and invert the whole thing so the rösti is now cooked side up.

4. Heat the remaining 2 tablespoons oil in the skillet and slide the rösti back in. Season with salt and pepper. Continue cooking, swirling and shaking the pan occasionally, until the rösti is deep golden brown and crisp on the second side, about 7 minutes longer. Slide the rösti into a cutting board. Serve immediately, with aioli or mayonnaise, or ketchup.

# POTATO HASH

**Hash is the kind of breakfast that happens when I plan
to go grocery shopping on a Friday night. This is
never a good idea. Let me give you an idea of how it works.**

**The Plan:** I wake up Friday morning fresh and
dewy-faced, ready for a full day of work, followed
by a trip to New York Mart for some produce, a
quick subway ride home, and a few hours of cooking. My wife gets home, we enjoy dinner, play a
couple rounds of online *Jeopardy!*, catch an episode
of *How I Met Your Mother*, and hit the sack early,
ready to face a hearty breakfast in the morning.

**The Reality:** I wake up Friday morning, barely over
a cold from earlier in the week, head in for a day
at work, get caught in meetings all morning before
finally getting to start my *real* work in the late afternoon, don't get as much done as I hoped, and say,
"Screw it, it's Friday, time for happy hour." Rather
than go grocery shopping, I get a cocktail, then realize that New York Mart is now closed, acknowl-

edge the grave error I've made in my meal planning,
and send down another cocktail to keep the first
one company. My wife ends up meeting me down-
town for another cocktail, followed by dinner out
(that's a bottle of wine and an after-dinner drink),
and since we've already made a night of it, we might
as well *really* make a night of it. Next thing I know,
it's noon on Saturday, the dog needs to be walked,
and I've got nothing but a few potatoes, a couple of
eggs, and some random leftovers in my pantry to
nurse us back to good health.

Thank god for hash, right?

Hash is the ultimate leftover-consumer. All you
need is a starchy root vegetable to form the base
(potatoes are the usual choice, but sweet potatoes or beets are great too), whatever leftovers you
have on hand—cooked meat, greens, vegetables,

whatever—a good cast-iron skillet, and a couple of eggs, and you've got the makings of a breakfast that will frighten *any* hangover into quiet submission. As I mention for my rösti recipes, the best way to get a good fluffy/crisp texture out of your potatoes is to boil them, dry them, and then fry them (see page 134). This allows the oil to penetrate a little deeper, increases the surface area by causing the potatoes to blister, and makes for a far crisper finished product. But who's got time for all that when there's a headache that needs tending to?

Instead, it's much easier to slice the potatoes, put them on a plate, and microwave them for the initial cooking step, as I do for rösti. This'll let you soften them and cook them through without having to worry about them getting waterlogged or too wet on their exterior, and what takes ten minutes in a pot takes under three minutes in the nuker. Once they are parcooked, I add the potatoes to a hot skillet to begin the crisping/charring process while I roughly chop up my vegetables—in this case, peppers and onions. In a hazy stupor, I've tried tossing the parcooked potatoes and other vegetables together before adding them to the skillet, but that is an exceedingly bad idea, resulting in burnt onions and none-too-crisp potatoes.

You need to give the taters a head start on the rest of the vegetables, only adding the veg once a crisp crust has developed.

By the way, don't ever feel like you're limited in what you can put into a good breakfast hash. Cabbages (like bok choy or Brussels sprouts) develop an awesomely sweet nutty, flavor as they char. Cured meats like pastrami or corned beef will crisp nicely, their fat flavoring the potatoes as they cook. Shallots and onions turn sweet and complex, while green vegetables like broccoli or asparagus get nicely charred and tender. By the time the potatoes are completely crisp, the peppers and onions in this one are perfectly tender and sweet, and the hangover has begun to emit a faint, high-pitched whimper of fear.

The nail in the coffin? A dash of hot sauce, which adds not only a touch of heat, but, more important, vinegar, to brighten things up.

All the awesome smells emanating from the skillet are enough to work my appetite up into a near frenzy, so a couple of eggs in the mix are a no-brainer. You can fry your eggs in a separate skillet, but it's much easier just to make a couple of wells in the hash, crack your eggs directly into the pan, and finish the whole thing off in the oven until the whites are just barely set but the yolks still runny. Start to finish, it takes under fifteen minutes, which means it's hot and on the table all before my wife is even back with the dog.

A great way to start your Saturday afternoon— bright and early.

Add your onions too soon and you end up with burnt onions.

# POTATO HASH
## WITH PEPPERS AND ONIONS

**NOTE:** The potatoes can also be parcooked on the stovetop instead of in the microwave. Cover them with cold salted water in a large saucepan and bring to a boil over high heat, then reduce to a simmer and cook until they are just barely undercooked. Drain and continue with step 2.

**SERVES 4**

1½ pounds russet (baking) potatoes (about 4 medium), peeled and cut into ½-inch cubes

3 tablespoons vegetable oil

1 small red bell pepper, thinly sliced

1 small green bell pepper, thinly sliced

1 small onion, thinly sliced

1 teaspoon Frank's RedHot or other hot sauce, or more to taste

Kosher salt and freshly ground black pepper

4 large eggs (optional)

1. If using the eggs, adjust an oven rack to the upper-middle position and preheat the oven to 400°F. Spread the potatoes on a large microwave-safe plate, cover with paper towels, and microwave on high until heated through but still slightly undercooked, 4 to 6 minutes.

2. Heat 2 tablespoons of the oil in a 12-inch cast-iron or nonstick skillet (or two 10-inch skillets) over high heat until lightly smoking. Add the potatoes and cook, stirring and tossing them occasionally, until well browned on about half the surfaces, about 5 minutes. Reduce the heat if the pan is smoking heavily.

3. Add the peppers and onions and cook, tossing and stirring occasionally, until all the vegetables are browned and charred in spots, about 4 minutes longer. Add the hot sauce and cook, stirring constantly, for 30 seconds. Season to taste with salt and pepper. If not using eggs, serve immediately.

4. If using eggs, make 4 wells in the potatoes and crack the eggs into them. Season the eggs with salt and pepper, then transfer the skillet to the oven. Cook until the whites are just set, about 3 minutes. Serve immediately.

### POTATO AND CORNED BEEF HASH

Replace the peppers and onions with 8 ounces leftover corned beef, shredded into bite-sized nuggets.

# KNIFE SKILLS:
## How to Cut a Potato

**Potatoes are heavy and unwieldy, with a lumpy round
shape that makes them hard to hold steady.**

This is bad news if you want even cuts—and all your fingers. The trick is to first slice off one side, creating a stable base to rest the potato on.

To dice a potato, start by peeling it with a Y-peeler, if peeling. Rinse under cold water and use the eye-remover to gouge out any eyes. Hold the potato firmly on a cutting board and slice a ¼- to ½-inch-thick slab off one side. Turn the potato cut side down and slice lengthwise into planks of even thickness. Working with a few planks at a time, stack them and slice lengthwise into even batons. Then, for dice, hold a few batons at a time and slice across them to create dice of even width. Cut potatoes will discolor, so cook immediately or store in cold water.

1

2

3

4

5

# BUTTERMILK PANCAKES

**They may be golden brown, crisp on the edges, and light and fluffy
in the center, but when you get right down to it, classic American pancakes
are not all that different from any leavened bread.**

Apart from its starch content, bread is basically just a ball of protein filled with gas (very much like my dog, in that respect). When flour is mixed with

---

My dog has a nasty habit of waiting until the elevator is completely full of people before silently releasing his gas, then staring innocently up at me as if to say, "I can't believe you just did that." I don't appreciate it.

liquid, two proteins naturally present in wheat, *glutenin* and *gliadin*, link together to form the resilient, stretchy protein matrix known as *gluten*. In leavened breads, air bubbles are formed in this matrix and expand, creating the familiar hole structure inside a loaf of bread (or a good pizza crust, for that matter).

With traditional or "slow" breads, that leavening

agent is a living fungus called yeast. As the yeast consumes sugars present in the flour, it releases carbon dioxide gas, forming thousands of teeny-tiny air pockets inside the dough and causing it to rise. Once you pop that dough into the oven, those air pockets heat up and further expand, and a phenomenon known as *oven spring* takes place. Finally, as the gluten and starches get hot enough, they set into a semisolid form, giving structure to the bread and turning it from wet and stretchy to dry and spongy.

The only problem with yeast? It takes a long, long time to work. Enter baking soda. Unrestricted by the protracted time frames of biological organisms, it relies instead on the quick chemical reaction between an acid and a base. Baking soda is pure sodium bicarbonate—an alkaline (aka basic) powder. When dissolved in liquid and combined with an acid, it rapidly reacts, breaking down into sodium, water, and carbon dioxide. Just as with yeasted breads, this carbon dioxide expands upon baking, leavening the gluten protein matrix. This type of chemically leavened bread is referred to as a quick bread, a broad category that includes everything from scones and biscuits to banana or zucchini bread and even pancakes.

Of course, for baking soda to work, a recipe needs to include a significant acidic ingredient. That's why you see so many classic recipes for buttermilk pancakes and buttermilk biscuits or cake recipes that contain vinegar. The buttermilk is not just a flavoring agent—it provides the necessary acid to react with the baking soda and leaven the bread. Around the middle of the nineteenth century, someone realized that rather than relying on the home cook to add an acidic ingredient to react with the baking soda, it'd be much simpler to add a powdered acid

directly to the baking soda itself, and baking powder was born. Composed of baking soda, a powdered acid, and a starch (to absorb moisture and prevent the acid or base from reacting prematurely), baking powder was marketed as the all-in-one solution for busy housewives. In its dry state, it's totally inert. But once you add a liquid, the powdered acid and base dissolve and react with each other, creating bubbles of carbon dioxide, without the need for an external acid source.

Neat, right? But hold on—there's more.

## Side Effects

The most interesting side effect of using baking soda in a recipe is that it affects browning in a major way. **The Maillard reaction**, named after Louise Camille Maillard, who first described its processes in the early twentieth century, is the set of reactions responsible for that beautiful brown crust on your steak and the deep color of a good loaf of bread. Aside from cosmetics, the reaction also produces hundreds of aromatic compounds that add an inimitable savoriness and complexity to foods.

As it turns out, the reaction occurs better in alkaline environments, which means that once you've added enough baking soda to neutralize the acid in a batter or dough, any extra you add will work to increase browning. So I made five batches of pancakes using identical batters consisting of flour, baking powder, egg, buttermilk, melted butter, salt, and sugar and varying amounts of baking soda, starting with none and increasing it by ⅛-teaspoon increments up to a full ½ teaspoon per batch. Each pancake was cooked on a preheated griddle for exactly 1½ minutes per side. The results very clearly demonstrate the browning effect of baking soda.

Baking soda affects pH, which in turn affects browning.

The pancake all the way on the left is inordinately acidic, due to the unneutralized buttermilk. It cooked up pale and bland. It was also under-risen, with a flat, dense texture. The one all the way around on the bottom, with a full ½ teaspoon of baking soda in the batter, had the opposite problem. It browned far too quickly, lending it an acrid burnt flavor tinged with the soapy chemical aftertaste of unneutralized baking soda. Interestingly enough, this pancake was also flat and dense—the large amount of baking soda reacted too violently when mixed into the batter. The carbon dioxide bubbles inflated too rapidly and, like an overfilled balloon, the pancake "popped," becoming dense and flaccid as it cooked.

This browning phenomenon isn't just limited to pancakes, of course. For example, cookie recipes routinely include baking soda to aid browning, even when there isn't an acid for it to react with.

## Double Bubble

If there's one major drawback with chemically leavened breads, it's that they need to be cooked pretty much immediately after the batter is mixed. Unlike a yeasted bread dough, which is low in moisture and kneaded until a tough, elastic gluten network forms to trap the massive amounts of carbon dioxide produced, a quick bread must be made with an extremely moist batter—baking powder simply doesn't produce enough gas to effectively leaven a thicker dough. Batters have relatively little gluten formation, meaning that they aren't all that great at trapping and holding bubbles. Once you mix a batter, your baking soda or baking powder immediately begins producing gas, and that gas almost immediately being trying to escape into the air. When working with quick breads, those who aren't into the whole brevity thing may run into difficulties.

Cook your pancakes immediately after mixing, and you get a light, tall, fluffy interior. Let the batter sit for half an hour, and you get a dense, gummy interior with few bubbles. But wait a minute, there are still *some* bubbles in there, right? Where did those come from?

Well, pretty much all baking powder is what is referred to as "double-acting." Just as the name indicates, it produces gas in two distinct phases. The first occurs as soon as you mix it with water; the second occurs only when it is heated (see "Experiment: Double-Acting Baking Powder," page 147).

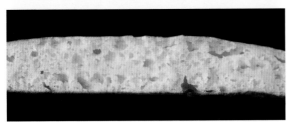

Double-acting baking powder gives you a second rise when cooked.

This second rise in the skillet makes for extra-light and fluffy pancakes.

## The Whites Are Light

So what if baking soda just isn't doing enough for you? How do you get your pancakes to stand even taller and lighter? I like to use a meringue—egg whites that have been whipped vigorously until they form a semisolid foam. Here's how it works:

- **Foam:** In the early phases of beating, the proteins in the egg whites—mostly *globulin* and *ovotransferrin*—begin to unfold. Like nerds at a *Star Wars* convention, they tend to gather together and bond in small groups. The whites start to incorporate a few bubbles and resemble sea foam.
- **Soft peaks:** As the whites are beaten, the groups of bonded egg proteins become more and more interconnected, eventually creating a continuous network of proteins that reinforce the walls of the bubbles you're creating. The whites begin to form soft peaks.
- **Stiff peaks:** As you continue to beat, the reinforced bubbles are broken into smaller and smaller bubbles, becoming so small that they are nearly invisible to the naked eye and thus the whites appear smooth and white, like shaving cream. When pulled into peaks, they remain stiff and solid.
- **Breakdown and weeping:** Keep going past the stiff-peak stage, and the proteins begin to bond so tightly with each other that they squeeze the moisture right out of the bubbles, resulting in a meringue that weeps and breaks. Acidic ingredients like cream of tartar or a touch of lemon juice can prevent egg white proteins from bonding too tightly, allowing you to form a foam that stays stable no matter how hard you beat it.

Add sugar and vanilla to the whites at the soft-peak stage, whip to stiff peaks, drop by the spoonful onto baking sheets, and bake at a low temperature, and you've got yourself classic meringue cookies. If you instead drizzle in a cooked sugar syrup toward the end of whipping, you'll end up with what's called an Italian meringue, a meringue that stays soft and supple even when browned—the kind of thing you'd want to top a lemon meringue pie with.

Here the use for meringue is much more simple: all you're going to do is fold it into the pancake batter. The extra air that the egg whites have incorporated expands as the pancakes cook, making them featherlight.

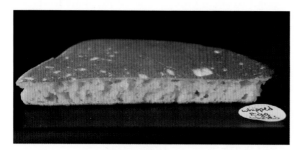

Whipped egg whites.

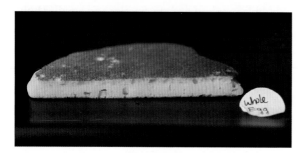

Plain eggs.

## Pancake Flavor

As far as flavoring buttermilk pancakes go, there are a few givens: Dairy fat, in the form of melted butter or milk, is essential. Not only does it add richness and flavor to the mix, but by coating the flour and limiting gluten development, it also ensures that your pancakes remain tender. Eggs help set the pancakes as they cook, as well as providing some extra lift. Buttermilk is obviously part of the equation, but I like my pancakes extra-tangy, and straight-up buttermilk just doesn't cut it for me. Increasing the quantity doesn't work—that just ends up throwing the liquid-to-solid ratio out of whack. Instead, I replace part of the buttermilk with a good amount of sour cream. It's both less moist than buttermilk and more sour, which allows me to add acidity without watering down the batter. If you don't have sour cream on hand, don't worry—the pancakes will still taste just fine with straight-up buttermilk.

# EXPERIMENT:
## Double-Acting Baking Powder

**Double-acting baking powder (the type sold in any supermarket) is designed to produce bubbles in two distinct phases: when it gets wet and then when it gets heated. You can see this for yourself.**

### Materials

- 1 teaspoon baking powder
- 1 tablespoon water

### Procedure

1. Combine the baking powder and water in a small bowl. You'll notice that the baking powder immediately starts bubbling and fizzing (if it doesn't, throw out your baking powder and buy a new can). This is the first reaction. After 30 seconds or so, all action will cease, and you'll end up with a still pool of chalky-looking liquid.
2. Now microwave that liquid for about 15 seconds to bring it up to 180°F. A second, vigorous batch of bubbling should occur. You may also notice the liquid thicken slightly.

### Results and Analysis

When the baking powder first gets wet, a reaction occurs between the sodium bicarbonate and one of the powdered acids, typically potassium bitartrate (aka cream of tartar), producing the first batch of bubbles. The second phase of the double act occurs only at higher temperatures (around 170° to 180°F), when a second powdered acid (typically sodium aluminum sulfate) reacts with the remaining sodium bicarbonate, producing another round of bubbles. The thickening action is a side effect of the starch used to keep the baking powder dry—it absorbs water and gelatinizes, thickening your liquid as it heats. Now isn't that *way* cooler than that baking soda volcano you built for your fourth-grade science fair?

# MIXING BATTER

It's important not to overmix a pancake batter. Just as with the batter on a good onion ring or piece of fried fish (see "Experiment: Gluten Development in Batter," page 899), the more vigorously you stir a

*continues*

batter, the more gluten develops, and the tougher it becomes. The result is underrisen or leathery pancakes. When mixing pancake batter, the goal is to do it as quickly as possible, just until it comes together, allowing a few lumps of dry flour to remain. Don't worry—they'll disappear as the pancakes cook.

## WHAT IS BUTTERMILK?

True buttermilk is the liquid whey left after cream has been churned to create butter. Traditionally this whey was allowed to ferment into a slightly thickened, sour liquid that would keep longer than fresh milk. These days, though, buttermilk is made from regular milk by dosing it with *Streptococcus lactis*, a bacteria that consumes *lactose*, the main sugar in milk, and produces lactic acid, which adds tartness to the buttermilk, as well as causing *casein*, the primary protein in milk, to curdle, thickening, or clabbering, the milk.

In some recipes, it's possible to substitute artificially clabbered milk—milk to which an acid like vineger or lemon juice has been added to thicken it—for buttermilk, but you'll always be left with a telltale flavor from the added acid. Much better is to substitute another soured dairy product. When I have no buttermilk on hand, I'll use yogurt, sour cream, or even crème fraîche diluted with milk.

| DAIRY PRODUCT | TO SUBSTITUTE FOR 1 CUP OF BUTTERMILK |
|---|---|
| Yogurt (full-fat or skim) | ⅔ cup yogurt whisked together with ⅓ cup milk |
| Sour Cream | ½ cup sour cream whisked together with ½ cup milk |
| Crème Fraîche | ½ cup crème fraîche whisked together with ½ cup milk |

## SUBSTITUTING BAKING SODA
## FOR BAKING POWDER

Baking soda is sodium bicarbonate. It reacts with liquid acids immediately upon contact to produce carbon dioxide. Carbon dioxide gets trapped within batters and expands upon baking, leavening your pancakes and other quick breads. Because baking soda reacts immediately, quick breads made with it must be baked or cooked right after mixing. And because of its alkalinity, baking soda can also hasten browning reactions, adding color (and thus flavor) to things like pancakes, cookies, and muffins.

Baking powder is sodium bicarbonate mixed with one or more of the powdered acids and a starch. It does not require another acid to activate it. As mentioned earlier, most baking powders are "double-acting," meaning they produce carbon dioxide once upon coming in contact with moisture and then again when heated. Because of this, baking powder–leavened goods are generally lighter and fluffier than those made with baking soda alone. This doesn't mean, however, that you can let a baking powder batter just sit around, expecting the second batch of bubbles to do all the leavening—the initial reaction is vitally important to the texture of your baked goods, and so these batters should be baked right away too.

Don't have baking powder on hand? It's quite simple to substitute with your own homemade mixture of baking soda, cornstarch, and cream of tartar. For every teaspoon of baking powder, use ¼ teaspoon baking soda, ½ teaspoon cream of tartar, and ¼ teaspoon cornstarch. But do bear in mind that your homemade mixture will not be double-acting, requiring you to be extra quick about getting your pancakes onto the griddle or your zucchini bread in the oven after mixing the batter.

# BASIC DRY PANCAKE MIX

Why buy a store-bought mix when homemade pancakes are so easy and so much better? You can use this mix immediately after putting it together or, better yet, do what I do: make a quadruple batch and store it in an airtight container in the pantry. That way, whenever you want to whip up a batch of pancakes, all you've got to do is add your wet ingredients, and you're ready to roll.

NOTE: This recipe can be scaled up to any size.

**MAKES ABOUT 2 CUPS, ENOUGH FOR APPROXIMATELY 16 PANCAKES**

10 ounces (2 cups) all-purpose flour

1 teaspoon baking powder

½ teaspoon baking soda

1 teaspoon kosher salt

1 tablespoon sugar

Combine all the ingredients in a medium bowl and whisk until homogeneous. Transfer to an airtight container. The mix will stay good for 3 months.

# LIGHT AND FLUFFY BUTTERMILK PANCAKES

**NOTE:** The sour cream can be replaced with more buttermilk.

**MAKES 16 PANCAKES, SERVING 4 TO 6**

1 recipe Basic Dry Pancake Mix (above)

2 large eggs, separated

1½ cups buttermilk

1 cup sour cream (see Note above)

4 tablespoons unsalted butter, melted

Butter or oil for cooking

Warm maple syrup and butter

1. Place the dry mix in a large bowl.

2. In a medium clean bowl, whisk the egg whites until stiff peaks form. In a large bowl, whisk the egg yolks, buttermilk, and sour cream until homogeneous. Slowly drizzle in the melted butter while whisking. Carefully fold in the egg whites with a rubber spatula until just combined. Pour the mixture over the dry mix and fold until just combined (there should still be plenty of lumps).

3. Heat a large heavy-bottomed nonstick skillet over medium heat for 5 minutes (or use an electric griddle). Add a small amount of butter or oil to the griddle and spread with a paper towel until no visible butter or oil remains. Use a ¼-cup dry measure to place 4 pancakes in the skillet and cook until bubbles start to appear on top and the bottoms are golden brown, about 2 minutes. Carefully flip the pancakes and cook on the second side until golden brown and completely set, about 2 minutes longer. Serve the pancakes immediately, or keep warm on a wire rack set on a rimmed baking sheet in a warm oven while you cook the remaining 3 batches. Serve with warm maple syrup and butter.

## BLUEBERRY PANCAKES

*Blueberries are much like peas in that the vast majority of the time, frozen are simply better for cooking with. They're picked ripe and flash-frozen immediately, leaving them sweet and flavorful. While a fresh blueberry may have a more pleasing texture, with plump flesh and skin that pops under your teeth, supermarket blueberries (particularly off-season berries) are completely lacking in flavor. Unless you're picking blueberries yourself or have access to a great local source, I'd suggest going with frozen in most applications, particularly cooked ones, where texture is not as much of an issue. The one drawback to be aware of: they leach color, staining everything in their path purple. The trick with pancakes is to add them after you've ladled the batter onto the griddle. You don't even have to scoop more batter on top of them—they'll get swaddled like little blue babies in the pancake batter.*

Sprinkle 1 to 2 tablespoons thawed frozen blueberries on top of each pancake as soon as you ladle them onto the griddle and proceed as directed.

# HOMEMADE RICOTTA

**News flash! Store-bought ricotta is almost invariably awful!**

True ricotta is made by adding acid to heated whey, typically the whey left over from the production of Pecorino Romano, although other kinds of whey are sometimes used. (*Ricotta* means "recooked," referring to the reheated whey.) The combination of heat and acid causes milk proteins (mainly casein) to bind together, trapping moisture and fat and forming soft, white curds. To make high-quality ricotta, these curds are then carefully removed from the whey (too much mechanical action can turn them rubbery) and allowed to drain, reducing their water content and concentrating their flavor and richness. The result is mind-blowingly simple yet decadent. Or at least it *should* be. The reality is that pretty much all mass-market ricotta producers don't bother to take the time to drain their cheese

properly. Instead, they load the stuff up with gums and stabilizers intended to keep the water (and thus their profits) from leaking out.

What you get is a gritty, gluey, rubbery paste. No thank you. Homemade ricotta, though, when made right, is creamy and tender with a mild, milky flavor and a slight tang from the acid used to curdle it. In fact, I like ricotta made at home with whole milk *better* than a traditional low-fat whey-produced ricotta. How do you like that?

With homemade ricotta, the single most important variable in terms of both flavor and texture is the acid you add to the mix.

- **Buttermilk** has many advocates, who claim it's the tastiest acid of choice. I had problems with it. In order to get the milk to curdle properly, I had to add buttermilk at nearly a 1:4 ratio, resulting in a final product with a very distinct sour flavor. It wasn't bad per se, but the flavor certainly limited its applications: I couldn't imagine stuffing it into ravioli, for instance. And the curd structure was also ever-so-slightly overdeveloped, giving the ricotta a sticky texture.

- **Distilled vinegar** gives the cleanest flavor, with soft, tender curds. Since bottled vinegar is always diluted to 5 percent acetic acid, using it is also the most consistent method. As long as your milk is fresh (older milk is more acidic than fresh milk, and thus requires less coagulant), you'll get identical results every time.

- **Lemon juice** also works very well, though I found that in some cases the amount I needed to use varied by about 25 percent, give or take. Most likely this is due to varying pH levels from lemon to lemon. Lemon juice gives the ricotta a very slight citrus tang that, while not as distinct as the buttermilk flavor, can be slightly off-putting in certain savory applications. On the other hand, it's wonderful for pancakes and blintzes, or feeding to your hard-working wife, drizzled with olive oil and sprinkled with sea salt, warm off a spoon.

Bottom line? For the most versatile ricotta, stick with vinegar. Use lemon juice when a lemon flavor is appropriate, and avoid buttermilk unless you're *really* into it.

## Draining Ricotta

To drain ricotta, place it in a fine-mesh strainer lined with cheesecloth (or a high-quality food-safe paper towel) set over a bowl. The final texture of ricotta can vary greatly depending on how well it's drained.

| DRAINING TIME | TEXTURE | BEST USES |
|---|---|---|
| Under 5 minutes | Extremely moist and creamy, like cottage cheese, with small, tender curds | Immediate consumption, while still warm. Try it drizzled with olive oil and sprinkled with sea salt and black pepper, or, for dessert, with honey and fruit. |
| 15 to 20 minutes | Small, tender curds with a cottage cheese–like consistency; moist and spreadable, but not runny | Moist, savory application, such as adding to a lasagna or topping a pizza, or mixing into your pancake batter |
| At least 2 hours, or up to overnight (refrigerated) | Large, dry, crumbly curds that can easily be molded into firm shapes | Cakes and pasta, like ricotta cheesecake or ricotta gnocchi |

# PASTEURIZED MILK

Unless you're buying your milk at a farm or squeezing it straight from the teats of your own herd, you're getting pasteurized milk—milk that's been heated in order to destroy bacteria and prolong its shelf life. There are three basic methods used to do this:

- **Regular pasteurized milk** has been heated to 161°F for around 20 seconds. This is the standard for most supermarket milks, which have a shelf life of a few weeks.
- **Ultra High Temperature pasteurized milk** has been heated much hotter—all the way up to 275°F—for 1 second. It is labeled UHT or "Ultra-Pasteurized" and has a shelf life of several months. Many organic milk producers use this method of pasteurization, as it allows their milks to sit in supermarket dairy cases for longer (organic milk often doesn't sell as quickly as regular milk). When packed into specially designed containers, UHT milk can actually keep, unrefrigerated, for months or even years.
- **Low-Temperature pasteurized milk** has been held at 145°F for 30 minutes. Many small farms pasteurize their milk with this method, as it doesn't produce the "cooked" flavor that UHT or regular pasteurized milk can have. The label generally doesn't indicate whether the milk is just pasteurized or if it's been low-temperature pasteurized, so unless you know the producer, chances are it's the former.

As far as their cooking qualities go, in most application, all of these types of milk will behave just about the same. For making ricotta, however, the higher the temperature the milk has been cooked to, the more breakdown you find in its proteins and sugars. For this reason, UHT milk tends to have a slightly sweeter flavor (complex carbohydrates are broken down into simpler, sweeter sugars during the pasteurization process). And UHT milks will not coagulate as well when making ricotta. I recommend standard pasteurized milk, Low-Temperature pasteurized milk, or, if you can get it, raw milk.

# FRESH RICOTTA
## IN 5 MINUTES OR LESS

**MAKES ABOUT 1 CUP**

4 cups whole milk

½ teaspoon table salt

¼ cup distilled white vinegar or
lemon juice (from 2 lemons)

1. Line a colander with four layers of cheesecloth or two layers of food-safe paper towels and set over a large bowl. Combine the milk, salt, and vinegar in a microwave-safe 2-quart liquid measure and microwave on high until lightly bubbling around the edges, 4 to 6 minutes; the milk should register about 165°F on an instant-read thermometer. Remove from the microwave and stir gently for 5 seconds. The milk should separate into solid white curds and translucent liquid whey. If not, microwave for 30 seconds longer and stir again. If necessary repeat until fully separated.

2. Using a slotted spoon or wire skimmer, transfer the curds to the prepared colander. Cover the exposed top with plastic wrap and allow to drain until the desired texture is reached. Leftover ricotta can be stored in a covered container in the refrigerator for up to 5 days.

**VARIATION**

You can make this recipe on the stovetop instead of the microwave. Heat the milk and vinegar mixture in a saucepan over medium-low heat, stirring constantly with a silicone spatula to prevent sticking or scorching until it reaches 165°F on an instant-read thermometer. Remove it from the heat and allow to rest until solid white curds form on the surface, about 2 minutes.

# WARM RICOTTA
## WITH OLIVE OIL AND LEMON ZEST

**SERVES 4**

1 cup Fresh Ricotta (above), just
made

2 tablespoons extra-virgin olive oil,
plus more for serving

2 teaspoons grated lemon zest
(from 1 lemon)

Flaky sea salt, like Maldon

Freshly ground black pepper

Place the ricotta in a serving bowl, drizzle with the olive oil, and sprinkle with the lemon zest and salt and pepper. Serve immediately with toast, passing extra olive oil.

# LEMON RICOTTA PANCAKES

**These are special-occasion pancakes. Serve them for brunch, at your own risk—you _will_ secure yourself the top seed as host for every brunch in the future.**

## MAKES 12 PANCAKES, SERVING 3 TO 4

½ cup buttermilk

1 cup Fresh Ricotta (page 154), drained for 30 minutes

2 tablespoons unsalted butter, melted and slightly cooled

2 large eggs

½ teaspoon vanilla extract

1 cup Basic Dry Pancake Mix (page 149)

2 teaspoons grated lemon zest (from 1 lemon)

Vegetable oil for cooking

Maple syrup

1. Whisk together the buttermilk, ricotta, melted butter, eggs, and vanilla extract in a medium bowl. Add the pancake mix and lemon zest and whisk until no dry flour remains (the mixture should remain lumpy—be careful not to overmix).

2. Heat ½ teaspoon oil in a 12-inch heavy-bottomed nonstick skillet over medium-high heat (or use an electric griddle) until it shimmers. Reduce the heat to medium and wipe out the skillet with a paper towel. Use a ¼-cup dry measure to scoop 4 pancakes into the pan and cook on the first side until bubbles start to appear on top and the bottoms are golden brown, 2 to 3 minutes. Flip the pancakes and cook until the second side is golden brown, about 2 minutes longer. Serve the pancakes immediately, or keep warm on a wire rack set on a baking sheet in a warm oven while you cook the remaining batches. Serve with maple syrup.

# WAFFLES

**Waffles are like the cool cousin of pancakes: a bit more complicated, a bit more interesting, and a bit crustier on the outside.**

But deep down, they're almost identical. When we're talking quick American-style waffles (as opposed to, say, a traditional slow-rising yeasted, chewy Belgian waffle), we're talking a chemically leavened batter, just like with pancakes. But try throwing your pancake batter into a waffle iron, and you will run into trouble. With pancakes, the steam evaporating as the pancakes cook has an easy escape route—you can see it coming out of the top of the pancakes as bubbles form. With a waffle trapped inside its metal cage, it's not so easy. Waffles made from pancake batter come out gummy, with a distinct lack of crispness.

But I wanted to be able to start my waffles with my basic pancake mix so that I wouldn't have to keep *two* mixes on hand in my pantry. I realized the solution had to be twofold: I needed extra leavening power to help the waffles rise in their constrained environment, and I needed a method to ensure that they got crisp faster and stayed crisp.

I first tried adding a bit of extra baking powder and baking soda when I mixed up my batter. It helped with the texture, but with too much chemical leavening, that soapy, metallic flavor started creeping in. I'd have to find a physical means to leaven my batter instead.

I was already adding a good deal of bubbles with my whipped egg whites—what if I were to add even more in the form of soda water? It's a trick that New Englanders have used forever: the beer in beer-battered fish is just as much about the leavening power of the bubbles as it is about the flavor of the ale. Even the Japanese use soda water to achieve an extra-light tempura. Using soda water does cut back a little on the flavor of the waffles, but it's not *too* noticeable, and it's a compromise I'm willing to make in the name of superior texture. A dash of vanilla (or orange liqueur, or even maple extract and bacon, if you'd like) adds plenty of flavor to keep you distracted. It's important that you use ice-cold club soda. Cold liquids retain carbonation better, and you want the batter to stay as bubbly as possible until it starts cooking. Club soda is superior to seltzer water in this case, because it contains sodium, which also helps it to retain its bubbles.

Crispness is all about dehydration and the setting of proteins, both things that are accomplished through heat and time. The key to extra-crisp waffles? Just cook them a little more slowly for a little bit longer. Not only does this lead to superior texture, it also serendipitously results in more even browning

# BASIC QUICK WAFFLES

**MAKES 8 SMALL ROUND WAFFLES, FOUR 4-WELL BELGIAN-STYLE WAFFLES, OR 4 LARGE SQUARE WAFFLES, SERVING 4**

1 recipe Basic Dry Pancake Mix (page 149)

2 large eggs

½ cup buttermilk

4 tablespoons unsalted butter, melted

1 cup ice-cold club soda

1 teaspoon vanilla extract

Butter or oil for the waffle iron

Maple syrup

1. Preheat an electric waffle iron, on the low-heat setting if you have the option, or a heat a stovetop waffle iron over medium-low heat. Place the dry mix in a large bowl.

2. In a medium bowl, whisk the egg whites until stiff peaks form. In a clean large bowl, whisk the egg yolks and buttermilk until homogeneous. Slowly drizzle in the butter while whisking. Carefully fold the egg whites into the yolk mixture with a rubber spatula until just combined. Fold in the soda water. Pour the mixture over the dry mix and fold until just combined (there should still be plenty of lumps).

3. If using a 7-inch round stovetop waffle iron, ladle ½ cup of batter into the iron and cook, flipping it occasionally, until the waffle is golden brown and crisp on both sides, about 8 minutes. If using a Belgian waffle iron, scoop ¼ cup batter into each well, close the iron, and immediately flip it, then continue cooking, turning occasionally, until the waffles are golden brown and crisp on both sides, about 10 minutes. If using an electric waffle iron, preheat and cook according to the manufacturer's instructions. Transfer to a plate, or keep warm on a rack on a baking sheet in a 200°F oven, and cook the remaining batches.

**ORANGE-SCENTED WAFFLES**

Replace the vanilla extract with 1 tablespoon orange liqueur, such as Grand Marnier, and add 1 teaspoon grated orange zest to the eggs and buttermilk in step 2.

**MAPLE BACON WAFFLES**

Replace the vanilla extract with maple extract (or 2 tablespoons maple syrup) and add 6 strips crisp bacon, crumbled, to the batter at the end of step 2.

# BUTTERMILK BISCUITS

**If my wife and I ever have identical twins, I'd like to name one Stanley
and the other Evil Stanley, for the purposes of scientific inquiry.**

We'll raise them exactly the same, but over time, Evil Stanley will undoubtedly begin to live up to his name because of a subtle difference in the way the world treats him. There is sure to be a tragic ending or two somewhere in the story. In the never-ending debate between nature versus nurture and their effect on the human mind, it's always fascinating to me to see how radically different the end results of seemingly similar starting cases can be.

So it is with pancakes and biscuits. Take a look at the ingredients lists, and they're nearly identical: flour, butter, baking powder, baking soda, and liquid

dairy. But one ends up fluffy, tender, and relatively flat, and the other ends up tall, flaky, and crisp. The difference is all in the details.

First off, biscuits are a dough, not a batter, which means that the ratio of flour to liquid is high enough that it can pull everything together into a cohesive ball that's soft but doesn't flow. Even more important is the way in which the butter is incorporated. With pancakes, the butter is melted and whisked into the batter, resulting in a sort of uniform tenderness. For great flaky biscuits, on the other hand, the butter is added cold and hard, and it's added before the liquid is. As you work the hard butter into the flour, you end up with a mealy mix comprised of small bits of butter coated in flour, some amount of a flour-and-butter paste, and some completely dry flour. Now add your liquid to this mix, and what happens? Well, the dry flour immediately begins to absorb water, forming gluten. Meanwhile, the flour suspended in the flour-butter paste doesn't absorb any water at all, and, of course, you've still got your clumps of 100-percent pure butter.

Kneading the dough will cause the small pockets of gluten to gradually link together into larger and larger networks. All the while, butter-coated flour and pure butter are suspended within these networks. As you roll the dough out, everything gets flattened and elongated. The gluten networks end up stretched into thin layers separated by butter and butter-coated flour.

Finally, as the biscuits bake, a couple things occur. First, the butter melts, lubricating the spaces between the thin gluten sheets. Next, moisture—from both the butter and the liquid added to the dough—begins to vaporize, forming bubbles that rapidly increase in volume and inflate the interstitial spaces between the gluten layers, causing them to separate. Meanwhile, remember there's also baking powder and baking soda involved. This causes the parts of the dough that are made up of flour and liquid to leaven and inflate, adding tenderness and making the texture of the biscuits lighter.

## Folding

One of the keys to ultratender biscuits is not all that different from making light pancakes: don't overmix. You want to knead the ingredients just until they come together. Overmixing can lead to excess gluten formation, which would make the biscuits tough. The other secret is to keep everything cold. If your dough warms up too much, the butter will begin to soften and become more evenly distributed in the dough. You want the butter in distinct pockets to help give the biscuits a varied, fluffy texture.

There are a couple ways to help achieve these goals. First is to incorporate the butter using a food processor. A food processor's rapidly spinning blade will make short work of the butter, with little time for it to heat up and begin to melt. The method by which you incorporate the buttermilk is also important. Some folks like to do it by hand, others in the food processor. I find that the absolute best way is with a flexible rubber spatula, gently folding the dough and pressing it onto itself in a large bowl. Not only does the folding motion minimize kneading (and thus gluten), it also causes the dough to form many layers that will separate as they bake, giving you the flakiness you're after.

For an extra boost of flakiness, I like to go one step further and make what's called a laminated pastry: pastry that has been folded over and over itself to form many layers. The doughs for classic French laminated pastries like puff pastry and croissants are folded until they form hundreds of layers. With my biscuit dough, I'm not quite so ambitious, but I've found that by rolling it out into a square and folding it into thirds in both directions, you create 9 distinct layers (3 × 3). Roll the resultant package out into a square again and repeat the process, and you've got yourself a whopping 81 layers (9 × 3 × 3)! How's that for flaky?

And guess what: a modern flaky American scone is really nothing more than a sweetened biscuit cut into a different shape. Master one, and you've mastered the other.

**1.** Combine flour, baking powder, baking soda, and salt in a food processor, then scatter with butter cubes. **2.** Pulse until butter is broken into ¼-inch pieces. **3.** Transfer to a large bowl and add buttermilk. **4.** Fold with a spatula. **5.** Transfer to a floured cutting board or work surface. **6.** Knead briefly and form into a rectangle. **7.** Roll into a 12-inch square. **8.** Fold the right third over to the center with a bench scraper.

**9.** Fold the left side over the right. **10.** Fold the top third down over the center. **11.** Fold the bottom third up over the top. **12.** Roll the smaller square out into a 12-inch square and scatter with grated cheese and scallions. **13.** Fold the right third over the center. **14.** Fold the left third over the center. **15.** Fold the top third down over the center. **16.** Fold the bottom third up over the center.

*continues*

**17.** Re-roll the dough into a 12-inch square. **18.** Cut six rounds out of the dough with a 4-inch biscuit cutter, gather the scraps, knead gently, re-roll, and cut out two more rounds. Place them on a parchment-lined baking sheet. **19.** Brush the top and sides of the biscuits with melted butter. **20.** Transfer to a preheated oven to bake. **21.** Rotate the biscuits halfway through baking. **22.** Let the biscuits cool for 5 minutes before serving. **23.** Try your best to resist.

# SUPER-FLAKY BUTTERMILK BISCUITS

**MAKES 8 BISCUITS**

½ cup buttermilk

½ cup sour cream

10 ounces (2 cups) unbleached all-purpose flour, plus additional for dusting

1 tablespoon baking powder

¼ teaspoon baking soda

1½ teaspoons kosher salt

8 tablespoons (1 stick) cold unsalted butter, cut into ¼-inch pats

2 tablespoons unsalted butter, melted

1. Adjust an oven rack to the middle position and preheat the oven to 425°F. Whisk together the buttermilk and sour cream in a small bowl.

2. In the bowl of a food processor, combine the flour, baking powder, baking soda, and salt and process until blended, about 2 seconds. Scatter the butter evenly over the flour and pulse until the mixture resembles coarse meal and the largest butter pieces are about ¼ inch at their widest. Transfer to a large bowl.

3. Add the buttermilk mixture to the flour mixture and fold with a rubber spatula until just combined. Transfer the dough to a floured work surface and knead until it just comes together, adding extra flour as necessary.

4. With a rolling pin, roll the dough into a 12-inch square. Using a bench scraper, fold the right third of the dough over the center, then fold the left third over so you end up with a 12-by-4-inch rectangle. Fold the top third down over the center, then fold the bottom third up so the whole thing is reduced to a 4-inch square. Press the square down and roll it out again into a 12-inch square. Repeat the folding process once more.

5. Roll the dough again into a 12-inch square. Cut six 4-inch rounds out of the dough with a floured biscuit cutter. Transfer the rounds to a parchment-lined baking sheet, spacing them about 1 inch apart. Form the dough scraps into a ball and knead gently two or three times, until smooth. Roll the dough out until it's large enough to cut out 2 more 4-inch rounds, and transfer to the baking sheet.

6. Brush the top of the biscuits with the melted butter and bake until golden brown and well risen, about 15 minutes, rotating the pan halfway through. Allow to cool for 5 minutes and serve.

**CHEDDAR CHEESE AND SCALLION BISCUITS**

In step 4, sprinkle 6 ounces grated cheddar cheese and ¼ cup sliced scallions over the 12-inch dough square before folding it the second time, and continue as directed. (See photos on pages 160–62.)

**BACON PARMESAN BISCUITS**

In step 4, sprinkle ½ cup crumbled cooked bacon and 2 ounces grated Parmigiano-Reggiano over the 12-inch dough square before folding it the second time, and continue as directed. Dust the biscuits with more grated Parmesan before baking.

*continues*

**FLAKY SCONES**

Add 2 tablespoons sugar to the dry ingredients in step 2. In step 4, if desired, scatter 1 cup frozen or chopped fresh fruit or berries over the dough before folding it the second time. In step 5, roll the dough into a 12-by-4-inch rectangle, then cut the rectangle into three 4-inch squares and cut each square into 2 triangles. Sprinkle the scones with an additional 2 tablespoons sugar and bake as directed.

# CREAMY SAUSAGE GRAVY

**White sauce, as classic French *béchamel* is known in the United States, is basically nothing more than milk that has been thickened with flour. There are a couple keys to making good white sauce: the first is to make sure that the flour is cooked. Raw flour tastes, well, raw. You want to cook the flour in butter until its raw aroma goes away and it takes on a very light golden color. After that, it's just a matter of slowly whisking in the milk. The more slowly you whisk it in, the smoother your sauce will be. As your white sauce heats, starch granules in the flour—which are like tiny water balloons filled with starch molecules—slowly absorb water from the milk, swelling up and eventually bursting, releasing starch molecules into the liquid. These starch molecules cross-link, thickening your sauce. White sauces need to be brought to a near boil to thicken fully.**

**Creamy sausage gravy is as simple as frying some good breakfast sausage (such as the kind you make yourself), then making a white sauce around it. I like mine nice and peppery. Serve immediately, on top of buttermilk or other savory biscuits.**

**MAKES ABOUT 3 CUPS, ENOUGH FOR 8 SERVINGS**

1 tablespoon unsalted butter

1 pound Maple-Sage Breakfast Sausage (page 507) or good-quality bulk sausage

1 small onion, finely chopped (about ⅔ cup)

2 tablespoons all-purpose flour

2 cups whole milk

Kosher salt and freshly ground black pepper

1. Heat the butter in a 10-inch heavy-bottomed nonstick skillet over medium-high heat until foamy. Add the sausage and cook, using a wooden spatula or spoon to break up the meat, until no longer pink, about 6 minutes. Add the onion and cook until softened, about 2 minutes.

2. Add the flour and cook, stirring constantly, until completely absorbed, about 1 minute. Gradually add half of the milk, whisking constantly, then whisk in the remaining milk and allow to come to a simmer, whisking constantly. Simmer, whisking, until thickened, about 3 minutes. Season to taste with salt and plenty of black pepper.

# EASY CREAM BISCUITS

**Don't want to bother with all that folding and shaping but still want tender, plush, light, buttery biscuits or scones for the brunch table? Cream biscuits or scones are the answer. Cream biscuits are to flaky biscuits what shortbread is to piecrust. That is, rather than getting the butter and flour to build up into flaky, irregular layers that separate upon baking, you simply add far more liquid fat (in the form of both melted butter and cream), enough to completely coat the flour. The result doesn't have the layers of a flaky biscuit, but it has all the tenderness and a uniquely fluffy texture of its own.**

**The best part? Just five ingredients (OK, six if you add sugar to make them into scones), one bowl, and fifteen minutes start to finish. Didn't I say these are easy?**

## MAKES 8 BISCUITS

10 ounces (2 cups) all-purpose flour

1 tablespoon baking powder

¾ teaspoon kosher salt

4 tablespoons unsalted butter, melted

1¼ cups heavy cream

1. Adjust an oven rack to the middle position and heat the oven to 425°F. Place the flour, baking powder, and salt in a large bowl and whisk to combine. Add 2 tablespoons of the melted butter and the cream and stir with a wooden spoon until a soft dough comes together.

2. Turn the dough out onto a floured surface and knead gently until it forms a cohesive ball. With a rolling pin, roll it into an approximate 8-inch square, ¾ inch thick. Use a 3-inch round biscuit cutter to cut out biscuits and place on a parchment-lined baking sheet, spacing them 1 inch apart. Gather up the scraps, reroll, and cut out more biscuits (you should end up with 8).

3. Brush the top of the biscuits with the remaining 2 tablespoons melted butter. Bake until the biscuits are golden brown and well risen, about 15 minutes, rotating the pan halfway through. Allow to cool for 5 minutes, and serve.

## CREAM SCONES

Add 3 tablespoons sugar to the dry mixture in step 1. Fold in ½ cup currants or raisins if desired.

# STICKY BUNS

**About once a year, whenever I feel that my marriage needs an artificial shot of undying devotion and true love, I'll wake my lovely wife up with the unmistakable scent of gooey sticky buns baking in the oven.**

I figure this act alone is enough to get me off the hook for a whole year's worth of minor marital infractions—or major ones, if I include a ramekin of orange-cream-cheese glaze for dipping on the side.

Seriously, these things are awesome. Awesome enough that I decided to include them to wrap up this chapter despite the fact that there's not all that much as far as "new" kitchen science goes here. Aside from minor tweaks to perfect the recipe, these sticky buns are pretty standard. But sometimes some minor tweaks is all it takes to perfect a standard.

## Rolling in Dough

Sticky buns are made with what is called an enriched dough, meaning that in addition to the flour, water-based liquid, salt, and leavener found in most doughs, you've also got fat—in this case, eggs and butter; ingredients like milk and yogurt provide both water and fat. The fat plays a vital role not only in the flavor of the buns, but also in their texture. In lean doughs

made without fat, gluten formation is exceptionally strong, because the flour proteins are easily able to come into direct contact with one another, rapidly forming a thick, sticky network of gluten. Because of this, lean doughs tend to have larger air bubbles trapped in them (stronger gluten means the dough can stretch longer and thinner before bursting), as well as a tougher, chewier structure. With enriched doughs, the fats act like a lubricant, preventing proteins from bonding too tightly.

Think of flour proteins as a group of hippie revelers forming a dance circle during a rare dry, sunny moment at Woodstock 1969. As they run into each other, they clasp each other's hands (as hippies are wont to do). Eventually, they're all linked together quite tightly. The circle can stretch out very far before any link breaks. Now let's imagine the same group of hippies in the same field, but this time in the pouring rain. If coated, as they are, with mud and water, clasping hands tightly becomes much more difficult. Perhaps small circles form here and there, but they are nowhere the size and strength of the dry circle. So it is with fats: they prevent large hippie circles of flour from forming in your dough, so to speak.

Because of this, enriched doughs tend to be more delicate than lean doughs, with a softer texture and smaller air bubbles. Of course, fats also add color and flavor to doughs. What fun would sticky buns be if they weren't golden and buttery?

There are recipes for sticky buns that use chemical leaveners like baking powder to induce a rapid rise, but this technique compromises flavor. Yeast is the only way to properly develop flavor and texture in a sticky bun. See, yeasts, like pretty much all living creatures, have a strong desire to procreate, and in order to do that, they must consume energy. This energy consumption comes in the form of sugars, which they digest and let off as both carbon dioxide and alcohol, along with numerous other aromatic compounds. It's the carbon dioxide getting trapped in the network of gluten formed by the flour that acts to leaven yeasted baked goods. The process takes time, however. There's only so much procreating a yeast can do, you know? A properly leavened sticky bun dough can take several hours to produce.

Well, *why can't I just add more yeast to start?* you might ask. The problem is that yeast has a flavor of its own, and it's not a particularly pleasant one. Start with a ton of yeast, and its slightly bitter, funky flavor will dominate the dough. The flavor of properly risen dough comes from the *by-products* of the yeast's actions: the complex array of aromatic chemicals that are produced as yeast slowly, slowly digests the sugars in the dough. For the best flavor, you must start with a relatively small amount of yeast and allow it plenty of time to perform its magic. This is as true for sticky buns as it is for pizza dough or baguettes.

If you've never made sticky buns, you'll probably find the process pretty much fun. You shape them by rolling a large piece of dough into a cylinder, then slicing it up to create shorter cylinders with a spiral pattern inside them. To keep these spiral layers separated from each other, a layer of butter and cinnamon sugar is spread over the flat sheet of dough before rolling it up.

There are really no two ways about it: sticky buns are a project involving multiple stages and plenty of time in the kitchen. (Hey, didn't I tell you I only make these about once a year?) But if my still-going-strong marriage is any indication, the results are definitely worth the effort.

# THE WORLD'S MOST AWESOME STICKY BUNS

**NOTES:** Being a night owl and a late riser, I like to prepare the buns and let them rise overnight in the fridge so all I have to do is bake them off in the morning. To do so, place the buns in the refrigerator immediately after covering them in step 6 and allow to rise for at least 6 hours, and up to 12. The next day, remove the buns from the fridge while the oven preheats, then proceed as directed.

For a nut-free version, the pecans can be omitted from the sauce.

## MAKES 12 STICKY BUNS

### For the Dough

3 large eggs

⅓ cup packed light brown sugar

¾ cup buttermilk

2 teaspoons kosher or 1 teaspoon table salt

2 teaspoons instant yeast

6 tablespoons unsalted butter, melted

20 ounces (4 cups) all-purpose flour, plus more for dusting

### For the Pecan-Caramel Sauce

4 tablespoons unsalted butter

⅔ cup packed light brown sugar

3 tablespoons buttermilk

4 ounces (about 1 cup) toasted pecans coarsely chopped

Pinch of kosher salt

### For the Filling

⅔ cup packed light brown sugar

1 tablespoon ground cinnamon

2 tablespoons unsalted butter, melted

### For the Orange–Cream Cheese Glaze (optional)

4 ounces cream cheese

¼ cup buttermilk

1½ cups confectioners' sugar

1 tablespoon grated orange zest (from 1 orange)

2 tablespoons fresh orange juice

Pinch of kosher salt

1. **Make the dough:** Whisk the eggs in a large bowl until homogeneous. Add the brown sugar, buttermilk, salt, yeast, and melted butter and whisk until homogeneous (the mixture may clump up a bit—this is OK). Add the flour and stir with a wooden spoon until a cohesive ball of dough forms.

2. Turn the dough out onto a lightly floured surface and knead for 2 minutes, or until completely homogeneous, smooth, and silky. Return to the bowl, cover with plastic wrap, and allow to rise at room temperature until roughly doubled in volume, about 2 hours.

3. **Make the pecan-caramel sauce:** Cook the butter and brown sugar in a small saucepan over medium-high heat, stirring occasionally, until the sugar is completely dissolved and the mixture is bubbling, about 2 minutes. Add the buttermilk, pecans, and salt and stir to combine, then pour the mixture evenly over the bottom of a 13-by-9-inch glass baking dish.

4. **Make the filling:** Combine the sugar and cinnamon in a small bowl and set aside.

5. **Roll out the dough:** Turn the dough out onto a floured surface and lightly flour it. Shape into a rough rectangle with your hands and then, using a rolling pin, roll into a rectangle about 16 inches long and 12 inches wide, with a short end toward you. Brush with the melted butter, leaving a 1-inch border along the top edge. Sprinkle with the cinnamon and sugar mixture and spread it with your hands until the buttered portion is evenly coated. Roll the dough up jelly-roll-style into a tight cylinder, using a bench scraper as necessary to assist you. Pinch the seam shut and turn the dough so that it's seam side down. Use your hands to even out its shape.

6. Use a sharp knife to cut the roll into 12 even slices: The easiest way to do this is to cut it in half, cut each half in half, and then cut each section into thirds. Nestle the 12 rolls with the swirl pattern facing up in the prepared baking dish, making sure the slices from the ends of the log go cut side down. Cover with plastic wrap and allow to rise until roughly doubled in volume, about 2 hours (for overnight instructions, see Note above). The rolls should be well puffed and pressed tightly against each other.

7. While the dough is rising, adjust an oven rack to the middle position and preheat the oven to 350°F. Transfer the baking dish to the oven and bake until the buns are golden brown and well puffed, about 30 minutes, rotating the dish once. Allow to rest for 5 minutes, then invert the buns onto a serving platter; scrape out any excess goo from the pan and spoon over the buns.

**8.** **Make the (optional) glaze:** Combine the cream cheese, buttermilk, confectioners' sugar, orange zest, orange juice, and salt in a small saucepan and cook over medium heat, whisking constantly, until simmering and homogeneous. Spoon half the glaze over the sticky buns, reserving the rest in a bowl to pass tableside. Serve immediately.

## HOT CHOCOLATE MIX

As a kid, I loved hot chocolate mix. I'd eat it straight from the packet, licking it greedily off my finger. I've never, however, liked it as it is meant to be served: mixed with hot water or milk. Thin and watery, too sweet or not sweet enough, with little chocolate flavor.

While real homemade hot chocolate is not *all* that hard to make (dissolve cocoa in butter, add chocolate and sugar, maybe some vanilla and/or bourbon, add milk, and whisk while heating), you can't deny the convenience of simply stirring a few tablespoons of powder into a cup of hot milk. So I decided to come up with a homemade recipe that

would match the convenience of a powder and beat it in terms of flavor and price.

To start, I tried simply grinding chocolate to a powder in the food processor (freezing it first makes this easy). While that made decent cups, by the time I'd added enough chocolate to the milk to get the flavor I wanted, the richness of the cocoa butter started to dominate, making drinking a full mug difficult.

I opted instead for a combination of 100%-cacao (unsweetened) chocolate, sugar, and Dutch-process cocoa. With these tweaks, my chocolate was tasting pretty good, but a few problems remained: It caked in the storage container overnight, making it hard to dissolve the next day. It broke when added to the milk, dispersing a fine layer of fat bubbles over the surface of the drink. And the result simply wasn't rich, thick, and creamy enough.

Many commercial mixes contain soy lecithin or dried milk proteins, both of which are intended to increase creaminess and help keep the milk fat, cocoa butter, and liquid nicely smooth and emulsified. I tried adding soy lecithin to my mix and it worked, but I decided against it (it's available in health food stores, but hardly a commonplace ingredient). Milk powder also helped with texture, but it left the chocolate with a distinct cooked-milk flavor—not right.

In the end, the simplest solution was to add some cornstarch to my mix. Not only did it prevent caking, it also thickened the milk, giving it a nice, smooth, creamy richness without adversely affecting flavor.

# HOMEMADE HOT CHOCOLATE MIX

**MAKES ENOUGH FOR ABOUT 18 TO 36 SERVINGS**

Two 4-ounce bars 100%-cacao baking chocolate

1 cup Dutch-process cocoa powder

1 cup sugar

2 tablespoons cornstarch

½ teaspoon kosher salt

1. Freeze the chocolate bars until completely frozen, about 10 minutes. Remove from the freezer, break into rough pieces, and place in the food processor, along with the cocoa powder, sugar, cornstarch, and salt. Process until completely powdered, about 1 minute. Transfer to an airtight container and store in cool, dark place for up to 3 months.

2. To make hot chocolate, add 1 to 2 tablespoons of the mixture, or more if desired, to 1 cup boiling milk and stir or whisk until combined. To thicken it further, return the pan to the heat and simmer for 30 seconds, until thick and smooth.

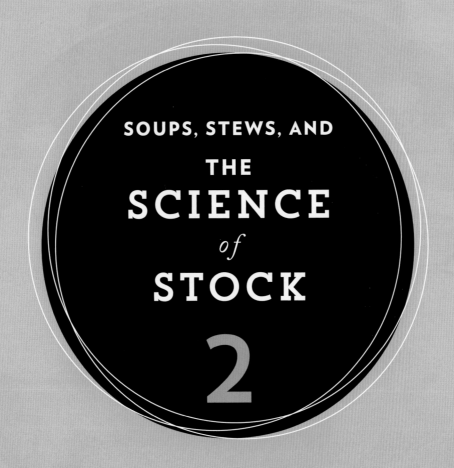

SOUPS, STEWS, AND
THE
SCIENCE
*of*
STOCK
2

Only the pure of heart can make a good soup.

# SOUPS, STEWS, AND THE
# SCIENCE *of* STOCK

## RECIPES IN THIS CHAPTER

MY WIFE
**HATES**
THE FACT
THAT OUR APARTMENT
ALWAYS SMELLS LIKE

FOOD.

**S**he treats the glorious scents of sizzling burgers and roasting chickens like enemy combatants, using guerilla tactics to hide jars of potpourri in places I'll never look—among the Russian literature, perhaps, or strategically disguised as one of the vacation souvenir knickknacks above her desk. As soon as I start a project in the kitchen, I wait the familiar swisssssh-clop of the window in the living room sliding open and the click-whir of the fan switching on, in her desperate attempts to preemptively ventilate.

That's why rainy days are my favorite. You can't open the windows during a thunderstorm, which ensures that the awesome aroma wafting from my giant pot of chili slowly simmering away on the stovetop saturates the curtains and carpets. And it's there to greet you every time you enter the apartment for at least a few weeks. It lives on in the bedsheets, ready to lull you to sleep like a warm glass of milk. It lingers on the shower curtain, greeting you every morning with its meaty, oniony aroma when you brush your teeth. My wife says I'm passive-aggressive. I tell her she's paranoid as I smile and heat up another bowl of chili.

This chapter is all about those wonderful, apartment-saturating, aromatic stews, soups, and braises—the kind of food so good that you check the weather report just *hoping* for a hurricane warning. And it all starts with stock.

# { STOCK }

A hundred years ago, when French chef Auguste Escoffier (perhaps the most august of chefs) codified classical French cuisine, cooking was based on the production and use of stock—the rich, savory liquid produced by simmering animal matter, bones, and vegetables in water for a long time. Meats were braised in it, vegetables were glazed with it, soups and stews were built on it, and it was reduced into rich sauces. Stock was made from chicken, duck, turkey, beef, veal, pork, sheep, you name it. If it had four legs or feathers, it's good bet that its bones and scraps would eventually find their way into a simmering pot.

These days, stock isn't quite as essential. Cooking is lighter, and many restaurants get by with just chicken stock. At home, I use chicken stock exclusively, and my wife has yet to complain that my food just isn't French enough. For many recipes, even a good canned broth will do just fine, though you want to make sure that it's low sodium so that you can control the salt level yourself. Most regular canned stocks or broths are too salty to reduce into a sauce.

There's still one place where a great stock is pretty much essential: **soup**. Like show dogs and children, soup can only be as good as the stock it's made from.

Unfortunately, as anyone who's ever worked in a restaurant can tell you, making stock is a slow business. It can take hours to extract flavor and break down the connective tissue from a pot of chicken bones and scraps. This isn't a problem when you're in the kitchen all day anyway: just keep a lazy eye on the huge stockpot on the back burner simmering away for six hours. But for a home cook? *Forget it.* A couple Sundays a year I'll give in and throw together a really traditional duck or veal stock, but for the other 363 days, I wanted to figure out a faster, better way.

## What's a Chicken?

As usual, I started with the basics, and in this case, the basics are a chicken. Once you've stripped away the feathers and the cluck, a chicken is actually a remarkably simple beast in culinary terms. Its matter can be divided into roughly four different parts:

- **Muscle** is what we think of as the meat on the chicken. It's the fleshy stuff that twitches and makes the bird go, and it can be further divided into two categories: **slow twitch** and **fast twitch**.
- **Slow-twitch** muscles are meant for sustained movement—i.e., the legs and thighs that keep the chicken standing, walking, and bending down or up. Because slow-twitch muscles are *aerobic* (they require oxygen to function), they are typically dense with capillaries carrying oxygen-rich red blood cells. That's why they appear to be darker.
- **Fast-twitch** muscles are used for short bursts of intense energy—they're the muscles that are found in chicken breasts, used to power the wings when a frightened chicken needs to escape from a dangerous situation. Because their activity is *anaerobic* (they don't require oxygen to twitch),

they tend to be less dense with capillaries, giving them their characteristic pale color.

Incidentally, the same distinction between fast- and slow-twitch muscles occurs in pretty much *all* animals, even humans. Ever wonder why a tuna is deep red while a cod is pale white? Tunas are made almost entirely of all-powerful slow-twitch muscles, which allow the fish to rapidly torpedo their way through the water for long periods of time. A cod moves only when he's eating or frightened.

- **Fat** provides insulation and energy storage for chickens. For humans, chicken fat just tastes delicious (provided it's cooked right). Fat is mainly found in large deposits around the legs and back of the bird, as well as in the skin. Contrary to popular belief, the skin of a chicken is *not* all fat—in fact, it's primarily made up of . . .

- **Connective tissue.** Composed of *collagen*, among other tissues, connective tissue is what keeps muscles attached to bones and bones attached to each other. In its natural state, it resembles a piece of yarn made of three separate strands that are tightly wound together, giving it lots of strength. Heat it up, and those strands unravel into *gelatin*, which can then form a loose matrix, giving stocks and sauces body and texture. Collagen is found everywhere, but it is particularly concentrated in the legs, wings, back, and skin of the bird. The older the animal, the more collagen there will be.

- **Bone** lends structure to the bird. Without bones, chickens would be little puddles of Jell-O, and not all that appetizing. Many cooks believe that bones are what give stock flavor; I'm skeptical (read on).

Depending on the part of the chicken you use, these parts are present in different ratios. To sum up, chicken legs are high in slow-twitch muscle,

have plenty of fat, and contain a good amount of connective tissue and bones. Breasts are almost completely fast-twitch muscle. Backs and carcasses have little meat of either kind but plenty of bone, connective tissue, and fat. Wings have the highest concentration of connective tissue of all, with a high proportion of fat and some bone.

To figure out exactly what each of these various tissues brings to the pot, I cooked a few batches of stock side by side: one made with just white meat, one made with just dark meat, one made with just bones, and one made with chicken carcasses, which have plenty of bones and connective tissue but relatively little meat.

Breasts.

Legs.

Backs.

After 4 hours of simmering, the meat-based stocks were flavorful (the one made with leg meat slightly more so than the breast) but had no body—

even when chilled to refrigerator temperature, these stocks remained a liquid, a sign that there was relatively little gelatin dissolved in the broth. The bones-only stock, as I suspected, was nearly flavorless, but it had a moderate amount of body. The stock made with carcasses was both flavorful and rich. This stock became a solid, rubber-like mass when chilled, due to the high amount of gelatin extracted in simmering. When sipped as a hot broth, it coated the mouth pleasantly, leaving the thin, sticky film on the lips characteristic of a good, rich broth.

A well-made stock should gel solid.

So carcasses are the way to go for the best balance between flavor and body. In a rare case of reverse economics, this also happens to be the cheapest way: you can accumulate carcasses by breaking down your own chickens (keep them in the freezer until you have enough to make a large batch of stock), or find them in most supermarkets at a bargain rate. But wings will do just fine if you can't get your hands on carcasses.

So we now know that for the optimum broth, we need two things: extraction of flavorful compounds from within muscle fibers (as indicated by the broths produced from chicken meat) and the extraction of gelatin from connective tissue to provide body. The question is, is there any way to speed things up a bit?

Well, I knew that chicken muscles look like long, thin tubes, and that extracting flavor from them is about slowly cooking them to extract their contents, much like squeezing a toothpaste tube. The degree to which those tubes are squeezed is depen-

dent upon the temperature to which the chicken is brought, but the rate at which those flavors come out is also dependent upon the distance they have to travel from the interior of the muscles to the stock. So, I wondered, would shortening the length of those tubes hasten the flavor extraction process?

I cooked three stocks side by side using chicken carcasses chopped to different degrees and found that indeed it *does* make a difference. Chopped carcasses gave up their flavor far faster than whole carcasses, and throwing roughly chopped chicken pieces into the food processor and finely grinding them worked even faster, producing a full-flavored broth in just about 45 minutes. It ain't pretty, but hey, it works!

Roughly chopped.

Finely chopped.

Pulverized.

But here's an interesting thing: although chopping the bones into pieces increased the rate of flavor extraction, it didn't have nearly as much impact on body development. Flavor extraction is a fast process—it's all about getting stuff out from inside the meat and dissolved in the water. Getting the gelatin out, on the other hand, requires not just extraction of the collagen; it is a chemical process that takes time, no matter how finely that collagen is chopped.

## TIME TO EXTRACT MAXIMUM FLAVOR AND BODY VERSUS DEGREE OF CHOPPING

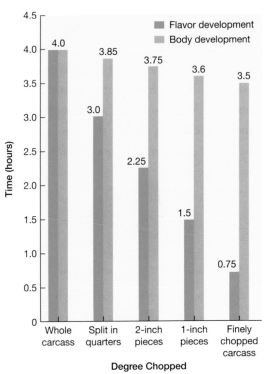

Remember how collagen looks like a twisted piece of yarn? Have you ever tried separating the separate strands in a piece of yarn? It's possible but time-consuming, and that's exactly what's going on in a pot of simmering connective tissue. At a 190° to 200°F simmer, it takes about 3 hours for 90 percent of the connective tissue to convert to gelatin, and then about an hour more for all of it to convert. Continue cooking past that point, however, and the gelatin itself starts breaking down, losing its thickening power. So the ideal chicken stock should be simmered for around 4 hours.

Luckily, chicken collagen isn't the *only* place where you can find gelatin. In fact, it's sitting right there in the supermarket:

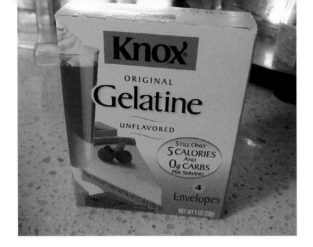

Packaged gelatin (the same stuff that makes Jell-O jiggly) is commercially available in powdered or sheet form. So all I had to do was take my 45-minute-simmered chopped-carcass broth and add commercial gelatin to it to produce a stock that was not only flavorful as one that had been simmered for hours, but just as rich!

So that's it, right? Stock in 45 minutes? Well, wait—what if we could make our stock taste even better than a traditionally made French-style stock? Not possible, you say? I'll prove it.

In a traditional French stock, clarity is valued above all else—it's fat and dissolved minerals and proteins (which we'll collectively call by their scientific name, "gunk") that make broth cloudy. If you keep a stock at a bare simmer, the fat rises to the surface in distinct bubbles that can be carefully skimmed off as it cooks, and the proteins coagulate into relatively large agglomerations that can be strained out.

But let your broth simmer vigorously, or—*mon dieu, non!*—actually come to a boil, and that gunk gets dispersed into millions of tiny droplets that simply can't be completely removed from the broth.

It can spell disaster in a fancy restaurant, where sauces and soups must be perfectly glossy and crystal clear, but do we really care about that at home? I, for one, will take flavor over appearance any day of the week, and fat is flavor.

A bonus to grinding up the chicken bones and scraps before making stock is that all the ground-up bits form a sort of floating raft that will collect stray proteins, minerals, and other gunk—exactly the same way that a French consommé is made. This helps solve part of the clarity problem. The rest of it, we can live with. Moral of the story: let your stock simmer away.

# EXPERIMENT:
## Fat = Flavor

It's not intuitive, but most of the flavorful compounds that give various types of meat its distinct flavor are not in the meat itself but in the fat that surrounds and runs through it. Don't believe me? Just think of this: why is most lean meat described as "tasting like chicken"? It's because without fat, meat has a very generic flavor. It *does* taste like a lean chicken breast. If you've got a food processor or meat grinder at home, you can even prove this to yourself with this experiment.

### Materials

- 12 ounces boneless lean beef (a cut like eye of round or well-trimmed sirloin will do well), cut into 1-inch cubes
- 1 ounce beef fat, trimmed from a steak (or just ask a butcher for some beef fat), cut into small pieces
- 1 ounce lamb fat trimmed from a lamb chop (or just ask the butcher for some lamb fat)
- 1 ounce bacon, cut into small pieces

### Procedure

1. Divide the beef into 3 portions and toss each portion with a different fat.
2. Using a food processor or meat grinder, grind the meat into hamburger (for more specific instructions on meat grinding, see page 485).
3. Form each portion into a patty and cook in a hot skillet or on the grill.
4. Taste the cooked patties.

### Results

How did the cooked patties taste? Most likely you found that the first patty tasted like a regular burger, while the second patty tasted like a lamb burger. The bacon-laced patty, well, you get the gist. Still unconvinced? Try the same thing replacing the lean beef with lean lamb. The patty with beef fat will *still* taste very much like a regular all-beef burger.

# DEFATTING A STOCK

In most situations, outside of fancy restaurants and fashion shows, a little extra fat is not necessarily a bad thing—go to any Japanese ramen house, for example, and they'll actually *add* fat to individual bowls of noodles for extra flavor—but there's a big difference between just enough and too much. A bit of fat emulsified into a broth and a few stray bubbles floating on the surface add richness and depth. A heavy slick adds nothing but a greasy flavor and mouth-coating oiliness. So, it's best to remove the slick of fat that forms on top of your stock as it cooks. But what's the best way?

If you've worked in a restaurant, you've probably been taught to be hyperaware of what your stock is doing at all times, carefully skimming any scum or fat that rises to the surface at regular intervals to keep the broth as clear as possible. But, again, that's a restaurant technique, for restaurant cuisine. When I make stock at home, I don't bother skimming it until it's finished. I strain the stock through a fine-mesh strainer into a fresh pot and let it rest for about 15 minutes, long enough to allow most of the fat and scum to rise to the surface so it can easily ladled off.

Even easier is to plan ahead and refrigerate the stock overnight; the fat will crystallize into an easy-to-remove layer that can be scraped off with a spoon, exposing the perfectly jelled stock underneath.

# STOCK, BROTH, *GLACE*, AND JUS

Strictly speaking, stock and broth are two distinct preparations. **Stock** is made by simmering bones, connective tissue, meat scraps, and vegetables in water. The connective tissue gives it body and a rich, unctuous mouthfeel. Reducing a stock by gently heating it to evaporate its water content concentrates both its flavorful compounds and its gelatin. Reduce it far enough, and it'll become viscous enough to form a

*continues*

coherent coating on food. At this stage, it's known as a *glace*, and it's exceedingly delicious. Just as Eskimos have many words for snow, the French have different words for reduced stock—*glace, glace de viande*, and *demi-glace*—according to how far the stock has been reduced.

**Broth** is made with meat and vegetables—no bones or connective tissue. It can be flavorful, but without collagen from connective tissue, it's about as thin as water. Yet while a classically trained chef may need to know this distinction, as a home cook, you don't have to worry about it. In this book, I use the word "stock" pretty much exclusively, because my Quick Chicken Stock recipe contains both bones and added gelatin. For the vast majority of home-cooked recipes, the two can be used interchangeably. The USDA makes no distinction in their labeling laws about what can be called "stock" or "broth." They state that the terms "may be used interchangeably as the resulting liquid from simmering meat and/or bones in water with seasonings." While some brands will choose to use one wording or another on their packaging, there is in fact no real difference in how they're manufactured.

**Jus** refers to the natural juices given off by a piece of meat that's being roasted. Often the jus will settle on the bottom of a roasting pan and evaporate while the meat cooks, leaving a coating of what is called *fond*—the delicious browned bits that form the base of pan sauces and gravies.

# FREEZING STOCK

There's no denying it—even a quick stock is a bit of a chore, and one that I don't want to go through every time I make a bowl of soup or need a quick pan sauce. Fortunately, stock freezes remarkably well. I keep my stock frozen in two different ways:

- **In ice cube trays.** Pour the stock into an ice cube tray, let it freeze completely, and then transfer the cubes to a zipper-lock freezer bag. You can pull out as much or as little as you need, and the cubes melt nice and fast. This is ideal for pan sauces where you don't need too much stock at a time.
- **In quart-sized Cryovac or freezer bags or plastic containers.** If you have a vacuum-sealer (such as a FoodSaver), a quart-sized bag is the ideal way to store stock. It freezes flat, so it takes up very little space in the freezer, and, better yet, defrosts under hot tap water in just a matter of minutes. If you don't have a vacuum-sealer, you can freeze stock in zipper-lock freezer bags. Make sure to squeeze out as much air as possible before sealing them, then lay them flat to freeze. Or, just use regular plastic deli containers.

The other trick to efficient and inexpensive stock production is to save all your chicken parts. Every time I break down a chicken, I toss the backs and wing tips into a gallon zipper-lock bag that I keep in the freezer. When the bag is full, I'm ready to make a batch of stock.

# QUICK CHICKEN STOCK

**NOTE:** This recipe also produces cooked chicken leg meat, which can be used in other recipes or saved for later use. Alternatively, replace the chicken legs with more backs and wingtips.

## MAKES 2 QUARTS

1 ounce (4 packets; about 3 tablespoons) unflavored gelatin

2 pounds skin-on chicken backs and wing tips, preferably from young chickens

2 pounds chicken legs

1 large onion, roughly chopped

1 large carrot, peeled and roughly chopped

2 stalks celery, roughly chopped

2 bay leaves

2 teaspoons whole black peppercorns

1 teaspoon fennel seeds

1 teaspoon coriander seeds

6 sprigs fresh parsley

1. Pour 4 cups water into a medium bowl and sprinkle with the gelatin. Set aside until the gelatin is hydrated, about 10 minutes.

2. Meanwhile, chop the chicken backs and wing tips with a cleaver into rough 2-inch pieces, or cut up with poultry shears. Working in two or three batches, transfer the chicken to a food processor and pulse until the texture roughly approximates ground chicken, stopping if a particularly hard bone gets stuck on the blade.

3. Transfer the ground chicken, chicken legs, onion, carrot, celery, bay leaves, peppercorns, fennel, coriander, and parsley to a large Dutch oven or stockpot and add cold water to just cover the ingredients, about 2 quarts. Add the hydrated gelatin and water and bring to a boil over high heat. Adjust the heat to maintain a brisk simmer, then skim off any foam and scum from the surface and discard. Cook for 45 minutes, adding water as necessary to keep the ingredients submerged. Remove from the heat and allow to cool for a few minutes.

4. Using a pair of tongs, transfer the chicken legs to a bowl and set aside to cool. Remove and discard any large pieces of bone and vegetables from the stock. Strain the stock through a fine-mesh strainer into a large bowl and discard the solids. Return the stock to the pot, bring to a rolling simmer, and reduce to 2 quarts, about 10 minutes.

5. Meanwhile, pick the meat off the chicken legs and reserve for another use (such as Chicken Vegetable Soup or Chicken and Dumplings; see page 191 or 237). Discard the bones and skin.

6. When the stock has finished reducing, allow it to rest until the excess fat and scum form a distinct layer on the surface, about 15 minutes, then skim off with a ladle—discard the fat or save it for another use. Alternatively, you can refrigerate the stock overnight and remove the solid fat from the top. The stock will keep in an airtight container in the refrigerator for up to 5 days, or it can be frozen for at least 3 months (see page 186).

# BASIC VEGETABLE STOCK

**MAKES 2 QUARTS**

1 ounce (4 packets; about 3
   tablespoons) unflavored gelatin

2 large onions, roughly chopped

2 large carrots, peeled and roughly
   chopped

4 stalks celery, roughly chopped

1 large leek, trimmed

8 ounces mushroom stems and
   scraps and/or whole mushrooms

2 Granny Smith apples, quartered

2 bay leaves

2 teaspoons whole black
   peppercorns

1 teaspoon fennel seeds

1 teaspoon coriander seeds

6 sprigs fresh parsley

1. Pour 4 cups of water into a medium bowl and sprinkle with the gelatin. Set aside until the gelatin is hydrated, about 10 minutes.

2. Meanwhile, put the onions, carrots, celery, leek, mushrooms, apples, bay leaves, peppercorns, fennel, coriander, and parsley in a large Dutch oven or stockpot and add cold water to just cover the ingredients. Add the hydrated gelatin and water and bring to a simmer over medium heat. Adjust the heat to maintain a slow, steady simmer and cook for 1 hour, adding hot water as necessary to keep the ingredients submerged.

3. Strain the stock through a fine-mesh strainer lined with cheesecloth into a large bowl and discard the solids. Return the stock to the pot, bring to a simmer, and reduce to 2 quarts, about 20 minutes. Remove from the heat. The stock will keep in an airtight container in the refrigerator for up to 5 days, or it can be frozen for up to 3 months (see page 186).

## CHICKEN SOUP WITH RICE (OR NOODLES)

As a hardcore Colombian from the mountains of Bogotá, my wife claims that she could live on soup alone. My naturally inquisitive mind is plagued by the strong urge to rigorously test this claim, but every time I decide to force her to embark on the first day of the rest of her liquid-diet-based life, the part of me that loves her intervenes and reminds me that life just wouldn't be as fun without her, soup or no. She, coming from a rice-eating country, prefers rice in her soup, while I, coming from New York, prefer the superior egg noodle. I have generations of Jewish grandmothers on my side, but she has the trump card of Maurice Sendak—though I do like to point out that had there been more words in the English language that rhyme with noodle, Sendak may well have written a different set of poems entirely. As it is, "In January, while I doodle/swirly patterns on my poodle" doesn't have the quite the same ring to it as the original.

That said, we *do* eat a lot of soup around here, and once you've got a great chicken stock to work with, you've won 99 percent of the battle for The Ulti-mate Vegetable Chicken Soup with Rice (or Noo-dles). The rest is as simple as adding the vegetables and rice or noodles. I like to start with a base of carrots, onions, and celery, then mix it up. Whatev-er's in season, whatever looks best—heck, whatever you can get your hands on and feel like adding—is the best strategy here. The key to remember is that different vegetables need to be prepped and cooked in different ways in order to maximize their soupworthiness.

## SOUP VEGETABLES: BEST STRATEGIES

Making any simple, broth-based vegetable soup is easy as long as you know how to prepare the various vegetables and when to add them. This chart describes how to deal with the most common soup vegetables.

| VEGETABLE | PREP | COOKING TIME |
|---|---|---|
| Carrot | Peel and cut into ½–inch chunks | 20 minutes |
| Cauliflower | Separate into florets, slice stems ¼ inch thick | 20 minutes |
| Celery | Peel and cut into ½-inch chunks | 20 minutes |
| Celery Root (celeriac) | Peel and cut into ½-inch chunks | 20 minutes |
| Jicama | Peel and cut into ½-inch chunks | 20 minutes |
| Kohlrabi | Peel and cut into ½-inch chunks | 20 minutes |
| Leek | Thinly slice or dice | 20 minutes |
| Onion | Thinly slice or dice | 20 minutes |
| Parsnip | Peel and cut into ½-inch chunks | 20 minutes |

*continues*

| VEGETABLE | PREP | COOKING TIME |
| --- | --- | --- |
| Potato | Peel and cut into ½-inch chunks | 20 minutes |
| Radish | Cut into ½-inch chunks | 20 minutes |
| Rutabaga | Peel and cut into ½-inch chunks | 20 minutes |
| Sweet Potato | Peel and cut into ½-inch chunks | 20 minutes |
| Asparagus | Cut into 1-inch lengths | 10 minutes |
| Bell Pepper | Cut into ½-inch chunks | 10 minutes |
| Broccoli | Separate into florets, slice stems ¼ inch thick | 10 minutes |
| Butternut Squash | Peel and cut into ½-inch chunks | 10 minutes |
| Cabbage | Slice ⅛ inch thick | 10 minutes |
| Collard Greens | Roughly chop leaves, cut stems into 1-inch segments | 10 minutes |
| Green Beans | Trim and cut into 1-inch pieces | 10 minutes |
| Kale | Roughly chop leaves, cut stems into 1-inch segments | 10 minutes |
| Summer Squash | Cut into ½-inch chunks | 10 minutes |
| Zucchini | Cut into ½-inch chunks | 10 minutes |
| Arugula | Remove tough stems | 5 minutes |
| Brussels Sprouts | Pull off leaves | 5 minutes |
| Chard | Roughly chop leaves, cut stems into 1-inch lengths | 5 minutes |
| Corn Kernels | Cut off cobs (separate into individual kernels if necessary) | 5 minutes |
| Lima Beans, frozen | None | 5 minutes |
| Peas, frozen | None | 5 minutes |

| VEGETABLE | PREP | COOKING TIME |
|-----------|------|--------------|
| Baby Spinach | None | 5 minutes |
| Curly Spinach | Roughly chop | 5 minutes |
| Watercress | Roughly chop | 5 minutes |

# THE ULTIMATE CHICKEN VEGETABLE SOUP
## WITH RICE (OR NOODLES)

**NOTES:** Instead of using the stock, you can simmer 4 chicken legs in 2 quarts low-sodium canned chicken broth for 30 minutes. Remove the legs, skim the fat from the broth, and add enough water to make 2 quarts. When the legs are cool enough to handle, pick off the meat, discarding the bones and skin, and reserve.

The carrots, celery, and onion are really just suggestions—feel free to use whatever vegetables you'd like (see the chart on pages 189–91), aiming for about 2½ cups total prepped vegetables.

**SERVES 4 TO 6**

1 recipe Quick Chicken Stock (page 187), including the picked leg meat (see Note above)

2 medium carrots, peeled and cut into medium dice (about 1 cup)

1 medium stalk celery, cut into medium dice (about ½ cup)

1 small onion, thinly sliced (about 1 cup)

½ cup long-grain white rice or 2 cups medium egg noodles

¼ cup chopped fresh parsley

2 cups kale torn into 1-inch pieces (about 6 leaves)

Kosher salt and freshly ground black pepper

1. Combine the chicken stock (reserve the meat for later), carrots, celery, onion, and rice, if using (not the noodles), in a Dutch oven and bring to a boil over high heat. Reduce to a simmer and cook until the vegetables are almost tender, about 15 minutes.

2. Add the greens and noodles, if using, and cook until all the vegetables and the rice (or noodles) are tender, about 5 minutes longer. Stir in the parsley and chicken meat and warm through. Season to taste with salt and pepper and serve.

## BEEF AND BARLEY STEW

When you grind as much meat as I do, you often find yourself with an extra pound or two of short ribs lying around. You could braise them in red wine and be all French, or you could cook them for a few days sous-vide like a modern restaurant, and both are great techniques. Sometimes, though, I'm in the mood for something simpler.

Short rib and barley stew is fantastic because it:

- is dumb easy to do,
- is made with pantry and fridge staples (aside from the short ribs),
- lasts for days and gets better with time,
- soothes the soul or warms the cockles of your heart, or, if you're really lucky, both at the same time, and
- tastes really, really good.

The method is pretty straightforward: sear the meat to add a bit of flavor, sauté the vegetables, and then simmer the whole thing down with a bit of Marmite, soy sauce, and tomato paste added for that umami kick.

Throughout the winter months, there are usually a few leaves of kale kicking around my fridge, since they make such a great salad (just marinate in olive oil and vinegar for a couple of hours—it stays crisp for days even after dressing! See page 825), and they go a long way in making this stew even tastier and heartier. If you prefer beef and barley soup to stew, just add some more stock at the end to thin it to the desired consistency.

# BEEF AND BARLEY STEW

**SERVES 4 TO 6**

2 pounds boneless beef short ribs,
cut into 1-inch chunks

Kosher salt and freshly ground
black pepper

2 tablespoons canola oil

2 medium carrots, peeled, split in
half lengthwise, and sliced into
½-inch pieces (about 1 cup)

2 medium stalks celery, split in half
lengthwise, and sliced into ½-inch
pieces (about 1 cup)

1 large onion, finely diced (about
1½ cups)

½ teaspoon Marmite

1 teaspoon soy sauce

2 medium cloves garlic, minced or
grated on a Microplane (about 2
teaspoons)

1 tablespoon tomato paste

4 cups homemade or low-sodium
canned chicken stock

One 14½-ounce can whole
tomatoes, drained and roughly
chopped

1 cup pearl barley

2 bay leaves

4 cups loosely packed roughly torn
kale leaves

1. Toss the short ribs in a large bowl with salt and pepper to coat. Heat the oil in a Dutch oven over high heat until smoking. Add the beef and cook, without moving it, until well browned on first side, about 5 minutes. Stir the beef and continue cooking, stirring occasionally, until browned all over, about 10 minutes total; reduce the heat if the bottom of the pot begins to scorch. Return the meat to the bowl and set aside.

2. Return the pot to medium-high heat and add the carrots, celery, and onion. Cook, stirring frequently, until the vegetables begin to brown, about 4 minutes. Add the Marmite, soy sauce, garlic, and tomato paste and cook, stirring, until fragrant, about 30 seconds.

3. Add the stock and scrape up the browned bits from the bottom of the pot with a wooden spoon. Add the tomatoes, barley, and bay leaves, then return the beef to the pot, increase the heat to high, and bring to a boil. Reduce to the lowest possible heat and cover the pot, leaving the lid slightly ajar. Cook, stirring occasionally, until the beef is completely tender and the barley is cooked through, about 2 hours.

4. Stir in the kale and cook, stirring constantly, until wilted, about 2 minutes. Season to taste with salt and pepper. Serve, or, for best flavor, cool and refrigerate in a sealed container for up to 5 days before reheating and serving.

# WHERE'S THE BEEF (STOCK)?

You may be wondering why I use chicken stock in my beef soup rather than beef stock, and the answer is simple: I'm lazy. Beef bones are large and it takes a long, long time to extract flavor and gelatin from them (restaurants will keep them simmering on a back burner all day). Chicken stock is quick, I usually have it on hand, and it has a nice neutral flavor that can easily pick up other flavors without overwhelming them. A beef stew made with chicken stock as the base will still be plenty beefy once it's done simmering.

What about store-bought broth? On most supermarket shelves, you'll find chicken, beef, and vegetable broth, all for about the same price. But that doesn't make sense, does it? If beef broth takes so much more effort—bigger bones, longer extraction times, more expensive meat—how can they sell it for the same price as chicken broth?

Here's the secret: *store-bought beef broth is not really beef broth*. There is actually very little beef in canned beef broth. Just like other humans, food manufacturers are lazy and concerned about their bottom line. Rather than spending all day simmering veal or beef bones, they opt to use natural and artificial flavorings. According to the USDA's labeling guidelines, beef or pork broth only has to have a Moisture Protein Ratio (MPR) of 135.1 to 1—that is, for every ounce of water, there is only 0.007 ounce of beef protein present. No wonder the broths don't taste much like meat!

To quickly and cheaply boost the flavor of canned beef broth, most manufacturers rely on yeast and vegetable extracts. While yeast extracts are great for adding a savory boost to your stews, I'd much rather have control over its addition myself (many of my recipes call for Marmite, a yeast extract). If you're gonna go canned, go with low-sodium chicken broth.

# SHOPPING FOR BROTH

Always buy chicken or vegetable broth, not beef (see "Where's the Beef (Stock)?," above). Here are a couple tips for what to look for in a canned broth:

- **Buy low-sodium broth.** This will allow you to adjust the seasoning to your taste instead of being tied to the (usually very high) salt levels of the canned broth.
- **Buy broth in resealable Tetra-packs, not cans.** Like many packaged foods, broth will start going bad as soon as you open it. A resealable Tetra-pack will help prolong its lifetime, allowing you to use as little or as much as you need while storing the rest in the fridge.

# HOW TO MAKE CANNED BEANS TASTE GOOD

I love beans and my wife loves soup, which makes winter a great time of the year for both of us. Well, not exactly. The reality is that my wife loves soup and I love whiskey. It's a would-be serendipitous situation for both parties involved, considering how often the spirit moves me to make soup when I'm tipsy. *I accidentally made soup again, dear,* I tell her. I say would-be, because my wife seems to prefer my sober soup to my drunk soup (even though the latter always tastes better to me). So this year I decided that I love beans almost as much as I love whiskey, and that I love my wife significantly more than I love either, and so I would trade in the bottle for the legume, and drunk soup for bean soup.

Of course, with the cold weather and lack of spirits, I was ready to get cracking right away. Problem is, dried beans are not exactly fast food, taking at best several hours, and at worst, a full day of soaking, followed by simmering. So I did what any sensible, sober man would do: I bought them canned. There are a couple of distinct advantages that canned beans have over dried. For one, their texture is pretty much always spot on. Bean canners have got the process down to an art, and you'd now be hard-pressed to open a can and find beans that were broken, chalky, hard, or anything short of perfectly creamy and intact—not always an easy thing to accomplish at home. For another thing, canned beans come with some nice, full-bodied liquid. Many recipes tell you to rinse the stuff off. That makes sense if you're making, say, a bean salad, but the liquid is awesome in soups, adding flavor and body to an otherwise thin broth.

There's only one real problem with canned beans: flavor.

With dried beans, you have the option of cooking your beans in any number of media—water, chicken stock, pork broth, dashi, a sweet molassesy tomato sauce—and adding whatever aromatics you like—onions, carrots, celery, bay leaves, thyme, pork fat—in order to get flavor built right into them. Canned beans, on the other hand, are designed to have a neutral flavor that will work moderately well in any dish but shine in none.

Luckily for us, there are means to getting a bit of flavor back into those guys.

Most simple canned bean soup recipes call for mixing the ingredients together, bringing the soup to a boil, and serving it immediately. There's no fault in doing that—canned beans, after all, are a convenience food. But what if I told you that by adding a ton of flavorful aromatic ingredients and a quick (15-minute) simmer, you could amplify the quality of your soup by an order of magnitude? *

OK, no dramatic drumroll there, I guess. Let's be honest: these are 30-minute bean soups. They aren't going to change your life the way, say, buying your best friend a lottery ticket that happens to be worth millions of dollars would, but they may well change your weeknight cooking routine.

---

\* I know, because I actually measured the soups' flavors by using a custom-built bean-o-matic flavor deductor.

# 30-MINUTE PASTA E FAGIOLI

Pasta e fagioli is the bean and pasta soup traditionally made with the leftovers from the Sunday gravy and was known to cause bouts of foolishness in Dean Martin. My version isn't packed with meat like nonna's undoubtedly was, but a ton of garlic, pancetta (you can use bacon, guanciale, or even crumbled sausage if you like), oregano, and a couple bay leaves add plenty of flavor to the mix.

**SERVES 4**

One 28-ounce can whole tomatoes

2 tablespoons extra-virgin olive oil, plus more for serving

1 tablespoon unsalted butter

3 ounces pancetta, finely chopped (optional)

1 medium onion, finely diced (about 1 cup)

6 medium cloves garlic, minced or grated on a Microplane (about 2 tablespoons)

½ teaspoon dried oregano

½ teaspoon red pepper flakes

4 cups homemade or low-sodium canned chicken stock

Two 15-ounce cans red kidney beans, with their liquid

2 bay leaves

1 cup small pasta, such as shells, ditali, or elbows

Kosher salt and freshly ground black pepper

2 tablespoons chopped fresh parsley

1. Pour the tomatoes into a medium bowl and squeeze each one through your fingers to break it up into small pieces (be careful—they can squirt). Set aside.

2. Heat the olive oil and butter in a large saucepan over medium-high heat until the butter is melted. If using pancetta, add it to the pan and cook, stirring constantly, until fragrant, about 2 minutes. Reduce the heat to medium, add the onion, garlic, oregano, and red pepper flakes, and cook, stirring, until the onion is fragrant and softened but not browned, about 3 minutes. Add the tomatoes, with their juice, the chicken stock, kidney beans, and bay leaves, bring to a boil over high heat, and then reduce to a bare simmer. Cook for 20 minutes, adding the pasta to the soup for the last 5 to 10 minutes (depending on the package directions).

3. Season the soup to taste with salt and pepper. Discard the bay leaves, stir in the parsley, and serve, drizzling each serving with olive oil.

# 30-MINUTE MINESTRONE

Minestrone is my soup of choice during the spring and early summer, when vegetables from the farmers' market are at their brightest and most flavorful. Some minestrone soups get cooked down for hours. I actually prefer my quick version, because it keeps the vegetables mildly crisp and fresh tasting. I always start with onion, carrot, celery, and canned tomatoes as my base, but on top of that, you can use the vegetables suggested here or go with anything from the table on pages 189–91. Just make sure to keep the total amount of extra vegetables at around 3 to 4 cups (not counting greens, which will cook down dramatically).

## SERVES 6 TO 8

2 tablespoons extra-virgin olive oil, plus more for serving

1 medium onion, finely diced (about 1 cup)

2 medium carrots, peeled and finely diced (about 1 cup)

2 stalks celery, finely diced (about 1 cup)

4 medium cloves garlic, minced or grated on a Microplane (about 4 teaspoons)

6 cups homemade or low-sodium canned chicken stock

1 cup diced canned tomatoes, with their juice

One 15-ounce can Roman (borlotti or cranberry) cannelini, or great northern beans, with their liquid

2 bay leaves

1 small zucchini, cut into ½-inch cubes or ½-inch half-moons (about ¾ cup)

1 small summer squash, cut into ½-inch cubes or ½-inch half-moons (about ¾ cup)

1 cup green beans cut into ½-inch segments

2 cups roughly chopped curly spinach or kale

1 cup small pasta, such as shells, ditali, or elbows

½ cup frozen peas

½ cup cherry tomatoes, cut in half

Kosher salt and freshly ground black pepper

¼ cup chopped fresh basil

1. Heat the olive oil in a large saucepan over medium-high heat until shimmering. Reduce the heat to medium, add the onion, carrots, celery, and garlic, and cook, stirring, until softened but not browned, about 3 minutes. Add the chicken stock, tomatoes and beans, with their liquid, and the bay leaves and bring to a boil over high heat, then reduce to a bare simmer. Cook for 20 minutes, adding the zucchini, squash, green beans, and spinach for the last 10 minutes, and the pasta for last 5 or 10 minutes (depending on the package directions).

2. Season the soup to taste with salt and pepper. Discard the bay leaves, add the peas, cherry tomatoes, and basil, and stir until the peas are thawed. Serve, drizzling each serving with olive oil.

# 30-MINUTE DON'T-CALL-IT-TUSCAN WHITE BEAN AND PARMESAN SOUP

Given the grim history of the usage of the word, I'm not positive that there's actually anything Tuscan about this bean soup, but it's delicious nevertheless. The key is plenty of rosemary and a hunk of rind from some good Parmigiano-Reggiano tossed in while it simmers. Much like simmering chicken bones, a Parmesan rind will add both flavor and body to the mix. The difference is that this takes only minutes to reach excellence. That, and plenty of good olive oil for drizzling.

**SERVES 4**

2 tablespoons extra-virgin olive oil, plus more for serving

1 medium onion, finely diced (about 1 cup)

2 medium carrots, peeled and finely diced (about 1 cup)

2 stalks celery, finely diced (about 1 cup)

4 medium cloves garlic, minced or grated on a Microplane (about 4 teaspoons)

½ teaspoon red pepper flakes

4 cups homemade or low-sodium canned chicken stock

Two 15-ounce cans cannellini or great northern beans, with their liquid

Four 6-inch sprigs rosemary, leaves removed and finely chopped, stems reserved

One 3- to 4-inch chunk Parmesan rind, plus grated Parmigiano-Reggiano for serving

2 bay leaves

3 to 4 cups roughly chopped kale or Swiss chard leaves

Kosher salt and freshly ground black pepper

1. Heat the olive oil in a large saucepan over medium-high heat until shimmering. Add the onions, carrots, and celery and cook, stirring, until softened but not browned, about 3 minutes. Add the garlic and red pepper flakes and cook, stirring, until fragrant, about 1 minute. Add the chicken stock, beans, with their liquid, the rosemary stems, Parmesan rind, and bay leaves, increase the heat to high, and bring to a boil. Reduce to a bare simmer, add the kale, cover, and cook for 15 minutes.

2. Discard the bay leaves and rosemary stems. Use an immersion blender to roughly puree some of the beans until the desired consistency is reached. Alternatively, transfer 2 cups of the soup to a regular blender or food processor and process until smooth, starting on low speed and gradually increasing to high, then return to the soup and stir to combine. Season to taste with salt and pepper.

3. Ladle into bowls, sprinkle with the chopped rosemary, drizzle with olive oil, and sprinkle with a grating of Parmigiano-Reggiano. Serve with toasted crusty bread.

# 30-MINUTE BLACK BEAN SOUP

**Black bean soup was a favorite of mine when I was a kid, and it still is. Cumin, garlic, and hot pepper flakes, along with onions and peppers, form the flavor base. The kicker is the canned chipotle pepper, which is everywhere these days, but no less delicious for it.**

## SERVES 4

1 tablespoon vegetable oil

2 green bell peppers, finely diced

1 large onion, finely diced

2 medium cloves garlic, minced or grated on a Microplane (about 2 teaspoons)

1 jalapeño or serrano pepper, seeded and finely chopped

1 teaspoon ground cumin

½ teaspoon red pepper flakes

1 chipotle chile packed in adobo, finely chopped, plus 1 tablespoon of the adobo sauce

4 cups homemade or low-sodium canned chicken stock

Two 15-ounce cans black beans, with their liquid

2 bay leaves

Kosher salt

For Serving (optional)

Roughly chopped fresh cilantro

Crema (Mexican-style sour cream)

Diced avocado

Diced red onion

1. Heat the oil in a large saucepan over medium-high heat until shimmering. Add the bell peppers and onions and cook, stirring frequently, until softened but not browned, about 3 minutes. Add the garlic, jalapeño, cumin, and red pepper flakes and cook, stirring, until fragrant, about 1 minute. Add the chipotle and adobo sauce and stir to combine. Add the chicken stock, beans, with their liquid, and the bay leaves, increase the heat to high, and bring to a boil. Reduce to a bare simmer, cover, and cook for 15 minutes.

2. Discard the bay leaves. Use an immersion blender to roughly puree some of the beans until the desired consistency is reached. Alternatively, transfer 2 cups of the soup to a blender or food processor and process until smooth, starting on low speed and gradually increasing to high, then return to the soup and stir to combine. Season to taste with salt.

3. Ladle the soup into serving bowls and serve, with cilantro leaves, sour cream, diced avocado, and/or diced red onion if desired.

# How to Make
# CREAMY Vegetable Soups
## WITHOUT A RECIPE

**When I was a totally green cook with my first serious restaurant job, working under Chef Jason Bond at what's now a landmark Boston restaurant, No. 9 Park, there were many moments when I learned a new technique or perfected an old one and said to myself, "Holy crap, I just made this?"**

But the very first was when Chef Bond taught me how to make a creamy chanterelle soup (read: Campbell's cream of mushroom soup on tasty, tasty crack), sweating aromatics, sautéing mushrooms, adding a good stock, and pureeing it all while emulsifying the mixture with fresh butter.

Like any great vegetable soup, the end result was something that tasted like a liquefied, purified, intensified version of itself—this soup tasted more like chanterelles than actual chanterelles. The magic lies in the way that aromatic ingredients can bring out other flavors, as well as the way in which liquids coat your mouth, giving more direct contact to your taste buds and olfactory sensors, and making for easier release of volatile compounds.

These days, there aren't too many vegetables in the world that I haven't made into a smooth, creamy soup, and there are even fewer that I've not loved,* but my experience has taught me something: that first process of making a chanterelle soup wasn't really just a recipe for chanterelle soup. It was a blueprint for making any creamy vegetable soup. You just need to break it down into its individual steps and figure out how to universalize them.

---

\* Wait a minute, something about that doesn't make sense. But you get what I mean.

Let's say, for instance, that I've never made a smooth carrot soup flavored with ginger and harissa, but I really like the idea. Here's how I'd go about it.

## Step 1: Prepare Your Main Ingredient

The simplest soups can be made by merely adding your main ingredients raw and simmering them in liquid later on. When preparing this type of soup, all you've got to do is get your main ingredient ready by peeling it (if necessary) and cutting it into moderately small pieces. The smaller you cut, the quicker your soup will cook down the line.

There are times when you may want to boost the flavor of a main ingredient by, say, roasting or browning it. This is an especially effective technique for sweet, dense vegetables like sweet potatoes and squashes, or brassicas like broccoli or cauliflower, all of which intensify in sweetness with some browning. To roast them, cut them into large chunks, toss them with some olive oil, salt, and pepper, set them in a baking sheet lined with aluminum foil or parchment paper, and roast in a 375°F oven until tender, with their edges tinged brown.

This works in two ways. First, the process of caramelization breaks down large sugars into smaller, sweeter ones. Second, enzymatic reactions that create simple sugars are accelerated with heat.

## Step 2: Choose Your Aromatics

Alliums—onions, leeks, shallots, garlic, and the like—are like the Best Supporting Actor of the soup pot. They're not there to steal the spotlight, but without them, your soup would be boring. Nearly every soup I make starts with either onions or leeks, along with some garlic or shallot (and sometimes all four!) cooked down in olive oil or butter.

Other firm vegetables such as diced carrots, bell peppers, celery, thinly sliced fennel, or ginger can work well in certain situations, but they tend to have a stronger impact on the finished flavor of the dish, so make sure that you really want them there. Make a carrot soup with just onions and it'll taste like carrot soup. Make a carrot soup with fennel or ginger, and it will taste like carrot-and-fennel soup or carrot-and-ginger soup.

## Step 3: Sweat or Brown Your Aromatics

Next big question: to sweat or to brown?

- **Sweating** is the process of slowly cooking chopped vegetables in a fat. You do it over moderate heat, and the goal is to get rid of some of the excess moisture within those vegetables, and to break down their cellular structure so that their flavor is released. With the case of alliums, there's another process going on: onion aroma is created when certain precursor molecules that exist within separate compartments in onion cells break out and combine with each other. Sweating an onion will break down cell walls, allowing this process to happen. The same holds true for garlic, shallots, and leeks.

- **Browning** starts out like sweating, but generally takes place over higher heat. Once excess liquid from vegetables has evaporated, the vegetables can begin to brown and caramelize, creating rich flavors, more sweet notes, and more complexity. You might think that more flavor is always better, and thus you should always brown your vegetables, but more often than not, this browning can be overpowering, making soups too sweet or competing too much with the subtler flavors of your main vegetable.

## Step 4: Add Second-Level Aromatics Like Spices and Pastes

After your aromatics have sweated or browned, the next phase is your secondary aromatics, and it's an optional stage that's often omitted. If you like very clean, pure-tasting soups, jump ahead. If you like playing with flavors and spices, then you'll have fun with this step.

These are things like ground spices (say, curry powder, ground cumin, or chili powder) and moist pastes (like tomato paste, harissa, or chopped chipotle peppers in adobo sauce). These types of ingredients benefit from a brief toasting or frying in hot oil, which alters some of their constituents into more complex, more aromatic products, as well as extracting fat-soluble flavors so that they disperse more evenly into the soup.

Because ground spices have such a high ratio of surface area to volume and most pastes have already been cooked, the process takes only a few moments —just until the spices start smelling fragrant.

## Step 5: Add Your Liquid

Your choice of liquid can have a big impact on the finished dish.

- **Chicken stock** is an easy fallback and always a good choice. It has a neutral, mild flavor that adds meatiness and savoriness to a dish without overwhelming any flavors. Likewise vegetable stock can bring similar complexity, though buyer beware: unlike store-bought chicken broth, most store-bought vegetable broths are vile. You're better off making your own.

- **Vegetable juice** is what you want if you value intensity of vegetable flavor over balance. Carrots cooked and pureed in carrot juice will taste insanely carroty. You can buy many vegetable juices at the supermarket these days, or juice your own with a home juicer. Mixing and matching a main ingredient with a different vegetable juice (like in my recipe for Roasted Squash and Raw Carrot Soup) can lead to great end results.
- **Dairy** such as milk or buttermilk is a good way to get yourself a heartier, creamier dish, though dairy fat does have the tendency to dull bright flavors. This is not necessarily a bad thing: dairy is the perfect foil for the intense flavor of broccoli in a creamy broccoli soup, or tomatoes in a cream of tomato soup, for instance.
- **Water** is a perfectly fine choice if the other options aren't available.

Whatever liquid you choose, don't use too much. Use just enough to cover your ingredients by an inch or so. You can always thin a thick soup out after blending, but reducing a pureed soup that's too thin is a much more difficult thing to do (if you don't want to risk burning it to the bottom of the pot).

After adding your liquid and main ingredient, bring the soup to a simmer and let it cook until the vegetables are just cooked through; you want them to be just tender enough to pierce with a knife with no resistance. For things like carrots, parsnips, and other root vegetables, you have a bit of leeway.

Overcooking won't be the end of the world. But for bright green vegetables like broccoli, asparagus, peas, string beans, or leafy greens, you want to make sure to stop cooking them before they start turning a drab green color—if a brightly colored soup is something you care about, that is.

## Step 6: Puree and Emulsify

Here's the fun part: pureeing. The smoothness of your final soup will depend on the tool you use.

- **A blender** will give you the smoothest result, due to its high speed and vortex action. When blending hot liquids, always hold the lid down with a kitchen towel, start the blender on low speed, and slowly bring it up to high. Unless you enjoy wearing hot soup.
- **An immersion blender** can give you a decently smooth result, depending on the power of your blender. It's by far the most convenient way to make soup, and it's a good choice if you're fine with a rustic, kind of chunky texture.
- **A food processor** should be your last choice. Because of its wide base and relatively low spinning rate, a food processor does more chopping than pureeing.

Whatever the pureeing method, I like to emulsify my soup with some fat during this stage—either butter or olive oil. This adds a rich texture to the soup.

Some recipes (including many of mine) will tell you to slowly drizzle in fat or add butter a knob at a time while the blender is running, which is a sure-fire way to get your fat to emulsify properly, but here's a secret: so long as you don't have the world's worst blender (and somebody out there does!), there's no real need to drizzle in the fat slowly. The vortex action of a blender is plenty powerful enough to emulsify the fat even if you just dump it all in at once.

If the ultimate in smoothness is your goal, finish off your pureed soup by using the bottom of a ladle to press it through a chinois or an ultra-fine-mesh strainer. The end results should be smoother than John Travolta strutting with a double-decker pizza slice.

## Step 7: Finish with Acid and Season

Seasoning is the final step just before plating and serving in any recipe. You can season as you go, but you never know if your soup has the right level of salt until you taste it in its final form. Now is the time to do that.

Equally important to bring out the best flavor in a recipe is acid. Because acidic ingredients quickly dull in flavor when cooked, it's best to add fresh acid right at the end, just before serving. For most vegetable-based dishes, lemon juice or lime juice is a great option, as their aroma complements vegetal flavors. Other good options would be a dash of cider vinegar, wine vinegar, or my favorite, sherry vinegar. The latter goes particularly well with soups made with plenty of extra-virgin olive oil.

## Step 8: Garnish and Serve

Your soup is essentially done at this stage, but a little garnish never hurt anybody. Here are some options:

- Flavorful oils, like walnut, pistachio, squash seed, or argan.
- Chopped fresh herbs or tender alliums, like parsley, tarragon, chives, or sliced scallions.
- Sautéed vegetables, like mushrooms, leeks, or garlic.
- Nuts, like almonds, hazelnuts, or pine nuts, toasted in olive oil or butter.
- Simple gremolata-style mixtures, like a blend of parsley, lemon zest, and grated garlic.
- Thinly sliced chilies.
- A drizzle of browned butter.
- Dairy products, like sour cream, crème fraîche, or heavy cream; plain or flavored with spices or pastes. Using a hint of the same spice you used earlier in step 4 can be a good way to boost flavor.

I think of the garnish as a final step to layer flavor and/or texture into the bowl.

## Step 9: Rinse and Repeat

Once you've got these eight basic steps down, you've got what it takes to start creating any number of creamy soups, combining any flavors you like. I'm not promising that every single combination of vegetables and aromatics will work out, but use this guide as a blueprint and you're well on your way to building the soup of your dreams. We all dream about soup, right?

# BOOM GOES THE BLENDER

**H**ere's a scenario that happened to me just last week: I was making a batch of tomato soup for my wife (who can't get enough of the stuff), and I'd just dumped the hot tomato mixture into the blender. As I reached for the On button, a tiny voice in the back of my mind said to me, *All of this has happened before, and all of it will happen again. Ask yourself this question: "Would an idiot do this?" If the answer is "yes," then do not do this thing.* Of course, I went ahead and turned on the blender anyway. The top popped off in a violent explosion, and my dog leaped behind the couch in fear as hot tomatoes splattered across the apartment with all the fury of a VEI-7* volcanic eruption. This kind of stuff happens because there are some mistakes I never learn from and, more important, because of thermodynamics and the physics of vapor formation.

See, there are a number of factors involved in the conversion of hot water to steam. Pressure is a big one. As we all know, steam takes up much more space than water. Because of this, if you apply enough pressure to a body of water, steam will not escape. When a big batch of tomato soup is sitting in a blender, there is significant pressure on all of the soup except the stuff at the surface. Steam escapes from the top alone, while the liquid below the surface sits there patiently waiting its turn. Switch on the blender, though, and suddenly you create tons of turbulence. A vortex is formed, the surface area of that body of liquid suddenly becomes much bigger, and steam is rapidly and violently produced. Moreover, the added exposure will also heat up the air in the headspace of the blender, causing it to expand. This rapid expansion causes the top of your blender to pop off and the hot tomatoes to go flying. BOOM!

So, how do you prevent it? There are two ways: First off, make sure that there's room for expansion in the blender. The hot steam needs an escape valve. The easiest thing to do is to remove the central plug in your blender top and cover the hole with a towel. This will allow expanding gases to escape but prevent liquids from flying out. The second, and easier, way to prevent blow-out is to start blending *very gently*. Start your blender at the lowest speed and slowly work your way up to the highest. Expansion will take place more slowly, allowing the gases plenty of time to escape, and enabling your kitchen and your dog to get away unharmed.

# STRAINING

**I**f "velvety-smooth" and "lump-free" are terms that you'd like to be able to use to describe your soups and sauces, a fine-mesh strainer is your tool of choice. But have you ever poured a batch of soup into a strainer set over a pot or bowl and waited for it to all pass through? Chances are, you ended up waiting a long, long time. As the soup passes through, the holes in the strainer get clogged.

*continues*

---

\* That's a volcanic eruption with devastating long-term effects on the surrounding area and profound short-term effects on the world. My apartment was devastated and my world was profoundly affected in the short term.

There are two solutions to this. The first is **The Rap**: Holding the soup-filled strainer above your container with one hand, rap a long heavy tool repeatedly against the edge of the strainer with your other hand (I use a heavy spatula or a honing steel). Soup should flow out of the bottom with each rap. For extra-thick or chunky soups, I use a second method, **The Spoon Press**: Holding the soup-filled strainer above your container with one hand, stir the contents using a large metal serving spoon, a ladle, or a rubber spatula, scraping the edge of the utensil against the mesh. This should force seeds, lumps, and other clog-inducing material out of the way so your soup can flow freely.

# QUICK TOMATO SOUP WITH GRILLED CHEESE

**Is there any rainy-day fare more classic than tomato soup and grilled cheese?**

It's quick, it's easy, it's filling, and it just takes you back to feeling like a kid again (if you ever left that stage in the first place).

Of course, there's the classic Campbell's tomato-soup-in-a-can, which, like a puppy that won't stop licking your face, is cloyingly cute and enjoyable up to a point, but sometimes what you want is the adult version. Luckily, making it is almost as simple. This version uses whole canned tomatoes pureed with a mixture of sautéed onions, oregano, and red pepper flakes, which gives it just the right amount of flavor and heat to bring balance to the dish. As with chili, adding a splash of liquor to the soup just before serving helps its aroma pop out of the bowl and into your nose, where it belongs, though it'll do fine without it. A drizzle of high-quality olive oil and a sprinkling of fresh herbs transform this childhood classic into a downright elegant lunch. Make sure you put on a tie and jacket before consuming.

A plain grilled cheese is great, but . . .

. . . adding some extra grated Parmesan
on the outside . . .

. . . gets you the cheesiest of grilled cheeses.

As for the grilled cheese? Everyone knows how to make grilled cheese, right? The key is plenty of butter and low-and-slow cooking. Cook it too fast, and your toast burns long before the cheese has had a chance to get even remotely gooey. That's how I'd been cooking my grilled cheeses for years—until my friend Adam Kuban showed me a better way. His trick: toast two slices of bread in butter, then add the cheese *to the toasted side* of one slice before closing the sandwich and proceeding as normal. The hot toasted bread not only adds buttery richness and extra-toasty flavor to the interior, it also helps get the cheese started on its way toward glorious gooeyness.

The cheese you use is really up to you, but I do admit that just like in my cheeseburgers, I like my grilled cheese with the meltiness of American. A good compromise is to use one slice of American for the goo factor and an extra slice of a more flavorful sharp cheddar or Swiss (or, if you want to be really fancy, Gruyère).

Finally, if you want to go full cheese on this one, I like to add a layer of Microplaned Parmigiano-Reggiano to the exterior of my sandwich. It crisps up like an Italian *frico* in the skillet, giving you an extra layer of cheesy crunch.

# 15-MINUTE PANTRY TOMATO SOUP

**NOTE:** I like to use high-quality canned tomatoes such as Muir Glen in this soup.

## SERVES 4

3 tablespoons unsalted butter

1 large onion, finely diced (about 1 ½ cups)

Pinch of red pepper flakes

½ teaspoon dried oregano

1 tablespoon all-purpose flour

Two 28-ounce cans whole tomatoes (see Note above), with their juice

½ cup whole milk or heavy cream

Kosher salt and freshly ground black pepper

2 tablespoons whiskey, vodka, or brandy (optional)

2 tablespoons extra-virgin olive oil

2 tablespoons chopped fresh herbs, such as parsley, basil, or chives (optional)

Extra-Cheesy Grilled Cheese Sandwiches (recipe follows)

1. Melt the butter in a medium saucepan over medium-high heat. Add the onions and cook, stirring frequently, until softened but not browned, 6 to 8 minutes. Add the pepper flakes and oregano and cook, stirring, until fragrant, about 30 seconds. Add the flour and cook, stirring, for 30 seconds. Add the tomatoes, with their juice, and stir, scraping the flour up off the bottom of the pan. Add the milk or cream and cook, stirring occasionally and breaking up the tomatoes with the spoon, until the whole thing comes to a boil. Reduce to a simmer and cook for 3 minutes.

2. Remove the soup from the heat and puree using an immersion blender. Or transfer to a standing blender, in batches if necessary, and puree, starting on low speed and gradually increasing to high, then return to the pan. Season the soup to taste with salt and pepper. Stir in the whiskey, if using, and bring to a simmer. Serve immediately, topping each serving with a generous drizzle of olive oil, a sprinkle of herbs, and if you like, a crack or two of freshly ground pepper, with the sandwiches alongside.

## Extra-Cheesy Grilled Cheese Sandwiches

*I usually use a slice each of American and cheddar cheese, forgoing a category B cheese. Serve half a sandwich per person with the soup, or double this recipe and make the sandwiches in two skillets. (If you have only one skillet, make the sandwiches in batches and keep the first batch warm in a low oven, on a rack on a baking sheet.)*

**NOTE:** Category A cheeses are good melting cheeses like American, cheddar, Jack, Fontina, young Swiss, Gruyère, Muenster, young provolone, and young Gouda, among others. Category B cheeses are strongly flavored grating cheeses like Parmigiano-Reggiano, Asiago, Pecorino, aged Manchego, and aged Gouda.

**MAKES 2 SANDWICHES**

2 tablespoons unsalted butter

4 slices high-quality white, whole wheat, or rye sandwich bread

4 ounces sliced Category A cheese (see Note above)

Kosher salt

½ ounce category B cheese, grated (optional; see Note above)

Brown mustard

1. Melt ½ tablespoon of the butter in a 12-inch stainless steel or cast-iron skillet over medium heat. Add 2 bread slices and swirl them around the pan with your hands until all the butter is absorbed. Cook, swirling the bread occasionally, until very lightly browned, about 1 minute. Remove to a cutting board, toasted side up, and immediately top with the sliced A cheese. Melt another ½ tablespoon butter in the skillet and toast the remaining 2 bread slices until lightly browned, then immediately press them toasted side down onto the cheese-topped slices to form sandwiches.

2. Melt another ½ tablespoon butter in the skillet and sprinkle with a pinch of salt. Reduce the heat to medium-low and place the sandwiches in the skillet. Swirl them around with your hands until all the butter is absorbed. Cook the sandwiches, swirling them with your hands and pressing down on them gently with a wide stiff spatula occasionally, until the bottoms are a deep, even golden brown, about 4 minutes. Remove the sandwiches to your cutting board with the spatula. Melt the remaining ½ tablespoon butter in the skillet, sprinkle with a bit of salt, and repeat the toasting procedure with the second side, until the sandwiches are golden brown on both sides and the cheese is thoroughly melted, about 4 minutes longer. Transfer the sandwiches to the cutting board.

3. If desired, spread the grated cheese evenly over a large plate. Press the sandwiches into the grated cheese, turning once and pressing on them until you get an even coating of cheese on both sides. Return the sandwiches to the skillet and cook until the grated cheese has melted and formed a golden brown crust, about 1 minute. Carefully flip the sandwiches and repeat with the second side. Transfer to the cutting board.

4. Slice the sandwiches in half on the diagonal, and serve immediately, with mustard and the tomato soup.

## CORN CHOWDER

My mother's corn chowder recipe involved a can of creamed corn, an equal amount of half-and-half, and a teaspoon of chicken bouillon. I loved that version growing up (and it's still a cornerstone of my little sister's recipe repertoire), but as I'm a semi–New Englander, chowder is a semisacred thing in my book, with a few hard-and-fast rules: All chowders contain dairy (don't give me none of that Manhattan clam chowder crap), most contain potatoes, and some contain pork—all traditional and inexpensive New England products. I *used* to make my corn chowder with bacon, the most readily available cured pork product at the supermarket, but I was never too happy with its dominating smoky flavor, so I switched over to unsmoked salt pork, which adds the characteristic porkiness without overpowering the sweet corn. And some days, when I'm trying to feel extra valorous or have simply let my freezer run empty, I'll forgo the pork altogether.

Most chowder recipes call for sweating some onions in butter, adding your corn kernels, potatoes, and dairy, and letting it cook down. As it cooks, the potatoes release some starch, thickening up the broth. None of this bothers me. What does bother me is what goes into the trash: the stripped corncobs.

Anyone else out there go for two or three rounds on their corn on the cob just to suck at the little bits of sweet milk left in the cob after you've eaten the kernels? Like the crispy fat around a rib bone, that's the tastiest part. Why would you want to throw it away? Instead, I use the corn-milking technique here: scraping out the milky liquid from the cobs with the back of a knife. By then infusing your base stock with both the scraped milk and empty corncobs (along with a few aromatics like coriander and fennel seed), you can vastly increase the corniness of the finished soup. (I mean that in a good way.)

It doesn't take long to infuse the stock—all of 10 minutes, which is just about enough time to sweat off your onions and corn kernels. Once you've got your corn-milk stock made, the rest is simple: simmer the onion-butter-stock-potato mix until the potatoes are tender, add some milk (I prefer it to cream, as the fattiness of cream can mask some of that sweet corn flavor), and then puree just enough of it to give the soup some body and help keep the butterfat properly emulsified into the mix.

The great thing about this stock-infusing technique is that it's totally adaptable. Sometimes I feel like making a smooth and sweet corn velouté, which I'll make like my chowder, but omitting the potatoes and cream and blending until completely smooth. And if you do like the flavor of bacon in your chowder, go for it—nothing's stopping you, except perhaps your cholesterol and your spouse.

I, fortunately, have a spouse who can be plied with corn soup when I really want to get my way. Might I suggest you try the same?

### How to Buy Corn

Want to know the secret to great chowder or corn on the cob? Great corn. It's as simple as that. The trick is getting the corn. After that, it's a cake walk.

The first time I tasted really great corn—one of those early food memories that made me realize food was more than just fuel—was on a second-grade field trip to an Upstate New York farm: me and the farmer on a tractor, the farmer grabbing an ear of corn as he drove by the field, shucking it, and handing it to me to taste. In my head I was thinking, "Holy Skeletor! I'd trade in my Battle-Armor He-Man for more of this!" which roughly translates to my current vocabulary as, "Holy f*&k, this tastes amazing!" (My eloquence has diminished significantly through the years.) Incredibly sweet, bright, and flavorful, it became the epitome of good corn in my mind, the corn that all corn since has tried to live up to—something that happens only rarely.

Because of the result of a happy mutation several hundred years ago, sweet corn has a far higher concentration of sugar in its kernels than regular old field corn (that's the stuff they feed animals with). But here's the hitch: as soon as the ear leaves the stalk, that sugar begins converting to starch. Within a single day of harvest, an ear of corn will lose up to 50 percent of its sugar when left at room temperature—and even more, up to 90 percent, when it's sitting out in the hot sun at the farmers' market.

Moral of the story. Buy your corn as fresh as possible (from the farmer if you can!), refrigerate it as soon as you can, and cook it the day you buy it.

# KNIFE SKILLS:
## How to Prepare Corn

**When selecting corn, look for ears that are tightly shut, with bright green leaves that show no signs of wilting.**

Squeeze the ears, particularly around their tips, to ensure that the kernels inside are full and juicy. A good ear of corn should have very little give and feel heavy for its weight.

Avoid ears of corn that have been preshucked or come packaged in plastic wrap. Any extra handling or packaging means that those ears are that much farther away from their original time of harvest.

The best way to store corn is not at all—don't buy it until the day you plan on eating it. If you must store it, keep it in its husks in the refrigerator's crisper drawer, but don't store it for more than a day, or you'll have starchy, flavorless corn on your hands. Instead, for longer-term storage, remove the kernels and then blanch them in boiling water for 1 minute, followed by a plunge into an ice water bath to chill them. Spread the blanched kernels out on a rimmed baking sheet and place in the freezer until fully frozen. Put the frozen kernels in a zipper-lock freezer bag and store them in the freezer for up to 3 months.

To remove the kernels from an ear of corn, first peel off the husk and silk and discard. Hold the cob in one hand and rest the end on the bottom of a large bowl. Hold your knife against the top of the ear, then cut downward, slicing the kernels as close to the cob as possible. They should fall neatly into the bowl. Repeat with the remaining kernels, rotating the cob as you go. Save the cobs if you want to extract their milk and make stock (see The Best Corn Chowder, page 212).

# THE BEST CORN CHOWDER

**NOTE**: Buy the absolute freshest corn you can find, and use it the day you bring it home.

**SERVES 6**

6 ears corn, husks and silks
　removed
6 cups homemade or low-sodium
　canned chicken stock
1 bay leaf
1 teaspoon fennel seeds
1 teaspoon coriander seeds
1 teaspoon whole black
　peppercorns
4 ounces salt pork or slab bacon,
　cut into ½-inch cubes (optional)
3 tablespoons unsalted butter
1 medium onion, finely diced
　(about 1 cup)
2 medium cloves garlic, minced or
　grated on a Microplane (about 2
　teaspoons)
1 to 2 russet (baking) potatoes,
　peeled and cut into ½-inch dice
　(about 1½ cups)
Kosher salt
2 cups whole milk or half-and-half
Freshly ground black pepper
Sugar (if necessary)
3 scallions, finely sliced

1. With a sharp knife, cut the kernels off the corn cobs. Reserve the cobs. Use the back of the knife to scrape the "corn milk" from the corncobs into a large saucepan. Break the cobs in half and add to the pan. Add the stock, bay leaf, fennel seeds, coriander seeds, and peppercorns and stir to combine. Bring to a boil over high heat, then reduce to just below a simmer and let steep for 10 minutes. Strain the stock through a fine-mesh strainer into a bowl; discard the cobs and spices.

2. While the stock infuses, heat the pork, if using, and butter in a 3-quart saucepan over medium-high heat until the butter melts. Add the onions, garlic, and corn kernels and cook, stirring frequently, until the pork has rendered its fat and the onions are softened, about 7 minutes. Reduce the heat if butter begins to brown.

3. Add the corn stock, potatoes, and 1 teaspoon salt, bring to a simmer, and simmer, stirring occasionally, until the potatoes are tender, about 10 minutes. Add the milk and stir to combine. The chowder will look broken, with melted butter floating on top. Use an immersion blender to blend the soup until the desired consistency is reached. Alternatively, transfer half of the soup to a regular blender and blend, starting on low speed and gradually increasing to high, until smooth, about 1 minute, then return to the remaining soup and stir well. Season to taste with salt, pepper, and sugar (with very fresh corn, sugar should not be necessary). Serve immediately, sprinkled with the sliced scallions.

# CREAMY BROCCOLI-PARMESAN SOUP

This creamy soup relies on the thickening and emulsifying power of a roux—cooked flour and butter—to give it a creamy consistency without the need for heavy cream, which can dull flavors. For a long, long time, drab army-green vegetables got a bad rap, but I'm trying to bring sexy back to thoroughly cooked broccoli (and green beans). There are definitely great things to be said about snappy, bright green stalks, but the flavor that develops when broccoli is cooked to well-done is unmatched by that of its al dente counterpart. A touch bitter, a hint of sulfur (in a good way), and a rich, grassy depth all emerge as the stalks soften.

The only downside here is that waiting for broccoli to soften this much can be a tedious process that takes up to an hour or more. But there's an old trick that the English use to make their traditional fish 'n' chips side of mushy peas: add some baking soda to the water. Baking soda raises the pH of the liquid, causing the pectin that holds the cells of the broccoli together to soften. Just a tiny pinch is enough to cut simmering time down by two-thirds.

To add some depth to the soup, I toss in a handful of anchovies (you can skip them for a vegetarian version), as well as a good amount of grated Parmesan, whose nutty tang plays nicely off the deep flavor of the broccoli. A handful of quick buttery croutons adds both texture and flavor.

5 tablespoons unsalted butter

1 medium onion, finely diced
(about 1 cup)

4 medium stalks celery, finely diced
(about 1 cup)

2 medium cloves garlic, minced or
grated on a Microplane (about 2
teaspoons)

4 anchovy fillets, finely chopped
(optional)

3 tablespoons all-purpose flour

2 cups milk

2 cups homemade or low-sodium
canned chicken stock or vegetable
stock, plus more if necessary

¼ teaspoon baking soda

12 cups broccoli florets, stems,
and stalks cut into 1-inch pieces
(about 1 large head)

3 ounces Parmigiano-Reggiano,
grated

2 tablespoons lemon juice (from 1
lemon)

Kosher salt and freshly ground
black pepper

4 slices hearty white sandwich
bread, crusts removed and cut
into ½-inch dice

1. Melt 3 tablespoons of the butter in a large Dutch oven or soup pot over medium-high heat. Add the onion, celery, and garlic and cook, stirring, until the vegetables are softened but not browned, about 5 minutes (reduce the heat if the butter begins to brown). Stir in anchovies, if using, and cook until fragrant, about 30 seconds.

2. Add the flour and cook, stirring constantly, until all the flour is absorbed, about 30 seconds. Stirring constantly, slowly pour in the milk, followed by the stock. Stir in the baking soda and broccoli florets and bring to a boil, then reduce the heat to maintain a simmer, cover, and cook, stirring occasionally, until the florets are completely tender and olive green, about 20 minutes.

3. Working in batches, transfer the mixture to a blender, add the Parmesan, and blend, starting on low speed and gradually increasing to high, until completely smooth, about 1 minute; add additional stock or water if necessary to thin to the desired consistency (I like mine thick). Pass through a fine-mesh strainer into a clean pot. (Alternatively, use an immersion blender to puree the soup directly in the original pot.) Whisk in the lemon juice and season the soup to taste with salt and pepper. Keep warm.

4. Melt the remaining 2 tablespoons butter in a large nonstick skillet over medium-high heat. When the foaming subsides, add the bread cubes and cook, tossing frequently, until golden brown on all sides, about 6 minutes. Season to taste with salt and pepper.

5. Serve the soup garnished with the croutons.

# INCORPORATING STARCHES

Have you ever tried adding flour or cornstarch directly to a hot soup in an attempt to thicken it, only to find that the starch clumps up into frustratingly impossible-to-destroy little balls? Here's the problem, and it has to do with the nature of the interaction between starch—a complex carbohydrate found in all sorts of plant matter, including flour—and water. Remember those little dinosaur-shaped sponges you'd get as a kid, which you'd drop into water, then wait for them to grow? That's exactly what starch molecules are like. When dry, they are tiny and shriveled. They can flow freely past each other. But expose them to water, and they start growing, getting bigger and bigger, until they eventually rub up against each other and bind, creating a water-resistant barrier. Are you starting to get the picture?

When a spoonful of flour or cornstarch lands on the surface of a pot of water or milk, the first parts to get wet are the starches on the outside of the granules, which rapidly expand, forming a waterproof seal. As you stir and submerge the clumps, a seal ends up forming around the entire clump, keeping the interior from getting wet.

So, how do you solve this problem? Two ways.

With a starch that doesn't need to be cooked before it is incorporated (such as cornstarch or potato starch), just dissolve the starch in a small amount of liquid to start. Starting with a smaller amount of liquid makes the mechanical stirring action of your spoon, fork, or whisk much more effective. Smaller amounts of liquid also get viscous more easily, making it simpler to bash up those pockets of dry starch. I use an equal volume of starch to liquid to start and stir it until homogeneous before adding the remaining liquid, or adding it to the rest of the liquid.

For starches that need to have their raw flavor cooked out of them, such as flour, start them in fat. Starch does not swell in fat, so by first combining flour with a fat like butter or oil and mixing it until homogeneous, you end up coating the individual starch granules, preventing them from swelling and sticking together when you first add the liquid. After you add it, the fat eventually melts away, so the starch is exposed and can be incorporated smoothly. This is the premise behind using a roux to thicken a soup or sauce.

Finally, remember that for starches to thicken properly, they must be brought to a complete boil to reach their optimal swelling size. You'll notice a soup thicken dramatically as it goes from just plain hot to actually boiling.

# HOW TO BUY BROCCOLI AND CAULIFLOWER

Shopping for broccoli and cauliflower is pretty much the same and, luckily, finding good specimens is not too difficult. These hearty members of the brassica family have a good shelf life and are firm enough that they don't bruise or break easily during storage and shipping. A head of broccoli should have tight florets that are an even dark green with greenish to purple buds. Cauliflower should be an even pale white;

*continues*

yellow or brown spots should be avoided, though if they are minor, you can simply trim them off. Look to the leaves as well, which should be tight around the base and appear bright pale green.

Once you get the broccoli or cauliflower home, keep it loosely wrapped in plastic or in a vegetable bag inside the crisper. It should stay good for at least a week. Once you've cut it into florets, it's best to use it as quickly as possible to prevent excess moisture loss, though florets can be stored in an airtight container or zipper-lock plastic bag with a damp paper towel placed in it for up to 5 days.

# KNIFE SKILLS:
## How to Cut Broccoli and Cauliflower

- **For Broccoli:** Trim the woody ends of the stalks and discard. Use the tip of your knife to cut off larger branches, then cut the florets off the stalks and trim to the desired shape and size. Cut any stubs off the stalks and discard. Peel the stalks, quarter them lengthwise, and slice into 1- to 2-inch lengths to cook along with the florets.

- **For Cauliflower:** Split the head in half through the center. Use the tip of a sharp knife to remove the hard central core as well as any green leaves around the base and discard. Break the cauliflower into large chunks with your hands, then use the tip of your knife to cut into florets of the desired size and shape.

1

2

3

4

5

6

7

# CREAMY MUSHROOM SOUP

**The key to great mushroom soup is to cook the mushrooms in butter long enough to drive off excess moisture and allow them to start to brown, deepening their flavor.**

**NOTE:** You can use plain button mushrooms, but for the best flavor, use a mix of mushrooms, such as button, portobello, shiitake, and/or other foraged or cultivated mushrooms.

## SERVES 6

2 pounds mushrooms (see Note above), cleaned and sliced ¼ inch thick (about 3 quarts)

4 tablespoons unsalted butter

1 large leek, white and pale green parts only, split in half and cut into ¼-inch-thick half-moons (about 1 cup)

1 medium onion, finely sliced (about 1 cup)

2 teaspoons fresh thyme leaves

3 tablespoons all-purpose flour

1 cup milk

4 cups homemade or low-sodium canned chicken stock, plus more if necessary

2 bay leaves

Kosher salt and freshly ground black pepper

1. Set aside 1 cup of the mushrooms. Melt 3 tablespoons of the butter in a large Dutch oven or soup pot over medium-high heat. Add the remaining mushrooms and cook, stirring occasionally, until they have given off their liquid and are beginning to brown, about 10 minutes. Add the leeks, onions, and half of the thyme and cook, stirring frequently, until the vegetables are softened, about 5 minutes.

2. Add the flour and cook, stirring constantly, until all the flour is absorbed, about 30 seconds. Stirring constantly, slowly pour in the milk, followed by the stock. Add the bay leaves and bring to a boil, then reduce the heat to maintain a simmer, cover, and cook, stirring occasionally, until the liquid is thickened and lightly reduced, about 10 minutes. Discard the bay leaves.

3. Working in batches, transfer the mixture to a blender and blend, starting on low speed and gradually increasing to high, until a rough puree forms, about 1 minute; add additional stock or water if necessary to thin to the desired consistency (I like mine thick). Pass through a fine-mesh strainer into a clean pot and season to taste with salt and pepper. (Alternatively, use an immersion blender to blend the soup directly in the original pot.) Keep hot.

4. Melt the remaining tablespoon of butter in a large nonstick skillet over medium-high heat. When the foaming subsides, add the reserved 1 cup mushrooms and cook, stirring and tossing frequently, until deep brown, about 8 minutes. Add the remaining thyme and season to taste with salt and pepper.

5. Serve the soup garnished with the sautéed mushrooms.

## WORKING WITH MUSHROOMS

**W**hen shopping for mushrooms of any sort, look for ones that don't have any soft or discolored spots on their caps, which can indicate decay. For mushrooms with gills (such as portobellos and shiitakes), examine the gills under the cap as well, as they'll often start to turn before the rest of the 'shroom. It's OK if the bottom of the stem is a little discolored, but it should not be overly dry, mushy, or starting to shred apart. As for dirt, it is no indication either way. Mushrooms grow in dirt, so it's inevitable you'll find some attached to them. Obviously, cleaner mushrooms are easier to work with, but a little dirt on the cap or clustered near the stem is no problem.

Once you get the mushrooms home, store them in a plastic bag with the top left open or in a perforated plastic container in the vegetable drawer of your refrigerator. Fresh mushrooms should keep for 3 to 5 days under optimal conditions.

### Can I Wash My 'Shrooms?

You've probably had folks tell you things like, "Mushrooms are basically living sponges. Don't get 'em wet, or they'll get soggy and you'll never cook them right." Those same people will recommend that you clean mushrooms by brushing them with a special mushroom brush (oh, please), or maybe with a damp paper towel. No wonder people don't like mushrooms—they're a pain in the butt to clean!

But is this level of fear really necessary? I tested out this theory by cooking a few batches of mushrooms side by side. One I cleaned meticulously with a damp paper towel. Another I cleaned under the tap and shook dry in a strainer. The last I cleaned under the tap and spun-dry in a salad spinner. I weighed all the batches before cooking and found that—hey, what do you know?—the washed-then-drained mushrooms gained only about 2 percent of their weight in water, while the washed-then-spun mushrooms gained about 1 percent. That's about 1½ teaspoons of water per pound, which in turn translates to an extra 15 to 30 seconds of cooking time.

What does this mean? It means that most of the water you add by washing mushrooms clings only to the surface. So long as you dry your mushrooms carefully before cooking, you can rinse them as much as you'd like. Cooking the spin-dried mushrooms side by side with the paper-toweled mushrooms confirmed this: they both cooked at exactly the same rate.

# KNIFE SKILLS:
## Slicing Mushrooms

### Button and Cremini Mushrooms

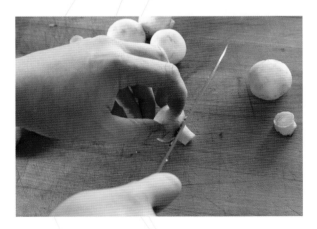 

White button and cremini mushrooms can be prepared the same way. Start by trimming off the bottoms of the stems, which can be woody or tough, and discard. Then hold each mushroom flat against the cutting board, with the stem side down, and cut into quarters for roasting or slice into thin strips for sautéing.

## Portobello Mushrooms

Portobello mushrooms are mature cremini. Start by trimming off the woody bottoms of the stems. Then, cradling each mushroom in your hand, use a spoon to remove the dark gills, which can discolor a dish and turn it muddy. The mushrooms can now be scored for roasting or grilling whole, or split in half and cut into thin slices for sautéing.

## Shiitake Mushrooms

Shiitakes have very rubbery stems which should be discarded. Use a paring knife to trim them off (removing them by hand can end up tearing the caps). Slice the caps into thin strips.

## PUMPKIN SOUP

Sometimes vegetables are so starchy on their own that no additional thickener or emulsifying agent is needed to create a creamy, ultrasmooth pureed soup. Sweet sugar pumpkin and squash are ideal candidates.

You *could* make a pumpkin soup just like the creamy broccoli soup on page 213 by simmering the cubed pumpkins in broth and pureeing the lot, but much, much tastier is to roast the pumpkin first. There's more than meets the eye to the process of roasting starchy vegetables like pumpkin (or, say, sweet potato)—it's not just about softening.

First off, roasting drives off some moisture, concentrating flavor. Second, there are enzymes naturally present in pumpkins, other squashes, and sweet potatoes that will aid in the conversion of starches

to sugars, intensifying their sweetness. While this process will occur naturally over time, slow-roasting hastens these reactions.

Finally, as the pumpkin roasts, liquid escapes from inside of it, making its way to the surface and bringing along some dissolved sugars with it. As the liquid evaporates, the sugars are left on the exposed surfaces of the pumpkin, where they begin to caramelize. The process of caramelization not only creates sweeter sugars, it also produces hundreds of varying flavor compounds that add depth to the finished soup.

I like to use a mix of pumpkins and squashes, splitting them, tossing them in olive oil, and then roasting them as slow as I have the patience for before scooping out their innards and pureeing them with stock and other flavorings.

# ROASTED PUMPKIN SOUP

**SERVES 6 TO 8**

4 pounds whole pumpkins and
winter squash, preferably a mix,
such as sugar, kabocha, delicata,
and acorn (about 2 to 3 medium
or small or 1 large)

2 tablespoons olive oil

Kosher salt and freshly ground
pepper

2 tablespoons unsalted butter

1 medium onion, finely sliced
(about 1 cup)

¼ teaspoon ground cinnamon
(optional)

¼ teaspoon ground nutmeg
(optional)

4 cups homemade or low-sodium
canned chicken stock plus more if
necessary

2 tablespoons maple syrup

1. Adjust an oven rack to the lower-middle position and preheat the oven to 350°F. Split the pumpkins and/or squash in half through the stem and use a large spoon to scoop out and discard the seeds. Transfer to a foil-lined rimmed baking sheet, cut side up. Rub all over with the olive oil and season with salt and pepper. Roast until the flesh is completely tender and shows no resistance when a sharp knife or cake tester is inserted into it, about 1 hour. Remove from the oven and allow to cool.

2. While the pumpkin is cooling, melt the butter in a large Dutch oven or soup pot over medium-high heat. Add the onion and cook, stirring frequently, until softened but not browned, about 4 minutes. Add the cinnamon and nutmeg, if using, and stir until fragrant, about 30 seconds. Add the chicken stock.

3. Using a large spoon, carefully scoop out the roasted pumpkin flesh and transfer it to the pot. Add enough water to barely cover the pumpkin and bring to a simmer.

4. Working in batches, transfer the mixture to a blender and blend, starting on low speed and gradually increasing to high, until completely smooth, about 1 minute, adding additional stock or water if necessary to thin to the desired consistency (I like mine thick). Pass through a fine-mesh strainer into a clean pot and reheat gently. Stir in the maple syrup, and season to taste with salt and pepper. Serve.

# Two PATHS to
## FRENCH ONION SOUP

What's French onion soup doing in a book of American food, you might ask? Here's the answer: my goal is to wean you off of those little packets of powdered brown stuff to show you that with science at your side, making *real* caramelized onions, and French onion soup, is not nearly as time-consuming or difficult as you think!

There is certainly no shortage of recipes for French onion soup out there, and the general method begins with the same basic technique: **cook down finely sliced onions over low, low heat** so their natural sugars slowly and evenly caramelize. Once the onions are completely broken down to a deep brown, jam-like consistency, just add stock, a splash of sherry, and a couple of aromatics, simmer it down, season with a bit of salt and pepper, and serve with cheesy croutons.

It's a simple process, and the results are infinitely better than any commercial version, but it's a major pain in the *cul*. All that slow caramelizing takes a good 3 to 4 hours of constant pot babysitting. Let it go just a bit too long or step away for 5 minutes, and you've burnt your onions, making the final product too bitter to use.

As with exercise and marriage, I often think about how great it'd be if there were a method that could deliver the exact same (or better!) results without the massive time commitment. The bad news? After several months (OK, *years*) of testing, and more than fifty pounds of onions later, I've discovered that there isn't really a perfect substitute for traditional caramelizing. The good news? You can get 90 percent of the way there in about 10 percent of the time. That's a pretty decent exchange rate.

Here are the basics.

## Seeking Sweetness

First, it's important to understand exactly what's going on when an onion browns.

Onions go from firm to soft to melting and golden brown as they cook.

- **Sweating** is the first stage of sautéing onions or other vegetables. As they slowly heat up, moisture from their interior (onions are roughly 75 percent water by weight, some other vegetables are even more watery) begins to evaporate, forcing its way out of the cells and causing them to rupture in the process. This breakdown of the cells is what causes vegetables to soften.

- **Enzymatic reactions** take place as the contents of the vegetable cells—a complex mix of sugars, proteins, and aromatic compounds (in the case of onions, mercaptans, disulfides, trisulfides, thiopenes, and other such long, no-reason-to-memorize chemicals)—are spilled out and begin to mix with each other.
- **Caramelization** begins to occur once most of the liquid has evaporated and the temperature of the onions starts creeping up into the 230°F-and-above zone. This reaction involves the oxidation of sugar, which breaks down and forms dozens of new compounds, adding color and depth of flavor to the onions.
- **Sweetening** of the onions also takes place. The large sugar molecule sucrose (aka white sugar) breaks down into the smaller monosaccharides glucose and fructose (the same two sugars that corn syrup is made of). Since one glucose molecule plus one fructose molecule is sweeter than a single sucrose molecule, the flavor of the caramelized sugars is actually *sweeter* than the sugar they started out as.
- **The Maillard reaction**, aka the browning reaction, also takes place at these temperatures. This is the same reaction that causes browning on your toast or your steak when you cook it (see page 292). The Maillard reaction is far more complex than caramelization, involving interactions among sugars, proteins, and enzymes. The products of the reaction number in the hundreds, and are still not fully identified.

In an ideal world, as the onions continue to cook, three things will happen at the same time: (1) the complete softening of the onions' cell structures, (2) maximum caramelization (i.e., as brown as you can get before bitter products begin to develop), and (3) maximum Maillard browning (with the same caveat as caramelization).

By enhancing these results, I should be able to speed up my overall process.

## Mission 1: Increase the Effects of Caramelization

The most obvious way to speed up caramelization is to add more raw ingredients, namely, sugar. The sugars in onions, as mentioned above, are glucose, fructose, and sucrose (a combination of one glucose and one fructose molecule)—exactly the same as the caramelization products of granulated sugar. So I tried cooking a touch of sugar in a dry skillet until it reached a deep golden brown, then adding the onions and tossing them to coat them in the hot caramel. It worked like a charm, shaving a good 4 to 5 minutes off my total cooking time and giving me sweeter, more deeply caramelized end results, without affecting the overall flavor profile of the finished product.

## Mission 2: Increase the Maillard Reaction

There are a number of things that affect the Maillard reaction, but the overriding factors are **temperature** and **pH**. At this point, I had no safe way to increase the temperature—just like with a steak, if you cook the onions too hot, the edges and outsides of each piece begin to burn before the interiors have a chance to release their chemicals. Low and slow is the only way to go.

On the other hand, I had a bit of control over the pH. In general, the higher the pH (i.e., the more basic or alkaline), the faster the Maillard reaction takes place. The key is moderation. While large amounts of baking soda dramatically increased the browning rate (by over 50 percent!), any more than ¼ teaspoon per pound of chopped onions proved to be too much—the metallic flavor of the baking soda took over.

I also noticed that the baking-powdered onions were much softer—not an undesirable trait for a soup. This is because pectin, the chemical glue that holds vegetable cells together, weakens at higher pH levels. Faster breakdown means faster release of chemicals, which means faster overall cooking.

## Mission 3: Increase the Heat

Back to the heat. As I mentioned, the problem with increasing the heat too much higher than medium-low is that the onions begin to cook unevenly. Some bits and edges will start to blacken long before other bits reach even the golden brown stage. Additionally, the sugars and proteins that get stuck to the bottom of the pot as the onions cook rapidly turn dark brown, because of their direct contact.

So, the question is, if you're cooking with high heat, what can you do to simultaneously even out the cooking all across the onions, remove the sticky browned gunk from the bottom of the pan, and regulate the overall temperature so that nothing burns? If you've ever made a pan sauce, the answer is so blindingly simple that I'm surprised it's not a completely common practice: just add water.

At first, adding water may seem counter-productive—it cools down the onions and the pot, forcing you to expend valuable energy heating it up and evaporating it. But here's the deal: both the browned patina on the bottom of the pot and the browned bits on the edges of the onions are made up of water-soluble sugar-based compounds that happen to be concentrated in a single area. Adding just a small amount of liquid to the pot at regular intervals means these compounds get dissolved and redistributed evenly throughout the onions and the pot. Even distribution leads to even cooking, which leads to no single part burning before the rest is cooked.

**So what does all this mean for your onions?** It means that you can cook them over a much higher flame (medium-high works well—even maximum heat is feasible, though it requires a little more attention), and every time they threaten to start burning, just add a couple tablespoons of water and you're smooth-sailing once again.

As I said, the flavor is not *quite* as deep and sweet as traditionally slow-cooked onions (sometimes there are simply no shortcuts to quality), but it's worlds better than anything you'll ever get out of a can, box, or packet—and it can go from start to finish and make it to the table in under 30 minutes. That's some seriously fast onion soup!

# FAST FRENCH ONION SOUP

**SERVES 4**

1 tablespoon sugar

5 pounds yellow onions (about 5 large), finely sliced (about 7½ cups)

2 tablespoons unsalted butter

¼ teaspoon baking powder

Kosher salt

¼ cup dry sherry

6 cups homemade or low-sodium canned chicken stock

2 bay leaves

6 to 8 sprigs fresh thyme

Freshly ground black pepper

1 baguette, sliced ½ inch thick and toasted

8 ounces Gruyère or Swiss cheese, grated

1. Pour the sugar into a large Dutch oven and cook over high heat, swirling the pot gently as the sugar melts, until it is completely liquid and a golden brown caramel. Add the onions and cook, stirring with a wooden spoon and tossing constantly until they are evenly coated in the caramel, about 30 seconds. Add the butter, baking powder, and 2 teaspoons salt and cook, stirring occasionally, until the onions are light golden brown and a brown coating has started to build up on the bottom of the pot, about 10 minutes.

2. Add 2 tablespoons water and scrape the browned coating from the bottom of the pot. Shake the pot to distribute the onions evenly over the bottom and cook, shaking occasionally, until the liquid evaporates and the browned coating starts to build up again, about 5 minutes. Add 2 more tablespoons water and repeat, allowing the coating to build up and scraping it off, then repeat two more times. By this point, the onions should be a deep brown. If not, continue the deglazing and stirring process until the desired color is reached.

3. Add the sherry, chicken stock, bay leaves, and thyme, bring to a boil, and reduce to a simmer. Simmer, uncovered, until the liquid is deeply flavored and slightly reduced, about 15 minutes. Season to taste with salt and pepper. Discard the bay leaves and thyme.

4. To serve, heat the broiler. Ladle the soup into four broilerproof bowls. Float the croutons on top and cover with the grated cheese. Broil until the cheese is melted, bubbly, and golden brown in spots. Serve immediately.

# ALL ABOUT ONIONS

In the mood for some chili? You're gonna need three cups of onions, medium dice. Making chicken stock? Two onions, large chunks, please. And what about onion soup? Yes, believe it or not, you'll need onions for that too.

No matter how you slice 'em, onions are used in a good 30 to 40 percent of any cook's savory-dish repertoire, if not more. They are the first thing you should learn how to cut when you pick up a knife, and, at least for me, still one of the most pleasurable foods to take a sharp blade to.

### What color onion should I be using?

There are four basic onion varieties available in most supermarkets: yellow, white, sweet (Vidalia or Walla Walla), and red. You may also occasionally see Spanish onions, which are larger, milder relatives of yellow onions. Although sweet onions have about 25 percent more sugar than standard onions, their flavor difference when raw has more to do with the amount of tear-inducing *lachrymators* they contain (see below). Yellow and white onions have more of these pungent compounds, but after cooking, they all but disappear.

For the most part, onions can be used interchangeably without catastrophic consequences (unless you consider red onions on a slider to be a catastrophe). But some onions are better suited for certain tasks than others.

- **Yellow onions** are the kitchen workhorse. They boast a good balance of sweetness and savoriness, though they can be quite pungent, and are best for cooked applications. If there is one onion you should never be without, this is it.

- **White onions** are extremely mild in flavor and have a distinct sweetness. When caramelized, they have a flat, one-dimensional flavor that can come across as cloying. They are best used raw or in soups.

*continues*

• **Sweet onions (Vidalia, Walla Walla, Maui, etc.)** cook similarly to yellow onions, but their mild pungency and sweetness are better enjoyed raw in preparations like chopped salads or fresh salsas, or sliced for sandwiches.

• **Red onions** are rarely used for cooking, as their pigment can turn an unappetizing blue with prolonged cooking, throwing off the color of your finished dish. Slightly more pungent than white or sweet onions, red onions are best used raw or in simple, quick-cooking applications, like on the grill or under the broiler.

• **Shallots** are the diminutive cousins of onions. They have a distinctly sweet and pungent flavor and are great both raw in salad dressing or cooked with other vegetables. Think of them more as onion seasoning than real onions, with the ability to provide onion flavor without overwhelming a dish.

**Does size matter?**
Ahh, the eternal question. The size of an onion has little bearing on flavor, though I prefer larger onions because I have to peel fewer of them to get the same volume of prepped onions.

**How do I tell the good onions from the bad?**
No matter what type of onions you choose, make sure that they are firm to the touch when you buy them. If they give even a little bit—particularly at the root or stem end—there's a good chance some of the interior layers may have begun to rot.

**Where's the best place to store them?**

Store onions in a cool, dry, dark place, never in a sealed container, which can trap moisture, leading to mold and rot. I keep mine in a Chinese bamboo steamer.

Half-used onions can be placed in a plastic bag in the refrigerator. Just use them within a few days.

**I've noticed that, like grandmothers and movie theaters, some onions smell more than others. Is there a way to know before I buy?**

How much an onion smells is largely dependent on how long it's been stored. The longer onions have been in storage (in some cases, up to months), the more pungent they'll be. Unfortunately, it's not always easy to tell, as they don't come with a date on the label. Generally, older onions have thicker, tougher skins, while newer onions will have thinner papery skins. But it's not like you have a choice anyway—markets don't offer "old onions" and "new onions."

The unfortunate answer is that with onions, you've got to play the hand you're dealt. But we've got a few tricks for dealing with them in our arsenal. Read on.

**What is it that makes onions smell, anyway?**

My favorite *Calvin and Hobbes* strip is the one where Calvin walks into the kitchen and sees his mom crying while cutting an onion. He walks away mumbling, "It must be hard to cook when you anthropomorphize all your vegetables." Classic. But there's a very real reason we cry when onions are cut into: defense.

Onions take up sulfur from the soil as they grow, storing it within larger molecules in their cells. Separately, they store an enzyme that catalyzes a reaction that breaks these larger molecules down into pungent, irritating sulfurous compounds. Only after the onion's cells are damaged by chopping or crushing do the precursors and enzyme mix, producing what are called *lachrymators*, the compounds that attack nerves in our eyes and nose, causing us to tear up and sneeze. Nature at its most defensive!

That's why an uncut onion will have very little aroma, but as soon as you slice it, the smell begins to permeate the room.

**Those lachrymators really get my tears flowing. Anything I can do to help it?**

There's no shortage of home remedies that are claimed to suppress or minimize tearing up: Light a fire (supposedly catalyzes some reaction that prevents lachrymators from forming—it doesn't work unless you are cutting your onion directly over or under the flame). Rinse the onion as you go (works OK, but wet hands and sharp knives don't mix). Place a piece of bread on the cutting board (does absolutely nothing). Suck on an ice cube or chew a toothpick (I can't even begin to fathom the rationale). Chill the onions in ice water for 10 minutes first (this works pretty well—the cold slows down enzymatic reactions). But of all the cures, there's only one that's really effective: just block your eyes. If you're a contact lens wearer, you've probably already noticed that onions don't really bother you. For the rest of you, ski goggles or swimming goggles are the way to go. Plus, they make you look really cool. Trust me.

*continues*

**Is there any way to get rid of that onion odor?**

Let's say you happen to have some extra-pungent onions (it happens to the best of us)—is there a way to tame them? I tried out a few different methods, from submerging them in cold water for times ranging from 10 minutes to 2 hours to chilling them to letting them air out on the counter.

Soaking the sliced onions in a container of cold water just led to onion-scented liquid in the container, without much of a decrease in the aroma in the onions themselves. Perhaps if I'd used an unreasonably small amount of onion in an unreasonably large container, the water would have diluted it more efficiently. Air-drying led to a milder aroma but also to dried-out onions and a papery texture.

The best method turned out to be the fast-est and easiest: just rinse away all those extra-pungent compounds under running water after you slice the onions—and not just that, but warm water. The speeds of chemical and physical reactions increase with temperature. Using warm water causes onions to release their volatile compounds faster—about 45 seconds is enough to rid even the most pungent onions of their kick.

But doesn't hot water turn the texture of an onion limp? No. Even if you use very hot tap water, it generally comes out at around 140° to 150°F or so, while pectin, the main carbohydrate "glue" that holds plant cells together, doesn't break down until around 183°F. There are other bits of the onion that, given enough time, will begin to soften at hot-tap-water temperatures, but it takes far longer than the 45-second rinse needed. Don't worry, your onions are safe.

# KNIFE SKILLS:
## Slicing and Dicing Onions

**The main question when it comes to slices is one of direction.**

If you call the stem and root ends of an onion its north and south poles, then an orbital slice looks like this:

. . . while a pole-to-pole slice looks like this:

At first glance, you may think, what's the big difference?

Let me answer your question with a question of my own: do you care about the flavor of what you put in your mouth? If the answer is no, then by all means slice your onions any which way. But if the answer is yes, consider this: onion cells are not perfectly symmetrical—they're longer in the pole-to-pole direction than in the orbital direct. Hence the direction you slice your onions will affect the number of cells you rupture and thus the amount of lachrymators that are formed. We know that some amount of this stuff is desirable: it makes your onions taste more oniony, your stews taste more meaty, your French onion soup sweeter. But too much can be overwhelming.

To see what difference it made, I split an onion in half, slicing each of the two halves in different ways, then placed the onion slices in identical covered containers and let them sit for 10 minutes on the counter before opening them and taking a whiff. There was no doubt that the orbitally sliced onion was stronger, giving off the powerful stench of White Castle dumpsters and bad dates.

When cooked into a recipe like a sauce or a soup, orbitally sliced onions also have an inferior texture—they come out tougher and wormy. With rare exception, I use pole-to-pole-sliced onions for all applications.

*continues*

**Dicing an onion is a task you're going to do many,
many, *many* times, so you'd better get used to it.**

- **THE BASIC PROCESS** always starts out the same: Place the onion on your cutting board and slice off the stem end (1), then put it on the face you just cut and split the onion in half (2). Peel the onion (3), and then from there . . .

- **FOR MEDIUM OR LARGE DICE:** Make 2 to 6 cuts in the onion half, running from pole to pole (4), leaving the root end intact to hold the onion together (5). Then make 2 to 6 perpendicular cuts to form large dice.

- **FOR SMALL DICE:** Make parallel cuts running pole to pole at ¼-inch intervals, leaving the root end intact. Then hold your knife horizontally and make a single slice about ¼ inch up from the base (6). Cut across the parallel cuts, using your curved knuckles as a guide for the knife (7). The onion should separate into fine dice (8). Discard the root end.

**QUICK TIP:** If you're working with a large volume of onions, to maximize efficiency, take every onion through each step before proceeding to the next step. In other words, peel all the onions before you start slicing any of them. Similarly, make all of your horizontal cuts before making your vertical cuts. It will keep your work space more organized, require fewer trips to the garbage can (or compost can), and make you look like a pro.

## CLASSIC FRENCH ONION SOUP

What about those lazy Sundays when you don't *mind* hanging around the kitchen for the several hours it takes to caramelize onions properly—what's the ideal method then?

Most recipes have you play the babysitter, cooking the onions slowly on the stovetop, stirring every few minutes. There are really two separate processes going on here. The onions are softening, releasing water and various dissolved sugars and other chemical compounds from inside their cells. Simultaneously, there's caramelization as those sugars are heated. Ideally, both of these things end up finishing at around the same time.

But here's what I wondered: could I divide the process into two distinct steps, first letting the onions fully soften and release their juices, *then* reducing those juices and browning them? If so, I should be able to save myself a bit of babysitting time by limiting my stirring to the final stage of cooking. To do it, I went back to a method I'd learned from Chef Jason Bond for making a sweet white onion puree back when I was a line cook at No. 9 Park: all you have to do is to cook onions thinly sliced pole-to-pole (they'll have a better texture when cooked) in butter in a heavy enameled cast-iron or stainless steel Dutch oven. Once they get going, throw on the lid, turn the heat down as low as possible, and let 'em sit there. As the onions heat, they give off liquid, some of which turns to steam, re-condenses on the roof of the pot, and then falls back down, keeping the onions moist as they soften.

After a couple hours (with just one or two stirs in the middle), the onions will be completely softened and have given up all the liquid and dissolved flavor compounds they're gonna give up. At this stage, it's a simple matter of reducing that sugary liquid over moderate heat until it's deeply caramelized and brown, using the browning-and-deglazing process I use for my quick caramelized onions.

The resultant soup is sweet, rich, and deeply complex.

# TRADITIONAL FRENCH ONION SOUP

**NOTE:** If your pot doesn't have a heavy tight-fitting lid, place a layer of aluminum foil over the pot, crimping the edges to seal tightly, then add the lid.

**SERVES 4**

4 tablespoons unsalted butter

5 pounds yellow onions (about 5 large), finely sliced (about 7½ cups)

Kosher salt

¼ cup dry sherry

6 cups homemade or low-sodium canned chicken stock

2 bay leaves

6 to 8 sprigs fresh thyme

Freshly ground black pepper

1 baguette, sliced ½ inch thick and toasted

8 ounces Gruyère or Swiss cheese, grated

1. Melt the butter in a large Dutch oven over medium heat. Add the onions and 1 teaspoon salt and cook, stirring frequently with a wooden spoon, until the onions have begun to soften and settled into the bottom of the pot, about 5 minutes. Cover the pot with a tight-fitting lid (see Note above), reduce the heat to the lowest setting, and cook, stirring every 45 minutes, until the onions are completely tender, about 2 hours.

2. Remove the lid and increase the heat to medium-high. Cook, stirring frequently, until the liquid has evaporated and a brown patina has started to form on the bottom of the pot, about 15 minutes. Add 2 tablespoons water and scrape the browned coating from the bottom of the pot. Shake the pot to distribute the onions evenly over the bottom and cook, shaking occasionally, until the liquid evaporates and the browned coating starts to build up again, about 5 minutes longer. Add 2 more tablespoons water and repeat, allowing the coating to build up and scraping it off, then repeat two more times. By this point, the onions should be a deep brown. If not, continue the deglazing and stirring process until the desired color is reached.

3. Add the sherry, chicken stock, bay leaves, and thyme, bring to a boil, and reduce to a simmer. Simmer, uncovered, until the liquid is deeply flavored and slightly reduced, about 15 minutes. Season to taste with salt and pepper. Discard the bay leaves and thyme.

4. To serve, heat the broiler. Ladle the soup into four broilerproof bowls. Float the croutons on top and cover with the grated cheese. Broil until the cheese is melted, bubbly, and golden brown in spots. Serve immediately.

# CHICKEN AND DUMPLINGS

**It's gonna be tough not to stick a joke about my late pup
Dumpling in here somewhere, but I'll try my best.**

Once you realize that chicken and dumplings is nothing more than chicken stock combined with biscuit dough, adding this dish to your repertoire is a snap. On page 187, you learned how to make an awesome chicken stock in record time, and we explored the science of biscuits in our breakfast chapter (see page 158), so the only question is: does a biscuit recipe need any modification to cook properly in the moist environment of a soup pot?

The answer, unfortunately, is yes. But not much. Regular biscuit dough tends to be really high in fat—a full 4 ounces of butter for every 10 ounces of flour. The flour can't form tough gluten sheets as readily as it otherwise would, because its proteins are lubricated by butter. In an oven, this is no problem. All you've got to do is get your biscuits on a baking sheet, and from there, don't touch 'em until they're baked and set. In the dynamic environment of a pot of soup, however, with bubbles simmering all around, condensation dripping from the ceiling, and pieces of chicken jostling it every which way, the delicate biscuit dough doesn't stand a chance: it's almost guaranteed to disintegrate, turning the broth sludgy and greasy.

The first step to modifying biscuit dough for dumplings is to reduce the fat. I found that 6 table-spoons, down from 8, was a good compromise, still leaving plenty of flavor but increasing stability. This introduced a new problem, though: with less fat, the dumplings were coming out a little dry and dense, tougher than they should have been. I tried increasing the amount of baking powder and baking soda, but neither one worked—the dumplings ended up with a strong chemical aftertaste. The easy solution? An egg.

Each of the two parts of an egg improves a dumpling dough in its own way. The fatty, protein-rich yolk replaces some of the fat that was lost when I cut back on the butter. But, unlike butterfat, which starts melting and leaking out of the dumplings at around 90°F, an egg yolk does the opposite, becoming firmer as it is heated. Emulsifying agents found in the yolk, like lecithin, also help ensure that the fat stays put inside the dumplings. The egg white in this case acts as a leavener. As the dumplings cook, their loose protein matrix begins to solidify, trapping bubbles of water, moist air, and carbon dioxide created by the baking powder and the baking soda/buttermilk reaction. As the dumplings continue to cook, this moist air expands, resulting in lightness and tenderness.

# CHICKEN AND DUMPLINGS

**NOTE:** Instead of using the chicken stock, you can simmer 4 chicken legs in 2 quarts low-sodium canned chicken broth for 30 minutes. Remove the legs, skim the fat from broth, and add enough water to make 2 quarts. When the legs are cool enough to handle, pick off the meat, discarding the bones and skin, and reserve.

**SERVES 4 TO 6**

1 recipe Quick Chicken Stock (page 187), including the picked leg meat (see Note above)

2 medium carrots, peeled and cut into medium dice (about 1 cup)

1 medium stalk celery, cut into medium dice (about ½ cup)

1 small onion, finely sliced (about 1 cup)

**For the Biscuit Dough**

¾ cup buttermilk

1 large egg

10 ounces (about 2 cups) unbleached all-purpose flour

1 teaspoon baking powder

¼ teaspoon baking soda

1½ teaspoons kosher salt, plus more for seasoning

4 tablespoons cold unsalted butter, cut into ¼-inch pats

¼ cup chopped fresh parsley

Freshly ground black pepper

1. Combine the chicken stock (not the meat), carrots, celery, and onion in a large Dutch oven and bring to a boil over high heat. Reduce to a simmer and cook until the vegetables are tender, about 20 minutes.

2. **Meanwhile, make the biscuits:** Whisk together the buttermilk and egg in a medium bowl.

3. In the bowl of a food processor, combine the flour, baking powder, baking soda, and salt and process until mixed, about 2 seconds. Scatter the butter evenly over the surface of the flour and pulse until the mixture resembles coarse meal and the largest butter pieces are about ¼ inch at their widest dimension. Transfer to a large bowl, add the buttermilk mixture, and fold with a rubber spatula until just combined. The dough will be a little shaggy and pretty sticky.

4. Stir the parsley and chicken meat into the stock. Season to taste with salt and pepper and bring to a simmer. Using a greased tablespoon measure, drop dumplings onto its surface, leaving a little space between them. Cover the pot, reduce the heat to low, and cook until the dumplings have doubled in volume and are cooked through (you can cut one open with a knife to peek, or insert a cake tester or toothpick—it should come out clean). Serve immediately.

# POT ROAST

**Give me a good American-style pot roast, in all its dripping, savory, messy, beefy glory, over a French *boeuf bourguignon* any day of the week.**

I love how the meat shreds under the slightest pressure from your fork. I love the rich onion-scented gravy, studded with flecks of beef debris. I love the tender carrots and potatoes, heavy with meat juices. For soul-satisfying cold-weather fare, it's about as good as it gets.

Pot roast is essentially a large piece of braised meat. Braising is the act of slowly cooking a piece of meat in a moist environment. The moisture can come from submerging it in liquid (in which case, it's technically called stewing) or by cooking it in a covered or partially covered vessel designed to trap moist air around the food. As the meat cooks at a low temperature in a moist environment, just as when making stock, the connective tissue, primarily made up of the protein collagen, slowly converts into gelatin. This is essential, because cooking also transforms meat in another important way: it drives out moisture—even if you cook it in a completely moist environment. Indeed, because water is such a great conductor of heat, beef boiled in 212°F water will actually get hotter and lose moisture faster than beef roasted in 212°F oven! But there are other reasons to keep liquid in your pot. First, it regulates the temperature, so that there's no chance of anything ever getting hotter than the boiling point of water. Second, it facilitates the transfer of flavors among different parts of the meat and the vegetables. Finally, what good's a pot roast without gravy?

All good meat recipes start with the right cut. For pot roast, any number of cuts high in connective tissue will do, but I prefer the chuck eye. It's beefy and has plenty of gelatin-rich connective tissue to keep it moist (see "Stewing Beef," page 241). After

browning the roast in a Dutch oven, I further bolster the flavor by adding a mirepoix of carrots, celery, and onion, browning them in the same pot (the moisture exuding from the vegetables will help deglaze the *fond* left behind by the browned beef). Time for the umami bombs: anchovies, Marmite, and soy make their way into pretty much every braised dish I make, for their savory glutamates.

While French-style braises may resort to rich reduced veal stocks to thicken the gravy, flour is the thickener of choice stateside. Next up, a bottle of wine. It's not especially traditional in an American pot roast, but, just like anchovies, Marmite, and soy sauce, wine is rich in savory glutamates, adding meatiness to the broth, as well as complex aroma and hint of acidity. Some chicken stock, a few peppercorns, and some sprigs of thyme and bay leaves round out my flavor profile.

After building my braising liquid, I put the beef back in, place a lid on the pot and set it in a 275°F oven to cook until tender, making sure to leave the lid slightly cracked. Why, you may ask? Temperature regulation. With a completely sealed lid, the water inside the pot rapidly reaches the boiling point—a temperature at which, given time, over 50 percent of the moisture in a piece of beef will be forced out. By keeping the lid ajar, you can keep the contents of the pot at around 185°F, even in a 275°F oven! (For a more complete explanation, see "Experiment: Boiling Water Under Cover," page 242, and "Stovetop Versus Oven," page 271). This lower temperature allows the collagen to slowly break down while still maintaining a good level of moisture inside the meat.

Slicing a pot roast after chilling overnight gives you uniform, tatter-free slices.

## Constant Heat Versus Constant Temperature

You may wonder why many braised recipes call for cooking in the oven instead of simmering on the stovetop. Here's the deal: a stovetop maintains a constant heat output, while an oven maintains a constant temperature. This means that on the stovetop, the same amount of heat energy is being transferred to the pot no matter how much stuff is inside it, and no matter how hot that stuff already is. A pot of stew that's barely simmering at the start of cooking over medium-low heat might be at a rapid boil toward the end of cooking, when some of the liquid has evaporated and the volume of the stew has diminished. In an oven, no matter how much or how little food there is in the pot, the temperature of the food remains the same. Moreover, an oven heats gently from all sides, while a burner focuses the heat on the bottom of the pot. Move all your braises to the oven, and you'll get better results—guaranteed.

When I made the pot roast, after about 3 hours the meat was at just the point I wanted it—tender enough that a knife or cake tester could slip in and out of it easily, but not so tender that it had lost its structure. (The broth, by the way, smelled awesome.) The problems arose when I tried to slice the meat. While hot, it was so tender that it was nearly an impossible task—it shredded and fell apart even with the sharpest knife and the gentlest touch. The best way to slice braised meat is to allow it to first cool completely.

Originally I thought it'd be best to allow the meat to cool down in the air of the kitchen instead of in the hot liquid inside the pot. But I tested identical halves of the same roast cooled in the air versus in the liquid and here's what I found:

## RETAINED WEIGHT VERSUS COOLING METHOD

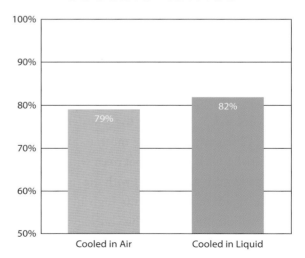

As you can see, the meat cooled in the liquid ended up with a good 3 percent more moisture than the one cooled in the air. This is because cooler meat can hold moisture more easily than hot meat. So, as the meat cools in the liquid, some of the liquid is reabsorbed. I also found that then allowing the pot roast to rest in the refrigerator for up to 5 days actually improved both flavor and texture, which means that for the ideal pot roast experience, you ought to cook the thing several days before you plan to eat it.

## STEWING BEEF

The best cuts of beef for stewing or braising have plenty of robust beefy flavor and lots of connective tissue to break down into rich gelatin. Here are a few of my favorites:

- **Chuck** comes from the shoulder of the steer and is intensely beefy, with a good deal of fat. The best chuck cuts for braising are the 7-bone and the chuck roll. I prefer the latter, which is a boneless cut that makes for easier slicing. Look for well-marbled pieces with a nice cylindrical shape for even cooking.
- **Brisket** comes from the steer's chest. A whole brisket contains two parts: the flat (also called "thin cut" or "lean") and the point (also called "deck" or "moist"). The flat is more commonly available in supermarkets, but if you can find the point, it's worth buying for its larger amount of connective tissue, fat, and flavor. Brisket is not quite as rich as chuck and has a distinctive metallic, grassy aroma.
- **Flap meat**, also known as sirloin tip, is generally sold as an inexpensive steak, but it's great stewed, which makes it one of the most versatile steaks around. A whole flap steak is a rectangular block of meat about 1½ inches thick, weighing about 2 to 3 pounds, with a very strong grain and plenty of fat. It has a deep, beefy flavor and a robust, ropy texture that holds up well to long cooking.
- **Round** comes from the rear leg and is available as many different cuts. Bottom round, with a flavor similar to chuck, is the best for braising, though its odd shape makes it a bit harder to handle. Eye of round is by far the leanest of all of the braising cuts and so has a tendency to dry out a bit. If you keep a careful eye on the temperature and make sure to remove it from the oven as soon as it is tender, it does make a decent lower-fat option (but where's the fun in that?).
- **Short ribs** are technically part of the chuck but are sold separately. They come in three forms: as hunks of meat on top of 6-inch sections of rib bones (called English-cut short ribs); as pieces of meat attached to three- to four-rib-bone cross-sections (flanken cut); and boneless. All three make great stew meat. Their abundant fat and connective tissue ensure that they'll be meltingly tender and rich when properly cooked.

# EXPERIMENT:
## Boiling Water Under Cover

**Does leaving the lid on or off really make much of a difference when cooking in an oven that's supposed to be maintaining a constant low temperature? Try this quick experiment to see for yourself.**

### Materials

- **Two identical pots filled halfway with water**
- **One lid**
- **An instant-read thermometer**

### Procedure

Preheat the oven to 275°F. Place both pots in it, one with the lid on and the other with the lid off. Let the water heat for 1 hour, then open the oven and immediately take the temperature of each pot of water.

### Results

The water in the lidded pot should be at around 210°F, while the water in the uncovered pot is probably closer to 185° to 190°F. Because of the cooling effect of evaporation (it takes a significant amount of energy for those water molecules to jump from the surface of the liquid—energy that they steal from the liquid itself, cooling it down), an open pot of stew in a 275°F oven will max out at around 185°F. Good news for you, because that's right in the optimal subsimmer stewing temperature zone.

Pop the lid on, and you cut down on the amount of evaporation. Less evaporation means a higher max temperature. In my simple test, putting the lid on increased the temperature in the pot by almost 25 degrees!

# ALL-AMERICAN POT ROAST
## WITH GRAVY

**SERVES 6 TO 8**

1 boneless chuck roast (about 5 pounds), pulled apart at the seam into 2 large chunks, excess fat and gristle trimmed

Kosher salt and freshly ground black pepper

2 tablespoons vegetable oil

4 anchovy fillets

2 medium cloves garlic, finely minced or grated on a Microplane (about 2 teaspoons)

1 teaspoon Marmite

1 tablespoon soy sauce

2 tablespoons tomato paste

2 large carrots, peeled and cut into 1- to 2-inch chunks

2 stalks celery, cut into 1-inch chunks

2 large onions, finely sliced (about 4 cups)

2 tablespoons all-purpose flour

1 bottle (750-ml) dry red wine

4 cups homemade or low-sodium canned chicken stock

¼-ounce (1 packet) unflavored gelatin

2 bay leaves

4 sprigs fresh thyme

1 pound russet baking potatoes (about 2 large), peeled and cut into 1- to 2-inch chunks

1. Adjust an oven rack to the lower-middle position and preheat the oven to 275°F. Pat the chuck roast dry and season it with salt and pepper. Tie kitchen twine tightly around each piece at 1-inch intervals to help it retain its shape.

2. Heat the oil in a large Dutch oven over high heat until lightly smoking. Add the chuck and cook, turning occasionally, until well browned on all sides, about 8 to 10 minutes. Transfer the beef to a large bowl.

3. Meanwhile, combine the anchovy fillets, garlic, Marmite, soy sauce, and tomato paste in a small bowl and mash with the back of a fork until a smooth, homogeneous paste is formed.

4. Return the pot to medium-high heat, add the carrots and celery, and cook, stirring frequently, until the vegetables begin to brown around the edges, about 5 minutes. Add the onions and cook, stirring frequently, until very soft and light golden brown, about 5 minutes. Add the anchovy mixture and cook, stirring, until fragrant, about 1 minute. Add the flour and cook, stirring, until no dry flour remains, about 1 minute. Increase the heat to high and, whisking constantly, slowly add the wine. Bring to a simmer and cook until the wine is reduced by half, about 15 minutes.

5. Meanwhile, pour the chicken stock into a large liquid measuring cup or a bowl and sprinkle the gelatin on top. Allow it to hydrate for 10 minutes.

6. Add the gelatin and chicken stock, bay leaves, and thyme to the Dutch oven, return the beef to the pot, and bring the liquid to a simmer. Cover, leaving the lid slightly cracked, place in the oven, and cook until the beef is completely tender (it should offer little to no resistance when you poke it with a cake tester or thin knife), 3 to 4 hours; add the potatoes to the pot about 45 minutes before the beef is done. Remove the pot from the oven and allow to cool for 1 hour.

7. Transfer the whole pot to the refrigerator and let rest at least overnight, or up to 5 nights.

8. When ready to serve, carefully remove the hardened layer of fat from the top of the cooking liquid and discard. Transfer the meat to a cutting board. Discard the bay leaves and thyme sprigs. Bring the liquid to a boil over high heat and reduce it until coats the back of a spoon but doesn't taste heavy. Season to taste with salt and pepper.

*continues*

9. Meanwhile, remove the twine from the beef and slice it against the grain into ½-inch-thick pieces. Place the pieces in overlapping layers in a 12-inch skillet and add a few ladles of sauce to moisten them. Cover the skillet and set over medium-low heat, shaking occasionally, until the meat is heated through, about 15 minutes.

10. Transfer the meat to warmed serving plates or a large platter and top with the cooked vegetables and more sauce. Serve immediately.

# GLUTAMATES, INOSINATES, AND THE UMAMI BOMBS

For many years, food scientists believed that our tongues were sensitive to four different basic tastes: sweetness, saltiness, sourness, and bitterness. Turns out that there's a fifth: *umami*. First discovered in Japan, it's best translated as "savory." It's the saliva-inducing qualities that, say, a good steak or a hunk of Parmesan cheese has in your mouth. Just as the sensation of sweetness is triggered by sugar, saltiness by salt, sourness by acid, and bitterness by a number of mildly poisonous classes of chemical, umami flavor is triggered by glutamates—essential amino acids found in many protein-rich foods. The key to getting many dishes to taste meatier—turkey burgers, chili, stew, soups, etc.—is to increase their level of glutamates.

Now, you can do this with powdered monosodium glutamate—a natural salt extracted from giant sea kelp—but some folks are squeamish about using it (I personally keep a little jar of it right next to my saltcellar). However, there are alternatives, namely, what I like to call the three umami bombs: Marmite, soy sauce, and anchovies.

If you've ever been to England, you've probably seen Marmite. It's that strange dark-brown goo with a sharp, salty, savory flavor that the Brits like to spread on toast in the morning. On toast, it's an acquired taste to be sure, but there's a lot of potential hiding in that little bottle. Made from a by-product of alcohol production, it's essentially concentrated yeast proteins, rich in both salt and glutamates. Soy sauce is made from fermented soybeans and acquires high levels of glutamates from both the amino-acid-rich soybeans themselves and the yeast and bacteria used to ferment them into the savory sauce. Anchovies and many other small oily fish are naturally rich in glutamates, and salting and aging them increases their concentration. Even many traditional recipes for rich meat-based stews will call for a few anchovies to help boost the dish's savory qualities.

There are many other ingredients that have high levels of glutamates—Worcestershire sauce, Parmigiano-Reggiano, powdered *dashi* (a Japanese broth made with sea kelp and bonito), to name a few—but they all have pretty strong flavor profiles. My three umami bombs are the only ones I know of that can increase meatiness while blending into the background, making the natural flavors of meat stronger without imposing their own will on the dish.

But which one to use? I made several batches of turkey burgers incorporating the ingredients directly into the grind. Turns out that adding all three together is exponentially more potent in triggering our meatiness-detectors than using any one on its own. Why does the combination work better than just one? Well, glutamates *are* the kings of meatiness, but there's a second chemical, *disodium inosinate*, that has been found to work in synergy with glutamic acid to increase the savoriness of food. In fact, in the August 2006 issue of the *Journal of Food Science*, researcher Shizuko Yamaguchi found that the synergistic effect of the two compounds is in fact quantifiable!

So, combine the two, and you get a synergistic effect, making each more powerful than if used on its own. Think of inosinates as the Robin to glutamate's Batman—they aren't necessary for the job, but they sure help an

*continues*

awful lot. ✳ By curing pork and fish, say, into prosciutto and Thai fish sauce, you can create a superconcentrated sources of inosinates. Anchovies actually have quite high levels of inosinate—more than they need to balance out their own levels of glutamate—while Marmite and soy don't have much at all. Combining the inosinate in the anchovies with the glutamate in the soy and Marmite makes for the ultimate combination.

## GLUTAMATE CONTENT
## IN COMMON INGREDIENTS

Many foods that we cook with every day are high in glutamates. Here's a chart of their relative glutamate content.

| | |
|---|---|
| Kombu (giant sea kelp) | 22,000 mg/100g |
| Parmigiano-Reggiano | 12,000 |
| Bonito | 2,850 |
| Sardines/Anchovies | 2,800 |
| Tomato Juice | 2,600 |
| Tomatoes | 1,400 |
| Pork | 1,220 |
| Beef | 1,070 |
| Chicken | 760 |
| Mushrooms | 670 |
| Soybeans | 660 |
| Carrots | 330 |

✳ Holy savory ground meat patties!

# SKILLET-BRAISED CHICKEN

**Quite a few other things may be as delicious as braised chicken
legs, but few things braise faster or easier, so, more often than not
when I'm in the mood for tender, moist, stewed meat, chicken it is.**

I rarely braise chicken using same recipe twice, but there are some common themes in all of them. The key to really great braised chicken is in the browning. You've got to brown the skin in the skillet until it's deeply golden brown and extraordinarily crisp, then make sure that the skin remains above the level of the liquid the entire time it's cooking so that the crispness remains. What you end up with is fall-off-the-bone-tender meat deeply flavored with sauce and the crisp skin of a perfectly roasted chicken.

You definitely want to use dark meat for this. It has more connective tissue, which slowly breaks down into gelatin as the chicken braises, lubricating the meat and adding some nice richness to the sauce. White meat will just dry out.

One of my favorite recipes uses a splash of white wine, a can of tomatoes, a handful of chopped capers and olives, and a good sprinkle of cilantro. If I'm in the mood, I substitute smoked paprika for the regular kind and sautéed peppers and onions for the tomatoes, omit the capers and olives, and finish it off with parsley and a splash of vinegar. Equally good is a version with more white wine, a couple cups of chicken stock, and plenty of parsley and pancetta. Yet another variation incorporates mushrooms, shallots, and slab bacon. Again, it's the technique that matters—the flavors are up to you.

These braises are even better flavorwise the second or third day, though you do lose some of the crispness in the chicken skin.

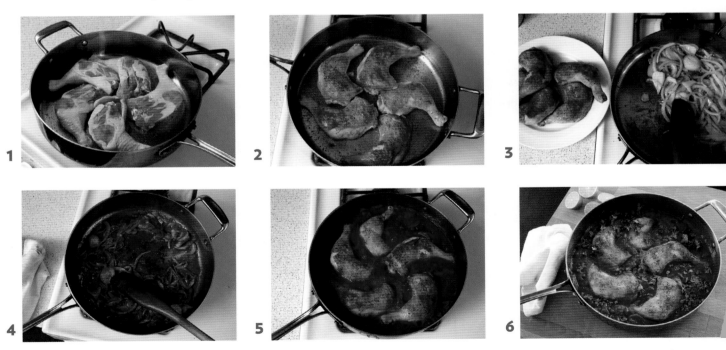

# EASY SKILLET-BRAISED CHICKEN
## WITH TOMATOES, OLIVES, AND CAPERS

**NOTE:** You can make this dish entirely on the stovetop by lowering the heat to the lowest setting when you add the chicken in step 4, covering the pan, and cooking until the chicken is tender, about 45 minutes.

### SERVES 4 TO 6

4 to 6 chicken leg quarters

Kosher salt and freshly ground
  black pepper

1 tablespoon vegetable oil

1 large onion, finely sliced (about
  1½ cups)

2 cloves garlic, finely sliced

1 tablespoon paprika

1 tablespoon ground cumin

1 cup dry white wine

One 28-ounce can whole tomatoes,
  drained and crushed by hand

½ cup homemade or low-sodium
  canned chicken stock

¼ cup capers, rinsed, drained, and
  roughly chopped

¼ cup chopped green or black
  olives

¼ cup fresh cilantro leaves

¼ cup lime juice (from 3 to 4 limes)

1. Adjust an oven rack to the lower-middle position and preheat the oven to 350°F. Lightly season the chicken legs with salt and pepper.

2. Heat the oil in a 12-inch ovenproof skillet or sauté pan over high heat until lightly smoking. Using tongs, carefully add the chicken pieces skin side down. Cover with a splatter screen or partially cover with a lid to prevent splattering and cook, without moving it, until the chicken is deep golden brown and the skin is crisp, about 4 minutes. Flip the chicken pieces and cook until the second side is golden brown, about 3 minutes longer. Transfer the chicken to a large plate and set aside.

3. Reduce the heat under the pan to medium-high, add the onions, and cook, using a wooden spoon to scrape up the browned bits from the bottom of the skillet and then stirring frequently, until completely softened and just starting to brown, about 4 minutes. Add the garlic and cook, stirring, until fragrant, about 30 seconds. Add the paprika and cumin and cook, stirring, until fragrant, about 1 minute longer.

4. Add the white wine and scrape up the browned bits from the bottom of the skillet. Add the tomatoes, chicken stock, capers, and olives and bring to a boil. Nestle the chicken pieces into the stock and vegetables so that only the skin is showing. Cover the pan and transfer to the oven. Cook for 20 minutes, then remove the lid and continue to cook until the chicken is falling-off-the-bone tender; and the sauce is rich, about 20 minutes longer.

5. Stir the cilantro and lime juice into the sauce and season to taste with salt and pepper. Serve immediately.

# EASY SKILLET-BRAISED CHICKEN
## WITH WHITE WINE, FENNEL, AND PANCETTA

**NOTE:** You can make this dish entirely on the stovetop by lowering the heat to the lowest setting when you add the chicken in step 4, covering the pan, and cooking until the chicken is tender, about 45 minutes.

**SERVES 4 TO 6**

4 to 6 chicken leg quarters

Kosher salt and freshly ground black pepper

1 tablespoon vegetable oil

3 ounces pancetta, diced

4 cloves garlic, finely sliced

1 large onion, finely sliced (about 1½ cups)

1 bulb fennel, trimmed and finely sliced (about 1½ cups)

1 large tomato, roughly chopped

1½ cups dry white wine

½ cup Pastis or Ricard

1 cup homemade or low-sodium canned chicken stock

1 bay leaf

¼ cup chopped fresh parsley

2 tablespoons unsalted butter

1 tablespoon lemon juice (from 1 lemon)

1. Adjust an oven rack to the lower-middle position and preheat the oven to 350°F. Lightly season the chicken legs with salt and pepper.

2. Heat the oil in a 12-inch ovenproof skillet or sauté pan over high heat until lightly smoking. Using tongs, carefully add the chicken pieces skin side down. Cover with a splatter screen or partially cover with a lid to prevent splattering and cook, without moving it, until the chicken is deep golden brown and the skin is crisp, about 4 minutes. Flip the chicken pieces and cook until the second side is golden brown, about 3 minutes longer. Transfer the chicken to a large plate and set aside.

3. Add the pancetta to the skillet and cook, stirring frequently, until lightly browned, about 3 minutes. Add the garlic and cook, stirring, until lightly browned, about 1 minute. Add the onions and fennel and cook, using a wooden spoon to scrape up the browned bits from the bottom of the skillet and then stirring frequently, until completely softened and just starting to brown, about 5 minutes.

4. Add the tomato, white wine, and Pastis, and scrape up the browned bits off from the bottom of the skillet. Add the chicken stock and bay leaf and bring to a boil. Nestle the chicken pieces into the stock and vegetables so that only the skin is showing. Cover the pan and transfer to the oven. Cook for 20 minutes, then remove the lid and continue to cook until the chicken is falling-off-the-bone tender and the sauce is rich, about 20 minutes longer. Transfer the chicken to a serving platter.

5. Stir the parsley, butter, and lemon juice into the sauce and season to taste with salt and pepper. Pour the sauce around the chicken and serve immediately.

# KNIFE SKILLS:
## How to Cut Fennel

### Fennel is a generally divisive vegetable.

Crisp, with a distinct anise flavor, it can be overpowering for some people. I prefer my fennel in small doses. Sliced superthin on a mandoline and tossed with citrus supremes and a nice lemony vinaigrette, it's a great winter salad that goes well with sausages, terrines, and other charcuterie. The frilly green fronds that grow out of the top are entirely edible and make a pretty garnish.

With fennel, just like the characters in an episode of *ThunderCats*, the good guys and the bad guys are easy to identify. Look for fennel bulbs that are pale green or white with no discoloration. The first thing you'll notice when fennel is past its prime is browning at the edges of the layers, so check there first. The layers should be tightly packed, and the fronds should be bright green and vigorous (1).

In its whole form, fennel will last for about a week in a loosely closed plastic bag in the vegetable crisper in the refrigerator, but once you cut it, it can brown rapidly, so chop it just before using.

To slice fennel, start by cutting off the thick stalks (they can be reserved for stock) (2). Standing the bulb on its base, split it in half (3). Use the tip of your knife to cut the core out of each half (4)—it should come out easily in a triangular-shaped wedge (5). Thinly slice the fennel from pole to pole (6).

For diced fennel, split and remove the core as for sliced fennel. Cut the fennel into thicker planks (7), then cut across them into dice (8).

1

3

4

5

6

7

8

# EASY SKILLET-BRAISED CHICKEN
## WITH PEPPERS AND ONIONS

**Serve this with egg noodles, spätzle, rice, or boiled potatoes.**

**NOTE:** You can make this dish completely on the stovetop by lowering the heat to the lowest possible setting when you add the chicken in step 4, covering the pan, and cooking until the chicken is tender, about 45 minutes.

### SERVES 4 TO 6

4 to 6 chicken leg quarters

Kosher salt and freshly ground
  black pepper

1 tablespoon vegetable oil

1 large onion, finely sliced (about
  1½ cups)

1 green bell pepper, finely sliced
  (about 1 cup)

1 red bell pepper, finely sliced
  (about 1 cup)

2 cloves garlic, finely sliced

1 tablespoon sweet Spanish smoked
  paprika (or regular paprika)

1 teaspoon dried marjoram (savory
  or oregano will also work just fine)

1 tablespoon all-purpose flour

1 cup dry white wine

3 cups homemade or low-sodium
  canned chicken stock

1 large tomato, peeled, seeded, and
  cut into ½-inch pieces

1. Adjust an oven rack to the lower-middle position and preheat the oven to 350°F. Lightly season the chicken legs with salt and pepper.

2. Heat the oil in a 12-inch skillet ovenproof or sauté pan over high heat until lightly smoking. Using tongs, carefully add the chicken pieces skin side down. Cover with a splatter screen or partially cover with a lid to prevent splattering and cook, without moving it, until the chicken is deep golden brown and the skin is crisp, about 4 minutes. Flip the chicken pieces and cook until the second side is golden brown, about 3 minutes longer. Transfer the chicken to a large plate and set aside.

3. Add the onions and peppers to the skillet and cook, using a wooden spoon to scrape up the browned bits from the bottom of the skillet and then stirring frequently, until completely softened and just starting to brown, about 4 minutes. Add the garlic and cook, stirring, until fragrant, about 30 seconds. Add the paprika, marjoram, and flour and cook, stirring, until fragrant, about 1 minute.

4. Add the white wine and scrape up the browned bits from the bottom of the skillet. Add the chicken stock and tomato and bring to a boil. Nestle the chicken pieces into the stock and vegetables so that only the skin is showing. Cover the pan and transfer to the oven. Cook for 20 minutes, then remove the lid and continue to cook until the chicken is falling-off-the-bone tender and the sauce is rich, about 20 minutes longer. Season the sauce to taste with salt and pepper and serve.

# EASY SKILLET-BRAISED CHICKEN
## WITH MUSHROOMS AND BACON

**NOTE:** You can make this dish entirely on the stovetop by lowering the heat to the lowest setting when you add the chicken in step 4, covering the pan, and cooking until the chicken is tender, about 45 minutes.

**SERVES 4 TO 6**

4 to 6 chicken leg quarters

Kosher salt and freshly ground
  black pepper

1 tablespoon vegetable oil

3 ounces slab bacon, diced

5 ounces button mushrooms,
  cleaned and sliced (about 2 cups)

1 large shallot, finely sliced (about
  ½ cup)

2 cloves garlic, finely sliced

2 teaspoons fresh thyme leaves

1 tablespoon unsalted butter

1 tablespoon all-purpose flour

1 cup dry white wine

3 cups homemade or low-sodium
  canned chicken stock

½ cup heavy cream

2 tablespoons fresh parsley leaves

1. Adjust an oven rack to the lower-middle position and preheat the oven to 350°F. Lightly season the chicken legs with salt and pepper.

2. Heat the oil in a 12-inch ovenproof skillet or sauté pan over high heat until lightly smoking. Using tongs, carefully add the chicken pieces skin side down. Cover with a splatter screen or partially cover with a lid to prevent splattering and cook, without moving it, until the chicken is deep golden brown and the skin is crisp, about 4 minutes. Flip the chicken pieces and cook until the second side is golden brown, about 3 minutes longer. Transfer the chicken to a large plate and set aside.

3. Add the bacon to the skillet and cook, stirring frequently, until lightly browned, about 3 minutes. Transfer to a plate. Add the mushrooms and cook, using a wooden spoon to scrape up the browned bits from the bottom of the pan and then stirring frequently, until the mushrooms give up all of their liquid and start to sizzle, about 8 minutes. Add the shallots and garlic and cook, stirring, until fragrant, about 1 minute. Add the thyme and butter and cook until the butter is melted. Add the flour and cook, stirring constantly, for 30 seconds.

4. Add the white wine and scrape up the browned bits from the bottom of the skillet. Add the chicken stock and bring to a boil. Nestle the chicken pieces into the stock and vegetables so that only the skin is showing. Cover the pan and transfer to the oven. Cook for 20 minutes, then remove the lid and continue to cook until chicken is falling-off-the-bone tender and the sauce is rich, about 20 minutes longer. Transfer the chicken to a serving platter.

5. Add the heavy cream to the sauce and cook, stirring, over high heat until thickened slightly, about 1 minutes. Season the sauce to taste with salt and pepper, stir in the parsley, and pour around the chicken. Serve.

# In SEARCH of
## THE ULTIMATE CHILI

Despite its Mexican origins, today's chili is a decidedly American dish. Even so, like a religion, the world of chili lovers is split into numerous factions, many of whom are prepared to fight to the death over what can and can't be used in "real" chili. Should it be made with ground beef or chunks? Are tomatoes allowed? Should we even mention beans? Since no one is going to agree on all of these factors, we're going to deal with not one, but *two* variations on beef chili—one a traditional Texas-style made with nothing but beef and chiles and the other the type most of us grew up on, with beans and tomatoes, including both a version made with short ribs and one made with ground beef.

Either way, there are a number of things we can all agree on about a good chili:

- It should have a rich, complex flavor that combines sweet, bitter, hot, fresh, and fruity elements in balance.
- It should have a robust, meaty, beefy flavor.
- If it contains beans, the beans should be tender, creamy, and intact.
- It should be bound together by a thick deep-red sauce.

To achieve these goals, I decided to break down chili into its distinct elements—the chiles, the beef, the beans, and the flavorings—and perfect each one before putting them all together in one big happy pot.

## The Chiles

I have bad memories of my chili-eating college days—when chili was made by adding a can of beans and a can of tomatoes to ground beef, then adding one of every spice on the rack (and too much cumin) and simmering it. The finished product inevitably had a totally unbalanced flavor with a powdery, gritty mouthfeel from all the dried spices.

Bottled chili powder is fine in a pinch, but to get the ultimate chili, my first plan of action was to ditch the powdered spices and premixed chili powders and go straight to the source: real dried chiles. Dried either in the sun, over a smoky fire (in the case of smoked chiles like chipotles), or—more commonly these days—in humidity- and wind-controlled rooms, dried chiles have remarkably complex flavors. Just like meat that is aged, as a chile dries, it loses moisture, concentrating its flavorful compounds inside each cell. These compounds come into closer contact with each other, allowing them to react and produce new flavors that weren't present in the fresh pepper.

Dried chiles are available in a baffling array, so to make my selection easier, I decided to taste every variety of whole chile I could find, taking note of both its spice level and its flavor profile. I noticed that most of them fell into one of four distinct categories: sweet and fresh, hot, rich and fruity, or smoky (see "Dried Chiles," page 258).

For my taste, a combination of chiles from the first three categories produced the most balanced mix (the smoky chiles tend to overpower other flavors). Even though chiles are dried, their flavor can dissipate with time, so it's important to get fresh dried chiles. They should have a leathery quality and still be flexible. If a chile splits or cracks when you bend it, move along to a different one. Dried chiles should be stored in a sealed container away from the light (I keep mine in zipper-lock bags in the pantry and used within about 6 months of purchase.

Toasting chiles develops flavor.

As with dried spices, the flavor of the chiles can be enhanced by toasting them dry (see "Whole Versus Ground Spices," page 257). This accomplishes two goals: First, the heat catalyzes reactions among individual compounds within the chili, creating new flavors. Second, the Maillard browning reaction takes place, resulting in hundreds of highly flavorful new compounds.

After toasting, I could go the traditional route and simply combine the chiles and grind them

into a powder, but I'm not one to bow to tradition. Instead, I found that by cooking them down in chicken stock and pureeing the moist chiles, I could create a completely grit-free paste for a concentrated flavor base for my chili. The best part? If I made a double or triple batch, I could freeze the puree in ice cube trays for long-term storage, giving me the convenience of jarred chili powder but vastly better texture and flavor.

I use a chile puree instead of powder for better flavor and texture.

## The Meat

Aside from beans, the meat is the biggest source of contention among chili lovers. Some (like my lovely wife) insist on ground beef, while others (like me) prefer larger stew-like chunks. More often than not, I begrudgingly let my wife have her way, which is why this time I was determined to fight for my own rights—or, at the very least, make her compromise her chili convictions.

After trying store-ground beef, home-ground beef, beef cut into 1-inch chunks, and beef roughly chopped by hand or in a food processor into a textured mix of ⅛- to ½-inch pieces, it was apparent that the last one was the winner. It provided little bits of nearly ground beef that added body and helped keep the chili (and my marriage) well bound while still providing enough large, chunkier pieces to give textural interest.

There are plenty of beef cuts that are great for stewing (see "Stewing Beef," page 241), but for my chili, I decided to go with robust, beefy short rib.

As anyone who's ever cooked ground beef knows, it's nearly impossible to properly brown a large pot of it. It's a simple matter of surface-area-to-volume ratio. Ground beef has tons of surface area liquid and fat can escape from. As soon as you start cooking it, liquid begins pooling in the bottom of the pot, submerging the meat and leaving it to gurgle and stew in its own gray-brown juices, which self-regulate its temperature to 212°F, far too low for flavorful browning to take place. Only after its juices have completely evaporated can any browning action occur. The sad truth? With ground (or, in our case, finely chopped) beef, you have to settle for either dry, gritty meat or no browned flavor.

Trying to brown ground meat is an exercise in futility.

But then I had a thought: why was I bothering trying to brown the beef after I chopped it? If browned flavor in the stew was what I was after, did it even matter *when* I browned the beef as long as it ended up getting browned? I grabbed another batch of short ribs, this time searing them in a hot

pan *before* removing the meat from the bone and chopping it to size.

The result? Chili with chopped-beef texture but deeply browned flavor. *Yippee ki-yay.*

Whole cuts brown rapidly.

Chopping whole cuts after browning gives you browned flavor and superior texture.

## The Beans

If you are from Texas, you may as well skip right to page 266. But if you're like me and believe beans are as integral to a great bowl of chili as the beef, if not more so, read on. To be honest, there's nothing wrong with using canned beans in chili. They are uniformly cooked, hold their shape well, and here the relative lack of flavor compared to cooked dried beans is not an issue. There are enough other flavors going on to compensate. But sometimes the urge to crack some culinary skulls and the desire for some food-science myth-busting is so strong that I can't resist. So, a quick diversion into the land of dried beans.

If you have worked for a chef or have a grandmother from Tuscany or a mother-in-law from South America, you may have at one point been told never to add salt to your beans until they were completely cooked, lest you prevent their tough skins from softening fully. In fact, in some restaurants I worked in, it was thought that overcooked beans *could actually be saved* by salting them after cooking. To think!

But how often have you actually cooked two batches of beans side by side, one soaked and cooked in salted water and the other soaked and cooked in plain water? Chances are, never. And now you'll never have to. I present to you the results of just such a test:

Salting beans can help tenderize their skins, preventing blowouts.

Both batches of beans were cooked just until they were fully softened, with none of the papery toughness of undercooked skins (about 2 hours for both batches, after an overnight soak). As you can clearly see, the unsalted beans (left) ended up absorbing too much water and blowing out long before their skins had properly softened, while the salted beans remained fully intact.

The problem? Magnesium and calcium, two ions found in bean skins that act kind of like buttresses,

supporting the skins' cell structure and keeping them firm. If you soak beans in salted water overnight, though, some of the sodium ions end up playing musical chairs with the calcium and magnesium, leaving you with skins that soften at the same rate as the beans' interiors.

The thing that *does* affect the cooking rate of beans is the pH level. Acidic environments tend to cause beans to seize up—which is why, for example, Boston baked beans, cooked in acidic molasses and tomato, can take as long as overnight to soften properly. Brining the beans in saltwater mitigates this effect to some degree, but the only way to be sure that your beans will cook properly in acidic stews (like, say, chili) is to soften them separately and add them to the pot later on.

So, where does the old anti-salting myth come from? Probably the same place most culinary myths come from: grandmothers, aunts, and chefs. Never trusted 'em, never will.

## The Flavorings

The chili-standard duo of cumin and coriander was a given, as were a couple of cloves. Their medicinal, mouth-numbing quality is a perfect balance for the spicy heat of the chiles, much as numbing Sichuan peppers can play off chiles in the Chinese flavor combination known as *ma-la* (numb-hot).

I also decided to give star anise a try, in a nod to British chef Heston Blumenthal and his treatment of Bolognese sauce. He found that, in moderation, star arise can boost the flavor of browned meats without making its presence known. He was right, as I quickly discovered. For maximum flavor, make sure to toast your spices whole before grinding them (see "Whole Versus Ground Spices," page 257).

All I needed now was the traditional combo of onion, garlic, and oregano, along with some fresh

chiles (for added heat and freshness) and tomatoes. I simmered everything together, added my cooked beans, simmered again, seasoned, and tasted.

So how'd it taste? Great. But not quite worthy of its "Best" title yet. It could still do with some more meatiness. It was time to reach into my Bat utility belt of culinary tricks for the one weapon that has yet to fail me, my umami bombs: Marmite, soy sauce, and anchovies.

These can increase the meatiness of nearly any dish involving ground meat or of stews (see "Glutamates, Inosinates, and the Umami Bombs," page 245). Adding a dab of each to my chile puree boosted my already-beefy short ribs to the farthest reaches of meatiness, a realm where seared skinless cows traipse across hills of ground beef, darting in and out of fields of skirt steak, stopping only to take sips of rivers overflowing with thick *glace de viand*.

I snapped out of my reverie with one thing on my mind: booze. Alcohol has a lower boiling point than water and, even more important, it can actually cause water to evaporate at a lower temperature. You see, water molecules are held loosely together like tiny magnets. When water and alcohol are mixed, each individual water molecule becomes farther away from the other water molecules, making it much easier for it to escape and vaporize. Since water- and alcohol-soluble aromatic molecules can only be detected by your nose if they escape into the air, it stands to reason that the more evaporation occurred, the more aromatic my chili would be.

I added a shot of liquor to my finished chili and gave it a side-by-side sniff test with a boozeless batch. No doubt about it, the alcohol improved its aromatic properties. After a thorough tasting of vodka, scotch, bourbon, and tequila in the name of good science, I came to the conclusion that in chili, *they're all good*.

## WHOLE VERSUS GROUND SPICES

You often hear chefs and recipe writers saying to use whole spices instead of ground and to toast your spices before using them, but you don't often hear about why. To figure out the answer, I made five batches of my Easy Weeknight Ground Beef Chili (page 261), using spices and chiles treated the following ways:

1. **Preground**, straight from the jar
2. **Preground**, toasted before adding
3. **Whole spices and dried chiles**, ground and used without toasting
4. **Whole spices and dried chiles**, ground and then toasted
5. **Whole spices**, toasted and then ground

*continues*

Every batch of chili made with whole spices and chiles—including those that weren't toasted—was superior in flavor to those made with preground spices. Of the two made with preground spices, the toasted version was slightly more complex, and of the three made with whole spices, the one with spices that were toasted before grinding was markedly superior to the one that used ground and then toasted spices. Why was this?

Well, toasting whole spices accomplishes two goals: First, it forces aromatic-compound–laden oils from deep within individual cells to the surface of the spice and the interstitial spaces between the cells. This makes it much easier to extract flavor when the spice is subsequently ground and incorporated into your food. Second, toasting also catalyzes a whole cascade of chemical reactions that produce hundreds flavorful by-products, greatly increasing the complexity of the spice.

When toasting preground spices, this latter reaction will definitely occur, but you've also got a problem: vaporization. The flavorful compounds inside spices are generally quite volatile—they desperately want to escape into the air and fly away. With whole spices, they remain relatively locked down: they can't escape very easily from their cellular prisons. With ground spices, on the other hand, there's nothing holding them back. They'll very rapidly fly into the air. You may have noticed that preground spices become far more aromatic as you toast them. Remember this—if you smell it while you're cooking, it will *not* be in your food when you serve it.

There are rare exceptions to the toast-before-grinding rule. Indian and Thai curries, for example, start with ground or pureed aromatics sautéed in fat. Because most of the aromatic compounds in spices are fat-soluble, they end up dissolved in the fat, flavoring the rest of the dish evenly and easily when other ingredients are added. But for the vast majority of applications, and *any time* you're going to toast your spices dry, make sure to do it before grinding them.

# DRIED CHILES

Dried chiles come in range of flavors and heat levels. To help make selection easier, I've broken them down into a few categories. The ideal chili should combine elements from several of them.

- **Sweet and fresh:** Distinct aromas reminiscent of red bell peppers and fresh tomatoes. These peppers include costeño, New Mexico (aka dried Anaheim, California, or Colorado), and choricero.
- **Hot:** Overwhelming heat. The best, like cascabels, also have some complexity, while others, like pequin or árbol, are all about heat and not much else.
- **Rich and fruity:** Distinct aromas of sun-dried tomatoes, raisins, chocolate, and coffee. Some of the best-known Mexican chiles, like ancho, mulatto, and pasilla, are in this category.
- **Smoky:** Some chiles, like chipotles (smoked dried jalapeños), are smoky because of the way they are dried. Others, like nora or guajillo, have a natural musty, charred-wood smokiness.

# CHILE PASTE

**NOTE**: Chile paste can be used in place of chili powder in any recipe. Use 2 tablespoons paste for every tablespoon powder.

### MAKES 2 TO 2½ CUPS

6 ancho, pasilla, or mulato chiles (about ½ ounce), seeded and torn into rough 1-inch pieces

3 New Mexico red, California, costeño, or choricero chiles (about ⅛ ounce), seeded and torn into rough 1-inch pieces

2 cascabel, árbol, or pequin chiles, seeded and torn in half

2 cups homemade or low-sodium canned chicken stock

1. Toast the dried chiles in a Dutch oven over medium-high heat, stirring frequently, until slightly darkened, with an intense roasted aroma, 2 to 5 minutes; turn down the heat if they begin to smoke. Add the chicken stock and simmer until the chiles have softened, 5 to 8 minutes.

2. Transfer the chiles and liquid to a blender and blend, starting on low speed and gradually increasing the speed to high and scraping down the sides as necessary, until a completely smooth puree if formed, about 2 minutes; add water if necessary if the mixture is too thick to blend. Let cool.

3. Freeze the chile paste in ice cube trays, putting 2 tablespoons into each well. Once the cubes are frozen, transfer to a zipper-lock freezer bag and store in the freezer for up to 1 year.

# THE BEST SHORT-RIB CHILI
## WITH BEANS

**Serve the chili with grated cheddar cheese, sour cream, chopped onions, scallions, sliced jalapeños, diced avocado, and/or chopped cilantro, along with corn chips or warmed tortillas.**

**NOTE**: Canned beans can be used in place of dried. Use three 15-ounce cans red kidney beans, drained, and add them at the beginning of step 5. Or omit the beans entirely.

**SERVES 8 TO 12**

5 pounds bone-in beef short ribs
(or 3 pounds boneless short ribs
or chuck), trimmed of silverskin
and excess fat

Kosher salt and freshly ground
black pepper

2 tablespoons vegetable oil

1 large yellow onion, finely diced
(about 1½ cups)

1 jalapeño or 2 serrano chiles, finely
chopped

4 medium cloves garlic, minced or
grated on a Microplane (about 4
teaspoons)

1 tablespoon dried oregano

1 cup Chile Paste (page 259) or ½
cup chili powder

4 cups homemade or low-sodium
canned chicken stock

4 anchovy fillets, mashed into a
paste with the back of a fork

1 teaspoon Marmite

1 tablespoon soy sauce

2 tablespoons tomato paste

2 tablespoons cumin seeds, toasted
and ground

2 teaspoons coriander seeds,
toasted and ground

1 tablespoon unsweetened cocoa
powder

2 to 3 tablespoons instant cornmeal
(such as Maseca)

2 bay leaves

1 pound dried red kidney beans,
soaked in salted water at room
temperature for at least 8 hours,
preferably overnight, and drained

1. Season the short ribs on all sides with salt and pepper. Heat the oil in a large Dutch oven over high heat until smoking. Add half of the short ribs and brown well on all sides (it may be necessary to brown the ribs in three batches, depending on size of the Dutch oven—do not crowd the pot), 8 to 12 minutes; reduce the heat if the fat begins to smoke excessively or the meat begins to burn. Transfer to a large platter. Repeat with remaining short ribs, browning them in the fat remaining in the Dutch oven.

2. Reduce the heat to medium, add the onion, and cook, scraping up the browned bits from the bottom of the pot with a wooden spoon and then stirring frequently, until softened but not browned, 6 to 8 minutes. Add the fresh chiles, garlic, and oregano and cook, stirring, until fragrant, about 1 minute. Add the chile paste and cook, stirring and scraping constantly, until it leaves a coating on the bottom of the pot, 2 to 4 minutes. Add the chicken stock and scrape up the browned bits from the bottom of the pot. Add the anchovies, Marmite, soy sauce, tomato paste, ground spices, cocoa, and cornmeal and whisk to combine. Keep warm over low heat.

3. Adjust an oven rack to the lower-middle position and preheat the oven to 225°F. Remove the meat from the bones, if using bone-in ribs, and reserve the bones. Chop all the meat into rough ¼- to ½-inch pieces (or finer or larger if you prefer). Add any accumulated meat juices from the the cutting board to the Dutch oven, then add the chopped beef, beef bones, if you have them, and bay leaves and bring to a simmer. Cover and place in the oven for 1 hour.

4. Meanwhile, transfer the drained beans to a pot and cover with water by 1 inch. Season to taste with salt. Bring to a boil over high heat, then reduce to a simmer and cook until the beans are nearly tender, about 45 minutes. Drain.

5. Remove the chili from the oven and add the tomatoes, vinegar, and beans. Return to the oven with the lid slightly ajar and cook until the beans and beef are tender and the stock is rich and slightly thickened, 1½ to 2 hours longer; add water if necessary to keep the beans and meat mostly submerged (a little protrusion is OK).

6. Using tongs, remove the bay leaves and bones (any meat still attached to the bones can be removed, chopped, and added to the chili if desired). Add the whiskey, if using, hot sauce, and brown sugar and stir to combine. Season to taste with salt, pepper, and/or vinegar.

One 28-ounce can crushed
   tomatoes
¼ cup cider vinegar, or more to
   taste
¼ cup whiskey, vodka, or brandy
   (optional)
2 tablespoons Frank's RedHot or
   other hot sauce
2 tablespoons dark brown sugar
Garnishes as desired (see the
   headnote)

7. Serve, or, for the best flavor, allow the chili to cool and refrigerate overnight, or up to 5 days in an airtight container, then reheat. Serve with
   some or all of the suggested garnishes and corn chips or tortillas.

# EASY WEEKNIGHT GROUND BEEF CHILI

**Sometimes even the best of us don't feel like going all out. Here's a much quicker weeknight chili that uses a few of the tricks learned from my Best Short-Rib Chili (page 259) and the 30-minute bean soup recipes (pages 196–200). I use ground beef here, which precludes real browning as an option, but a couple of smoky chipotle chiles added to the mix lend a similar sort of deep complexity. Serve the chili with grated cheddar cheese, sour cream, chopped onions, scallions, sliced jalapeños, diced avocado, and/or chopped cilantro, along with corn chips or warmed tortillas.**

**SERVES 4 TO 6**

4 tablespoons unsalted butter

2 medium onions, grated on the large holes of a box grater (about 1½ cups)

2 large cloves garlic, minced or grated on a Microplane (about 4 teaspoons)

1 teaspoon dried oregano

Kosher salt

2 chipotle chiles packed in adobo, finely chopped

2 anchovy fillets, mashed to a paste with the back of a fork

½ cup Chile Paste (page 259) or ¼ cup chili powder

1 tablespoon ground cumin

½ cup tomato paste

2 pounds boneless ground chuck

One 28-ounce can whole tomatoes, drained and chopped into ½-inch pieces

One 15-ounce can red kidney beans, drained

1 cup homemade or low-sodium canned chicken stock, or water

2 to 3 tablespoons instant cornmeal (such as Maseca)

2 tablespoons whiskey, vodka, or brandy (optional)

Freshly ground black pepper

Garnishes as desired (see the headnote)

1. Melt the butter in a large Dutch oven over medium-high heat. Add the onions, garlic, oregano, and a pinch of salt and cook, stirring frequently until the onions are light golden brown, about 5 minutes. Add the chipotles, anchovies, chile paste, and cumin and cook, stirring, until aromatic, about 1 minute. Add the tomato paste and cook, stirring until homogeneous, about 1 minute.

2. Add the ground beef and cook, using a wooden spoon to break up the beef into pieces and stirring frequently, until no longer pink (do not try to brown the beef), about 5 minutes. Add the tomatoes, beans, stock, and cornmeal and stir to combine. Bring to a boil, reduce to a simmer, and cook, stirring occasionally, until the flavors have developed and the chili is thickened, about 30 minutes.

3. Stir in the whiskey, if using. Serve with some or all of the suggested garnishes, along with corn chips or tortillas.

# VEGETARIAN CHILI

## Why does vegetarian chili get such a bum rap?

I mean, there's the obvious: chili is a divisive issue, especially among those who love chili. But hey, guess what? Beans can taste good in chili. Tomatoes can taste good in chili. Heck, even pork and tomatillos can taste good in chili. So why shouldn't we be able to make a completely meatless version that tastes great as well?

I've seen a few decent vegetarian chilis in my lifetime, but for some reason, they all seem to fall into the "30 minutes or less" camp. That in and of itself is not a bad thing—vegetarian chili as a general rule doesn't need to be cooked as long as meat-based chilis because vegetables, especially canned beans, tenderize faster than meat—but long, slow cooking nets you benefits in the flavor development. Fast chili recipes are inevitably not quite as rich and complex as you'd like them to be. My goal was to create a 100-percent vegetarian (actually, it's vegan) chili that has all of the deep flavor, textural contrast, and rib-sticking richness that the best chili should have.

First things first: faux meat is not in the picture. I want my vegetarian chili to celebrate vegetables and legumes, not to try and imitate a meaty chili. With that out of the way, we'll move on to the second thing: great chili has to start with great chiles. That's what it's all about. I've seen recipes calling for just a couple tablespoons of prefab chili powder for an entire pot of beans and tomatoes. The only way to achieve great flavor is to blend up the chiles yourself, starting with whole dried chiles. And we've already got a great recipe for a complex chile paste devised for the meat chili, so why not use it here as well?

Next up, the beans. For me, a great chili has to show some character and diversity. You don't want completely uniform beans in every bite, you want a range of textures. Here's where you've got to make some creative choices.

Many vegetarian chiles take the kitchen-sink, big-car-compensation approach: *Hey, we can't use beef,* is the apparent thought process, *so let's throw in every damn type of bean and vegetable imaginable.* That method definitely gets you textural as well as flavor variety, but it can become a bit too jumbled. Better to make a couple of well-balanced choices and focus on perfecting them.

Kidney beans are a must in my chilis; I grew up with kidney beans in my chili, and I will continue to enjoy them in my chili. You, on the other hand, are free to substitute whatever type of bean you want. There's certainly something to be said for dried beans, and I do sometimes opt to brine dried beans overnight to make chili 100 percent from scratch, but canned beans are a sure thing. They're never over- or undercooked, they're never bloated or busted. They are lacking in the flavor department, but with a good simmer in a very flavorful liquid, you can easily make up for this (see "How to Make Canned Beans Taste Good," page 195). And the great thing is that the liquid base for chili is naturally low in pH (both the chiles and the tomatoes are acidic), and beans and vegetables soften very slowly in acidic liquid. This means you can simmer your canned beans for a significant period of time in your chili before they really start to break down.

But what about more texture? I tried using a mixture of kidney beans and other smaller beans and grains (chickpeas, flageolets, barley), but the real key turned out to be using the food processor. By pulsing a couple cans of chickpeas in the food processor, I was able to roughly chop them into a mixture of big chunks and tiny pieces. Adding this to my chili gave it great body and a ton of textural contrast.

## Amping Up Flavor

The key to rich flavor in vegetable chili is two-fold: first, a long simmer during which water is driven off so that flavors are concentrated and various volatile compounds break down and then recombine to add complexity, and second, a good source of glutamic acid, the chemical responsible for the flavor we recognize as savory, sometimes called umami. So, time to reach for the umami bombs again (see page 245). Anchovies are out of the picture for obvious reasons, but a touch of Marmite and soy sauce adds a ton of richness to the chili. Other than that, the flavor base is pretty straightforward: onions sweated in a little vegetable oil, garlic, oregano (the dried stuff is fine for long-cooking applications like this), and a couple of canned chipotle chiles in adobo sauce to add a touch of smokiness and heat, compensating nicely for the lack of browned beef.

Finally, as we saw with our meaty Short-Rib Chili (page 259), there are certain aromas that are carried well with steam, while others are actually carried better via vaporized alcohol. My chili has got plenty of liquid in it, so the steam bit's covered. Adding a couple shots of booze just before serving takes care of the rest. I like bourbon or whiskey, because I've usually got it around, but Cognac, tequila, or vodka will work well. Just make sure that it's at least 80 proof (40 percent alcohol by volume)—and unsweetened.

The truth of the matter is that the key to great vegetarian chili is to completely forget that you're working on a vegetarian chili. Chili greatness lies in the careful layering of real chiles, textual contrast in each bite, and a rich, thick consistency packed with savory flavor. Whether it's made with beans, beef, pork, or ground yak hearts, for that matter, if you get the basics right, you're off to a good start.

# THE BEST VEGETARIAN BEAN CHILI

**Serve with grated cheddar cheese, sour cream, chopped onions, scallions, sliced jalapeños, diced avocado, and/or chopped cilantro, along with corn chips or warmed tortillas.**

### SERVES 6 TO 8

Two 14-ounce cans chickpeas (with their liquid)

One 28-ounce can whole tomatoes

1 cup Chile Paste (page 259) made with water instead of chicken stock

2 chipotle chiles in adobo sauce, finely chopped, plus 2 tablespoons of the sauce

2 tablespoons vegetable oil

1 large onion, finely diced (about 1½ cups)

3 medium cloves garlic, minced or grated on a Microplane (about 1 tablespoon)

tablespoons ground cumin

2 teaspoons dried oregano

1 tablespoon soy sauce

1 teaspoon Marmite

Two 14-ounce cans red kidney beans, drained and liquid reserved

2 tablespoons vodka, bourbon, tequila, or Cognac

Kosher salt

2 to 3 tablespoons instant cornmeal (such as Maseca)

Garnishes as desired (see the headnote)

1½ tablespoons ground cumin

1. Drain the chickpeas, reserving the liquid in a medium bowl. Transfer the chickpeas to a food processor and pulse until just roughly chopped, about three 1-second pulses. Set aside.

2. Add the tomatoes, with their juice, to the chickpea liquid and use your hands to break the tomatoes up into rough chunks, about ½ inch each. Add the chile paste and chipotles, along with their sauce, and stir to combine.

3. Heat the oil in a Dutch oven over medium-high heat until shimmering. Add the onion and cook, stirring frequently, until softened but not browned, about 4 minutes. Add the garlic, cumin, and oregano and cook, stirringly, until fragrant, about 30 seconds. Add the soy sauce and Marmite and cook, stirringly, until fragrant, about 30 seconds. Add the tomato mixture and stir to combine.

4. Stir in the chickpeas and kidney beans. If necessary, add some of reserved bean liquid so the beans are just barely submerged. Bring to a boil over high heat, reduce to a bare simmer, and cook, stirring occasionally, until thick and rich, about 1½ hours; add more bean liquid as necessary if the chili becomes too thick or sticks to the bottom of the pot.

5. Add the vodka and stir to combine. Season to taste with salt and whisk in the cornmeal in a slow, steady stream until the desired thickness is reached. Serve, or, for best results, allow the chili to cool and refrigerate at least overnight, or up to a week, then reheat to serve.

6. Serve with some or all of the suggested garnishes, along with corn chips or tortillas.

# TEXAS CHILI CON CARNE

**OK, folks, this is it.**

The real deal. For all of you Texans who turned your noses up or rolled your eyes at my bean-filled chili recipe, I will make it up to you. This recipe is real-deal chili con carne, old-school Texas-style. What does that mean? First of all, absolutely no beans. And no tomatoes. Indeed, there's very little that goes into the pot other than beef and chiles (and plenty of both!). That doesn't mean there aren't a few things to discuss, however. Let's get to it.

## The Meat

The original chili was made with dried beef pounded together with suet and dried chiles into a sort of pemmican-like dry mix intended to last a long time and be quick and nutritious for cowboys to rehydrate and stew up out on the range. These days, we've got refrigerators and fresh meat. So we

use them. What we're looking for here is a meat that's good for stewing—that is, rich in connective tissue and fat and high in flavor.

In general, beef falls across a spectrum of tenderness, with the relatively bland but tender cuts on one end and the very flavorful but tough cuts on the other. These cuts generally correspond with the muscles that the steers use least to most during their life: So, on the far left side would be relatively unworked muscles like tenderloin or loin cuts (strip steak, porterhouse, etc.)—very tender but relatively flavorless. On the other end of the spectrum are hardworking muscles like short ribs, shin, oxtail, and chuck (shoulder). Chuck is the ideal stew cut, with great flavor, a good amount of fat, and plenty of connective tissue in one well-balanced package. As the meat slowly cooks down in flavorful liquid, all of that connective tissue—mostly composed of the protein collagen—breaks down into rich gelatin, which is what gives good stewed beef its luxurious texture.

## Cut and Sear

The chili of my youth was made with ground beef, which is key if you don't want to spend the time to properly stew your meat. Grinding beef shortens its fibers, making it far more tender, and ground beef chili can be ready to eat in under an hour. But that's not what we're after here. Real Texas chili is made with big chunks of meat and requires long, slow stewing. I played around with a few different sizes and settled on 2-inch chunks (they shrink to about an inch and a half after cooking). I like having to shred a large cube of beef apart with my spoon before eating it, if only to remind myself how perfectly tender the meat has become.

As for searing, as we know, there's always a trade-off. Searing helps develop nice browned flavors via the Maillard reaction, but it also results in tougher, drier meat. At the high temperatures required for browning, meat muscle fibers contract greatly and expel so much liquid that even after a long simmer in the pot, the edges of the meat cubes are relatively dry. I vastly prefer the softer texture of unseared meat.

The solution? You could sear large chunks of meat as I did for the Short-Rib Chili with Beans (page 259), but here's another solution: brown only half the meat. You develop plenty of browned flavor but retain good texture in the rest of the meat. Worried that the flavor will be concentrated only in the meat that you sear? Don't sweat it. Most of those flavorful compounds are water-soluble, meaning that there's plenty of time for them to dissolve and distribute themselves throughout the chili as it cooks.

We already know how to make the most of chiles, so I won't bang your head against it again. OK, maybe one last time:

1. Use fresh dried chiles, not chili powder.
2. Toast the chiles, then simmer them in liquid and, finally, blend them to prevent any grittiness in the final product.

And . . . that's about it. Beef, chiles, and time are all it takes. I occasionally add an onion and perhaps a few cloves of garlic that I sauté in the pot after braising the beef. If I'm feeling feisty, I may also add a few spices from the rack: cumin, cinnamon, allspice, a bit of dried oregano—all are good in small quantities, but totally optional (Texans, please don't kill me!).

The only question left is how to stew the meat. Ideally, you want to cook the meat at as low a temperature as possible (to avoid causing undue muscle fiber contraction) while still softening its connective tissues. The easiest way to to this is to use a very large, heavy pot with plenty of surface area for evaporation (this helps limit the chili's maximum temperature) and to use as low a flame as possible on the stovetop or, better yet, put the pot into a low-temperature (200° to 250°F is good) oven, which will heat more gently and evenly than a burner.

Leaving the lid slightly ajar reduces vapor pressure on the surface of the stew, which can also limit its upper temperature. With a heavy lid, stew temperatures can push up to 212°F. Leave that same lid slightly cracked, and your stew will stay closer to 190° or even 180°F—much better. Even slow-cooked meat can be overcooked, so you want to carefully monitor your chili and pull it off the heat just when the meat becomes tender. This usually takes 2½ to 3 hours.

Then you can leave the chili as is, but I like to thicken mine with a bit of cornmeal.

Like any good marriage, the marriage between beef and chili gets better and more intimate with time. Let the chili sit overnight in the fridge and it'll taste even better the next day. I promise, it's worth the wait. Meaty? Check. Hot, rich, complex chili flavor? Check. And that's really all Texas chili needs. A sprinkle of cilantro, sliced scallions, and perhaps some grated cheese (I like cotija, but Jack, Colby, and cheddar will all work fine) make for good accompaniments. As do warm tortillas or corn chips. As does some good beer or whisky. And, fine, if you'd like, you can go ahead and add a can of beans. Just don't tell anyone I told you to.

# KNIFE SKILLS:
## Trimming Chuck

**Chuck is a versatile and inexpensive muscle, but it's also got a lot of big swaths of fat and connective tissue. When ground for burgers, this is not such a bad thing, but nobody wants to bite into a big chunk of soft fat in their chili, so it's important to trim it off.**

- **ROTATE THE ROAST** until you find its major seam (depending on exactly what part of the chuck you get, it may have more than one). The meat should pull apart easily at this seam; if necessary, use the tip of a sharp butcher's or chef's knife to nick any stubborn connective tissue.

- **TO TRIM A WHOLE CHUCK ROAST**, start by dividing it in half. This will make it less clumsy to maneuver on your cutting board.

- **ONCE YOU'VE SEPARATED** the chuck into single large-muscle groups, all of the fat and connective tissue should be exposed. Use your sharp knife to trim them, then discard and cut the chuck into pieces as desired.

# REAL TEXAS CHILI CON CARNE

**Serve with grated cheese, sour cream, chopped onions, chopped scallions, sliced jalapeños, diced avocado, and/or chopped cilantro, along with corn chips or warmed tortillas.**

### SERVES 6 TO 8

4 pounds boneless beef chuck, trimmed of gristle and excess fat and cut into 2-inch chunks

Kosher salt and freshly ground black pepper

2 tablespoons vegetable oil

1 large onion, finely diced

4 medium cloves garlic, minced or grated on a Microplane (about 4 teaspoons)

1 tablespoon ground cumin

½ teaspoon ground cinnamon (optional)

¼ teaspoon ground allspice (optional)

2 teaspoons dried oregano

1 cup Chile Paste (page 259)

2 quarts homemade or low-sodium canned chicken stock

2 to 3 tablespoons instant cornmeal (such as Maseca)

Garnishes as desired (see the headnote)

1. Season half of the meat with salt and pepper. Heat the oil in a large Dutch oven over high heat until smoking. Add the seasoned meat and cook, without moving it, until well browned on bottom, about 6 minutes. Transfer the meat to a large bowl, add the uncooked meat, and set aside.

2. Return the Dutch oven to medium-high heat, add the onion, and cook, stirring frequently, until softened but not browned, about 6 minutes. Add the garlic, cumin, cinnamon and allspice, if using, and oregano and cook, stirring, until fragrant, about 1 minute.

3. Add the meat, along with the chile paste and chicken stock, and stir to combine. Bring to a boil over high heat, then reduce to a simmer, cover, leaving the lid just barely ajar, and cook, stirring occasionally, until the meat is completely tender, 2½ to 3 hours. (Alternatively, the chili can be cooked in a 200° to 250°F oven with the lid of the Dutch oven slightly ajar.)

4. Season the liquid to taste with salt and pepper. Whisk in the cornmeal in a slow, steady stream until the desired thickness is reached. Serve, or, for best results, allow the chili to cool overnight, then reheat the next day to serve. Serve with some or all of the suggested garnishes, along with corn chips or tortillas.

# CHILE VERDE

**We all know that chili is thick, rich, spicy, meaty, complex, and red, right?**

But what about green chili, that equally complex, fresher, porkier cousin common to many Southwestern states? The most basic and hardcore version of New Mexico chile verde is made by simmering rich cuts of pork in a thick stew of roasted Hatch chiles, onions, garlic, salt, and little else. As the meat is braised until tender, the broth picks up the flavor of the melted pork fat along with the uniquely sweet and bitter taste of the chiles, made smoky from roasting until nearly blackened.

Grown in the town of Hatch (population: ~2,000) in southern New Mexico, these chiles provide a complex backbone that few other single ingredients can. I don't spend much time in New Mexico—I have a thing about heat and dream catchers—and fresh Hatch chilies rarely make their way to the Northeast, which leaves me with canned or frozen chiles. But neither of these two roast particularly well, and that smoky char is the best part of green chili. Luckily, I also don't place much credence in authenticity. I'll settle for delicious. I live far enough away from New Mexico that, hopefully, I'll be able to see the dust trail roused up by the violently inclined green chili fanatics approaching and beat a hasty retreat.

## The Peppers

It's certainly possible to get some form of Hatch chile to your door, no matter where you live. The internet is rife with Hatch chile distributors, promising to ship you authentic canned or frozen peppers straight from the source. The problem with these is not really a texture or flavor problem—it's that canned or previously frozen chiles are downright impossible to char properly before stewing them.

They're simply too wet. And preroasted canned or frozen chiles don't have the deep flavor of home-roasted chilies, which left me with one option: find some suitable substitutes.

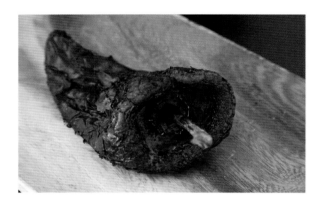

Poblano peppers were an obvious place to start. They're readily available, and they have a deep, earthy flavor. To add a bit of brightness and some of those characteristic bitter notes to the mix, I also added a few cubanelle peppers. A couple of jalapeños, with their heat and grassiness, rounded things out. If you want to go for full-on smoke, you can light a fire in an outdoor grill and roast your chiles over the glowing embers until completely blackened all over. For us apartment dwellers, roasting them over an open gas flame or under a broiler works just as well. The goal here is total carbonization of the exterior. As it heats, the liquid just under the surface converts to steam, forcing the skin outward and away from the flesh. This small area of air and water vapor just under the skin insulates the flesh underneath, preventing it from burning. After the peppers are completely blackened, the loosened skin slips right off, leaving the flesh clean and uncharred but infused with deep smoky flavor from the charred skin.

## The Tomatillos

Even among the less zealous chiliheads out there, tomatillos are a point of contention. Although tomatillos are a member of the same family as tomatoes (though completely different from unripe green tomatoes), it's actually most closely related to gooseberries. The flavors of the two—dominated by a citrus-like tartness with a distinct, savory finish—are remarkably similar as well, though gooseberries tend to be sweeter. One of the great things about tomatillos is that they are also quite high in pectin, the sugar-based jelling agent that is the primary thickener in most jellies. If you include tomatillos, you don't really need any other thickener in your chili (many classic tomatillo-less recipes call for flour or another starch), and the tartness it brings to the party is a welcome flavor addition as well. I char mine under the broiler to maximize that smoky aroma.

It's in recipes like these when the importance of every detail really comes out: Charring every surface of the chiles to maximize smokiness. Carefully monitoring your roasting tomatillos so that they char and soften while still retaining some of their fresh acidity. Sometimes the process of slowly, deliberately building flavors is just as rewarding as the finished dish.

## The Pork

Red chili is all about beef, but with green chili, pork is king. I tried a few different cuts, including sirloin, belly, country-style ribs, and shoulder. The shoulder and sirloin fared the best, maintaining moisture and favor throughout the cooking. The belly was simply too fatty, and while parts of the ribs were great, other sections cut nearer to the lean loin of the pig were much too dry. Pork shoulder requires a bit more work than sirloin to get ready for stewing (boning, trimming away excess fat), but it's significantly cheaper, which gives it an edge in my book.

In my previous chili experiments, I'd found that browning small pieces of meat is a very inefficient method, and that it's much better to brown whole cuts and cut them up afterward. With shoulder, that's a little more difficult—you basically have to dismantle the whole thing to clean and bone it properly. My solution? Use the same method I use for the Texas chili con carne: rather than browning all of the pork, brown only half of it, but allow the pot to develop a rich, deep brown *fond* before adding the onions and the rest of the pork. The coloring built up by the first batch of pork is more than adequate to give the finished dish a rich, meaty flavor. On top of that, the tender texture of the unbrowned pork is far superior to that of the browned stuff.

## Stovetop Versus Oven

The only question remaining was how to cook the dish. Oftentimes, for short-simmered sauces, I'll do 'em directly on the stovetop, just keeping an eye on them as they cook to prevent them from burning. However, for braised dishes that need to be cooked for upward of 3 or 4 hours, the oven shows a couple of distinct advantages.

On the stovetop, the stew cooks only from the

bottom, which can lead to food burning on the bottom of the pot if you aren't careful. An oven mitigates this by heating from all sides at the same time. Moreover, a gas or electric flame set at a certain heat level is a *constant-energy-output* system, meaning that at any given time, it is adding energy to the pot above it at a set rate. An oven, on the other hand, is a *constant-temperature* system. That is, it's got a thermostat that controls the temperature of the air inside, adding energy only as needed in order to keep the temperature in the same basic range. That means that whether you are cooking a giant pot of stew or the Derek Zoolander Stew for Ants, it'll cook at the same rate from beginning to end. For these reasons, it's better to use the oven for long, thick braises. (For more on this, see "Pot Roast," page 239.)

Here's another question we touched on earlier (see page 242): lid on or lid off? If you believe the classical wisdom, you want your lid on as tightly as possible, in order to preserve moisture. And more moisture in the pot means more moisture in the meat, right? Unfortunately, that's not really how braising works. Basically, when you are braising, there are two counteracting forces that you need to balance.

Collagen breakdown—the conversion of tough connective tissue into soft gelatin—begins slowly at around 140°F and increases at an exponential rate as the temperature goes up. Pork shoulder cooked at 140°F might take 2 days to fully soften, while at 180°F, the time is cut down to a few hours. On the other hand, muscle fibers tighten and squeeze out moisture as they are heated, beginning at around 130°F and getting worse as the temperature rises. Unlike collagen breakdown, which takes both time and heat to take place, the muscle squeezing happens almost instantaneously—meat that has been heated to 180°F for even one second will be wrung dry.

Like an unstoppable force meeting an immovable object, it's nearly impossible to achieve collagen breakdown without simultaneously squeezing muscle fibers. The good news: the gelatin created by collagen breakdown goes a long way to mitigating the drying effects of tightening muscle fibers. But the real key to a well-cooked braise is to cook it at a low temperature so that the meat doesn't enter the so-tight-even-gelatin-can't-save-it range.

Now we all know that water boils at 212°F, right? But, interestingly enough, even in a 250°F oven, you can greatly affect the temperature of the water inside a pot by allowing evaporation or not. This is because evaporation, the act of converting water into steam, takes so much energy—it takes more than five hundred times as much energy to convert one gram of water into steam as it does to raise that same gram of water's temperature by one degree!

Check out this graph demonstrating the braising temperature of a pot with the lid sealed versus that with the lid left slightly ajar:

## BRAISING TEMPERATURE VERSUS TIME

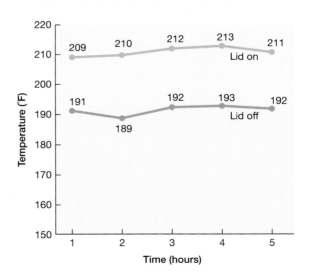

With the lid closed, the liquid inside the pot hovers around boiling temperature, sometimes even rising above it (the pressure from the tightly sealed heavy lid allows the liquid to go above its regular boiling point, just as in a pressure cooker). So the meat ends up overcooking, and by the time it is tender, it's also relatively dry and stringy. The uncovered pot, on the other hand, stays a good 20 degrees lower, keeping the meat inside at a temperature far closer to the ideal. The meat gets tender and retains moisture as it cooks, giving us a far juicer, more tender end result.

Don't you just hate that? You try your hardest to be cool and just write about the joys of slow-cooking pork, then a graph has to go and sneak its way into the works. Graphs are always doing that to me. My apologies. (It is The Food Lab, after all.) Anyhow, at least now you understand exactly why your pork chile verde is so deeply flavored, succulent, and complex.

# CHILE VERDE
## WITH PORK

**Serve with diced onions, sour cream, grated cheese, chopped cilantro, and lime wedges, along with warm tortillas.**

**SERVES 4 TO 6**

3 pounds trimmed boneless pork shoulder, cut into 1-inch cubes

Kosher salt

5 poblano peppers

5 cubanelle peppers

2 pounds tomatillos (about 15 medium), husks removed, rinsed, and patted dry

6 cloves garlic, not peeled

2 jalapeño peppers, stems removed and split lengthwise in half

3 tablespoons vegetable oil

2 cups loosely packed fresh cilantro leaves

1 large onion, finely diced (about 1½ cups)

1 tablespoon ground cumin

4 cups homemade or low-sodium canned chicken stock

Garnishes as desired (see the headnote)

1. Toss the pork with 2 tablespoons salt in a large bowl until thoroughly coated. Set aside at room temperature for 1 hour.

2. Meanwhile, roast the poblano and cubanelle peppers by placing them directly over the flame of a gas burner, turning occasionally, until deeply charred on all surfaces, about 10 minutes. Or, if you don't have a gas burner, you can achieve similar results under the broiler or on an outdoor grill. Place the roasted peppers in a bowl, cover with a large plate, and let steam for 5 minutes.

3. Peel the peppers under cool running water, discard the seeds and stems, and pat dry, then roughly chop. Transfer to a food processor.

4. Preheat the broiler to high. Toss the tomatillos, garlic, and jalapeños with 1 tablespoon of the vegetable oil and 1 teaspoon salt. Transfer to rimmed baking sheet lined with foil and broil, turning once halfway through cooking, until charred, blistered, and just softened, about 10 minutes. Transfer to the food processor, along with any juices. Add half of the cilantro and pulse the mixture until roughly pureed but not smooth, about 8 to 10 short pulses. Season to taste with salt and pepper.

5. Adjust an oven rack to the middle position and preheat the oven to 225°F. Heat the remaining 2 tablespoons oil in large Dutch oven over high heat until smoking. Add half of the pork and cook, without moving it, until well browned, about 3 minutes. Stir and continue cooking, stirring occasionally, until well browned on all sides. Add the remaining pork and the onion and cook, stirring frequently and scraping up any browned bits from the bottom of the pot, until the onions are softened, about 4 minutes. Add the cumin and cook, stirring until fragrant, about 1 minute.

6. Add the chicken stock and pureed chile mixture and stir to combine. Bring to a boil, cover, and transfer to the oven, leaving the lid slightly ajar. Cook until the pork shreds easily with a fork, about 3 hours.

7. Return the pot to the stovetop and skim off and discard any fat. Adjust the liquid to the desired consistency by adding water or boiling it to reduce. Stir in the remaining cilantro and season to taste with salt. Serve, or for best results, let the chili cool and refrigerate overnight, then reheat to serve. Serve with some or all of the suggested garnishes, along with warm tortillas.

# EASY HAM, BEAN, AND KALE STEW

**Once I realized what a difference soaking beans in salted water overnight makes in terms of their texture, I ended up cooking more beans than I could possibly hope to shake a stick at. Here's another of my favorite bean-based dishes. Serve it with sliced crusty bread that's been brushed with olive oil and toasted.**

## SERVES 8 TO 10

1 pound dried white beans, such as great northern, cannellini, or Tarbais, soaked in salted water at room temperature for at least 8 hours, preferably overnight, and drained

1 pound leftover smoked ham bones, scraps, and/or meat

3 quarts homemade or low-sodium canned chicken stock, or water

1 large onion, split in half

1 large clove garlic

3 bay leaves

Kosher salt

1 large bunch kale, stems trimmed (about 2 quarts loosely packed)

Freshly ground black pepper

Extra-virgin olive oil

Sherry vinegar

1. Combine the beans, ham, stock, onion, garlic, bay leaves, and 1 teaspoon salt in a large Dutch oven and bring to a boil over high heat. Reduce the heat to medium-low and simmer until the beans are completely tender, about 45 minutes, adding water as necessary to keep the beans covered. Remove from the heat, remove the ham, and set aside. Discard the onion, garlic, and bay leaves.

2. When the ham is cool enough to handle, shred the meat into small pieces and return to the pot; discard the bones. Add the kale to the pot, bring to a vigorous simmer, and cook, stirring occasionally, until some beans have broken down completely, the liquid is reduced to a thick stew-like consistency, and the kale is completely tender, about 30 minutes.

3. Season the beans to taste with salt and pepper. Serve in shallow bowls, drizzling each one with plenty of extra-virgin olive oil and a sprinkle of sherry vinegar. Serve.

STEAKS, CHOPS,
CHICKEN, FISH, AND

THE

SCIENCE

*of*

FAST-COOKING
FOODS

3

A steak is mostly water, but it's the fat that counts.

# STEAKS, CHOPS, CHICKEN, FISH, AND THE
# SCIENCE *of* FAST-COOKING FOODS

## RECIPES IN THIS CHAPTER

**QUESTION: WHAT DO MCDONALD'S AND THE MOST EXPENSIVE STEAK HOUSE IN NEW YORK HAVE IN COMMON?**

**A**NSWER: They're both fast-food restaurants.

Wait, what? How so? Well, both specialize in *fast* cooking methods. That is, cooking methods designed simply to bring cuts of meat to the desired serving temperature while perhaps adding some nice browned flavor to the exterior. Fast-cooking methods include pan-searing, grilling, broiling, and sautéing, and they are reserved for tender cuts of meat like prime rib, pork chops, chicken breasts, beef steaks, lamb chops, and, yes, hamburgers. They differ from slow-cooking methods, which are designed not only to heat foods, but also to heat them gently enough and for long enough to allow connective tissues to slowly break down. Slow-cooking methods include true barbecuing (i.e., smoking) and braising. Some cooking methods, such as roasting, steaming, or simmering, can be used for any cut of meat, though the temperature and time period you apply them for will vary depending on what you are cooking. We'll get to those slower methods in later chapters.

Steak houses, just like fast food restaurants, are all about speed. Unlike other types of restaurants, where many dishes and their components are prepared in advance and finished to order, steak houses start pretty much solely with raw ingredients, slapping them on the hot grill or under a mammoth broiler only after you place your order. Grilling, pan-searing, and broiling are among the most ancient of cooking methods, and they bring out the primal urge in all of us. How many of you can resist a hot rib-eye steak fresh out of the pan with a crisp, well-charred crust and a juicy medium-rare center, dripping with juices? Go on, raise your hands.

I thought so.

Many people see a meal at a steak house as the ultimate night out. The way to celebrate that big promotion or graduation. A means to reaffirm your dominance in the food chain. A once-a-year splurge to be taken with plenty of red wine, creamed spinach, hash browns, and brandy, and a time when neither your significant other nor your doctor can say peep. This is *my* life, *my* day, *my* meat, you silently tell yourself as you bang your fat steak knife on the table, toppling that neat tower of asparagus into its pool of creamy hollandaise.

But here's a little secret: there's nothing those steak houses do that you can't do better at home, and cheaper. All it takes is a bit of science, know-how, and practice.

In this chapter, we'll not only talk about how to roast the tenderest, crispest-on-the-outside, juiciest-on-the-inside, melt-in-your-mouthiest prime rib roast to feed a whole crowd of hungry carnivores, but we'll also address the "fast" part of fast-cooking methods. That is, steaks, chops, chicken breasts, and other proteins that can go from fridge to table in thirty minutes or less. Are you ready?

# The FOOD LAB's Complete Guide
## TO BEEF STEAKS

I've cooked a lot of beef in my life, from steak for two to entire 25-pound Prime-grade, grass-fed, dry-aged, grain-finished, well-marbled 7-rib racks. Beautiful hunks of meat that have left my apartment permanently perfumed with the sweet, musky scent of crisp beef fat and my mind permanently stained with the insatiable desire to taste that beef again. This is not an altogether unpleasant state of affairs to be in in my line of work.

But wait—*Prime-grade, grain-finished, marbled? What do all these terms mean?* you cry. And, more important, *Why should I care?*

Here are the answers to every question you've ever had or might ever have about beef roasts and steaks.

**What exactly are the differences between a roast and steak? Do they come from different parts of the cow?** Put most simply, a roast is a large piece of meat—generally at least 2 inches thick—that is cooked in the oven and sliced before serving. A steak is a thinner piece of meat—2 inches thick or under—intended to be cooked and served as is. Practically speaking, there's not much difference between the two other than size. Both roasts and steaks are cooked via fast-cooking methods. That is, when cooking them, the intent is to get the meat up to a specific final temperature and then serve it, as opposed to slow-cooking methods like braising, which require you to hold meat at a given temperature long enough that connective tissues can break down. Because of this, cuts used for roasts and steaks must come from parts of the cow that are relatively tender to begin with. In many cases, the cuts of beef used for roasts and steaks overlap. For instance, a rib-eye or Delmonico steak is essentially simply a single-bone rib roast, and a tenderloin steak or filet mignon is a steak-sized slice of a tenderloin roast or Chatêaubriand.

## The Four High-End Steaks You Should Know

**I'm looking to buy some good steak. What do I need to know?**
As mentioned above, the difference between a steak and roast essentially comes down to size. Any good roast can be cut into individual steaks. While cheaper cuts like sirloin, flank, and skirt, and cheffy cuts like hanger and flatiron, are becoming increasingly popular and available these days, the kings of the steak house are still those cuts that come from the *longissimus dorsi* and the *psoas major* muscles. The longissimus dorsi are a pair of long, tender muscles that run down either side of the spine of the steer, outside the ribs, all the way from the neck to the hip. The psoas major are a pair of shorter muscles that start about two-thirds of the way down the steer's spine and run on the opposite side of the ribs to the longissimus—the inside. Commonly referred to as the filet mignon or tenderloin, the psoas major are by far the tenderest meat on the steer. That, coupled with their small size, makes them

the most expensive cut (that whole supply-and-demand thing, you know?).

Out of these two muscles come a number of different steaks. The chart on page 284 shows what you'll find at the typical butcher shop.

**Why would I want to eat steaks from these muscles?**
The tenderness of a steak is inversely related to the amount of work that the muscle does during the steer's lifetime. So, as relatively unused muscles, the longissimus dorsi (commonly referred to as the loin or backstrap) and the psoas major are extremely tender, making them ideal candidates for steak (and also quite expensive). The former has an advantage over the latter in that it contains a generous amount of fat, both in large swaths around the central eye of meat and, more important, within the muscle itself in a web-like network known as marbling.

**Why is marbling important?**
Mainly because it lubricates the muscle fibers. At room or fridge temperature, the fat is solid, but when cooked, it melts, helping muscle fibers slip around each other more easily as you chew, resulting in more tender, juicier meat. Marbling is also important because most of the flavor in red meats comes from the fat. Indeed, there are studies in which tasters fed portions of lean beef and lean lamb were unable to identify them correctly but were easily able to do so when given a portion with fat. Fatty beef just tastes beefier.

## Grading

**What do the labels from the government on my beef telling me that it's "Prime" or "Choice" mean?**
The USDA grades beef into eight categories: Prime, Choice, Select, Standard, Commercial,

Utility, Cutter, and Canner. The top three are the only ones you are likely to see fresh at the supermarket; the rest are used for packaged foods and other products (if you ever see a steak house advertising "100% Utility-Grade Beef!" run, screaming).

Prime-grade boneless rib-eye.

- **Prime-Grade Beef** is the USDA's highest designation. It comes from younger cattle (under 42 months of age) and is highly marbled, with firm flesh. Less than 2 percent of the beef produced in the United States gets this designation, and the vast majority of it goes to steak houses and fancy hotels. If you happen to find some at your butcher, make sure to thank him.
- **Choice-Grade Beef** has less marbling and tenderness than Prime beef. If cooked right, though, it'll still be plenty juicy and flavorful. For low-fat cuts like tenderloin or certain sirloin steaks, you can expect the meat quality to be nearly indistinguishable from that of Prime beef. This is the standard option at high-end supermarkets.
- **Select-Grade Beef** is much leaner than Prime or Choice but still tender and high in quality. Its main drawback is its relative lack of marbling, so it will not be as juicy or flavorful as higher grades.
- **Standard- and Commercial-Grade Beef** can be found in some supermarkets but will be sold as "ungraded." It's often the choice for generic store-brand beef, and it shows very little marbling and is markedly tougher than other grades. Avoid it.
- **Utility, Cutter, and Canner** are almost never sold retail. You can get them in the form of beef sticks, jerky, or preformed burgers, or in the fillings of things like frozen burritos or sausages.

**Prime-grade beef is so expensive and difficult to find. Is it really worth seeking out?**
Good question. I held a blind beef tasting pitting Choice-grade beef against Prime, cooking both in the exact same manner and to the same temperature (oh, the horrors I put up with in the name of science!). Of the eight tasters present, there was an overwhelming and unanimous preference for the Prime-grade beef, though the Choice was still quite tasty.

Prime generally costs about 25 percent more per pound than Choice, which is a hefty chunk of change if you're feeding a crowd. In my household, though, steak night is a rare occasion, and one that I save up for.

**What about this "Kobe beef" I keep hearing about?**
Kobe beef is a type of high-quality, well-marbled beef named after a professional basketball player. Wait. Strike that, reverse it.

True Kobe beef comes from Tajima-breed Wagyu cows. This is a breed that was originally a work animal used to plow rice fields in the mountainous Hyogo prefecture of Japan. Once beef started to become more popular in Japan, some people noticed that meat from these cows had an unusually high level of marbling and a distinct, delicate flavor. The cows were consequently carefully bred to maximize these characteristics. The result is beef that is absolutely stunning in the degree of marbling it has, far outstripping even USDA Prime beef.

| NAME OF STEAK | TENDERNESS (on a scale of 1 to 10) | FLAVOR (on a scale of 1 to 10) | ALSO SOLD AS | WHERE IT'S CUT FROM |
|---|---|---|---|---|
| Rib-Eye | 7/10 | 9/10 | Beauty steak, market steak, Delmonico steak, Spencer steak, Scotch fillet, entrecôte | The front end of the longissimus dorsi, from the rib primal of the steer. The farther toward the head of the steer you get, the more of the spinalis muscle you'll find in your steak—that's the cap of meat that wraps around the fatter end of the steak. |
| Strip | 8/10 | 7/10 | New York Strip, Kansas City strip, top sirloin (which has nothing to do with the sirloin primal) | The longissimus dorsi muscle, in the short loin primal (that's the primal just behind the ribs). |
| Tenderloin | 10/10 | 2/10 | Fillet, filet mignon, Châteaubriand (when cut as a center-cut roast serving 2 or more, tournedos (when cut from the smaller tapered section of the tenderloin closest to the rib primal) | The central section of the psoas major muscle in the short loin primal. |
| T-Bone | Combination of tenderloin and strip | Combination of tenderloin and strip | Porterhouse (when tenderloin section is 1½ inches or wider) | The T-bone is a two-for-one cut—it's comprised of a piece of tenderloin and a piece of strip, separated by a T-shaped bone. The regular T-bone is cut from the front end of the short loin primal, just after the tenderloin starts, giving it a smallish piece of tenderloin (between ½ and 1½ inches wide); porterhouse steak is cut from farther back and has a section of tenderloin at least 1½ inches wide. |

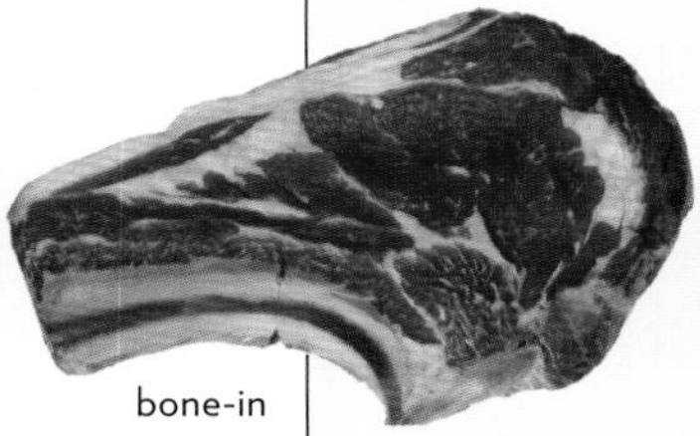

bone-in

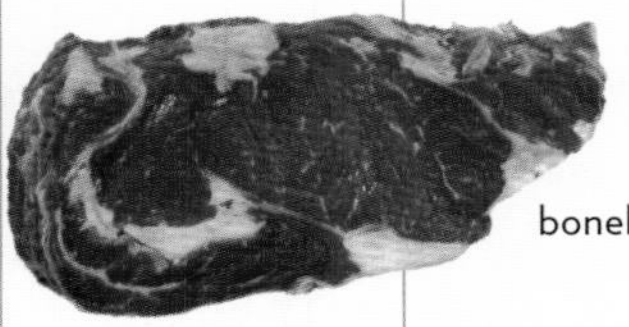

boneless

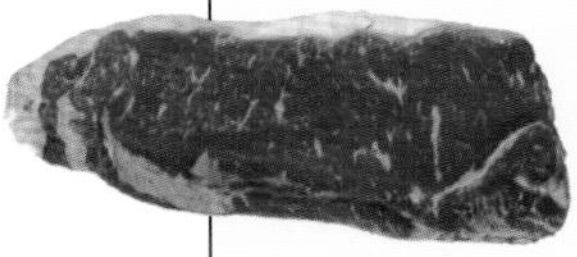

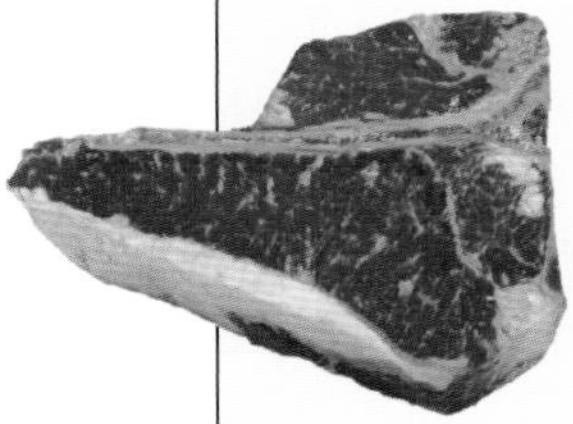

| WHAT IT TASTES LIKE | THE BEST WAY TO COOK IT |
|---|---|
| Fat is where a lot of the distinctive flavor of beef comes from, making the rib-eye, which is highly marbled with a large swath of fat separating the longissimus from the spinalis, one of the richest, beefiest cut available. The central eye of the meat tends to be smooth-textured with a finer grain than a strip steak, while the spinalis section has a looser grain and more fat. Many people (myself included) consider the spinalis to be the absolute tastiest quick-cooking cut on the cow. | Panfrying, grilling, broiling. Because its copious fat is likely to cause flare-ups, grilling can be a bit tricky. Have a lid ready, and stand by with tongs in case you need to rapidly spring into action and retrieve it from the depths of a fireball. This is my favorite cut for pan-searing. |
| A tight texture with a definite grain means strip steaks are moderately tender but still have a bit of chew. Good marbling and a strong beefy flavor. Not as robust as a rib-eye steak, but much easier to trim, with no large pockets of fat, making it an easy-to-cook, easy-to-eat cut. A favorite of steak houses. | Panfrying, grilling, broiling. It's easier to grill than rib-eye, as less fat means less flare-ups and less burning. |
| Extremely tender, with an almost buttery texture. Very low in fat, and correspondingly low in flavor. Unless you are looking for a low-fat cut or prize tenderness above all else, you're better off with one of the other less-expensive cuts. | Panfrying or grilling. Because it's so low in fat, and fat conducts heat more slowly than muscle, tenderloin tends to cook much faster than other steaks and is thus more prone to drying out. Panfrying in oil and finishing by basting it with butter helps add some richness, as does wrapping it in bacon before grilling (a common approach). Even better is to purchase and roast or grill-roast it whole as a Châteaubriand—less surface area means less moisture loss—and then cut into steaks. Because of its mild flavor, it's often paired with flavorful sauces or compound butters. |
| The strip section tastes like strip and the tenderloin tastes like tenderloin. | Grilling, broiling. Because of the irregularly shaped bone, pan-searing is extremely difficult with a T-bone. As the meat cooks, it tends to shrink down a bit, so the bone ends up protruding, preventing the meat from getting good contact with the pan surface and inhibiting browning. You're much better off grilling or broiling it. When grilling or broiling, make sure you position the steak so that the tenderloin is farther away from the heat source than the strip is, to promote even cooking. |

But Kobe beef is difficult to come by even in Japan, and depending on the import laws at the moment, it is impossible (or at least illegal) to find true Kobe beef in the United States. What you're far more likely to see is "Kobe-style" beef, most of which comes from domestic animals crossbred from Wagyu and Angus cattle. Meat from our domestic "Kobe" beef tends to be leaner, darker, and stronger in flavor than its Japanese counterpart, an artifact of their Angus lineage and their American-style grass and grain feed. Good American Kobe-style beef is usually the most expensive beef on the market.

Incidentally, if you ever see someone offering "Kobe burgers," kindly refrain from ordering them. Kobe beef is prized for its marbling, tenderness, and subtle flavor. Burgers already have plenty of fat and tenderness because of their grind, and subtle flavor is *not* what you want in a hamburger. It's a marketing gimmick, pure and simple. And yes, Kobe Bryant really is named after the beef.

## Color and Size

**How come when I buy beef sometimes it appears purple and sometimes deep red? Should I select one over the other?**

Here's the reason: it has to do with the conversion of one of the muscle pigments, *myoglobin*, and its exposure to oxygen. Immediately after being cut, meat is a dark purplish color—the color of myoglobin. Soon oxygen begins to interact with the iron in myoglobin, converting it to *oxymyoglobin*, which has a bright cherry-red color. Have you ever noticed how when you cut into a rare steak in an oxygen-rich environment (like your house) it starts out dark, then "blooms" into redness? Now try the same thing in the vacuum of outer space. See the difference? So, even though the bright red color is the one most associated with freshness, it's really got nothing to do with it—purple meat can be just as fresh. You are particularly likely to notice this dark color in vacuum-sealed meat.

Eventually enzymes present in the meat will cause both myoglobin and oxymyoglobin to lose an electron, forming a pigment called *metmyoglobin*. It's got a dirty brown/gray/green color. While it doesn't necessarily indicate spoilage, it does mean that the beef has been sitting around for a while.

**Do you mean to tell me that the color of red meat doesn't come from blood?**
Precisely that. The beef you buy in the supermarket contains little to no blood, which is drained out immediately after slaughter. Blood contains a pigment very similar to oxymyoglobin, called *hemoglobin*. So, next time your friend orders his beef "bloody rare," you can correct him, saying, "Don't you mean rare enough that the myoglobin pigment in the muscle has not yet had a chance to break down?"

Say that, then duck. People who eat bloody-red beef tend to have anger management issues.

## Labels: Natural, Grass-fed, and Organic

Labeling laws in the United States are confusing at the very least and in many cases worthless. And bear in mind that it's not in the best interest of the vast majority of beef producers to make the labeling any clearer—the less the consumer knows about how meat gets on their table, the better. The majority of cattle in this country are raised on pasture for most of their lives, though when pastured, their diets are supplemented with corn and other grains. They are almost all finished for the last few months in high-density feedlots where they are fed a grain-based (mostly corn and soy) diet in order to promote the marbling and fattiness we find so appealing. They are routinely treated with prophylactic antibiotics, not just to stave off diseases, but also to promote faster growth. Regular cattle are not particularly happy animals for the last couple months of their lives.

Fortunately, there are alternatives at the supermarket. Here are a few labels you might see and what they mean:

- **"Natural"** means basically nothing. There's no enforcement, there are no rules. It's basically an honor system on the producer's part, and no third party checks it.
- **"Naturally Raised,"** on the other hand, does mean something. As of 2009, the label ensures that the animals are free of growth promotants and antibiotics (except coccidiostats for parasites) and that they were never fed animal by-products. What this means in practical terms is that you can be assured that the meat is free of any antibiotic residue and that the cattle were raised in clean, relatively uncramped environments designed to prevent them from ever needing antibiotics.
- **"Organic Beef"** is certified and inspected by the government, and the animals must be fed completely organic feed grains and be antibiotic- and hormone-free. They must also have access to pasture, though in reality "access" could be a single patch of grass on the far side of a large dirt feedlot. Organic cattle are also subject to stricter enforcement in terms of humane treatment. More recent legislation mandates that at least 30 percent of their dry-matter intake needs to come from pasture for 120 days out of the year. That's good news.
- **"Grass-Fed"** cattle must, at some point in their lives, have been raised on a diet of grass. They do not necessarily receive a 100-percent grass diet, nor are they necessarily finished on grass. Most "grass-fed" cattle are fed grain for their last few weeks to fatten them. The very definition of what "grass" means is also open to debate: many producers want to include young cornstalks under the "grass" umbrella, effectively diluting the label.

I generally choose well-marbled Organic or Naturally Raised beef if possible, from specific ranches that I know do a good job with their cows. Next time you're in the supermarket, take a look at the labels, write down the names of the producers, and look them up on the internet. You'll be amazed at what information a little Googling can get you.

**Is grass-fed beef really healthier than grain-fed?**
Many studies indicate that it is. It's certainly healthier for the cow, a ruminant animal whose digestive system has evolved to break down grass. Then again, even grain-finished cows are only finished on grain for a few months before they are slaughtered—hardly enough time to put them at risk of serious health problems, so I'm frankly not too concerned about that argument. According to Marion Nestle, a professor of nutrition and public health at NYU, grass-fed cows tend to have lower levels of *E. coli*, as well as lower levels of dangerous bacteria in their feces, and require fewer antibiotics, making them, all in all, safer to consume. They also tend to have higher levels of omega-3 fatty acids (that's the healthy stuff), along with higher levels of transconjugated linoleic acids (CLAs).

**Wait—aren't CLAs a trans-fat, and aren't trans-fats bad for you?**

Yes indeedy. The very same trans-fats that occur in artificially hydrogenated fats are present in the meat of all ruminants, like cattle, sheep, and goat. However, there is some indication that CLAs from cows are actually healthier than the artificial trans-fats in hydrogenated oils, though it's unclear whether it is real nutritionists who think this or perhaps just the folks from the Beef Council. As with most matters related to nutrition, the literature seems to indicate that nobody really knows what the heck is going on.

**I'm getting tired of all this talk about health and nutrition. What about flavor?**

Grass-fed beef in general has a more robust, slightly gamier flavor than grain-fed beef, and if it's grain-finished, it can be just as tender and moist. Personally, I try and find beef from a producer who has allowed the cattle to graze for the majority of their lives on pasture, then finished them with a supplemental grain. Most of mine comes from Creekstone Farms, which supplies a lot of the beef at Pat LaFrieda, New York's most well-known meat purveyor. Your best bet is to get to know your butcher and talk to him about what you want. How you balance flavor, nutrition, and ethics is up to you.

## Buying Tenderloin Steaks

**Is it true that in tenderloin steaks, marbling and aging are not as much of a factor?**

Yep, it's true. Tenderloin is one of the leanest cuts of beef, so even Prime-grade tenderloin won't have an excessive amount of fat. It's known more for its tenderness than for its flavor. In fact, when purchasing tenderloin, I normally don't bother paying for anything beyond Choice grade. Proper aging of tenderloin is also nearly impossible, for the simple fact that it does not have enough surrounding fat for it to be dry-aged without going rancid or drying out. Tenderloin that's labeled "aged" is almost guaranteed to have been wet-aged—that is, aged in a Cryovac bag, which will improve tenderness, but not flavor.

When buying tenderloin steaks, there are two ways to go about it. You can buy already butchered steaks, which are almost invariably too thin to cook properly, or worse, unevenly shaped and sized. Much better is ask your butcher for a whole two-pound center-cut roast, also called a Châteaubriand—this is a tenderloin from which the skinny and fat ends have been trimmed off. You can then cut it into perfectly even steaks yourself when you take it home. This is enough meat to feed four Americans or a half dozen Europeans. For more on this, check out "How to Trim a Whole Beef Tenderloin," page 317.

## Reasonably Sized Pan-Seared Steaks

You'll understand the title of this section when you get to "Unreasonably Large Pan-Seared Steaks," page 297.

Chicken might be more popular than beef at the supermarket these days (we'll get to quick-cooking chicken recipes here too), but we're still a nation of beef eaters. Is there anything that strikes us on a more basic, carnal, primal level than a perfectly marbled hunk of medium-rare beef with a juicy rosy-pink center and a deep, dark, crisply browned

crust? Bacon and sex, perhaps (in that order), but that's it. It's why we fork over top dollar at steak houses on a nightly basis. But as I mentioned at the beginning of the chapter, they're not doing anything in those kitchens that you can't do yourself at home. You just need to know two things: how to buy a good steak and how to cook it.

As for buying, we've already covered all the basics. Here's a quick recap of what you are looking for:

- **Well-marbled meat.** If you buy conventional meat, look for Prime or at the very least Choice grade. If you prefer Organic or Grass-Fed, look for plenty of intramuscular fat.
- **Fresh meat** that's been cleanly butchered. If the display case that's in front of the customers looks messy, imagine what it looks like back in the meat-cutting room.
- **Aged steaks,** if you can afford them.

Unless you like your meat well-done, you're also best off buying thick-cut steaks—that is, at least 1½ inches thick—so that you have plenty of time to develop a nice sear on the outside before the interior has a chance to overcook. It's better to buy one bigger thick-cut steak and serve two people perfectly cooked meat than to buy two thinner steaks and serve two people overcooked meat.

Congratulations—now that you've got a great steak in your kitchen, you've won 80 percent of the battle. The only thing left to do is not mess it up. Here are some commonly asked questions about cooking steak and the answers.

## When should I salt my steak?

Read a half dozen cookbooks or listen to a half dozen celebrity chefs, and you're likely to hear at least as many different responses about when you should salt your meat. Some claim that salting immediately before putting it in the pan is best. Others opt not to salt the meat at all, instead salting the pan and placing the meat on top of the salt. Still others insist on salting up to a few days in advance. Who's right?

To test this, I bought myself six thick-cut bone-in rib-eyes (I love the smile butchers get in their eyes when you do this) and salted them at different 10-minute intervals before searing them one at a time in a hot skillet—so the last steak went into the pan immediately after salting, while the first steak went in a full 50 minutes after salting. All of the steaks were allowed to stand at room temperature for the full 50 minutes, ensuring that they were all at the same starting temperature when cooking began. The results? The steak that was salted immediately before cooking and those that were salted 40 or 50 minutes ahead turned out far better than those that were salted at any point in between. What was up with those steaks?

Here's what happens:

- Immediately after salting, the salt rests on the surface of the meat, undissolved. All the steak's juices are still inside the muscle fibers. So searing at this stage results in a clean, hard sear.
- Within 3 or 4 minutes, the salt, through the process of osmosis, will begin to draw out liquid from the beef, and this liquid beads up on the surface of the meat. Try to sear at this point, and you waste valuable heat energy simply evaporating this pooled liquid. The pan temperature drops, your sear is not as hard, and crust development

and flavor-building Maillard browning reactions are inhibited.

- At around 10 to 15 minutes, the brine formed by the salt dissolving in the meat's juices will begin to break down the muscle structure of the beef, causing it to become much more absorptive. Then the brine begins to slowly work its way back into the meat.
- After 40 minutes, most of the liquid has been reabsorbed into the meat. A small degree of evaporation has also occurred, causing the meat to be ever-so-slightly more concentrated in flavor.

Not only that, but I found that even once the liquid has been reabsorbed, it doesn't stop there. As the meat continues to rest past 40 minutes, the salt will slowly work its way deeper and deeper into its muscle structure, giving you built-in seasoning beyond just the outer surface you get if cooking right after salting.

The absolute best steak I had was one that I salted on both sides and allowed to rest, uncovered, on a rack in the refrigerator overnight. It appeared to dry out slightly, but that was only superficial—the amount of drying that occurs with overnight rest (about 5-percent moisture loss) is negligible compared to the amount of moisture driven off during cooking (upward of 20 percent, or even more in the hard-seared edges). And after cooking, the steak that had been salted and rested overnight actually ended up with 2-percent *more* moisture than one that was salted and cooked immediately, due to the beef's increased ability to retain water as the salt loosened the muscle structure.

Also, as the salt makes its way back into the meat in the longer-salted steaks, the meat becomes a deeper color. That's because the dissolved proteins scatter light differently than they did when they were still whole.

**Moral of the story**: If you've got the time, salt your meat for at least 40 minutes, and up to overnight, before cooking. If you haven't got 40 minutes, it's better to season immediately before cooking. Cooking the steak anywhere between 3 and 40 minutes after salting is the worst way to do it.

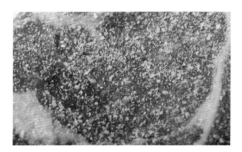

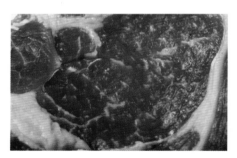
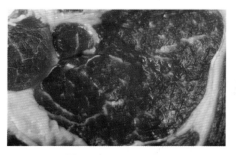

The stages of salting.

**What about salting even further in advance? Any benefit there?**

Actually, yes. Salting your steak up to three days before you plan to cook it while letting it rest uncovered on a rack set in a rimmed baking sheet in the fridge is called dry-brining, and it will improve your steak in three very noticeable ways.

First, as the steak sits, the salt will slowly continue to work its way into the meat, seasoning it more deeply. Second, as that salt works its way in, it will continue to break down muscle proteins, allowing your steak to retain more moisture when you cook it. This leads to juicier steaks. Third, and most important, resting your steak uncovered will allow the very outer edges to dry out a bit. This may sound like a bad thing, but keep this in mind: that moisture is all moisture you're going to drive out when searing your steak *anyway*. Getting it out of the way before your steak hits the pan means that you get more efficient browning. Better browning = better flavor, and faster browning = less overcooked meat under the surface of the steak.

**What exactly is searing, and does it really "seal in juices"?**

Since the mid-nineteenth century up until recently, it was believed (and is still believed by many) that searing a piece of meat—that is, exposing it to extremely high heat in order to rapidly cook its exterior—would cauterize the pores in its surface, thereby sealing it and reducing moisture loss. Let me say straight off the bat: **the theory is false**, and it's exceedingly simple to test. Here's what you do:

1. Take two identical steaks and measure their raw weight.

2. Cook one steak by first searing it in a hot, hot skillet, then transferring it to a 275°F oven to cook through to an internal temperature of 125°F. Remove it from the oven and allow it to rest for 10 minutes (during which time its internal temperature should rise by about 5 degrees, to 130°F, then fall again). Weigh this steak and note the amount of weight loss (which translates to fat and moisture loss).

3. Now cook the second steak by first placing it in the oven until it reaches an internal temperature of around 115°F, then transfer it to a blazing-hot skillet and cook, flipping occasionally, until it's nicely browned and has reached an internal temperature of 125°F. Allow it to rest, same as the first steak, then weigh it and note the amount of weight loss.

Both steaks have been exposed to searing and roasting and both steaks have been cooked to the exact same final temperature. The only difference is the order in which the operations were carried out. Now, if there were any truth at all to the "searing locks in juices" theory, we'd expect that the steak that was seared-then-roasted will have retained more juices than the steak that was roasted-then-seared. In reality, however, both steaks lose a fairly similar amount of juice—and, if you repeat the experiment, in most cases, the roasted-then-seared steak will actually stay juicier.

This is due to the fact that a cold steak going into a hot skillet takes longer to sear than a steak warm from the oven going into a hot skillet does. The extreme heat of a skillet is great for developing browned flavors, but it also causes muscle proteins to contract violently, squeezing out juices. For juicer steaks, the less time you spend at high heat, the better.

**So why do we bother searing at all?**

It's simple: flavor. The high heat of searing triggers the cascade of chemical reactions known as the Maillard reaction. I've referred to this elsewhere, but here's a quick recap.

Named after Louis-Camille Maillard, the scientist who discovered it, the Maillard reaction is the complex series of chemical reactions that causes foods to brown. It's often confused with caramelization ("That steak has a beautifully caramelized crust!"), but in fact the two reactions are distinct. Caramelization occurs when sugars are heated, while the Maillard reaction occurs when sugars *and* proteins are heated. The places you're most likely to see it are when searing or roasting meat (meat contains natural sugars), when baking bread or making a piece of toast (flour contains carbohydrates built with sugars and proteins), or when roasting coffee beans.

Although Maillard reactions can occur at relatively low temperatures, they are glacially slow until your food reaches around 350°F. That's why boiled foods, which have an upper limit of 212°F (determined by the boiling point of water), will never brown. With high-temperature searing, frying, or roasting, however, browning is abundant. First a carbohydrate reacts with an amino acid (the building blocks of protein), which then continues to react, forming literally hundreds of by-products that will in turn react with each other to form still more by-products. To this day, the exact set of reactions that occurs when Maillard browning takes place has not been fully mapped out or understood. What we *do* understand is this: it's darn delicious. Not only does it increase the savoriness of foods, but it also adds complexity and a depth of flavor not present in raw foods or foods cooked at too low a temperature.

*That's* why your steak tastes so much meat-ier when it's properly browned, and that's why for most people, the outer crust of the meat is the tastiest part.

**I've read that letting your meat come to room temperature before cooking gives you better results. Any truth in that?**

Let's break this down one issue at a time. First, the internal temperature. While it's true that slowly bringing a steak up to its final serving temperature will promote more even cooking, the reality is that letting it rest at room temperature accomplishes almost nothing.

To test this, I pulled a single 15-ounce New York strip steak out of the refrigerator, cut it in half, placed half back in the fridge, and the other half on a ceramic plate on the counter. The steak started at 38°F and the ambient air in my kitchen was at 70°F. I then took temperature readings of its core every ten minutes. After the first 20 minutes—the time that many chefs and books will recommend you let a steak rest at room temperature—the center of the steak had risen to a whopping 39.8°F. Not even a full two degrees. So I let it go longer. 30 minutes. 50 minutes. 1 hour and 20 minutes. After 1 hour and 50 minutes, the steak was up to 49.6°F in the center—still colder than the cold water that comes out of my tap in the summer, and only about 13 percent closer to its target temperature of a medium-rare 130°F than the steak in the fridge.

After two hours, I decided I'd reached the limit of what is practical, and had gone far beyond what any book or chef recommends, so I cooked the two steaks side by side. For the sake of this test, I cooked them directly over hot coals until seared, then shifted them over to the cool side to finish. Not only did they come up to their final temperature at nearly the same time (I was aiming for 130°F) but also they showed the same relative evenness of cooking, and they both seared at the same rate.

Long story short: pulling your steaks out early is a waste of time.

### What's the best fat to use?

Now we're ready to start cooking. But before the steak hits the pan, we've got to add some fat. When searing, fat accomplishes two goals. First, it prevents things from sticking to the pan by providing a lubricating layer between the meat and the hot metal. Did you know that meat proteins actually form a chemical bond at the molecular level with metal when they are heated in contact with it? Proper preheating and using fat will help prevent this from happening. Second, fat conducts the heat evenly over the entire bottom surface of the steak. It may not look so from afar, but the surface of a piece of meat is very bumpy, and these bumps only get exaggerated as the meat is heated and starts shrinking and buckling. Without oil, only tiny bits of the steak will actually come in direct contact with the hot pan, and you will end up with a spotty sear—almost burnt in some spots and gray in others. You need to use at least enough oil to conduct heat to the portions of the steak that are not in direct contact with the metal.

But what's the best medium to sear in? Butter or oil? And, if I'm using oil, which one? Some claim that a mixture of both is best, using the rationale that butter alone has too low a smoke point (see "The Smoke Points of Common Oils," page 860)—it begins to burn and turn black at temperatures too low to properly sear meat. Somehow cutting the butter with a bit of oil is supposed to raise this smoke point. Unfortunately, that's not true. When we say that "butter is burnt," we're not really talking about the butter as a whole—we're talking spe-

Butter browns rapidly as it gets to searing temperatures.

cifically about the milk proteins in butter, the little white specks you see when you melt it. It's these milk proteins that burn when you get them too hot, and believe me, they couldn't care less whether they're being burnt in butterfat or in oil. Either way, they burn.

What all this means is that the best cooking medium for a steak is plain old oil. At least to start. Adding butter to the pan a minute or two before you finish cooking is not a bad idea. This is just long enough to allow a buttery flavor and texture to coat the meat, but not so long that the butter will burn excessively, producing acrid undertones.

There are some who keep bottles and bottles of various oils on hand for different cooking projects. I limit my oils to a more reasonable three. One is a high-quality extra-virgin olive oil that I use for flavoring dishes. One is the peanut oil I use for deep-frying (see Chapter 9), and the third is the canola oil I use for pretty much all of my other cooking

---

\* Butter has a good amount of saturated fat, which gives it a richer, thicker texture than vegetable oil (for more on saturated versus unsaturated fats, see page 855).

projects. Canola oil has a reasonably high smoke point, making it great for searing, but, more important, it has a very neutral flavor and is inexpensive, with neither the "corniness" that comes with corn oil or the high prices that come with safflower, grapeseed, and many other oils.

### How often should I be flipping my steak?

There's a problem when it comes to cooking steak, and it has to do with your two conflicting goals. You see, for most folks, the ideal internal temperature of a finished steak is around 130°F—medium-rare. This is the stage at which it's rosy pink, tender, and juicy. But you also want a deep-brown, crisp, crackly crust, a by-product of the Maillard reaction.

A few years ago, food scientist Harold McGee published an article in the *New York Times* that mentioned an interesting technique: multiple flips. It goes against all classical and backyard wisdom— we all *know* you should only flip a steak (or a burger, for that matter) once, right? I mean, how can you even *ask* that question? Well, I've always been of the mind that if an answer exists, and, clearly, there is an answer to this, the question is worth asking. Fortunately, the question is one that's fairly straightforward to test.

Those on the "one flip" side claim "more even cooking" and "better flavor development" as the selling points of the method. Curiously, the few people on the "multiple flips" side claim the exact same benefits from multiple flips, adding "shortened cooking time" to the mix. So, who's right?

I cooked a few different steaks to the same internal temperature of 130°F. One was flipped just once, another was flipped every minute, a third every 30 seconds, and a fourth every 15 seconds. Interestingly, the steak that was flipped every 30 seconds reached the desired temperature the fastest of all four, fol-

lowed by the one flipped every 15 seconds, followed by the one flipped every minute, and finally by the one flipped just once. The fastest-cooking steak took about 2 minutes less than the slowest-cooking one.

Then I served them up to a few friends I had over for dinner, asking them to tell me which ones had the nicest crusts, which were the most evenly cooked, and which were the tastiest. From the outside, they had trouble telling them apart—the amount of browning seemed pretty close to equal. However, once we cut into them, the differences were more apparent: the steaks flipped only once had distinct bands of overcooked meat at the edges, while the ones flipped multiple times were more evenly cooked. It wasn't enough of a difference to make anyone say that one steak was particularly *bad* per se—they all got devoured—but it was enough to prove that those in the "only flip it once!" camp have no basis in reality to back up their claims.

Steak flipped just once (*left*) versus multiple times (*right*).

More evenly cooked meat in a shorter period of time seems pretty win-win to me. If cooking via the single-flip method, when you flip the steak over, the second side will be barely any warmer than it was when the steak first went into the skillet. Your cooking is only halfway done. If you add more flips, on the other hand, what you are essentially doing is approximating cooking both sides of the steak simultaneously. Neat, right?

Incidentally, the steak flipped every 15 seconds took longer than the steak flipped every 30 seconds

*From left*: 120°, 130°, 140°, 150°, and 160°F internal temperature.

because it spent too much time in the air above the pan rather than in direct contact with the pan itself.

**Moral of the story:** All you supple-wristed crazy flippers out there, don't worry, you're doing the right thing. And for all you single-flippers? Well, you can keep doing what you're doing and it probably won't hurt your steaks none, but lighten up a bit, will ya?

### What is "carry-over cooking," and how does it affect how I cook my meat?

We know that meat cooks from the outside in, right? So, at any given moment, the exterior layers of your steak are hotter than the very center, where we take our temperature reading for doneness. Once you pull the steak out of the pan, though, heat energy from the outer layers of meat has two places it can go: out or in.

Most of that energy will dissipate into the air as the steak rests. But some of it will continue traveling into the meat. The result is that after you take a steak out of the pan or off the grill, its internal temperature will continue to rise. The amount it rises is determined by a number of factors, but the overriding one is the size of the steak or roast. A thin steak—say, an inch thick or less—will barely rise a couple of degrees, but a big, fat 1½- to 2-incher can rise a good 5 degrees as it rests. A prime rib roast can rise by as much as 10 degrees.

That's why it's always a good idea to take your meat off the heat *before* it reaches the final temperature you'd like it cooked to. (See "The Importance of Resting Meat," page 306.)

Speaking of temperature . . .

### How do I know when my beef is done?

While temperature is really a matter of personal taste, I wanted to lay out some actual data on temperature versus eating quality. So I cooked five Prime-grade New York strips to temperatures ranging from 120° to 160°F and fed them to a group of a dozen tasters. The chart below represents percentage of overall weight loss (i.e., moisture loss) that each steak experienced while cooking.

- **120°F (rare):** Bright red and slippery in the interior. At this stage, the meat fibrils (which resemble bundles of juice-filled straws) have yet to expel much moisture, so, in theory, this should be the juiciest steak. However, because of the softness of the meat, chewing causes the fibrils to push past each other instead of bursting and releasing their moisture, giving the sensation of slipperiness, or mushiness, rather than juiciness. Additionally, the abundant intramuscular fat has yet to soften and render.

### MOISTURE LOSS VERSUS FINAL COOKING TEMPERATURE IN BEEF

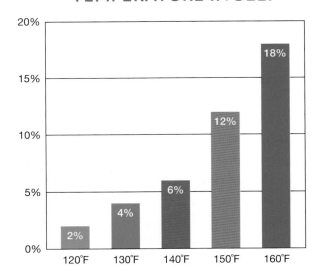

- **130°F (medium-rare):** The meat has begun to turn pink and is significantly firmer. Moisture loss is still minimal, at around 4 percent. The intramuscular fat has begun to render, which not only lubricates the meat, making it taste juicier and more tender, but also delivers fat-soluble flavor compounds to the tongue and palate—beef at this temperature tastes significantly "beefier" than beef at 120°F. When tasting the steaks blind, even self-proclaimed rare-meat lovers preferred this one, making it the most popular choice.
- **140°F (medium):** Solid rosy pink and quite firm to the touch. With more than a 6-percent moisture loss, the meat is still moist but verging on dry. Prolonged chewing results in the familiar "sawdust" texture of overcooked meat. But the fat is fully rendered at this stage, delivering plenty of beefy flavor. This was the second most popular choice.
- **150°F (medium-well):** Still pink but verging on gray. At this stage, the muscle fibrils have contracted heavily, causing the moisture loss to jump precipitously—up to 12 percent. Definite dryness in the mouth, with a chewy, fibrous texture. The fat has fully rendered and begun to collect outside the steak, carrying away flavor with it.
- **160°F (well-done):** Dry, gray, and lifeless. Moisture loss is up to 18 percent, and the fat is completely rendered. What once was cow is now dust.

So, as far as temperature goes, my strong recommendation is to stick within the 130° to 140°F range. To all you hard-core carnivores out there who insist on cooking your well-marbled, Prime-grade steaks rare, you are doing yourself a disservice: unless it renders and softens, the fat in a well-marbled piece of meat is worthless. You may as well be eating lean, Choice-, or Select-grade beef.

And as for people who cook their beef well-done, well, let's just say that you have a special place in my heart right next to *Star Wars* Episode I and that kid who stapled my arm to the table in second grade.

**Conclusion:** For most people, 130° to 140°F is best.

**I've heard folks say I should never stick a fork in my steak to flip it. Any truth in that?**
Watch a Johnsonville Brat commercial, and you'll be told that poking with a fork is one of the cardinal sins of sausage cookery, and they're right: a sausage has an impermeable casing for a reason—to keep all of those rendered fats and juices right in there with the meat. Pierce it, create holes, and you'll see a fountain of golden juices spring forth, like out of a kid after a long car ride. A steak, on the other hand, has no such casing to protect it—so, is it OK to poke or not?

I cooked two steaks of known weight side by side. The first I carefully turned with tongs each time. The second, I used a *fourchette de cuisine* (that's what French cooks call those two-pronged kitchen forks) completely indiscriminately, mercilessly (though not excessively) poking the steak this way and that as I flipped it. Afterward, I weighed both steaks again. The result? Exactly the same weight loss.

Poking with a fork to turn the steak is a completely risk-free move.

Poking a steak with a fork leads to negligible moisture loss.

The thing is, with steaks, moisture loss is due to one thing: muscle fibers tightening, due to the application of heat, and squeezing out their liquid. Unless you manage to completely pierce or slash those muscle fibers, the moisture they lose is directly proportional to the temperature to which you cook your steak. A fork is simply not sharp enough to harm muscle fibers in any significant way. Yes, you'll see a minuscule amount of juices seeping their way out of the fork holes, but it's a negligible amount. Indeed, that's why the many-bladed tenderizing tool known as a Jaccard is able to tenderize meat without causing it to lose any excess moisture—it separates muscle fibers, but it doesn't actually cut them or open them up.

What about that most-shunned of techniques, the old cut-and-peek? Surely slashing a cooking steak open with a knife and looking inside is going to have a detrimental effect on it, right? Well, yes and no. Yes, a knife actually severs muscle fibers, allowing them to leak their contents to the outside world. But the amount of moisture loss is very minimal. Cut-and-peek too many times, though, and you run the risk of shredding your steak. In reality, one or two peeks won't be detectable in the end product.

But there's a bigger problem with the cut-and-peek method: it's not accurate. Because of the fact that juices get squeezed out quickly from hot meat, when you cut into the center of a still-hot steak (like one that's still sitting in the pan), it'll appear to be much rarer than it really is. If you continue to cook your steak until it appears right by the cut-and-peek method, it'll be overcooked by the time you actually eat it. Remember that thick steaks continue to rise in temperature even after being removed from the pan.

What does that mean? It means that if you haven't yet, you should go out and get yourself a good digital thermometer, duh!

## Unreasonably Large Pan-Seared Steaks

With really big steaks, say 1½ inches thick or greater, you run into another problem: it's nearly impossible to cook them through to the center in a skillet without burning the outside. Conventional restaurant kitchen technique is that the best thing to do in this situation is to sear the steaks first in a hot skillet, then pop them into a hot oven to finish cooking through to the center. And this method works. But there are better ways to do it. See, that method is designed for use in a kitchen where order and expediency are the immediate goals of the line cook. As soon as an order for a steak comes in, the easiest course of action is to sear it, then throw it in the oven and forget about it until it's done, so you can focus on other things, say, plating the half dozen orders for chicken that table twelve is waiting for. At home, we don't have the same urgency—we have more time to plan and more time to execute.

As it turns out, a better way to cook a thick steak is to start it in a hot pan and then *turn down the heat.* You want to time it so that the optimal level of browning is achieved just as the center of the meat reaches the desired final temperature. How do you brown using moderate heat? First of all, use some butter. The milk proteins in butter brown naturally, giving the steak a jump start. Second, baste the steak. By spooning the hot fat over the meat as it cooks, you allow both sides to get the browning effects of the butter while simultaneously shortening its cooking time. For more on butter-basted steaks, see the recipe on page 312.

Alternatively, you can go radical and use my technique for cooking your steak in a beer cooler (wait, what?—skip ahead to page 392, and you'll see).

# DRY-AGING BEEF

Sometimes I get e-mails from readers that say something along the lines of, "You said one thing in article X, and then a couple years later, in article Y, you said almost the complete opposite. What gives? Don't you believe in science, and doesn't science deal in facts?"

There's only one kind of science that isn't open to contradicting itself: the bad kind. Science by definition needs to be open to considering and accepting contradictory evidence and redefining "facts." Heck, if new theories weren't allowed to be formed or conclusions debunked with further experimentation, we'd still believe in crazy things like spontaneous generation, static universes, and even that searing meat seals in juices. And then where would we be?

I bring this up because I once went through great pains to test and explain precisely why you cannot

dry-age meat at home—no way, nohow. But now I'm going to explain to you exactly how you can dry-age at home, how relatively simple it is, and how it can vastly improve the eating quality of your steaks and roasts to the point that they are better than what you can buy at even the best gourmet supermarket.

Now before you call up the National Committee of Good Science and send them to confiscate my calculator (by which I mean my head), let me explain that I still stand by 100 percent of what I wrote earlier: if you are starting with individual steaks, dry-aging at home is not feasible. When I tried dry-aging steaks at home, blind tasting showed that between the first day and the seventh day of such aging, there was absolutely zero perceptible improvement in the eating quality of the steaks.

But we all know that individual steaks is not how meat is dry-aged by professionals, right? They start with whole sub-primals—large cuts of meat with bones and fat caps fully intact, aging them uncovered in temperature-, humidity,- and airflow-controlled rooms designed to allow them to age for weeks or months without rotting. The question is, how can we do this ourselves at home?

I got my hands on 80 pounds of Prime-grade bone-in, fat-cap-intact beef ribs to get my answers. Over the course of two months or so, I aged them in close to a dozen different ways in order to determine what works, what doesn't, and what matters. Here's what I found.

## The Purpose of Aging

### How does aging work?

Good question! First, a brief rundown on why you might want to age meat. Conventional wisdom cites three specific goals when dry-aging meat, all of which contribute toward improving its flavor or texture.

Moisture loss is said to be a major factor. A dry-aged piece of beef can drop up to 30 percent of its initial mass through water loss, which concentrates its flavor. At least that's the theory. But is it true? [Cue dramatic foreshadowing music.]

**Tenderization** occurs when enzymes naturally present in the meat act to break down some of the tougher muscle fibers and connective tissues. A well-aged steak should be noticeably more tender than a fresh steak. But is it?

**Flavor change** is caused by numerous processes, including enzymatic and bacterial action, along with the oxidation of fat and other fat-like molecules. Properly dry-aged meat will develop deeply beefy, nutty, and almost cheese-like aromas.

### But is aged meat really better than fresh meat?

It depends. I had a panel of tasters test meats aged to various degrees and rank them in overall preference, tenderness, and funkiness. Almost everybody who tasted meat that had been aged for a couple of weeks—the time by which some degree of tenderization has occurred but seriously funky flavor has yet to develop—preferred it to completely fresh meat.

But folks were more mixed about meat aged longer than that. Many preferred the more complex, cheese-like flavors that develop with meat aged for between 30 and 45 days. Some even liked the really funky flavors that developed in 45- to 60-day meat. Where you lie on that spectrum is a matter of personal taste. I personally prefer meat aged for 60 days, but beyond that, it gets a little too strong for me.

### OK, I'm sold. But why would I want to do it at home when I can order it online or from my butcher?

Two reasons: First, bragging rights. How awesome is that dinner party where you can say to your friends, "Like this beef? I aged it for 8 weeks myself," gonna be?

Second, it saves you money. Lots of money. Aging meat takes time and space, and time and space cost money. That cost gets passed on to the consumer. Well-aged meat costs anywhere from 50 to 100 percent more than an equivalent piece of fresh meat. At home, so long as you are willing to give up a corner of your fridge or you have a spare mini-fridge, the extra costs are minimal.

You may have heard that in addition to the time and space required, much of the cost of aged meat comes down to the amount of meat that is wasted—that is, the meat that dries out and needs to be trimmed off. This is not as big a factor as you'd think, and we'll find out why soon.

## Selecting Meat to Age

**What cut of meat should I buy for aging?**

To age meat properly, you need a large cut that is best cooked with quick-cooking methods. This makes the standard steak house cuts—New York strip, rib, and porterhouse—the ideal cuts for aging. (See page 282 for more on the four high-end steaks you should know.) The easiest to find whole (and my personal favorite) is the rib.

**What's the minimum size I'll need for proper aging? Can I age a single steak?**

Nope, unfortunately you can't age individual steaks. You could wrap the steaks in cheesecloth or paper towels, set them on a rack, and leave them in the fridge for about a week, but during that time, no detectable texture or flavor changes will occur. Try to age them even longer, and (assuming they don't start rotting) the meat gets so dried out as to be completely inedible. After trimming away the desiccated and slightly moldy bits (perfectly normal for dry-aged meat), you are left with a sliver of meat about a half-centimeter thick, impossible to cook to anything lower than well-done, making your effective yield a big fat zero.

The simple truth is that in order to dry-age, you need large cuts of meat, and you need to age them in open air.

**So, of these larger cuts of meat, what should I look for?**

Rib sections come in several different forms, each with its own numerical designation.

- **The 103** is the most intact. It's an entire rib section (with ribs 6 through 12 of the steer), along with a significant portion of the short ribs, the chine bone completely intact and a large flap of fat and meat (called "lifter meat," not to be con-

fused with the coveted *spinalis dorsis*) covering the meaty side. It's unlikely you'll find this cut even if you ask the butcher for it.

- **The 107** has been trimmed somewhat, with the short ribs cut short, some (but not all) of the chine bone sawed off, and the outer cartilage removed. This is how rib sections are commonly sold to retail butchers and supermarkets, where they can be further broken down.

- **The 109A** is considered ready to roast and serve. It has had the chine bone nearly completely sawed off and the lifter meat removed. The fat cap is then put back in place.

- **The 109 Export** is essentially identical to the 109A, except it does not have the fat cap. This is the cut you'll see on your Christmas table or at that fancy-pants hotel buffet. The meat on this cut is only minimally protected on the outside.

I aged a 107, a 109A, and a 109 Export in a mini-fridge (this one from Avanti) set at 40°F, in which I had placed a small computer fan to allow for air circulation (I had to cut a small notch in the sealing strip around the door so the fan's cord could pass through), simulating a dry-aging room on a small scale. I made no attempt to regulate the humidity, which bounced around from 30 and 80 percent (higher at the beginning, lower as the aging progressed).

I found that the more protection you have, the better your final yield. Why does exterior protection matter when aging meat? Because when you dry-age meat for any length of time that's enough to make a difference, the exterior layers get completely desiccated and so must be trimmed away. The less protected the "good" meat, the more of it you'll have to throw in the trash.

Provided you start with a 109A or another cut

with the fat cap intact, your yield will amount to basically the equivalent of a regular roast. If you imagine your prime rib as a long cylinder, the only meat you'll end up losing is at either end. The fat cap and bones will completely protect the sides.

## What Causes Flavor Changes During Dry-Aging?

**So aged meat doesn't really lose much moisture. But wait a minute, haven't I read that aged steaks can lose up to 30 percent of their weight in water? Isn't that one of the reasons why aged steak is so expensive?**

Don't believe everything you read. That 30 percent figure is deceptive at best and an outright lie at worst. Yes, it's true that if you dry-age an untrimmed, bone-on, fat-cap-intact prime rib, you'll end up losing about 30 percent of its total weight over the course of 21 to 30 days or so. What "they" don't tell you is that the weight is almost exclusively lost from the outer layers—that is, the portion of the meat that would be trimmed off anyway, regardless of whether it is aged or not.

Has it never struck you as just a little bit odd that the aged rib-eye steaks in the butcher's display case aren't 30 percent smaller than the fresh rib-eyes on the display? Or that aged bone-in steaks are not stretching and pulling away from their bones? (I mean, surely the bones aren't shrinking as well, are they?)

The fact is, with the exception of the cut faces that will need to be trimmed off, the edible portion of an aged prime rib is pretty much identical to that of a fresh prime rib.

**OK, let's say I'm now convinced about that. Does that mean that the whole idea that "the meat flavor is concentrated" in an aged steak because of dehydration is also false?**

I'm afraid so. It's a great idea in theory, but the facts don't support it.

First, there's simple visual inspection: a trimmed steak cut from an aged piece of beef is pretty much the same size as a trimmed steak cut from a fresh piece of beef.

Next, I measured the density of beef aged to various degrees against completely fresh meat. To do this, I cut out chunks of meat of identical weights from the centers of rib eyes aged to various degrees, making sure to avoid any large swaths of fat. I then submerged each chunk of meat in water and measured its displacement. What I found was that meat aged for 21 days displaced about 4 percent less liquid than completely fresh meat—a slight increase, but not much. Meat aged all the way up to 60 days displaced a total of 5 percent less—showing that the vast majority of moisture loss occurs in the first three weeks.

What's more, once the meat was cooked, these differences in density completely disappeared. That is, the less aged the meat was, the more moisture it expelled. Why is this? One of the side effects of aging is the breakdown of meat proteins and connective tissue. This makes the meat more tender, as well as causing it to contract less as it cooks. Less contraction = less moisture loss.

When all was said and done, in many cases, the meat that was 100 percent fresh ended up losing more liquid than dry-aged meat.

Finally, a simple taste test was the nail in the coffin: meat dry-aged for 21 days (the period during which the largest change in density of the interior meat occurs) was indistinguishable from fresh meat in terms of flavor. The improvements were only in texture. It wasn't until between the 30- and 60-day marks that noticeable changes in flavor occurred, and during that time period, there was essentially no change in internal density. Thus, moisture loss is not tied to flavor change.

**Why does meat being aged stop losing moisture after the first few weeks?**
It's a matter of permeability. As meat loses moisture, its muscle fibers get more and more closely packed, making it more and more difficult for moisture under the surface to escape. After the first few weeks, the outer layer of meat is so tight and tough that it is virtually impermeable to moisture loss.

**If it's not moisture loss, what factors do affect the flavor of aged beef?**
A couple of things: The first is enzymatic breakdown of muscle proteins into shorter fragments, which alters their flavor in desirable ways. But this effect is completely secondary to the far more important change that occurs when fat is exposed to oxygen—it's the oxidation of fat as well as bacterial action on the surfaces of the meat that cause the most profound flavor change, the funkiness you get in meat that has been aged for over 30 days.

It's true, though, that much of this funky flavor is concentrated on the outermost portions of the meat—the parts that largely get trimmed away—and for this reason, if you want to get the most out of your aged meat, it's essential that you serve it with the bone attached (not the fat cap, which should be completely removed and discarded). The outer parts of the bones will hold tons of oxidized fat and funky meat. The aromas from this meat reach your nose as you're eating, altering your entire experience. Lovers of aged steak also prize the *spinalis*—the outer cap of meat on a rib-eye—for its richer, more highly aged flavor.

## Aging Setup

**What sort of setup do I need for aging steak at home? Is it relatively simple?**
It's very simple and requires virtually no special equipment. There are just a few things you'll need:

- **Fridge space.** The best thing is to use a dedicated mini-fridge, so that the meat smells don't permeate other food. It can get a little . . . powerful. The mini-fridge I kept by my desk when testing aged meat would fill the office with the aroma of aging meat if I peeked inside it for even a moment. Similarly, aged meat can pick up aromas from your refrigerator. Unless your refrigerator is odor-free, a mini-fridge is the way to go.
- **A fan.** To promote drying of the surface and even aging, you want to stick a fan inside your fridge to keep air circulating. This works in much the same way as a convection oven, promoting more even cooling and humidity. I used a slim computer fan I ordered online for about $30.
- **A rack.** The meat must be elevated on a rack. I tried aging pieces of meat on a plate and directly on the floor of the fridge. Bad idea. The part in contact with the plate or refrigerator floor didn't dehydrate properly and ended up rotting. Aging on a wire rack or directly on a wire fridge shelf with a rimmed baking sheet underneath to catch drips is the way to go.
- **Time.** Patience, little grasshopper. You will be rewarded with the steak of your dreams for your patience.

**But what about humidity? I hear humidity needs to be kept [high, or low, or medium, or nonexistent, or etc.]. What should it be and how can I control it?**
I aged meat in fridges kept at relative humidities ranging from 30 to 80 percent as well as in fridges

that fluctuated wildly with no controls. Guess what? All of them produced excellent aged beef.

And it makes sense. As noted above, after the first couple of weeks, the outer layers of the beef become all but impenetrable to moisture. So, it really doesn't make much difference how humid or dry the environment is, because the interior meat is protected. That's good news for home dry-agers!

## Timing

**OK, I'm nearly convinced. How long should I age my meat?**

I had tasters taste steaks aged for various lengths of time. In order to ensure that all the steaks were fairly ranked and that differences in actual cooking were minimized, I cooked them in a sous-vide water bath to 127°F before finishing them with a cast-iron pan/torch combo. The steaks were tasted completely blind. The final results were largely a matter of personal preference, but here's a rough guide to what happens over the course of 60 days of aging:

- **14 days or less:** Not much point. No change in flavor, very little detectable change in tenderness. Few people preferred this steak.
- **14 to 28 days:** The steak is noticeably more tender, particularly toward the higher end of this range. Still no major changes in flavor. This is about the age of the steaks at your average high-end steak house.
- **28 to 45 days:** Some real funkiness starts to manifest itself. At 45 days, there are distinct notes of blue or cheddar cheese, and the meat is considerably moister and juicier. Most tasters preferred the 45-day-aged steak above all others.
- **45 to 60 days:** Extremely intense flavors emerge. A handful of tasters enjoyed the richness of this long-aged meat, though some found it a little too much to handle after more than a bite or two. It is rare that you will find any restaurant serving a steak this aged.

## OK, JUST GIVE ME THE TL/DR VERSION. HOW DO I AGE MY STEAK?

- **Step 1:** Buy a prime rib. Make sure that it is bone-in, preferably with the chine bone still attached, and the complete fat cap intact. If you are buying from a butcher shop, ask them not to trim it at all. A decent butcher will not charge you full price then, since they are making money by selling you that extra fat and bone.
- **Step 2:** Place the meat on a rack in a fridge, preferably a dedicated mini-fridge in which you've stuck a desk fan or small cabinet fan set to low (with a small notch cut in the door seal for the cord). Set the temperature to between 36° and 40°F.

*continues*

- **Step 3**: Wait. Wait for anywhere between 4 and 8 weeks, turning the meat occasionally to promote even aging. It'll start to smell. This is normal.
- **Step 4**: Trim. For a step-by-step guide to the process, see "Knife Skills: Trimming Aged Beef," page 305.
- **Step 5**: Cook.
- **Step 6**: ???
- **Step 7**: Profit.

**What about wet-aging? What is it, and does it work?** Wet-aging is simple: put your beef in a Cryovac bag and let it sit on the refrigerator shelf (or, more likely, on a refrigerated truck as it gets shipped across the country) for a few weeks. Then tell your customers that it's aged, sell it at a premium.

The problem is that wet-aging is nothing like dry-aging. For starters, there is no oxidation of fat in wet-aging, which means that there is no development of funky flavors. A minimal amount of flavor change will occur through enzymatic reactions, but these are, well, minimal. Additionally, wet-aging prevents the drainage of excess serum and meat juices. Tasters often describe wet-aged meat as tasting "sour" or "serumy."

Wet aging can result in the same tenderizing and moisture-retaining benefits as dry aging, but that's

about it. In reality, wet-aging is a product of laziness and money-grubbing. It's easy to let that Cryovacked bag of beef from the distributor sit around for a week before opening the bag and calling it "aged." When you are being sold "aged" meat, be sure to ask whether it's been dry-aged or wet-aged. If the butcher doesn't know the answer or is unwilling to share, it's best to assume the worst.

The other drawback to wet-aging is that it can't be done for as long as dry-aging. It seems counterintuitive, considering that a wet-aged hunk of meat is largely protected by the outside environment. But if even a smidge of harmful anaerobic bacteria makes its way into that bag, the meat will rot inside its cover, though giving no indication that it's done so until you open it up.

# KNIFE SKILLS:
## Trimming Aged Beef

**STEP 1: PEEL OFF THE FAT**
Start by peeling off the outer fat cap. Since it was already removed once during butchering, this should be a pretty simple process.

**STEP 2: START TRIMMING**
Trim off the outer fat layers. The goal is to remove as little meat as possible, so work in thin slivers, going deeper and deeper, and then stopping as soon as the meat and fat look fresh. If the meat is a little slippery, use a clean kitchen towel to get a better grip.

**STEP 3: ALMOST THERE!**
Keep trimming the outer surfaces until only clean white fat and red meat are showing. Follow up by trimming the dried-out layer from the cut surfaces. You may need to fiddle around a bit to get the meat off the bones there, depending on how the beef was butchered.

**STEP 4: READY TO ROAST!**
Trimmed and ready to cook as a roast. To cut into individual steaks for cooking, read on.

**STEP 5: STEAK!**
Carefully slice through the meat between the bones. The only difficult part will be around the chine bone—you'll need to trim around it before cutting it off and discarding it. You'll end up with thick steaks, to serve about 2 people each.

# THE IMPORTANCE OF RESTING MEAT

- **Thought process of prehistoric man:** Start large fire. Cook large steak over large fire. Rip steak from fire with bare hands, bite into it, and allow succulent juices to dribble down chin. Howl at moon and chase mammoths.
- **Thought process of modern man:** Start large fire. Cook large steak over large fire. Rip steak from fire with bare hands, allow steak to rest in a warm place undisturbed for 10 minutes. Bite into it, and allow succulent juices to dribble down throat. Discuss latest Woody Allen movie with civilized friends while secretly wishing you could howl at moon and chase mammoths.

If there's one cooking mistake that regular folks make more often than any other, it's not properly resting meat before serving it. *You mean I have to wait before I can tuck into that perfectly charred rib-eye?* Unfortunately, yes.

Here's why:

This is a picture of a steak that was cooked in a skillet to medium-rare (an internal temperature of 125°F). The steak was then placed on a cutting board and immediately sliced in half, whereupon a deluge of juices started flooding out and onto the board. The result? Steak that is less than optimally juicy and flavorful. This tragedy can be easily avoided by allowing your steak to rest before slicing.

I've always been told that this deluge happens because as one surface of the meat hits the hot pan (or grill), the juices in that surface are forced toward the center, increasing the concentration of moisture in the middle of the steak, and then, when the steak is flipped over, the same thing happens on the other side. The center of the steak becomes supersaturated with liquid—there's more liquid in there than it can hold on to—so when you slice it open, all that extra liquid pours out. By resting the steak, you allow the liquid that was forced out of the edges and into the center time to migrate back out to the edges.

Seems to make sense, right? Imagine a steak as a big bundle of straws, representing the muscle fibers, each straw filled with liquid. As the meat cooks, the straws start to change shape, becoming narrower and putting pressure on the liquid inside. Since the meat cooks from the outside in, the straws are pinched more tightly at their edges and slightly less tightly in their centers. So far, so good. Logically, if the edges are pinched more tightly than the center, liquid will get forced toward the middle, right? Well, here's the problem: water is not compressible. In other words, if you have a two-liter bottle filled to the brim with water, it is (nearly) physically impossible to force more water into that bottle without changing the size of that bottle. Same thing with a steak.

Unless we are somehow stretching the centers of the muscle fibers to make them physically wider, there is no way to force more liquid into them. You can easily prove that the muscle fibers are not getting wider by measuring the circumference of the center of a raw steak versus a cooked one. If liquid were being forced into the center, the circumference should grow. It doesn't—it may appear to bulge, but that is only because the edges shrink, giving the illusion of a wider center. In fact, the exact opposite is the case. Since the center of a medium-rare steak is coming up to 125°F, it too is shrinking—and forcing liquid out. Where does all that liquid go? The only place it can: out of the ends of the straws, or the surface of the steak. That sizzling noise you hear as a steak cooks? That's the sound of moisture escaping and evaporating.

## Give That Theory a Rest

So why does an unrested steak expel more juices than a rested one? Turns out that it all has to do with temperature.

We already know that the width of the muscle fibers is directly related to the temperature to which the meat is cooked, and to a degree, this change in shape is irreversible. A piece of meat that is cooked to 180°F will never be able to hold on to as much liquid as it could in its raw state. But once the meat has cooled slightly, its structure relaxes—the muscle fibers widen up slightly again, allowing them to once again hold on to more liquid. At the same time, as the juices inside the steak cool, proteins and other dissolved solids cause them to thicken up a bit. Have you ever noticed that if you leave pan drippings from a roast to sit overnight, they are almost jelly-like? This thickening helps prevent those juices from flowing out of the steak too rapidly when you slice it.

I cooked a half dozen steaks all to an internal temperature of 130°F, then sliced one open every 2½ minutes to see how much juice would leak out. Here's what happened:

- **After no resting:** The meat at the exterior of the steak (the parts that were closest to the pan) is well over 200°F. At this temperature range, the muscle fibers are pinched tightly shut, preventing them from holding on to any moisture. The center of the steak is at 125°F. While it can hold on to some of its juices at this temperature, cutting the meat fibers open is like slitting the side of a soda bottle: some juice might stay in there (mostly through surface tension), but liquid is going to spill out.
- **After 5 minutes of resting:** The outermost layers of meat are down to around 145°F and the center of the steak is still at 125°F. At this stage, the muscle fibers have relaxed a bit, stretching open a little wider. This stretching motion creates a pressure differential between the center of the muscle fibers and the ends, pulling some of the liquid out from the middle toward the edges. As a result, there is less liquid in the center of the steak. Cut it open now, and some of the liquid will spill out, but far less than before.
- **After 10 minutes of resting:** The edges of the steak have cooled all the way down to around 125°F, allowing them to suck up even more liquid from the center of the steak. What's more, the center of the

*continues*

steak has cooled down to around 120°F, causing it to widen slightly. Cut the meat open at this stage, and the liquid will be so evenly distributed throughout the steak that surface tension is enough to keep it from spilling out onto the plate.

The difference is dramatic. Look back at the unrested steak, then take a look at this one:

With the unrested steak, all those delicious succulent juices are all over the plate. With the rested steak, everything stays inside, right where it belongs.

But wait a minute—how do we know that those juices really are staying inside the rested steaks? Is it not possible that in the 10 minutes you allowed it to rest that the liquid has simply evaporated, leaving you with a steak that is equally unmoist? To prove this is not the case, all you need to do is weigh the steaks before and after cooking. Aside from a minimal amount of weight loss due to rendered fat, the vast majority of weight loss comes from juices that are forced out of the meat. When cooked to 130°F, a steak loses around 12 percent of its weight during cooking. Cut it open immediately, and you lose an additional 9 percent. But allow it to rest, and you can keep the additional weight loss down to around 2 percent.

Resting is not just for steaks, by the way. At a fundamental level, pretty much *all* meat behaves the same way, whether it's a 30-pound standing rib roast or a 6-ounce chicken breast. The only differences are that just as cooking times are different for different-sized pieces of meat, so are resting times. By far the easiest and most foolproof way to test if your meat has rested long enough is the same way you can tell if your meat is cooked properly: with a thermometer.

Ideally, no matter how well-done you've cooked your meat, you want to allow it to cool until the very center is about 5 degrees below its maximum temperature. So for a medium-rare 130°F steak, you should allow it to cool to at least 125°F in the center before serving. At this stage, the muscle fibers have relaxed enough and the juices have thickened enough that you should have no problem with losing juices. With in a 1½-inch-thick steak or a whole chicken breast, this translates to around 10 minutes. For a prime rib, it may take as long as 45 minutes.

Afraid your steak will lose its crust as it rests? Easy solution: Reheat its pan drippings (or melt a panful of butter if you cooked the steak on the grill) and pour, smoking hot, over the steak just before serving.

## The Rules for Pan-Seared Steak

Let me explain. No, there is too much. Let me sum up:

1. **Dry your steak and season liberally at least 45 minutes before cooking.** A wet steak will not brown properly, as energy from the pan will go into evaporating the excess moisture rather than browning the steak properly. Salting your steak and letting it rest will draw out some moisture at first, but eventually, as muscle fibers break down, that moisture will be sucked back into the meat, leaving you with a well-seasoned, perfectly dry surface.

2. **Room temperature? Don't bother.** For better results, place your steak on a rack set in a rimmed baking sheet in the fridge for up to 3 days. I've found that even after resting at room temperature for 2 hours, when cooked side by side with a steak straight from the fridge, the results are completely indistinguishable. Don't bother.

3. **Use the heaviest pan you've got.** The heavier the pan, the more energy it can retain, and the more efficiently your steak will sear. A heavy cast-iron pan is my top choice for searing steak.

4. **Manage your temperature.** The goal is to get the steak perfectly browned just as it reaches its target internal temperature. For a normal-sized steak, say an inch thick or so, this means using a hot, hot pan. For a thick steak, use a more moderate temperature and baste the meat.

5. **Don't crowd the pan.** Too many cold steaks can make even the hottest pan cool too fast to cook effectively. For best results, make sure that your steaks have at least an inch of room around them on all sides. When cooking a large number of steaks, use multiple skillets, cook in batches, or, better yet, take things outdoors to the grill.

6. **Flip as often as you'd like.** Flipping multiple times not only gets your meat to cook a little faster, it also makes it cook more evenly and develop a crust just as nicely as a single flip will. That said, the difference is minimal, so if you don't want to flip every 30 seconds, don't sweat it.

7. **If using high heat for a normal-sized steak, don't add butter and aromatics until close to the end.** Butter contains proteins that can aid in browning, but if you add it too early on in the process, it can burn, turning bitter and acrid. Start your steak in oil and add the butter only for the last few minutes of cooking. If you'd like, add some aromatics, like thyme or rosemary stems, bay leaves, smashed garlic cloves, or sliced shallots at the same time. For a thicker steak that is cooked over more moderate heat, the butter can be added a little earlier.

8. **Get the edges!** Well, assuming you like your steak as thick as I do, there's a significant edge that sees little to no direct action during the entire searing process, but that edge is often the fattiest, most delicious part of the steak. It deserves love just as much as the next guy. Pick up your steak with tongs and get those edges seared!

9. **Rest before serving.** For maximum juiciness, it's important to let your meat rest for at least a few minutes after cooking. This allows muscle proteins to relax and the meat juices to thicken slightly so that they stay in place until the bite of steak reaches your mouth.

# THE LEIDENFROST EFFECT, AND HOW TO TELL IF YOUR PAN IS PREHEATED

Pop quiz: I've got two identical pans. One is maintained at 300°F on a burner and the other is maintained at 400°F. I then add a half ounce of water to each pan and time how long it takes for the water to evaporate. How much faster will the water in the 400°F pan evaporate than the water in the 300°F pan?

A. About ten times as fast.

B. At one-and-a-third times the rate.

C. At almost the same rate.

D. None of the above—I've already seen through your trick question.

You got it. The water in the 400°F pan will actually take longer to evaporate. In fact, when I performed this test at home, it took nearly ten times as long for the water in the hotter pan to vaporize. This seems contrary to pretty much everything we've learned so far, doesn't it? I mean, hotter pan = more energy, and more energy = faster evaporation, right?

The principle demonstrated by my test was first observed by Johann Gottlob Leidenfrost, an eighteenth-century German doctor. Turns out that if you give a drop of water on a pan enough energy, the steam that it produces will be pressed out so forcefully that it will actually lift the water droplet clear off the surface of the pan. Because the water is no longer in direct contact with the pan and is insulated by this layer of steam, the transfer of energy between the pan and the water becomes quite inefficient, so the water takes a long time to evaporate.

The very center of this skillet is still relatively cool, resulting in water that just bubbles as it sits there. The edges, however, are hot enough to induce the Leidenfrost effect, causing the entire blob of water to form a cohesive unit that elevates itself above the surface of the pan.

Here's a closer look at a Leidenfrost-ified water droplet:

This effect can be very useful in the kitchen as a means for judging how hot a pan is if you don't own an exceedingly sexy infrared instant-read thermometer like I do. Drop a bead of water on a pan while heating it. If it stays on the surface and evaporates rapidly, your pan is under 350°F or so—a suboptimal temperature for most sautéing and searing. If the pan is hot enough for the Leidenfrost effect to kick in, the water will form distinct drops that skid and scoot over the surface of the metal, taking quite a while to evaporate. Your pan is hot enough to cook in.

# QUICK AND EASY PAN-SEARED STEAKS

**NOTE**: For best results, let the steaks rest for at least 45 minutes at room temperature or up to 3 days, uncovered on a rack set in a rimmed baking sheet, in the refrigerator after seasoning them.

## SERVES 4

Two 1-pound boneless or bone-in
   rib-eye or strip steaks, 1 to 1½
   inches thick

Kosher salt and freshly ground
   black pepper

2 tablespoons vegetable oil

2 tablespoons unsalted butter

4 sprigs fresh thyme

2 large shallots, finely sliced (about
   ½ cup)

1. Carefully pat the steaks dry with paper towels. Season generously with salt and pepper on all sides.

2. Heat the oil in a 12-inch cast-iron or stainless steel skillet over high heat until smoking. Carefully add the steaks and cook, occasionally flipping them over, until both sides have developed a light brown crust (if the oil starts to burn or smokes incessantly, reduce the heat to medium-low), about 6 minutes.

3. Add the butter, thyme, and shallots to the pan and continue to cook—reducing heat if smoking is excessive—turning the steaks frequently, until they are deep brown on both sides and the center of the steaks registers 120°F for medium-rare, or 130°F for medium, on an instant-read thermometer, about 5 minutes longer. Transfer the steaks to a large plate, tent with foil, and allow to rest for 5 minutes before slicing in half and serving one half-pound steak per person.

4. Meanwhile, make a pan sauce if you like (see pages 318–21). Or serve the steaks with a compound butter (page 326), Foolproof Béarnaise (page 322), or Dijon mustard as desired.

# BUTTER-BASTED PAN-SEARED THICK-CUT STEAKS

**NOTES**: This recipe is designed for very large bone-in steaks, at least 1½ to 2½ inches thick and weighing 1½ to 2 pounds. Porterhouse, T-bone, rib-eye, and New York strip will all work well. Do not use tenderloin steaks, as they are likely to overcook.

For best results, let the steak rest for at least 45 minutes at room temperature or up to 3 days, uncovered, on a rack set in a rimmed baking sheet, in the refrigerator after seasoning it.

## SERVES 2 OR 3

One 1½- to 2-pound bone-in
  T-bone, porterhouse, strip, or rib-
  eye steak, 1½ to 2½ inches thick
Kosher salt and freshly ground
  black pepper
¼ cup vegetable or canola oil
3 tablespoons unsalted butter
6 sprigs fresh thyme or rosemary
  (optional)
2 large shallots, finely sliced (about
  ½ cup; optional)

1. Carefully pat the steak dry with paper towels. Season liberally on all sides (including the edges) with salt and pepper.

2. Heat the oil in a 12-inch cast-iron skillet over high heat until just beginning to smoke. Carefully add the steak, reduce the heat to medium-high, and cook, flipping the meat frequently, until a pale golden brown crust starts to develop, about 4 minutes.

3. Add the butter and herbs and shallots, if using, to the skillet and continue to cook, flipping the steak occasionally and basting it with the foaming butter, shallots, and thyme until an instant-read thermometer inserted into thickest part of the steak, away from the bone, registers 120°F for medium-rare, or 130°F for medium, 4 to 8 minutes longer. To baste, tilt the pan slightly so that butter collects by the handle and use a spoon to pick up the butter, and pour it over the steak, aiming at the light spots. If the butter begins to smoke excessively or the steak begins to burn, reduce the heat to medium. When the steak is done, transfer it to a rack set in a rimmed baking sheet. Let rest for 5 to 10 minutes. Reheat pan drippings until smoking, and pour over steak to re-crisp.

4. Carve the steak and serve.

# A NEW WAY TO COOK TENDERLOIN STEAKS

**Sure, the tenderloin is the most tender, buttery-smooth cut of meat on the whole steer, but it is bland, plain and simple.**

Because of its lack of fat, it's also extremely unforgiving when it comes to cooking. Fat plays two roles in a steak as it cooks: First, it's an insulator. Energy doesn't get transferred as efficiently through fat as through lean meat, which means that the more fat a steak has, the longer it takes to cook—and the wider the window of time during which it is perfectly cooked. A fatty rib-eye steak might have a good 45-second window during which you can pull it off the heat and have it be perfectly medium-rare. A tenderloin, on the other hand, goes from underdone to overdone in a matter of seconds. Not only that, but fat also gives you a nice buffer zone for overdoneness. Because fat lubricates and flavors meat, a nicely marbled steak will still taste pretty good even if it's slightly overcooked. Not so for a tenderloin, which turns pasty and chalky when cooked even a shade beyond medium.

All this is to say that it requires quite a bit of skill and patience to cook a tenderloin steak properly—at least it does if you're doing it the traditional way.

The last time I overcooked a tenderloin, I thought to myself, as I often do, "Shouldn't there be a much easier, more foolproof way to do this?"

Indeed there is.

The problem is that the high unidirectional heat of a skillet or grill makes not overcooking a tenderloin steak a very tough task indeed. So I first considered slow-roasting the steaks in a relatively low-temperature 275°F oven until perfectly cooked to medium-rare, followed by a hard sear in a skillet to crisp up the edges and brown them. That worked reasonably well, but the window of time for perfectly cooked steaks was still a matter of moments. *So how do I increase that window?* I thought to myself. *Why not just cook the meat as a single large roast*, then cut it into steaks? Because of its more limited surface area, a whole roast is far easier to cook evenly than individual steaks, especially when you consider that even with the most careful butchering, not all steaks are going to be of an equal size and shape, making it nearly impossible to cook them all to the exact same

degree of doneness. A larger roast also has a much bigger window for perfectly cooked meat by sheer virtue of the fact that it cooks more slowly.

I fired off another round, this time cooking a 2-pound tenderloin roast whole until it reached about 20 to 30 degrees below my desired final temperature of 130°F. After removing it from the oven, I sliced it into four evenly sized steaks, gently flattening each one, then seared them in a hot skillet with oil and finished them with butter. What resulted were steaks that were perfectly cooked from edge to edge, with a beautifully browned crisp crust—far better cooked than I'd ever managed using the traditional method. More even cooking was a happy by-product of the method. A steak cooked in the traditional way with high heat from the get-go will end up with a good amount of overcooked meat toward the exterior—raw steaks have to sit in a hot skillet for a good amount of time as they develop a good sear, and all the while, they're slowly overcooking. But a slow-roasted steak seared right at the end spends relatively little time in the hot skillet, resulting in more evenly cooked meat throughout.

Take a look at these two steaks below: the one on the left was cooked using the traditional hot-skillet approach, while the one on the right was cooked as part of a whole roast, then portioned into steaks, followed by a sear. Both have the exact same internal temperature, but there is far more perfectly cooked rosy meat in the roasted-then-seared steak than in the traditional steak.

Traditional searing leaves a gray ring.

My method gives you perfectly even cooking.

I know which one I'd rather eat.

OK, I hear some of you skeptics: is this *really* the way you cook steak at home all the time? No, of course not. It takes longer, and sometimes even *I* don't have extra time in the kitchen. If I'm in a rush, cooking prebutchered steaks using the method outlined in the Quick and Easy Pan-Seared Steaks recipe (page 311) works just as well for tenderloin—just be extra, extra, extra careful with that thermometer.

# PERFECT TENDERLOIN STEAKS

**NOTE:** A center-cut tenderloin roast is also referred to as a Châteaubriand. Ask your butcher for a 2-pound center-cut tenderloin roast, or trim one yourself (see page 317).

**SERVES 4**

One 2-pound center-cut tenderloin
  roast
Kosher salt and freshly ground
  black pepper
1 tablespoon vegetable or canola oil
1 tablespoon unsalted butter

1. Adjust an oven rack to the center position and preheat the oven to 275°F. Season the tenderloin liberally on all sides with salt and pepper. Place on a wire rack set on a rimmed baking sheet and roast until an instant-read thermometer inserted into the center of the meat registers 100°F for medium-rare (about 45 minutes), or 110°F for medium (about 50 minutes). Transfer to a cutting board (the roast will appear gray and uncooked at this stage).

2. Slice the roast into 4 even steaks, pat dry with paper towels, and season the cut surfaces with salt and pepper. Heat the oil and butter in a 12-inch cast-iron skillet over high heat until the butter is browned and lightly smoking. Add the steaks and cook until crusty on the bottom, about 1 minute. Flip the steaks with tongs and cook on the second side until crusty, about 1 minute longer. If the oil and butter are beginning to burn or smoke too heavily, reduce the heat. Flip the steaks onto one of their sides and cook, turning occasionally, until browned all over, about 1 minute longer. Transfer the steaks to a cutting board, tent with foil, and let rest for 5 minutes.

3. Serve the steaks with a pan sauce (see pages 318–21), compound butter (page 326), Foolproof Béarnaise (page 322), or Dijon mustard as desired.

# KNIFE SKILLS:
## How to Trim a Whole Beef Tenderloin

**Tenderloin steaks are expensive, but buying a whole untrimmed tenderloin can help you save some money. Not only that, you get some nice beef scrap to use for soups, burgers, or dog food in the process. Here's how to do it.**

- **TRIM OFF THE SILVERSKIN.** Use a sharp boning knife to remove the silverskin—the tough membrane that surrounds muscles—a small strip at a time (1). Insert the tip of the knife under the silverskin and, using your free hand to hold the tenderloin steady, cut away, trying to remove as little meat as possible (2). Cut in one direction first, then flip the knife over, grasp the end of the strip of silverskin you just removed, and cut back in the opposite direction to remove a full strip (3). Repeat until all the silverskin is removed.

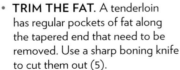

- **SEPARATE THE GRISTLE.** Tenderloins have a long "chain" of fat and gristle that runs along one side. It's quite easy to remove. Start by separating the chain with your hands, gently pulling on it to pry the meat apart at the natural seam. Then use just the tip of your knife to cut through any bits of tough connective tissue or membrane (4).

- **TRIM THE FAT.** A tenderloin has regular pockets of fat along the tapered end that need to be removed. Use a sharp boning knife to cut them out (5).

- **TRIM THE LARGE END.** The large end of the tenderloin has some fat and connective tissue hiding in its folds. Use the tip of the knife to slide under and carefully remove these (6).

- **CUT THE TENDERLOIN INTO SECTIONS.** The tenderloin can be roasted whole at this point—just fold the tapered end back on itself and tie it in place to get an even thickness along the whole length of the tenderloin. Or cut off the tapered end and the fatter end and save them for another use (7); this will yield you a center-cut tenderloin roast, which is what you want for a perfect-presentation roast or for steaks (8).

# PAN SAUCES

**You may notice after cooking your steak in a skillet that you're left with a brownish residue on the bottom of the pan. Don't scrub it out!**

You see, as meat cooks and shrinks, it exudes juices that are loaded with proteins. As these juices evaporate, the proteins end up on the bottom of the pan, where they eventually stick and brown—exactly like the browned proteins on the surface of your steak. The French, who seem to have a fancy-sounding word for everything, call this *fond*, as in *fondation*, or foundation, because it's the flavor base that all good pan sauces are built on. Here in America the technical term is tasty-brown-gunk-on-the-bottom-of-the-pan.

Pan sauces are made by *deglazing* (fancy word for "pouring liquid into the hot pan") the pan, usually with wine or stock. By rapidly reducing this liquid, then adding a couple of aromatics and finishing it off with a knob of butter, you've got yourself a quick and easy sauce that cooks in just about the same amount of time it takes for your steak to rest properly. Built-in timer!

Here are a couple of simple pan sauces. The toughest part of a pan sauce comes at the very end, when typically a bit of cold butter is whisked into the sauce to thicken it, give it some richness and body, and mellow out its flavor. The French call this step *monter au beurre*, which translates roughly to, "Please, Mrs. Cow, make my sauce extra-smooth and delicious." Or something like that. It's not a hard process, but if you aren't careful, the sauce can break, with greasy pools of butterfat floating on the surface of a thin, watery sauce. You don't want this to happen. The easiest way to prevent it? Add just a touch of flour to the skillet before adding the liquid. The starch in the flour will absorb liquid, which will cause it to expand and thicken the sauce and then keep your butter well emulsified when you get around to adding it.

# SIMPLE RED-WINE
## PAN SAUCE

**Use a decent-quality dry red wine. This recipe works just as well with lamb chops. If you'd like to use it for chicken or pork, substitute a dry white wine or white vermouth for the red wine.**

**SERVES 4**

1 medium shallot, finely minced
    (about ¼ cup)

4 tablespoons unsalted butter

1 teaspoon all-purpose flour

1 cup homemade or low-sodium
    canned chicken stock

1 cup dry red wine

1 tablespoon Dijon mustard

1 tablespoon minced fresh parsley

1 teaspoon fresh lemon juice

Kosher salt and freshly ground
    black pepper

1. After cooking the steaks, pour off the excess fat from the pan and return it to medium heat. Add the shallot and cook, stirring constantly with a wooden spoon, until softened, about 1 minute. Add 1 tablespoon of the butter and the flour and cook, stirring, for 30 seconds. Slowly whisk in the stock, wine, and mustard. Scrape up the browned bits on the bottom of the pan with the spoon, increase the heat to high, and simmer the stock until reduced to 1 cup, about 5 minutes.

2. Off the heat, whisk in the parsley, lemon juice, and the remaining 3 tablespoons butter. Season with salt and pepper to taste. Pour over your steaks and serve immediately.

# PORCINI-VERMOUTH
## PAN SAUCE

**This recipe works equally well with beef, pork, and chicken.**

### SERVES 4

½ ounce (about ¾ cup) dried
  porcini mushrooms

1½ cups homemade or low-sodium
  canned chicken stock

1 large shallot, minced (about ¼
  cup)

4 tablespoons unsalted butter

1 teaspoon all-purpose flour

1 teaspoon soy sauce

½ cup dry vermouth

1 teaspoon tomato paste

1 teaspoon fresh lemon juice

1 teaspoon chopped fresh thyme

Kosher salt and freshly ground
  black pepper

1. Before cooking the steaks, rinse the porcini in a large bowl of cold water to remove dirt and sand. Lift the porcini from the bowl and transfer to a microwave-safe 1-quart measure. Add the chicken stock and microwave on high for 1 minute. Set aside in a warm spot while you cook the steaks.

2. After cooking the steaks, pour off the excess fat from the pan and set the pan aside. Pour the porcini/chicken broth through a fine-mesh strainer into a small bowl, pressing on the porcini with a spoon to extract as much liquid as possible; reserve the liquid. Chop the porcini into rough ½- to ¼-inch pieces and return to the soaking liquid.

3. Return the steak pan to medium heat, add the shallot, and cook, stirring constantly with a wooden spoon, until softened, about 1 minute. Add 1 tablespoon of butter and the flour and cook, stirring, for 30 seconds. Slowly whisk in the soy sauce and vermouth, then whisk in the porcini and their liquid. Scrape up the browned bits on the bottom of the pan with the spoon, then whisk in the tomato paste. Increase the heat to high and simmer the sauce until reduced to 1 cup.

4. Off the heat, whisk in the lemon juice, thyme, and the remaining 3 tablespoons butter. Season with salt and pepper to taste. Pour over your steaks and serve immediately.

# SMOKY ORANGE-CHIPOTLE
## PAN SAUCE

**This recipe works equally well for steak, chicken, and pork.**

**NOTE:** To get strips of citrus zest, remove the zest with a peeler, being careful not to get much white pith, then thinly slice with a knife.

**SERVES 4**

1 medium shallot, finely minced (about ¼ cup)

4 tablespoons unsalted butter

1 teaspoon all-purpose flour

2 cups homemade or low-sodium canned chicken stock

A dozen thin strips of orange zest (from 1 orange; see Note above)

¼ cup orange juice

2 chipotle chiles packed in adobo sauce, finely chopped, plus 1 tablespoon of the sauce

2 teaspoons fresh lime juice (from 1 lime)

1 tablespoon minced fresh cilantro

Kosher salt and freshly ground black pepper

1. After cooking the steaks, discard the excess fat from the pan and return it to medium heat. Add the shallot and cook, stirring constantly with a wooden spoon, until softened, about 1 minute. Add 1 tablespoon of the butter and the flour and cook, stirring, for 30 seconds. Slowly whisk in the stock, orange zest, juice, chiles, and adobo sauce and stir to combine. Scrape up the browned bits on the bottom of the pan with the spoon, then increase the heat to high and simmer the stock until reduced to 1 cup, about 5 minutes.

2. Off the heat, whisk in the lime juice, cilantro, and the remaining 3 tablespoons butter. Season with salt and pepper to taste. Pour over your steaks and serve immediately.

# BÉARNAISE SAUCE

**Béarnaise is the ultimate creamy accompaniment to steak.**

Find your tenderloin a bit lacking in fat and flavor? Never fear, béarnaise sauce will come to the rescue! If you've already learned how to make Foolproof Hollandaise Sauce (page 111), then congratulations—you know how to make foolproof béarnaise as well. The two are nearly identical, the only difference being the liquid element. While hollandaise is an emulsion of butterfat, egg yolks, and lemon juice, béarnaise replaces the lemon juice with a tarragon-and-shallot-scented vinegar-and-white-wine reduction. Everything else is *exactly* the same.

# FOOLPROOF BÉARNAISE

## MAKES ABOUT 1 CUP

1 cup dry white wine

½ cup white wine vinegar

2 medium shallots, thinly sliced (about ½ cup)

6 sprigs fresh tarragon, leaves removed and minced (about 2 tablespoons), stems reserved

3 large egg yolks

½ pound (2 sticks) unsalted butter, cut into rough tablespoon-sized chunks

Kosher salt

1. Bring the wine, vinegar, shallots, and tarragon stems to a boil in a small saucepan over medium-high heat and cook until the mixture is reduced to about 1½ tablespoons and syrupy. Strain through a fine-mesh strainer into a small bowl.

### TO MAKE BÉARNAISE WITH AN IMMERSION BLENDER

2. Add the egg yolks and wine reduction to the blender cup (or a cup that will just barely hold the head of your blender).

3. Melt the butter in a small saucepan over medium-low heat and continue to heat until the butter just begins to bubble and registers 180° to 190°F on an instant-read thermometer. Transfer it to a liquid measuring cup, leaving the thin layer of whitish liquid in the pan (discard it).

4. Insert the head of the blender into the bottom of the cup and run the blender. Slowly pour in the melted butter. You should see the sauce begin to form at the bottom of the cup. As the sauce forms, slowly pull the head of the blender up to incorporate more melted butter, until all the butter is incorporated and the sauce has the consistency of heavy cream. Season to taste with salt and stir in the chopped tarragon. Transfer to a serving bowl or small saucepan, cover, and keep in a warm spot (not directly over heat!) until ready to serve.

### TO MAKE BÉARNAISE IN A STANDARD BLENDER OR FOOD PROCESSOR

2. Add the egg yolks and wine reduction to the blender or food processor and blend on medium speed until smooth, about 10 seconds.

3. Melt the butter in a small saucepan over medium-low heat and continue to heat until the butter just begins to bubble and registers 180° to 190°F on an instant-read thermometer.

4. With the blender running on medium speed, slowly drizzle in the butter over the course of 1 minute, stopping to scrape down the sides as necessary and leaving the thin layer of whitish liquid in the bottom of the pan (discard it). The sauce should be smooth, with the consistency of heavy cream. Season to taste with salt and stir in the chopped tarragon. Transfer to a serving bowl or small saucepan, cover, and keep in a warm spot (not directly over heat!) until ready to serve.

# GRILLED RIB-EYE (OR T-BONE, OR PORTERHOUSE, OR STRIP STEAK) FOR TWO

**Grilled steaks and pan-seared steaks are two completely different beasts flavorwise, but as far as cooking technique goes, there are only a few minor distinctions.**

For one thing, the heat you can get out of charcoal briquettes (or, better yet for searing, real hardwood coals) is far greater than what you can get out of a home stovetop range, leading to superior charring, as well as the singeing of the dripping beef fat, which gives grilled beef its characteristic smoky, ever-so-slightly acrid (in a good way) flavor. It's a flavor you simply can't get from a stovetop or even a gas grill, both of which burn significantly cooler than coal.

Going thick is always a good idea on the grill (steaks at least an inch and a half thick)—it's the best way to guarantee that you'll get plenty of good crust development while being able to maintain a nice, expansive medium-rare center. But ultra-thick Flintstones-sized double-cut bone-in big-enough-to-serve-two-fully-grown-Thundercats rib-eye steaks (commonly referred to as cowboy chops) require a bit of extra care when cooking. Their thickness makes it all too easy to end up with a burnt exterior and cold, raw middle.

Just as with roasting a large standing rib roast, the very best way to guarantee that you maximize that medium-rare center—you want to see pink from edge to edge—while still getting a nicely charred crust is to first cook the steak over a very gentle low heat before finishing it over ripping-hot heat to sear its surface. It's better to do it in this order rather than searing first and then cooking through, because a prewarmed steak will sear much faster, minimizing the amount of overcooked meat under the surface (and we all know now that searing does not lock in juices, right?)

## Step 1: Season

Season all sides of the steak well with kosher salt and black pepper at least 45 minutes before cooking, and up to overnight. At first the salt will draw moisture out of the surface of the meat, but then it will create a brine with this extracted liquid that will dissolve some of the meat proteins, allowing them to reabsorb the liquid, and the salt along with it. The result is deeper flavoring and a more tender texture. Letting the steak rest for up to 3 days uncovered on a wire rack set in a rimmed baking sheet in the fridge will further improve it.

## Step 2: Cook over Indirect Heat

Build a two-zone indirect fire with at least a full chimney's worth of coals banked entirely to one side of the grill. Or, if using a gas grill, heat one set of burners to high and leave the rest off. Cook the meat on the cooler side of the grill with the lid on, flipping the steak every 5 minutes or so, until it reaches 10 degrees less than the desired finished temperature (115°F after this step for medium-rare, 125°F for medium) on an instant-read thermometer. For a really thick steak, this can take up to half an hour or so.

## Step 3: Sear

Once you're within 10 degrees of your final serving temperature, transfer the steak to the hot side of the grill and leave the lid open; this will supply the coals with plenty of oxygen and allow them to burn hotter. Sear the steak, turning frequently, until it has built up a significant charred crust and reads 125°F for medium-rare, 130°F for medium. If you don't like the ultracharred taste of singed fat, keep a squirt bottle filled with water nearby to put out any flare-ups (I personally like the flavor).

## Step 4: Rest

Transfer the steak to a cutting board and allow it to rest for 10 minutes. During this time, its internal temperature should climb up to its maximum, then drop down again by a couple of degrees. You want to serve it after the temperature has peaked at 130°F and dropped back down to 128°F for medium-rare, or 140°F and then 138°F for medium.

## Step 5: Serve

After sufficient resting, carve the steak and serve immediately. Actually, with a bone-in steak this large, I like to serve it whole, allowing guests to cut hunks off for themselves. A 2-pound bone-in steak will serve at least two very hungry people, more likely three. This is rich stuff!

# PERFECT GRILLED STEAK
## FOR TWO

**NOTES**: For best results, use a bone-in steak, but a boneless steak can also be used; it should weigh about 1 pound. A New York strip, T-bone, or porterhouse steak can be used in place of the rib-eye.

For best results, let the steak rest for at least 45 minutes at room temperature, or up to 3 days, loosely covered, in the refrigerator after seasoning it.

**SERVES 2**

One 1½-pound bone-in rib-eye
   steak, at least 2 inches thick
Kosher salt and freshly ground
   black pepper

1. Season the steak liberally on all sides with salt and pepper. Set on a plate.

2. Light a chimneyful of charcoal. When all the charcoal is covered with gray ash, pour out and arrange on one side of a charcoal grill. Set the grilling grate in place, cover the grill, and allow to heat for 5 minutes. Or, if using a gas grill, heat one set of burners to high and leave the rest off. Clean and oil the grilling grate.

3. Place the steak on the cooler side of the grill, cover, and cook with all of the vents open, flipping the steak and taking the temperature with an instant-read thermometer every few minutes until it registers 115°F for medium-rare, or 125°F for medium, 10 to 15 minutes. Transfer to a cutting board and let rest for 2 minutes; leave the grill lid open. The added oxygen flow should get the coals burning very hot.

4. Transfer the steak to the hot side of the grill and cook, flipping frequently, until a deep char has developed and the internal temperature registers 125°F for medium-rare, or 135°F for medium, about 3 minutes. Transfer to a cutting board and allow to rest until the internal temperature peaks and then drops back down to 128°F for medium-rare, or 138°F for medium, about 10 minutes, then carve and serve.

## COMPOUND BUTTER

Even simpler than making a pan sauce is to serve your steaks with a compound butter, made by adding aromatics to softened butter. Place a disk or a dollop on each hot steak so that it can slowly melt, essentially emulsifying itself with the meat juices into a luxurious sauce. The great thing about compound butters is that you can make them in advance, wrap them into in a few layers of plastic wrap, and freeze them, then pull them out whenever you need them.

Pan sauces can also benefit from compound butters. Rather than swirling regular butter into the sauce to finish it, use a bit of compound butter to add complexity and flavor.

# MASTER RECIPE FOR COMPOUND BUTTER

**MAKES 4 TO 6 OUNCES**

8 tablespoons (1 stick) unsalted
   butter, at room temperature
Any of the Compound Butter
   Seasonings (recipes follow)

1. Combine the butter and seasonings in a medium bowl, using a fork to mash the butter until it's homogeneous.
2. Lay a 12-inch-long piece of plastic wrap on the work surface. Transfer the butter to the lower quarter of the plastic wrap, attempting to get it as close to a log shape as possible, then carefully roll the butter up in the plastic wrap to form a log. Twist the ends to tighten them. Refrigerate until hardened before using, or wrap tightly in foil, place in a zipper-lock freezer bag, and freeze for up to 6 months. To serve, slice off as much frozen butter as you need and let it soften at room temperature for 30 minutes.

## Compound Butter Seasonings

### Lemon-Parsley Butter Seasoning

2 tablespoons finely chopped fresh parsley

2 teaspoons grated lemon zest (from 1 lemon)

1 tablespoon fresh lemon juice

1 medium clove garlic, minced or grated on a Microplane (about 1 teaspoon)

Kosher salt to taste

### Blue-Cheese Butter Seasoning

4 ounces Gorgonzola, Roquefort, or Stilton cheese, softened

1 teaspoon Worcestershire sauce

1 small shallot, finely minced (about 2 tablespoons)

### Garlic-Chili Butter Seasoning

2 medium cloves garlic, minced or grated on a Microplane (about 2 teaspoons)

1 teaspoon chili powder

1 serrano or ½ jalapeño pepper, finely chopped

¼ teaspoon cayenne pepper

½ teaspoon ground cumin

2 teaspoons fresh lime juice (from 1 lime)

2 tablespoons finely minced fresh cilantro

Kosher salt to taste

# MARINATED Steak
## FOR THE GRILL OR THE PAN

We've talked about the expensive, ultra-tender cuts of beef. Now we move on to my favorites: the butcher's cuts. Those cuts of relatively inexpensive beef that require just a bit more care and attention to cook right, but reward you with incredible flavor. One of the keys to getting there? Proper marinating.

Before we go further, let's get one thing straight: marinades will not rescue poorly cooked or bland meat. After testing hundreds of marinade variations on all sorts of meat, I've found that the best marinades share three common ingredients: oil, acid, and a salty liquid, preferably a protease (more on those later).

### Key to Great Marinades #1: Oil

Oil is essential for three purposes. First, it emulsifies the marinade, making it thicker and tackier, causing it to stick more efficiently to the meat.

Second, many flavorful compounds—like those in onions, garlic, and many spices—are oil soluble. With a fat-based medium coating the meat, you get better, more even flavor distribution. Finally, the oil helps the meat cook more evenly, providing a buffer between the heat of the grill and the surface of the meat. Omitting it detracts from all three of these qualities.

### Key to Great Marinades #2: Acid

I used to think that acid was essential in a marinade for tenderizing purposes, and it's true—acid can slightly tenderize tough connective tissue in meat. Unfortunately, excessive acid can also start to chemically "cook" meat, denaturing its protein and causing it to firm up and eventually turn chalky (think: ceviche). If you're going to use acid in a marinade, it's best to go with no more than equal parts acid and oil, and limit exposure time to under 10 hours to prevent the meat from getting chalky. You may be surprised to learn that despite their reputation, marinades do not actually penetrate particularly far into meat—even after the course of a night, a marinade will penetrate no farther than a millimeter or two, and that penetration rate slows down the longer you marinate. So really, a marinade's effects are largely limited to the surface of the meat.

### Key to Great Marinades #3: Salt and Proteases

The final ingredient in a good marinade is a salty liquid. The muscle protein myosin will dissolve in a salty liquid, leaving the meat with a looser texture and a better ability to retain moisture. Want to do even better than just salt? Consider adding a protease to your marinade as well; That's an enzyme that breaks down proteins. Soy sauce is a great choice.

### Bonus: Aromatics

Aromatics are mainly a surface treatment, but they can still be quite powerful. Garlic, shallots, dried spices, herbs, or chiles are all good things to experiment with.

### HOW TO MARINATE

The goal with marinating is to maximize contact between the meat and the marinade. To do this, marinate your meat in a plastic zipper-lock bag with all the air squeezed out (I do this by leaving a small air hole along one edge of the zipper lock, squeezing all the air towards it, then sealing it at the last moment before juices start leaking out), or, even better, seal the steaks in a Cryovac-style bag with a vacuum sealer.

Timing-wise, you should marinate for at least 1 and up to 12 hours. Less, and the marinade simply doesn't stick as well. More, and the meat will start to get a bit too mushy and chalky around the edges, having a slightly cooked appearance from any acids or proteases present.

## THE SIX INEXPENSIVE STEAKS YOU SHOULD KNOW

There are dozens of cuts sold in supermarkets as cheap steak options, but here are my six favorites. These are the pieces of the steer that chefs love to use because not only are they less expensive, but they've got *character*.

The high-end steaks are all cut from the same general region of the steer—along the ribs and spine. Why? Because the muscles in that area—the longissimus dorsi and the psoas major—do little to no work during the steer's lifetime. They are large, tender, and remarkably easy to cut into big, juicy, meaty steaks.

The so-called butcher's steaks, on the other hand, come from all over the steer, and they're not always quite as easy to extract. Many of them are whole muscles that must be trimmed just so to be tender enough and large enough to cook as steaks. There are also not many of them on a steer. For every 20 pounds of rib-eye and T-bone steaks you can get from a steer, for example, you'll get perhaps 1 to 2 pounds hanger steak.

These butcher's cuts tend to be more flavorful because of the work the muscles do, but because they're not as marketable to the general public

Flank steak.

and require a bit more skill to cook and serve, they remain much cheaper than their mainstream counterparts. **This is good news for you!**

Here are the six steaks worth knowing. Some of them—like flank—are edging up in price to the not-so-inexpensive range, but no matter where you shop, you're bound to find one of them at a reasonable price.

Short ribs.

Hanger steak.

Tri-tip steak.

Skirt steak.

Flap meat (sirloin tips).

| NAME | TENDERNESS (on a scale of 1 to 10) | FLAVOR (on a scale of 1 to 10) | WHAT IT TASTES LIKE | THE BEST WAY TO COOK IT |
|---|---|---|---|---|
| **Hanger Steak** | 7/10 (when sliced against the grain) | 8/10 | This is my go-to butcher's cut. If properly trimmed, it'll come in 8- to 10-inch-long strips about 2 inches wide. Half-trimmed hanger steak will come as a larger piece with a strip of connective tissue running down the center that needs to be trimmed off. Hanger steak has kind of a coarse, squishy texture when rare, but when cooked to medium-rare to medium, it acquires a firm juiciness. | Pan-searing and grilling. Best cooked to at least medium-rare or medium (the texture is quite unpleasant any rarer). This makes it an excellent candidate for cooking sous-vide in a beer cooler (see page 392). |
| **Skirt Steak** | 6/10 | 7/10 | Skirt steak is sold as "inside" or "outside" skirt, depending on precisely where it's cut from; see page 333. Also sold as "fajita steak," it's got a tough membrane attached to one side of it that is usually peeled off before it's sold (if you find it with the membrane still attached, grip it with a kitchen towel and peel it away). It's one of the more flavorful cuts, with a distinct gaminess and plenty of fat. Like hanger steak, it's best cooked to medium and thinly sliced. Because of its thinness, it can be difficult to gauge doneness properly, so it may require some practice. | Grilling. Skirt is very thin, so high heat is essential to get it nicely browned before it overcooks in the center. To slice it, first cut it crosswise into 3- to 4-inch-long segments, then cut each segment lengthwise (against the grain) into thin strips. |
| **Flank Steak** | 7/10 | 5/10 | Once an inexpensive cut and a great alternative to the premium steaks, flank steak now often runs to nearly as much as a strip steak. A wide, flat, rectangular piece of beef with a moderately beefy flavor, plenty of juice, and a strong grain, it's a favorite for the grill. | Grilling. After resting, flank steak must be thinly sliced against the grain. Start by splitting it in half lengthwise (with the grain), then cutting each strip crosswise (against the grain) into thin pieces. |

| NAME | TENDERNESS (on a scale of 1 to 10) | FLAVOR (on a scale of 1 to 10) | WHAT IT TASTES LIKE | THE BEST WAY TO COOK IT |
|---|---|---|---|---|
| Short Ribs | 6/10 | 10/10 | This is my little secret (OK, the Argentines and Koreans know about it too). Most people consider the short rib to be merely a braising cut, but it makes a superbly beefy steak as well. The key is to find large short ribs with pieces of meat at least 1½ inches square and several inches long. If it's grilled or pan-seared to medium and sliced ultrathin against the grain, it's hard to think of a more flavorful steak. | Grilling, pan-searing. Short ribs are very high in fat, so they take longer to cook than most other steaks. Slice ultrathin against the grain before serving. |
| Tri-Tip Steak | 5/10 | 4/10 | Texturally, tri-tip resembles a flat-cut brisket, though it doesn't have nearly as much outside fat. Flavorwise, I'd peg it closer to an eye-round roast. It's not huge on beef flavor or fat, so it's generally a good idea to season it generously and serve it with a flavorful sauce. Because of its uneven tapered shape, tri-tip is a good choice when you've got guests who like meat at varying degrees of doneness. | Grilling. It's a bland cut, so the flavor a grill imparts can be a boon. |
| Flap Meat (Sirloin Tips) | 6/10 | 6/10 | A whole flap is a rectangular block of meat about 1½ inches thick, weighing about 2 to 3 pounds. Flap has a very strong grain and plenty of fat, with a rich, beefy flavor that works well in a range of applications. Butchers will often mislabel other lesser cuts as "sirloin tips" or "tip steaks." Look for, or ask your butcher for, the entire muscle to ensure you are getting the right thing. | Excellent for a wide variety of purposes: grilling, pan-searing, or even simmering in soups or stews; great for kebabs. If grilled or pan-seared, flap steak should be sliced with the grain into thick slabs and then against the grain into thin strips. |

## Hanger Steak

Hanger is like that indie band that hasn't quite hit Top 40 mainstream status yet but is big enough that everybody and their mother has heard about it. Most have even given it a try. For a long, long time, hanger wasn't even sold to the general public, reserved mostly for ground beef, or taken home by the butcher (earning it the nickname "butcher's steak"). If you traveled in France, you would have seen it on bistro menus as the *onglet*—a popular cut for steak frites. In the United States, however, your chances of running into it were much slimmer.

Then, sometime around the late 1990s or early aughts, chefs caught wind of it and it started appearing on menus of American bistros and fancy restaurants alike. Chefs liked it because it offered the full, beefy flavor and richness of more expensive cuts of meat like rib-eye or strip steak without the hefty price tag. These days, hanger steak has become so popular that it's no longer as cheap as it used to be (after all, there are only two on each steer, and they aren't particularly large), but it still comes in at around half to a third the price of a typical high-end steak at the supermarket.

**Also sold as:** Butcher's steak, hangar (this is an incorrect spelling but appears frequently), *arrachera* (Mexico), *fajitas arracheras* (south Texas), bistro steak, *onglet* (France).

**Where it's cut from:** From the plate section of the cow (the front of the belly); it "hangs" off the cow's diaphragm, hence the name.

**Shopping:** Hanger steak can be found in a few different forms in the market. Straight from the steer, it comes as two rather large, loose-grained muscles stuck together with a ton of connective tissue and silverskin surrounding it. If you're lucky, you'll have a good butcher who knows how to break it down into two well-trimmed steaks. Each one will be about a foot long and weigh in at 8 to 10 ounces, with a triangular cross section.

I've seen these individual steaks butterflied into wider, thinner steaks, supposedly to make for more even cooking. Really, though, a butterflied hanger steak is too thin to cook to medium-rare while still developing a nice crust, so I'd avoid them. Instead, stick to a regular trimmed steak or buy it untrimmed and do it yourself.

**Trimming:** You'll want to start by removing all the silverskin and excess fat from the exterior with a sharp, sharp boning knife: Slide the tip of the knife under the silverskin, grab the skin with your free hand, and then carefully pull the knife under it, taking off as little meat as you can. Eventually you'll end up with a piece of meat with two muscles attached by a thick sinew that runs down their center.

Cut the steak in half along that sinew to separate it into two individual steaks, then trim each one, and you're ready to cook.

**Cooking:** There are a several ways to cook hanger, both indoors and out, but no matter where you cook it, you want to make sure that you cook it to medium-rare or medium, no more, no less. Unlike, say, a rib-eye steak, which will still be pretty tender and juicy at medium and beyond, a hanger steak has a very coarse texture with a distinct grain running through it. Anywhere beyond medium, and it gets too rubbery to chew.

Undercook it, on the other hand, and you get meat that is mushy and slippery. Rare hanger steak is simply not the same as rare tenderloin, rib-eye, or strip steak. Use a thermometer, and cook it to the sweet spot between 125° and 130°F (this gives it some leeway to rise in temperature as it rests).

Very high heat is essential. Hanger steaks are relatively thin, and you want to give them a nice char before they get a chance to start overcooking. On the grill, I'll pile up a full chimney of coals under one side of the grill grate and cook the steaks full blast from start to finish, flipping occasionally. On the stovetop, use a pan cast-iron pan and go for smoking-hot high heat. Hanger steaks work well with marinades.

Finally, hanger steaks are good candidates for sous-vide cooking (see page 392), as it guarantees that they cook evenly all the way through. Cook them in a 125° to 130°F water bath, then finish them off with the highest-possible heat on a grill or the stovetop.

**Slicing and serving:** Like any meat, hanger steaks should be allowed to rest for a few minutes after cooking. Let it rest, then slice against the grain and serve. When properly cooked and sliced, a hanger steak is every bit as tender as a rib-eye.

## Skirt Steak

Of all the inexpensive cuts on the cow, skirt is probably the greatest dollar-to-flavor value there is. Riddled with plenty of buttery, beefy fat with a deep, rich flavor and a tender, juicy texture, it's a tough cut not to like. Indeed, I'd say that its flavor is even better than the rib-eye and far superior to a relatively bland tenderloin or New York strip.

That is, it's a tough cut not to like if you've had it cooked and sliced properly. All too often, you head out to a midrate taqueria where the skirt steak (known as *asfajitas*—"little belts"—in Mexico) sits around in piles on the edges of the griddle, slowly overcooking and turning from tender, juicy, steak-fit-for-a-king into your typical tough, leathery, livery-tasting taco stuffing.

Equally bad is the uncle who throws it onto a too-cool grill, forgets to rest it, and then slices it improperly, reducing it to inedibly tough rubber bands. Do *not* be this uncle. Your family may still love you, but they certainly won't like you.

**Also sold as:** Fajita meat.

**Where it's cut from:** The outside skirt is the diaphragm muscle of the cow. The inside skirt is part of the flank.

**Shopping:** The outside skirt is the traditional cut for fajitas, but it is generally sold only to restaurants; you'd be hard-pressed to find it retail. It comes with a tough membrane attached to it, which needs to be trimmed before it can be cooked.

Inside skirt is part of the flank, and it is the more widely available form of skirt steak. It generally comes with the membrane removed, making trimming an easy job at home. All you've got to do is remove some of the excess fat from the exterior, and you're good to go.

Trimming: Use a sharp knife, and try to take off the fat without digging into the meat. Some fat will remain inside the steak—this is good. It'll render as the meat grills, basting it as it cooks, giving the steak that much more richness and adding to its intense beefy, buttery flavor.

Cooking: There's a single rule when it comes to cooking skirt steak: intense, unrelenting, high heat. Forget starting low and slow or cooking sous-vide. Skirt steak should be cooked over the highest-possible heat from start to finish, and here's why: With a normal thick steak, if you were to cook it over intense heat the entire time, you'd end up with meat that burnt to a crisp on the outside before the center reached the appropriate medium-rare. Skirt steak has the opposite problem: it's so thin that unless you cook it over maximum heat, it'll be overcooked before you get any chance to develop a good sear on the exterior.

I like to light up an entire chimney of coals, pile them all under one side of the grill, and then add a few more coals on top just for good measure. As soon as those coals on top are hot, throw the suckers on and cook them with a single flip.* If you have

---

\* I know I've said to flip your steaks multiple times in the past, but the faster cooking this produces ends up overcooking your skirt.

hardwood coals, now's the time to break 'em out. They burn faster and hotter than briquettes, making them the ideal choice for grilling skirt steak.

Like hanger and other loose-textured cuts, skirt takes well to rubs and marinades. At the very least, use plenty of salt and pepper.

Slicing and serving: Take a look at the way the grain flows. Slice the steak with the grain to divide it into 3- to 4-inch-long segments, then rotate each one of those 90 degrees and slice into thin strips.

**Flank Steak**

It's hard to think of a cut of meat that is more conducive to cooking for a crowd than flank steak. It's got a robust, beefy flavor and a pleasantly tender texture, with a bit of good chew. It comes in large regular shapes that make cooking, slicing, and serving easy, and the steaks are thin enough that they'll cook through in a matter of minutes but thick enough that you can still get a nice medium-rare center.

Flank steaks are pretty versatile as far as cooking method goes, though they can be a bit unwieldy indoors because of their size. The best way by far is on the grill. With their large surface area, the steaks are made for picking up nice char and smoky flavors, and the types of dishes they can be transformed into, such as fajitas or steak salads, are perfect for al fresco dining.

Also sold as: Stir-fry beef (it's usually sliced in this instance).

Where it's cut from: The flank, which is on the steer's belly, toward the rear end.

Shopping: Once an inexpensive cut and a great alternative to premium steaks, flank steak can run nearly as much as a strip steak these days. When shopping for it, look for an even deep red color with a fair amount of fine fat running along the length of the muscle. Poorly butchered flank steak will either have a thin membrane still attached to parts of it or have had that membrane removed so aggressively that the surface of the meat has been shredded. Look for smooth-textured pieces without nicks or gouges.

A standard whole flank steak can weigh anywhere between 2 and 4 pounds. Plan on cooking a pound of flank steak for every 3 diners, or a pound and a half if your friends are as hungry as mine usually are.

Trimming: Flank steak is so popular that most butchers sell it trimmed and ready to cook.

Cooking: Flank has a close-grained texture that makes it suitable for serving anywhere between rare and medium. It can be tough to fit into a skillet without cutting it into smaller pieces, but it's great on the grill. Cook as you would a hanger steak—

high heat, flipping it occasionally, until it develops a good crust on both sides. If it begins to burn before the center reaches the desired temperature, transfer it to the cooler side of the grill to finish more gently.

Flank steak doesn't sop up marinades quite as readily as the more loose-textured hanger or skirt steak, but it's still worth marinating it for an added flavor boost.

Slicing and serving: Use a long, thin sharp carving knife or chef's knife to slice the steak against the grain. Holding the knife at a shallow angle will get you pieces that are a bit wider, which makes for a better presentation.

## Short Ribs

The Koreans and the Argentineans know something that we don't: short rib is the best cut of meat for grilling. In Korean restaurants, it's on the menu as *kalbi*. At most of them, you'll find the short ribs cut flanken-style—that is, thin slices cut across the ribs so you see a few rib-bone cross sections in each slice. At fancier restaurants, you'll find the ribs served as a single bone each, the meat carefully butterflied so that it stretches out into a long, thin strip.

In Argentina, the cut is known as *asado de tira*, and it's served thick-cut, grilled on an open fire, and drizzled with herb-oil-and-vinegar-based chimichurri sauce. More intensely beefy than a strip steak, more well marbled than a rib-eye, far more flavorful than a tenderloin, and thicker and meatier than a skirt or hanger steak, there's nothing—and I mean *nothing*—better on the grill than short ribs.

**Also sold as:** *Kalbi* (Korean), Jacob's Ladder (UK, when cut across the bones), *asado de tira* (Argentina).

**Where it's cut from:** The ribs (duh). Although short ribs can be cut numerous ways, they generally come from the area of the ribs a bit farther down toward the belly than rib steaks or strip steaks (which come from closer to the back). When cut into long slabs with bones about 6 to 8 inches in length, short ribs are referred to as "English cut." When sliced across the bones so that each slice has four to five short sections of bone, they are known as "flanken-style."

**Shopping:** Like any meat, short ribs can vary in quality. The very best come from high up on the ribs, close to where rib-eye steaks are cut from. The top 6 inches or so is what you're looking for. With ribs cut from this region, you'll find a bone about 6 inches long, 1½ inches wide, and ½ inch thick, with a slab of meat sitting on top of it about an inch high.

Some less-scrupulous butchers will sell sections cut from much lower down on the rib as short ribs. You'll recognize these by the skimpy amount of meat on them. Don't bother with them, they won't work at all (unless you've got a couple of hungry dogs). Look for meaty ribs with plenty of marbleing.

Either English- or flanken-style will work just fine on the grill, but I prefer to buy English-cut ribs. This affords me the possibility to remove the meat from the bone in one relatively thick steak. If

you can manage to find boneless short ribs, all the better. Simply slice them into individual steaks and they're ready to cook—no waste.

**Trimming:** If the ribs have a big cap of fat, trim it to about ⅛ inch. There should be no need to remove any sort of silverskin or connective tissue. English-cut short ribs can be cut off the bone as well. Save the bones for stock (or the dogs!).

**Cooking:** Because short ribs have such a high fat content—they are unforgivably rich—they're a relatively foolproof cut to work with. The intramuscular fat acts as an insulator, which means that they cook a bit more slowly, giving you a larger window of time to pull them off the grill at the desired level of doneness. I treat my short ribs much as I would a high-end Japanese Wagyu-style steak. That is, whether you like your regular steaks rare or well-done, I very strongly suggest cooking your short ribs to medium-rare—about 130°F. Any cooler than that, and the intramuscular fat will remain solid and waxy, rather than turning unctuous and juicy. Much hotter, and the fat will start running out copiously, making your ribs tough and dry.

Short ribs cook best over a moderately hot, not blazing-hot fire. Like all things, fat has a tendency to burn when it gets too hot. If you were to cook your ribs over an inferno, that dripping fat would vaporize, leaving a foul-tasting sooty deposit on the surface of your meat. You want to have the short ribs cooked through to the center exactly when the exterior becomes deep brown and crusty. They can also be cooked like a steak in a hot cast-iron pan.

I prefer my short ribs the Argentine way: seasoned with nothing but salt and served with chimichurri sauce.

**Slicing and serving:** Short ribs are a bit tougher than premium cuts of meat, so, once again, slicing thinly against the grain before serving (or instructing your diners to do so) is the way to go.

## Tri-Tip Steak

If you're not from Santa Maria, California, you may not have heard of tri-tip, the large, tender, triangular muscle cut from the bottom sirloin of a steer. If you are from Santa Maria, you can bet your bowl of pinquito beans that you've had more than your share of the cut.

It's the primary cut used for Santa Maria–style barbecue, a regional barbecue style that's not well known outside of central California and that, by some standards, wouldn't qualify as "real" barbecue

at all. See, Santa Maria–style barbecue is technically a fast-cooking method—that is, the meat is cooked over an open pit burning with red oak just until medium-rare. No low-and-slow smoking, no breakdown of connective tissue (which, actually, tri-tip is very low in), no fancy barbecue sauces. Just seasoned beef, grilled, sliced, and served with a bowl of native pinquito beans, a tomato salsa, and buttery garlic bread.

Sounds pretty darn good to me.

**Also sold as:** Santa Maria steak; Newport steak (when cut into individual steaks); *aguillote baronne* (France); *punta de anca*, *punta de Solomo*, or *colita de cuadril* (Latin America); *maminha* (Brazil).

**Where it's cut from:** The bottom sirloin, from the muscle group that controls the steer's back legs (it is the muscle that applies force to the steer's kneecaps).

**Shopping:** There's not much to watch out for here—tri-tips are pretty consistent. If you have a choice between Prime and Choice grades, I'd go with Prime—this is a case where you're going to want all the fat you can get, because tri-tip is a generally lean cut that is prone to drying out.

**Trimming:** A tri-tip should need no trimming, but if there is any silverskin on the surface, use a thin sharp knife to remove it.

**Cooking:** Tri-tip is not huge on beef flavor or fat, so it's a generally good idea to season it generously or serve it with a flavorful sauce. It has a tapered shape that makes it cook unevenly. If you have diners who prefer more-well-done meat, give them slices from the thinner, tapered end.

Traditional Santa Maria–style barbecue calls for

salt, black pepper, and perhaps a bit of garlic rubbed onto the meat before cooking. Personally, I like a somewhat heartier spice rub with paprika, a bit of cumin, cayenne, and some brown sugar. The flavor really helps boost the meat when sliced and served.

Just as when cooking a big fat steak, the key to even cooking, juicy meat, and a nice crust is to start the sucker over the cooler side of the grill with the lid on (you can add some soaked wood chunks to the coals—tri-tip takes well to smoke) and cook it to within 5 to 10 degrees of its target pull temperature (which is 5 degrees lower than its target final temperature). For medium-rare, that's about 115° to 120°F. After that, slide it on over to the hot side and cook it until nicely charred all around.

Slicing and serving: Pull it off, let it rest for about 10 minutes (it is, after all, a big cut of meat), and then thinly slice against the grain with a sharp knife.

You can serve it with whatever kind of sauce you like. The traditionalist in me says to go with a tomato-and-celery-studded Santa Maria–Style Salsa (page 348); the who-cares-as-long-as-it-tastes-good-person in me often tells the traditionalist to shut the hell up and serve it with any of the sauces in this chapter.

## Flap Meat

I first knew flap meat by its local New England name of sirloin tips. Go to any old-school dive or tavern with a menu, and you're bound to run into them, cut into cubes, stuck on a skewer, and grilled over an open coal fire, just like they do at Santarpio's in East Boston. When grilled right, flap meat is tender and juicy, and has a robust beefy flavor that a lot of other cuts for kebabs lack. That, and it's cheap. Not just cheaper-than-tenderloin-but-still-kinda-expensive cheap, but actually cheap.

It wasn't until I moved back to New York City that I realized that nobody outside New England knows what sirloin tip is, and it wasn't until even later that I realized that the "faux hanger" and "flap meat" that butchers around here sell are in fact the exact same cut of beef, just left whole rather than sliced into tips.

Of all the inexpensive cuts of beef, flap meat is one of the most versatile. It takes great to fast-cooking methods like grilling or searing. It's excellent cooked whole and then sliced into thin strips. It can't be beat cubed and put on skewers. It has a coarse texture that grabs onto marinades and seasonings. It's even great as a slow-cooked braise, where it falls apart into tender shreds, like a Cuban *ropa vieja*.

Also sold as: Faux hanger, *bavette* (France), sirloin tip (New England).

Where it's cut from: The bottom sirloin butt—the same general region the tri-tip comes from.

Shopping: Flap meat comes in several forms, depending on where you live, but it's pretty much always delivered to the supermarket or butcher as a whole cut of meat. So, if you live in an area where selling it in strips or cubes is the norm (such as New England), instead ask the butcher to sell you a whole trimmed flap. This gives you more options when you get it home. As it is a relatively lean cut, there's not really any need to spring for Prime-grade flap; the Choice stuff will taste just as good and costs less.

**Trimming:** Flap meat generally requires very little trimming; remove any silverskin.

**Cooking:** More so than any other cut I know of, flap meat is pretty terrible when it's cooked rare. You can see for yourself when it's still raw: this is some mushy-ass meat. Only by cooking it to medium-rare or medium can you get it firm enough to not squish around in your mouth as you chew it.

Flap meat works particularly well on the grill. It doesn't require the extreme heat of skirt steak, and it doesn't have the fat flare-up problems of short ribs, which makes it pretty simple to cook. Just build a hot fire on one side of the grill, lay on the flap (after seasoning it, of course), and flip every minute or so until it gets to at least 125°F at its thickest part.

**Slicing and serving:** Flap meat has an extremely coarse grain with an obvious direction: it runs crosswise all the way down the steak. This makes it hard to cut against the grain into thin bite-sized pieces (you'd end up with strips sliced lengthwise). Instead, the best thing to do it first divide it into three or four pieces, slicing with the grain, then to rotate each of those pieces 90 degrees and thinly slice them against the grain.

Since its shape, thickness, and proclivity for marinades makes flap similar to flank steak, you can use them pretty much interchangeably. Think of it as flank steak's tastier, sexier cousin.

## GOING AGAINST THE GRAIN

As you know, meat is made up of bundles of long muscle fibers that are laid out parallel to one another. Take a close look at your meat, and you'll see that, just like wood, it's got a grain. In some muscles, like the loin (where strip and rib-eye steaks come from) or tenderloin (aka fillet), that grain is very fine: the muscle fiber bundles are thin enough that they don't form a significant grain. Cuts from weak muscles like these will be soft and tender no matter how you slice 'em. Inexpensive cuts, like skirt steak, hanger steak (shown above), or flank, have thicker muscle fiber bundles with a clearly defined grain.

These fibers are tough cookies—they have to be. Their job is to move the moving parts of an animal that is much much bigger than you. Try to tear a single muscle fiber by stretching it along its length, and you'll have a pretty hard time. But pulling individual muscle fibers apart from one another is relatively easy. So, before putting a piece of flank, hanger, or skirt steak in your mouth, the goal should be to shorten those muscle fibers as much as possible with the help of a sharp knife. If you cut with your knife parallel to the grain, you end up with long muscle fibers that are tough for your teeth to tear through. Slicing the meat thinly against the grain, however, delivers very short pieces of muscle fiber that are barely held together. Take a look below at the difference between slicing with the grain (on the left) and slicing against the grain (on the right):

Really, that's about all you need to know, so you have full permission to skip the rest of this sidebar right now. But for those of you who, like me, had the greatest geometry teacher in the world in ninth grade and were thus instilled with a preternatural desire to draw triangles and measure stuff, well, in the words of Mr. Sturm, get your gas masks, because we are climbing Mount Elegance, and the air up there is quite thin!

Quantitatively, how much of an effect does the way you slice a steak actually have? Let's set up some definitions:

- Let w be the distance you move the knife between slices (i.e., the width of a slice).
- Let m be the length of the meat fibers in each slice.
- Finally, let $\theta$ be the angle between the knife blade and the meat fibers.

Given a bit of high school trigonometry, you can quickly come up with the following formula:

$$m = w \, / \sin(\theta)$$

So, what are the implications of this? Well, if our goal is to minimize the length of the meat fibers (m), then we need to maximize $\sin(\theta)$. When the meat is cut into 0.5-inch strips at a 90-degree angle to the direction of the meat fibers, $\sin(\theta)$ is equal to 1 (i.e., maximized) and the meat fibers are exactly as long as the slice is wide, i.e., 0.5 inch. Cut them at a 45-degree angle instead, and while their width is still 0.5 inch, the length of the meat fibers has reached 0.707 inch (that's the square root of 0.5, for all you nerds out there who get excited over 45°–45°–90° right triangles). That's an increase of almost 50 percent! Now take it to the extreme: if you were to cut perfectly parallel to the meat fibers, then $\sin(\theta)$ will be equal to 0 and, according to the unbreakable laws of mathematics, your meat fibers would stretch all the way into infinity. That's one big cow!

Moral of the story: Cut your meat perpendicular to the grain to maximize tenderness. This applies to all types of meat: beef, lamb, pork, turkey, buffalo, bison, mammoth—you name it. If it's got muscles, the direction you slice matters.

# PAN-SEARED HANGER STEAK
## WITH HERB AND GARLIC MARINADE

**This recipe is for an oil-based marinade designed to allow the flavors of the herbs and garlic to season the meat evenly. As such, it won't do much harm to let it sit for longer than the six or so hours I recommend for a regular marinade. In fact, it will take a little longer for the flavors to transfer, as they first need to come out of the herbs into the oil, and then from the oil to the meat. Leaving the herbs and spices whole allows you to wipe them off before searing (or grilling).**

**NOTE**: Hanger steak can be replaced with flank, skirt, or flap.

### SERVES 4

**For the Marinade**

¼ cup extra-virgin olive oil

4 cloves garlic, split in half and gently smashed with the side of a knife

1 tablespoon whole black peppercorns

1 teaspoon fennel seeds

1 teaspoon coriander seeds

½ cup roughly chopped fresh parsley leaves and stems

4 sprigs fresh thyme, roughly chopped

2 tablespoons soy sauce

Kosher salt

2 pounds hanger steak, cut into four 8-ounce steaks

1 teaspoon vegetable oil

2 tablespoons unsalted butter

1. **To make the marinade:** Combine the olive oil, garlic, peppercorns, fennel seeds, coriander seeds, parsley, thyme, and soy sauce in a small bowl and whisk well. Season to taste with salt. Place the hanger steaks in a gallon-sized zipper-lock freezer bag, add the marinade, press out the air, and seal the bag. Allow the meat to marinate in the refrigerator, turning occasionally, for at least 6 hours, and up to 24.

2. Remove the meat from the marinade and wipe off the excess, including any whole spices, herbs, and garlic cloves. Heat the vegetable oil in a 12-inch cast-iron or heavy-bottomed stainless steel skillet over high heat until smoking. Carefully add the steaks and cook, flipping them frequently, until both sides have developed a light brown crust (if the oil starts to burn or smokes incessantly, reduce the heat to medium-low), about 4 minutes.

3. Add the butter and continue to cook, turning the steaks frequently and reducing the heat if the pan is smoking excessively, until they are deep brown on both sides and the center registers 125°F for medium-rare, or 135°F for medium, on an instant-read thermometer, about 5 minutes longer. Transfer the steaks to a large plate and allow to rest for 5 minutes before slicing and serving.

4. Meanwhile, make a pan sauce if desired (see pages 318–21), or serve the steaks with a compound butter (page 326) or Dijon mustard.

# STEAK HOUSE–STYLE GRILLED MARINATED FLANK STEAK

**This marinade is my version of an A-1–style Worcestershire-based steak sauce, flavored with anchovy, soy sauce, Marmite, and brown sugar. It's ultrasavory and a little bit sweet. The flank steak can be replaced with hanger, skirt, or flap meat.**

## SERVES 4

### For the Marinade

½ cup Worcestershire sauce

¼ cup soy sauce

3 tablespoons brown sugar

4 anchovy fillets

2 teaspoons Marmite (optional)

2 cloves garlic, split in half and gently smashed with the side of a knife

2 tablespoons Dijon mustard

2 tablespoons tomato paste

½ cup vegetable oil

2 tablespoons chopped fresh chives

1 medium shallot, minced (about 2 tablespoons)

1 flank steak (about 2 pounds)

1. **To make the marinade:** Combine the Worcestershire sauce, soy sauce, brown sugar, anchovies, Marmite, if using, garlic, mustard, tomato paste, and vegetable oil in a blender and blend until smooth, creamy, and emulsified. Transfer one-third of the marinade to a small bowl, add the chives and shallots, and stir to combine; refrigerate. Place the flank steak in a gallon-sized zipper-lock freezer bag, add the remaining marinade, press out the air, and seal the bag. Allow the meat to marinate in the refrigerator, turning occasionally, for at least 1 hour, and up to 12 hours.

2. Remove the steak from the marinade and pat dry with paper towels. Ignite a large chimneyful of coals and wait until they're covered in gray ash, then spread evenly over one half of the grill. Put the cooking grate in place, cover, and allow the grill to preheat for 5 minutes. Or, if using a gas grill, heat one set of burners to high and leave the rest off. Scrape the cooking grate clean.

3. Place the flank steak on the hot side of the grill and cook until well charred on the first side, about 3 minutes. Flip and continue to cook until the second side is well charred, about 3 minutes longer. Transfer the steak to the cooler side of the grill, cover, and cook until the center registers 125°F for medium-rare, or 135°F for medium, on an instant-read thermometer. Transfer to a cutting board and allow to rest for at least 5 minutes, then carve the steak and serve, passing the reserved marinade at the table.

# SPICY THAI-STYLE FLANK STEAK SALAD

**This is probably my favorite beef marinade of all time. Sweet, spicy, sour, and salty, it encompasses the four basic flavors of Thai cuisine. While there's no salt added to the recipe, the fish sauce and soy sauce are plenty salty.**

**NOTES**: You can also cook the meat indoors in a grill pan or in a large skillet—split the steak crosswise in half so you can fit it in the pan. Flank steak can be replaced with hanger, skirt, or flap meat.

## SERVES 4

### For the Marinade

½ cup packed brown sugar

¼ cup water

3 tablespoons Asian fish sauce

2 tablespoons soy sauce

⅓ cup lime juice (from 3 to 4 limes)

2 medium cloves garlic, minced or grated on a Microplane (about 2 teaspoons)

1 tablespoon Thai chile powder or red pepper flakes

¼ cup vegetable oil

1 flank steak (about 2 pounds)

### For the Salad

½ cup loosely packed fresh mint leaves

½ cup loosely packed fresh cilantro leaves

½ cup loosely packed fresh basil leaves

4 medium shallots, finely sliced (about 1 cup)

1 small cucumber, cut into ½-inch chunks

1 to 2 cups mung bean sprouts, rinsed and dried

1. **To make the marinade:** Combine the sugar and water in a small saucepan and heat over medium heat, stirring, until the sugar is completely dissolved. Transfer to a small bowl. Add the fish sauce, soy sauce, lime juice, garlic, and chile powder and stir to combine. Transfer half the marinade to a small container and reserve. Add the oil to the remaining marinade and whisk to combine. Place the flank steak in a gallon-sized zipper-lock freezer bag, add the marinade, press out the air, and seal the bag. Allow the meat to marinate in the refrigerator, turning occasionally, for at least 1 hour, and up to 12 hours.

2. Remove the steak from the marinade and pat dry with paper towels. Ignite a large chimneyful of coals and wait until they're covered in gray ash, then spread evenly over one half of the grill grate. Put the cooking grate in place, cover, and allow the grill to preheat for 5 minutes. Or, if using a gas grill, heat one set of burners to high and leave the rest off. Scrape the cooking grates clean.

3. Place the flank steak on the hot side of the grill and cook until well charred on the first side, about 3 minutes. Flip the steak and continue to cook until the second side is well charred, about 3 minutes longer. Transfer the steak to the cooler side of the grill, cover, and cook until the center of the steak registers 125°F for medium-rare, or 135°F for medium, on an instant-read thermometer, about 5 minutes longer. Transfer to a cutting board, tent with foil, and allow to rest for at least 5 minutes.

4. Thinly slice the beef against the grain and transfer to a large bowl. Add the herbs, shallots, cucumber, bean sprouts, and reserved marinade and toss to combine. Serve immediately.

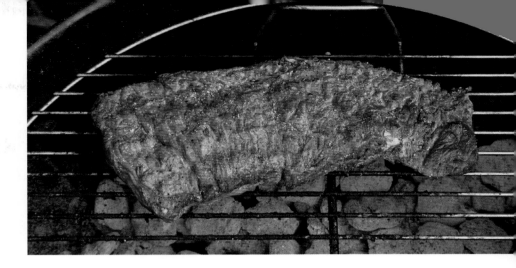

# GRILLED FLAP MEAT (STEAK TIPS)
## WITH HONEY-MUSTARD MARINADE

**Flap meat can be bought as a single large piece of meat, but it's more commonly served cut into strips labeled "sirloin tip" or "steak tips." Either will do for this recipe. If you buy it whole, divide it into individual serving portions before marinating. Flap can be replaced with hanger, flank, or skirt steak.**

**SERVES 4**

For the Marinade

½ cup soy sauce

¼ cup honey

¼ cup Dijon mustard

2 medium cloves garlic, minced or grated on a Microplane (about 2 teaspoons)

½ cup vegetable oil

Kosher salt and freshly ground black pepper

2 pounds flap meat

1. **To make the marinade:** Combine the soy sauce, honey, mustard, and garlic in a medium bowl and whisk, then slowly add the oil, whisking constantly. Transfer half of the marinade to a small container and reserve. Place the meat a gallon-sized zipper-lock freezer bag, add the remaining marinade, press out the air, and seal the bag. Allow the meat to marinate in the refrigerator, turning occasionally, for at least 1 hour, and up to 12 hours.

2. Remove the meat from the marinade and pat dry with paper towels. Ignite a large chimneyful of coals and wait until they're covered in gray ash, then spread evenly over one half of the grill grate. Put the cooking grate in place, cover, and allow the grill to preheat for 5 minutes. Or, if using a gas grill, heat one set of burners to high heat and leave the rest off. Scrape the cooking grates clean.

3. Place the meat on the hot side of the grill and cook until well charred on the first side, about 3 minutes. Flip the meat and continue to cook until the second side is well charred, about 3 minutes longer. Transfer the meat to the cooler side of the grill, cover, and cook until the center of the meat registers 125°F for medium-rare, or 135°F for medium, on an instant-read thermometer, about 5 minutes longer. Transfer to a cutting board, tent with foil, and allow to rest for at least 5 minutes.

4. Thinly slice the meat against the grain, drizzle with the reserved marinade, and serve.

# GRILLED MARINATED SHORT RIBS
## WITH CHIMICHURRI

**Like the marinade for Hanger Steak (page 342), this oil-based marinade benefits from an extended marination time.**

### SERVES 4

½ cup extra-virgin olive oil

3 medium cloves garlic, minced or grated on a Microplane (about 1 tablespoon)

1 small shallot, minced (about 1 tablespoon)

½ teaspoon red pepper flakes

2 tablespoons minced fresh oregano or 2 teaspoons dried oregano

4 pounds flanken-style or English-cut short ribs

Kosher salt and freshly ground black pepper

1 recipe Chimichurri Sauce (page 395)

1. Combine the olive oil, garlic, shallot, pepper flakes, and oregano in a large bowl. Add the short ribs to the large bowl, season with plenty of salt and pepper, and toss until thoroughly coated. Transfer the short ribs to two gallon-sized zipper-lock freezer bags, press out the air, and seal the bags. Allow the meat to marinate in the refrigerator, turning occasionally, for at least 6 hour, and up to 24 hours.

2. Ignite a chimneyful of coals and wait until they're covered with gray ash, then spread evenly on one half of the grill grate. Set the cooking grate in place, cover the grill, and allow to preheat for 5 minutes. Or, if using a gas grill, heat one set of burners to high and leave the rest off. Clean and oil the grilling grate.

3. Wipe the excess marinade off the short ribs and place on the hot side of the grill. Cook, turning frequently, until they are charred on all sides and the center of the meat registers 125°F on an instant-read thermometer, 8 to 10 minutes. If the fire flares up, transfer the short ribs to the cooler side of grill and cover the grill until the flames die down, then continue to cook on the hot side. Transfer the ribs to a cutting board, tent with foil, and let rest for 5 minutes, then serve with the chimichurri.

# SANTA MARIA–STYLE GRILLED TRI-TIP

**NOTE**: If desired, add a few chunks of oak that have been soaked in water for 30 minutes to the coals. The tri-tip can be replaced by a boneless top sirloin roast.

**SERVES 4 TO 6**

1 tri-tip roast (about 2½ pounds)

4 medium cloves garlic, minced or grated on a Microplane (about 4 teaspoons)

Kosher salt and freshly ground black pepper

1 recipe Santa Maria–Style Salsa (recipe follows)

1. Light a chimneyful of coals and wait until they're covered with gray ash, then spread evenly on one half of the grill grate. Set the cooking grate in place, cover the grill, and allow to preheat for 5 minutes. Or, if using a gas grill, heat one set of burners to high and leave the others off. Clean and oil the grilling grate.

2. Rub the roast all over with the garlic and season well with salt and pepper. Place on the cooler side of grill, cover, and cook, turning and flipping occasionally, until into the thickest part of the roast registers 105° to 110°F on an instant-read thermometer, 20 to 30 minutes. If using a charcoal grill and the coals are no longer hot, remove the roast from the grill, add another 4 cups of coals, and wait 5 minutes for them to heat up, then return the meat to the hot side of the grill.

3. Continue to cook, turning the roast regularly, until it is well charred on the exterior and the center registers 120° to 125°F, 5 to 8 minutes longer. Transfer to a cutting board and let rest for 10 minutes, then thinly slice the meat and serve with the salsa.

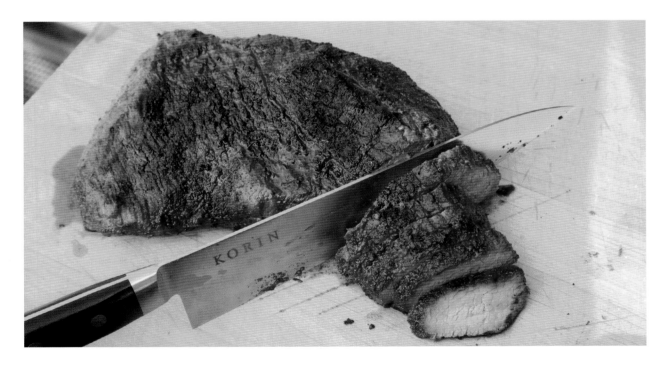

## Santa Maria–Style Salsa

**MAKES ABOUT 4 CUPS**

2 large ripe tomatoes, diced (about 2½ cups)

Kosher salt

1 stalk celery, peeled and diced (about ½ cup)

4 scallions, chopped (about ½ cup)

1 California chile (poblano or Hatch can be substituted), diced (about ½ cup)

¼ cup chopped fresh cilantro

2 medium cloves garlic, minced or grated on a Microplane (about 2 teaspoons)

1 tablespoon red wine vinegar

1 teaspoon Worcestershire sauce

1. Combine the tomatoes with ½ teaspoon salt in a colander and toss to coat. Set the colander in the sink and let sit for 30 minutes.

2. Combine the drained tomatoes, celery, scallions, chile, cilantro, garlic, vinegar, and Worcestershire in a medium bowl and toss. Season to taste with salt. Let sit at room temperature for at least 1 hour before serving. Refrigerate leftovers in a sealed container for up to 5 days.

# THE BEST FAJITAS

**I'm not particularly proud of my time spent working
at the kinds of cheesy chain restaurants you'd find next to the
Victoria's Secret at the mall, or perhaps in Times Square.**

But aside from making me shun any writer that uses the phrase "X to perfection," it did teach me one valuable lesson: People looooooove meat served on a sizzling platter. It was a well-known phenomenon: If a waiter sold one order of Extreme Fajitas™ to a table in their section, a half dozen more orders would quickly follow.

The meat itself should be ultra-juicy, with an overwhelming, almost buttery beefiness—this is skirt steak, after all, the butteriest of all beef—accented by a marinade that is slightly sweet, very savory, and packed with lime and chili. While fajitas are traditionally made with outside skirt—part of the diaphragm muscle of the steer—the cut is pretty much unavailable unless you work for a restaurant that special orders it. At the butcher or meat counter, you're far more likely to find inside skirt, which will do us just fine.

## GRILLED SKIRT STEAK FAJITAS

**NOTE:** If skirt is unavailable, substitute with hanger or sirloin flap (also sold as sirloin tip in New England—it's different from sirloin steak). Flank steak can also be used. For best flavor, grind your own chile powder from a mix of equal parts ancho and guajillo chiles.

## SERVES 4 TO 6

½ cup soy sauce

½ cup lime juice (from 6 to 8 limes)

½ cup canola oil

¼ cup packed brown sugar

2 teaspoons ground cumin seed

2 teaspoons freshly ground black pepper

1 tablespoon chile powder (see the headnote)

3 medium cloves garlic, finely minced (about 1 tablespoon)

2 pounds trimmed skirt steak (about 1 whole steak), cut crosswise into 5- to 6-inch pieces (see page 333 for trimming instructions)

1 large red bell pepper, stemmed, seeded, and cut into ½-inch-wide strips

1 large yellow bell pepper, stemmed, seeded, and cut into ½-inch-wide strips

1 large green bell pepper, stemmed, seeded, and cut into ½-inch-wide strips

1 white or yellow onion, cut into ½-inch slices

12 to 16 fresh flour or corn tortillas, hot

1 recipe Pico de Gallo, for serving, if desired (recipe follows)

Guacamole, sour cream, shredded cheese, and salsa, for serving, if desired

1. Combine the soy sauce, lime juice, canola oil, brown sugar, cumin, black pepper, chile powder, and garlic in a medium bowl and whisk to combine. Transfer ½ cup of the marinade to a large bowl and set aside. Place the steaks in a gallon-sized zipper-lock bag and add the remaining marinade. Seal the bag, squeezing out as much air as possible. Massage the bag until the meat is fully coated in marinade. Lay flat in the refrigerator, turning every couple of hours for at least 3 hours and up to 10.

2. While the steak marinates, toss the peppers and onion in bowl with the reserved ½ cup marinade. Refrigerate until ready to use.

3. When ready to cook, remove the steaks from the marinade, wipe off any excess, and transfer the steaks to a large plate. Light one chimneyful of charcoal. When all the charcoal is lit and covered with gray ash, pour out and arrange the coals on one side of the charcoal grate. Set a cooking grate in place, cover the grill, and allow it to preheat for 5 minutes. Clean and oil the grilling grate.

4. Place a large cast-iron skillet over the cooler side of the grill. Transfer the steaks to the hot side of the grill. Cover and cook for 1 minute. Flip the steaks, cover, and cook for another minute. Continue cooking in this manner—flipping and covering—until the steaks are well-charred and an instant-read thermometer inserted into the center registers 115° to 120°F for medium-rare or 125° to 130°F for medium. Transfer the steaks to a large plate, tent with foil, and allow to rest for 10 to 15 minutes.

5. Meanwhile, transfer the cast-iron skillet to the hot side of the grill and allow it to preheat for 2 minutes. Add the pepper and onion mix and cook, stirring occasionally, until the vegetables are softened and beginning to char in spots, about 10 minutes. When the vegetables are cooked, transfer the steaks to a cutting board and pour any accumulated juices from the plate into the skillet with the vegetables. Toss to coat.

6. Transfer the vegetables to a warm serving platter. Thinly slice the meat against the grain and transfer to the platter with the vegetables. Serve immediately with hot tortillas, pico de gallo, guacamole, and other condiments as desired.

## Classic Pico de Gallo

**NOTE:** Use the ripest tomatoes you can find. In the off-season, this generally means smaller plum, Roma, or cherry tomatoes.

**MAKES ABOUT 4 CUPS**

1½ pounds ripe tomatoes, cut into ¼- to ½-inch dice (about 3 cups, see Note above)

Kosher salt

½ large white onion, finely diced (about ¾ cup)

1 to 2 serrano or jalapeño chiles, finely diced (seeds and membranes removed for a milder salsa)

½ cup finely chopped fresh cilantro leaves

1 tablespoon lime juice from 1 lime

1. Season the tomatoes with 1 teaspoon salt and toss to combine. Transfer to a fine mesh strainer or colander set in a bowl and allow to drain for 20 to 30 minutes. Discard the liquid.

2. Combine the drained tomatoes with the onion, chiles, cilantro, and lime juice. Toss to combine and season to taste with salt. Pico de gallo can be stored for up to 3 days in a sealed container in the refrigerator.

Pan-Seared Pork Chops with Brandied Cherries (page 355).

# PAN-SEARED PORK CHOPS

**Now that we've perfected pan-searing
beef, how hard could pork be?
The answer: not very.**

Indeed, the only difference between cooking a fat steak and cooking a fat pork chop lies in how to select it and the final temperature. In the old days (i.e., before the 1990s), people in this country believed that all pork had to be cooked to at least 165°F in order to rid it of pesky worms that could infect you. These days, our pork is just as safe as our beef, so you can cook it to medium or even medium-rare with confidence. I like my pork best when it's a rosy pink, 135° to 140°F.

When selecting pork chops, you're likely to find four options at the butcher. They're all cut from the loin of the pig.

| NAME | DESCRIPTION |
|------|-------------|
| **Blade-End Chop** | Cut from the front of the loin near the shoulder, it contains several muscle groups, all divided by swaths of fat. This is my favorite cut, because it is self-basting. It's very hard to get a dry blade chop! |
| **Rib Chop** | Cut from farther back along the loin, this is the picture-perfect chop you'll find at fancy restaurants. It's got a large, smooth eye of meat and is easy to cook evenly. However, that eye of meat is relatively low in fat, making it easy to overcook, and relatively dry. |
| **Center-Cut Chop** | The equivalent of a T-bone steak, a center-cut pork chop contains part of both the loin and the tenderloin. It's impressive looking, but I find it very hard to cook evenly. The tenderloin cooks faster than the loin, and the bone gets in the way, preventing you from getting good contact with the pan. |
| **Sirloin Chop** | Cut from near the back end of the loin, this chop contains tons of muscles from the hip and ham. Compared to the other chops, it's tough and not particularly tasty, and I don't buy it very often. |

Pork has a particular affinity for sweet sauces, so when serving it, I usually like to make a fruit preserve–based pan sauce (page 355), or perhaps one flavored with maple syrup (page 356) or even apples and cider (page 356).

# ENHANCED PORK

Unlike beef, which has a ton of internal fat, most pork these days is relatively lean, making it very easy to overcook. One way to fix this problem is by brining it (see page 358). Just as with turkey and chicken, a soak in a saltwater solution can cause the meat's protein structure to loosen up, allowing it to retain more moisture as it cooks, though, just as with poultry, it'll also dilute the flavor. Pork producers figured this out a while ago, and many of them started injecting their pork with a briny solution. The vast majority of industrially raised pork in this country is now sold in this form, known as "enhanced" pork. To see if your pork has been enhanced, check the label: if it is, it'll say something like "with up to 10% sodium solution."

Seems convenient, right? Prebrined meat? Unfortunately, the brine solution sits in the meat for far too long. Rather than tasting simply juicy, most enhanced pork borders on spongy, with an odd ham-like texture. I much prefer to buy unenhanced natural pork and dry-brine it myself, allowing me to control exactly how much liquid it retains. Oftentimes I don't brine it at all, knowing that if I cook them carefully and monitor the temperature, the chops'll still end up nice and juicy.

**To dry-brine pork chops,** season well on all surfaces with kosher salt. Place on a wire rack set in a rimmed baking sheet and refrigerate, uncovered, for at least 45 minutes and up to 3 days.

# BUCKLE UP!

Pork chops have a tendency to buckle as they cook, making it hard to maintain even contact with the pan. This happens when the layer of fat around the chop's exterior shrinks faster than the meat inside, squeezing it and causing it to buckle. To prevent this from happening, score the fat in two or three places with a sharp knife. Your chops won't look as flawless, but they'll cook a heck of a lot more evenly.

# BASIC PAN-SEARED PORK CHOPS

**Just as with beef, it's better to cook pork bone-in. While the bone won't add flavor to your meat, it does act as an insulator, and there is less exposed surface area with a bone-in chop, which helps it to retain more moisture as it cooks. For best results, season the chops, place them on a wire rack set in a rimmed baking sheet, and refrigerate, uncovered, for at least 45 minutes and up to 3 days.**

## SERVES 4

Four 6- to 8-ounce bone-in pork chops (blade-end or rib), about 1 inch thick, brined if desired (see page 354)

Kosher salt and freshly ground black pepper

1 tablespoon vegetable oil

1 tablespoon unsalted butter

1 teaspoon minced fresh thyme (optional)

1. Pat the pork chops dry and season with salt (omit the salt if the chops were brined) and pepper. Heat the oil in a 12-inch cast-iron or heavy-bottomed stainless steel skillet over high heat until smoking. Carefully add the chops and cook, flipping them frequently, until both sides have developed a light brown crust (if the oil starts to burn or smokes incessantly, reduce the heat to medium-low), about 5 minutes.

2. Add the butter and thyme, if using, and continue to cook, turning the chops frequently, until they are deep brown on both sides and an instant-read thermometer inserted in the center registers 135°F for medium, about 5 minutes longer. Transfer the chops to a platter, and allow to rest for 5 minutes before serving.

## PAN-SEARED PORK CHOPS WITH BRANDIED CHERRIES

½ cup brandy

½ cup dried cherries or pitted sweet or sour fresh cherries

2 tablespoons sugar

1 tablespoon unsalted butter

1 tablespoon balsamic vinegar

Kosher salt and freshly ground black pepper

Combine the brandy and cherries in a small bowl and set aside. Cook the chops as directed. While the chops are resting, add the cherry mixture and the sugar to the skillet, return to medium-high heat, and cook, scraping up the browned bits from the bottom of the pan, until the brandy is reduced to about ¼ cup and the mixture is slightly syrupy. (Be careful, the brandy may ignite.) Off the heat, whisk in the butter and vinegar. Season to taste with salt and pepper. Pour over the chops and serve.

*continues*

## MAPLE-MUSTARD-GLAZED PAN-SEARED PORK CHOPS

½ cup maple syrup

¼ cup molasses

1 tablespoon bourbon

2 tablespoons whole-grain mustard

Kosher salt and freshly ground
   black pepper

Whisk together the syrup, molasses, bourbon, and mustard in a small bowl and set aside. Cook the chops as directed. While the chops are resting, pour off the fat from the skillet. Add the maple syrup mixture, return to medium-high heat, and cook, scraping up the browned bits from the bottom of the pan, until the mixture is reduced by half and syrupy. Season to taste with salt and pepper. Return the chops to the skillet and turn to coat with the glaze. Serve immediately, pouring the extra glaze on top.

## PAN-SEARED PORK CHOPS WITH APPLE AND CIDER SAUCE

½ cup cider vinegar

½ cup apple cider

½ cup packed dark brown sugar

Pinch of ground cinnamon

Pinch of ground cloves

2 tablespoons unsalted butter

1 Granny Smith apple, peeled,
   cored, and cut into ½-inch cubes
   (about 1 cup)

Kosher salt and freshly ground
   black pepper

Whisk together the vinegar, cider, brown sugar, cinnamon, and cloves; set aside. Cook the chops as directed. While the chops are resting, add the butter and diced apples to the pan, return to medium-high heat, and cook, stirring constantly, until the apples are browned and softened, about 3 minutes. Add the vinegar mixture and reduce until syrupy, about 4 minutes. Season to taste with salt and pepper. Pour the apples and sauce over the chops and serve immediately.

# PAN-SEARED CHICKEN PARTS

**There's a terrible problem sweeping this country, an insidious one that has managed to work its way into nearly every household, with no distinctions among race, gender, or class.**

I'm talking about dry chicken breasts, and it's time for us to Just Say No. Luckily, there's an easy program to help you do it, and all it requires is a small investment: a good instant-read thermometer.

As with many problems in life, the root causes of dry chicken are noble, and they stem from the government's—more specifically, the Food and Drug Administration's—recommendation to cook chicken breasts to 165°F. Just like beef and pork, chicken meat tightens as you cook it, and by the time it's reached 165°F, it's irretrievably, irrevocably dry. With chicken legs, this is not so much of a problem. Because of their large amounts of fat and connective tissue, you can cook a chicken leg all the way to 180°F or even 190°F and still get some semblance of succulence. Chicken breasts, though, with their large, roundish shape and total lack of fat, can't handle temperatures much above 145°F (for a discussion on chicken safety, see page 360).

So, what's the best way to cook a chicken? If we learned anything from pan-searing steaks and pork chops, it would seem to be to flip the chicken repeatedly as it cooks. Well, I tried it and quickly learned that, a chicken is not a cow, and there are key differences between the structure of a chicken breast and that of a steak that make flip-cooking unfeasible—namely, the skin. Luckily for us, the skin provides some amazing benefits that allow us to cook chicken much more easily than we can cook beef and pork. Now, now, I know some odd folks don't like eating chicken skin (surely its most delicious feature!), but I'm here to tell you that regardless of whether you end up eating it or pushing it to the side of your plate, it should *stay on the chicken while you cook it.*

Here's the thing: without skin, what happens when you try to pan-sear a chicken breast? The meat at the exterior dries out, turning stringy and leathery, and not at all pleasant. Try using the multiple-flip method with a boneless, skinless chicken breast, and you'll end up leaving half the chicken on the bottom of the pan as you flip it. There are few things in life I hate more than skinless, boneless chicken breasts. Rule Numero Uno for great chicken is to start by buying skin-on, bone-in chicken. Leaving the skin on the chicken not only prevents the breast meat from overcooking at the surface, it also allows you to let the chicken cook for a longer time on one side without the danger of uneven cooking you'd get with steaks or chops. Its because chicken skin, with its abundance of fat, is a natural insulator. Think about it: fat, through millions of years of evolution, has been *designed* to help regulate temperature. Its purpose is to even out abrupt changes in temperature so that animals can move relatively freely between cold and hot environments without dying of shock. Perhaps it was never intended to see a situation as extreme as the refrigerator to the skillet, but it performs its role admirably nevertheless.

With a layer of skin on one side and an insulating bone on the other, it becomes exponentially easier to cook chicken evenly. All you've got to do is cook it skin side down first, in a hot skillet (don't try and move it until it releases itself freely), then flip it over and toss it into a moderately hot oven to finish. So long as you're careful about monitoring its temperature and allow it to rest, never again will you have to deal with anything other than moist, juicy, tender meat. And, just like with pork chops, brining your chicken before cooking can help it to become even juicier, though it's a largely unnecessary step, provided you're careful with your thermometer, and it has the same drawback as when brining pork—the chicken flavor gets diluted with water. I see it more as a safeguard against overcooking.

## BRINING MEAT: THE BIG TRADE-OFF

Cold, hard fact time: all meat dries out and toughens as it cooks, particularly in the very hot zones on its exterior. Yet we want the center to cook through. How does one heat the center without cooking the exterior to dry oblivion? Enter **brining**, the process in which a lean cut of meat (like turkey, chicken breast, or pork) is soaked in a saltwater solution to help it retain moisture during cooking. Sure, sure—this is nothing new. The Scandinavians and Chinese have been extolling the virtues of brining for millennia, but is it worth it? What are the trade-offs? Before we jump on the bandwagon, consider a few simple queries: namely, what does it do, how does it work, and should I bother?

### Why Brine?
Let's start with what brining actually accomplishes.

Time to break out the science. I started with a dozen nearly identical chicken breasts. Three of them were cooked as is. Three were soaked overnight in a 6-percent solution of saltwater (about ½ cup Diamond

Crystal kosher salt, or ¼ cup table salt, per quart of water) before cooking. Three were salted and left to sit overnight before cooking (a technique sometimes referred to as dry-brining), and the final three were soaked overnight in just water before cooking. I measured the weight of the chicken breasts (and consequently the amount of moisture lost) at each stage of the process.

All twelve breasts were cooked simultaneously in a 275°F oven until they reached an internal temperature of 150°F. Here's what happened:

## WEIGHT LOSS IN COOKED CHICKEN BREASTS

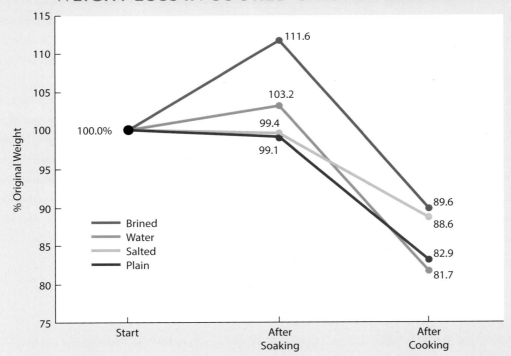

*continues*

|          | STARTING WEIGHT | AFTER SOAKING | AFTER COOKING |
|----------|-----------------|---------------|---------------|
| Plain    | 100%            | 99.1%         | 82.9%         |
| Brined   | 100%            | 111.6%        | 89.6%         |
| Salted   | 100%            | 99.4%         | 88.6%         |
| Water    | 100%            | 103.2%        | 81.7%         |

As you can see, while a plain chicken breast lost about 17 percent of its moisture during cooking, a brined breast lost only 10 percent. Salted breasts came in just under brined breasts, at 11 percent. Soaked in plain water, the breasts gained about 3 percent in weight prior to cooking, but all of that extra water came right back out—water-soaked breasts fared no better than plain breasts.

From this data, we know that salt is doing something to help retain moisture, whether applied through a brine or simply rubbed on the surface of the breasts. How does it accomplish this? It's the same as salting a steak before you cook it: it's in the shape of the proteins. In their natural state, muscle cells are tightly bound within long protein sheaths—this doesn't leave much room for extra water to collect in the meat. But as anyone who has ever made sausages or cured meats knows, salt has a powerful effect on muscles (see page 496). A salt solution will effectively denature (read: unravel) the proteins that make up the sheath around the muscle bundles. In their loosened denatured state, you can fit more water into those muscles than in their natural state. Even better, the denatured proteins in the sheaths contract far less as they cook, squeezing out much less moisture.

So, which method is better: brining or extended salting? From the chart alone, you'd guess brining; the meat retains an entire extra percentage point of moisture. But is this all good news? I can hear you all now saying, no more dry pork, chicken, or turkey? Sign me up! Not so fast. There's a major trade-off when it comes to brining, and it's got to do with flavor. You see, while your meat may end up juicier, remember that much of the juice it's now holding on to is nothing more than tap water. This can have a pronounced effect on the flavor of the meat. With salting, on the other hand, all of the juices in there naturally occur in the meat.

I've repeated this test numerous times with everything from turkey to pork chops and always come to the same conclusion: salting and resting your meat is superior in every way to brining.

## CHICKEN TEMPERATURE AND SAFETY

Take a look at the USDA's basic cooking guidelines, and you'll see that they recommend cooking foods to higher temperatures than anyone in their right mind would want to eat them at. They recommend cooking all pork, beef, and lamb to at least 145°F—well into the medium to medium-well range. Poultry has it even worse. The USDA recommends that *all* poultry, whether ground or in whole parts, be cooked to at least 165°F. It's no wonder that most folks think chicken is a dry meat.

The USDA likes to play it safe, and their guidelines for safety are intended to be simple to understand and foolproof at the expense of being accurate. The rules are designed such that anybody from the burger-flipper at Wendy's to the most amateur home cook can understand and grasp them, ensuring food safety across the board. I, on the other hand, prefer to put a little more faith in my readers' intelligence.

The fact of the matter is that bacteria are not quite as simple as the "alive at 164°F, dead at 165°F" the guidelines would have you believe them to be. Rather, a number of factors, including free moisture, fat content, the levels of dissolved solids like salt or sugar, and temperature, can all affect bacterial growth and decay in different ways.

The USDA knows this, of course, and if you delve deep enough into their guides, you'll find some useful charts that begin to shed some light on the subject of meat safety. Below is a graph I put together by drawing data from their chart on obtaining a 7 log reduction in salmonella in chicken.

## MINIMUM SAFE COOKING TIME VERSUS TEMPERATURE FOR CHICKEN

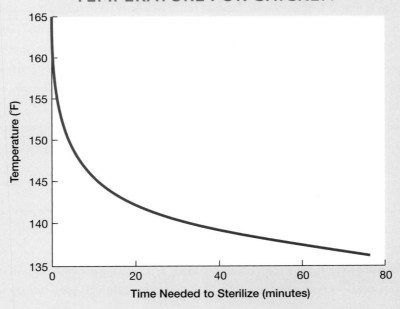

What you find is that there isn't a simple temperature limit that defines when chicken is safe to consume; rather, it's a combination of temperature and time. The line on the graph above essentially represents how long a piece of chicken must be held at a specific temperature in order to be considered safe to consume.

*continues*

---

※A 7 log reduction would mean destroying 99.99999% of the bacteria present—enough to give you a good margin for food safety.

| DEGREES FAHRENHEIT | TIME FOR CHICKEN | TIME FOR TURKEY | TIME FOR BEEF |
| --- | --- | --- | --- |
| 135 | 63.3 minutes | 64 minutes | 37 minutes |
| 140 | 25.2 minutes | 28.1 minutes | 12 minutes |
| 145 | 8.4 minutes | 10.5 minutes | 4 minutes |
| 150 | 2.7 minutes | 3.8 minutes | 72 seconds |
| 155 | 44.2 seconds | 1.2 minutes | 23 seconds |
| 160 | 13.7 seconds | 25.6 seconds | Instantaneous |
| 165 | Instantaneous | Instantaneous | Instantaneous |

So, at 165°F, the chicken is safe pretty much instantaneously. This is why the USDA sets 165°F as the lower limit for their general guidelines. At 155°F, the same bacterial reduction takes 44.2 seconds to occur in chicken. At 150°F, 2.7 minutes: 2.7 minutes! That's all!

What this means is that as long as chicken stays at 150°F or higher for at least 2.7 minutes, it is as safe to eat as chicken that has been cooked to 165°F. I monitored a few pieces of chicken that I cooked to 150°F before removing them from the heat and found that their temperature increased to about 153°F during the first few minutes and stayed at well above 150°F for a good 6 minutes before they started to cool down again. Even chicken cooked to 145°F can be safe if you let it rest. It easily maintains its temperature for the requisite 8.4 minutes.*

My mom often comments on how incredibly moist my roast chicken is, believing that it's some secret technique or marinade I'm using to get it that way. Want to know the secret? *Just don't overcook it.*

## What Temperature Should I Cook Chicken To?

Fact: 145°F chicken is not for everybody. Some folks used to chicken having a certain texture may find it a little too wet and soft. Here's a quick guide to chicken temperatures:

- **140°F**: Pinkish-tinged and almost translucent; extremely soft, with the texture of a warm steak; fleshy.
- **145°F**: Pale, pale pink but completely opaque; very juicy, a little soft. This is my favorite temperature for chicken.
- **150°F**: White and opaque, juicy, and firm.
- **155°F**: White and opaque, starting to turn a little bit stringy; bordering on dry.
- **160°F and higher**: Dry, stringy, and chalky.

---

* A modern sous-vide–style cooking apparatus allows you to cook and hold chicken and other meats at even lower temperatures with completely safety. You can approximate the results at home with a beer cooler. See page 389 for details.

# KNIFE SKILLS:
## How to Break Down a Chicken

**If there's one knife skill that can save you money *and* make you look cool at the same time, it's breaking down a chicken.**

Consider that boneless breasts often cost around three times more per pound than a whole chicken does. For the same price as a two-pack of breasts, you can buy a whole chicken, which comes with those same breasts, plus two legs and a back. And, wait for it—if you're really lucky, you'll get a free liver, heart, and gizzard thrown in to sweeten the deal! Of course, if you don't know how to break down the chicken, this is not too useful. But just follow the pictures, here and you'll be breaking down chickens like the pros.

**STEP 1: THE TOOLS**
You'll need a sharp chef's knife and a pair of poultry shears or a cleaver. Extra coolness points if you've got the cleaver.

**STEP 2: SPREAD 'EM**
Grab the chicken by the drumstick and pull the leg out from the body until the skin is stretched taut.

**STEP 3: THE FIRST INCISION**
Start the operation by cutting through the skin between the leg and the body. Don't cut too deep—just through the skin. No matter what Cat Stevens says, the first cut should be the shallowest.

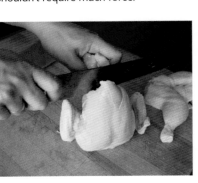

**STEP 4: POP THE JOINT**
Grab the leg in one hand and twist it down away from the body until the ball joint pops out of the socket. This shouldn't require much force.

**STEP 5: REMOVE THE THIGH**
Use your chef's knife to completely remove the leg by cutting through the joint you just exposed, making sure to get the little nugget of meat that sits closest to the chicken's spine (this is called the oyster and should be fought over at the table).

**STEP 6: THE OTHER FOOT**
Repeat steps 2 through 5 on the other leg.

**STEP 7: CRACK THE BACK**
Hold the chicken by the back and position it vertically on your cutting board, with the butt end pointing up. Use your chef's knife to cut through the skin and cartilage between the breast and the back until you get through the first or second ribs.

*continues*

## STEP 8: BREAK OUT THE CLEAVER

Switch over to your cleaver and continue cutting through the ribs, using short, firm strokes. Alternatively, use poultry shears to cut the through the ribs on both sides.

## STEP 9: CUT THROUGH THE SHOULDERS

Use the tip of the cleaver to cut through the shoulder bones on either side, or use the poultry shears.

## STEP 10: HALFWAY DONE!

The backbone should now be completely separated from the whole breast. Save it for stock.

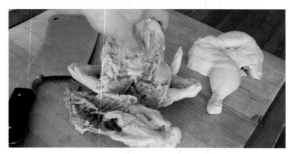

## STEP 11: SPLIT THE BREAST

To split the breast, cut through either side of the breastbone until you hit the sternum. Using your free hand, press down firmly on the blade until it cracks through the bone.

## STEP 12: 4-PIECE CHICKEN

A standard 4-piece (quartered) chicken is two breast halves (with or without the wings attached) and two leg quarters. The back is an added bonus. To break down the chicken into 8 or 10 pieces, keep going.

## STEP 13: FIND THE BALL

Use your fingertip to locate the ball joint between the thigh and drumstick.

## STEP 14: CUT AT THE JOINT

Cut through the joint with your chef's knife, separating the thigh from the drumstick. Repeat with the second leg.

## STEP 15: DIVIDE THE BREASTS

Cut each breast crosswise in half by pressing down on your knife blade with your free hand until you crack through the breastbone.

## STEP 16: FINISHED

Eight pieces of chicken, ready for stewing, braising, pan-roasting, or frying. Cut the wings off at the joint, and you've got a ten-piece chicken.

# PAN-ROASTED CHICKEN PARTS

**SERVES 4**

One 4-pound chicken, cut into
8 serving portions, or 3 pounds
bone-in, skin-on mixed chicken
pieces (thighs, drumsticks, breast
quarters), brined or dry-brined if
desired (see pages 358 and 579)

Kosher salt and freshly ground
black pepper

2 teaspoons vegetable oil

1. Adjust an oven rack to the center position and preheat the oven to 350°F. Season the chicken generously with salt (omit the salt if using brined chicken) and pepper.

2. Heat the oil in a 12-inch heavy-bottomed stainless steel skillet or sauté pan over medium-high heat until wisps of smoke appear. Swirl the oil to coat the pan, then remove from the heat and carefully add the chicken, skin side down. Return to the heat and cook, without moving the chicken, until the skin is a deep golden brown, about 5 minutes. Flip the chicken and continue to cook until the second side is lightly golden, about 3 minutes.

3. Transfer the skillet to the oven and roast until the thickest part of the chicken breast registers 150°F on an instant-read thermometer and the thighs and drumsticks register 175°F (remove the pieces to a plate as they reach their temperature and cover loosely with aluminum foil). If desired, make a pan sauce (see pages 368–70) while the chicken rests for 10 minutes. Serve.

## QUICK CHICKEN CUTLETS

There are times in life—say, fifteen minutes before that Mr. Wizard marathon is about to start—when even simple pan-roasted chicken parts take too long to cook. At these moments, wise home cooks call on that great savior of the quick dinner: chicken cutlets. Made by splitting a boneless chicken breast horizontally in half and gently pounding the resulting pieces to a thickness of about ¼ inch, they cook in less time than it takes to boil an egg. Once you've prepped the chicken, you've got all of 3 minutes before it's ready to eat, 10 minutes if you want to add a pan sauce (I would).

Most supermarkets carry cutlets, but you can make your own starting with whole boneless, skinless chicken breasts; see "How to Prepare Chicken Cutlets," page 367. Because of the cutlets' thinness, you've *got* to use very high heat when cooking them lest they overcook before they have a chance to brown and build up flavor. I like to dredge my cutlets in a little bit of flour before adding them to the skillet. The thin coat of flour browns more efficiently than plain chicken does, giving you color more quickly while offering some amount of protection to the meat.

My initial thought when cooking chicken cutlets was that both sides should be cooked for about the same amount of time—after all, I wanted even browning, right? But after trying this method over and over, I found that it simply wasn't possible to brown both sides without really overcooking the meat. So, why not just brown one side *extra* well? It worked like a charm. I placed the cutlets in a skillet with hot oil and cooked them without moving until the first side was well browned. At this stage, the cutlets were already almost cooked through. All they needed was a quick kiss of the flame on the

second side—about 30 seconds' worth—and they were finished, ready to rest and serve.

# 3-MINUTE CHICKEN CUTLETS

**SERVES 4**

4 boneless, skinless chicken breast
  halves, cut horizontally in half and
  pounded to a ¼-inch thickness
  (see below), brined or dry-brined
  if desired (see pages 358 and 579)

Kosher salt and freshly ground
  black pepper

1 cup all-purpose flour

3 tablespoons vegetable oil

1. Season the chicken cutlets with salt (omit the salt if using brined chicken) and pepper and dredge lightly in the flour, shaking off the excess. Transfer to a plate or cutting board.
2. Heat 1½ tablespoons of the oil in a 12-inch heavy-bottomed stainless steel skillet or sauté pan over high heat until smoking. Add half of the cutlets and cook, without moving them, until well browned on bottom, about 2½ minutes. Carefully flip the cutlets and cook on the second side until cooked through, about 30 seconds longer. Transfer to a large plate and tent with aluminum foil. Add the remaining 1½ tablespoons oil to the pan, heat until smoking, and cook the remaining cutlets. Let rest for 5 minutes before serving.
3. If desired, make a pan sauce (see pages 368–70) while the chicken rests.

## KNIFE SKILLS:
## How to Prepare Chicken Cutlets

**The most difficult step in preparing chicken cutlets is the cutting: it requires a sharp knife and a little practice.**

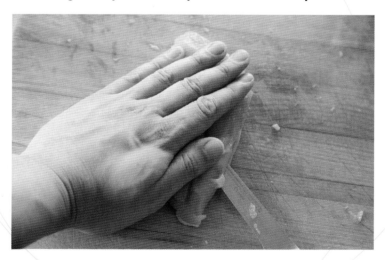

If you're still a little green in the kitchen, you'll probably make a few holes in your chicken breasts before you get the hang of it—no worries, they'll taste just as good. Once you've got the breast split, pounding is fun and easy. The key is not to pound too hard. Hard pounding can make holes in your meat and it gives you less control over the final thickness. Easy does it, OK?

A dedicated meat pounder is a good investment if you make tons of chicken cutlets or, say, chicken-fried steak. But if you're only an occasional cutlet eater, the bottom of a heavy skillet will do just fine.

Start by placing a boneless, skinless chicken breast, with the tenderloin removed, on your cutting board, parallel and close to the edge of the counter. Hold the chicken with the palm of one hand and, with the other, use a sharp knife to make a horizontal slice all the way through the chicken. Do not saw back and forth—if you need to take second stroke to get all the way through, open up the cut chicken, reset the knife, close the chicken around it, hold it in place, and take another stroke.

Once you've split the chicken, working with one piece at a time, place the chicken between two sheets of plastic wrap or inside a gallon-sized zipper-lock plastic bag with its sides cut open. Gently pound the chicken with a meat pounder or a skillet until it's an even ¼ inch thick. Repeat with the remaining cutlet(s).

## SAUCES FOR PAN-ROASTED CHICKEN PARTS OR CHICKEN CUTLETS

# LEMON CAPER
## PAN SAUCE

**SERVES 4**

1 large shallot, minced (about ¼
   cup)
4 tablespoons unsalted butter
1 teaspoon all-purpose flour
1½ cups dry white wine
3 tablespoons capers, rinsed,
   drained, and roughly chopped
3 tablespoons lemon juice (from 2
   lemons)
2 tablespoons minced fresh parsley
Kosher salt and freshly ground
   black pepper

1. After cooking the chicken, discard the excess fat from the pan, return it to medium heat, and add the shallot. Cook, stirring with a wooden spoon, until softened, about 1 minute. Add 1 tablespoon of the butter and the flour and cook, stirring constantly, for 30 seconds. Slowly whisk in the wine and capers. Scrape up any browned bits from the bottom of the pan with the spoon, increase the heat to high, and simmer the liquid until reduced to 1 cup, about 5 minutes.

2. Off the heat, stir in the lemon juice, parsley, and the remaining 3 table-spoons butter. Season with salt and pepper to taste. Pour over the rested chicken and serve immediately.

# MUSHROOM-MARSALA
## PAN SAUCE

**SERVES 4**

4 tablespoons unsalted butter

8 ounces button mushrooms, cleaned and finely sliced (about 4 cups)

1 large shallot, minced (about ½ cup)

1 teaspoon fresh thyme leaves

1 teaspoon all-purpose flour

1 tablespoon tomato paste

1 teaspoon soy sauce

1½ cups sweet Marsala

1 tablespoon lemon juice (from 1 lemon)

Kosher salt and freshly ground black pepper

1. After cooking the chicken, discard all but 2 tablespoons of the fat from the skillet. Return to high heat, add 1 tablespoon of the butter and the mushrooms, and cook, stirring frequently with a wooden spoon, until the mushrooms are well browned, about 6 minutes. Add the shallots and thyme and cook, stirring, for 1 minute. Add the flour and tomato paste and cook, stirring constantly, for 30 seconds. Slowly whisk in the soy sauce and wine. Scrape up any browned bits from the bottom of the pan with the spoon and simmer the liquid until reduced to 1 cup, about 5 minutes.

2. Off the heat, whisk in the lemon juice and the remaining 3 tablespoons butter. Season with salt and pepper to taste. Pour over the rested chicken and serve immediately.

# BRANDY-CREAM
## PAN SAUCE

**SERVES 4**

1 large shallot, minced (about ¼
  cup)

4 tablespoons unsalted butter

1 teaspoon all-purpose flour

1 cup homemade or low-sodium
  canned chicken stock

½ cup brandy

1 tablespoon whole-grain mustard

¼ cup heavy cream

1 tablespoon fresh lemon juice
  (from 1 lemon)

2 tablespoons minced fresh parsley

Kosher salt and freshly ground
  black pepper

1. After cooking the chicken, discard the fat from the pan, return it to medium heat, and add the shallot. Cook, stirring with a wooden spoon, until softened, about 1 minute. Add 1 tablespoon of the butter and the flour and cook, stirring constantly, for 30 seconds. Slowly whisk in the chicken stock and brandy. Scrape up any browned bits from the bottom of the pan with the spoon. Add the mustard and cream, whisk to combine, and simmer the liquid until reduced to 1 cup, about 5 minutes.

2. Off the heat, whisk in the lemon juice, parsley, and the remaining 3 tablespoons butter. Season with salt and pepper to taste. Pour over the rested chicken and serve immediately.

# SALMON THAT DOESN'T STINK

**I hate salmon.**

With a passion. Chalky, dry, smelly, slimy-skinned, the worst of the worst when it comes to fish.

At least, that's what I would've said about a decade ago, when the only salmon I had tasted was the overpoached stuff at buffets or overcooked specimens at restaurants that, frankly, didn't know what they were doing. I don't know if I was running in the right circles, but it seemed de rigueur in my youth for everyone to cook salmon to a shade just past well-done. We didn't exit these culinary dark ages until sometime in the 1990s, by which time my bias against the fish had been firmly established.

It wasn't until I started cooking in nice restaurants (the kind that I could never afford to go to as a civilian) that I realized that it wasn't the *salmon* that was at fault, but rather the cook. Properly cooked salmon is amazing, whether it's got crisp, crackly skin that can rival the best roast chicken's or tender, moist, flavorful meat that melts on your tongue like butter (or sometimes both!). There's a reason after all, why salmon is the most popular fresh fish in this country, and why it's the fish I've chosen to work mainly with here.

That said, the techniques I'm going to discuss apply to any thick, robust fish fillets, such as halibut, red snapper, or sea bass.

## PAN-SEARING SALMON

There's an unholy trinity of fates that can befall salmon. If you've ever cooked salmon, these are probably all too familiar sights:

### The Picked Scab

Flaky bits of salmon flesh that get stuck to the pan as it cooks. Not only does it make the cooked fillet look like a pockmarked teenage crater-face, it also makes the pan a bitch and a half to clean when you're done.

### The Leatherhead

There may be something soft and tender underneath, but the dried-out, stringy, crusty, downright malicious exterior is all you can pay attention to.

## The Bloomin' 'Bumin

That's right. The white gunk that gets squeezed out of the layers of salmon flesh, oozy and unattractive like a popped pimple (what's up with the blemish similes today?). It's not just that it's something you don't want to eat, it's also a pretty surefire indication that the salmon has been overcooked beyond repair.

Luckily, the first two problems can be solved relatively easily.

### A CLOSER LOOK

How many of you have been intimate with a salmon fillet? Raise your hands.

*I thought so.* Well, it's time you made the leap. When was the last time you looked closely at the cross section of a salmon fillet—I mean *really* closely?

Well, here's what you'd see:

Starting from the top, we've got:

- **Pale-orange/red flesh.** This is the bulk of the matter, and if you get your salmon fillets skinless, then it's basically all you're left with. Depending on the species of salmon, the color can vary from a deep, dark red to a paler orange-pink. We'll talk more in a moment about salmon flesh's cooking characteristics. Right above the meat, you'll find a layer of . . .
- **Subcutaneous Fat.** Depending on the species, the time of year, the availability of food, and a number of other factors, the thickness of the fat layer may vary, but all salmon have it. It serves both as an energy store for the fish and as a means of insulating its body from wide temperature changes between the ocean waters and the river it swims to during the spawning season.
- **Skin.** Some fish have thick, leathery skin. Salmon skin is some of the nicest around, very similar in thickness and texture to chicken skin, making it ideal for cooking.

It's these second two layers—the subcutaneous fat and the skin—that are of interest to us here. We know that the role of that fat is to insulate the salmon against rapid temperature changes, so why not harness that feature in our cooking method? Just like all meats, the texture of salmon flesh changes as a direct result of the temperature it is raised to.

- **At 110°F and below,** the flesh is essentially raw. Translucent and deep orange or red, it has the soft, fleshy texture of good sashimi.
- **At 110° to 125°F,** the salmon is medium-rare. The connective tissue between layers of flesh has begun to weaken, and if you insert a cake tester or toothpick into the fillet, it should slide in and out

with no resistance. The meat is relatively opaque but still juicy and moist, without any chalkiness or fibrousness.

- **At 125° to 140°F**, we are beginning to enter medium to well-done territory. Flakiness will increase, and a chalky texture will start to develop, though it won't be extreme. Albumin will start to get expelled from between the contracting muscle fibers and will begin to coagulate in unattractive white clumps on the exterior of the salmon. In the early stages of this clumping, your salmon is still rescuable, if you stop cooking it IMMEDIATELY.
- **At 140°F and above**, the salmon has reached its limit. From here on out, it's just going to get chalkier, drier, and more unattractive. This is what salmon that sits in the steam table at the cafeteria looks like, and probably why you didn't like salmon as a kid.

So, your goal is really to keep as much of the salmon below the 140°F temperature range (and, preferably, closer to the 125°F range) as possible. To do this, make sure to always cook salmon skin on if you're pan-roasting, even if you plan on serving it skinless.* By cooking salmon with the skin on, you can alleviate any sort of overcooking problems in the outer layers of flesh. The insulative subcutaneous fat acts as a heat barrier, transmitting heat to the interior flesh very, very slowly. This slow heat transfer means that skin-on salmon cooks much more evenly and gently than skinless salmon. The skin fulfills the exact same role that a batter or breading supplies on a piece of fried chicken or a tem-

---

*\* Why do you want to serve your salmon skinless? is a question most often followed up by, I see, and do you find pleasure anywhere in your life?*

pura shrimp—a buffer to slow down heat transfer and provide a crisp element while keeping the flesh underneath from overcooking.

You may ask, *but what about the other side of the fillet? A salmon fillet only has skin on one side, right?* And right you are. We still have the potential problem of overcooking the skinless side. The solution? *Just cook it through almost entirely with the skin side down.* French chefs who want to sound lofty like to call this unilateral cooking—cooking from one side only. Personally, I cheat just a bit, flipping the salmon over for the last 15 seconds or so, just to firm up the second side. But cooking salmon skin-on does lead to a few other possible problems that need to be dealt with.

## THE WOES OF SKIN-ON FILLETS

First off, if you aren't careful, you *still* get the leaking albumin problem with skin-on salmon fillets. Even worse is this guy:

Yep, don't tell me that hasn't happened to you. At its worst, the skin gets solidly fused to the skillet and the meat ends up completely separating from the skin as it cooks. This is not a terrible thing if you don't plan on eating the skin, and indeed, if you want to serve a skinless fillet, it's the best way to do it: cook the fillet skin-on, then slide a thin spatula in between the skin and the flesh to separate them.

At best, you end up with something that looks like this:

It's not the end of the world—what's left of the skin is still relatively crisp, and the flesh underneath *may* be perfectly cooked, but it's certainly not the kind of thing that's gonna impress the mother-in-law.

After cooking my way through several pounds of salmon fillets at various temperature ranges, I found that the key to getting the skin to stay intact is serendipitously the same method that gets you the most evenly cooked, moistest, tenderest salmon. And guess what? This technique will work for *any* firm-fleshed thick fish fillet, such as halibut, bass, grouper, or snapper. I've broken it down into a few easy steps.

### Perfect Fish Tip 1: Preheat the Oil

Know why that fish likes to stick to the metal pan? It's not just a matter of being, well, sticky, it's actually a chemical bond that occurs between the fish

and the pan at a molecular level. This happens with all meat. With land-dwelling animals like beef or pork, though, it's not quite as big a deal. The robust flesh of land animals sticks to itself better than it sticks to a pan. The worst you'll get is a deposit of browned proteins that have been expelled from the meat as it cooks.

With tender fish, on the other hand, it's very easy for it to stick to the pan better than it sticks to itself. Rather than lifting cleanly, it tears. The key to preventing this is to make sure that the skin heats up as fast as possible. With enough hot fat in the pan, the skin will have heated up, causing its proteins to tighten and coagulate, before it comes into direct contact with the hot metal. This prevents it from forming a strong molecular bond with the metal and makes subsequent flipping easier.

### Perfect Fish Tip 2: Drier Is Better

Nothing cools down oil faster than wet stuff being added to it. Rather than working toward searing your meat, the energy from the hot oil ends up getting used to evaporate excess moisture. The drier your meat before it goes in the pan, the better. I press my fish fillets firmly between paper towels before I transfer them to the pan (skin side down).

### Perfect Fish Tip 3: Hold On Tight—She's a Fighter!

Fish skin shrinks as it cooks due to proteins tightening and water and fat being driven off. What happens when this occurs? Like a bimetal strip, the fish fillet will begin to curl up on itself. This curvature can cause uneven cooking—the edges of the skin in tight contact with the pan will end up overcooking and burning while the central regions that are lifted off the pan will barely cook at all. This is not an ideal state of affairs.

To counteract this problem, I use a thin flexible metal fish spatula to hold the fillets firmly in place as their bottom sides cook. This is of utmost importance during the first minute or two of cooking. After that, the fillets' shape will be set and they will continue to cook evenly.

### Perfect Fish Tip 4: Slow and Steady

In order to get perfectly crisp skin, three things need to happen simultaneously: fat needs to render out, water needs to evaporate, and proteins need to set. Cook too hot, and your water will evaporate too fast. The temperature rapidly climbs, and the proteins set and start to burn before the fat has had a chance to render out properly. You'll end up with fish skin that is slightly crisp in spotty patches on the surface while still gelatinous, greasy, and fatty underneath. Preheating the oil over a relatively high heat to prevent the sticking problems explained in Tip 1, then immediately reducing the heat once the fish is added solves the problem. The result is shatteringly crisp, perfectly rendered, brown crunchy skin, just like the best pan-seared chicken.

And here's another bird to kill with that same stone: slow and steady cooking also leads to a more evenly cooked finished result. Say good-bye to coagulated albumin!

### Perfect Fish Tip 5: Don't Flip Until It's Ready!

If there's one most important trick I've learned about pan-roasting foods in all my years cooking, it's this one: *never force your food out of the pan.* It'll come on its own when it's good and ready. I use that flexible metal fish spatula for all of my flipping, but I find that if after some very gentle prying the fish doesn't release, it's not ready yet. Let it continue cooking, and once the skin is completely rendered and crisp, it should detach from the pan quite easily.

### Perfect Fish Tip 6: Break Out the Thermometer

Do as I say, not as I do.

You knew this one was coming, right? If you've been paying attention, you'll know I'm a huge fan of the Thermapen instant-read thermometer. Get yourself one and you will never, repeat *never*, overcook your fish again. That is, unless you are like me and stop to take photos while the fish is in the skillet, thereby letting it come all the way up to 137°F when you were aiming for 120°F.

### Perfect Fish Tip 7: Just a Kiss on the Backside

Just a repeat from what I said earlier: cook your fish almost all the way through skin side down, as that is the side that's insulated. The other side need just the briefest kiss from the flame.

FOLLOW ALL these tips, and you should end up with something that looks like the photo below: brown, crisp, crackling skin, with no greasy, gelatinous fat underneath. A thin layer of ever-so-slightly flaky meat underneath that, followed by a wide expanse of tender, juicy, not-the-least-bit-chalky flesh, and a central core with a creamy, buttery texture, bordering on sashimi-esque. *This* is how fish should look and taste.

# HOW TO BUY, STORE,
## AND PREPARE FISH
## FOR COOKING

There's a reason that fish is far more prone to spoilage than the flesh of land animals. Spoilage occurs via two means: the breakdown of cells through the action of enzymes naturally present in meat and the proliferation of bacteria. Both of these things occur at faster rates as the temperature increases.

Now, land-based, warm-blooded animals like cows, chickens, and pigs are used to living in warm environments and they have metabolisms to match. Take their meat and refrigerate it, and enzymatic and bacterial action slows down to a crawl. Fish, on the other hand, are designed to operate in the low temperatures of ocean waters. Some species of Arctic or deepwater fish spend most of their lives in water that is just barely above freezing. Compared to that, the 40°F of an average refrigerator is positively balmy. So fish-related enzymatic action will occur just fine in your fridge or the fish market's display case.

That's why it's so hard to find great fish on a regular basis, and why you should cook fish as soon as possible after buying it.

## Buying

Here are some tips for buying fish:

- **Look at the fish market itself**. Is everything clean and orderly? Is the fish displayed with care and kept on top of and under ice at all times? If your fishmonger looks careless with his wares even in front of the customer, chances are you're not getting goods worth paying for.
- **Look at the fish**. Does it look fresh? Whole fish should have shiny scales, perfectly clear eyes, and bright red gills. Bloodshot or cloudy eyes are an early sign of spoilage. Fillets and steaks should look shiny, fresh, and moist.
- **Smell the fish**. Fresh fish should not have a "fishy" aroma, it should just smell faintly briny. Any hint of ammonia is a bad sign.
- **Poke the fish**. If they'll let you, that is. Fresh fish has resilient flesh that springs back when you gently poke it. If the flesh is mushy or your finger leaves an indentation, leave the fish behind.

*continues*

---

\* Why, you might ask, does meat have enzymes in it designed to destroy it? It's because our bodies (and that cow's body, and that cod's body) are constantly growing new cells and retiring old ones. It's a natural part of the life cycle, and our bodies have to have the means to destroy old cells as they grow out of date—that's what the cell-destroying enzymes are for. One of the results of our bodies losing the ability to destroy old cells while new cells are being created is called cancer.

## Storing

As with asparagus or corn, the best way to store fish is not to. You should plan on buying it just before you're going to take it home, put it in the fridge as soon as you get there, and cook it within a few hours.

If you must buy your fish the day before or even 2 days before, keep it as cold as possible. With very clean hands, rinse the fish thoroughly under cold running water to wash away any surface bacteria, then carefully pat it dry with paper towels. Transfer the fish to a zipper-lock plastic bag, squeeze out as much air as possible, and seal the bag.

Place a frozen ice pack in the bottom of a baking dish and place the bag with the fish directly on top of it. Place a second ice pack on top of the fish. (You can also use zipper-lock bags of ice in place of the ice packs.) Store the fish for up to 2 days, replacing the ice packs as they thaw.

## Boning Fillets

More often than not, fish fillets you buy from the supermarket or fish market still have a few bones left in them. To remove them, you'll need a pair of sturdy tweezers.✳ Gently run your fingers back and forth along each fillet until you locate the ends of the bones. They'll feel like firm little bumps in the flesh. Use your fingertip to press the flesh right around the end of the bone to expose it a little bit. Grasp the exposed end of the bone firmly with the tip of the tweezers. Hold your free hand flat against the fish as near the bone site as possible and extract the bone by pulling it in the direction it's pointing, to minimize damage to the flesh. Discard the bone and repeat until the fish is clean.

## Skinning Fillets

I like to cook my fish skin on even if I'm going to serve it skinless, as the skin helps insulate the fish from the heat of the pan or oven, giving you more evenly cooked flesh. But if you'd like to remove the skin from a fish fillet, here's how it's done:

---

✳There are tweezers made specifically for this task. My favorite brand is Global. Their fish tweezers are precise, sturdy, and easy to grip.

- **Step 1: First incision.** Use a sharp chef's knife to cut through the flesh right at the very edge of the fillet, exposing the skin.
- **Step 2: Grab on tight.** Grab the exposed skin, using a clean kitchen towel or a sturdy paper towel to get a grip.
- **Step 3: Slide the knife in.** Slide the knife under the flesh, angling the blade down toward the skin. Slowly take the skin off by very gently moving the knife back and forth as you pull the fish back by the skin. The fish should move more than the knife.
- **Step 4: Finish her off.** Keep going, gently moving the knife back and forth in a sawing motion as you pull the skin, until it is completely detached from the fillet. Trim off any small bits you may have missed.

# SAMPLIN' SALMON

There was a time not long ago when salmon was salmon: it was the pink fish that skinny people ordered at restaurants or fancy ladies in French hats picked on at a high-class buffet. These days, diners are a little more aware of what's out there, or are at least aware that there *are* options when it comes to salmon. Here's a quick guide to what you might find in the market:

- **King salmon**, also known as chinook, is the largest salmon species and one of the most popular at the fish counter. In the wild, these salmon can grow to over a hundred pounds and live for several years, making them prized among game fisherman. Their large, thick fillets make for relatively easy cooking, though they are not the most flavorful species. Farm-raised king salmon tend to be smaller, with a bit more intramuscular fat, giving them more richness.
- **Coho** are far smaller than king salmon, with denser, brighter, more flavorful flesh. With relatively little intramuscular fat and a very fine texture, the fish is great for cured preparations such as gravlax.
- **Sockeye salmon** get their name from a Halkomelem word from the indigenous people of British Columbia—nothing to do with either socks or eyes. Known for their deep red flesh and full flavor, they are quite small, which makes them difficult to cook—their thinner fillets are prone to overcooking.
- **Arctic char** are . . . not salmon. But they have a similar reddish-orange flesh, colored by the carotenoid pigments they get from feasting on small shellfish. Their flavor and cooking qualities are quite similar to sockeye salmon, though they tend to be a little fattier.

In general, I prefer larger, fattier king salmon for high-heat cooking methods like pan-roasting, and I find that more flavorful coho and sockeye take well to slower cooking methods, like poaching. The thickness and higher fat content of king salmon fillets offer a little more protection from overcooking or drying out, both things that salmon is prone to do in the high heat of a pan or an oven.

# ULTRA-CRISP-SKINNED PAN-ROASTED FISH FILLETS

**NOTE:** This recipe will work with any skin-on firm-fleshed fish fillets, such as salmon, snapper, grouper, or bass. I prefer salmon rare to medium-rare; white-fleshed fish should be cooked at least to medium.

**SERVES 4**

4 skin-on fish fillets (about 6 ounces each)

Kosher salt and freshly ground black pepper

2 tablespoons vegetable or canola oil

1. Press the fillets between paper towels to dry all surfaces thoroughly. Season on both sides with salt and pepper. Heat the oil in a large heavy-bottomed stainless steel skillet over medium-high heat until shimmering. And the fish fillets skin side down, immediately reduce the heat to medium-low, and cook, pressing gently on the back of the fillets with a flexible metal fish spatula to ensure good contact between skin and pan for the first minute. Then continue cooking until the skin has rendered its fat and is crisp, about 5 minutes longer. If the skin shows resistance when you attempt to lift the fish with a spatula, allow it to continue to cook until it lifts easily.

2. Flip the fish and cook on the second side until an instant-read thermometer inserted into the thickest part registers 120°F for medium-rare, or 130°F for medium, about 1 minute longer. Transfer the fish to a paper-towel-lined plate and allow to rest for 5 minutes before serving.

**RELISHES AND SAUCES FOR PAN-ROASTED FISH**

# BASIL-CAPER RELISH

**MAKES ABOUT ⅔ CUP**

2 tablespoons capers, rinsed, drained, and roughly chopped

2 tablespoons chopped kalamata or Taggiasche olives

1 small shallot, minced (about 1 tablespoon)

1 Thai bird or serrano pepper, seeded and chopped

½ cup chopped fresh basil

2 scallions, finely sliced

3 anchovy fillets, finely chopped

1 tablespoon fresh lemon juice (from 1 lemon)

1 tablespoon balsamic vinegar

1 teaspoon honey

⅓ cup extra-virgin olive oil

Kosher salt and freshly ground black pepper

Combine the capers, olives, shallot, chile pepper, basil, scallions, anchovies, lemon juice, vinegar, and honey in a small bowl. Whisking constantly, add the olive oil in a thin, steady stream. Season to taste with salt and pepper. Serve spooned over pan-roasted fish.

# CHERRY TOMATO–SHALLOT RELISH

**MAKES ABOUT 2 CUPS**

2 cups cherry tomatoes, sliced into
  quarters

1 shallot, finely sliced (about ¼
  cup)

2 tablespoons chopped fresh
  parsley

1 tablespoon red wine vinegar or
  balsamic vinegar

3 tablespoons extra-virgin olive oil

Kosher salt and freshly ground
  black pepper

Combine the tomatoes, shallot, parsley, vinegar, and olive oil in a small bowl. Season to taste with salt and pepper. Serve spooned over pan-roasted fish.

# DILL–LEMON CRÈME FRAÎCHE

**MAKES ABOUT 1 CUP**

¾ cup crème fraîche

1 teaspoon grated lemon zest (from
  1 lemon)

2 tablespoons fresh lemon juice

2 tablespoons finely minced fresh
  dill

1 tablespoon capers, rinsed,
  drained, and minced

Kosher salt and freshly ground
  black pepper

Combine the crème fraîche, lemon zest, lemon juice, dill, and capers in a small bowl. Season to taste with salt and pepper. Serve with pan-roasted fish.

# BASIC TARTAR SAUCE

**MAKES ABOUT 1 CUP**

¾ cup mayonnaise

2 tablespoons sweet pickle relish

2 tablespoons capers, rinsed,
  drained, and minced

1 teaspoon sugar

1 small shallot, minced or grated
  on a Microplane (about 1
  tablespoon)

1 teaspoon freshly ground black
  pepper

1 teaspoon distilled or white wine
  vinegar

Combine all the ingredients in a small bowl. Cover and let stand for at least 15 minutes before serving. Serve with pan-roasted fish.

# Beer COOLERS, Plastic BAGS,
## AND THE SCIENCE OF SOUS-VIDE

There's been a small revolution going on in restaurant kitchens since the early 2000s. It's changed everything from the way line cooks cook and chefs conceive dishes and menus to the way fast food chains maintain consistency and organize their workflow. I'm talking sous-vide, from the French for "under vacuum," the cooking method wherein food is placed in a vacuum-sealed pouch and cooked in a temperature-controlled water bath. The technique was first introduced to the public in the 1970s at Michel Troisgros' eponymous restaurant in Roanne, France, but it wasn't until early in this century, when chefs gained access to very precise, laboratory-grade equipment that it became both practical and possible to implement on a large scale.

You may be thinking: "OK, interesting, but I'm a home cook, and I couldn't tell a water circulator from a rotary evaporator—what's this got to do with me?" You'll just have to trust that I'll get there in a moment.

According to famed British chef Heston Blumenthal of The Fat Duck, outside London, "Sous-vide cooking is the single greatest advancement in cooking technology in decades," and he's not the only one who thinks so. Everyone from Thomas Keller of New York's Per Se and California's The French Laundry to your local Chipotle Mexican Grill is serving food cooked sous-vide.

Here's what's great about it. Recall the problem of temperature gradients developing in meat? To recap, food cooks from the outside in, which means that the outer layers are going to be hotter than the very center. Thus, cooked foods develop an internal bull's-eye pattern: perfectly cooked at the very center and increasingly overcooked as you move out to the edges.

So, for example, imagine you're starting with a steak that's a consistent 40°F through and through. Place it in a 500°F pan, and the outer layers will almost immediately reach around 212°F, the temperature at which the internal moisture at the surface of the steak starts to evaporate. Eventually all the moisture will dissipate and the temperature of the outer layers of the steak will continue to increase. It's quite easy for those outer layers to achieve temperatures in excess of 200°F (that's beyond the well-done 160°F stage for steak) before the core temperature has even begun to shift. By the time the center reaches 130°F (medium-rare), the outer layers are hopelessly overcooked.

Now imagine cooking the same steak in a constant 130°F environment. Sure, it'll take much longer for the center to get up to 130°F, but it'll get there eventually and, in the meantime, the outer layers have no chance of overcooking.

That's precisely what sous-vide cooking is all about. If you place the meat in an airtight vacuum-sealed pouch and submerge it in a temperature-controlled water bath, the water very efficiently transfers heat energy to the steak while maintaining a very precise temperature. The result is meat that's cooked evenly from edge to edge.

And because the water bath is maintained at the final serving temperature of the meat, there's absolutely no way you can overcook it. Need to walk the dog? No problem—your steak will be waiting. Forgot to add the fabric softener? Take your time. The steak will still be there, exactly as you left it. This makes hosting dinner parties extraordinarily easy.

Because the Maillard browning reaction doesn't take place at standard sous-vide temperatures, most recipes call for searing meat in a hot skillet to add color and flavor after removing it from the bag.

In addition, meat naturally contains enzymes called *cathepsins* that slowly break down tough muscle tissue at first and then work faster and faster as the temperature increases. Giving the meat extra time in the lower temperature range means the cathepsins work overtime, making an already tender steak even more tender. And tender meat is not just about texture—the more loosely packed the muscle fibers, the less they contract during cooking and the fewer juices they expel, making the slow-cooked meat more juicy as well.

The advantage for restaurant chefs is obvious. Even at the best restaurants in the world, when using traditional cooking methods, the occasional over- or undercooked protein can be a problem. The seasoned line cooks at Peter Luger's, who've been turning-and-burning steaks since before vegans existed, still produce the occasional slightly-too-well-done porterhouse. But with sous-vide cooking, even a monkey with a toupee can produce perfectly cooked proteins without fail: Chicken with a juiciness the Colonel's wife only dreamed of. Salmon so tender it melts if you look at it too hard. The kind of double-thick pork chops that would've made me break out a celebratory PBR midservice when I was still a line cook. We're talking *perfect* food here.

Of course, there's a catch, and it's a big one: a typical water circulator will set you back about $1,000. Even the cheaper home versions now on the market (like the Sansaire and the Anova) are at least a couple hundred dollars, all in.

In fact, there's a whole legion of people out there on the internet who've devoted considerable time and resources into figuring out ways to put together a cheaper sous-vide setup. These fall into two categories:

- **Category 1: The rice-cooker, aquarium-bubbler, PID-controller method.** It's accurate, but it requires a fair amount of DIY know-how and costs a couple hundred dollars to hack together.
- **Category 2: The David Chang pot-of-water-on-the-stove, fiddle-with-the-heat-as-necessary method.** This is less accurate, and it requires you to hover around the stove for the entire cooking time.

Convinced that there was a faster, easier, cheaper, and more foolproof way to achieve the same results, I started poking around. Essentially, in order to create a low-temperature water cooker, all you need to do is keep a large body of water at the same temperature for a couple of hours: so, a well-insulated box should do the trick. And, fortunately, there's already a tool in pretty much every home that's designed *precisely* for the purpose of keeping large volumes of food—or liquid—at a stable temperature: the beer cooler.

Here's how it works: A beer cooler can keep things cool because it is a two-walled plastic chamber with air space between the walls. The air space acts as an insulator, preventing thermal energy from the outside from reaching the cold food on the inside. Of course, insulators work both ways—the cooler is just as good at keeping thermal energy on the inside from escaping to the outside. Once you realize that a beer cooler is just as good at keeping hot things hot as it is at keeping cold things cold, the rest is easy: Fill your cooler with water just a couple degrees higher than the temperature you'd like to cook your food at (to account for temperature loss when you add the cold food), seal your food in a zipper-lock freezer bag, drop it in, and close your cooler. Leave the food there until it is cooked. It's as simple as that.

## But What About the Vacuum?

Surely vacuum-sealing the food is an essential step in effective sous-vide cooking? As it turns out, not really. The main reason to vacuum-seal food is that any air bubbles trapped inside the plastic bag can act as an insulator, preventing the food from cooking evenly. As long as you can get all the bubbles out of your bag, there's no reason it shouldn't work just as well as a bag sealed with a vacuum-sealer and there's an easy way to do just that. This technique was demonstrated for me by Dave Arnold, an instructor at the French Culinary Institute and contributor to the Cooking Issues blog (www .cookingissues.com):

1. Place your food inside a zipper-lock freezer bag and seal it, leaving the last inch unsealed.

2. Holding on to the bag, slowly lower it into a large pot, cooler, or sinkful of water, using your hands to release any trapped air bubbles as you go.

3. Continue lowering it until just the very top edge of the bag is above the surface (don't allow any water to leak inside).

4. Seal the bag. You now have your food inside a completely airless bag.

A note about nomenclature: While the term "sous-vide" technically refers to the vacuum-sealing portion of the cooking process, in common usage these days, it's come to mean the act of cooking food in a temperature-controlled water bath—so that's how I'm going to use it, even though my cooler method actually makes no use of a vacuum at all. Deal with it.

So does the beer cooler method really work? I pitted my beer cooler/freezer bag method (total cost: $21.90) against a SousVide Supreme/Food-Saver combo (total cost: $569.98), comparing their performance in four categories:

1. The ability to cook proteins to a precise temperature all the way from the edges to the center.

2. The ability to hold cooked foods at serving temperature for several hours without any loss of quality. (Low temperatures and a sealed bag prevent overcooking or loss of moisture from cooked foods. This is an invaluable asset, allowing a line cook, or a harried spouse, to serve hot food at a moment's notice, without having to worry about precise timing.)

3. The ability to tenderize tough pieces of meat. Traditional braises use relatively high temperatures—180°F or so—to tenderize tougher cuts. But at these temperatures, the muscle fibers will expel quite a bit of their juices. With sous-vide cookery, much lower temperatures (say around 140°F) are applied for much longer periods of time—sometimes up to 72 hours. The result is extremely tender meat with no loss of juiciness. It's particularly effective for cuts like beef chuck or short rib.

4. The ability to cook vegetables without loss of flavor. Vegetables cooked in vacuum-sealed

pouches naturally soften in their own juices. In some cases, this can be overpowering (ever try sous-vide celery root?), but in others, the results can be downright extraordinary. Sous-vide carrots taste more like a carrot than any carrot you've ever tasted.

Before I even began, though, I threw in the towel as far as categories 3 and 4 go. There's no way my beer cooler would stay warm for the requisite 24 hours. Previous testing had shown me that it loses about 1 degree per hour when it's in the 140° to 150°F range.

Vegetables presented an even bigger problem. Pectin, the tough glue that keeps vegetable cells connected, doesn't begin to break down until 183°F. Even after only 15 minutes, a beer cooler filled with water this hot will have cooled by several degrees—it just doesn't work. So for the time being, it looks like if prolonged (2 hours+) or relatively hot (160°F+) cooking is among your requirements for a sous-vide cooker, you're going to have to spring for the real deal.

On the other hand, I'd easily argue that categories 1 and 2 are in fact the primary use of a sous-vide machine—particularly for a home cook. A quick Google search of the types of recipes home cooks have been playing around with confirms this.

Confident, I then moved on to the field tests, cooking steaks to 125°F and chicken to 140°F (sound like a salmonella trap?—we'll get to food safety in a minute). In both cases, the results were completely indistinguishable from each other.

Here's the coolest part: it just so happens that the hot water from my tap comes out at 135°F—the perfect temperature for cooking steak. What luck! The beer cooler is more easily transportable than a professional water circulator, and it doesn't require an electric outlet. So, last summer, I was able to start cooking a 2-pound dry-aged rib-eye in my kitchen, carry the whole cooler out to my deck 2 hours later, slap the beef on a blazing-hot grill for 30 seconds on each side just to mark it and brown the exterior, and then enjoy the most perfectly cooked meat that's ever come off my Weber. Anywhere you have access to hot water and a cooler, you can cook sous-vide. Think of the possibilities. Hotel rooms. Backyards. Boats. Movie theaters.

## ON FOOD SAFETY

Anyone who has taken a ServSafe food-handling course has heard of the "danger zone": the temperature range between 40° and 140°F where bacteria supposedly multiply at accelerated rates. According to ServSafe directives, no food can stay in that zone for longer than 4 hours total.

Of course, this is strictly absurd. Imagine throwing out a ripe Camembert just as it is finally approaching its optimal serving temperature because it's been on the cheese board for a couple of hours. Or think of throwing out *jamóns Iberico* or *prosciutto* or even a good old country ham just 4 hours into its months-long curing process, all of which takes place in this so-called danger zone. Small fortunes' worth of dry-aged beef would have to be chucked in the bin to comply with these draconian regulations.

ServSafe rules, along with the rules set by the U.S. Department of Agriculture, are intended to absolutely eliminate the possibility of food-borne illness—they are designed to have a large margin for error, as well as to be simple to understand at the expense of accuracy. In reality, any number of factors, including salt level, sugar level, and fat level, as well as water content, can affect how rapidly food will become unfit to consume. Not only that, but temperature and time have a much more nuanced effect on food safety than we are led to believe.

When we talk about fresh food—particularly meat—posing a health hazard, what we're really talking about is bacterial content and the toxins they can produce. As meat sits, bacteria present on its surface will begin to breed and multiply, eventually growing to a dangerous level. Below 38°F or so—fridge temperature—the bacteria are lethargic, multiplying very, very slowly. Take the meat all the way down to freezing temperatures, and the water necessary for the basic life functions of a bacterium turn into ice, making it unavailable to them. That's why frozen meat can last for months, even years, if properly sealed.

But when meat gets warmer, the bacteria become more and more active, and they will continue to do so until they've gotten so hot that they die. This kill-temperature can vary from bacterium to bacterium, but in general it's around 120°F, with the very hardiest (*Bacillus cereus*) finally kicking the bucket at around 131°F.

*Ah,* you're thinking to yourself, *so I only need to cook my meat to 131°F for it to be safe.* Well, yes and no. Just like cooking, destroying bacteria—the process of pasteurization—takes both temperature *and* time to accomplish. (See "Chicken Temperature and Safety," page 360.)

With a temperature-controlled water bath, you have the ability not only to cook chicken to lower temperatures, but, more important, to hold it there until it's completely safe to consume.

What does this mean for a home cook? It means that you no longer have to put up with dry 165°F chicken.

A chicken breast cooked sous-vide to 140°F and held for 25 minutes is just as safe as chicken cooked to 165°F, and incomparably moister and more tender. It glistens with moisture as you cut it. It practically oozes juices as you chew. Equally stunning results can be achieved with pork, which will submit to your fork like butter.

With some foods, the final serving temperature you're after is actually *below* the level at which bacteria begin to be actively killed off. Rare salmon, or steak, is around 120°F, for example. For foods like these, you must be very careful not to let them sit for too long. To be safe, I don't hold my medium-rare steaks or salmon in the cooler any longer than 3 hours. After that, dinner becomes a game of Russian roulette. And never, ever cook food in the cooler, let it cool, and reheat it. This is absolutely inviting illness or worse.

With that somewhat dry lesson out of the way, let's move on to the actual techniques—the fun part.

## COOKING IN YOUR COOLER

These recipes require an accurate thermometer as well as a beer cooler with at least a 2.5-gallon capacity and a tight-fitting lid. Some coolers retain heat better than others. Heat retention can be further improved by draping several towels over the cooler during cooking. Leaving it in a warm spot also helps—I leave mine in direct sunlight on a warm day or in a warm corner of the kitchen indoors. Here are the basic steps for cooking in your cooler:

1. Season the food generously on all sides with salt and pepper. Place in a single layer in gallon-sized zipper-lock freezer bags along with any aromatics or rub. Squeeze out as much air from the bags as possible with your hands and close them, leaving 1 inch unsealed. (See photos on page 386.)

2. Heat at least 2 gallons of water to the designated temperature, using an instant-read thermometer to ensure accuracy (the hot water from your tap may be hot enough, without having to heat it on the stovetop). Pour the water into the cooler.

3. One at a time, slowly submerge each bag of food in the water until only the unsealed edge is exposed. Any remaining air should have been forced out of the bag as it was submerged. Seal the bag completely.

4. Close the cooler, drape it with a few towels, and set it in a warm spot for the specified cooking time, checking the temperature of the water every 30 minutes or so and topping it up with boiling water as necessary to maintain it at within 3 or 4 degrees of the desired final temperature (with a very good cooler, this may be unnecessary).

5. Remove the food from the bags and sear in hot fat, on a grill, or with a blowtorch to trigger the Maillard reaction and add textural contrast to the food.

# SEARING

As I mentioned earlier, sous-vide cooking is deficient in one key aspect: it doesn't brown your meat. The browning reactions that give your meat those wonderful crusty, roasted aromas only take place to a significant degree at temperatures well above 300°F, a good 170 degrees hotter than normal sous-vide cooking temperatures—which means that you still need to pull out the sauté pan to finish it. The key is to sear the meat as quickly as possible, to prevent overcooking. Get your skillet or grill ready and ripping hot, dry the meat thoroughly before adding it to the pan (wet meat will cool the pan down faster than dry meat), and leave it in the pan only long enough to color.

There's some debate over whether or not meat should be preseared before bagging and cooking it sous-vide. The idea is that the flavors created by a presear will penetrate the meat as it cooks in the bag, giving it a deeper, roastier flavor. I cooked a few steaks side by side to see if this was true.

Visually, there's not much distinction. Flavorwise? In a blind tasting, tasters were split across the board over which one they preferred, and when asked to identify which steak was which, they fared no better than with blind chance.

**Conclusion:** Don't bother with the presear—you develop plenty of flavor with just a single, post-water-bath sear.

And what is the best method of searing? These are three that I employ often.

## Pan-Searing

**Advantages:** Easy and done indoors. Drippings stay in contact with meat.
**Disadvantages:** Without an ultrapowerful burner, crust can take a while to develop, leading to slightly overcooked meat underneath.

## Grilling

**Advantages:** High heat leads to fast char and good flavor development.
**Disadvantages:** Requires an outdoor grill. Steak loses moisture and flavor to drippings.

## Torching

**Advantages:** Very high heat makes it easy to char. You look badass doing it.
**Disadvantages:** Charring can be uneven, resulting in some blackened bits before the rest of the steak has even browned. If you're not careful, it can also leave your steak tasting like uncombusted fuel.

So, if none of these three methods is perfect on its own, why settle for just one? For steaks and chops, by combining pan-searing and torching into one hybrid technique, you can avoid all the disadvantages of either one alone.

I started by first searing one side of a steak in smoking-hot oil and butter (the browned butter solids help kickstart browning reactions). As soon as the browning started, I flipped the steak over and immediately started cooking that top surface with the full blast of a propane torch. The layer of oil and butter clinging to its surface helped to distribute the heat of the flame evenly, leading to excellent, all-over browning and charring and creating an unbeatable steak house broiler–quality crust in record time. Finally, I flipped the steak back over and torched the second side.

What about the problem of the uncombusted propane leaving its telltale aroma? Turns out to not be a problem in this case. Because of the heat of the skillet and the increased convection caused by the shifting heat of the pan, the propane gets plenty of oxygen and heat, allowing it to fully combust, leaving behind nothing but sweet, succulent, charred beef.

# SOUS-VIDE AND RESTING

We all know that it's important to rest your meat before serving. It gives time for the juices inside to settle and thicken slightly, preventing them from leaking out excessively when you cut into the steak.

There are, as it turns out, some downsides to resting: Namely, your steak will have a more significant, crusty, snappy, sizzling crust when it's fresh off the burner than after it's rested. This more appetizing crust will subsequently lead to greater production of saliva, which in turn will lead to a juicier sensation in your mouth when you chew the steak—or so the theory goes. There is something very appealing about that sizzling crust you get just as the steak comes off the heat, and I have to restrain myself while letting it rest.

Cooking a steak sous-vide can help a great deal in this department: because there is no large temperature gradient built up inside, you do not need to wait for temperatures to normalize—they're normalized to begin with! The only gradient that builds is during the brief searing stage, which only requires a few moments of resting. Still, this is enough time for some of that crackling crust to disappear.

Wouldn't it be great if there was a single technique that gave us the best of both worlds? Well, fortunately for us, there is.

The trick is to allow the steak to rest normally, and then, just before serving, reheat its fatty pan drippings until they're smoking hot and pour them over the steak. The steak will sizzle and crisp, while the interior stays perfectly well-rested and juicy. Adding some aromatics to those pan drippings is never a bad idea, and re-collecting them after pouring them over the steak and serving them alongside the steak in a little heated pitcher gives you a built-in sauce right there. This method will work for *any* steak or chop, not just those cooked sous-vide.

I propose that the resters and nonresters of the world now unite over some juicy, crusty, sizzling steaks to celebrate.

## FATS AND AROMATICS

A few quick tests proved that aromatics added to the sous-vide bag *do* indeed flavor the meat as it cooks. But would adding a fat like oil or butter to the bag help as well? For this test, the aromatics consisted of 3 sprigs of thyme and a sliced garlic clove. I placed just the steak, thyme, and garlic in one bag, and I placed the same ingredients along with 2 tablespoons butter in a second bag.

My hope was that as the butter melted, it would pick up all the fat-soluble flavor molecules from the garlic and thyme, helping to distribute their fragrance evenly over the meat and further enhancing its flavor. I also included a nonaromaticized steak as a control.

The results were a shock: tasters unanimously picked the nonbutter version as the most aromatic. Some even had trouble telling the difference between the buttered version and the version with no aromatics at all! How could this be? I looked at the sous-vide bags the steaks had been cooked in and had my answer: the bag with the buttered steak contained a large amount of highly aromatic melted butter. Turns out that rather than traveling into the meat, where I wanted it to go, the aroma ended up in the butter, getting thrown away along with the bag.

**Conclusion:** Aromatics are fine, but leave out the butter if you want to maximize their flavor.

### COOKING STEAK IN A COOLER

Cooking steak in a cooler is the absolute best, most foolproof way to ensure that you get yours to exactly the level of doneness that you like. As we've already discussed, the degree of doneness is largely a personal matter, but if you are the kind who likes your steaks rare, I strongly suggest keeping an open mind about taking your steak a shade more toward well-done. At a rare 120°F, the beef fat is still relatively solid—it hasn't begun to melt into the surrounding meat, which means that for all intents and purposes, it may as well not be there. In my book, undercooking a heavily marbled rib-eye or expensive Japanese-style Wagyu steak is just as much of a crime as overcooking it.

In an impromptu blindfolded tasting I held at a dinner party, I found that even most self-proclaimed rare steak lovers actually preferred the texture and flavor of medium-rare (130°F) or even medium steaks (140°F) when they couldn't see the color of the meat they were eating. Try it out, and see what you think.

One more thing: cooking sous-vide is ideal for cheaper "butcher's cuts" of beef. The expensive steak house cuts—New York strip, rib-eye, porterhouse, T-bone, and filet (tenderloin)—have historically been prized for their extreme tenderness, not particularly for their flavor. But more flavorful cuts like hanger, blade, or flatiron steak are much more difficult to cook correctly—even a tad over- or undercooked, and you're left with a tough, stringy, chewy mess. Cooked properly, though, these can be every bit as tender as the more expensive cuts, and with more flavor to boot!

That's why those cuts are commonly referred to as "butcher" or "chef" cuts—chefs and butchers love them because they are cheap and, with proper preparation, delicious. With a sous-vide setup, *anyone* can properly cook these tricky cuts. Like most fatty cuts of beef, these are at their best when cooked to medium-rare or medium. Do you want to pay $7 per pound for a hanger steak that is just as tender as and tastes much better than a $16-per-pound strip steak? **Yes, please!**

# COOLER-COOKED RIB-EYE STEAKS
## WITH SHALLOTS, GARLIC, AND THYME

**SERVES 4**

2 bone-in 2-inch-thick, dry-aged rib-eye steaks, 2 to 3 pounds total

Kosher salt and freshly ground black pepper

2 tablespoons canola oil

4 tablespoons unsalted butter

1 medium shallot, sliced

4 cloves garlic, smashed

4 sprigs fresh thyme or rosemary

1. Cook the steaks in your cooler as directed on page 389, using 133°F water for medium-rare or 143°F water for medium, for at least 1 hour, and up to 3 hours.

2. Remove steaks from bags and pat dry with paper towels. Heat the canola oil and butter in a 12-inch cast-iron or stainless steel skillet over high heat until the butter browns and begins to smoke. Add the steaks and cook, without moving them, for 30 seconds. Flip the steaks and cook on the second side for 1 minute. Meanwhile, if desired, blast the top sides with a propane torch set on the hottest setting. Flip the steaks and torch second side until well browned and charred in spots, about 30 seconds longer. Using tongs, place one steak on top of the other, then stand them on one edge, and hold against the hot pan to crisp up the fat; continue all the way around the edges. Transfer to a wire rack set on a rimmed baking sheet and allow to rest for 5 minutes. Set the pan aside.

3. When the steaks have rested, add the shallot, garlic, and herbs to the skillet, increase the heat to high, and cook until aromatic and lightly smoking, about 30 seconds.

4. Pour the hot pan drippings over the steaks. Transfer the steaks to a large serving plate and transfer the drippings from the baking sheet to a small warmed pitcher. Serve immediately, with the pitcher of drippings on the side.

# COOLER-COOKED HANGER STEAK
## WITH CHIMICHURRI

**SERVES 4**

4 hanger steaks, 8 ounces each

Kosher salt and freshly ground black pepper

12 sprigs fresh thyme (optional)

2 cloves garlic (optional)

2 shallots, thinly sliced (optional)

2 tablespoons canola oil

1 recipe Chimichurri Sauce (recipe follows)

1. Cook the steaks in your cooler as directed on page 389, using 133°F water for medium-rare, or 143°F water for medium, for at least 45 minutes, and up to 3 hours.

2. After removing the steaks from the bags, discard the aromatics, and pat the steaks dry with paper towels. Heat the oil in a 12-inch heavy-bottomed stainless steel or cast-iron skillet over high heat until heavily smoking. Add the steaks to the skillet and sear until well browned on all sides, turning with tongs, about 2 minutes total.

3. Transfer the steaks to a large plate, tent with foil, and allow to rest for 5 minutes, then serve with the chimichurri.

### Chimichurri Sauce

**MAKES ABOUT 1 CUP OF SAUCE**

¼ cup finely chopped fresh parsley

¼ cup finely chopped fresh cilantro

2 teaspoons finely minced fresh oregano

4 medium cloves garlic, minced or grated on a Microplane (about 4 teaspoons)

½ teaspoon red pepper flakes

¼ cup red wine vinegar

½ cup extra-virgin olive oil

Kosher salt to taste

Whisk together all of the ingredients in a small bowl. Let stand at room temperature for at least 1 hour before serving. The chimichurri can be stored in a sealed container in the refrigerator for up to a week.

# COOLER-COOKED LAMB RACK
## WITH SALSA VERDE

Lamb rack is one of the most delicious—and most expensive—cuts of meat at the butcher, making it the ideal candidate for cooking in your cooler. This is a cut you *really* don't want to mess up.

This recipe has you finish the lamb in a hot skillet to crisp up and brown the exterior, but you can just as easily finish it off on a hot charcoal grill. Bring the cooler out to the backyard or deck with you while you polish off a couple of cocktails, then transfer the lamb from the cooler to the hot grill. Chow time!

NOTES: I prefer leaving extra fat and breast meat attached to the ribs for gnawing and bone sucking, but, it can be removed (frenched) if desired. Make sure you buy the racks with the chine bone removed to make it easier to cut between chops after cooking. Most prepackaged lamb will already have the bone removed; if buying lamb from the butcher's case, ask your butcher to do this for you—it's impossible to do at home without a hacksaw.

**SERVES 4**

Two 8-rib lamb racks, 3 to 4 pounds
total

Kosher salt and freshly ground
black pepper

8 sprigs fresh rosemary or thyme
(optional)

2 medium shallots, roughly
chopped (optional)

4 medium cloves garlic, roughly
chopped (optional)

1 tablespoon vegetable oil

1 tablespoon butter

1 recipe Salsa Verde (recipe follows)

1. Cook the lamb in your cooler as directed on page 389, using 133°F water for medium-rare, or 143°F water for medium, for at least 45 minutes, and up to 3 hours.

2. After removing the lamb from the bags, discard any aromatics and pat the lamb dry with paper towels. Heat the oil and butter in a 12-inch heavy-bottomed stainless steel or cast-iron skillet over high heat until smoking. Add the lamb to the skillet fat side down and sear until well browned on all sides, turning with tongs, about 3 minutes total (for particularly large racks, sear the lamb in two batches, tenting the first batch with foil after searing to retain the heat).

3. Transfer the seared lamb to a platter, tent with foil, and allow to rest for 5 minutes, then carve and serve with the salsa verde.

## Salsa Verde

**MAKES ABOUT 1 CUP**

¼ cup capers, drained

¼ cup finely minced fresh parsley

¾ cup extra-virgin olive oil

1 tablespoon honey

4 anchovy fillets, finely minced

1 medium clove garlic, minced or
grated on a Microplane (about 1
teaspoon)

Kosher salt and freshly ground
black pepper

Press the capers between a double layer of paper towels to remove excess moisture. Finely chop and transfer to a large bowl. Add parsley, olive oil, honey, anchovies, and garlic and whisk to combine. Season to taste with salt and pepper.

# COOLER-COOKED CHICKEN
## WITH LEMON OR SUN-DRIED-TOMATO VINAIGRETTE

**If you're used to dry, well-done chicken, cooking it via this method will be an absolute revelation. Who knew that chicken could be so incredibly succulent and moist?**

**It's up to you whether or not you want the chicken skin on or off, but I prefer to keep the skin on because I love crispy skin, and browning chicken without skin will *always* leave you with a stringy layer, no matter**

what you do. **The skin provides insulation that allows you to brown the chicken without sacrificing texture. With traditional cooking methods, leaving chicken breasts on the bone helps slow down the cooking process, resulting in more evenly cooked meat. With sous-vide cooking, however, the process is already so slow that I found little difference between chicken cooked bone-in versus bone-off.**

**The last drawback to cooking chicken in a water bath is that you develop very little** *fond*—the flavorful **bits that stick to your pan when you sear meat—this makes forming a pan sauce impossible. The good news is that the reason those flavorful bits aren't on the pan is that they stay stuck to the chicken where they belong. A simple squeeze of lemon or a vinaigrette will do the chicken just fine.**

**SERVES 4**

4 skin-on boneless chicken breasts,
  6 to 8 ounces each
Kosher salt and freshly ground
  black pepper
1 tablespoon canola or vegetable oil
1 lemon, cut into wedges, or
  1 recipe Sun-Dried-Tomato
  Vinaigrette (recipe follows)

1. Cook the chicken in your cooler as directed on page 389, using 148°F water, for at least 1 hour 35 minutes, and up to 3 hours.

2. After removing the chicken from the bags, pat it dry with paper towels. Heat the oil in a 12-inch heavy-bottomed stainless steel or cast-iron skillet over high heat until shimmering. Add the chicken to the skillet skin side down and sear until the skin is crisp, about 3 minutes.

3. Transfer to a serving platter and serve with the lemon wedges (or vinaigrette).

## Sun-Dried-Tomato Vinaigrette

**MAKES ABOUT 1 CUP**

½ cup oil-packed sun-dried
  tomatoes, drained and chopped
  into ¼-inch pieces, plus 2
  tablespoons of the oil
1 jalapeño pepper, finely minced, or
  less to taste
1 teaspoon honey
½ teaspoon soy sauce
½ teaspoon Frank's RedHot or
  other hot sauce (optional)
2 teaspoons lemon juice (from 1
  lemon)
1 tablespoon minced fresh mint
1 medium shallot, finely minced
  (about ¼ cup)

Whisk together all ingredients in a medium bowl. The vinaigrette can be stored in a sealed container for up to 3 days.

# COOLER-COOKED BRATS AND BEER

**Sausages are forgiving, but as anyone who has ever been to a family reunion can tell you, it *is* possible to overcook a brat. Just as with any other meat, the hotter it gets, the tighter the muscles squeeze, and the more juices come out. This recipe takes advantage of sous-vide in two ways: First, your brats come out perfectly cooked and totally juicy. Second, by adding a liquid cooking medium to your bag, you can flavor them as they cook. Beer does just fine.**

**SERVES 4 TO 6**

2 pounds store-bought bratwurst (about 8 links) or Bratwurst-Style Sausage (page 505)

2 cups pilsner-style beer

1 tablespoon vegetable oil (if pan-searing)

1 tablespoon butter (if pan-searing)

8 torpedo rolls or other buns of your choice, split and toasted

Condiments as desired

1. Cook the brats and beer in your cooler as directed on page 389, using 143°F water, for at least 45 minutes and up to 3 hours.
2. Remove the brats from the bag and discard the beer. Carefully dry with paper towels. Cook on a hot charcoal or gas grill, turning occasionally, until well browned, about 1 minute per side. Alternatively, heat the vegetable oil and butter in a large skillet over medium-high heat until the foaming subsides and cook the brats, turning occasionally, until well browned on all sides, about 2 minutes.
3. Serve in the toasted rolls, with condiments as desired.

# COOLER-COOKED PORK CHOPS
## WITH BARBECUE SAUCE

**Ever wonder how that fancy restaurant you went to for your last anniversary got that enormous pork chop for two so perfectly tender and flavorful throughout? Chances are it was cooked in a water bath.**

**Most of the time, I prefer simple seasonings. Just salt and pepper on a perfectly cooked piece of meat will do me fine. But sometimes you're in the mood for something more exciting. Rubbing the chops with a spice rub and finishing them on the grill with a slathering of sweet and tangy barbecue sauce is my answer for those days.**

**SERVES 4**

2 teaspoons chili powder

1½ tablespoons dark brown sugar

1½ teaspoons kosher salt

⅛ teaspoon cayenne pepper

½ teaspoon ground coriander

¼ teaspoon ground fennel

½ teaspoon ground black pepper

4 bone-in pork rib chops, about an inch thick (10 ounces each)

About 1½ cups of your favorite sweet barbecue sauce

1. Combine the chili powder, 1 tablespoon of the brown sugar, the salt, cayenne, coriander, fennel, and pepper in a small bowl and mix well. Rub the mixture evenly all over the pork chops. Cook in your cooler as directed on page 389, using 143°F water, for at least 45 minutes and up to 3 hours.

2. Preheat a grill to high heat. Remove the pork chops from their bags and pat dry with paper towels. Brush the top of each chop with 1 tablespoon of the barbecue sauce, transfer to the grill, sauced side down, and cook for 1 minute, brushing the top with more sauce. Flip the chops and cook for 1 minute longer. Transfer the chops to a plate and brush both sides with more sauce. Cover with foil and allow to rest for 5 minutes.

3. Remove the foil, brush the chops with more sauce, and serve, passing the remaining sauce at the table.

# COOLER-COOKED OLIVE OIL–POACHED SALMON

**If you are used to salmon that is firm and opaque, you may want to skip this. But for those of you who love the flavor of sashimi-quality salmon, give this method a shot. Salmon cooked to 120°F has an extremely delicate, almost custard-like texture and translucent flesh that literally melts in your mouth. Adding some extra-virgin olive oil to the bag gently scents the exterior of the fish.**

**SERVES 4**

4 skinless center-cut salmon fillets, about 6 ounces each

Kosher salt and freshly ground black pepper

¼ cup extra-virgin olive oil

Extra-virgin olive oil for drizzling or 1 recipe Grapefruit Vinaigrette (recipe follows)

1. Season the salmon with salt and pepper. Cook the the salmon and olive oil as directed on page 389, using 120°F water, for at least 20 minutes, and up to 1 hour.

2. Carefully remove the salmon from the bag (it will be very fragile) and pat it dry with paper towels. Transfer to a platter and serve, drizzling with more olive oil or the grapefruit vinaigrette.

### Grapefruit Vinaigrette

**MAKES ABOUT ½ CUP**

1 ruby red grapefruit, rind and
  pith removed, cut into ¼-inch
  suprêmes (see "How to Cut
  Citrus Suprêmes," page 770),
  juice reserved separately

1 tablespoon whole-grain mustard

1 teaspoon honey

1 tablespoon finely minced fresh
  parsley, basil, or tarragon, or a mix

¼ cup extra-virgin olive oil

Kosher salt and freshly ground
  black pepper

Combine the grapefruit juice, mustard, honey, and herbs in a small bowl. Whisking constantly, slowly drizzle in the olive oil until an emulsion is formed. Season to taste with salt and pepper. Add the grapefruit suprêmes to the bowl and stir.

# COOLER-COOKED CHEESEBURGERS

**Can a burger really benefit from sous-vide cooking?**

You bet it can—at least when it's a fat pub-style burger. Many chefs avoid cooking burgers sous-vide because the vacuum-sealing process can compress meat, turning burgers dense and tough. With the water-dipping bag-sealing method, however, this is not a problem.

We are still left with one question: what's the best way to sear the burger post-cooking?

Deep-frying is the method favored by Dave Arnold at the French Culinary Institute (he also adds clarified butter to the bags when cooking burgers, but I'm not such a fan of the flavor). It creates a wicked crispy crust, with the advantage that the crust forms evenly on all sides of the patty—top, bottom, and all the way around. It's also splatterless and won't smoke out your apartment the way searing can. The problem with it is that deep-frying oil temperature maxes out at around 400°F—even lower if you consider that the temperature of the oil immediately surrounding a piece of frying food is significantly lower than the rest of the pot. This brings us back to the same old problem with traditional cooking methods: By the time a decent crust has formed, a good ⅛ to ¼ inch of meat has overcooked and turned leathery around the edges.

Pan-searing does not have that problem. As long as you use a heavy-bottomed pan and let it preheat until it's really, *really* hot, your burger will form a deep brown crust in well under 45 seconds or so, preventing that leathery skin from forming. Pan-searing also produces better flavors in the crust, due to the higher heat and charring.

Of course, searing is a slightly messier affair, so which method wins in the end? It's all about personal preference. Deep-frying produces a superior crusty texture without creating a mess in the kitchen, but pan-searing offers better flavor development and better internal texture. As I don't mind cleaning the kitchen, my vote goes to pan-searing.

# SOUS-VIDE CHEESEBURGERS

**SERVES 4**

1½ pounds freshly ground beef

Kosher salt and freshly ground black pepper

2 tablespoons vegetable oil, if pan-searing, 2 quarts if deep frying

4 slices cheese (I recommend American or cheddar)

4 soft hamburger buns, lightly toasted

Condiments as desired

1. Divide the meat into four 6-ounce portions and gently shape each one into a patty 4 inches wide by approximately ¾ inch thick. Season generously with salt and pepper. Place the patties in individual sandwich-sized zipper-lock freezer bags, seal and cook as directed on page 389, using 123°F water for rare, 133°F for medium-rare, or 143°F for medium, for at least 30 minutes, and up to 3 hours.

**IF PAN-SEARING**

2. Remove the patties from the bags and carefully dry on paper towels. Season again with salt and pepper. Heat the vegetable oil in a 12-inch heavy-bottomed cast-iron or stainless steel skillet over high heat until it begins to smoke. Add the patties and cook until well browned on the first side, about 45 seconds. Flip the patties, add the cheese, and cook until the second side is well browned, about 45 seconds longer. Place the patties on the buns, top with condiments as desired, and serve.

**IF DEEP FRYING**

2. Heat the oil in a large wok or Dutch oven to 400°F. Season the patties again with salt and pepper. Carefully lower the patties into the hot oil using a metal spider. Fry until deep brown, about 2 minutes. Transfer to a paper-towel-lined plate and immediately top with cheese. Place the patties on the buns, top with condiments as desired, and serve.

BLANCHING,
SEARING, BRAISING,
GLAZING, ROASTING, AND

THE
# SCIENCE
*of*
# VEGETABLES

## 4

My favorite vegetable is entirely dependent on my mood and the season.

# BLANCHING, SEARING, BRAISING, GLAZING, ROASTING, AND THE
## SCIENCE *of* VEGETABLES

## RECIPES IN THIS CHAPTER

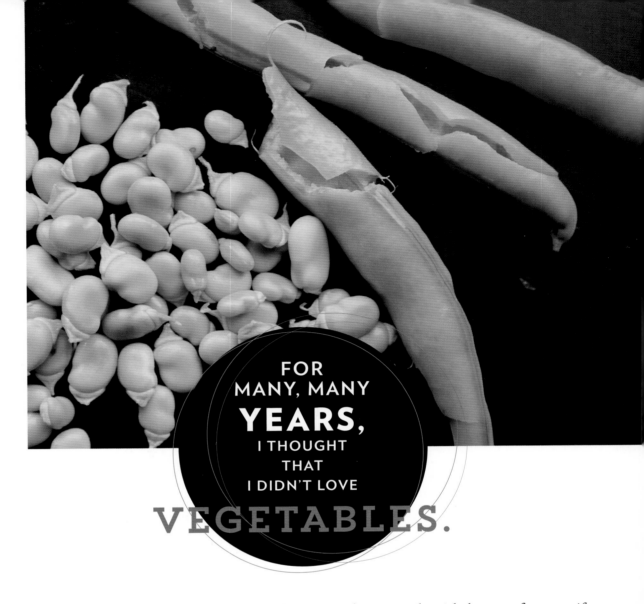

FOR
MANY, MANY
**YEARS,**
I THOUGHT
THAT
I DIDN'T LOVE

VEGETABLES.

I mean, some of them were OK. I dug salads, carrot sticks were cool, artichokes were fun to eat if only for the novelty. Asparagus was even good when it was still green and crunchy and dipped into Kewpie mayonnaise. But for the most part, vegetables were pushed over to the "choke down with a cup of water when my mom's not looking" corner of the plate. I now realize that my dislike of vegetables is entirely my mother's fault (sorry to break it to you, Ma). See, kids don't dislike broccoli. They dislike *mushy* broccoli. They don't dislike Brussels sprouts. They dislike Brussels sprouts that smell like farts and have the texture of old cheese. Ah, if only my mom had known how to properly roast or sear a Brussels sprout when I was a kid, I could've had a good couple extra decades of fine vegetable dining!

In this chapter, we'll discuss five basic techniques for cooking vegetables—blanching/steaming, searing/sautéing, braising, glazing, and roasting/broiling—the ins and outs of how they work, and the best times to use them. With a little luck, I'll turn every single one of you into a vegetable believer.

# THE FIVE BASIC VEGETABLE-COOKING TECHNIQUES
## AND WHAT THEY'RE GOOD FOR

| | BLANCHING/ STEAMING | SEARING/ SAUTÉING | BRAISING | GLAZING | ROASTING/ BROILING |
|---|---|---|---|---|---|
| Artichokes | X | X (when small) | X (when small) | | X (when small) |
| Asparagus | X | X | X | | X |
| Beets | X | | X | X | X |
| Bell Peppers | | X | X | | X |
| Bitter Greens (e.g., radicchio, endive) | | | X | | X |
| Bok Choy | X | X | | | X |
| Broccoli | X | X | X | | X |
| Broccoli Rabe | X | X | | X | |
| Brussels Sprouts | X | X | X | X | X |
| Cabbage | X | X | X | | X |
| Carrots | X | X | X | X | X |
| Cauliflower | X | X | | | X |
| Celery | X | X | | X | |
| Corn | X | X | | | X |
| Eggplant | | X | | | X |
| Green Beans | X | X | X | | X |
| Hearty Greens (e.g., kale, chard) | | X | X | | X |

*continues*

| | BLANCHING/ STEAMING | SEARING/ SAUTÉING | BRAISING | GLAZING | ROASTING/ BROILING |
|---|---|---|---|---|---|
| Leeks | | X | X | | X |
| Mushrooms | | X | | X | X |
| Onions | | X | X | X | X |
| Parsnips | X | X | | X | X |
| Peas | X | X | | X | |
| Radishes | | | | X | |
| Salsify | X | X | X | X | X |
| Scallions | X | X | | | |
| Spinach | X | X | X | | |
| Tomatoes | | | | | X |
| Turnips | X | | | X | |
| Zucchini | X | X | | | X |

# Essential VEGETABLE Technique #1:
## BLANCHING/STEAMING

Is there anything more beautiful on the plate than emerald-green, perfectly tender-crisp stalks of blanched asparagus? What are the tricks to achieving vegetables that both look *and* taste fantastic?

We've all read enough books by big-name chefs to know that when you cook a green vegetable, you're supposed to use a large pot of heavily salted water. But why? And is volume the only thing that matters? What about lid-on versus lid-off? Does pH play a role? And do I *really* need to drop my green vegetable into a large bath of ice water immediately after cooking? I gathered up a few pounds of vegetables and headed into the kitchen to answer these questions.

### Volume and pH

I cooked a half pound of green beans in amounts of water ranging from 2 cups all the way up to a full

2 gallons, noting both the temperature of the water and the time it took to cook the beans to tenderness in each case. Some trends immediately started to surface. No matter how much water I started with, the energy needed to bring the pot back to a boil after I added the beans was the amount of energy it took to raise the ½ pound of green beans up to 212°F, which meant that every single pot returned to a boil at around the same rate (for more on this, see "The Best Way to Cook Pasta," page 674). On the other hand, the temperatures of the pots with very little water dropped much more precipitously when I added the beans to them. Not only did the beans cooked in less water take longer to tenderize, they also ended up cooking to a drab army green instead of the brilliant shamrock green of those cooked in a gallon of water or more.

To understand why this is the case, let's take a look at the outside of a green vegetable. Like all living matter, vegetables are composed of many individual cells. With vegetables, these cells are bound in place with *pectin*, a glue-like carbohydrate-based molecule. Within vegetable cells, there are various pigments, enzymes, and aromatic compounds. Green vegetables in particular contain *chloroplasts*, the tiny organelles (that's a small organ) responsible for converting sunlight into energy through the use of the pigment *chlorophyll*, which is responsible for giving green vegetables their bright green color.

Between all of the plant cells are trapped tiny pockets of gases that scatter light waves, partially obscuring your view of the vegetable's bright green pigments. As soon as you plunge it into boiling water, though, those gases escape and expand, and the unhindered view of its pigments makes the vegetable appear suddenly much greener. At the same time, an enemy from within—an enzyme called *chlorophyllase*—is working to destroy the vibrant green color by altering the shape of the chlorophyll. Chlorophyllase is most active at temperatures below 170°F, and it is destroyed at around 190°F. *That's* the reason why a large pot of water is necessary. With a small pot of water, the vegetables spend too much time under that 170°F cutoff, giving the chlorophyllase a head start in dulling their appearance. With large pots of boiling water that never drop below 190°F, the chlorophyllase is rapidly destroyed before it has a chance to get to work on the chlorophyll.

Even with the cholorphyllase out of commission, a boiling green vegetable will eventually begin to turn drab as heat causes irreversible changes in its structure. These changes are exacerbated by acidic conditions—even a few teaspoons of lemon juice or vinegar in a large pot of boiling water can cause green vegetables to rapidly turn dull when cooked in it. This is the second reason why vegetables should be cooked in plenty of water. As they cook, they release their naturally acidic contents into the cooking medium, acidifying it and hastening browning. Using a large volume of water dilutes this acidity. Similarly, vegetables should be cooked with the lid off, to encourage the partial evaporation of any of these acidic compounds. *Ah*, you're thinking to yourself, *if acid is the enemy of green vegetables, why not add a pinch of baking soda to the water to keep them bright and green?* And you'd be right—baking soda *does* keep vegetables greener. Unfortunately, it also hastens the breakdown of their cells, causing them to turn mushy while imparting a soapy aftertaste.

Beans cooked in plain water, acidic water, and baking soda water.

Seems like in this case, the pro-big-potters are right. It's the only way to achieve vegetables that are simultaneously bright green and tender-crisp.

**Shocking!**

Once your vegetables are cooked perfectly, the question remaining is how to prevent them from *over*cooking. At the restaurants I've worked in, we'd plunge them into a huge bowl of ice water and leave them there until completely chilled. But in my home kitchen, I tested the ice-water method side by side with two other methods: running the drained vegetables under cool tap water and simply leaving them in a bowl at room temperature. Both the ice-water vegetables and the cool-running-water veg came out identically, so clearly the ice is overkill—cold water will do just fine. Surprisingly though, it turned out that even when the vegetables are simply placed in a bowl and left on the counter, the ones around the edges lose heat to the air fast enough to prevent overcooking. It's only the vegetables in the center that end up mushy and dark green. So, as long as you spread the vegetables out in a single layer—say, on a rimmed baking sheet—you don't even really need the cold water. I still use it just for the sake of convenience, but it's good to know this in a pinch.

Whatever cooling method you use, it is of vital importance that you dry your vegetables in a salad spinner or with a clean kitchen towel before adding them to a salad. That is, unless you like watered-down salad.

BLANCHING IS the gateway to so much more than keeping green vegetables green, however. In common cook's parlance, blanching is the act of dropping vegetables into a large pot of boiling salted water and lightly cooking them. Most often, the vegetables are then used in another recipe, whether it's sweet peas that are lightly blanched in salted water, then tossed in a skillet with some butter, or green beans that are blanched until almost completely soft before being stirred into a creamy mushroom sauce and baked in a casserole. It's a softening step for vegetables, allowing you better control over the texture of the finished dish, rather than trying to cook all the ingredients in one step or one pot. It's also a valuable organizational and time-saving tool for dinner parties or holidays.

Remember this: anytime blanching a vegetable is part of a recipe, you can *always* cool that vegetable, dry it carefully, and finish the rest of the recipe later. This means that a recipe for, say, a broccoli or cauliflower gratin that calls for blanching the stalks, then covering them with a cheese sauce and baking can actually be broken down into two distinct steps that don't have to be done with one immediately following the other. Heck, you could boil your broccoli on Monday, then toss it with your cheese sauce and bake it on Thursday if you'd like. This kind of flexibility in a recipe makes planning and executing far simpler.

I've included steaming in this section as well, since it essentially accomplishes the same goal as blanching.

## BUTTERED SNAP PEAS

Here is blanching at its absolute simplest. You boil your snap peas just until barely tender, then add them to a buttery sauce (or just butter and lemon juice) you have waiting in a skillet. Toss to combine over high heat, and serve. Of course, as I mentioned above, you can blanch the peas ahead of time, cool them down, and serve them whenever you'd like by preparing the sauce, adding the peas straight from the fridge, and tossing until heated through.

Want to make the recipe even easier? Skip the snap peas and use regular old frozen green peas. Because freezing vegetables disrupts their cell structure in much the same way that blanching does, if you use frozen peas, there's no need to even boil them beforehand. Simply thaw them under running water, drain, and add them to the buttery sauce in the skillet.

## PICKING PEAS

Peas are one of the few vegetables that are almost always better frozen than fresh. Why is that? Peas start losing flavor and sweetness the minute their pods are picked off the vine. There's a noticeable difference in the texture and flavor of a pea that's been off the vine for even as little as six hours or so and one that's just-picked. This means that when you see "fresh" peas at a supermarket or even at most farmers' markets, they are likely to have been off the vine for several days, slowly becoming less sweet and more starchy. Frozen peas, on the other hand, are individually frozen very soon after they are picked, ensuring that their sugars and creamy texture are intact. Small, spherical peas are also the absolute ideal shape for rapid freezing and thawing, which helps them to maintain good texture throughout the process by minimizing the stress on the cells that can be caused by large-ice-crystal formation during slow freezes. Unless you have an extremely reputable farmer you're buying your peas from in the spring, stick to the frozen version.

# HOT BUTTERED SNAP PEAS
## WITH LEMON AND MINT

**SERVES 4**

2 tablespoons unsalted butter

1 teaspoon lemon juice (from 1 lemon)

Kosher salt

1 pound sugar snap peas, strings and ends removed

2 tablespoons chopped fresh mint

1 teaspoon grated lemon zest

Freshly ground black pepper

1. Add the butter and lemon juice to a 12-inch heavy-bottomed stainless steel skillet and set aside.

2. Bring 4 quarts water and ¼ cup salt to a boil in a Dutch oven over high heat. Add the snap peas and cook until bright green and tender but still with a bit of bite, about 3 minutes. Drain the peas and add to skillet with the butter and lemon juice. Set over high heat and cook, tossing and stirring, until the butter is completely melted and the snap peas are coated. Stir in the mint and lemon zest. Season with salt and pepper to taste and serve.

# HOT BUTTERED SNAP PEAS
## WITH SCALLIONS AND HAM

**SERVES 4**

2 teaspoons olive oil

4 ounces baked ham, cut into ¼- to ½-inch cubes

6 scallions, whites sliced into ¼-inch segments, greens finely sliced (reserve separately)

2 tablespoons unsalted butter

Kosher salt

1 pound sugar snap peas, strings and ends removed

1 teaspoon grated lemon zest (from 1 lemon)

1 teaspoon lemon juice

Freshly ground black pepper

1. Heat the oil in a 12-inch heavy-bottomed stainless steel skillet over medium-high heat until shimmering. Add the ham and scallion whites and cook, stirring and tossing occasionally, until the ham is just starting to turn golden brown, about 3 minutes. Remove from the heat, add the butter, and set aside.

2. Bring 4 quarts water and ¼ cup salt to a boil in a Dutch oven over high heat. Add the snap peas and cook until bright green and tender but still with a bit of bite, about 3 minutes. Drain the peas and add to the skillet. Set over high heat and cook, tossing and stirring, until the butter is completely melted and the snap peas are coated. Stir in the scallion greens and lemon zest and juice. Season with salt and pepper to taste and serve.

# HOT BUTTERED SNAP PEAS
## WITH LEEKS AND BASIL

**SERVES 4**

1 large leek, white part only, split in
    half and cut into ¼-inch slices

2 tablespoons unsalted butter

Kosher salt

1 pound sugar snap peas, strings
    and ends removed

2 tablespoons chopped fresh basil

1 teaspoon grated lemon zest (from
    1 lemon)

1 teaspoon lemon juice

Freshly ground black pepper

1. Place the leeks and butter in 12-inch heavy-bottomed stainless steel skillet over medium heat and cook, stirring frequently, until the leeks are tender but not browned, about 5 minutes; reduce the heat if the butter or leeks start to brown. Remove from the heat and set aside.

2. Bring 4 quarts water and ¼ cup salt to a boil in a Dutch oven over high heat. Add the snap peas and cook until bright green and tender but still with a bit of bite, about 3 minutes. Drain the peas and add to the skillet. Set over high heat and cook, tossing and stirring, until the snap peas are coated. Stir in the basil and lemon zest and juice. Season with salt and pepper to taste and serve.

# HOT BUTTERED PEAS

**SERVES 4**

1 pound (about 3 cups) frozen peas

2 tablespoons unsalted butter

1 teaspoon grated lemon zest (from
    1 lemon)

1 teaspoon fresh lemon juice

Kosher salt and freshly ground
    black pepper

1. Place the peas in a colander and run under hot water until thawed and warmed, about 4 minutes; drain well.

2. Melt the butter in a large saucepan over medium-high heat. Add the peas and toss to coat. Cook, stirring, until heated through, about 2 minutes. Stir in the lemon zest and juice. Season with salt and pepper to taste and serve.

# HOT BUTTERED PEAS
## WITH BACON, SHALLOTS, AND TARRAGON

**If you were to cook the butter along with the bacon from the beginning, its proteins would all brown, altering their structure and emulsifying properties. Stirring the cold butter in with the peas (which act as a temperature regulator) ensures a nice, creamy, glaze-y coating in the finished dish.**

**SERVES 4**

1 pound (about 3 cups) frozen peas

2 slices thick-cut bacon, cut into
½-inch-wide strips

1 medium shallot, finely sliced
(about ¼ cup)

2 tablespoons unsalted butter

2 tablespoons finely chopped fresh
tarragon

1 teaspoon grated lemon zest (from
1 lemon)

1 teaspoon lemon juice

Kosher salt and freshly ground
black pepper

1. Place the peas in a colander and run under hot water until thawed and warmed, about 4 minutes.

2. Cook the bacon in a large saucepan over medium heat, stirring frequently, until it has rendered its fat and is crisp. Add the shallot and cook, stirring frequently, until softened, about 3 minutes. Add the peas and butter and toss to coat, then cook, stirring, until heated through, about 2 minutes. Stir in the tarragon and lemon zest and juice. Season with salt and pepper to taste and serve.

# HOT BUTTERED PEAS
## WITH PROSCIUTTO, PINE NUTS, AND GARLIC

**SERVES 4**

1 pound (about 3 cups) frozen peas

2 teaspoons vegetable oil

3 ounces thinly sliced prosciutto,
cut into thin ribbons

½ cup pine nuts

2 medium cloves garlic, minced or
grated on a Microplane (about 2
teaspoons)

2 tablespoons unsalted butter

2 tablespoons finely chopped fresh
parsley

1 teaspoon grated lemon zest (from
1 lemon)

1 teaspoon lemon juice

Kosher salt and freshly ground
black pepper

1. Place the peas in a colander and run under hot water until thawed and warmed, about 4 minutes.

2. Heat the oil in a large saucepan over medium-high heat until shimmering. Add the prosciutto and cook, stirring frequently, until just beginning to crisp, about 3 minutes. Add the pine nuts and cook, stirring constantly, until lightly toasted, about 2 minutes. Add the garlic and cook, stirring, until fragrant, about 30 seconds. Add the peas and butter and toss to coat, then cook, stirring, until heated through, about 2 minutes. Stir in the parsley and lemon zest and juice. Season with salt and pepper to taste and serve.

# KNIFE SKILLS:
## How to Chiffonade Basil and Other Herbs

*Chiffonade* **is a French chef's term for "cut into ribbons."**

The key to cutting herbs into a chiffonade efficiently is to stack the leaves—that way, you can cut through many at once. For basil or other leafy herbs like mint, sage, and parsley, start by placing one especially large leaf bottom side up on a cutting board (1), then stack more leaves on top (2). Once you have a stack of about 10, roll the leaves up into a tight cylinder (3) and use a sharp knife to slice them as thin as possible (4). You will end up with a shower of herb ribbons (5).

# UPGRADED GREEN BEAN CASSEROLE

This recipe relies on blanching as a parcooking technique more than the buttered snap peas do. In the snap pea recipes, you could potentially cook the peas from start to finish in the skillet. It'd take a bit of skill to do and the end results wouldn't be that great, but it's doable. This recipe, on the other hand, would not be possible without blanching. Put raw green beans in a casserole, and they simply won't cook through. The classic Campbell's green bean casserole is a staple on many American tables, particularly around the holidays, but there are some easy ways you can improve on the out-of-the-can version.

Here's the deal: If the only thing you do is substitute fresh green beans for the canned variety, you're giving your casserole a major upgrade. But substitute your own mushroom sauce made out of real mushrooms (using chicken stock and a splash of soy add a big umami boost to the dish) instead of canned cream of mushroom soup and top the whole thing off with some crisply fried shallots, and you can proudly say goodbye to Sandra Lee and proclaim that semi-homemade is a thing of the past.

My fried shallots were inspired by Thai-style fried shallots, something that you should have on hand in your kitchen all the time. I make mine in batches of a couple pounds (to cook more than what's called for in this recipe, increase the amount of oil so that the shallots are just barely poking out of the surface). Add them to sandwiches or soups, use as a garnish for cooked meats, or just eat 'em straight out of the jar. Sometimes I forget to hide the jar and come home to find the sweet smell of shallots on my wife's breath. She, of course, blames the dog. As far as I know, the dog has yet to figure out how to leave a keyboard covered with greasy little fingerprints.

NOTE: Homemade fried shallots rock, but you can also get them prefried in Thai or Vietnamese markets.

**SERVES 6 TO 8**

1½ pounds button mushrooms, cleaned

2 teaspoons soy sauce

2 teaspoons lemon juice (from 1 lemon)

2 cups homemade or low-sodium canned chicken stock

1½ cups heavy cream

1 recipe Fried Shallots (recipe follows), plus 2 tablespoons of the strained oil

2 tablespoons unsalted butter

2 medium cloves garlic, finely minced or grated on a Microplane (about 2 teaspoons)

¼ cup all-purpose flour

Kosher salt and freshly ground black pepper

2 pounds green beans, ends trimmed and cut into 2-inch segments

1. Smash the mushrooms under the bottom or a large skillet until broken into ¼- to ½-inch pieces. Roughly chop into ⅛- to ¼-inch pieces. Set aside.

2. Combine the soy sauce, lemon juice, chicken stock, and cream in a 1-quart liquid measure or medium bowl. Set aside.

3. Heat the shallot oil and butter in 12-inch nonstick skillet over high heat until the butter melts and the foaming subsides. Add the mushrooms and cook, stirring occasionally, until their liquid has evaporated and the mushrooms begin to sizzle, 6 to 10 minutes. Reduce the heat to medium-high, add the garlic, and cook, stirring, until fragrant, about 30 seconds. Add the flour and cook, stirring constantly, until light golden blond, 1 to 2 minutes. Whisking constantly, add the stock and cream mixture. Bring to a boil, whisking, then reduce to a simmer, and cook, whisking, until the mixture has a consistency somewhere between pancake batter and heavy cream, about 5 minutes. Season to taste with salt and pepper and set aside.

4. Adjust an oven rack to the lower-middle position and preheat the oven to 350°F. Bring 4 quarts water and ¼ cup kosher salt to boil in a Dutch oven over high heat. Fill a large bowl with 4 cups ice cubes and 2 quarts water. Add the green beans to the boiling water and cook until tender but still green, about 7 minutes. Drain and immediately transfer to the ice water to cool completely; drain well.

5. Combine the green beans, mushroom sauce, and 1 cup of the fried shallots in a bowl. Transfer to a 9-by-13-inch rectangular casserole or 10-by-14-inch oval casserole. Bake until hot and bubbly, 15 to 20 minutes. Top with the remaining fried shallots and serve.

1

2

3

4

5

## Fried Shallots

**MAKES ABOUT 2 CUPS**

1 pound shallots, sliced ⅛ inch
  thick, preferably on a mandoline
2 cups canola oil
Kosher salt

1. Line a rimmed baking sheet with 6 layers of paper towels. Combine the shallots and oil in a wok or medium nonstick saucepan; the shallots should barely stick up above the oil. Place over high heat and cook, stirring frequently, until the shallots are completely soft, about 20 minutes. Then continue to cook, stirring constantly, until the shallots are light golden brown, about 8 minutes. Immediately drain in a fine-mesh strainer set over a heatproof bowl or saucepan; set the shallot oil aside.

2. Transfer the fried shallots to the paper towels. Then lift up one end of the top layer of towels and roll the shallots off onto the second one. Blot with the first towel to absorb excess oil, then repeat, transferring the shallots from one layer of paper towels to the next, until only one layer remains. Season well with salt and allow to cool completely, about 45 minutes.

3. Once they are cooled, transfer the shallots to an airtight container and store at room temperature for up to 3 months. The cooled and strained shallot oil can be used for salad dressing or stir-fries.

### FRIED GARLIC

Replace the shallots with peeled whole garlic cloves. Place the garlic in the bowl of a food processor and pulse 8 to 10 times, scraping down the sides and redistributing the garlic as necessary, until it is chopped into pieces no larger than ⅛ inch across. Cook as directed—the garlic may cook a little faster, but do not let it get beyond golden brown, or it will become very bitter. The key is to drain the garlic about 15 to 20 seconds before you think it is completely done, as it will continue to cook after draining. It may take a couple of trials to get the exact timing down, but the results are worth it.

# CHEESY BROCCOLI OR CAULIFLOWER CASSEROLE

At its core, a broccoli or cauliflower casserole is not really much different from macaroni and cheese, and, indeed, you can make it exactly the same way, simply replacing the parboiled or soaked pasta (see page 744) with blanched cauliflower or broccoli. Note that broccoli florets tend to gather and hold on to water, which can cause them to become waterlogged after blanching or to overcook in the casserole. To avoid this, transfer the broccoli to a rimmed baking sheet immediately after blanching, so that the water has a chance to evaporate and escape.

A hand blender delivers extra-glossy cheese sauce.

Vegetables won't absorb as much liquid as pasta will, so you want the cheese sauce to be a little thicker than it would be for mac and cheese. To create a smoothly flowing sauce, I use a mix of evaporated and whole milk (evaporated milk has extra proteins in it that help keep the sauce nice and emulsified), along with a bit of flour to thicken and gelatin to add creaminess. Garlic, a dash of hot sauce, and some ground mustard add flavor. A blender will make things extra-smooth. (For more on cheese sauce, see page 714.)

Once the broccoli or cauliflower is blanched and the sauce is made, all you need to do to complete the dish is to combine them, throw them in a casserole, and top them with a layer of buttery bread crumbs that will brown and crisp as the casserole bakes.

**NOTE**: Use a good melting cheese like American, Cheddar, Jack, Fontina, young Swiss, Gruyère, Muenster, young provolone, and/or young Gouda, among others (see "Cheese Chart," pages 717–21).

## SERVES 6 TO 8

3 slices hearty sandwich bread, crusts removed and torn into chunks

1 small shallot, finely minced

2 tablespoons chopped fresh parsley

4 tablespoons unsalted butter

Kosher salt

2 pounds broccoli or cauliflower (or a mix), cut into bite-sized florets

½ cup whole milk

¼ ounce (1 packet) unflavored gelatin

2 medium cloves garlic, minced or grated on a Microplane (about 2 teaspoons)

1 tablespoon all-purpose flour

One 12-ounce can evaporated whole milk

1 teaspoon Frank's RedHot or other hot sauce, or more to taste

½ teaspoon ground mustard

8 ounces cheese (see Note above), grated

1. Adjust an oven rack to the middle position and preheat the oven to 400°F. Combine the bread, shallot, parsley, 1 tablespoon of the butter, and a pinch of salt in the bowl of a food processor and pulse until coarse bread crumbs have formed and no large chunks of butter remain. Set aside.

2. Bring 4 quarts water and ¼ cup salt to a boil in a Dutch oven over high heat. Add the broccoli (or cauliflower) and cook until just tender but still with a bit of bite, about 3 minutes. Drain, spread on a rimmed baking sheet, and set aside.

3. Place the whole milk in a small bowl and sprinkle the gelatin evenly over the top. Set aside to soften.

4. Melt the remaining 3 tablespoons butter in a large saucepan over medium-high heat. Add the garlic and cook, stirring, just until fragrant, about 30 seconds. Add the flour and cook, stirring constantly, until light golden blond, about 2 minutes. Whisking constantly, slowly add the evaporated milk, followed by the gelatin mixture. Whisk in the hot sauce and mustard and bring to a boil over medium-high heat, whisking occasionally to prevent the bottom from burning. Remove from the heat, add the cheese all at once, and whisk until fully melted and smooth. If a smoother sauce is desired, blend with an immersion blender or in a standing blender. Season to taste with salt and more hot sauce if desired.

5. Stir the broccoli into cheese sauce, then transfer to a 13-by-9-inch baking dish or 10-by-14-inch oval casserole. Scatter the bread crumbs evenly over the top. Bake until golden brown and bubbling, about 25 minutes, rotating the dish once halfway through cooking.

## THE ULTIMATE STEAMER: THE MICROWAVE

There are few kitchen tools more feared, more maligned, and more misunderstood than the microwave. It's understandable. You put your food in a box, the box shoots out some invisible rays, and all of a sudden, your food is cooked and hot. Must be magic, right?

The reality is far more benign. A microwave works by sending out long waves of electromagnetic radiation that create an oscillating magnetic field inside the chamber. Because water molecules are polar—that is, they are sort of like tiny magnets with a positive end and a negative end—the oscillating magnetic field causes them to rapidly jostle up and down. It's the friction that this jostling water creates that in turn heats up your food. That's why a microwave won't affect objects that don't contain water or some other magnetic molecules.

But hold on, back up. Electromagnetic radiation. Isn't that like, all dangerous and stuff? Well, sure, certain types of it *are* dangerous. But EM radiation (as we'll refer to it from now on) comes in many forms. Indeed, the very light that you see coming from the sun, from a flashlight, or from the quiet glow of your iPad is a form of EM radiation. It just happens to be of a wavelength that your eyes can detect. (That's right—your head has radiation detectors built right into it.) Radio waves are another form of EM radiation. The X-rays a doctor shoots at your chest when you accidentally swallow a lobster whole when eating too fast are a more dangerous form of EM radiation. Even the heat from an oven or a red-hot poker is EM radiation. It's everywhere, but again, not all radiation is dangerous. Microwaves fall squarely into the "nondangerous" category. At least, so long as you don't try and stick yourself behind the shielded door of the cooker.

That said, the microwave has some severe limitations as a cooking tool. For one, it's nearly impossible to brown foods in it in a satisfactory way. Meat cooks fast but comes out flaccid and anemic looking. What it *is* good for is heating up liquids, reheating leftovers when you don't care too much about exterior texture, and steaming vegetables. The last is the use I put it to most.

See, just as with brief blanching, the goal of steaming is to very lightly cook vegetables just until they lose their raw taste, but not until they begin to turn mushy. Since a microwave very efficiently and rapidly uses the liquid inside the vegetables themselves to heat them from the inside, you can micro-steam vegetables in a matter of minutes.

# MICRO-STEAMING VEGETABLES

**To micro-steam, lay vegetables in a single layer on a microwave-safe plate. Cover them with a triple layer of damp paper towels. Microwave on high heat until the vegetables are tender, 2½ to 6 minutes, depending on the power of your microwave.**

| VEGETABLE | HOW TO PREPARE | SPECIAL INSTUCTIONS |
|---|---|---|
| Asparagus | Peel the fibrous skin off the stalks, starting from just below the head, if desired. Leave whole. | |
| Bok Choy | Remove the core, separate the individual leaves, and wash carefully to remove dirt. | |
| Broccoli | Cut off the florets in 1-inch pieces. Peel the tough stalks and slice ¼ to ½ inch thick. | May take longer than other vegetables. |
| Broccoli Rabe | Wash carefully, then remove any tough sections from the stalks. | |
| Brussels Sprouts | Cut in half or peel off individual leaves. | |
| Cauliflower | Cut off florets in 1-inch pieces. | May take longer than other vegetables. |
| Celery | Peel the exterior, then cut on a bias into ¼- to ½-inch slices. | |
| Corn | Microwave in the husk. Alternatively, remove the kernels from cobs and microwave. | For corn in the husk, microwave for about 1½ minutes per ear. For kernels, microwave uncovered in a bowl, stopping every 30 seconds to stir, until hot. |
| Green Beans | Trim the ends. | |
| Frozen Peas | Use straight from the freezer. | Microwave uncovered in a bowl, stopping every 30 seconds to stir, until hot. |
| Spinach | Trim tough stems, wash, and dry. | Microwave one-third the amount called for at a time in a large bowl, checking and tossing every 30 seconds, until wilted. |
| Zucchini | Cut into ¼- to ½-inch disks. | |

# MICRO-STEAMED ASPARAGUS
## WITH HOLLANDAISE OR MAYONNAISE

**NOTE**: Hot asparagus and hollandaise are a natural pair, as are cold asparagus and hollandaise's colder cousin, mayonnaise. That said, if you want to make matters even simpler, a bit of compound butter (page 326) melted over hot spears makes a fine side dish: even a simple drizzle of extra-virgin olive oil and a squeeze of lemon will do.

**SERVES 2 OR 3**

8 ounces asparagus, bottom 1½ inches trimmed off, stalks peeled if desired

Kosher salt and freshly ground black pepper

1 Foolproof Hollandaise Sauce (page 111) or Foolproof Homemade Mayonnaise (page 807), for serving

1. Place the asparagus in an even layer on a large microwave-safe plate and season with salt and pepper. Lay a triple layer of damp paper towels on top of the asparagus, completely covering them, or cover with a clean damp kitchen towel. Microwave on high power until the asparagus is bright green and tender but still crisp, 2½ to 6 minutes, depending on the power of your microwave.

2. Remove the asparagus, arrange on a warmed serving platter, and serve with the hollandaise. Alternatively, run the asparagus under cold running water until chilled, dry carefully, and serve with mayonnaise.

# MICRO-STEAMED ASPARAGUS
## WITH POACHED EGG AND WALNUT VINAIGRETTE

**SERVES 2**

8 ounces asparagus, tough bottoms trimmed off, stalks peeled if desired

Kosher salt and freshly ground black pepper

2 Perfect Poached Eggs (page 107), kept warm in a bowl of hot water

¼ cup Walnut Vinaigrette (recipe follows)

1. Place the asparagus in an even layer on a large microwave-safe plate and season with salt and pepper. Lay a triple layer of damp paper towels on top of the asparagus, completely covering them, or cover with a clean damp kitchen towel. Microwave on high power until the asparagus is bright green and tender but still crisp, 2½ to 6 minutes, depending on the power of your microwave.

2. Remove the asparagus and arrange on warmed serving plates. Top with the poached eggs, drizzle with the vinaigrette, and serve.

## Walnut Vinaigrette

*Use on roasted vegetables like beets or sweet potatoes or on robust bitter greens such as radicchio, endive, or frisée.*

**MAKES ABOUT 1½ CUPS**

2 ounces (about ½ cup) walnuts, toasted and roughly chopped

3 tablespoons sherry vinegar

1 tablespoon water

1 tablespoon Dijon mustard

1 tablespoon honey

1 small shallot, minced or grated on a Microplane (about 1 tablespoon)

½ cup extra-virgin olive oil

¼ cup canola oil

1 teaspoon walnut oil (optional)

½ teaspoon kosher salt

¼ teaspoon freshly ground black pepper

Combine the walnuts, vinegar, water, mustard, honey, and shallots in a medium bowl and whisk together. To keep it from moving, set the bowl on top of a heavy medium saucepan lined with a dish towel. Whisking constantly, slowly drizzle in the olive and canola oil; the dressing should emulsify and thicken significantly. Whisk in the walnut oil, if using. Season with the salt and pepper. The dressing will keep in a sealed container in the fridge for up to 2 weeks: shake vigorously before using.

# ALL ABOUT ASPARAGUS

**Y**ou may be wondering why asparagus gets its own gigantic sidebar all to itself. It's because (a) I really love it and (b) among the vegetables that are always available from the supermarket, there are very few that show as drastic a difference in flavor between the imported year-round stuff and the fresh-from-the-earth spring variety. Asparagus contains a lot of sugar at the moment of harvest, but as it sits around waiting to be cooked, this sugar quite rapidly begins to form starch molecules, turning tender, sweet stalks bland and starchy.

**How do I pick the best asparagus at the market?**

Whether you choose bright green spears, mild white stalks (which are grown underground to prevent chlorophyll development), or one of the purple varieties, you should look for the same things: firm, crisp stalks with tight, fully closed budding tips. As asparagus ages, the petals on the tips will slowly open up, dry out, or fall off. Asparagus should appear moist but not wet, fresh-cut, and bright, not dry or woody.

Your best bet for good asparagus is at a farmers' market or direct from a farm. Unless your supermarkets are far, far better than mine, the asparagus you get in them, even during peak season, have been out of the earth for far too long to let their flavor shine. Unless it comes direct from a farmer, asparagus's point of origin is always written either on a label or rubber band affixed to the bunch. Do me, your farmer, and your taste buds a favor: if you live in New England, don't buy Peruvian asparagus in the middle of May.

**What about thickness? Does it make a difference in flavor?**

Asparagus comes in all sizes, from slim pencil-wide stalks to big fat ones as thick as your thumb, but believe it or not, size has nothing to do with age. Asparagus grows from an underground crown, from which scores of stalks shoot forth. It takes about three seasons for this crown to begin producing edible stalks, and after that, it'll continue to produce stalks for at least a couple of decades. It's the age and variety of the crown that determines the thickness of a stalk. A farmer can't simply wait for a thin stalk to grow into a thicker one—that won't happen until a few seasons later. While both can be fantastic, I generally choose one size over the other depending on how I'm going to cook it (or, more likely, I choose my cooking method based on the size of asparagus I happen to pick up at the farmers' market).

- **Spears about ⅓ inch thick or thinner** tend to be more intense in flavor and less watery. They're also a little bit tougher and snappier, due to their higher ratio of fibrous skin to softer interior. This makes them ideal for blanching and serving hot or cold, stir-frying, or even just eating raw as a snack. Higher-heat methods like broiling or grilling tend to dry them out a little too much, though if you like that charred asparagus flavor, you might still consider cooking them with these methods.

- **Spears thicker than ⅓ inch** are considerably more tender than thinner stalks, but they can get a little watery if you steam or boil them. High-heat cooking methods like grilling, broiling, stir-frying, and pan-searing are best, allow-ing you to get them nice and caramelized on their exterior while still maintaining a bit of bite. I also use large spears for braising.

**I've heard that I need to snap off the bottoms of the stalks and that somehow the stalk will tell me where it's supposed to snap. Any truth in this crazy story?**

Depending on its age, the bottom part of a stalk of asparagus can get unpleasantly woody or fibrous and so usually needs to be trimmed. But what's the best way to do this? Traditional wisdom does tell you that the "foolproof" way is to simply grasp the stalk at both ends and snap it, the asparagus magically breaking exactly where it needs to. This question is often debated, and most people come down on the side of snapping. But is it really the best method?

After some pretty extensive testing, I've come to realize that it's all a bunch of hokum. Indeed, depending on exactly how you apply force to the stalk, you can get it to snap pretty much anywhere along its length, even when your hands are in the exact same position. Check this out:

I snapped every one of these stalks with my hands, holding each one of them at the exact

*continues*

same point, and was still able to make them break wherever I wanted them to—quite easily, I might add. Far better is to line them up, determine where most of the stalks appear to become woody (the stalk will begin to fade to white at that point), and then slice them all at once with a knife, picking out any outliers and trimming them as necessary on a case-by-case basis.

### Should I peel my asparagus?

Even when the stalks are properly trimmed, the outer layers can have a fibrousness that is apparently upsetting to the palates of people who do things like write Michelin guides. Use a

standard Y-peeler to shave the skin off, starting right below the tip. The flavorful trimmings can be added to your vegetable or chicken stock or blanched and pureed into a soup or sauce.

### What's the best way to store asparagus?

The best way to store asparagus is not to. As I said, its flavor dramatically diminishes over time, so the sooner you get it in the pan and into your belly, the better. If you absolutely must store asparagus, treat it like you would a bunch of flowers.* Trim the ends and stand the stalks up in a cup or glass of water, loosely cover the tips with a plastic bag to prevent evaporation, and set the whole thing in the fridge. Some people recommend adding salt or sugar to the water, but I've never been able to detect any difference in flavor if you do this. Don't bother.

---

\* Which in fact asparagus isn't, despite what you may have read. Asparagus "flowers"—the parts you eat—are actually modified stem structures. True asparagus flowers are six-tepaled (not to be confused with petaled, though they are, in fact, also six-petaled), bell-shaped affairs with poisonous red berries. What does tepaled mean? Look it up!

# MICRO-STEAMED GREEN BEANS
## WITH OLIVES AND ALMONDS

**SERVES 2 OR 3**

8 ounces green beans, ends trimmed

Kosher salt and freshly ground black pepper

¼ cup chopped kalamata or other olives

2 tablespoons slivered or sliced almonds, toasted

1 tablespoon extra-virgin olive oil

2 tablespoons chopped fresh parsley

1 teaspoon lemon juice (from 1 lemon)

1. Place the green beans in an even layer on a large microwave-safe plate and season with salt and pepper. Lay a triple layer of damp paper towels on top of the beans, completely covering them, or cover with a clean damp kitchen towel. Microwave on high power until the beans are bright green and tender but still crisp, 2½ to 6 minutes, depending on the power of your microwave. Transfer to a large bowl.
2. Toss the beans with the olives, almonds, olive oil, parsley, and lemon juice. Season to taste with salt and pepper and serve.

# MICRO-STEAMED CORN
## WITH GARLIC-CHILI BUTTER

**Corn is pretty much custom-made for the microwave. The husks trap in steam as it heats up, effectively creating a steam chamber inside, cooking the kernels quickly and efficiently. All you need for an ear of microwaved corn is a quick buttery spread you can whip up as it cooks.**

**SERVES 4**

4 ears corn in the husk

3 tablespoons unsalted butter, at room temperature

1 small clove garlic, minced or grated on a Microplane (about ½ teaspoon)

2 tablespoons sliced scallions

1 teaspoon chili powder

Kosher salt and freshly ground black pepper

1. Place the corn on microwave-safe plate and microwave on high power until cooked through, about 6 minutes.
2. Meanwhile, combine the butter, garlic, scallions, and chili powder in a small bowl and mash with a fork until smooth. Season to taste with salt and pepper.
3. Husk the corn and remove the silks. Rub with the butter and serve immediately.

# Essential VEGETABLE Technique #2:
## SEARING/SAUTÉING

Though searing and sautéing are similar in appearance and setup—both require the use of a skillet and some fat—there are some subtle differences between the two.

- **With searing**, the goal is to brown the exterior of the vegetables in order to promote the development of the complex flavors created by the Maillard reaction. These reactions take place in the realm above 350°F or so. Cooking with very little movement helps improve this browning, as it allows energy from the skillet and the hot oil to be transferred to the same part of a piece of food for a long period of time. High heat is the order of the day here.

- **With sautéing**, the goal is to cook a whole mess of small bits of food evenly. Oftentimes sautéing is the first step in a much longer recipe, as when you soften onions in olive oil as the first step to a biscuit gravy or pasta sauce. Other times, sautéing is the only step needed to take a food from raw to table-ready.

## BRUSSELS SPROUTS WORTH EATING

Brassicas—that is, cabbage, broccoli, Brussels sprouts, cauliflower, and the like—are particular well suited for searing. They are rich in the sulfurous compounds known as *glucosinolates*, and even the slightest bit of overcooking can cause these chemicals to break down, producing the foul-smelling odor of long-cooked cabbage or sprouts. It's what I always imagined Charlie Bucket's house smelled like. The effect is less pronounced in broccoli and cauliflower, but still rather unpleasant. At the same time those sulfurous compounds are being released, more desirable, distinctively sharp mustard-like compounds are being actively destroyed by an enzyme within the vegetable. This enzyme isn't deactivated until it reaches around 180°F or so. In order to get the best flavor from brassicas, your goal is to get them above this 180°F mark as quickly as possible. Searing is one of the best ways to accomplish this,

Let's talk Brussels sprouts, the much-maligned vegetable that happens to be my favorite. I love them any number of ways—shredded into crispy hash, roasted in a hot oven, braised with white wine—but certainly my favorite way is seared in a ripping-hot skillet that not only heats them rapidly, but also chars the leaves a bit, developing the sort of rich, sweet, nutty flavors that only the best brassica-tinted dreams are made of.

Want an even better suggestion? Sear them in pork fat. When picking the appropriate pork product for your sprouts, anything fatty and cured will do, really. It's just a matter of personal taste. Slab bacon works well. I cut it into strips that you can call *lardons* if you're fancy or French, or both. Equally good, or maybe even better, would be *guanciale*—

salty cured pork jowl. I've done this with dry-cured Spanish chorizo as well, which may be my favorite fat to use. No matter what, before you start cooking the sprouts, you've got to render the fat out of the bacon or pork.

You could try doing that in a dry skillet, but air is a notoriously poor conductor of heat, which means that only the part of the bacon in direct contact with the pan is really heating up. Much better is to start with a little water in the skillet—just enough to cover the bacon. If you blast the heat, the water quickly evaporates, all the while heating the bacon and getting the rendering process started. By the time the water is gone, enough fat will have melted out (though I like to add a bit of vegetable or olive oil as extra insurance) that the bacon should be able to finish cooking quickly and evenly, crisping up far better than it would on its own.

Once you've got your supply of rendered pork fat, it's time to cook the sprouts themselves. If you're doing an unusually large amount, you can always jack up the oven to maximum temperature, toss the sprouts with the pork fat, and roast them until charred (oven-sear them, if you will). By splitting the sprouts in half, you increase their surface area and also give them a stable surface to sit upon while searing. This helps maximize the delicious charring that gives sprouts the nuttiness and charm that makes them worth eating.

After they're charred in the bacon or other pork fat, I season the sprouts with plenty of salt and pepper (I don't like to do it beforehand, because I find the salt from the bacon fat penetrates the sprouts as they cook, making it hard to judge the salt level), then toss them together with the crisp bacon. If you're feeling extra-plucky, you can go for a full half-and-half bacon-to-sprout ratio. Trust me, you'll be popping them like Scrumdiddlyumptious candy bars.

The other great way to cook sprouts is to go the whole nine yards and finely shred them with a knife them before charring. The cooking procedure for sprouts split in half and shredded sprouts is exactly the same.

# KNIFE SKILLS:
## How to Prepare Brussels Sprouts

**Getting Brussels sprouts ready to cook is an easy task, but it's a fiddly one, since each little head needs to be trimmed individually (1). For me, the first step in cleaning the sprouts is to ask around if anybody wants to help make dinner. The job is much easier when you delegate wisely.**

Here's how to tell your volunteer to do it: Start by slicing off the bottom of each sprout (2), then peel off the outer layer of leaves until you get to the tightly packed, pale green core (3). The outer leaves are a little rubbery and can be discarded. The sprouts could be cooked as is, or for better surface area for caramelization, split in half (4). To shred sprouts for hash (5), place each half sprout on a cutting board, cut side down, and carefully shred the sprout with a sharp chef's knife (6). Prepared sprouts can be stored in a zipper-lock plastic bag in the refrigerator for up to 3 days.

1

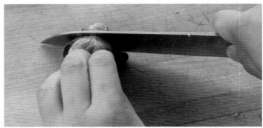
2

3

4

5

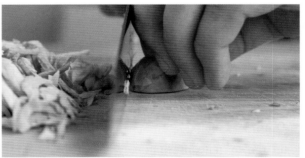
6

# SEARED BRUSSELS SPROUTS
## WITH BACON

**NOTE:** To prepare these for a crowd, double or triple the ingredients. Preheat the oven to 500°F and proceed through step 1. Toss the sprouts and rendered bacon fat together in a large bowl, then transfer to one or two rimmed baking sheets and roast until cooked through and well charred, about 10 minutes. Toss the sprouts with the reserved bacon and serve.

### SERVES 4 TO 6

8 ounces slab bacon, cut into ½-by-¼-inch-thick strips

3 tablespoons olive oil

2 pounds small Brussels sprouts, bottoms trimmed, outer leaves discarded, and split in half

Kosher salt and freshly ground black pepper

1. Put the bacon and olive oil in a 12-inch nonstick skillet, add enough water to barely cover the bacon, and cook over medium-high heat, stirring to separate the bacon, until the water completely evaporates. Continue to cook, stirring and tossing occasionally, until the bacon is crisp on all sides, about 8 minutes longer. Transfer to a fine-mesh strainer set over a large heatproof bowl to drain, then set the bacon aside.

2. Toss the sprouts with the bacon fat until well coated. Wipe out the skillet, add 1 teaspoon of the bacon fat, and heat over high heat until smoking. Add as many sprouts as fit in single layer, cut side down, and cook, without moving, until deeply charred, about 3 minutes. Flip and cook until the second side is charred and the sprouts are tender-crisp, about 3 minutes longer. Season to taste with salt and pepper and transfer to a serving bowl. Repeat with the remaining sprouts.

3. Add the bacon to the serving bowl and toss to combine. Serve immediately.

# PAN-SEARED ZUCCHINI AND CORN

Can anyone else simply not wait for corn season to start every year? I get so impatient that I'll visit the farm stand or supermarket every day toward the start of the season, waiting for the first ears of the local stuff to hit the market. The best way to eat corn is, of course, as *elotes* or *esquites*, the Mexican street snack of grilled corn with cheese, chile, lime, and cream. Or maybe the best way is to bury it underground with a heap of seaweed, chouriço, lobsters, and clams for a traditional New England clambake. Oh no, wait. It's gotta be corn chowder, or, *oh oh*, I know—sautéed with butter and beans into savory succotash. OK, so it's good pretty much no matter how you cook it. Here's another one to add to your arsenal: kernels charred in a skillet. If you separate the kernels from the ear and cook them over crazy-high heat in a skillet with a bit of oil, they get deeply charred while still retaining a sweet bite. It's my go-to method when cooking corn indoors, because it allows you to capture some of the smoky, complex, sweet flavor of grilled corn.

It's a fantastic way to make a quick side for chicken, pork, or seafood dishes, but it's also great as the centerpiece of a simple vegetable-based meal (you could add some crisp cooked bacon and cook the corn in the bacon fat if you want to up your meat factor). Add some hard-seared cubed zucchini, a bit of onion and garlic, hot chile, and a big squeeze of lime, and you've got a snack that's good enough to eat out of a bowl.

It's very easy to accidentally overcook watery zucchini. I've done it too many times. I impatiently throw it into the skillet before it's hot enough. What happens after a few seconds of a pathetic attempt at searing is that the zucchini gives off liquid, which rapidly reduces the temperature of the pan. It's a vicious cycle. At lower temperatures, the zucchini doesn't cook fast enough, instead releasing more moisture and cooling the pan further. What you end up with is a mushy, colorless pile. The key to great seared zucchini is high, high heat, allowing the pan to preheat fully before adding the squash, and not crowding the pan.

**SERVES 4**

3 tablespoons vegetable oil

4 ears corn, shucked and kernels removed (about 3 cups kernels)

1 medium onion, finely diced (about 1 cup)

1 jalapeño pepper, seeds and ribs removed and finely minced

2 medium cloves garlic, minced or grated on a Microplane (about 2 teaspoons)

2 cups zucchini cut into ½-inch cubes (about 2 medium)

¼ cup chopped fresh basil or parsley

3 tablespoons lemon juice (from 2 lemons)

2 tablespoons extra-virgin olive oil

Kosher salt and freshly ground black pepper

Freshly grated Parmigiano-Reggiano (optional)

1. Heat half of the oil in a 12-inch stainless steel or cast-iron skillet over high heat until smoking. Add the corn, toss once or twice, and cook, without stirring, until charred on the first side, about 2 minutes. Toss and stir the corn and repeat until charred on a second side, about 2 minutes longer. Continue tossing and cooking until well charred all over, about 10 minutes total. Add the onion and jalapeño and cook, tossing and stirring, until softened, about 1 minute. Add the garlic and cook, stirring, until fragrant, about 1 minute. Transfer to a bowl.

2. Rinse out the pan (there might be corn sugars burnt onto the bottom, but they should come off easily with water), carefully dry, and return to high heat. Add the remaining oil and heat until smoking. Add the zucchini and cook, without stirring, until well charred on the first side, about 2 minutes. Toss the zucchini and char on a second side, about 2 minutes longer. Toss and char once more, then transfer to the bowl with the corn. Fold in the basil, lemon juice, and olive oil and season to taste with salt and pepper. Serve immediately, sprinkled with Parmesan if desired.

# MEXICAN STREET CORN SALAD

*Elotes*, the on-the-cob version of Mexican street corn, is a staple on my balcony grill over the summer. It's about as easy and inexpensive a dish as you can think of, and there is nothing—really, nothing—that'll get snatched up and eaten as fast as a hot plate of it.

I usually count on making at least an ear and a half per person. To speed things up, I'll keep a big bowl of the coating mixture—that's garlicky mayonnaise, crumbled Cotija cheese (crumbled feta or grated Romano also works well), chopped cilantro, lime juice, and a pinch of chili powder—at the ready so that as soon as my corn comes off the grill, all nice, hot, and charred, it gets a dunk in the sauce, then a pass-off to a waiting mouth. That first bite of hot charred corn, when the cheesy sauce inevitably gets smeared all over your cheeks, just tastes of summer to me. Delicious, fat-smothered summer.

But there are times when a more . . . demure approach must be taken. When there are prim-and-proper aunts or brand-new ties involved, for instance. For those occasions, I go for *esquites*, the spoon-ready version of *elotes*. Rather then slathering the ears of corn with the sauce, you slice the kernels off after cooking and toss them together with the sauce ingredients into a sort of hot salad that's decorous enough to consume with impunity in mixed company.

I tend to make *esquites* when I don't want to bother firing up the grill, because truth be told, it's just as tasty and easy to make indoors as it is out. The key for cooking *esquites* indoors is to remove the kernels from the cobs before you cook them, then cook them in a ripping-hot wok (you can use a skillet, though it's a bit messier), letting them sit in place until the sugars caramelize and you get a deep, dark char before tossing and letting them char again. When it's done right, a few kernels will jump and pop just like popcorn. I've had kernels leap clear across the kitchen on occasion. A careful eye and a splatter guard will protect you from any corn-kernel mortar fire.

**Once the corn is charred, toss it, still hot, with the remaining ingredients. It can be served straightaway, but it is just as good at room temperature, making it an ideal picnic dish.**

2 tablespoons vegetable oil

4 ears corn, shucked and kernels
  removed (about 3 cups kernels)

Kosher salt

2 tablespoons mayonnaise

2 ounces Cotija or feta cheese,
  finely crumbled, or 2 ounces
  Pecorino Romano, grated

½ cup finely sliced scallion greens

½ cup fresh cilantro leaves, finely
  chopped

1 jalapeño pepper, seeded and
  finely chopped

1 to 2 medium cloves garlic, minced
  or grated on a Microplane (1 to 2
  teaspoons)

1 tablespoon lime juice (from 1
  lime)

Chili powder or red pepper flakes

1.  Heat the oil in a large wok or nonstick skillet over high heat until shimmering. Add the corn kernels, season to taste with salt, toss once or twice, and cook, without stirring, until charred on the first side, about 2 minutes. Toss the corn, stir, and repeat until charred on a second side, about 2 minutes longer. Continue tossing and charring until the corn is well charred all over, about 10 minutes total. Transfer to a large bowl.

2.  Add the mayonnaise, cheese, scallions, cilantro, jalapeño, garlic, lime juice, and chili powder to taste and toss to combine. Taste and adjust the seasoning with salt and more chili powder if desired. Serve hot or at room temperature.

# PAN-ROASTED PEARL ONIONS

Not quite searing, not quite sautéing, this technique develops the same deep flavors of well-browned seared vegetables, but it does it very slowly, giving the onions plenty of time to soften and absorb some of the butter. It's amazing how sweet and nutty they become as they cook. While you can certainly serve the onions on their own, they make a great addition to other vegetable dishes. When pearl onions or, even better, cipollini, their flat Italian cousins, are in season, you'll find a supply of pan-roasted onions in my refrigerator, ready to be reheated and added to a pan of sautéed peas or green beans or stirred into a beef stew. When I'm feeling up to the task, I'll actually start with fresh pearl onions and peel them. It's a time-consuming task, but I find it relaxing. Other times, I get lazy and just go with frozen pearl onions. Honestly, they work really well in this dish.

NOTES: Frozen onions can be used in place of peeled fresh ones. Thaw in a bowl of warm water, then spin-dry in a salad spinner. Cippolini onions can also be substituted.

To store the cooked onions, transfer to a plate and allow to cool completely, then transfer to a sealed container and refrigerate for up to 1 week. Add to other sautéed vegetable dishes as desired.

### SERVES 4

3 tablespoons unsalted butter

1 pound fresh pearl or cippolini onions, peeled (see Note above)

Kosher salt and freshly ground black pepper

Melt the butter in a 12-inch heavy-bottomed stainless steel or cast iron skillet over medium heat. Add the pearl onions and reduce the heat to low. The onions should be at a steady sizzle with small bubbles forming under them—adjust the heat as necessary. Cook, shaking and turning the onions every 7 to 10 minutes, until completely tender and well browned on all surfaces, about 30 minutes. Season to taste with salt and pepper. Serve, or reserve for another use as desired (see Note above).

# KNIFE SKILLS:
## How to Peel Pearl Onions

**The easiest way to peel pearl onions if you've got a *huge* batch
of them is to cut off the tops and bottoms, then plunge them into
boiling water for about 1 minute, immediately followed by an
ice bath. The skins should slip right off.**

For a more reasonably sized batch of onions, it's faster to simply peel them by hand rather than to wait for a pot of water to boil (1). Start by cutting off both ends with a sharp knife (2). Then, holding an onion in one hand, use a paring knife to lightly score just the outermost layer (3). Alternatively, stand the onion on one of its cut surfaces on your cutting board and use the tip of the knife to score it. Once scored, the outer layer should be easy to remove (4).

# GARLICKY SAUTÉED SPINACH

**This is just about the fastest vegetable side dish you can throw together. It's got four ingredients (six if you want to be a stickler and count the salt and pepper), and it quite literally cooks in five minutes. Because spinach is so moist, there's no need to add any liquid to the pan—it releases plenty on its own as it wilts. I'd love to be able to tell you that I've got some sort of magic food-sciencey trick that'll make your sautéed spinach *much better than anyone else's*, but this is a case where the traditional method is so quick, easy, and perfect that I can't find a single fault in its process or its end result.**

Swiss chard will work as well.

**NOTE**: This recipe will work equally well with trimmed Swiss chard or young kale leaves (just cut out the thicker portions of the stems). Do not use baby spinach—it will become too wet while cooking.

## SERVES 4

2 tablespoons extra-virgin olive oil

4 medium cloves garlic, finely sliced

2 pounds curly spinach, washed, drained, and tough stems removed

Kosher salt and freshly ground black pepper

Fresh lemon juice

Heat the olive oil in a 12-inch sauté pan over high heat until shimmering. Add the garlic and cook, stirring constantly, until fragrant and just beginning to brown, about 1 minute. Immediately add half of the spinach and cook, stirring and tossing, until lightly wilted. Add the remaining spinach and cook, using tongs to turn the spinach every 30 seconds or so, until it is completely wilted and the excess liquid has mostly evaporated, about 4 minutes. Season to taste with salt, pepper, and lemon juice and serve.

## PAN-ROASTED MUSHROOMS

When you first add mushrooms to a hot skillet, everything seems to be going fine, but then disaster strikes. Your mushrooms start leaking copious amounts of liquid into the pan, steaming instead of sautéing. You begin to panic. *Am I ruining them? Are they going to be ready for dinner? Why does this always have to happen to me?!?*

Never fear! Unlike meat, which will overcook and get tough if you let it steam in a skillet, mushrooms will stay tender no matter *how* long they sit there. Once the water has evaporated, they'll start sizzling again, developing both color and flavor.

You can season your mushrooms with nothing but salt and pepper, but I find that a small splash of soy sauce helps bring out their savoriness.

# PAN-ROASTED MUSHROOMS
## WITH THYME AND SHALLOTS

**SERVES 4 TO 6**

1 tablespoon vegetable oil

1½ pounds button or cremini mushrooms, cleaned and sliced into quarters

1 medium shallot, finely minced (about ¼ cup)

2 medium cloves garlic, minced or grated on a Microplane (about 2 teaspoons)

2 teaspoons fresh thyme leaves

2 teaspoons soy sauce

1 teaspoon fresh lemon juice

2 tablespoons unsalted butter

Kosher salt and freshly ground black pepper

1. Heat the oil in a large nonstick skillet over high heat until shimmering. Add the mushrooms and cook, tossing and stirring frequently, until they have given up their liquid and the liquid has completely evaporated, about 8 minutes. Continue cooking, tossing and stirring frequently, until the mushrooms are deep brown, about 10 minutes longer.

2. Add the shallot, garlic, and thyme and toss to distribute evenly. Cook, stirring and tossing, until fragrant, about 30 seconds. Remove from the heat, add the soy sauce, lemon juice, and butter, and toss until the butter is melted. Season to taste with salt and pepper. Serve immediately.

## CRYO-BLANCHING AND SAUTÉING

Cryo-blanching is a technique that was developed by my friends Alex Talbot and Aki Kamozawa over at the blog ideasinfood.com (they have a book of the same name). I've been following their work ever since I was a lowly line cook at Clio, the very restaurant where the couple honed their culinary chops. The technique sounds simple: rapidly freeze vegetables, then thaw and cook. But the concept is pretty brilliant. As I mentioned earlier in the discussion on peas (see page 411), freezing vegetables actually causes many of the same reactions as blanching does, namely, helping cells to break down and internal gases to escape. As the vegetables freeze, ice crystals forming within their cells will puncture cell walls, weakening their structure. After thawing, what you end up with is a vegetable that is partially softened but still has bright, fresh flavor with a bit of crunch remaining. Eat them as is, and you won't be all that happy—their texture tends to be a little . . . flaccid. But if you sauté them after thawing to soften them just the slightest bit more, you'll end up with vegetables with perfect color, perfect texture, and the brightest, freshest flavor you've ever had from a sautéed vegetable.

The other beauty of the technique, of course, is that you can store your vegetables pretty much indefinitely in the freezer, requiring just a half hour or so to let them thaw at room temperature.

### Preparing Vegetables for Cryo-Blanching

Essential to good cryo-blanching is rapid freezing. This means two things: first, you must use vegetables with a small cross section, like green beans, asparagus, or peas, and second, you must freeze them rapidly in a single layer. If you've got a vacuum-sealer (such as a FoodSaver), you can arrange your vegetables in a single layer in the bag before sealing them, then toss the bag directly into the freezer. Alternatively, lay your vegetables out in a single layer on a rimmed baking sheet and place them in the freezer, uncovered. Once they are completely frozen (give it a few hours to be safe), transfer them to a zipper-lock freezer bag, squeeze out any excess air, seal, and return to the freezer. They should be good for at least a few months and can be cooked directly from frozen.

# CRYO-BLANCHED GREEN BEANS
## WITH FRIED GARLIC

**NOTE**: Follow the instructions for preparing vegetables for cryo-blanching above. The fresh beans can also be blanched in a 4 quarts of salted water for 3 minutes and drained.

**SERVES 4 TO 6**

1 tablespoon olive oil, plus more for drizzling

1 pound cryo-blanched trimmed green beans (see Note above), thawed and dried with paper towels

2 cloves garlic, thinly sliced

Kosher salt and freshly ground black pepper

2 tablespoons Fried Garlic (page 419)

1. Heat the olive oil in a 12-inch skillet over medium-high heat until shimmering. Add the beans and cook, without stirring, until lightly blistered on the first side, about 1 minute. Add the sliced garlic and to cook, stirring and tossing, until the garlic is light golden brown. Season to taste with salt and pepper.

2. Serve sprinkled with the fried garlic.

# Essential VEGETABLE Technique #3:
## BRAISING

Braising is a slow-cooking process you see most often applied to tough cuts of meat (see All-American Pot Roast on page 243, for instance). It's a process in which meat is first seared in hot oil (dry heat), then slow-cooked in a pot with liquid (moist heat). The result is meat that has the flavor that comes with good browning but becomes completely fork-tender as the connective tissues slowly break down. Vegetables also take well to braising, and the technique is almost identical, with two key differences: **First is temperature.** In order to be fully tenderized, vegetables must be cooked to at least 183°F, the temperature at which pectin, the intracellular glue that holds them together, begins to break down. That means that while with meats it's preferable to keep the liquid at below a simmer, with vegetables, you can simmer away without fear of them toughening or drying out. **Second is time.** Vegetables cook much faster than meats do. While a beef stew can take upward of 3 hours to tenderize the meat, most braised vegetables will be as tender as they'll ever become in 20 minutes or less. This is good news for you.

For a while in the late 1990s and early 2000s, it became bafflingly fashionable to serve vegetables that were still essentially raw. Foodies and their ilk called them "al dente" and proclaimed that any green bean that was not perfectly emerald green with a hearty crunch and raw freshness in the center was not worth gracing their lips. I call BS on them. Indeed, I'd take a nice pot of rib-sticking green beans stewed with bacon grease until completely-tender-bordering-on-mushy any day over their crunchy sautéed counterpart. OK, perhaps that's going a bit too far, but in the winter? No side dish could be better. Ditto for asparagus. The recipe below is my absolute favorite way to prepare asparagus, but one that was looked down upon for many years. *Why would you want to eat drab green vegetables?* people would ask. Because they taste as awesome as MacGyver was cool, that's why. I sear my stalks in a bit of oil first to develop flavor, then deglaze the pan with water or stock, add a big knob of butter, put a lid on the whole thing, and let the asparagus cook in the liquid as it reduces. By the time the stalks are tender, your stock and butter will have emulsified into a slick, stalk-coating sauce that adds richness and sweet flavor to each bite. It's awesome.

# BRAISED ASPARAGUS

**This was a technique that my old mentor, Ken Oringer, taught me at his restaurant Clio in Boston. I firmly believed that army-green asparagus was a bad thing until I tasted these tender, buttery spears. I've been a convert ever since.**

### SERVES 4

2 tablespoons vegetable oil

1 pound asparagus, tough bottoms trimmed, stalks peeled if desired

Kosher salt and freshly ground black pepper

1 cup homemade or low-sodium canned chicken or vegetable stock

3 tablespoons unsalted butter

1 teaspoon lemon juice (from 1 lemon)

1. Heat the oil in a 12-inch sauté pan over high heat until lightly smoking. Add the asparagus in as close to a single layer as possible, season with salt and pepper, and cook, without moving it, until lightly browned on the first side, about 1½ minutes. Shake the pan and cook until the asparagus is browned again, 1½ minutes longer.

2. Add the stock and butter to the pan, immediately cover it, and cook until the asparagus is completely tender and the stock and butter have emulsified and reduced to a shiny glaze, 7 to 10 minutes. If the stock completely evaporates and butter starts to burn before the asparagus is cooked through, top up with a few tablespoons of water. Stir in the lemon juice and serve immediately.

# BRAISED STRING BEANS
## WITH BACON

My good friend and fellow food writer Meredith Smith is from Kentucky, where braised greasy beans (yes, that's what the bean variety is called) are a way of life. True greasy bean pods have thick strings and big beans inside that take about an hour to soften properly. They're also impossible to find round my parts. Fortunately, braised string beans capture plenty of that slow-cooked beany flavor, and they're available pretty much anywhere. (Please don't tell Meredith I'm using regular string beans. She'll kill me.)

   If anyone ever tries to tell you that green beans absolutely *must* be perfectly al dente and snappy to be worth eating, just shove a handful of these in their mouth and they'll clam right up.

**NOTE:** For best results, use slab bacon, but sliced bacon can be used in its place: Cut the bacon crosswise into ½-inch-wide strips.

**SERVES 4 TO 6**

8 ounces slab bacon, cut into
   1-by-½-inch-thick lardons
1 tablespoon vegetable oil
3 medium cloves garlic, thinly sliced
½ teaspoon red pepper flakes
½ cup homemade or low-sodium
   canned chicken stock
1½ pounds string beans, ends
   trimmed
¼ cup cider vinegar
1 tablespoon sugar
Kosher salt and freshly ground
   black pepper

1. Place the bacon and oil in a large saucepan, add ½ cup water, and bring to a simmer over medium-high heat, then cook until the water is evaporated and the bacon is crisp and well rendered, about 10 minutes.

2. Add the garlic and pepper flakes and cook, stirring, until fragrant, about 30 seconds. Add the chicken stock and scrape up any browned bits from the bottom of the pan. Add the beans, 2 tablespoons of the vinegar, and the sugar, reduce the heat to medium-low, cover, and cook, stirring occasionally, until the beans are completely soft and cooked through, about 1 hour; add water as necessary if the pan begins to dry out and the beans start to sizzle.

3. Stir in the remaining 2 tablespoons vinegar, season to taste with salt and pepper, and serve.

# BRAISED LEEKS
## WITH THYME AND LEMON ZEST

Leeks are the archetypal wingman. They disappear into stews and soups and gently flavor sautéed vegetables. They melt into sauces and hide out in stir-fries. I mean, they even play second fiddle to potatoes, for god's sake—that's potato-leek soup, not leek-potato soup.

Well, Mr. Leek, today, opportunity knocks.

When braised, leeks retain their subtle aroma but acquire a completely tender, almost meaty texture as they slowly break down and absorb liquid. To heighten their sweetness a bit, it's best to caramelize the cut surfaces of the split leeks in hot oil before adding the liquid. Why the cut surface? Because it's easier to lay them flat that way. Even though only the edges will acquire any color, everything's cool—those wonderful browned compounds are water-soluble, which means that after you add your liquid, many of them will dissolve and spread throughout the dish.

You can use any type of liquid you'd like, but I like to use chicken stock, with a few nuggets of good butter. The butter keeps things lubricated while the leeks cook, and it will add richness to the sauce the leeks and stock form as they slowly braise in the oven.

**NOTE:** For best results, look for leeks about 1 inch in diameter.

**SERVES 4 TO 6**

8 medium leeks (see Note above)
   white and light green parts only
2 tablespoons extra-virgin olive oil
2 tablespoons unsalted butter
6 sprigs fresh thyme
1 cup homemade or low-sodium
   canned chicken stock
Kosher salt and freshly ground
   black pepper
2 teaspoons grated lemon zest
   (from 1 lemon)
2 teaspoons lemon juice

1. Adjust an oven rack to the center position and preheat the oven to 375°F. Split the leeks lengthwise in half, leaving the root end attached, and discard any tough outer layers. Rinse under cold running water and pat dry with paper towels.

2. Heat the olive oil in a 12-inch heavy-bottomed skillet over medium-high heat until shimmering. Add half of the leeks cut side down and cook, without moving them, until well browned on the cut side, about 4 minutes. Transfer cut side up to a 13-by-9-inch baking dish. Sear the remaining leeks and add to the baking dish.

3. Dot the top of leeks with the butter, scatter the thyme over the top, and pour the chicken stock over the leeks. Cover tightly with foil, place in the oven, and cook until the leeks are completely tender, about 30 minutes; remove the foil for the last 10 minutes of braising. Remove from the oven and allow to cool for 5 minutes.

4. Discard the thyme stems, season to taste with salt and pepper, and transfer the leeks to a serving platter. Stir the lemon zest and juice into the pan juices and pour on top of the leeks. Serve immediately.

# QUICK CHICKPEA AND SPINACH STEW
## WITH GINGER

Restaurant fare is both complex and time-consuming to make—that's why you pay a lot for it. But bringing some of that flavor home needn't be either.

This chickpea and spinach stew is based on the *garbanzos con espinacas* that I used to make with Chef John Critchley at Toro in Boston. It's about as classic a Spanish bar snack as there ever was, and you'll find it all over Spain, flavored with everything from smoky chorizo and rich morcilla (blood sausage) to simpler preparations served with nothing but a spritz of bright sherry vinegar.

At the restaurant, we'd painstakingly make vegetable stock, brine dried beans, sweat aromatics, braise spinach, and crush olives under the hooves of real live Spanish *burros* to scatter over the finished dish. At least, we did most of that stuff. Painstakingly tasty is how I'd describe that kind of food. At home, I'm happy to take a couple of shortcuts.

This version, which ends up somewhere between a soup and a stew, relies on canned chickpeas and their liquid for body, but giving them a bit of a simmer with some aromatics—garlic, onion, bay leaf, and smoked paprika—adds a ton of flavor to them. (Remember the lessons on canned beans in Chapter 2?) The unique part is the bit of ginger added to the pureed tomatoes. It's not enough to make itself obvious, but it's just enough to add a bit of complex heat to the saucy backbone of the dish.

This is great served hot in a bowl as is, but to be honest, I actually like it better on the second day, served room temperature on top of slices of dark toast drizzled with olive oil. Perfect fare for when you want to act all cool, sophisticated, and *suave* at that Spanish wine tasting you're going to host. Or something like that.

One 28-ounce can whole tomatoes

One 1-inch knob ginger, peeled

¼ cup extra-virgin olive oil, plus
more for serving

1 medium onion, finely sliced

4 cloves garlic, finely sliced

1 teaspoon sweet or hot Spanish
smoked paprika

12 ounces spinach, trimmed,
washed, drained, and roughly
chopped

Two 14-ounce cans chickpeas, with
their liquid

2 bay leaves

2 teaspoons soy sauce

Kosher salt

Sherry vinegar (optional)

1. Drain the tomatoes in a strainer set over a medium bowl. Transfer the liquid and half of the tomatoes to a blender, add the ginger, and blend on high speed until completely pureed. Set aside. Roughly chop the remaining tomatoes and set aside separately.

2. Heat the olive oil in a 12-inch skillet over high heat until shimmering. Add the onion, garlic, and paprika and cook, stirring frequently, until the onion is softened and very slightly browned. Add the tomato-ginger puree and stir to combine. Add the spinach a handful at a time, allowing each handful to wilt before adding the next. Reduce the heat to medium and simmer, stirring occasionally, until the spinach is completely tender, about 10 minutes.

3. Add the chopped tomatoes, chickpeas, with their liquid, bay leaves, and soy sauce and bring to a boil over high heat. Reduce to a bare simmer and cook, stirring occasionally, until thickened into a thick stew, about 30 minutes.

4. Season to taste with salt and serve immediately, drizzling with extra virgin olive oil and, if desired, a few drops of sherry vinegar.

# THE ULTIMATE CREAMED SPINACH

**The key to great creamed spinach is time. Low, slow cooking to slowly reduce the spinach's juices, along with the creamy béchamel sauce, into a rich, thick coating with a near-pudding-like texture. I like to finish mine with a bit of crème fraîche for a bit of added creaminess and acidity. For the absolute ultimate holiday side dish, make your crème fraîche from scratch (see page 123) and broil the whole shebang topped with Parmigiano-Reggiano.**

**NOTE:** Trimmed kale or chard leaves will work just as well here.

**SERVES 4**

3 tablespoons unsalted butter

2 medium shallots, finely minced (about ½ cup)

2 medium cloves garlic, minced or grated on a Microplane (about 2 teaspoons)

2 pounds curly spinach, trimmed, washed, and drained

1 tablespoon all-purpose flour

1½ cups heavy cream

½ cup whole milk

¼ teaspoon freshly grated nutmeg

¼ cup crème fraîche

Kosher salt and freshly ground black pepper

2 ounces Parmigiano-Reggiano, finely grated (about 1 cup; optional)

1. Melt the butter in a Dutch oven or large saucepan over medium-high heat. Add the shallots and garlic and cook, stirring, until softened, about 2 minutes. Add the spinach in four batches, turning each batch with tongs or a rubber spatula and allowing it to wilt before adding the next.

2. Add the flour and cook, stirring constantly, until no dry flour remains. Slowly stir in the heavy cream and milk. Bring to a simmer, stirring, then reduce the heat to the lowest possible setting and cook, stirring occasionally, until the spinach is completely softened and the sauce has thickened, about 1½ hours.

3. Stir in the nutmeg and crème fraîche and season to taste with salt and pepper. Remove from the heat.

4. If desired, preheat the broiler to high. Transfer the creamed spinach to a 1-quart oval or round casserole dish and top it with the cheese. Broil until the spinach is bubbly and the cheese has formed a well-browned crust, about 2 minutes. Serve.

# Essential VEGETABLE Technique #4:
## GLAZING

This is the ultimate all-in-one-pan technique for vegetables, and it works particularly well for those vegetables that don't get hurt by long cooking: carrots, parsnips, onions, turnips, radishes—hearty vegetables of all sorts, in fact. The traditional French technique involves cooking vegetables in a skillet with butter, chicken stock, sugar, and salt, slowly simmering them so that by the time the vegetables are cooked through, the stock has reduced and emulsified with the butter into a shiny, flavorful glaze that coats the vegetables. It's a simple technique, but it's far from foolproof, particularly in the home kitchen. In a restaurant on a high-output burner, the stock boils extremely rapidly and this bubbling action helps it to emulsify with the butter in the pan quite easily. With lazy bubbling on a home stove, however, it's much more difficult to get a stable emulsion.

On top of that, the gelatin in good chicken stock really helps stabilize the emulsion. Use a thin store-bought stock, vegetable stock, or plain water, and your emulsion becomes doubly hard to form. The solution? A bit of "artificial" thickening: the tiniest bit of cornstarch—½ teaspoon for a full 4 to 6 servings of vegetables—is enough to easily stabilize the coating without turning it gloppy or thick.

# GLAZED CARROTS
## WITH ALMONDS

**SERVES 4 TO 6**

1½ pounds large carrots, peeled
and cut into 1-inch disks or faux
tournée (see "Knife Skills: How to
Cut Carrots," page 454)

½ teaspoon cornstarch

1½ cups homemade or low-sodium
canned chicken stock

2 tablespoons unsalted butter

2 tablespoons sugar

Kosher salt

2 tablespoons finely minced fresh
parsley or chives

¼ cup toasted slivered almonds

½ teaspoon lemon zest and 1
teaspoon lemon juice (from 1
lemon)

Freshly ground black pepper

1. Toss the carrots with the cornstarch in a 12-inch heavy-bottomed stainless steel skillet until no clumps of starch remain. Add the stock, butter, sugar, and 1 teaspoon salt and bring to a boil over high heat, then reduce to a simmer and cook, stirring occasionally, until the carrots are almost tender (they should show little resistance when poked with a cake tester or the tip of a knife), about 10 minutes.

2. Increase the heat to high and boil, tossing occasionally, until the sauce is reduced to a shiny glaze. Remove from the heat, add the herbs, almonds, lemon zest and lemon juice, and toss to combine. Season to taste with salt and pepper and serve.

### GLAZED PEARL ONIONS
Replace the carrots with peeled pearl or cipollini onions and omit the almonds.

### GLAZED TURNIPS, RADISHES, OR RUTABAGA
Replace the carrots with small radishes or turnips, split into quarters or halves, or large turnips, radishes, and/or rutabagas, cut into 1-inch chunks. Omit the almonds.

# KNIFE SKILLS:
## How to Cut Carrots

**Carrots come in odd shapes and sizes, but cutting them uniformly is essential to cooking evenly. Here's how to do it.**

- **TO START**, peel the carrots using a Y-peeler (1), making single strokes all the way from the head to the tip to remove the skin in as few pieces as possible—this will help keep the shape of the carrot more uniform. Next, trim off the blunt end of the carrot (2), as well as the tip if it is still dirty. For rough chunks, cut the carrot as is. Otherwise, read on.

1

2

3

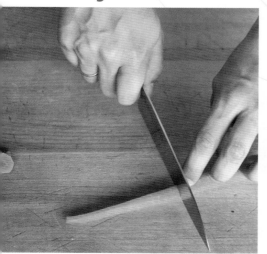

- **DICE** of various sizes are the most common way to cut carrots. Large dice can be nice in stews, while medium or small dice are more suited for soups, hearty sauces like Bolognese, and chunky chopped salads. To cut, split the carrot crosswise in half (3), then split each half lengthwise into quarters (4). Hold the batons together and cut into dice of the desired size (5).

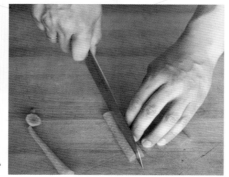

4

5

- **FAUX TOURNÉ** is what I do when I'm too lazy to make a true tourné, which means all the time. The fancy seven-sided football that you get with a true tourné wastes carrot, looks pretentious, and is entirely unnecessary. This method, also called roll-cutting or oblique-cutting, is supersimple, and it produces attractive evenly shaped tapered pieces perfect for glazing or adding to more refined stews. Hold a peeled carrot on a cutting board parallel to the bottom edge of the board. Slice 1 inch of the thick end off at a 45-degree angle (6). Roll the carrot forward so that it rotates 90 degrees and cut another 1-inch segment off at a 45-degree angle (7). Repeat rolling and slicing until the entire carrot is cut (8). This method works equally well with parsnips.

6

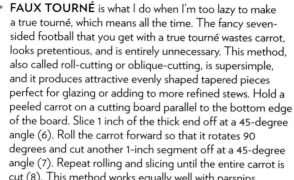

7

8

- **JULIENNE AND BRUNOISE** are what you use when you want to start getting really fancy. These look great when cooked with fish en papillote, for example, and also work well in stir-fries and sauté. To finely julienne a carrot, start by cutting it into 3-inch lengths. Then slice off an ⅛-inch strip from one side of each segment to create a stable base for the carrot to sit on (9). (Use the scraps for stocks or compost, or eat immediately.) Set each piece of carrot on its trimmed side and carefully slice it lengthwise into thin planks (10). Stack two or three of the planks and slice into even julienne (11). To cut brunoise, take a bunch of your julienne and cut crosswise into tiny even cubes (12). Getting these perfect is one of the most satisfying things you can do with a knife in the kitchen.

9

10

11

12

# Essential VEGETABLE Technique #5:
## ROASTING/BROILING

On the surface, roasting, grilling, or broiling vegetables may not appear to be too similar to searing vegetables in a skillet, but the goal is the same: to caramelize and brown the exterior of the vegetable while leaving a bit of fresh crunch in the center. It's dead simple to do under the broiler. Take stalks of asparagus, for example. The high heat of the broiler rapidly starts caramelizing and charring the asparagus's sugars while allowing the bulk of it to remain crisp and sweet. Charred + sweet + crisp + easy = huge win for all home cooks. In order to make sure that the asparagus cooks relatively easily and doesn't dry out under the broiler, it's essential to toss the spears with a thin coating of oil before they hit the heat. Oil is not only a better heat-distribution medium than the naked air, but by filling in all the microscopic nooks and crannies left behind by evaporating moisture, it also keeps vegetables from turning shriveled or leathery. You can get fancy by drizzling melted herb butter or lemon juice or sprinkling grated cheese on broiled or grilled vegetables after cooking, but in all honesty, the best way to eat them is straightaway with your fingers.

The key to using the oven is to use crazy high heat to maximize caramelization while still maintaining a pleasant crispness. The best way to do this is to use a heavy rimmed baking sheet and preheat it for at least 10 minutes or so on the bottom rack in a 500°F oven. Toss your vegetables with a bit of olive oil, salt, and pepper, and throw them onto the pan. The vegetables should sizzle and start browning as soon as they hit the pan. Alternatively, place the vegetables on a rimmed baking sheet a few inches away from the broiler element heated to high. In just a few minutes, you're good to go. Sprinkle with some lemon or a nice sharp grating cheese. Parm works, Pecorino is better, Cotija is just plain cool.

## POST-ASPARAGUS STINKY URINE DISORDER

One last thing you may notice after your asparagus has been eaten: That haunting smell—haunting as in it comes back and surprises you long after you thought it was gone—is caused by S-methyl thioacrylate and S-methyl 3-(methylthio)thiopropionate, chemicals identified in 1975 at the University of California at San Diego. It's not known exactly why some people seem unable to digest them, but it is known that the degree of "Post-Asparagus Stinky Urine Disorder" (a term I just coined, PASUD for short) is related to your genealogy. Fewer than half of Britons, apparently, suffer from it, while almost 100 percent of the French do. I know which country I'd rather be in for sporting matches during asparagus season. But the real kicker is that it turns out that not only do some people not produce the odor, some people *cannot smell it*, which means that in these self-reported studies, it's unclear whether or not the folks who claimed that they don't have asparagus pee actually just don't know that they do have it. It's not easy to find volunteers to check for it either.

# BROILED ASPARAGUS
## WITH PARMESAN BREAD CRUMBS

**NOTE:** Pecorino, Cotija, or any other hard grating cheese can be substituted for the Parmigiano.

**SERVES 4 TO 6**

2 slices white sandwich bread, crusts removed and torn into rough chunks

2 tablespoons unsalted butter

Kosher salt and freshly ground black pepper

1 ounce Parmigiano-Reggiano, finely grated (about ½ cup)

1 pound asparagus, tough bottoms trimmed, stalks peeled if desired

1 tablespoon extra-virgin olive oil

1 lemon, cut into wedges

1. Preheat the broiler to high. Add the bread, 1 tablespoon of the butter, a pinch of salt, and a few grinds of pepper to a food processor and pulse until coarse crumbs are formed.

2. Melt the remaining tablespoon of butter in a 12-inch skillet over medium-high heat. Add the bread crumbs and cook, tossing and stirring occasionally, until golden brown and crisp, about 5 minutes. Transfer to a medium bowl, add the Parmesan, and toss to combine. Season to taste with salt and pepper if necessary and set aside.

3. Arrange the asparagus in a single layer in a foil-lined broiler pan or rimmed baking sheet. Drizzle with the oil and shake to coat evenly. Sprinkle with salt and pepper. Broil 2 inches from the heating element until tender and well charred, about 8 minutes. Sprinkle with the bread crumbs and serve with the lemon wedges.

# ROASTED BRUSSELS SPROUTS
## AND SHALLOTS

**NOTE:** This recipe is easily doubled to serve a large crowd. Use two rimmed baking sheets with the oven racks in the lower- and upper-middle positions. Rotate and swap the trays halfway through cooking.

### SERVES 4

1 pound Brussels sprouts, bottoms
    trimmed, outer leaves discarded,
    and split in half
4 ounces shallots, finely sliced
2 tablespoons extra-virgin olive oil
Kosher salt and freshly ground
    black pepper
1 tablespoon balsamic vinegar

1. Adjust an oven rack to the lower-middle position, place a heavy rimmed baking sheet on the rack, and preheat the oven to 500°F. Toss the Brussels sprouts, shallots, and oil in a large bowl. Season generously with salt and pepper.

2. Carefully remove the baking sheet from the oven, using oven mitts or a folded dish towel. Transfer the Brussels sprouts to the pan (they should immediately begin to sizzle) and return to the oven. Roast until the sprouts are completely tender and well charred, about 15 minutes, stirring and shaking the pan once or twice. Remove from the oven and drizzle the sprouts with the balsamic vinegar. Stir to coat and serve.

# ROASTED BROCCOLI
## WITH GARLIC-ANCHOVY BREAD CRUMBS

**Broccoli becomes incredibly sweet and nutty, with a really concentrated flavor, when roasted at a high temperature. It's one of my favorite ways to eat it. Tossing the roasted broccoli with a bit of anchovy-scented bread crumbs adds another dimension to the dish.**

**NOTE:** The broccoli stalks can be peeled and roasted too if desired. Cut them on a bias into ½-inch slices.

**SERVES 4**

1½ pounds broccoli, cut into 3- to 4-inch-long florets, each floret cut lengthwise in half

3 tablespoons extra-virgin olive oil

Kosher salt and freshly ground black pepper

1 teaspoon fresh lemon juice

1 recipe Garlic-Anchovy Bread Crumbs (recipe follows; optional)

1. Adjust an oven rack to the lower-middle position, place a heavy rimmed baking sheet on the rack, and preheat the oven to 500°F. Toss the broccoli with the oil in a large bowl. Season generously with salt and pepper.

2. Carefully remove the baking sheet from the oven, using oven mitts or a folded dish towel. Transfer the broccoli florets to the pan, shaking it so that most end up flat side down (they should immediately begin to sizzle) and return to the oven. Roast until the broccoli is tender and well charred, about 10 minutes, stirring and shaking the pan once or twice. Sprinkle the broccoli with the lemon juice and the anchovy bread crumbs if desired and serve.

## Garlic-Anchovy Bread Crumbs

*These bread crumbs can be used to add texture and flavor to any roasted, grilled, or broiled vegetable dish. They also go great with pasta.*

**NOTE:** The bread can also be dried by placing it directly on a rack in a preheated 275°F oven for about 20 minutes, turning once halfway through.

**MAKES ABOUT 1 CUP**

3 slices hearty sandwich bread, dried overnight on the counter (see Note above)

3 tablespoons unsalted butter

2 medium cloves garlic, minced or grated on a Microplane (about 2 teaspoons)

4 anchovy fillets, finely chopped

2 tablespoons finely chopped fresh parsley

Kosher salt and freshly ground black pepper

1. Tear the bread into rough 2-inch pieces and pulse in a food processor until coarse crumbs are formed.

2. Melt the butter in a 12-inch heavy-bottomed skillet over medium heat. Add the garlic and anchovies, reduce the heat to medium-low, and cook, stirring occasionally, until the garlic is lightly browned and the anchovies have dissolved, about 6 minutes. Add the bread crumbs, increase the heat to medium-high, and cook, stirring and tossing constantly, until the bread crumbs are well toasted and browned, about 3 minutes. Transfer to a plate and let cool. Toss with parsley and season with salt and pepper. Cooled bread crumbs can be stored in a sealed container at room temperature for up to a week.

# ROASTED CAULIFLOWER
## WITH PINE NUT, RAISIN, AND CAPER VINAIGRETTE

**Like broccoli, roasted cauliflower gets sweet and nutty and is great on its own with just a drizzle of really good olive oil or lemon, but I like to turn mine into a warm salad with a more elaborate vinaigrette. This one is inspired by a dish my friend Einat Admony occasionally serves at her awesome restaurant Balaboosta, with a toasted pine nut vinaigrette made with raisins, capers, and a touch of honey.**

**SERVES 4**

1 head cauliflower, trimmed and cut
    into 8 wedges
6 tablespoons extra-virgin olive oil
Kosher salt and freshly ground
    black pepper
1 tablespoon sherry vinegar
1 tablespoon honey
2 tablespoons capers, rinsed,
    drained, and roughly chopped
¼ cup toasted pine nuts
¼ cup raisins
2 tablespoons finely chopped fresh
    parsley leaves

1. Adjust an oven rack to middle position, place a heavy rimmed baking sheet on the rack, and preheat the oven to 500°F. Toss the cauliflower with 3 tablespoons olive oil. Season generously with salt and pepper.

2. Carefully remove the baking sheet from the oven, using oven mitts or a folded dish towel. Transfer the cauliflower wedges to the pan and return to the oven. Roast until the cauliflower is tender and deeply browned on both sides, about 20 minutes total, flipping the cauliflower with a thin metal spatula halfway through roasting.

3. While the cauliflower roasts, combine the remaining 3 tablespoons olive oil and the vinegar, honey, capers, pine nuts, raisins, and parsley in a medium bowl and season to taste with salt and pepper.

4. Transfer the cooked cauliflower to a serving plate and spoon dressing on top. Serve immediately.

# ROASTED MUSHROOMS

The problem with mushrooms, of course, is that they are so watery and spongy. Roasting them on a large baking sheet is an ideal way to cook them, as it gives them ample room for evaporation. Unlike other vegetables, where the goal is to minimize the amount of time they spend in the oven, when roasting mushrooms, you want to leave them in until nearly all their moisture is driven out so that they can brown properly. A well-roasted mushroom will end up about half its original size and a quarter of its original weight.

**SERVES 4**

1½ pounds white button or
  cremini mushrooms, cleaned and
  quartered

2 tablespoons extra-virgin olive oil

Kosher salt and freshly ground
  black pepper

6 to 8 sprigs fresh thyme

1. Adjust an oven rack to the middle position and preheat the oven to 400°F. Toss the mushrooms with the olive oil in a large bowl and season to taste with salt and pepper.

2. Transfer to a rimmed baking sheet lined with parchment paper. Scatter the thyme sprigs evenly over the top. Roast until the mushrooms have exuded all of their liquid, the liquid has evaporated, and the mushrooms are well browned and flavorful, 30 to 45 minutes. Discard the thyme sprigs, transfer the mushrooms to a serving bowl, and serve.

## BETTER ROASTED SWEET POTATOES

OK, so sweet potatoes are sweet, but they're not *that* sweet, right? I mean, sure, you could go with maple syrup or honey and marshmallows on top, but I wouldn't wish one of those monstrous casseroles on my worst enemy, let alone my own family. Much better are *really well-roasted* sweet potatoes. At their best, they're creamy, flavorful, and sweet, with a slightly crisp, caramelized crust. Too often though, roasted sweet potatoes end up mealy, starchy, and bland. How can the same vegetable produce such distinctly different results? How does one get a sweet potato to really live up to its name?

Here's the deal: **starch is made from sugar**. More precisely, starch is a *polysaccharide*, which means that it's a large molecule consisting of many smaller sugar molecules (in the case of sweet potatoes, *glucose*). The thing about sugar, though, is that unless it's broken down to relatively simple forms, it doesn't taste sweet to us. Your tongue simply doesn't recognize it. It helps to imagine sugar molecules as a bunch of circus midgets (OK, "little people," if you will). When they're all standing in a row, it's easy for us to identify them as midgets. But stack them up on one another and throw a trench coat over 'em, and they're effectively hidden.

Now, sweet potatoes contain plenty of starch molecules. The goal when roasting them is to try and break down as many of the starch molecules as possible into sweet-tasting maltose (a sugar consisting of two glucose molecules): pull off that trench coat and knock the little-person stack down. We do this with the help of an enzyme naturally present in the potatoes that is active between the range of 135° and 170°F. Essentially, the longer a sweet potato spends in that zone between 135° and 170°F, the sweeter it becomes. To test this, I cooked three

batches of potatoes. The first I popped directly into a 350°F oven and baked until tender. The second I parcooked in a temperature-controlled water bath at 150°F for 1 hour before baking. The last I parcooked in the water bath overnight before baking.

More sugary sweet potatoes brown faster.

You can immediately see that the parcooked potatoes browned better, indicating a higher sugar content that allowed them to caramelize faster. The color difference was also reflected in the flavor: the parcooked potatoes were significantly sweeter and more flavorful than the plain roasted potatoes, which were starchy and bland. Interestingly, the hour-long-parcooked potatoes were nearly as flavorful as the overnight potatoes, which means you've really got to cook them for only an hour at 150°F.

If you've got a temperature-controlled water bath, the path to better sweet potatoes is an obvious one. Just bag your potatoes, cook 'em as long as you'd like at 150°F (any higher, and I found they softened too much before cooking), and then pop them into in the oven while your turkey is resting.

For the rest of you, there are a couple of options. You could always go the beer-cooler sous-vide route outlined on page 384. It's cheap and effective, and it will easily hold the proper temperature for the requisite hour. Just put your potatoes in a zipper-lock bag with the air squeezed out, then put them into a cooler filled with water at 150°F, close the lid, wait an hour, and you're ready to roast.

The good thing about sweet potatoes is that they're less finicky than, say, a steak, which means that you don't have to worry about getting the temperature exactly right. In fact, as long as your water's above 135° and below 170°F, it'll have a positive effect on their sweetness.

Don't want to use the cooler? Here's an even easier way to do it: Bring 3 quarts of water to a boil in a large pot. Add 4 cups of room-temperature water. This should bring your water down to around 175°F. Add a few pounds of sliced or diced sweet potatoes to that water, and it'll come down to well within the requisite range. Pop a lid on the pot, put it in a warm part of your kitchen, and leave it there for a couple hours, then simply drain the potatoes and roast at your leisure. Your mouth will thank you, as will your family.

## I YAM WHAT I YAM

So, as all of you highly educated eaters probably already know, the thing that we call a yam in the United States is not a yam. A true yam is a gigantic starchy, sticky root from a large grass-like plant native to Africa. These days, yams are mostly found in Africa, South America, and the Pacific Islands. Very rarely do they make their way here.

The things we call yams are in fact a type of sweet potato, a different plant entirely. Sweet potatoes come in a few different varieties, but they can basically be broken down into two groups, which behave differently when cooked.

- **Dry sweet potatoes**, like white-fleshed American sweet potatoes or Okinawan purple potatoes, are less sweet and starchier than moist sweet potatoes. They turn fluffy when cooked, and in many recipes can be a good substitute for regular potatoes (albeit with a flavor all their own).
- **Moist sweet potatoes**, like garnet or ruby yams, are the more commonly available variety in the United States. They have a higher water content and sugar content than dry sweet potatoes, and they cook up creamy and rich rather than fluffy. They are the ones called for in most recipes.

# EXTRA-SWEET ROASTED SWEET POTATOES

**SERVES 6 TO 8**

5 pounds sweet potatoes, peeled
    and cut into ½-inch disks or 1-inch
    chunks

¼ cup olive oil

Kosher salt and freshly ground
    black pepper

1. Bring 3 quarts of water to a boil in a large pot over high heat. Remove from the heat and add 4 cups of room-temperature (70°F) water, then immediately add the potatoes and cover the pot. Place in a warm spot and let stand for at least 1 hour, and up to 3 hours.

2. When ready to roast the potatoes, adjust the oven racks to the upper- and lower-middle positions and preheat the oven to 400°F. Drain the sweet potatoes and transfer to a large bowl. Toss with the olive oil and season to taste with salt and pepper.

3. Spread the sweet potatoes on two rimmed baking sheets and roast until the bottoms are browned, about 30 minutes. Carefully flip the potatoes with a thin offset spatula and roast until the second side is browned and potatoes are tender, about 20 minutes longer. Serve immediately.

## { POTATOES }

Despite the fact that they qualify as vegetables under certain definitions, potatoes are so vastly different from the other veggies in this chapter that they deserve a section all to themselves. There are few vegetables you find in the supermarket that are as varied as potatoes. They come in myriad shapes and sizes, but don't let all that variety fool you. When it comes to cooking qualities, potatoes can basically be broken down into three categories: high-starch, medium-starch, and low-starch (also referred to as waxy). While most any potato will technically work in any recipe, the end results can be vastly different. French fries made with red potatoes are creamy and dense in the center, as opposed to the fluffy fries made from russets. Mashed potatoes made with medium-starch Yukon Golds will be buttery, thick, and intense, quite unlike the light, almost-aerated mash you get with russets. Waxy red potatoes will hold their shape in soups and stews even after they've been cooked to oblivion, while russets will absorb flavors and eventually fall apart.

I prefer russet potatoes in nearly every application for the reason that they tend to absorb flavors better than any other variety. Also, I'm too lazy to keep more than one variety of potato around. I'll occasionally buy smaller bags of yellow fingerling or red creamer potatoes, or pick up some baskets of flavorful tiny new potatoes from the farmers' market if I feel like making something different to go with my meal. But that's just me. Other people may like to have a full range in their pantry. In my wife's native Colombia, the national dish, *ajiaco*, is a soup made with several different varieties of potato. The starchier spuds disintegrate, thickening the broth, while the waxier ones stay firm. You end up with an awesome textural mix. This soup alone is reason enough to make me consider expanding my potato pantry.

Here's a chart to help you decide which potatoes are right for a given application:

| APPLICATIONS | HIGH-STARCH POTATOES (Russets, Idaho) | MEDIUM-STARCH POTATOES (Yukon Golds, Yellow Finn, Bintje, round white potatoes) | LOW-STARCH/WAXY POTATOES (Red Bliss, most varieties of fingerlings, Huckleberry) |
|---|---|---|---|
| Baking Whole | Best: potatoes turn out fluffy and moist with thick, crisp-chewy skin. | Not recommended; wet and sticky results. | Not recommended; very sticky interiors. |
| Roasting in Pieces | Very good: crisp exterior, fluffy middles. | Creamier and less crisp than russets. | These take some work to get them crisp but have intensely potatoey, creamy interiors if done properly. |
| Salad | Best if you want optimum dressing-flavor absorption; potatoes tend to break down a bit. | Halfway between high-starch and waxy potatoes; will absorb a bit of dressing and maintain shape fairly well. | Best if you prefer firm chunks of potatoes covered with but not necessarily penetrated by dressing. |
| Mashing | Best for light, fluffy mash. | Best for dense, creamy mash. | Not recommended except for niche applications like French-style *pommes puree* (where butter and cream make up half of the volume). |
| Boiling/ Steaming | Not recommended except as a parcooking step for other recipes (salad, fries, hash)—potatoes tend to disintegrate. | Best for lightly creamy potatoes with buttery flavor. | Best for solid, substantial chunks of potatoes with very little breakdown and a waxy, firm texture. |
| Hash and Pancakes | Best: crisp, crunchy edges and tender, fluffy interiors. | Not recommended | Not recommended |
| Soups and Chowder | Best for starchy thickened soups where you don't want discrete chunks of potato. | Not recommended | Best for thinner soups or chowders where distinct chunks of potato are desired. |
| French Fries | Best: maximum crispness with a fluffy interior | Not recommended | Not recommended |

# THIS IS THE ULTIMATE POTATO CASSEROLE

**It was the middle of a Tuesday night when I awoke,
the moon still high in the night sky, turned around, shook my
wife awake and said, "Adri, Adri, you must awake! I
just had an idea and I must peel some potatoes. Make haste!"**

She gave me the usual "You interrupted my favorite activity—sleeping—to ask me to do some menial kitchen work? How about this idea: you go do it yourself, then make yourself comfortable on the couch" look before rolling back over and nodding off. Sometimes I just don't understand her.

Nevertheless, I went into the kitchen and got to work on the first batch of what would end up being my favorite potato recipe in years.

Here's the idea: We all love potato gratin, correct? The creamy, layered casserole of potatoes and cream with a crisp browned top. We also love Hasselback potatoes, those cute side dishes achieved by making thin, parallel slices almost all the way through a potato, stuffing butter and cheese in between them, and roasting them until golden brown and crisp. What if I were to combine the two concepts into one gloriously crispy, creamy, crunchy, cheesy casserole?

The dish starts out just like most potato gratins: sliced potatoes. If you've got an inexpensive Japanese mandoline slicer, then it's a snap. I tried it with both peeled and unpeeled potatoes and preferred the cleaner crunch you get from peeled potatoes. From there, it progresses like a standard gratin: I mix heavy cream, grated cheese (I use Comté and Parmesan), fresh thyme leaves, salt, and pepper in a bowl, then I add the potatoes and toss them all together. This step is worth taking your time with: it's crucial that every single potato slice gets coated on all sides with the mixture. That means prying or sliding apart all the slices of potato that are stuck together and dipping them into the fatty mixture.

Here's where we turn things on their head: rather than stacking sliced potatoes horizontally like in a traditional gratin, I pack them into a greased casserole dish standing on their edges, working my way around the perimeter of the dish and packing it all in tightly. Because potatoes vary wildly in shape, you end up with tons of little nubby bits sticking out all over the top surface.

As the potatoes cook, the cream eventually starts to boil, simmering up and over the tops of the potatoes, basting them as they roast, aiding further in preventing them from getting leathery. A final layer of cheese added halfway through the uncovered stage of cooking adds a layer of flavor to the final casserole.

During the final stages of cooking, the cream eventually loses enough moisture that it breaks, releasing its butterfat, which coats and then gets slowly absorbed into the potatoes as they continue to lose water content. Milk proteins in the cream and the cheese coagulate, creating little pockets of curd-like tenderness between slices. The final dish is nothing short of glorious. Look at it. I mean, look at it. Every bite has a combination of crisp-but-moist upper potato ridges and rich and creamy potatoes underneath with a cheese underscoring the whole affair.*

It's so good that I've decided to make this only in the middle of the night while my wife is fast asleep and finish it all myself, picking at the crispy cheesy bits around the edges of the casserole dish in the wee hours of the morning, leaving just enough to hint at the glory that it once contained. What dear? What's the lingering aroma of garlic and thyme in the air, you ask? Ah, well, you snooze you lose, hon.

---

* I strongly believe that more affairs ought to be underscored with cheese.

# CHEESY HASSELBACK POTATO GRATIN

**Because of variation in the shape of potatoes, the amount of potato that will fit into a single casserole dish varies. Longer, thinner potatoes will fill a dish more than shorter, rounder potatoes. When purchasing potatoes, buy a few extra in order to fill the dish if necessary. Depending on the exact shape and size of the potatoes and the casserole dish, you may not need all of the cream mixture.**

**SERVES 8**

3 ounces finely grated Comté or Gruyère cheese

2 ounces finely grated Parmigiano-Reggiano

2 cups heavy cream

2 medium cloves garlic, minced

1 tablespoon fresh thyme leaves, roughly chopped

Kosher salt and freshly ground black pepper

4 to 4½ pounds russet potatoes, peeled and sliced ⅛ inch thick on a mandoline slicer (7 to 8 medium, see the headnote)

2 tablespoons unsalted butter

1. Adjust the oven rack to the middle position and preheat the oven to 400°F. Combine the cheeses in a large bowl. Transfer ⅓ of the cheese mixture to a separate bowl and set aside. Add the cream, garlic, and thyme to the cheese mixture. Season generously with salt and pepper. Add the potato slices and toss with your hands until every slice is coated with the cream mixture, making sure to separate any slices that are sticking together to get the cream mixture in between them.

2. Grease a 2-quart casserole dish with the butter. Pick up a handful of potatoes, organizing them into a neat stack, and lay them in the casserole dish with their edges aligned vertically. Continue placing potatoes in the dish, working around the perimeter and into the center until all the potatoes have been added. The potatoes should be very tightly packed. If necessary, slice an additional potato, coat it with the cream mixture, and add it to the casserole. Pour the excess cheese/cream mixture evenly over the potatoes until the mixture comes halfway up the sides of the casserole. You may not need all the excess liquid.

3. Cover the dish tightly with foil and transfer to the oven. Bake for 30 minutes. Remove the foil and continue baking until the top is pale golden brown, about 30 minutes longer. Carefully remove from the oven, sprinkle with the remaining cheese, and return to the oven. Bake until deep golden brown and crisp on top, about 30 minutes longer. Remove from the oven, let rest for a few minutes, and serve.

# MASHED POTATOES

**Mashed potatoes are a particularly divisive topic in my family.**

See, I like mine rich and perfectly smooth, with plenty of butter and heavy cream, lots of black pepper, and maybe some chives if I'm dressing to impress (I usually am). Somewhere between a dish on its own and a sauce, the mash should have the consistency of a pudding, slowly working its way across a tilted plate. I like to pick up a piece of turkey and swirl it in my gravy-covered potatoes so that they coat it, their buttery richness working into the cracks in the meat. Sounds good, right? Who could possibly want it any other way? *My sister. That's who.*

For Pico (yes, that's her real name), mashed potatoes should be fluffy and thick enough to stand up under their own weight, *Close Encounters of the Third Kind*–style. The kind of mashed potatoes that can hold their own on the plate. The kind that you want to turn into a TV commercial with a pat of butter slowly melting on top. I'm talking smooth but light and fluffy. So how do you arrive at such two different results with the same starting ingredients?

It's all got to do with starch.

## The Starch

For our purposes, potatoes can be thought of as basically three different things. First, there are the *cells*, the little microscopic bubbles that all living things are made from. These cells are held together with *pectin*, a sort of natural plant glue, and the *walls* of the cells are where the starch is concentrated.

The starch molecules—a type of carbohydrate—come bundled up in tight granules. As potatoes cook, the pectin breaks down and individual cells expand and separate, releasing starch granules into the outside environment. These starch granules absorb water like little balloons, eventually popping and releasing sticky starch molecules. The concentration of released starch that makes its way into the final mashed potatoes to a large degree determines their consistency.

To put it simply: for lighter, fluffier potatoes, the goal is to incorporate as little starch as possible in the final product.

- **Potato type** (see pages 465–66) plays a huge role in this. Mealy russet potatoes have cells that readily fall apart from each other, meaning you don't have to cook them or work them too hard to get them to a relatively smooth consistency. Less working means fewer burst starch granules, which means fluffier mashed potatoes. Waxier Yukon Golds or Red Bliss require longer cooking, and they must be worked fairly hard to separate their cells, making for creamier mashed potatoes.
- **The mashing method** can drastically alter your end results. Carefully pressing the potatoes through a tamis (drum sieve), ricer, or food mill will separate the cells with minimal shearing action to break up the starch. Throw potatoes in a food processor, and an avalanche of starch gets released, turning your potatoes into the consistency of melted mozzarella cheese (there are some recipes, like

Potatoes whipped in the food processor become gummy.

the famously elastic *pommes aligot*, that require you to beat your potatoes into stretchy oblivion). Whipping the potatoes with an electric mixer will develop some starchiness but still keep the potatoes creamy.

- **Soaking and/or rinsing** the potatoes can help you reduce the amount of starch that remains on them. Cutting potatoes into smaller pieces before cooking and rinsing them under cold water will wash away much of the excess starch. But there is a downside to soaking—it rinses away some of the enzymes necessary to properly break down pectin. Soak your potatoes for too long or cut them too small before soaking, and they'll never soften, no matter how long you boil them.

So, just by knowing these factors, we should be able to determine the best way to get either style of mashed potatoes.

### Cream of the Crop

Ultracreamy mashed potatoes are really more of a French thing than an American one, and if you want to be fancy, you can call them *pommes puree*. The goal is to get them superrich but not heavy or leaden. This requires some careful cooking to allow just enough starch to be released to give the potatoes the right texture, but not so much that they turn gluey. The best way I've found to do this is to boil medium-starch potatoes (like Yukon Golds) just until tender enough to be pierced with a cake tester or paring knife with no resistance. Starting them in cold water helps them cook more evenly, as well as strengthening some of their pectin, which keeps them from totally falling apart in the water.

I tried several methods for pureeing the potatoes, including pressing them through a tamis (lots of work), throwing them straight into a stand mixer

(they never get smooth), and using the food processor (really, really bad idea). The best and easiest method is to just pass them through a ricer into the bowl of a stand mixer. You don't even need to peel them. After that, I whisk them on high speed with the paddle attachment, adding melted butter, heavy cream, salt, and pepper. If I'm not serving vegetarians, I also like to add a bit of chicken stock, which gives the puree an intensely savory quality (don't give away the secret).

On to the second variety.

### Fluff Enough?

Getting potatoes light and fluffy is a little bit trickier. One thing is clear: you want to start with mealy russets that will fall apart with minimal prodding and release starch in an easy-to-rinse-off manner. At first I thought that simply rinsing away as much starch as possible before cooking would be the key, and to test, I made three batches of potatoes. The first I cut into large chunks, the second into 1-inch dice, and the last I grated on the large holes of a box grater. I rinsed all three batches under cold water until the liquid ran clear. By collecting the drained milky liquid from each batch of potatoes and comparing it, it was easy to see that the grated potatoes released far more starch than either of the other types. Let's see how it translates down the line.

Turns out that another weird phenomenon occurs when you try to cook grated and rinsed potatoes: they simply don't soften. I boiled the grated potatoes for a full 45 minutes to no avail. Even after forcing them through a ricer, pebbly, hard bits remained. What the heck was going on?

It's got to do with that pesky pectin. When exposed to calcium ions, pectin cross-links, forming stronger bonds that are resistant even to prolonged cooking. As it happens, potato cells are *full* of calcium ions just waiting to burst out. By grating the taters, I ended up releasing so much calcium that the pectin got strengthened to a point where it *never* softened. The other two batches—the large chunks and the smaller dice—both formed a moderately fluffy mash, but to get the potatoes even

fluffier, I found that rinsing the potatoes of excess starch both before *and* after cooking was the key. A quick pass through the ricer, and a little bit of lubrication provided by some butter and whole milk gently stirred in with a rubber spatula, and my sister's potatoes were ready for sculpting.

With both potato cooking methods in order, my sister and I can finally get back to fighting over really *important* things, like who gets to play the guitar part on Beatles Rock Band.

# ULTRA-FLUFFY MASHED POTATOES

**SERVES 6 TO 8**

4 pounds russet (baking) potatoes

2 cups whole milk

12 tablespoons (1½ sticks) unsalted
butter, cut into ½-inch pats, at
room temperature

Kosher salt and freshly ground
black pepper

1. Peel the potatoes and cut into rough 1- to 2-inch chunks. Transfer to a colander and rinse under cold water until the water runs clear.

2. Bring 4 quarts of water to a boil in a Dutch oven or stockpot over high heat. Add the potatoes and cook until completely tender when pierced with the tip of a knife, about 15 minutes.

3. Meanwhile, heat the milk and butter in a small saucepan over medium heat, stirring occasionally, until the butter is melted.

4. Drain the potatoes in a colander and rinse under hot running water for 30 seconds to wash away excess starch. Set a ricer or food mill over the empty pot and pass the potatoes through. Add the milk and butter and fold gently with a rubber spatula to combine. Season to taste with salt and pepper and reheat as necessary, stirring constantly. Keep warm until ready to serve.

# RICH AND CREAMY MASHED POTATOES

**SERVES 6 TO 8**

4 pounds Yukon Gold potatoes, scrubbed

½ pound (2 sticks) unsalted butter, melted

2 cups heavy cream, or as needed

Up to 1 cup homemade or low-sodium canned chicken stock (optional)

Kosher salt and freshly ground black pepper

1. Place the potatoes in a Dutch oven or stockpot, cover with cold water, and bring to a boil over high heat. Reduce to a simmer and cook until the potatoes are tender; a paring knife should pierce them with no effort.

2. Carefully peel potatoes under cool running water—the skin should slip off easily—and transfer to a large bowl. Pass the potatoes through a ricer or food mill into the bowl of a stand mixer. Add the melted butter and half of the cream. Whip on low speed, using the paddle attachment, until the cream and butter are incorporated, about 30 seconds. Increase the speed to high and whip until smooth and creamy, about 1 minute. Adjust to the desired consistency with more cream and/or chicken stock. Season to taste with salt and pepper and reheat as necessary, stirring constantly. Keep warm until ready to serve.

# SUPER-CRISP ROASTED POTATOES

How often do you get roasted potatoes that look like they're going to be awesomely crisp only to find that rather than crispness, all you've got is a papery (or worse, leathery) skin on the exterior?

Roasting potatoes is not quite as easy as roasting most other vegetables. See, with roast potatoes, we've got a different set of goals than when, say, roasting Brussels sprouts. First off, we want the potatoes to be cooked through all the way to the center. Fluffy and moist is what we're after. Second, we want the exterior to be extremely crisp. We're talking crisper-than-a-French-fry crisp. Simply toss a potato coated with a bit of oil in the oven, and what you end up with is a potato with a paper-thin sheath of crispness around its exterior that very rapidly softens and turns leathery as internal moisture seeps through it.

So, what makes a potato crisp? The answer is building up a dehydrated layer of gelatinized starch on the exterior of the spud, much like when you fry a French fry (see page 904). To do this, you've got to parcook them, allowing their starches to soften and expand, and then recrystallize by cooling them a bit.

Boiling potatoes then roughing them up before roasting adds surface area and extra crunch.

Our secondary goal is to increase their surface area. Craggy, uneven surfaces crisp up a lot better than smooth ones. Luckily, we can kill two birds with one stone here. By boiling your potatoes before you roast them, not only do you ensure that their outer layers of starch are properly gelatinized, you also soften their exteriors enough that they get a bit battered and bruised when you toss them around with oil before roasting them. This creates a sort of potato-oil paste that adds plenty of surface area and acts almost like a batter for fried foods, creating an extra layer of crispness as the potatoes roast.

Take a look at these potatoes, roasted side by side in the same oven. One batch was started raw, another was boiled first, and the third was boiled, then roughed. The difference is obvious.

Next question: What's the best fat to use? People often tout the awesomeness of duck fat with potatoes, and for good reason: it tastes awesome. Duck fat has a distinct richness and aroma that gets

absorbed very easily into the surface of a spud. On top of that, it's got plenty of saturated fat and a high smoke point, which makes it an ideal medium for crisping up fried or roasted foods. (In general, the higher the saturated fat content of an oil, the more efficiently it'll crisp foods.) Can't get duck fat? Well, turkey fat or chicken fat collected from roasted birds will do just fine.

Bacon fat and rendered lard are also fine choices, as is just about any sort of animal-derived fat.

If you must, extra-virgin olive oil will certainly do admirably well, though you won't get quite the same level of crispness you'd get with an animal fat.

Once your potatoes are tossed in fat and seasoned well, all you've got to do is roast them in an extremely hot oven until they crisp up. I roast mine on an unlined heavy rimmed baking sheet (they have a tendency to stick to foil). The key is to make sure you let the undersides crisp up completely before you even attempt to lift or flip them. If the potatoes don't come off relatively easily, you run the risk of breaking off the tops, leaving the crisp bottoms cemented to the bottom of the pan. This is not an ideal situation.

MORAL OF THE STORY: Your potatoes will release themselves from the pan when they're good and ready. Don't force them.

**SERVES 4 TO 6**

3 pounds russet (baking) potatoes, scrubbed and cut into 1- to 2-inch cubes

Kosher salt

¼ cup duck fat, bacon fat, or olive oil

Freshly ground black pepper

Chopped fresh herbs, such as thyme, rosemary, parsley, and/or chives

1. Adjust the oven racks to the upper- and lower-middle positions and preheat the oven to 450°F. Place the potatoes in a large saucepan and cover with cold water by 1 inch. Season generously with salt. Bring to a boil over high heat, reduce to a simmer, and cook until the potatoes are just barely cooked through, about 10 minutes (a knife or cake tester inserted into a potato should meet little resistance). Drain and transfer to a large bowl.

2. Add the fat and a few generous grinds of pepper to the hot potatoes and toss well; the potatoes should end up with a thin coating of potato-fat paste. Spray two rimmed baking sheets with nonstick cooking spray (or coat with a thin layer of oil). Transfer the potatoes to the baking sheets and roast until the bottoms are crisp, rotating the pans halfway through cooking, about 25 minutes. Test the potatoes by trying to pry one or two pieces off the baking sheet with a stiff metal spatula. If they don't come off easily, roast for additional 3-minute increments until they do.

3. Flip the potatoes with the spatula, making sure to get all the crisped bits off the bottom, then continue to roast until golden brown and crisp all over, about 25 more minutes. Transfer to a serving bowl, season to taste, and toss with chopped herbs.

# CRISPY SEMI-SMASHED NEW POTATOES

The idea for this dish came from one of my wife's favorite foods: deep-fried plantains. To make the Colombian dish of *patacones*, you deep-fry plantains until soft, then smash 'em into disks and fry them again. Works for plantains, why not potatoes?

The first time I made these potatoes, my wife walked in, took one look at what I was doing, and said, "You're making *patacones* out of potatoes? Weirdo." Of course, she ate every last one.

The key to these potatoes is the same as with my crisp roast potatoes, but it's even easier and takes place 100 percent on the stovetop. Rather than frying them the way you would with plantains, I start by boiling new potatoes (you can use any color or variety) in their skins until completely tender. Then it's a simple matter of smashing them with a skillet so their guts start to spill out (the better to crisp!) and browning them in hot fat until crisp. They come out as crisp as French fries and incredibly creamy and flavorful within.

**SERVES 4 TO 6**

1½ pounds new potatoes, scrubbed

Kosher salt

¼ cup vegetable or canola oil or duck fat

2 tablespoons chopped fresh chives, parsley, or other herbs

Freshly ground black pepper

1. Place the potatoes in a large saucepan and cover with cold water by 1 inch. Season heavily with salt. Bring to a boil over high heat, then reduce the heat and simmer until the potatoes are cooked through, about 20 minutes. Drain and let cool slightly.

2. Adjust an oven rack to the middle position and preheat the oven to 250°F. Place one potato on a cutting board and, using the bottom of a skillet, smash it gently to a thickness of about ½ inch. Use a spatula to scrape the smashed potato off the cutting board and onto a large plate. Repeat with the remaining potatoes.

3. Heat the oil in a large nonstick skillet over medium heat until shimmering. Add half of the potatoes and cook occasionally shaking the pan gently, until golden brown and crisp on the bottoms, about 8 minutes; reduce the heat if the potatoes threaten to burn. Flip the potatoes and brown the second side, about 8 minutes longer. Transfer to a rimmed baking sheet and keep warm in the oven while you cook the remaining potatoes.

4. When all the potatoes are cooked, transfer to a bowl. Add the herbs and toss to combine, season to taste with salt and pepper, and serve.

# BUTTERY, ONIONY CHARRED HASH BROWNS

I've got an Irish friend who for the longest time didn't eat anything but brown and white food (I'm sure you know the type): fried chicken, steak, cheeseburgers, white bread, grilled cheese sandwiches, potatoes in all forms, etc. He's since evolved to appreciate the odd orange food and has even forayed into the realm of green recently (baby steps), but brown and white are still where his heart lies. This is the dish I created for his birthday one year in the heyday of his brown era. It's essentially a cross between a steak house–style hash brown—that is, a potato cake made by pressing boiled potatoes into a skillet, with some onions for good measure, and then frying—and mashed potatoes. It starts off like normal steak house hash browns, but rather than simply frying the mass of potatoes just once, you let the whole thing cook until it's almost burnt, saving it at the last minute by breaking up the crust and folding it into the center, then repeat the process two more times.

What you end up with is a pile of creamy, textured potatoes riddled throughout with sweet caramelized onions and intense browned flavors from the repeated crisping. They are not like any other potatoes you've ever had, I promise. (I mean that in a good way.)

**SERVES 4 TO 6**

2½ pounds russet (baking) potatoes, peeled and cut into 1- to 2-inch cubes

Kosher salt

8 tablespoons (1 stick) unsalted butter

2 tablespoons olive oil

2 large onions, thinly sliced (about 3 cups)

Freshly ground black pepper

1. Place the potatoes in a large saucepan and cover with cold water by 1 inch. Season generously with salt. Bring to a boil over high heat and boil until the potatoes are completely tender but not falling apart, 8 to 10 minutes; a cake tester or fork should pierce the potatoes with no effort. Drain the potatoes, transfer to a large bowl, and set aside.

2. Heat 4 tablespoons of the butter and the olive oil in a 12-inch heavy-bottomed nonstick or cast-iron skillet over medium-high heat until the butter melts. Add the onions and cook, stirring frequently, until softened and beginning to brown, 8 to 10 minutes. Using tongs or a slotted spoon, transfer the onions to the bowl with the potatoes, leaving the fat in the skillet. Fold the potatoes and onions together and season to taste with salt and pepper.

3. Add 2 tablespoons butter to the skillet and heat over high heat, swirling until the foaming subsides and the butter starts to brown. Add the potato mixture to the skillet and press down with a silicone spatula until an even cake is formed. Reduce the heat to medium-high and cook, shaking the pan gently every minute or so, until the bottom is completely browned and verging on charred, about 5 minutes. Use the spatula to lift and flip the potatoes in sections, slicing and folding to incorporate the browned sections into the interior. Press the cake into the skillet again and repeat the browning and folding steps two or more times: you should end up with a potato cake that is riddled with crisp bits of potato and sweet onions.

4. Push the potatoes to one side, add the remaining 2 tablespoons butter, and melt it. Then lift the potatoes over the butter and spread and press into an even layer. Cook until the cake is completely crisp on the bottom, about 5 minutes, shaking the skillet every minute or so to prevent sticking. Remove the skillet from the heat and place an inverted plate over the top. Carefully flip the potatoes out onto the plate, so that the crisp side is facing up, and serve (or keep warm in a low oven until ready to serve).

BALLS, LOAVES,
LINKS, BURGERS, AND

# THE
# SCIENCE
*of*
# GROUND
# MEAT

# 5

I'm a cook by trade but a grinder by nature.

# BALLS, LOAVES, LINKS, BURGERS, AND THE
# SCIENCE *of* GROUND MEAT

## RECIPES IN THIS CHAPTER

HAVE **YOU** EVER
NOTICED
. . .

that many times, a group of things that *seem* pretty similar at first glance upon close inspection turns out to be nothing of the sort? I'm talking about things like *Star Trek* versus *Star Wars* fans (*Star Trek* is science fiction, *Star Wars* is fantasy, dammit!), computer geeks versus physics geeks (hint: physics geeks wear shoes more often).

And so it is with meat loaf, sausages, and hamburgers. I mean, all three are made of seasoned and cooked ground meat. How different can they be? Plenty different! On page 544, we're going to discuss the travesty of sticking a small patty-shaped meat loaf in a bun and calling it a hamburger. What you've got there is not a hamburger, it's a meat loaf sandwich.

Just so we're all on the same page here, let's go through a few definitions. These are the three great pillars that support ground-meat cuisine and lift it to magnificent heights. Don't worry if you don't quite understand everything yet—all will be revealed.

- **A hamburger** is a patty of pure ground beef with no salt, seasonings, flavorings, or additives of any kind mixed into it. It can be cooked any number of ways, but key to its production is that salt and pepper *should go only on the exterior*, and we'll see why later. Its texture should be loose, tender, and juicy.
- **A sausage** is ground meat to which enough salt has been added (about 1.5 percent of the weight of the meat) that the protein *myosin* has dissolved, allowing it to subsequently bind the meat together when the raw mixture is stirred. The mixture can be flavored with spices, vegetables, herbs, or any number of other additives, but the keys to its production are salt and meat. Its texture should be springy, snappy, and juicy.
- **A meat loaf** and **meatballs** are made of seasoned ground meat to which bread crumbs, eggs, and/or dairy products have been added in order to discourage the cross-linking of meat proteins and produce a more tender finished product. As with sausages, an array of other flavorings can be added. Its texture should be tender and moist

So, how do you get three such different finished products from what amounts to basically the same set of ingredients? That's what we're going to explore in this chapter.

# The FOOD LAB's Guide
## TO GRINDING YOUR OWN MEAT

I'm a cook by trade but a grinder by nature. Nothing pleases me more than the careful, controlled deconstruction and reconstruction of what nature has so carefully put together.

Grinding your own meat at home is a uniquely satisfying experience. There's something profoundly beautiful to me about watching the chopped meat fall out of the grinder into the bowl, deep red interspersed with discrete creamy-white bits of fat. I like the way it starts out free and pebbly as you pick it up but comes together in your hands to form a burger patty. And I like how salt can make the grind sticky, ready to be beaten into a juicy sausage or a tender meat loaf.

I've never met anyone who's decided to go back to using store-bought ground beef after having tried beef ground fresh at home. Once you grind, you never rewind. Why should you grind? Four reasons:

- **It's safer.** Packaged ground beef can contain meat from hundreds, even thousands, of animals, and not necessarily from the nicest bits either. This means that you've got to be extra careful when cooking with packaged ground meat—chances of contamination are higher.
- **Better flavor.** Unless you've got a really great butcher, you're stuck with whatever ground beef the supermarket has on hand. Usually this is no more specific than knowing the fat content. Even when it's labeled with cuts like chuck, round, or sirloin, there's no guarantee that you're not just getting scraps. Grinding at home allows you to control the flavor of your grind, along with the fat content.
- **Better texture.** Preground meat sits in its packaging, slowly being compressed and oxidizing. And it's often ground much finer than is ideal for the

perfect burger. Grinding it fresh lets you keep it nice and loose, improving both moisture level and texture after cooking.

- **It's cool.** Those who grind their own beef for their sausages and burgers get instant street cred in my book. You can't help but look badass.

If you are still asking the question "Why should I grind my own meat?" instead of "How do I grind my own meat?" it's possible that you're simply hopeless. For the rest of you, read on.

## GRINDER, PROCESSOR, OR HAND-CHOPPING?

As a home cook, you have five different options for getting ground beef for your burgers: buying preground beef from the store, having your butcher grind meat for you fresh, grinding at home with a dedicated meat grinder or stand mixer attachment, grinding at home in the food processor, or hand-chopping with a knife. There are advantages and drawbacks to each method. Store-bought beef, for example, is easy, but, as I noted above, you have little control over flavor or texture. The food processor provides excellent results if you don't have a meat grinder, but it can take a little bit of planning.

To find the best method for grinding beef, I gathered a couple of pounds of beef, ground using five methods: store-bought 80% lean ground chuck, chuck ground fresh by the butcher; chuck ground in the meat grinder, chuck chopped in a food processor, and chuck finely chopped by hand. I tasted the results of each method cooked into two types of burgers, griddled and grilled, as well as made into sausages and meatballs. I analyzed all of the results based on texture, flavor, and ease of preparation. Here are the results.

| METHOD OF GRINDING | TEXTURE | COOKING NOTES/BEST USES | EASE OF USE |
|---|---|---|---|
| **Preground from the Supermarket** | Compact and dense | It's fine in a pinch, but don't expect a memorable burger. It's best for smashed burgers (see page 549) and thin patties for grilling or pan-searing, where tight texture is not as big a drawback. If buying preground beef, look for stuff ground in-store rather than packed in vacuum-sealed containers offsite, which have inferior texture and inconsistent flavor. | Very easy—open the package and go. But it's more difficult to shape patties properly because the meat is tightly compressed. |

| METHOD OF GRINDING | TEXTURE | COOKING NOTES/BEST USES | EASE OF USE |
|---|---|---|---|
| **Ground Fresh from the Butcher** | Can vary from butcher to butcher. Your best bet is to ask for a coarse single grind through a ¼-inch plate. | Far superior to packaged ground beef, with the added benefit that you can request specific cuts to be ground. Be careful not to compress the meat as you transport it, and use it soon after purchase for optimal texture. | Provided you've got a great butcher, this is as easy as buying preground meat. |
| **Dedicated Meat Grinder or Stand Mixer Attachment** | Very loose and airy, with plenty of nooks and crannies for crisping/collecting juices. It gives an even grind that is ideal for sausage. | Best all-around method, particularly excellent for loose pan-seared patties or for larger grilled burgers that will stay tender and juicy as you cook them. | Once you have the grinder or attachment, it's very easy. Cleanup can be a bit of a pain (getting meat and fat out of an L-shaped tube ain't easy), but my dishwasher takes care of most of the hard work. |
| **Food Processor** | A little uneven—some large chunks mixed with very fine bits. | Unless you've got a brand-new processor blade, some amount of smearing (see page 498) will occur, even when your meat is partially frozen before grinding. Not as good as a grinder, but far better than store-bought ground beef. | Can be a little annoying: freezing the meat adds an extra step to the process, as does working in batches. Cleanup can also be a chore. |
| **Hand-Chopped** | You can chop the meat as fine or as coarse as you like, but no matter what, you're going to end up with a good degree of variation in the texture. This is a good thing for burgers. They end up with nice tiny bits of meat and fat for crust formation but retain enough larger pieces that each bite has a few steak-like moments with a touch of chew. Very satisfying indeed. Not ideal for sausage. | When I've got the time, this is my new go-to method, particularly for thicker burgers where that steak-like quality of the chunky beef really makes for an interesting finished texture. | There's no denying it: this process is time consuming, at least three or four times as slow as a grinder or processor. On the bright side, it's good exercise, and cleanup is a snap. |

## BASIC TIPS FOR GRINDING MEAT

There are a few keys to great ground meat that apply no matter what method you employ to grind it.

- **Start with good meat.** Just as you can't make a great sandwich on terrible bread, you can't make good ground meat out of poor meat. And, of course, you can't make a great burger or meat loaf out of poor ground meat. Start with whole cuts from a reputable butcher or supermarket, and select your cuts to optimize fat content and flavor.

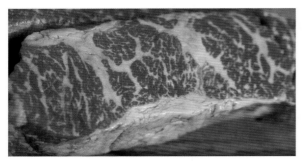

Abundant marbling makes for juicier burgers and meatballs.

- **Keep everything cold.** This is the single most important thing when it comes to grinding. Warm meat will smear (as opposed to chop), the fat will leak out, and it will result in a cooked texture similar to papier-mâché—pulpy and dry. Ugh. Place the grinder and all of its parts (or the food processor bowl and blade) in the freezer for at least one hour before grinding (I store my grinder in the freezer). Keep your meat well chilled right until ready to grind. If you are making sausage that will require several grinds, chill the meat in between grinds to ensure perfect texture. If using a food processor, freeze the meat for 15 minutes before attempting to grind.

- **Watch for smearing.** Keep an eye on the meat as it comes out of the grinder or flies around the food processor bowl. Ideally, it'll come out of each die hole in discrete little pieces. You should be able to clearly identify fat and meat. If it starts coming out as one mass, looks wet, and collects on the surface of the die, you are in trouble. Similarly, if it smears around the bowl of the food processor or chunks of fat ride along on the blade, it's getting too warm. Remove the meat, rechill it, and try again.

- **Keep your blades sharp.** In a meat grinder, the blade is the only part of your grinder that should ever need much care or attention. A dull blade will smear meat. Luckily, the blade and plate should actually get better and better with repeated use. The metal grinds down microscopically each time you use it, so the contact between the blade and the plate gets tighter and tighter. Nothing grinds as smoothly as a well-taken-care-of, well-used grinder. You will occasionally need to get your blades resharpened if they've gotten way too dull; once a year or so for a moderately-well-used grinder is more than enough. Or simply buy a few replacement blades. They can usually be had for a few bucks. Similarly, if your food processor blade gets too dull, it should be replaced. If chopping by hand, always use a sharp, sharp cleaver.

- **Keep everything clean.** Part of the reason to grind your own meat is to make it safer. Working in a messy or unsanitary environment will throw all that out the window. Make sure to start with a clean cutting board when trimming your meat, wash your hands and your knife carefully as you work, avoid cross contamination, and don't distract yourself with other tasks when grinding meat, no matter what method you use. If you use a grinder, allowing meat to dry and stick to the blade or inside the feed tube is a good way to get yourself sick. Remove and wash all parts of the grinder well between grinds.

# HOW TO BUY, USE, AND CARE FOR A MEAT GRINDER

There's really not much to it when it comes to using a meat grinder. Basically all you've got to do is assemble the grinder with the plate you desire, take your trimmed meat (grinders hate sinew and connective tissues, so make sure to trim it all out), feed it into the hopper, turn the grinder on (if using a stand mixer attachment, a relatively fast speed is the way to go—I've found that about 6 to 8 on the KitchenAid produces the best results), or start grinding with a manual model, and press the meat through. Ground meat, simple as that!

Here's a basic guide on how to select, use, and maintain a grinder.

## The Parts

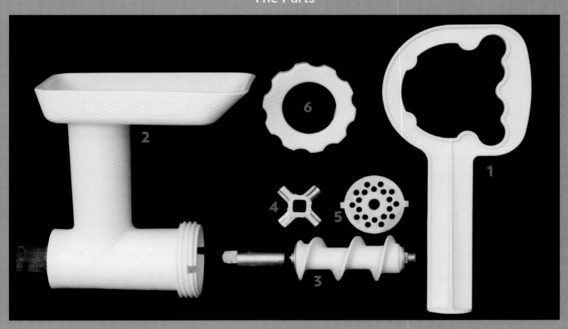

All meat grinders consist of the same basic parts:

- **The pusher (1) and hopper (2)** are where you add the cubes of meat. The pusher is used to force the meat down the feed tube and to keep things moving. Usually there is a tray located on top of the feed tube where extra meat can be set before being pushed into the grinder. The larger this tray, the more convenient it is to grind large batches of meat.
- **The screw or auger (3)** is the main working part of the grinder. It steadily pushes meat down the shaft toward the blade.
- **The blade (4) and plate (5)** are what do the actual grinding. The blade is a small cross-shaped piece with a sharp edge on each arm that rotates against the plate (also called the die). The plate is a flat

*continues*

piece of metal with holes cut into it. As the screw forces the meat into these holes, the blade cuts it into a fine mince. The size of the holes determines the fineness of the final grind.

- **The cover (6)** is used to keep the blade and plate in place as the meat is ground.

Although the basic parts are the same, you have a number of options when it comes to buying a meat grinder. The good news is, none of them is bad.

## Manual Grinders

Manual grinders are the cheapest way to get good-quality freshly ground meat at home and are a great option for casual grinders who don't own a stand mixer. You have two choices: If you've got a nice woodworking table or countertop and are planning to do a lot of grinding, a bolt-mounted meat grinder is the way to go. At less than $40, one of these should last you a near lifetime of grinding, provided you care for the working parts properly. For an even cheaper, if slightly less sturdy option, at $29.95, a clamp-mounted model allows you to work on any tabletop you'd like, though sometimes getting the pieces to fit together properly can be a pain. Still, one of these will still grind your meat well.

## Stand Mixer Attachments

The next level up is for anyone who owns a stand mixer. All of the major brands, including KitchenAid, Viking, and Cuisinart, have their own attachments. I use the plastic KitchenAid grinder attachment at home. The great thing about buying a meat grinder attachment is that you already know that the hardest-working part of your grinder—the motor—is a workhorse that can power through even the toughest grinding projects.

You are, obviously, stuck buying the attachment for whatever brand mixer you own, but none of them are that bad. While both the Cuisinart and the Viking feature all-metal parts, which can stay chilled for longer than those in the plastic KitchenAid model, they are also three times as expensive. Stand mixer attachments are a great choice if you make a lot of sausage. You can grind the meat directly into the mixer bowl, then attach the bowl to the machine and immediately start mixing it with the paddle to develop the protein.

## Stand-Alone Grinders

I don't know many home cooks outside of those who do a lot of hunting who have a need for a stand-alone grinder. Although these usually come with a wider assortment of plates and a bigger feed tube and screw shaft, the motor is only as good as the price you pay for it. Cheaper models will work no better than stand mixer attachments, and more expensive models are only necessary if you plan on doing a whole lot of grinding. I grind way more meat than the average cook, and my KitchenAid attachment has yet to fail me.

The one distinct advantage that stand-alone grinders have is that most of them have a reverse function—a real time-saver if you are trying to chop especially troublesome meat with lots of connective tissue that gets caught in the blade.

## HOW TO GRIND IN A MEAT GRINDER

Here are the basic steps:

- **Chill your grinder.** Place all the parts in the freezer until thoroughly chilled.
- **Trim your meat.** Start with whole cuts and carefully trim them of any excess sinew. A certain amount of fat is OK—even desirable—but make sure your ratio of fat to lean is on point. I generally aim for around 20 percent fat to 80 percent lean, plus or minus 5 to 10 percentage points, depending on the application.
- **Cut your meat into 1- to 2-inch cubes and chill them.** Keeping everything ice-cold will make sure that the fat stays firm and easily choppable.

- **Grind from large to small die.** If you need an extra-fine grind for certain types of sausages, make sure to grind your meat twice, chilling it between batches: once through a ¼-inch die, then a second time through the smaller die. This will help prevent smearing and will give you a more even grind and a better textured sausage in the end.
- **If it starts to smear, stop immediately.** The meat should come out in clean, discrete pieces. If it starts to come out as a solid mass or extrude unevenly, you've got some sinews or other gunk stuck in the blade. Stop, remove the blade, clean it carefully, and start again.
- **Use a paper towel to get the last bits of meat out.** Once you're almost done with your batch of meat, you'll find that the last few cubes will have trouble getting through the die on their own. To push them through, crumple up a paper towel and feed it into the tube. It'll push the meat out but will not get pushed through the grinder itself. As an added bonus, it'll clean your tube as it goes.

## HOW TO GRIND IN A FOOD PROCESSOR

Here are the basic steps:

- **Chill your processor bowl and blade.** Place the bowl and blade in the freezer for at least 15 minutes before grinding.
- **Trim your meat.** Start with whole cuts, carefully trimmed of any excess sinew. A certain amount of fat is OK—even desirable—but make sure your ratio of fat to lean is on point. I generally aim for around 20 percent fat to 80 percent lean, plus or minus 5 to 10 percentage points, depending on the application.
- **Cut your meat into 1- to 2-inch cubes and partially freeze them.** You want your meat to be as close to frozen as it can possibly be without actually making it so hard that the processor will not be able to cut through it. I cut the meat into cubes, place them on a plate, and throw them in the

freezer for about 15 minutes, until they start to feel firm around the edges.

- **Set up the processor and add some of the meat.** Trying to grind too fast will cause some meat to ride around the blade, making for an uneven grind. Do not try and grind more than half a pound (225 grams) of meat at a time in a 10- or 11-cup food processor.

- **Pulse to chop.** Don't let the processor run. Rather, pulse it in rapid bursts to allow the large chunks of meat to settle back down to the bottom of the bowl, where the blade will cut into them. Pulse until you get your desired grind size—generally 10 to 12 fast pulses will get you a reasonable grind for burgers, while 8 to 10 is what you want for chili and stews.
- **Empty the bowl and repeat.** Repeat until all the meat is ground.

## HOW TO CHOP MEAT BY HAND

Here are the basic steps:

- **Trim your meat.** Start with whole cuts, carefully trimmed of any excess sinew. A certain amount of fat is OK—even desirable—but make sure your ratio of fat to lean is on point. I generally aim for around 20 percent fat to 80 percent lean, plus or minus 5 to 10 percentage points, depending on the application.

- **Slice your meat thin.** Use a sharp chef's knife or carving knife to slice the meat into thin, thin slices. Stack the slices, then cut them into thin strips. Finally, rotate the strips 90 degrees and cut them into small pieces.
- **Use a cleaver.** Once you have small pieces of meat, use a cleaver to reduce the pieces to the desired texture, using the weight of the cleaver to do most of the work for you.

## SAUSAGE = MEAT PERFECTED

The story of the sausage is a humble one. Sausages were originally created as a means to expand the useful portions of an animal. In the days before refrigeration, easy transportation, and canned dog food, when an animal was slaughtered, it was essential that none of it went to waste. People discovered that by chopping up the less desirable parts of the animal, adding salt to them, and stuffing them into various other bits of the animal (say, the intestine or stomach), they could create a food that not only lasted longer than fresh meat but also was in fact quite tasty in its own way. Thus sausages were born. As with many other foods—like duck confit, fruit jams, or beef jerky—what was once a means of preservation, an act of necessity, is still practiced today because the end result is so freaking delicious.

I mean, is there anything more humble yet so perfect as a well-made sausage? Evolution has done a pretty darn good job of producing delicious hunks of meat for us to enjoy, but with the sausage, we do Mother Nature one better: the right fat content, the right sodium content, proper seasoning, and perfect texture are built right in. How many pork chops can claim that? And to think that sausages are also one of the most inexpensive forms of meat on the market!

Thing is, most people are familiar only with store-bought sausages or, if they're lucky, those made by a good local butcher. And many of these can be quite good. Even some nationally available brands do decent work. But the real joy of sausages is their completely customizable nature. Who says that sweet and hot Italian are the only types worth eating (much as I love 'em)? Once you understand the basic principles of sausage making and have a handle on the technique, there's no limit to your creativity. You like the flavor of juniper and cinnamon with venison? Go ahead and make a warm-spiced venison sausage. Broccoli rabe and cheese with chicken? No problem: Or how about just some garlic and parsley with pork? Yep, you can do that too.

# THE CRAFT OF THE CURE

Y ou remember that scene in the teen movie when the shy, nerdy girl in the glasses that none of the boys really wants to talk to gets a makeover, trades in the specs for contacts, throws on a dress, and is suddenly the hottest one at the prom? Well, that never happens in real life. But how's this for an awesome transformation: turning the tough, sinewy back leg of a pig into silky, rich, sweet prosciutto, or the fatty, squishy belly into smoky, crisp strips of bacon? That is the craft of curing, and it's one of the pinnacles of cuisine.

There's often confusion as to exactly what "curing" means when it comes to meat. Curing is the act of preserving meat or fish through the use of one of three methods: chemical cures, smoking, or dehydration/fermentation. It is these three processes that are responsible for the incredible array of flavors you find in preserved meats from around the world. In France, it's called *charcuterie*. In Italy, it's *salumi* (incidentally, *salami* is a specific type of *salumi*). Whatever you want to call it, it's the craft of preserving meat—of taking those bits that nobody really wanted to eat and turning them into the best-tasting stuff around. In this book, I've included recipes for the simplest form of curing, letting meat and salt mingle overnight—and you'll be amazed at the transformation that meat undergoes in even that short amount of time. But the world of curing goes far beyond the scope of these recipes. Here's quick overview of what you'll find out there:

- **Chemical cures** involve the use of salt, sugar, nitrates, and/or nitrites. The idea is that by adding a high-enough concentration of these chemicals (yep, salt and sugar are chemicals), you create an environment that is inhospitable to bacterial growth. Salt and sugar we know about. As for the other two, sodium nitrate and potassium nitrate both break down into nitrites as a sausage ages, and these nitrites prevent the growth of certain types of bacteria (mainly the botulism-causing bacteria *Clostridium botulinum*). They play an important role in flavor development as well as helping cured meats to retain a pink hue even after being cooked. Their effect on human health is debated, but it's probably a good idea to keep your intake of them to a reasonable level. None of the recipes in this book call for nitrates or nitrites, as none of the sausages are dry-cured. These days, you may see bacon, salami, and other meats that have traditionally been cured with nitrates marketed as "uncured" and "nitrate free," but it's a bit of a misnomer: they are still cured with salt and often with natural sources of nitrates like celery extract.

- **Smoking meat** with a wood fire also aids in its preservation. During combustion, nitrogen dioxide is formed and released from the smoldering wood. This gas reacts with the water on the surface of a piece of meat to form nitric acid. It's this acid that inhibits the growth of bacteria. Smoke also creates compounds that prevent the oxidation of fat. A smoked pork belly (aka bacon) will go rancid much more slowly than a fresh or simply chemically cured pork belly. Thinner pieces of smoked meat, like jerky or thin hot dogs, can become completely penetrated with nitric acid and other compounds formed in the smoking process, while in larger cuts of meat—say pastrami—these will only penetrate the outer ¼ inch of so of meat. Incidentally, nitric acid prevents the breakdown of muscle pigments during cooking, which is what creates the bright pink "smoke ring" you see on a properly smoked brisket or rack of ribs.

- **Dehydration/fermentation** is the oldest form of curing, and it's almost exclusively applied in conjunction with a chemical cure. Bacteria need water to survive, so fully or partially dehydrated meat will keep for far longer than fresh meat. Many sausages, such as Italian *salami*, *soppressata*, and French *saucisson sec* are dried by hanging them in the open air, during which time friendly yeasts and bacteria partially break down the meat to create a whole host of wonderful, sweet, pungent, and funky flavors. European hams and American country hams lose a great deal of moisture through hanging as well. By slicing meat thin, you can increase its surface area, creating more places for internal moisture to escape—beef jerky and Native American pemmican are made by drying thin strips of seasoned meat and fat.

Some cured meats go through only one or two of these processes. Most fresh sausages go through a chemical cure, with or without a smoking step. Lox is made by chemically curing salmon overnight. Cold-smoke it, and you've got smoked salmon. Bacon is chemically cured and smoked but minimally dried. Italian *pancetta* (cured pork belly) and *guanciale* (cured pork jowl) are cured with salt and nitrates but not smoked and only minimally dried. Some products, like, say, Austrian *speck* or all-American Slim Jims, go through all three processes, making them quite long lasting indeed.

If you want to experiment with your own smoking or dehydrating or move beyond the basic sausage recipes in this book to more complicated emulsified sausages (think hot dogs, bologna, or *mortadella*), I'd suggest checking out Michael Ruhlman and Brian Polcyn's authoritative books on the subject, *Charcuterie* and *Salumi*.

Properly mixed sausage meat should be glossy and sticky looking.

## Sausage: Meat, Fat, and Salt

Flavorings are all well and good, but there are really only three ingredients that are required to make sausage: meat, fat, and salt.

- **Meat,** by which I mean lean muscle mass, makes up the vast majority of a sausage, and the type you use is important. Remember the general rule that the more a muscle is used during the lifespan of an animal, the tougher but more flavorful it will become—so there's an inverse relation between flavor and tenderness. While relatively bland cuts like loin and tenderloin are prized for their tenderness as steaks, chops, or roasts, for sausage making, the tougher cuts are desirable. Grinding renders tenderness a moot point, so you might as well go for the most flavorful cuts you can get. Pork butt (shoulder), with its plentiful marbling and connective tissue, is superior to pork loin or tenderloin. Chicken thighs work better than chicken breasts. Beef short ribs or brisket makes better sausage than strip steak. All of this is good for us, because the ideal sausage cuts are the ones that happen to be the cheapest.

- **Fat** should not be feared. Every great sausage that's ever been made has been great in part because it is dripping in fat—at least 20 percent of it by weight. Fat adds juiciness and mouth-coating richness to a sausage. Fat is where most of the flavor in meat comes from (try eating a bite of beef steak cooked in lamb fat—you'll swear you were eating a lamb chop). Making a sausage without fat is like having brunch without cocktails. Where's the fun in that? If you're scared of fat, just eat less sausage. One bite of a properly fatty sausage is infinitely more enjoyable than any amount of a dry low-fat version.

- **Salt** is by far the most important ingredient in sausage. A sausage simply *cannot* be made without the right amount of salt. The name itself comes from the Latin root for salt (*sal*, hence Spanish *salchicha* or Italian *salciccia*). Without salt, a sausage will not bind properly. It will be mealy and mushy instead of snappy and springy. It will be dry and bland instead of juicy and flavorful.

The most basic recipe for a sausage is as follows: Start with a mixture of 1- to 2-inch chunks of meat and fat in a 4:1 ratio (for extra-juicy sausages, you can use up to 30 percent fat). Add 1 to 2 percent of the total weight of the meat in salt and allow the mixture to sit overnight in the fridge. The next day, grind the meat in a well-chilled meat grinder or food processor. Mix it together carefully with your hands or with the paddle attachment on the mixer. Pack into casings if desired, or shape and cook immediately. That's it. Simple, right? So, how does the alchemy work?

## A Sausage Worth Its Salt

My mother and I are constantly at odds when it comes to food. I tend to like things saltier than the average person, while she can't abide even a pinch of salt. We usually end up compromising somewhere in the middle, but there is one food my mother will never get to enjoy in a low-salt variant: sausage. As anyone who's ever tried to make a sausage without salt will tell you, it simply does not work.

To prove this, I ground two batches of pork, both cut from the same shoulder. The first was seasoned with salt weighed out at 2 percent of the total weight of the meat (you *must* use a scale when making sausage!) and allowed to rest for 8 hours in the refrigerator. The latter was left completely unseasoned. I ground both of them in a meat grinder and formed them into balls. Even before cooking, there was a drastic difference in texture. While the salted meat stuck firmly to itself in a tight ball, the unsalted meat was messy—mushy even.

I then poached them in 180°F water until they reached an internal temperature of 160°F, before cutting them in half. Check out what happened.

You can plainly see that while the salted sausage on the left held together, with a smooth, resilient texture, the unsalted sausage on the right completely crumbled, in much the same way that an overcooked burger will do.

I recorded the weight of each sausage before and after cooking. The sausage that was salted lost only 20 percent of the amount of moisture that the unsalted sausage lost during cooking—as we learned in the lesson on brining (see page 358), salted meat is better at retaining moisture than unsalted meat. Flavorwise, the difference was undeniable: the sausage salted overnight was vastly, unequivocally, night-and-day superior, with a juicy, snappy mouth feel, to its mealy unsalted brother.

## Basic Anatomy

Despite the fact that ground meat looks like, well, ground meat, it's actually got a fairly complex structure. In their whole form, muscle fibers resemble thick bundles of telephone cables, where each individual wire inside the bundle is a single juice-filled strand (known as a *fibril*) constructed of proteins. Chop up these cable bundles, as you do when you grind meat, and you end up with a whole bunch of shorter bundles. Shorter, but still intact: the protein strands are still held tightly within.

When salt is applied to the meat, at first some of the juices contained within the muscle fibrils are drawn out through the process of *osmosis*. That's the tendency for a solution to travel across a permeable membrane in the direction of lower solute concentration to higher concentration (translation: when there's lot of salt outside of a meat cell and not much inside, water from within the cell will travel out to try and even out the concentration of solutes on the outside and inside). The salt then dissolves in these juices, creating a briny liquid. Certain meat proteins, namely *myosin*, will partially dissolve in the presence of this brine.

Salted meat at 4 hours and 0 hours.

Essentially, the bundles of telephone wires become looser, their ends fraying out. You can see this happening when you let a chunk of salted meat sit. The exterior gets darker and darker as the proteins dissolve. This makes it far easier for the proteins to then cross-link when you knead the ground meat. Indeed, just by touching salted ground meat versus regular ground meat, you can instantly tell the difference: the salted meat is much stickier.

## SAUSAGE WEIGHT LOSS AT 160°F

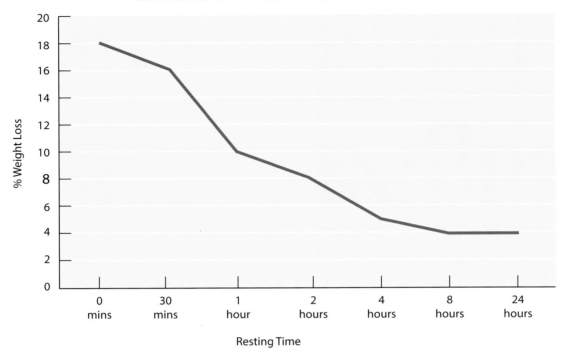

It's this dissolved protein that subsequently cross-links with other proteins to cause a sausage to tighten up, giving it its springy, resilient texture. On top of this, with some of their proteins loosened up by the brine, the muscle fibers are able to hold on to much more moisture than before (see "Experiment: Meat, Salt, and Time," page 502).

### In Good Time

What about timing? Does it matter how long you let your salted meat rest? To test this, I divided one pork shoulder into eight different test batches. The first batch was left unsalted. The remainder were salted for intervals of time ranging from 24 hours all the way down to immediately before grinding. I then cooked all the different ground meats in vacuum-sealed bags in a 160°F water bath and weighed them after draining them.

As you can see in the chart above, there's a pretty clear advantage to letting your meat rest before grinding and forming sausages. A wait of 2 hours saves you half of the juices that would otherwise be lost, while 4 hours saves you a full 75 percent. Not bad. Beyond 8 hours or so, the changes become incremental, shaving off a half percentage point or two before finally maxing out at around a 3.6-percent moisture-loss level after a few days of salting.

### Fear the Smear

A meat grinder is the best way to grind meat for sausage (though a food processor will do just fine), but it's not as simple as throwing chunks of salted meat into the hopper. Before you start grinding, you must remember the most important rule in sausage making: everything must stay cold at all times. Why, you ask? Let me warn you that the picture you are about to see is not pretty. I'm almost afraid to show it to you.

That is sausage meat that was ground in a warm meat grinder. Appetizing, right? The problem is with the fat. As fat gets warmer, it gets softer. Rather than cutting cleanly through the grinder, warm fat will smear, turning your fine grind into a mushy paste. Try cooking up this paste, and rather than staying in your sausage in nice juicy little pockets like good fat should, the smeared fat will flow out of your meat in vast rivulets, leaving nothing behind but dry, mealy, lean meat.

To prevent this from happening, it is vitally important that both the meat and the grinder be cold. If the meat for my sausage has not come straight out of the coldest part of my fridge, I spread it on a rimmed baking sheet and place it in the freezer for 15 minutes before attempting to grind it.

Equally important is the proper mixing of fat, meat, and seasoning. Because well-ground meat is chopped, the tiny individual pieces of meat don't actually get much of a chance to rub up against one another and develop strong protein bonds through the dissolved myosin. So, after your meat is ground, it needs to be kneaded, precisely like a ball of dough. The metaphor is especially apt: the purpose of kneading dough is to build up bonds between flour proteins to create better structure, while the purpose of blending sausage meat is to build up bonds between meat proteins to create better structure. That it also helps the flavorings to become more evenly distributed is an added bonus.

The absolute easiest way to do this is in a stand mixer using the paddle attachment, though a large metal bowl and bare hands are much more fun.

However you do it, just make sure to work quickly. You'll need to mix the sausage meat for at least a minute or two, until it is sticky and tacky instead of loose or crumbly, but if it threatens to get too warm, throw it in the freezer for a few minutes to cool it off—you've come too far to ruin your sausage now!

## Flavoring

For the simplest sausages, all you need is pork shoulder and salt. But the beauty of a sausage is that, unlike a solid piece of meat, you can build your flavorings directly into it. And now that you know the keys to good texture and moisture, you are free to flavor your sausages however the heck you'd like. You can even make them with whatever meat you'd like, though here's a word of warning: pork fat is the best fat to use for sausages, even when using other meats. Highly saturated fats like beef or lamb become waxy and will form an unpleasant coating inside your mouth as they cool. On the other end of the spectrum, chicken or duck fat is almost completely liquid at room temperature. Only pork fat is firm but not waxy at room temperature, and it has a relatively neutral flavor, making it the ideal fat for any number of sausages. Whenever I make, say, a nice lamb merguez (a North African sausage flavored with harissa), I'll combine my lamb with some pork back fat or even nice fatty bacon (making sure to adjust the salt level accordingly to account for the salt in the bacon). Same for my venison sausage, which I make from the organic proceeds of my annual hunting trip.

I usually try to limit flavoring ingredients to about 2 percent of the total weight of the sausage, often far less. I've included a number of recipes in this chapter to get you started and to provide basic guidelines for some of the more popular seasonings, but really, your imagination and your palate are the only things restricting you in this realm. Experiment a bit and see what you can come up with.

If you're afraid to inadvertently ruin an entire batch of sausage meat with overzealous seasoning, start by scaling down: season just a tiny portion of your meat, then cook a small patty of it in a skillet or in the microwave (I usually just throw a quarter-sized patty on a microwave-safe plate and zap it for about 15 seconds, until it's cooked through). This'll allow you to taste and adjust the seasoning as necessary before any full-on commitment.

## To Stuff or Not to Stuff

There's no two ways about it: stuffing sausages is time-consuming. If I want to make a big batch of casing-stuffed sausage, I'll generally devote at least a couple hours of leisurely kitchen time to the task—and that's coming from someone who practiced stuffing sausages in a professional setting for many years. Complete amateurs can expect to mess up a at least a few times, ending up with burst casings, unevenly shaped sausages, and meat-smeared clothes before getting it right.

Most meat grinders and attachments come with funnels designed for stuffing sausages. They will work in a pinch but can be a real headache to use. The main problem is that they don't push the meat through forcefully enough, so stuffing the sausages takes five or ten times longer than it should. All the while, the meat is slowly warming up. I've had better luck stuffing sausage with a pastry bag (this requires two people—one to squeeze the bag, the other to pull the casing off the end as the meat comes out), but if you're really serious about sausage making, you'll want a piston-based stuffer that pushes the meat out with a lever rather than trying to force it

out with a screw. The result is faster, tighter sausages with fewer air bubbles.

Most good butchers will sell you hog or lamb casings packed in salt. To use them, rinse them inside and out with cold water and let them soak in a bowl of cold water for at least half an hour. Then open up one end of a casing and thread it over the end of the sausage stuffer, leaving about 6 inches hanging off the end. Slowly extrude the sausage meat, using one hand to guide the casing off the end of the stuffer. When stuffing sausages, you want the casing to be full but not too tight. Get it too tight, and your sausage will burst when you try and twist it off into links or when the casing shrinks during cooking.

Once you've extruded the meat into the casing, tie off each end with a piece of butcher's twine or knot it, then twist it into individual links of whatever size you'd like. Tie a small piece of twine around each joint to keep the links tightly sealed.

Don't want to bother with all this fuss? Don't worry, you don't have to. Casings are great because they create a natural cooking vessel for your sausage, as well as add plenty of texture and snap, but well-made sausage can do just fine on its own, formed into patties or logs. When shaping casing-less sausage, keep a bowl of cold water nearby to moisten your hands. Wet hands make handling sausage much simpler.

Below you'll find a few recipes for basic sausage seasoning mixes. While most of the recipes in this book are American in origin or at least seen through an American lens, I've included a few sausage recipes from around the world, because once you know how to make good sausage, it seems such a waste not to get a variety of simple recipes under your belt. After that, we'll talk about cooking them.

If using casings, all sausages should be stuffed into hog casings unless otherwise noted.

## USE THE METRIC SYSTEM!

You may notice that while many of the recipes in this book are written in standard U.S. measures (that is, cups, pounds, and ounces), you'll occasionally see me giving metric units (grams and liters). Why the inconsistency?

While imperial units are the preferred unit in the United States and work perfectly well for recipes in which precision is not of utmost importance (say, making a quick pan sauce or a batch of scrambled eggs), for certain recipes, generally baking and charcuterie, a little too much salt or a bit too much water can spell the difference between success and failure. It's at times like this that the metric system excels. Why? Several reasons.

- **It's more precise.** Right off the bat, your basic unit of weight in the metric system—the gram—is more precise than your basic unit of weight in the imperial system—the ounce. This means that measuring small amounts on a scale is more accurate.
- **It's easier to work out percentages.** Say you have a sausage recipe that calls for 1.5 percent salt by weight and 2 pounds of meat. (I picked a nice round number to make it even easier.) How much salt should you add? The math ain't easy, is it? Even when you *do* arrive at the right answer—0.48 ounce—how are you supposed to measure a strange number like that? With the base-10 metric system, figuring out percentages is essentially built in. Got 100 grams of meat? That's 1.5 grams of salt. 1,000 grams of meat? 15 grams of salt. 200 grams of meat? 3 grams of salt. See how easy that is?
- **It makes scaling a snap.** Sometimes you might feel like making a huge batch of bread dough or a small batch of sausage. With the imperial system and its 16 ounces to a pound, it's not easy to scale even a basic recipe up or down. With the metric system, all of your units can be scaled up and down with ease. Doubled, halved, tripled, no matter what, they consist of the base-10 math that we're used to working with.

# EXPERIMENT:
## Meat, Salt, and Time

**So I've come out and said that good sausage can't be made without salt and that the salt takes some time to work its magic. But you don't have to take my word for it, prove it to yourself. Here's how:**

## Materials

- 1 pound boneless pork shoulder, cut into 1-inch chunks
- Kosher salt

- A meat grinder or food processor, grinder parts or processor bowl well chilled

## Procedure

Divide your meat mixture into thirds and place each portion into a zipper-lock plastic bag. Add 0.1 ounce kosher salt (about 1 teaspoon) to one bag and toss until the salt is evenly distributed. Seal all three bags and refrigerate overnight.

The next day, add 0.1 ounce salt to one of the unsalted bags and toss the meat to coat it evenly. Then immediately grind all three batches of meat, one after the other, and mix each one separately in a bowl.

Weigh out 1 ounce of meat from each batch and form it into a small patty, then fry all three side by side in a skillet. Weigh the patties after cooking and take note of their weight loss. Finally, taste, taking note of both texture and flavor.

## Results

Even before beginning cooking, you'll notice that the meat that was salted the night before is significantly stickier. This is the dissolved myosin cross-linking and creating a tighter protein matrix. Because salt also loosens muscle fibers and allows them to retain more moisture, you'll find that the batch that was salted overnight will retain between 10 to 20 percent more moisture than the one salted just before cooking or the unsalted one.

When you taste the patties side by side you'll see that the one that was salted overnight has a springy, juicy texture and is quite flavorful. The one salted just before grinding will also be flavorful but will have a looser texture and have lost more juices as it cooked. Finally, the one with no salt will be bland and have a loose mushy or crumbly texture, more like hamburger than sausage. Not only does salt flavor meat, but in the case of sausage, it enhances its texture by dissolving muscle proteins, allowing them to cross-link and provide resilience and spring to the mix, as well as loosening muscle fibers to allow them to retain more moisture as they cook.

**Moral of the story:** Salt your sausage meat at least a night in advance.

# BASIC HOMEMADE SAUSAGE

**There are four basic steps to making any type of sausage:**

1. **Salt and season**. Add aromatic flavorings and salt to the chunks of meat and fat. As discussed on page 496, salt is absolutely essential and should be measured out to between 1 and 2 percent of the total weight of the meat and fat.

2. **Rest**. Resting allows time for the salt to break down some of the meat proteins, getting them ready to bind to each other when the meat is ground and blended. A proper rest should be between 12 and 24 hours.

3. **Grind**. Pass the rested seasoned meat through a grinder.

4. **Blend**. Blend the sausage meat together, preferably in a stand mixer with a paddle attachment. This step will help dissolved proteins to cross-link with each other, allowing them to form a sticky matrix (much as gluten develops in kneaded bread dough) that will trap in moisture and fat and give the sausage resilience and bounce.

*continues*

This is the simplest sausage you can make and frankly is quite boring—it's intended to be merely a base to which you can add flavorings, either those I suggest or any that suit your fancy. Just remember, the important factors are fat and salt content. As long as those remain the same and you take care to cure, grind, and mix properly, you can add flavorings to your heart's content without affecting the great texture of your sausage.

**NOTE:** To make sausage with preground pork, combine the salt and any other seasonings with the pork and allow to sit in a sealed container in the refrigerator for 12 to 24 hours. The next day, massage the meat for 5 minutes by hand or mix in a stand mixer fitted with the paddle attachment at medium speed until homogeneous and tacky, about 2 minutes. You can replace a few ounces of the shoulder meat with back fat or cubed bacon if you want to increase the fat content.

### MAKES 1 KILOGRAM (ABOUT 2 POUNDS 3 OUNCES)

1 kilogram (about 2 pounds 3 ounces) pork shoulder with at least 20% fat content, cut into rough 1-inch chunks

15 grams (about 0.5 ounce/1½ tablespoons) kosher salt

Seasoning as desired (recipes follow)

1. Combine the meat, salt, and seasonings in a large bowl and toss with clean hands until homogeneous. Transfer to a gallon-sized zipper-lock bag and refrigerate for at least 12 hours, and up to 24 hours.

**TO GRIND WITH A STAND MIXER ATTACHMENT**

2. Place the feed tube, shaft, plates, die, and blade of the grinder in the freezer for at least 1 hour. Then attach the grinder to your mixer and fit it with the ¼-inch die. On medium speed, grind the pork mixture into the mixer bowl. Feed a crumpled paper towel through the grinder to force out any remaining bits of sausage. With the paddle attachment, mix the sausage on medium-low speed until homogeneous and tacky, about 2 minutes. Shape and cook as desired.

**TO GRIND WITH A FOOD PROCESSOR**

2. Place the processor bowl and blade in the freezer for 15 minutes. Working with 200 grams (about 7 ounces) at a time, place the seasoned pork in the bowl of the food processor and pulse until finely ground, about 15 short pulses. Transfer to a large bowl. Repeat until all the pork is ground. Using clean hands, knead the pork until it is homogeneous and tacky, about 5 minutes. Shape and cook as desired.

3. Uncooked sausage will keep for up to 5 days in the refrigerator.

### Seasoning Mix for Garlic Sausage

*This makes a basic sausage, great on its own, but I especially love it served with simmered French lentils.*

3 medium cloves garlic, minced or grated on a Microplane (about 1 tablespoon)
2 teaspoons freshly ground black pepper

### Seasoning Mix for Sweet or Hot Italian Sausage

*This is a flavoring blend for classic red-sauce–Italian-joint sweet or hot sausage. This is the sausage you want to cook with your broccoli rabe (page 696), to scatter in juicy chunks over pizza, or to grill and serve with peppers and onions in hot crisp hoagie rolls at your next cookout.*

2 medium cloves garlic, minced or grated on a Microplane (about 2 teaspoons)
2 tablespoons fennel seeds
1 teaspoon dried oregano
¼ teaspoon freshly grated nutmeg
1 teaspoon freshly ground black pepper
1 tablespoon red wine vinegar
2 tablespoons red pepper flakes (for hot sausage)

### Seasoning Mix for Bratwurst-Style Sausage

*Use this blend for the classic German sausage that is a natural with sauerkraut on the grill or seared and served with hot grainy mustard.*

3 medium cloves garlic, minced or grated on a Microplane (about 1 tablespoon)
1½ teaspoons freshly grated nutmeg
½ teaspoon ground ginger
1 teaspoon freshly ground black pepper
½ cup sour cream or crème fraîche

*continues*

## Seasoning Mix for Mexican Chorizo

*Unlike its dry-cured Spanish counterpart, Mexican chorizo is a fresh sausage, seasoned with warm spices and vinegar. Tart and hot, it's great as a taco filling, on top of nachos, cooked and stirred into a cheese dip, or simmered, sliced, and served with beans.*

2 medium cloves garlic, minced or grated on a Microplane (about 2 teaspoons)

3 tablespoons red wine vinegar

1 tablespoon paprika

½ teaspoon cayenne pepper

¼ teaspoon ground cinnamon

¼ teaspoon ground cloves

1 teaspoon ground cumin

1 teaspoon dried oregano

½ teaspoon freshly ground black pepper

## Seasoning Mix for Merguez-Style Lamb Sausage

*Merguez is a North African sausage that is traditionally made with lamb. To use lamb, substitute 710 grams (25 ounces) trimmed lamb shoulder plus 200 grams (7 ounces) pork fat (belly or back fat) in place of the pork shoulder. Harissa is a North African spice paste available in specialty grocers or online. I use DEA brand, which is more vegetal and less overtly spicy than some. Merguez is traditionally stuffed into slender lamb casings, but it works great as a free-form sausage.*

Free-form merguez on the grill.

**NOTE:** Sumac is the ground dried berries of a flowering plant indigenous to Africa and North America; it has a tart, lemony flavor.

3 medium cloves garlic, minced or grated on a Microplane (about 1 tablespoon)

3 to 4 tablespoons harissa (or more, depending on how spicy you like it)

1 teaspoon dried oregano or 1 tablespoon minced fresh oregano

2 teaspoon fennel seeds

1 tablespoon sumac (optional; see Note above)

1 teaspoon freshly ground black pepper

# MAPLE-SAGE BREAKFAST SAUSAGE

This is a classic breakfast sausage with a sweet and savory blend of maple syrup, sage, and two types of pepper. You'll never have to say the words "Jimmy Dean" again. Great as patties in a breakfast sandwich, as links to dip into your eggs, or crumbled into white gravy (see page 164) to slather over your biscuits. Breakfast sausage should be stuffed into lamb casings or formed by hand. Starting the sausage off with bacon not only adds some sweet smoky flavor, it also ensures a more cohesive structure in the finished product, since bacon is already cured.

NOTE: This recipe can also be made with preground pork. Increase the pork to 2 pounds, the salt to 0.32 ounce (about 2¾ teaspoons), and omit the bacon. Combine all the ingredients and allow to rest in the refrigerator for at least 1 hour, preferably overnight.

## MAKES 1 KILOGRAM (ABOUT 2 POUNDS 3 OUNCES)

680 grams (about 1½ pounds) pork shoulder, trimmed and cut into 1-inch cubes (see Note above)

320 grams (about 11 ounces) slab bacon, cut into 1-inch cubes

15 grams (about 0.5 ounce/1½ tablespoons) kosher salt

2 medium cloves garlic, minced or grated on a Microplane (about 2 teaspoons)

2 tablespoons maple syrup

1 teaspoon red pepper flakes

2 teaspoons ground sage

½ teaspoon dried marjoram

1 teaspoon freshly ground black pepper

Lamb casings (optional)

1. Combine the meat, salt, garlic, syrup, and seasonings in a large bowl and toss with clean hands until homogeneous. Transfer to a gallon-sized zipper-lock bag and refrigerate for at least 12 hours, and up to 24 hours.

### TO GRIND WITH A STAND MIXER ATTACHMENT

2. Place all the parts of the grinder in the freezer for at least 1 hour prior. Attach the grinder to your mixer and fit it with the ¼-inch die. On medium speed, grind the pork mixture into the mixer bowl. Feed crumpled a paper towel through the grinder to force out any remaining bits of sausage. With the paddle attachment, mix the sausage on medium-low speed until homogeneous and tacky, about 2 minutes. Shape and cook as desired.

### TO GRIND WITH A FOOD PROCESSOR

2. Place the processor bowl and blade in the freezer for 15 minutes. Working with 200 grams (about 7 ounces) at a time, place the seasoned pork in the bowl of the food processor and pulse until finely ground, about 15 short pulses. Transfer to a large bowl. Repeat until all the pork is ground. Using clean hands, knead the pork until it is homogeneous and tacky, about 5 minutes. Shape and cook as desired.

3. Uncooked sausage will keep for up to 5 days in the refrigerator.

# { COOKING SAUSAGES }

I'm sure you have been to at least a couple of those weekend warrior cookouts. You know, the kind where the host builds a gigantic fire made up of 75-percent lighter fluid in the grill, barely waits for it to die down, and then throws on a few bratwursts to cook, haphazardly flipping them to and fro with a big fork and keeping a lazy eye on them as he attends to the more important matter of cold beer. By the time the dessicated blackened carcasses come off the grill, the only consolation is that the beer has at the very least deadened your senses to the point that you can manage to choke them down with plenty of mustard and ketchup to lubricate them.

OK, maybe a bit of an exaggeration, but people seem to be under the impression that a sausage is less delicate, less prone to overcooking than, say, a good steak. And this is true—to a degree. Because it is cured with salt, a sausage essentially comes pre-brined. When cooked to a given temperature, it'll retain more moisture than unsalted meat cooked to the same temperature. But that doesn't mean you shouldn't treat it with care. How you cook your sausage depends on whether or not it's stuffed into a casing or formed by hand.

## Cooking Sausage Links Indoors

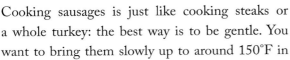

Cooking sausages is just like cooking steaks or a whole turkey: the best way is to be gentle. You want to bring them slowly up to around 150°F in

the center. Indoors, it's possible to do this in a dry skillet with a lid over low heat or in the oven, but here's the most foolproof method to cook your sausages: Place your sausages in a pan or a pot full of cold water and place it over a medium-high burner. Allow the water come up to a bare simmer, then turn off the heat. Let the sausages poach until they reach an internal temperature of 140° to 150°F. Of course, you don't have to cook your sausages in water. If you plan on serving them with a flavorful sauce or other accompaniment—say, sauerkraut or a spicy tomato sauce—then you can simmer them directly in that liquid for better flavor exchange.

Now, you could eat the sausages as is, and they'd be totally delicious, but usually you want to get some good browning on the exterior. If you're indoors, do this in a hot skillet with melted butter or oil. Add the hot cooked sausages and sauté them briefly, just until both sides are browned. You should be able to do this in just a few minutes, during which time the interior temperature of the sausage will rise to around 160°F, resulting a sausages that are perfectly cooked both inside and out.

## Cooking Sausage Links on the Grill

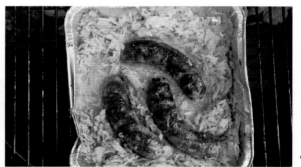

How does this method translate to the grill? There are a couple of options. The simplest is to build a two-zone indirect fire with all the coals placed under half of the grill (or half the burners shut off). Place the sausages on the cooler side of the grill, cover it, and cook until the sausages reach 140° to 150°F. Then transfer them over to the hotter side for some last-minute browning.

An even better but slightly more involved method is to partially fill an aluminum foil pan with moist, flavorful ingredients—say, sauerkraut and its liquid or some sliced onions and apples with a cup of beer—place the sausages in it, and set it on the hot side of the grill. As the contents come to a simmer and steam, the sausages will cook through gently. Once they reach 140° to 150°F, slide the pan over to the cooler side of the grill, sear the sausages directly over the coals, and then return them to the pan to stay warm until ready to serve. While grilling gently is far better than grilling hard and fast, the pan-grill method is even better, with a moisture-loss savings of about 50 percent compared to high-heat grilling.

## WEIGHT LOSS BY PERCENTAGE OF STARTING WEIGHT

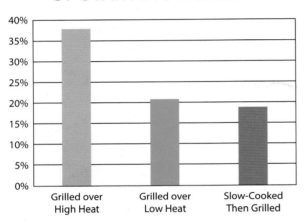

Just as with any meat, it's important to allow sausages to rest after cooking in order to prevent juices from escaping when you cut them. If all went well, your grilled sausages will be juicy and moist from the edges to the center and fairly evenly cooked, with a reasonable amount of char on the outer surface and a faint pink smoke ring around the edges.

### Cooking Free-Form Sausages

Free-form sausages are even easier to cook than links, though not quite as exciting. Because you can form them into thin patties or slender logs, cooking them evenly is not as difficult. A patty or a thin log can be cooked just like a hamburger, bearing in mind that you want it to come up to 160°F in the center to avoid mushiness. Because sausages almost always have a high fat content, you must be careful to keep an eye out for flare-ups if cooking on a grill over direct heat. The easiest way to deal with a flare-up is to just move the sausages away until the flames die down. Covering the grill will also help reduce flare-ups by starving the coals of oxygen.

For fat log-shaped free-form sausages, you can gently lower them into barely simmering water and poach until cooked through, then finish off in a hot skillet with butter or oil, or on the grill, just like an encased sausage.

# THE WRONG WAY TO GRILL SAUSAGES

Here are a few things that might go wrong when you grill sausages.

### 1. The Chest-Burster

A burnt and busted-open casing, sooty flavor, juices lost to the grill gods.

**What happened:** This is what happens when you throw a sausage over the highest-possible heat. Just like other meats, sausages contract as they cook, and in proportion to what temperature they are cooked to. Cook a sausage over high heat, and the casing and the outer layers will quickly get very hot, causing them to contract a great deal. Meanwhile, the raw meat in the center has not contracted at all.

What happens next is sort of like what happens to the Incredible Hulk, but instead of the Hulk growing faster than his clothes, imagine his clothes shrinking in proportion to his body. The casing and outer layers crack and burst open. Liquefied fat and meat juices expelled from the center pour out onto your fire, causing it to flare up, leaving a sooty deposit all over your sausage. The result is an acrid-tasting sausage with a dry, juiceless center.

*continues*

## 2. The Two-Fer

An inedible raw center and a burnt, cracked exterior.
Simultaneously overcooked and undercooked. Not good.

What happened: Another product of high-heat cooking. This time you wised up and put the sausage on a moderately hot grill. But you still end up with a sausage that cooks too fast, overcooking the exterior before the center has a chance to come up to temperature.

## 3. The Unloved Grandmother

OK, so this time you take it to the opposite extreme, cooking the sausage the entire way through on the cooler side of the grill. You get a tiny bit of browning, with no bursting at all, and when you pull it off the heat, it looks plump and juicy as can be—but within moments, it deflates like a sad balloon, a wrinkled, dry shell of its former self.

What happened: With not enough heat, by the time you get any significant browning on the exterior, the interior layers have already overcooked. Steam and expanded muscle tissue will give the sausage a plump appearance while it's still hot, but as soon as it comes off the grill and cools slightly, it shrivels up.

# GARLIC SAUSAGE
## WITH LENTILS

**These sausages are cooked with the simmer-then-fry technique to get them perfectly done both inside and out.**

### SERVES 4 TO 6

4 tablespoons unsalted butter

1 medium onion, minced (about 1 cup)

1 small carrot, peeled and cut into fine dice (about ½ cup)

2 medium cloves garlic, minced or grated on a Microplane (about 2 teaspoons)

8 ounces (about 1 cup) French Puy lentils

2 cups homemade or low-sodium canned chicken stock

Kosher salt and freshly ground black pepper

1 recipe Garlic Sausage (page 505), stuffed into casings (6 to 8 links), or about 2 pounds store-bought sausages

¼ cup minced fresh parsley

2 tablespoons extra-virgin olive oil

1 tablespoon red wine vinegar

1. Melt 3 tablespoons of the butter in a large saucepan over medium-high heat. Add the onion and carrot and cook, stirring frequently, until softened but not browned, about 4 minutes. Add the garlic and cook, stirring, until fragrant, about 30 seconds. Add the lentils and chicken stock and bring to a simmer, then reduce the heat to low, cover, and cook, stirring occasionally, until the lentils are fully tender, about 45 minutes. Remove the lid and season to taste with salt and pepper. Keep warm over low heat to evaporate any remaining liquid, stirring occasionally.

2. Meanwhile, place the sausages in a 12-inch skillet and cover with water. Bring to a bare simmer over high heat, cover, remove from the heat, and let poach until an instant-read thermometer inserted into the thickest part of the sausages registers 140° to 145°F, about 10 minutes.

3. Drain the sausages and set aside. Return the skillet to medium-high heat. Add the remaining tablespoon of butter and heat until the foaming subsides and the butter begins to brown. Lower the heat to medium, add the sausages, and cook, turning occasionally with tongs, until well browned on both sides, about 5 minutes. Transfer to a cutting board, tent with foil, and allow to rest for 10 minutes. By this point, the lentils should be ready.

4. Stir most of the parsley, olive oil, and vinegar into the lentils, reserving some for garnish. Slice the sausages on a bias if desired. Serve the lentils topped with the sausages, sprinkled with the remaining parsley and drizzled with olive oil and vinegar.

# GRILLED ITALIAN SAUSAGE
## WITH ONIONS AND PEPPERS

**This is the simplest method of cooking sausages on the grill: indirect heat to cook through, followed by direct heat to color the exterior.**

### SERVES 4 TO 6

1 recipe Sweet or Hot Italian Sausage (page 505), stuffed into casings (6 to 8 links), or about 2 pounds store-bought sausages

2 large onions, cut into ½-inch rounds

3 bell peppers (mixed colors), split into quarters, cores and seeds removed

2 tablespoons olive oil

Kosher salt and freshly ground black pepper

4 to 6 crusty hoagie rolls

Wooden skewers

1. Ignite a large chimneyful of charcoal. When the briquettes are coated in gray ash, pour out and distribute evenly over one side of a charcoal grill. Set the cooking grate in place. Or, if using a gas grill, heat one set of burners to high and leave the others off. Scrape the grill grates clean.

2. Place the sausages on the cooler side of the grill, cover, and cook until an instant-read thermometer inserted into the thickest part of the sausages registers 140° to 145°F, about 15 minutes. Uncover the grill, move the sausages to the hot part of the fire, and cook, turning occasionally with tongs, until browned on all sides, about 2 minutes. Transfer to a cutting a board or large plate, tent with foil, and allow to rest for 10 minutes.

3. Meanwhile, skewer each onion horizontally with a wooden skewer to keep the rings in place. Skewer the peppers. Brush the onions and peppers with the olive oil and season with salt and pepper. Place on the hot side of the grill and cook, flipping occasionally, until tender, about 10 minutes. Transfer to a cutting board and discard the skewers.

4. Split the onion disks in half and cut the peppers into strips. Toss to combine. Meanwhile, toast the rolls on the grill until lightly charred, about 3 minutes.

5. Serve the sausages in the rolls with the onions and peppers.

# GRILLED OR PAN-ROASTED MERGUEZ
## WITH YOGURT, MINT, AND MOROCCAN SALAD

**Because of their slender girth, merguez sausages can be cooked directly over a hot grill fire or in a skillet with no parcooking required.**

**NOTE:** You can also make this recipe with bulk sausage. Form the meat into cylinders around wooden or metal skewers.

### SERVES 4 TO 6

1 large tomato, cut into ½-inch dice

1 large cucumber, peeled, halved lengthwise, seeded, and cut into ½-inch dice

1 small red onion, thinly sliced

Kosher salt and freshly ground black pepper

1 cup full-fat yogurt, preferably Greek-style

1 tablespoon lemon juice (from 1 lemon)

¼ cup minced fresh mint

1 recipe Merguez-Style Lamb Sausage (page 506), stuffed into casings (12 to 16 links), or about 2 pounds store-bought sausages

1 tablespoon vegetable oil (if pan-roasting)

8 to 12 pita breads or Grilled Flatbreads (page 517 or 518)

1. Combine the tomato, cucumber, and onion in a medium bowl. Season to taste with salt and pepper and allow to rest at room temperature for 45 minutes.

2. Meanwhile, combine the yogurt, lemon juice, and mint in a small bowl and whisk together. Season to taste with salt and pepper and refrigerate until ready to use.

#### TO COOK ON THE GRILL

3. Ignite a chimneyful of charcoal. When most of the briquettes are coated in gray ash, pour out and distribute evenly over one side of the grill. Set the cooking grate in place, cover the grill, and preheat for 5 minutes. Or, if using a gas grill, heat it to high. Scrape the grill grates clean. Cook the sausages directly over the hot part of the fire, turning occasionally with tongs, until an instant-read thermometer inserted into the thickest part of the sausages registers 150°F, about 8 minutes. Transfer to cutting board, tent with foil, and allow to rest for 5 minutes.

4. If using pita bread, toast on the grill, about 20 seconds per side. Stack on a plate and cover with a clean dish towel.

#### TO COOK ON THE STOVETOP

3. Heat the oil in a large nonstick or cast-iron skillet over medium heat until shimmering. Add the sausages and cook, turning occasionally with tongs, until an instant-read thermometer inserted into the thickest part of the sausages registers 150°F, about 8 minutes. Transfer to cutting board, tent with foil, and allow to rest for 5 minutes.

4. If using pita bread, wipe the skillet clean with a paper towel and heat over medium heat. Toast each pita bread until warm, about 20 seconds per side. Stack on a plate and cover with a clean dish towel.

5. To serve, drain the tomato salad. Serve the sausages with the warm bread, tomato salad, and yogurt sauce.

# EASY GRILLED NAAN-STYLE FLATBREAD

**Like all yeast-raised breads, this one takes a little planning to allow time for it to rise properly. However, the actual cooking process is a snap. For an even easier version, see Even Easier Grilled Flatbread (page 518), which uses baking powder as its leavener.**

## MAKES 12 BREADS

### For the Dough

**600 grams (about 21 ounces/4 cups) bread flour**

**7 grams (1 packet/about 2 teaspoons) instant or rapid-rise yeast**

**12 grams (about 0.4 ounce/2½ teaspoons) kosher salt**

**24 grams (about 0.8 ounce/5 teaspoons) sugar**

**360 grams (12.75 ounces/1½ cups plus 1 tablespoon) full-fat yogurt or whole milk, or as necessary**

**8 tablespoons (1 stick) unsalted butter, melted**

### TO MAKE THE DOUGH

1. Combine the flour, yeast, salt, and sugar in the bowl of stand mixer and whisk until combined. Add the yogurt and knead with the dough hook on low speed until the dough comes together into a smooth ball; it should stick slightly to the bottom of the bowl as it is kneaded (add a little more yogurt or milk if necessary). Continue to knead until slightly elastic, about 5 minutes. Cover tightly with plastic wrap and allow to rise at room temperature until roughly doubled in volume, about 2 hours.

2. Turn the dough out onto a floured work surface. Using a bench scraper or a knife, cut it into 12 even pieces. Roll each piece into a ball, then place on a well-floured surface (leaving a few inches of space between them) and cover with a floured cloth. (Alternatively, place each ball of dough in a covered pint-sized deli container.) Allow to rise at room temperature until doubled in volume, about 2 hours.

### TO COOK ON THE GRILL

3. Ignite a chimneyful of coals. When the coals are coated in gray ash, pour out and spread evenly over one side of the grill. Set the grill grate in place, cover, and allow to preheat for 5 minutes. Or, if using a gas grill, heat it to high. Scrape the grill grate clean.

*continues*

4. Working with one ball of dough at a time, with your hands, stretch the dough into an oblong roughly 10 inches long and 6 inches across (or use a rolling pin). Once you have two or three pieces stretched, lay them out on the grill (above the coals) and cook, without moving them, until the tops are bubbled and blistered and the bottoms are charred in spots and light golden brown, 30 seconds to 1 minute. Flip with a large spatula, a pizza peel, or tongs and cook until the second side is charred and browned, another 30 seconds to 1 minute. Remove from the grill and immediately brush with melted butter. Transfer to a large plate and cover with a clean dish towel while you cook the remaining breads, stacking them as you make them.

### TO COOK ON THE STOVETOP

3. Heat a large ridged grill pan over medium-high heat for at least 10 minutes.
4. Meanwhile, with your hands, stretch one ball of dough into an oblong roughly 10 inches long and 6 inches across (or use a rolling pin). Lay it in the grill pan and cook, without moving it, until the top is covered in bubbles and the bottom is charred along the grill marks and pale brown across the rest of its surface, 1 to 1½ minutes. Carefully flip, using a metal spatula or tongs, and cook until the second side is charred and browned, another 1 to 1½ minutes. Remove from the pan and immediately brush with melted butter. Transfer to a large plate and cover with a clean dish towel while you shape and cook the remaining breads, stacking them as you make them.

# EVEN EASIER GRILLED FLATBREAD

**This flatbread can be cooked start to finish in about 30 minutes on the grill or stovetop.**

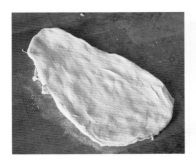  

For the Dough

600 grams (about 21 ounces/4 cups) bread flour

10 grams (about 0.35 ounce/1 tablespoon) baking powder

12 grams (about 0.4 ounce/2½ teaspoons) kosher salt

24 grams (about 0.8 ounce/5 teaspoons) sugar

360 grams (12.75 ounces/1½ cups plus 1 tablespoon) full-fat yogurt or whole milk, or as needed

8 tablespoons (1 stick) unsalted butter, melted

## TO MAKE THE DOUGH

1. Combine the flour, baking powder, salt, and sugar in the bowl of a stand mixer and whisk until combined. Add the yogurt and knead with the dough hook on low speed until the dough comes together into a smooth ball; it should stick slightly to the bottom of the bowl as it is kneaded (add a little more yogurt or milk if necessary). Continue to knead until lightly elastic, about 5 minutes.

2. Turn the dough out onto a floured work surface. Using a bench scraper or a knife, cut it into 12 even pieces. Roll each piece into a ball and cover with a clean kitchen towel or plastic wrap.

## TO COOK ON THE GRILL

3. Ignite a chimneyful of coals. When the coals are coated in gray ash, pour out and spread evenly over one side of the grill. Set the grill grate in place, cover, and allow to preheat for 5 minutes. Or, if using a gas grill, heat it to high. Scrape the grill grates clean.

4. Working with one ball of dough at a time, roll the dough into an oblong roughly 10 inches long and 6 inches across. Once you have two or three pieces rolled out, lay them on the grill (above the coals) and cook, without moving them, until the bottoms are charred in spots and light golden brown, 30 seconds to 1 minute. Flip with a large spatula, pizza peel, or tongs and cook until the second side is charred and browned. Remove from the grill and immediately brush with melted butter. Transfer to a large plate and cover with a clean dish towel while you cook the remaining breads, stacking the breads as you make them.

## TO COOK ON THE STOVETOP

3. Heat a large ridged grill pan over medium-high heat for at least 10 minutes.

4. Meanwhile, working with one ball of dough at a time, roll the dough into an oblong roughly 10 inches long and 6 inches across. Lay it in the grill pan and cook, without moving it, until the top is covered in bubbles and the bottom is charred along the grill marks and pale brown across the rest of its surface, 1 to 1½ minutes. Carefully flip, using a metal spatula or tongs, and cook until the second side is charred and browned, another 1 to 1½ minutes. Remove from the pan and immediately brush with melted butter. Transfer to a large plate and cover with a clean dish towel while you shape and cook the remaining breads, stacking the breads as you make them.

# GRILLED OR PAN-ROASTED BRATWURST
## WITH BEER, MUSTARD, AND SAUERKRAUT

**Either method produces great results for these beer-cooked brats, allowing the flavors of the sausage and beer to mingle and marry.**

### SERVES 4 TO 6

One 1-pound package sauerkraut,
  with its liquid

A few sprigs fresh thyme

½ cup lager-style beer

2 pounds Bratwurst-Style Sausage
  (page 505), stuffed into casings.
  (6 to 8 links), or about 2 pounds
  store-bought bratwurst

Spicy whole-grain mustard

1 tablespoon vegetable oil (if
  cooking on the stovetop)

Crusty hoagie rolls, toasted

### TO COOK ON THE GRILL

1. Ignite a chimneyful of charcoal. When the briquettes are coated in gray ash, pour out and distribute evenly over one side of the grill. Or, if using a gas grill, heat one set of burners to high and leave the others off. Clean the grilling grate.

2. Place the sauerkraut, thyme, and beer in a disposable aluminum foil pan and nestle the sausages in it. Place the pan on the hot side of the grill and cook until simmering, about 4 minutes. Slide to the cooler side of the grill, cover the grill, with the vented part of the lid positioned over the sausages, and cook, with all the vents open, until an instant-read thermometer inserted into the thickest part of the sausages registers 140° to 145ºF, about 15 minutes, turning once halfway through.

3. Remove the lid. Using tongs, remove the sausages from the pan and place on the hot side of grill. Cook, turning occasionally, until well browned and crisp, about 3 minutes total. Return to the pan and allow to rest, uncovered, for 10 minutes, then serve with the toasted rolls.

### TO COOK ON THE STOVETOP

1. Combine the sauerkraut, thyme, and beer in a 12-inch sauté pan and stir well. Nestle the sausages in the mixture and bring to a simmer over medium heat. Reduce the heat to the lowest setting, cover, and cook, turning the sausages occasionally, until an instant-read thermometer inserted into the thickest part of the sausages registers 140° to 145°F, about 12 minutes. Remove the sausages from the pan; keep the sauerkraut mixture warm.

2. Heat the oil in a large nonstick or cast-iron skillet over medium-high heat until shimmering. Add the sausages and cook, turning occasionally; until browned on all sides, about 5 minutes. Return to the warm sauerkraut mixture and allow to rest for 10 minutes, then serve with the toasted rolls.

# GRILLED OR PAN-ROASTED MEXICAN CHORIZO
## WITH SPICY TOMATO-CAPER SAUCE

**Spicy Mexican chorizo combines perfectly with a quick tomato and caper sauce. You can serve this in buns if you'd like, but it'd go equally well scooped on top of some steamed rice or, sliced up, with tortillas.**

### SERVES 4 TO 6

2 tablespoons vegetable oil (only 1 tablespoon if cooking on the grill)

1 large onion, finely sliced (about 1½ cups)

1 tablespoon chili powder

2 teaspoons ground cumin

One 28-ounce can whole tomatoes, drained and broken up by hand into rough chunks

¼ cup capers, rinsed, drained, and roughly chopped

¼ chopped green or black olives

¼ cup chopped fresh cilantro, plus more for garnish

Kosher salt and freshly ground black pepper

1 recipe Mexican Chorizo (page 506), stuffed into casings (6 to 8 links), or about 2 pounds store-bought sausages

Lime wedges (optional)

### TO COOK ON THE GRILL

1. Heat 1 tablespoon vegetable oil in a large saucepan over medium-high heat until shimmering. Add the onion and cook, stirring frequently, until softened, about 4 minutes. Add the chili powder and cumin and cook, stirring, until fragrant, about 1 minute. Add the tomatoes, capers, olives, and cilantro, season to taste with salt and pepper, and remove from the heat. Set aside.

2. Light a chimneyful of charcoal. When the coals are covered with gray ash, pour out and spread evenly over one side of the grill. Set the cooking grate in place, cover the grill, and allow to preheat for 5 minutes. Or, if using a gas grill, heat one set of burners to high and leave the others off. Clean and oil the grilling grate.

3. Transfer the tomato mixture to 10-inch square disposable aluminum foil pan. Nestle the sausages into the mixture. Place the tray on the hot side of the grill and cook until simmering, about 4 minutes. Slide to the cooler side of the grill, cover, with the vents over the sausages, and cook, with all the vented part of the lid positioned open, until an instant-read thermometer inserted into the thickest part of the sausages registers 140° to 145°F, about 15 minutes, turning once halfway through.

4. Remove the lid. Using tongs, remove the sausages from the sauce and place on the hot side of the grill. Cook, turning occasionally, until well browned and crisp, about 3 minutes. Return to the sauce and allow to rest for 10 minutes, then serve, garnished with cilantro and lime wedges if desired.

*continues*

Grilled Mexican Chorizo with Spicy Tomato-Caper Sauce.

**TO COOK ON THE STOVETOP**

1. Heat 1 tablespoon of the vegetable oil in a 12-inch sauté pan over medium-high heat until shimmering. Add the onion and cook, stirring frequently, until softened, about 4 minutes. Add the chili powder and cumin and cook, stirring, until fragrant, about 1 minute. Add the tomatoes, capers, olives, and cilantro and season to taste with salt and pepper.

2. Nestle the sausages into the mixture and reduce the heat to medium. Bring to a simmer, then reduce the heat to the lowest setting, cover, and cook, turning the sausages occasionally, until an instant-read thermometer inserted into the thickest part of the sausages registers 140° to 145°F, about 12 minutes. Remove the sausages from the pan; keep the tomato sauce warm.

3. Heat the remaining tablespoon of oil in a large nonstick or cast-iron skillet over medium-high heat until shimmering. Add the sausages and cook, turning occasionally until browned on all sides, about 5 minutes. Return to the tomato sauce and allow to rest for 10 minutes, then serve, garnished with cilantro and lime wedges if desired.

# GRILLED OR PAN-ROASTED HOT DOGS
## WITH SAUERKRAUT

You may think of hot dogs as your fallback, never-fail, always-OK-but-never-great backyard tubesteak, but a great hot dog can be so much more than that. Just ask anyone from **New Jersey**. Good hot dogs are first and foremost about the dog itself, and try as you might, you're *never* gonna make a hot dog at home as good as you can get from a professional hot-doggery.

Whether you like salty-smoky **New York**–style all-beef franks, or the German-style beef-and-pork franks you find in **Michigan**, or even a neon-red **Red Hot** from the North Country, there's one thing for certain: the best dogs have natural casings. Without the casing, a hot dog has no snap. With no snap, well, . . . what's the point?

How can you tell if a hot dog has a natural casing? There are several things to look for:

- **The label**. Most packages will be labeled "skinless" or "natural casings."
- **Curvature**. A hot dog with a natural casing will have a slight curve to it even before you cook it. This is because lambs' intestines—the casing of choice for hot dogs—are not symmetrical. Dogs with artificial cellulose casings or, even worse, no casing at all will be straight as an arrow.
- **A nipple at either end**. Examine the ends of the hot dogs. If you see a little nipple like the knot in a balloon, you've got yourself a natural-casing frank. If you see a pinched-star-shaped pattern, you're looking at the mark left by a cellulose casing being pinched shut.

**SERVES 4 TO 6**

One 1-pound package sauerkraut, with its juices

8 to 12 natural-casing all-beef hot dogs (such as Boar's Head, Sabrett, or Dietz & Watson)

1 tablespoon canola oil (if cooking indoors)

8 or 12 hot dog buns

Brown mustard

## TO COOK ON THE GRILL

1. Light a chimneyful of charcoal. When the coals are covered with gray ash, pour out and spread evenly over on one side of the grill. Set the cooking grate in place, cover the grill, and allow to preheat for 5 minutes. Or, if using a gas grill, heat one set of burners to high and leave the others off. Clean and oil the grilling grate.

2. Place the sauerkraut and juices in a 10-inch square disposable aluminum foil pan. Nestle the hot dogs into the sauerkraut. Place the tray on the hot side of the grill and cook until simmering, about 4 minutes. Slide to the cooler side of the grill, cover the grill, with the vents positioned over the hot dogs, and cook, with all the vents open, until the hot dogs are heated through, about 10 minutes, turning once halfway through.

3. Remove the lid. Using tongs, remove the hot dogs from the sauerkraut and place on the hot side of the grill. Cook, turning occasionally, until well browned and crisp, about 3 minutes. Return to the sauerkraut.

4. Meanwhile, toast the buns on the grill if desired. Serve the hot dogs and sauerkraut with the buns and mustard.

## TO COOK ON THE STOVETOP

1. Place the sauerkraut and juices in a 12-inch sauté pan, nestle the hot dogs into the sauerkraut, and bring to a simmer over medium heat. Reduce the heat to the lowest setting, cover, and cook, turning the dogs occasionally, until heated through, about 8 minutes. Remove the hot dogs from the pan; keep the sauerkraut warm.

2. Heat the oil in a large nonstick or cast-iron skillet over medium heat until shimmering. Add the hot dogs and cook, turning occasionally, until crisp on all sides, about 3 minutes. Return to the sauerkraut.

3. Meanwhile, toast the buns under a hot broiler if desired. Serve the hot dogs and sauerkraut with the buns and mustard.

# ALL-AMERICAN MEAT LOAF

In his 1958 classic, *365 Ways to Cook Hamburger* (at least, it *should've* been a classic), along with hundreds of recipes for burgers, sauces, soups, meatballs, and casseroles, Doyne Nickerson offers no fewer than seventy recipes for meat loaf. Seventy! A different loaf every night for over two months! Ten loaves apiece for every man, woman, and child on the cast of *Full House*! (Another classic.) Amongst this litany are such colorful offerings as Chili Hot Top Meat Loaf (it's flipped upside down and glazed with Heinz Chili Sauce), Sunshine Meat Loaf (that'd be a loaf topped with ketchup-filled peach halves), and two—count 'em, two—variations on Banana Meat Loaf (one with green bananas mashed into the meat, the other topped with bacon and ripe banana).

With such a varied and prolific precedent set, you may be disappointed to find out that I offer but a single, lonely recipe for plain old all-American meat loaf and not even one recipe that combines ground beef with bananas. But while Nickerson is unparalleled in his prolificacy, I plan on besting him in thoroughness.

You see, Americans are proud of their meat loaf, and rightfully so. It's one of our national dishes and deserves a place up on the pedestal, rubbing shoulders with the likes of hamburgers, barbecue, and hot dogs. I mean, it's a loaf made out of meat. What could be more decadent-yet-comforting than that?

The very best meat loaf should be tender and moist, with a distinctly soft but never mushy texture. "Velvety" and "rich" should come to mind when tasting it, tender enough to slice with a fork but firm enough to pick up that bite without it breaking. It should be a sponge for moisture, oozing juices when you eat it but not leaving a puddle on your plate. It should be deeply rich and meaty in flavor and savory, with just a hint of vegetable undertones to complement and lighten the slice. But make no mistake: meat loaf is about the meat. And, of course, it needs to reheat well for sandwiches.

We already know quite a bit about how ground meat behaves from our adventures with sausage, and we've learned the benefits of grinding your own meat (or, at the very least, having it ground fresh at the butcher). From those starting blocks, arriving at perfect meat loaf is just a short skip and a jump away.

All-beef loaf.

## Meaty Matters

*From left*: ground beef, veal, and pork.

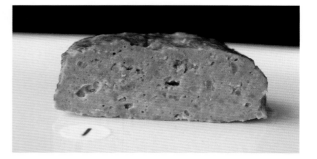

All-pork loaf.

All-veal loaf.

Let's start with the very basics. Anyone who's been to a supermarket has seen those plastic wrapped trays labeled "meat loaf mix," which contain a combination of pork, beef, and veal. Why the mix? What does each of these meats bring to the table? To find out, I made several identical meat loaves using a very simple mix of meat and a few sautéed vegetables (carrots, onions, celery). Each loaf was cooked in a vacuum-sealed plastic bag in a water bath set at precisely 145°F. That way, I was certain that each batch was cooked identically. For my first test, I cooked three loaves: 100-percent beef, 100-percent pork, and 100-percent veal.

After more tastings, including an exclusionary test (beef and pork alone, beef and veal alone, and pork and veal alone), and combining all three, a few things became obvious. Pure beef cooked in meat loaf form loses quite a bit of moisture and acquires a coarse, gritty texture and slightly livery flavor. Pork has a much milder flavor and more fattiness, with a less coarse, softer texture. Compared to beef and pork, veal loses very little moisture at all, and it has a tender, almost gelatinous texture when cooked. However, it's completely lacking in flavor. Why do three different meats cooked in the same manner return such different results?

Well, pigs and cattle differ mostly in their fast-twitch versus slow-twitch muscles (see page 179). Cows are large animals that spend most of their time walking around and grazing, requiring plenty of long, sustained effort from their muscles, which

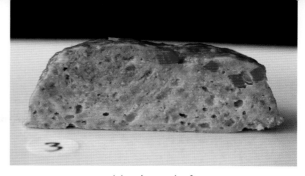

Mixed-meat loaf.

eventually turns them coarse, and flavorful, with a deep red color—a by-product of the oxygenation necessary for them to perform work. Pigs are smaller and less active for sustained periods of time. You may see them trot over to the trough to gorge, but they spend the bulk of their time lying in the mud or in the shade to keep cool. Consequently, their dark slow-twitch muscles are less developed. Instead, you'll find plenty of paler, more fine-textured fast-twitch muscle, as well as a good deal of stored fat. As mentioned earlier, pork fat is also softer than beef fat making it more pleasant to eat at normal serving temperatures. So, by combining beef and pork, you end up with a mix that has the great flavor of beef but an improved texture and softer fat from the pork.

Then what does veal bring to the mix?

The difference between veal and beef is a little more subtle, having to do with the age of the animal. When a cow (or almost any mammal, for that matter) is born, its muscles are not very well developed. Its fat is soft and malleable, its muscles pale and mild-flavored, with a high proportion of soluble collagen, the connective protein that transforms into gelatin as it cooks. It's the underdeveloped musculature that gives veal its tenderness, but it's the gelatin that lends ground veal its ability to retain moisture. How does this work? It helps to think of gelatin molecules as individual links in a very fine wire mesh and individual molecules of water as tiny water balloons. As the collagen is converted to gelatin inside a meat loaf as it cooks, these molecules of gelatin gradually link up with each other, forming a net that traps water molecules, preventing them from escaping. It's this same quality of gelatin that allows you to turn several cups worth of water into a quivering Jell-O mold with just a few tablespoons of powdered gelatin.

Thus the mix. Beef provides robust flavor, pork provides a good amount of tender fat, and veal provides plenty of gelatin to help retain moisture: The mixture provides the optimum balance of flavor, texture, fat content, and moisture-retaining ability. *Or does it?*

## Say No to Veal

Here's the problem with veal: it doesn't taste like much. Sure, it adds gelatin to the mix, but it dilutes the meaty flavor at the same time. It can also be a bit of a pain to seek out (I have to travel all the way to the supermarket by my mom's apartment to get it, which means I've got to visit my mother every time I want veal; this can be problematic). I'd seen a few recipes that suggested replacing the veal with powdered gelatin—an ingredient I always keep in my pantry. I made a couple loaves side by side, one with an equal mix of ground chuck, ground pork, and ground veal, and the second one with a mix of ground chuck, ground pork, and a couple tablespoons of unflavored gelatin hydrated in a bit of chicken stock and cooked until dissolved (I made sure to add the same amount of chicken stock to the first loaf as well). Texturewise, both loaves proved to be moist and tender. Flavorwise, the no-veal loaf had a clear advantage.

Gelatin it is.

## Meat Loaf Binders and Extenders

So, up to now, what we've essentially got is something that's halfway between a burger and a sausage. It's got the basic fat content of a burger, with the key difference being that the salt is mixed right into the meat rather than just seasoning the exterior. We all know what happens when you add salt to meat before mixing it, right? It causes the meat to become sticky and bind with itself as the salt slowly dissolves muscle proteins. But this is not a good thing for meat loaf, where tenderness and a loose, velvety texture are desired above all. We can mitigate those effects by adding the salt immediately before mixing and only mixing as much as necessary, but there are better ways to improve texture—namely, with binders and additives. Let's look at the most common ones to determine what role they play.

**Eggs** are an ingredient in nearly every meat loaf, and they have two distinct roles. Egg yolks, which are mostly water but contain a good amount of protein and fat, add flavor, richness, and moisture. They also help bind the meat together and get the loaf to set in a stable form without the need to overwork the meat. Egg whites have even more water in them, are devoid of any fat at all, and have a very mild flavor. Their main role is to add extra loose proteins to the mix to assist the egg yolks in their quest to add structure without overworking the meat or adding toughness. We'll definitely include them.

**Milk and other dairy products**, like heavy cream and buttermilk, contain both water and fat, adding two types of moisture to our meat loaf. There's a long-held theory that milk can tenderize ground meat, and this is the reason often cited for cooking ground meat in milk to make a Bolognese-style ragù. I'm pretty skeptical about this. Milk is mainly water, with some milk fat and a few proteins thrown in. What could cause it to tenderize meat?

Some sources claim that adding milk limits the cooking temperature to 212°F (the temperature at which water boils), which keeps meat from overcooking. What? Limiting temperature to 212°F? What good does that do? Meat toughens at temperatures a good 70 to 75 degrees below this threshold. Besides, plain old water (which is abundant in the meat and all the vegetables you add to meat loaf) will perform that function just as well. Indeed, cooking three batches of meat side by side, one simmered in milk, one simmered in water, and one allowed to simmer in its own juices, left me with three batches of meat that were equally tough. Fact of the matter is, milk does *not* tenderize meat. The only way to guarantee tender meat is *not to overcook it*. And that's a simple matter of using a thermometer when you bake the meat loaf.

Milk does, on the other hand, add moisture and fat and is worth including for that fact alone. Heavy cream works better. Better still is buttermilk, which has a unique tang that adds depth and complexity to the finished dish.

**Bread crumbs** may, at first glance, seem like an unnecessary extender—something added just to stretch your meat a little bit further—but they are perhaps the most important ingredient of all when it comes to improving the texture of a meat loaf. Aside from absorbing and retaining some moisture as the meat loaf cooks, they physically impede the meat proteins from rubbing up too closely to one another, minimizing the amount of cross-linkage and thus dramatically increasing tenderness. In many ways, the physical structure of a meat loaf is much like the structure of an emulsified sauce stabilized with starch. In the latter case, starch acts like a bouncer, keeping fats from coalescing, while in the former, bread crumbs do the job, keeping meat proteins apart. I found that using crumbs from fresh bread slices ground in the food processor provided better moisture and binding capabilities than dried bread crumbs.

Finally, **mushrooms**, while not necessarily a standard meat loaf ingredient, are an invaluable addition. Why do I include them under binders and extenders rather than lump them in with the aromatics? Because they act much more like bread crumbs than they do like, say, onions. Mushrooms are extremely porous and are full of flavorful liquid. At the same time, they are soft and spongy. Just like bread crumbs, they prevent the meat proteins from interlocking, increasing tenderness while simultaneously adding flavor as they slowly release their liquid. In fact, they're so much like bread that I treat them exactly the same way—grind them in the food processor and add them to the raw mix, no parcooking necessary at all!

So, to summarize, we have the following chart:

| INGREDIENT | EFFECT | HOW TO INCORPORATE |
|---|---|---|
| Egg Yolks | Add richness and moisture, help bind the meat and bread to lend structure without toughness. | Add to the meat mixture. |
| Egg Whites | Bind meat and bread to lend structure without toughness (more effective at binding than egg yolks). | Add to meat mixture. |
| Bread Crumbs | Help retain moisture and physically impede meat proteins from cross-linking, increasing tenderness. | Moisten with milk or stock to create a panade (a mixture of bread and a liquid). Add to meat mixture. |
| Milk (or other liquid dairy) | Adds moisture and tenderizes | Use to soak bread crumbs. |
| Gelatin | Increases the capacity to retain moisture as meat loaf cooks. | Bloom in chicken stock, cook to dissolve, and add to meat mixture (or use to moisten bread crumbs). |
| Mushrooms (chopped) | Physically impede meat proteins from cross-linking, increasing tenderness while simultaneously adding flavor. | Add to meat mixture. |
| Salt | If added too early, it can cause meat proteins to dissolve and cross-link, creating a bouncier, firmer texture. | Add to meat mixture just before mixing and cook immediately. |

## The Key to Great Flavor: A Concentrated Flavor Base

With the meat mix and the texture of the loaf squared away, I shifted my focus to flavorings.

The base of carrots, onions, and celery made sense to me—the three vegetables are a classic addition to meat dishes and sauces for a reason—but when they are simply diced and added to the meat mix, their texture doesn't quite work in meat loaf; I found it interfered with the velvetyness I desired. How to deal with this? Easy, just chop them finer and soften them. I used the food processor (already on my countertop to make the bread crumbs and chop mushrooms) to chop them into small pieces before sautéing them in butter until tender, adding a touch of garlic and Spanish paprika as well.

We've got the vegetables in there, now for a few ingredients to up the meaty backbone of the loaf, namely deploying my trusty umami bombs: anchovies, Marmite, and soy sauce. All three of these ingredients are rich in glutamates and inosinates, chemical compounds that trigger signals that tell our brains we're eating something savory and meaty. They make the meat loaf taste meatier without imparting a distinct flavor of their own. After sautéing all the ingredients for my flavor base together— the vegetables and the umami bombs—I added some chicken stock and buttermilk, along with soft-

ened gelatin, and reduced to a concentrated liquid simply bursting with flavor.

Mixing this flavor base into my meat produced a mixture wetter than any other meat loaf mix I'd seen. This led to a moister end product (that retained moisture with the help of the gelatin), but it proved problematic when shaping the loaf. I *could* bake it in a loaf pan, but I prefer making free-form loaves on a baking sheet to maximize surface area for flavorful browning or glazing. The solution was to use a hybrid method. I packed my meat loaf mix into a loaf pan, covered it with foil, and then inverted the whole thing onto a rimmed baking sheet, spreading out the foil so that I now had a foil-lined baking sheet with an inverted meat loaf and loaf pan on top of it. I baked this way for about half an hour—just long enough to set its shape— and then used a spatula and kitchen towels to lift off the pan. The result was a perfectly loaf-shaped meat loaf (just right for slicing into sandwiches), with all the advantages of a free-form loaf and its extra surface area.

You can leave your meat loaf completely undressed, but I kind of like the old-fashioned, low-brow sweet vinegariness of a ketchup-and-brown-sugar glaze. Draping the loaf in bacon wouldn't do any harm either. I still haven't tried topping my loaf with bananas as Mr. Nickerson so helpfully suggested.

As he could tell you, though, the beauty of meat loaf lies in the almost infinite ways in which it can be customized. So long as your ratio of meat to binders is correct, the sky's the limit as to what you can do. I sometimes add chopped pickles or briny olives. Pine nuts or almonds also add texture and flavor. My mother—who, I believed for a long time, looked for ways to hide raisins where you'd least expect them–would probably enjoy some raisins in her loaf. I'm not one to judge.

# ALL-AMERICAN MEAT LOAF

*continues*

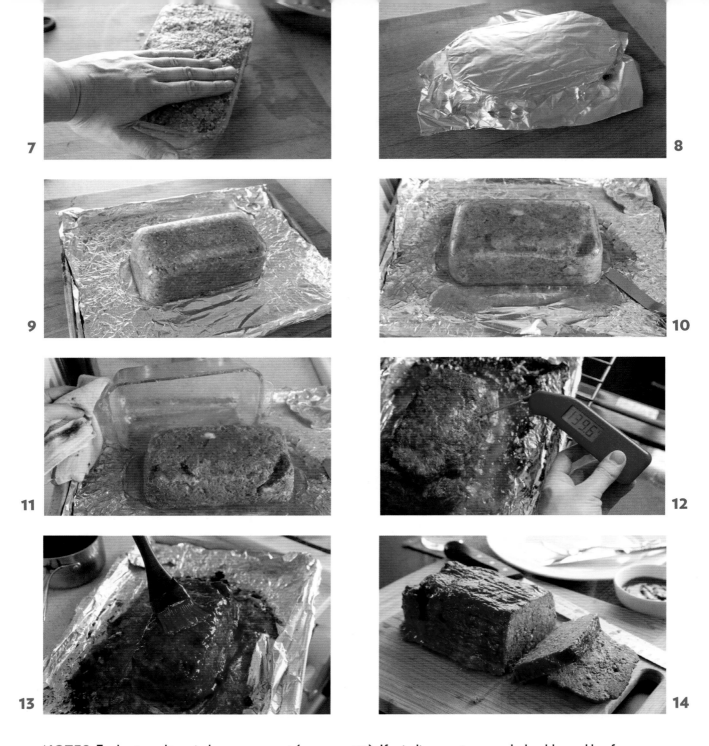

7

8

9

10

11

12

13

14

**NOTES:** For best results, grind your own meat (see page 485). If grinding meat, use pork shoulder and beef chuck (or a mix of short rib meat and brisket). Keep your hands well moistened when forming the loaf, to prevent sticking.

If you don't require or desire a perfect loaf shape, the meat loaf can be formed free-form on a foil-lined rimmed baking sheet without using a loaf pan, though it will sag a bit and come out only a couple of inches tall. Cooking instructions are the same.

½ cup homemade or low-sodium
  canned chicken stock

¼ cup buttermilk

½ ounce (2 packets; about 1½
  tablespoons) unflavored gelatin

2 slices high-quality white sandwich
  bread, crusts removed and torn
  into rough pieces

4 ounces button or cremini
  mushrooms, cleaned

3 anchovy fillets

½ teaspoon Marmite

2 teaspoons soy sauce

1 teaspoon paprika

2 cloves garlic, roughly chopped
  (about 2 teaspoons)

1 small onion, roughly chopped
  (about ¾ cup)

1 small carrot, peeled and roughly
  chopped (about ½ cup)

1 stalk celery, roughly chopped
  (about ½ cup)

2 tablespoons unsalted butter

12 ounces freshly ground pork (see
  Note above)

1¼ pounds freshly ground beef (see
  Note above)

2 large eggs

4 ounces cheddar, provolone,
  Monterey Jack, or Muenster
  cheese, finely grated (about 1 cup)

¼ cup finely minced fresh parsley

Kosher salt and freshly ground black
  pepper

1. Combine the chicken stock and buttermilk in a liquid measuring cup and sprinkle the gelatin evenly over the top. Set aside.

2. Place the bread and mushrooms in a food processor and pulse until finely chopped. Transfer to a large bowl and set aside.

3. Add the anchovies, Marmite, soy sauce, paprika, and garlic to the processor bowl and pulse until reduced to a fine paste, scraping down the sides of the bowl as necessary. Add the onion, carrot, and celery and pulse until finely chopped but not pureed.

4. Heat the butter in a 10-inch nonstick skillet over medium-high heat until foaming. Add the chopped vegetable mixture and cook, stirring and tossing frequently, until it is softened and most of the liquid has evaporated, about 5 minutes; the mixture should start to darken a bit. Stir in the buttermilk mixture, bring to a simmer, and cook until reduced by half, about 10 minutes. Transfer to the bowl with the mushrooms and bread, stir thoroughly to combine, and let stand until cool enough to handle, about 10 minutes.

5. Add the meat mixture to the bowl, along with the eggs, cheese, parsley, 1 tablespoon salt, and 1 teaspoon pepper. With clean hands, mix gently until everything is thoroughly combined and homogeneous; it will be fairly loose. Pull off a teaspoon-sized portion of the mixture, place it on a microwave-safe plate, and microwave it on high power until cooked through, about 15 seconds. Taste the cooked piece for seasoning and add more salt and/or pepper as desired.

6. Transfer the mixture to a 9-by-5-inch loaf pan, being sure that no air bubbles get trapped underneath. (You may have some extra mix, depending on the capacity of your pan; this can be cooked in a ramekin or free-form next to the loaf.) Tear off a sheet of heavy-duty aluminum foil large enough to line a rimmed baking sheet and use it to tightly cover the meat loaf, crimping it around the edges of the pan. Refrigerate the meat loaf while the oven preheats. (The meat loaf can be refrigerated for up to 2 days.)

7. Adjust an oven rack to the lower-middle position and preheat the oven to 350°F. When the oven is hot, remove the meat loaf from the refrigerator and, without removing the foil cover, carefully invert it onto the rimmed baking sheet. Loosen the foil and spread it out, leaving the pan on top of the meat loaf (see Note above). Fold up the edges of the foil to trap the liquid that escapes from the meat loaf while baking. Bake until just beginning to set (the top should feel firm to the touch), about 30 minutes.

*continues*

### For the Glaze

¾ cup ketchup

¼ cup packed brown sugar

½ cup cider vinegar

½ teaspoon freshly ground black
  pepper

Mustard or ketchup (optional)

8. Use a thin metal spatula to lift an edge of the inverted loaf pan, jiggling it until it slides off the meat loaf easily, and use oven mitts or a folded kitchen towel to remove the pan, leaving the meat loaf on the center of the foil. Return to the oven and bake until the center of the meat loaf registers 140°F on an instant-read thermometer, about 40 minutes longer. There will be quite a bit of exuded juices; this is OK. Remove from the oven and let rest for 15 minutes. Increase the oven temperature to 500°F.

9. **Meanwhile, make the glaze:** Combine the ketchup, brown sugar, vinegar, and pepper in a small saucepan and cook over medium-high heat, whisking occasionally, until the sugar is melted and the mixture is homogeneous, about 2 minutes. Remove from the heat.

10. Use a brush to apply some glaze to the meat loaf in a thin, even layer, then return it to the oven and bake for 3 minutes. Glaze again and bake for 3 minutes longer. Glaze one more time and bake until the glaze is beginning to bubble and is a deep burnished brown, about 4 minutes longer. Remove from the oven and allow to rest for 15 minutes. Slice and serve with any extra glaze and mustard or ketchup as desired.

# LEFTOVER MEAT LOAF SANDWICH

**There's nothing better than a meat loaf sandwich. Heck, sometimes I'll make meat loaf just so that I can reheat it for sandwiches the next day!**

**SERVES 1**

1 to 2 slices leftover All-American
   Meat Loaf (page 531)

1 slice American, cheddar, Swiss, or
   Monterey Jack cheese

1 hamburger bun, toasted

Toppings and condiments as
   desired, such as mustard, ketchup,
   and pickles

Preheat the broiler to high. Place the meat loaf on a foil-lined broiler pan and broil until just starting to crisp on the edges, about 5 minutes. Top with the cheese and broil until melted, about 1 minute longer. Transfer to the toasted bun, dress as desired, and eat.

## TESTING RAW MEAT MIXES FOR SEASONING

You may have noticed a problem: it's always a good idea to taste your food for seasoning (that'd be salt and pepper) as you progress in a recipe, but it's not feasible to taste raw meat mixes until they're cooked. How do you know your meat loaf (or sausage, or stuffing; etc.) has enough salt in it before you pack it into a pan or bake it?

There are two quick and easy ways. First is to simply pull off a small amount and throw it into a hot skillet, cooking it like a miniature hamburger patty. Even faster is to throw a bit on a microwave-safe plate. A teaspoon-sized amount will cook through in 10 to 15 seconds. Then taste and adjust your mix before committing to cooking it.

# ITALIAN MEATBALLS WITH TOMATO SAUCE

**Here's the great thing about cooking: once you learn the basic *whys* and *hows* of the craft, they can be applied to almost any situation.**

Take classic Italian-American meatballs, for instance. We learned the ins and outs of meat loaf on page 525 and of sausages on page 493. With that knowledge, we basically know all there is to know technique-wise about those meatballs. In essence, Italian-American meatballs are nothing more than tiny spherical meat loaves with a few sausage-like properties, simmered in a rich, meaty tomato sauce. (If you're Italian, don't tell your grandmother I said that.) Sure, they don't usually have the same set of aromatics as meat loaf, but in terms of basic technique, they are nearly identical. So anyone who's ever made meat loaf knows how to make meatballs, and vice versa. If only all of life were this easy! Rather than the onions, carrots, and celery I use in my meat loaf mix, I limit my meatballs to a simpler flavoring blend of garlic, parsley, and Parmesan cheese, making them even easier to form.

## Meatball Texture

On the spectrum of sausage to meat loaf, a meatball falls much closer to the meat loaf end, but there should be *one* sausage-like characteristic about them—namely a bit of bounce to their bite. How to go about getting it?

My first thought was the salt. I know that if I salt my sausage meat and let it rest before mixing it, it gives the sausages a bouncier, tighter texture. It works exactly the same way for meatballs. I made two batches of meatballs side by side. One was mixed right after seasoning, the other was left to sit for 30 minutes before mixing. The batch that rested had a markedly more cohesive texture, with a nice resilient bounce. I also tried mechanically kneading the balls as I would a sausage mix with the paddle attachment on my stand mixer, but the result was too much springiness. Hand-forming is the way to go.

## Flavor Exchange

Next question: how to cook the meatballs once they're formed? While a large meat loaf can cook just fine in the oven, developing some nice browning on its exterior as it slowly cooks through, this is not possible with meatballs. Because of their small size, they end up hopelessly overcooked by the time any browning occurs. That's why meatballs are traditionally cooked with a two-stage process: frying and simmering. A quick fry in a hot skillet will brown their exterior, adding texture and flavor (shallow-frying in a layer of oil gets you a more even layer of browning than sautéing), while simmering them in a pot of sauce will not only allow them to cook through to the center, but also add plenty of meaty flavor to the sauce. A little give and take, if you will.

For the sauce, I kept things simple, going with a basic marinara flavored with oregano, red pepper flakes, and garlic, all sautéed in a mixture of oil and butter (for more on marinara sauce, see page 693).

But there's one problem that more astute readers might have caught—we have two separate but conflicting goals here. Long simmering is good for the sauce—it helps build in meaty flavor—but it's bad for the meatballs—they end up hopelessly overcooked. Those long-cooked pasta sauces where the meatballs simmer on a back burner all day may *sound* like a good idea, a romantic idea, even, but it's not the best way to go if tender meatballs are your goal. A meatball's interior shouldn't ever get too far north of 160°F, which means about 10 minutes of gentle simmering *at the maximum*. But if you're not simmering your meatballs in the sauce, how are

you ever going to develop good rich meaty flavor in there? Ten minutes is not nearly enough time!

The solution? Take a few raw meatballs and brown them in a Dutch oven to build a flavorful base for the sauce from the get-go. Allowing the sauce to simmer for an hour with the mashed-up meat gives it plenty of flavor. Then, after browning the rest of the balls in hot oil, all that's left to do is drop them into the meaty sauce and simmer just until they're cooked through. I like these meatballs so much that sometimes I don't even bother with the spaghetti.

## TENDER ITALIAN MEATBALLS
### WITH RICH TOMATO SAUCE

*continues*

## SERVES 6 TO 8

### For the Meatballs

1 pound freshly ground chuck or
  lamb

1 pound freshly ground pork
  shoulder

Kosher salt

1 cup buttermilk

¼ ounce (1 packet) unflavored
  gelatin

4 slices high-quality white sandwich
  bread, crusts removed

2 teaspoons soy sauce

½ teaspoon Marmite

4 to 6 anchovy fillets, mashed to
  a paste with the back of a fork
  (about 1 tablespoon; reserve half
  the paste for the sauce)

2 large eggs

3 ounces Parmigiano-Reggiano,
  finely grated (about 1½ loosely
  packed cups)

6 medium cloves garlic, minced or
  grated on a Microplane (about
  2 tablespoons; reserve half the
  garlic for the sauce)

½ cup chopped fresh parsley

### For the Sauce

¼ cup extra-virgin olive oil

4 tablespoons unsalted butter

1 large onion, finely diced (about 2
  cups)

Reserved anchovy paste and garlic
  from above

1½ teaspoons dried oregano

1 teaspoon red pepper flakes

1. **To make the meatballs:** In a large bowl, combine the ground beef and ground pork with 1 tablespoon salt. Mix thoroughly with your hands, then set aside at room temperature for 30 minutes.

2. Meanwhile, pour the buttermilk into a large bowl and sprinkle the gelatin over it. Allow to hydrate for 10 minutes. Add the bread and allow to soak for 10 minutes, turning the bread occasionally until completely saturated.

3. After the meat has rested, add the soy sauce, Marmite, half of the anchovy paste, the eggs, cheese, half the garlic, and the parsley to the bowl. Add the bread and buttermilk mixture and mix gently with your hands until well combined; do not knead excessively. Pull off a teaspoon-sized portion of the mixture, place it on a microwave-safe plate, and microwave it on high power until cooked through, about 15 seconds. Taste the cooked piece for seasoning, and add more salt to the meat mixture as desired.

4. Using wet hands or a #40 scoop, form the meat mixture into meatballs about 1½ inches in diameter (the balls should slightly overfill the scoop, about 3 tablespoons per meatball—you should get 28 to 32 balls), placing them on a large plate as you go. Refrigerate.

5. **To make the sauce:** Heat the olive oil in a Dutch oven over medium-high heat until shimmering. Add 4 raw meatballs and mash them against the bottom of the pot, breaking them up. Let the meat cook, without stirring, until well browned on the bottom, about 3 minutes. Add the butter and onions and scrape up the browned bits from the bottom with a wooden spoon. Cook, stirring occasionally, until the onions are mostly softened and translucent, about 3 minutes.

6. Add the remaining garlic, the oregano, pepper flakes, and the remaining anchovy paste and cook, stirring and breaking up the meat, until fragrant, about 1 minute. Add the tomatoes and bring to a boil, then reduce the heat to maintain a simmer. Partially cover and cook until the sauce is thick and rich, about 1 hour.

7. Meanwhile, heat the vegetable oil in a 10-inch nonstick or cast-iron skillet over medium-high heat until it reaches 350°F (a meatball should sizzle vigorously when you dip the edge of it into the hot oil). Carefully transfer one-third of the remaining meatballs to the pan. The temperature will drop to around 300°F—adjust the flame to maintain this temperature (the oil should continue to sizzle vigorously but not smoke)

Three 28-ounce cans whole
  tomatoes, crushed by hand or
  with a potato masher into rough
  ½-inch chunks

1½ cups vegetable or canola oil
¼ cup chopped fresh basil
Freshly grated Parmigiano-
  Reggiano for serving

and cook until the meatballs are well browned on the first side, 1 to 2 minutes. Carefully flip them with a small offset spatula or a fork and cook on the second side until well browned, about 3 to 4 minutes longer. Using tongs, transfer the meatballs to a paper-towel-lined plate. Repeat with the remaining two batches of meatballs, allowing the oil to return to 350°F before adding each batch. Set the meatballs aside. (Discard the oil or strain and save for another use.)

8. After the sauce has cooked for 1 hour, add the meatballs and simmer for 10 minutes longer. Season with salt to taste, stir in the basil, and serve, with grated Parmigiano-Reggiano (and pasta if desired).

# PORK MEATBALLS
## WITH MUSHROOM CREAM SAUCE

**SERVES 4 TO 6**

For the Meatballs

1 pound freshly ground pork
   shoulder

Kosher salt

¼ cup buttermilk or heavy cream

¼ ounce (1 packet) unflavored
   gelatin

2 slices high-quality white sandwich
   bread, crusts removed

1 teaspoon soy sauce

½ teaspoon Marmite

2 anchovy fillets, mashed to a paste
   with the back of a fork

3 medium cloves garlic, minced or
   grated on a Microplane (about
   1 tablespoon; reserve about 2
   teaspoons for the sauce)

1 teaspoon red pepper flakes

1 large egg

1 tablespoon sugar

½ teaspoon ground fennel

Freshly ground black pepper

2 cups vegetable oil

1. **To make the meatballs:** In a large bowl, combine the ground pork with the salt. Mix thoroughly with your hands, then set aside at room temperature for 30 minutes.

2. Meanwhile, pour the buttermilk into a large bowl and sprinkle the gelatin over it. Allow to hydrate for 10 minutes. Add the bread and allow to soak for 10 minutes, turning the bread occasionally until completely saturated.

3. After the meat has rested, add the soy sauce, Marmite, anchovy paste, 1 teaspoon of the garlic, the red pepper flakes, egg, sugar, fennel, and pepper to taste to the bowl. Add the bread and buttermilk mixture and mix gently with your hands until well combined; do not overknead. Pull off a teaspoon-sized portion of the mixture, place it on a microwave-safe plate, and microwave it on high power until cooked through, about 15 seconds. Taste for seasoning, and add more salt and pepper to the meat mixture as desired.

4. Using wet hands, form the mixture into meatballs about 1 inch in diameter (a generous tablespoon per meatballs—you should get roughly 30 meatballs), placing them on a large plate as you go.

5. Heat the vegetable oil in a 10-inch nonstick or cast-iron skillet over medium-high heat until it reaches 350°F (a meatball should sizzle vigorously when you dip the edge of it into the hot oil). Set aside 4 meatballs and carefully transfer half of the remaining meatballs to the pan. The temperature will drop to around 300°F—adjust the flame to maintain this temperature (the oil should continue to sizzle vigorously but not smoke) and cook until the meatballs are well browned on the first side, 1 to 2 minutes. Carefully flip them with a small offset spatula or a fork and cook on the second side until well browned, about 3 to 4 minutes longer. Using tongs, transfer the meatballs to a paper-towel-lined plate. Repeat with the remaining meatballs, allowing the oil to return to 350°F before adding the second batch. Set the meatballs aside (discard the oil or strain and save for another use).

## For the Sauce

3 tablespoons unsalted butter

8 ounces button mushrooms, cleaned and sliced

1 small onion, finely diced (about 1 cup)

Reserved garlic from above

1 tablespoon all-purpose flour

1½ cups homemade or low-sodium canned chicken stock

1 teaspoon soy sauce

½ cup heavy cream

2 teaspoons sugar

Kosher salt and freshly ground black pepper

1 teaspoon fresh lemon juice

1 teaspoon fresh thyme leaves

6. **To make the sauce:** Return the skillet to medium-high heat, add 2 tablespoons of the butter, and heat until the foaming subsides. Add the mushrooms and cook, stirring occasionally, until they give up their liquid and start to brown, about 8 minutes. Push the mushrooms to one side and add the remaining tablespoon of butter to the center of the skillet. Add the reserved 4 raw meatballs and mash with a wooden spoon, then cook, stirring to break up the meat, until it is no longer pink, about 1 minute.

7. Add the onions, stir the meat, mushrooms, and onions together, and cook, stirring occasionally, until the onions are softened, about 3 minutes. Add the remaining garlic and cook, stirring, until fragrant, about 30 seconds. Add the flour and cook, stirring constantly, for 30 seconds. Slowly pour in the chicken stock, scraping up any browned bits from the bottom of the pan. Add the soy sauce, heavy cream, and sugar, stir to combine, and bring to a simmer.

8. Add the remaining meatballs to the sauce and cook, stirring and turning them occasionally, until they are cooked through and the sauce has thickened to the consistency of heavy cream, about 5 minutes. Season with salt and pepper to taste, stir in the lemon juice and thyme, and serve.

# MY WIFE GETS HOME EACH NIGHT AND GREETS ME WITH A KISS—

. . . an innocent gesture that I know is merely a breath test in disguise. I notice her taking a short, sharp inhale as her nose gets close to my face, searching for a whiff of that unmistakable perfume, that salty aroma of deceit and disloyalty. I try to hold my breath, but it's too late.

She confronts me: "You've been cooking burgers again, haven't you?"

Some would call my love of burgers an obsession. Others, a mental illness (hello, Mom), and still others—namely, my wife—a source of constant sorrow. A couple years ago, she forced us to move to a new apartment because the glorious smell of burgers and grilled onions had managed to permeate even the very walls. We may have to move again soon.

I love my wife, but burgers are my mistress.

# FIVE RULES FOR BETTER BURGERS

Anybody can make a decent burger—that's its beauty—but it takes a bit of know-how to take a burger from decent to shockingly, mind-blowingly good. Here are the five most basic rules to get you there.

### 1. Choose your beef wisely, and grind it yourself.

Even more so than with sausages, meat loaf, or meatballs, burgers really shine with freshly ground beef. I can't possibly emphasize this enough. Buying store-bought ground beef is a crap shoot. You're never quite sure when it was ground, what part of the cow it came from, or even how many different cows are in the package. Not to mention baddies like *E. coli*, rough handling, and tight shrink-wrap packaging that can lead to leaden patties.

Freshly ground beef has a looseness, juiciness, and flavor that will make every store-bought ground beef burger you've ever eaten want to hang its head in shame. Grinding it yourself also allows you to select exactly which cuts go into your patties, so you can fine-tune your burgers to your personal taste. If you've never done it, the task may seem daunting at first, but it's really not that difficult (see page 485 for tips on grinding meat).

If you do decide to go with store-ground beef, look for ground chuck that's at least 20-percent fat (it'll be labeled 80/20, 80% lean, 20% fat). If possible, ask the butcher to coarsely grind a fresh batch for you.

### 2. Don't futz around with the meat.

Despite outward appearances, ground meat is not inert. From the moment you lay your hands on it, it is changing dynamically, reacting to every knead, every sprinkle of salt, and every change in temperature. Working the meat unduly will cause proteins to cross-link with each other like tiny strips of Velcro, making your finished burgers denser and tighter with every manhandling of the grind.

For the most tender burgers, grind your meat fresh and form your patties as tenderly as possible. For griddled patties with superior nooks and crannies for cheese-catching, I sometimes like to grind my meat directly onto a baking sheet and gently coax it into patties without ever picking it up until just before I cook it. Superb.

Just check out the difference when a loose patty and compressed patty are cooked side by side:

*continues*

A compressed patty has a firmer texture and far fewer pockets to hold on to rendered juices and cheesy goo. The unmanhandled patty, on the other hand, has a loose texture that retains juices better (even at medium-well to well-done, it's insanely, chin-drippingly juicy), has far more surface area for browning, and, as you can see, has plenty of nooks and crannies brimming with melted cheese and juices.

Additionally, adding junk like onions, herbs, eggs, bread crumbs, or *anything* to your ground meat not only forces your to overhandle the mix, but instantly relegates your burgers into the "meat loaf sandwich" category. You've spent a long time carefully selecting and grinding your beef. Let it speak for itself.

There is one exception to this rule: the smashed burger. You may have heard that once a burger hits the grill or griddle, you should never, under any circumstances, press on it, lest you squeeze out valuable juices. Well, yes and no. True, once the patty has heated to the point where its fat has begun to liquefy, squeezing it will wring it out like a sponge. On the other hand, you can feel free to press on the burger at any point before this stage. In fact, in the case of smashed burgers—a style with recent strong representation from chains like Shake Shack and Smashburger—you actually *want* to press on it—and hard—right when it starts cooking. More on these later.

### 3. Season liberally, but do not salt the beef until the patties are formed.

Burgers shouldn't be confused with health food, or even everyday food. A great burger is a once-in-a-while treat, and if I'm going to treat myself, I want it to taste *awesome*. Without enough salt and pepper, even the most carefully selected and ground meat blend will taste flat—you're better off dining with the King or the Clown, who, despite their significant shortcomings, at least understand the benefits of a little sodium chloride.

A word of warning: salt your beef only after the patties are formed. Salt will dissolve muscle proteins, which subsequently cross-link, turning your burgers from moist and tender to sausage-like and springy (which is desirable for actual sausages; see page 496). The effect is quite dramatic.

The two burgers shown below were made with identical cuts of beef, formed in the exact same way, and cooked in the same skillet to the same internal temperature. The only difference? The one on the left was salted only on its exterior, after forming the patties, while the one on the right had salt mixed into the meat prior to forming the patties.

This tighter texture translates not just to an oddly smooth texture, but also to a firmer, tougher patty as well. To demonstrate this, I dropped a Dutch oven on top of half of each patty from a fixed height, taking note of how widely they splattered. Check out the results. Again, the regular patty is on the left, the salted patty is on the right:

Patties formed from unsalted (*left*) and salted (*right*) hamburger meat.

Now do you get it? Leave the salt alone until after the patties have been formed.

As for the pepper, I like to be liberal with it. A burger without pepper is like a bath without bubbles. Sure, it'll get the job done, but where's the fun in it?

## 4. Love your bun.

This is good general life advice that is particularly appropriate for burgers. Buns come in all shapes, sizes, densities, and styles. Make sure you've got the right one for the job at hand and treat it with the dignity and respect it deserves. I can't count the number of times I've seen an otherwise decent hamburger marred by an inappropriate bun selection.

For smaller, thinner patties, like the diner-style griddled burger, soft, sturdy, and slightly sweet Martin's Potato Rolls set the benchmark, although any soft, squishy, supermarket bun will do. (I like Arnold or Pepperidge Farm buns in the absence of Martin's. Sesame seeds are up to you). And, for the love of science, butter and toast that bun before you put your burger in it! Sliders, with their thin slip of a patty, require the softest of the soft supermarket buns, steamed through and designed to completely melt away as you eat. The generic store brand will almost always work.

A bigger pub-style burger can overwhelm a soft bun with juices, soaking through and dissolving the base

*continues*

before the burger even hits your table. Toasting the bun can mitigate some of these effects, but for the most part, you're better off selecting a sturdier roll or, if you've got one nearby, a custom burger bun from an artisan bakery. Brioche has its adherents, but I prefer my buns to be a little less buttery and sweet, so as not to compete with the flavor of the beef. Do avoid anything with an overly chewy crumb or a tough crust; a tough bun will force the burger patty to squeeze out of the back as you bit into it, a dreaded condition known to those in the industry as "backslide." Your bun should *always* be more tender than your burger.

## 5. Cook hot.

Except for extremely rare cases (as with, say, sliders or steamed burgers), the goal when cooking a burger is to maximize crispy crust formation and browning on the exterior. The best way to do this is to cook your burgers as hot as you can: high heat and a ripping-hot cast-iron or heavy stainless steel skillet if you're cooking on the stovetop (do not use nonstick—nonstick coatings will produce toxic vapors when heated hot enough to sear a burger), or a well-preheated grill with burners on full blast or a massive amount of coals. If it looks like the burgers are beginning to burn before the center reaches the temperature you want, lower the heat or move the burgers to a cooler part of the grill. (Or, better, start them cool and finish hot.)

As for doneness in thick burgers, sure, you can be all macho and try and gauge a burger's doneness by poking at it with your finger, or you can just use that awesome instant-read thermometer you bought (see page 565).

Temperatures for burgers are exactly the same as for steaks, though burgers cook significantly faster. Here's a rough guide:

- **120°F and below** for rare (red/raw in the center)
- **130°F** for medium-rare (pink and warm)
- **140°F** for medium (totally pink, starting to dry out)
- **150°F** for medium-well (grayish pink, significantly drier)
- **160°F and above** for well-done (completely gray, very little moisture)

# SMASHING BURGERS VERSUS SMASHED BURGERS

**"Never, ever press down on your burger!"**

How many times have you read that in a book or heard a TV chef say it? "It squeezes the juices out!" they cry. "It turns your lunch into a hockey puck!" they scream. You've heard it so many times you can't help but believe it's true, right? Well, OK, Mr. Smarty-Chef, I'll believe you, but first you must answer me these questions three:

- **Question the first:** One of my favorite burgers in New York—the one that folks'll stand in line for an hour to get—is smashed. How does the Shake Shack burger still retain such abundant juiciness?
- **Question the second:** The SmashBurger chain of burger joints has built its reputation on its smashing technique. Have all of its fans (which are legion) been deluded into enjoying the flavor of hockey pucks?
- **Question the third:** I just had what was the finest burger I've tasted in recent memory at Off-Site Kitchen in Dallas, where—guess what?—the burgers are smashed. What gives?

Now, these questions are largely rhetorical. Anybody who's been making burgers for a while knows the answer: not smashing your burgers is always—sometimes—only sort-of occasionally good advice. So, when is it OK to smash your burgers and when is it not? First, let's consider the advantages of smashing a burger.

## In Crust We Trust

There's really only one reason to smash a burger and it's the reason that all three of the burgers I mention above (as well as countless others) taste so good:

the Maillard reaction. It's what creates the crust on your steak or burger, the golden brown color on your toast, and the complex, pleasing aromas and flavors that accompany that browning. It's the smell of a steak house and fresh bread from the oven. And it's the smell of a good burger joint. It doesn't just make meat taste good, it actually makes it taste more meaty.

Most of these browning reactions don't take place until foods are heated to at least 300°F or so, and they are greatly accelerated at temperatures higher than that, so if maximizing browning is your goal when cooking a burger (and it should be!), then it's plain to see why smashing a burger can improve its flavor: It increases the surface area directly in contact with the hot metal, resulting in more browning.

While it's true that given enough time you can brown even a nonsmashed burger, there are a couple problems: If the heat is too high, the browning will be uneven—at worst, the bits of meat directly in contact with the skillet or griddle will burn before the bits elevated above it can even begin to brown. With lower heat, you can get more even browning, but it takes longer—long enough that your burger will end up overcooking in the middle (and overcooking is the real path to dry burgers).

Smashing allows you to get a deep brown crust before the interior overcooks, even with a relatively small patty.

### The Juice Is Loose

So, when is it *not* a good idea to smash? We all have a pretty good idea of this, but I cooked through a couple dozen burgers, smashing them at various stages in order to make sure. The results? If you don't want to lose juices, you must smash within the first 30 seconds of cooking.

When ground beef is cold, its fat is still solid and its juices are still held firmly in place inside small chopped-up segments of muscle fibers. That's the reason why you can push and press on raw ground meat without squeezing out too much liquid, and the reason why you can smash a burger during the initial phase of cooking without fear of losing moisture.

But what happens as the meat warms up?

When you look at a burger under a microscope, you see what basically amounts to an interconnected network of proteins interspersed with fat and water-based liquids. Like all meats, as a burger cooks, this protein network tightens, squeezing out liquids. Simultaneously, the fat begins to render and liquefy, allowing it to be squeezed out right along with the juices.

In a properly formed burger—one that is made with meat that's been ground right, kept chilled, and minimally handled while shaping—the protein matrix is relatively loose. Even once fat has been liquefied and juices have been squeezed out of the protein network, they can remain trapped in the patty, only getting released when you bite into the burger, in much the same way that liquids can be trapped in a sponge and only released when it is squeezed. But press down on a burger during this phase, and the juices come gushing out into the skillet or onto your coals. You're left with what amounts to a meat patty with the texture of a sponge that's been run through a wringer.

All burgers will lose weight as you cook them—it's not possible to hold on to all the liquefied fat and exuded juices. In my testing, 4-ounce burgers that started as round pucks and were smashed down to ½-inch thickness any time before 30 seconds still lost a little over 20 percent of their weight during cooking. But this is comparable to burgers of the same weight and thickness that were cooked with no smashing at all. Both burgers tasted quite juicy, while the smashed burger (obviously) had better flavor.

## SMASH TIME VERSUS FINAL WEIGHT

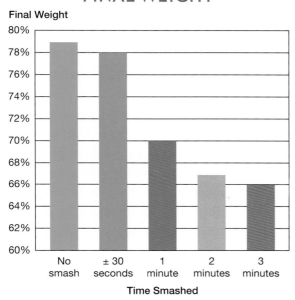

Once you start smashing after the 1-minute mark, that's when juices *really* start to flow and you end up with a dramatically drier burger—a good 50-percent more moisture is lost in a burger smashed after 1 minute versus one smashed within 30 seconds. Move into the territory of double or even triple smashing—that is, smashing once at the beginning, then getting impatient and smashing again during the middle and latter phases of cooking—and a burger can easily lose half of its weight to the evil griddle gods. I've seen more than one short-order cook with a backup of orders resort to this dastardly method, and not once have I ever taken more than one bite of a burger that's been subjected to it.

## FOUR RULES FOR SMASHING SUCCESS

If you know the basic rules for burgers (see page 545), making a smashed burger is simple.

## Rule 1: Use a Stiff, Sturdy Spatula

No flexible spatulas or cheap plastic ones here. You need a heavy-gauge stainless steel spatula with a fully riveted handle.

## Rule 2: Use a Heavy Stainless Steel or Cast-Iron Skillet

The goal is steady, even heat, so you want to use a relatively thick pan and allow it to preheat for long enough that there are no hot or cool spots. I let my pan preheat over medium heat for a few minutes, pumping it up to high just before I add the meat.

## Rule 3: Smash Early and Smash Firmly

Form anywhere from 2 to 5 ounces of meat into a puck about 2 inches high, season liberally with salt and pepper, place it in the preheated skillet, and smash it with the spatula, using a second spatula to add pressure if necessary. Then cook, without moving it, until a deep-brown crust develops. This'll take about a minute and a half.

## Rule 4: Leave No Crust Behind

The whole goal of smashing is to develop a nice browned crust, so it's important that you scrape it all up intact when you flip the burger. Once again, a sturdy metal spatula is your friend. I find that turning the spatula upside down to help scrape the crust off is pretty effective. If your crust is properly developed and your burger properly smashed, it should spend very little time on its second side— just enough to finish cooking through and to allow the cheese (if added) to melt, 30 seconds or so.

AND, WELL, that's it. So simple, so fast, so freaking delicious. The great thing about smashing burgers is that it's so efficient at developing good flavor that even using store-bought ground beef (*gasp!*) will result in a darn tasty burger. Smashing is my go-to method when the mood for a burger strikes and I don't feel like lugging out the grinder.

# CLASSIC DINER-STYLE SMASHED CHEESEBURGERS

**NOTE**: My toppings of choice are thinly sliced raw onions, pickles, and Fry Sauce (page 553), but feel free to use whatever you'd like.

**SERVES 2 TO 4**

1 pound freshly ground beef (store-
   bought or home-ground)

4 soft hamburger buns, preferably
   Martin's Potato Sandwich Rolls

2 tablespoons unsalted butter,
   melted

Vegetable oil

Kosher salt and freshly ground
   black pepper

1 small onion, finely sliced
   (optional)

4 deli-cut slices American cheese

Toppings as desired

1. Divide the meat into 4 even portions and form into pucks about 2 inches high and 2½ inches wide. Refrigerate until ready to use.

2. Open the buns but do not split the hinges. Brush lightly with the butter, then toast under a hot broiler or in toaster oven until golden brown, about 1 minute. Set aside.

3. Using a wadded-up paper towel, rub the inside of a 12-inch heavy-bottomed stainless steel or cast-iron skillet with vegetable oil, then heat over medium-high heat until just beginning to smoke. Season the beef pucks on the top with salt and pepper, then place seasoned side down in the skillet. Using a wide heavy spatula, press down on each one until it is roughly 4 to 4½ inches in diameter and ½ inch thick; it helps to use a second spatula to apply pressure on the first one. Season the tops with salt and pepper. Cook, without moving the burgers, until a golden brown crust develops on the bottom, about 1½ minutes. Use the edge of the spatula to carefully scrape up and flip the patties one at a time, making sure to get all the browned bits. If using onions, add to the tops of the burgers, then cover each with a cheese slice. Continue to cook until the patties are the desired doneness—about 30 seconds longer for medium-rare.

4. Top the buns and/or patties as desired, transfer the patties to the buns, close the burgers, and serve.

# FRY SAUCE

**Go to any burger joint in the Midwest and ask for fry sauce, and you'll get a little tub of pink, creamy goo to dip your fries in or slather on your burger. At its most basic, it's a mix of mayo and ketchup. I like to liven mine up with a few spices and some pickle juice.**

**MAKES ABOUT ⅔ CUP**

½ cup mayonnaise, preferably
   homemade (page 807)

2 tablespoons ketchup

1 tablespoon yellow mustard

1 tablespoon kosher dill pickle juice

1 teaspoon sugar

Pinch of cayenne pepper

Combine all the ingredients in a bowl and whisk until smooth. The sauce will keep in a covered container in the fridge for up to 2 weeks.

# BIG, FAT, JUICY GRILLED BURGERS

**Crusty on the outside, medium-rare and juicy in the middle, a good grilled pub-style or backyard burger is what most of us think of when we think of the archetypal hamburger, whether that's accurate or not.**

Because of their hefty girth, it's relatively simple to develop a good crust on these burgers before the interior dries out, but there are a few important steps that separate a good pub-style burger from a perfect one. As usual, the most important is to use freshly ground beef, but it doesn't stop there.

## Battling the Bulge

I'm about to show you some images that may frighten you. Those amongst you with a fragile constitution may choose to avert your eyes. You've probably all seen this before, and it's not a pretty picture. This is what happens when a poorly shaped burger hits the grill.

Here's a list of the basic symptoms fat burgers tend to suffer from:

- **Soggy Bottoms.** This occurs when the eater is forced to squeeze the bun together in order to compress the patty to a mouth-friendly girth. Juices squeeze out and saturate the bottom bun. High chance of TBF (Total Bun Failure).
- **Bun Gap.** There's a large gap between the edges of the bun and the burger, requiring the eater to take several meat-deficient bites despite having carefully measured and sized the patty before cooking.

A burger with incorrect bun-to-meat ratio.

- **Thickness Approaching Width.** The burger bulges in the center, leaving the eater with a shape that

A bulging burger that overwhelms its bun.

is awkward for both the hands and the mouth. In extreme cases, burgers may reach near-spherical proportions.
- **Dry Matter.** This symptom is too gruesome to show in photographs. It occurs when the griller notices that the burger is beginning to acquire a golf-ball-like shape halfway through cooking. He responds by pressing down on it with the back of a spatula. Fat and juices fall into the flames and ignite. The result is a flat patty, squeezed dry, and singed on the outside with a heavy deposit of black soot from the burnt fat.

Here's why *all* of these problems occur. As we know, when meat cooks, it contracts. With very thin patties that cook through quickly, this contraction is fairly even—they shrink equally from all directions, remaining relatively flat. With a thick burger, however, while the edges seem to get smaller and smaller, the center bulges farther and farther out. Why?

It's all got to do with the amount of rare meat left in the center of the burger. Since a burger cooks from its edges as well as its flat faces, the sides of the burger tend to cook much faster than the middle—so a large amount of uncontracted rare meat remains in the center. At the same time, the edges contract not just in terms of thickness, but also circumferentially. This tightening action is a lot like a belt being cinched around a fat waist that has nowhere to go but up and over—the center of the patty is squeezed out.

The simple solution? Just compensate for the bulging and shrinking before you start cooking the patties by making the patties slightly wider than your buns and using your fingertips to create a slight depression in the center of each patty. The burgers will end up flat, and you and your guests

To form a burger patty that cooks up flat . . .

. . . make a shallow indentation in the center.

The finished patty will flatten as it cooks.

won't be forced to squish them into place as you eat, losing valuable juices.

## Getting Even

After the dreaded bulge, the second most gruesome fate of a pub burger is poor cooking, which in the worst cases results in an exterior that's charred beyond recognition with a still-cold center. It's exciting to see flames leaping up around the patties, but it's not doing you any favors in the flavor department. Those flames are from fat dripping out of the patties, hitting the hot coals below, and

vaporizing, leaving a sooty, acrid deposit on the surface of your burgers. Pressing down on the patties as they cook exacerbates the problem.

The easiest way to deal with it is to cook in stages, using a two-level fire—that's a fire where all of the coals are banked under one side of the grill (or in the case of a gas grill, one set of burners is turned to high, the others are shut off).

A two-level fire is the best way to cook a thick burger.

What this allows you to do is cook over high heat to get the sear you need (if the coals or flames start to flare up, just cover the grill with the lid until the flames subside) and grill over lower heat to gently cook the burger through. Conventional cooking wisdom, as we know, tells us that you should sear first and finish on the cool side. But! This methodology is all based on the false premise that searing first helps lock in juices, something which we now know to be a fallacy. In fact, after running a few side-by-side tests, I found that *doing the exact opposite* results in more evenly cooked meat: Start your burgers on the cool side of the grill, using a thermometer to check for doneness, and then transfer them over to the hot side once they're within 10 degrees of your desired final serving temperature. By doing this, you warm the meat to the point that

Big flames look cool but are bad for flavor.

a deep, well-charred crust can develop over the high heat in about half the time it would take for it to do so if you slapped the burger on there raw. Minimizing time spent over high heat maximizes even cooking, resulting in a better, tastier burger.

## It's OK to Flip Out

Backyard burger chefs seem to have extraordinarily strong opinions on the issue of flipping, and how often it should be done, but we've already seen that,

with steaks (see page 294), it doesn't really make a difference how often you flip—indeed, multiple flips will yield a marginally better finished product in less time. Some quick tests confirmed that the same is equally true for burgers: multiple flipping is A-OK. But there's one good reason why you should consider easing up on the flipping toward the end: the cheese. A good grilled cheeseburger should have the cheese melted completely by the heat of the grill. Once you flip that burger and top it, there's no going back, so you'd better be sure that the top side is seared before you drape it in cheese.

Not only that, but grills are inherently more cumbersome than stovetops. Who wants to—or can—flip a grillful of burgers constantly? Just know that the next time you come across one of those backyard grill-Nazis who absolutely *insists* that one flip is the way to go, just smile, nod, and let him cook the way he wants to. *Rule one of grilling is: Never question the guy with the spatula.*

But do make sure to quietly revel in your superior knowledge, and maybe make fun of him behind his back.

# PUB-STYLE THICK AND JUICY CHEESEBURGERS

**SERVES 4**

2 pounds freshly ground beef
  (store-bought or home-ground)

Kosher salt and freshly ground
  black pepper

4 deli-cut slices American or
  cheddar cheese

4 hamburger buns, toasted

Condiments and toppings as
  desired

1. Divide the meat into four 8-ounce portions and gently shape each one into a patty 4½ inches wide by approximately ¾ inch thick. Place on a flat surface and create a dimple in the center of each patty by pushing down with three or four fingers: the dimple should be about ¼ inch deep and 3 inches across. Season the burgers generously with salt and pepper. Refrigerate until ready to use.

**TO COOK ON THE GRILL**

2. Ignite a large chimneyful of charcoal. When the briquettes are coated in gray ash, pour out and spread evenly over one side of the grill. Place the cooking grate in place. Or, if using a gas grill, heat one set of burners to high and leave the others off. Scrape the cooking grate clean.

3. Place the burgers on the cool side of the grill, as far from the heat as possible, cover, and cook until an instant-read thermometer inserted in the center of the burgers reads 110°F for medium-rare, or 120°F for medium, 10 to 15 minutes.

4. Uncover the grill and transfer the burgers to the hot side. Cook until the first side is well charred, about 1 minute. Flip the burgers and add the cheese. Continue to cook until the second side of the burgers is charred and the cheese is melted, about 1 minute longer. The center of the burgers should read 130°F for medium-rare, or 140°F for medium. Assemble the burgers, topping them as desired, and serve.

## TO COOK UNDER THE BROILER

2. Preheat the broiler to high. Place the patties on a foil-lined broiler pan and position it so that tops of patties are 2½ to 3 inches below the broiler element. Broil until the tops are well browned and beginning to char, about 3 minutes. Flip the patties and broil until an instant-read thermometer inserted in the center reads 130°F for medium-rare, or 140°F for medium, about 3 minutes longer.

3. Top each patty with a slice of cheese and place back under the broiler for about 25 seconds to melt it. Place on the toasted buns, topping the burgers as desired, and serve.

The perfect patty: flat, juicy, and slightly bigger than the bun after cooking.

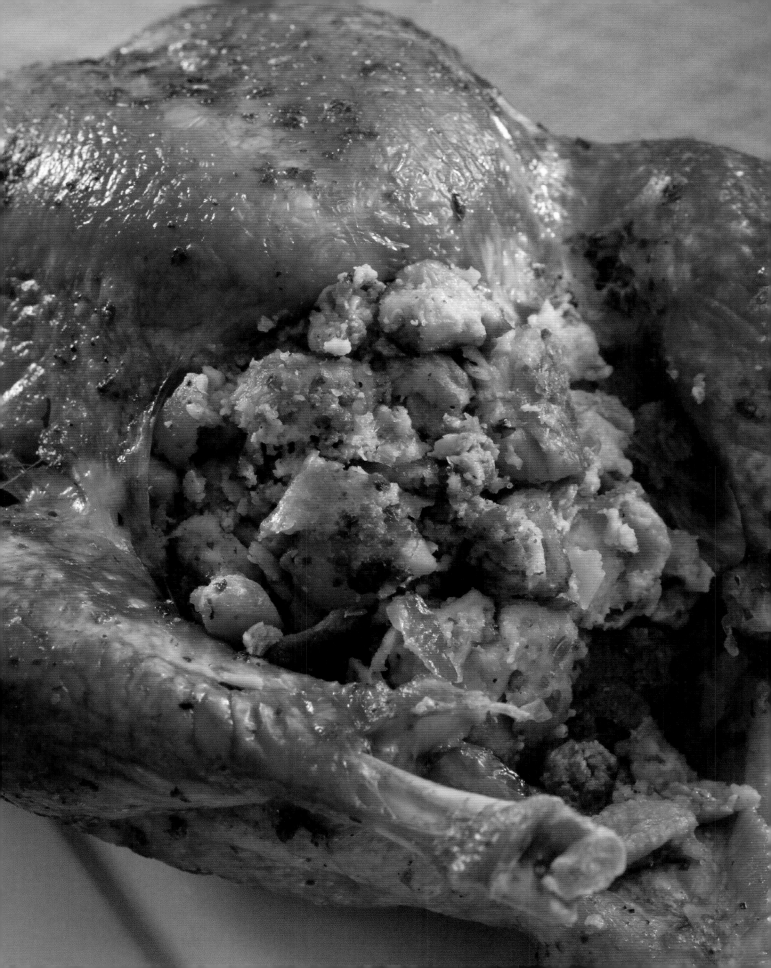

CHICKENS, TURKEYS,
PRIME RIB, AND

# THE
# SCIENCE
## *of*
# ROASTS

6

# CHICKENS, TURKEYS, PRIME RIB, AND THE
# SCIENCE *of* ROASTS

## RECIPES IN THIS CHAPTER

. . . **buy a digital instant-read thermometer.** You will never over- or undercook a piece of meat again. Do it. No more excuses, no more timing charts or poking with your fingers. Just buy a good thermometer, and don't look back. I guarantee you will not regret it.

With that out of the way, on to the introduction.

F or most Americans (and many others around the world), a roast is part and parcel with holiday meals. What would Thanksgiving be without its golden brown crisp-skinned centerpiece, or Christmas without its rosy-centered prime rib or shiny glazed ham?

But roasts aren't just for the holidays. I can think of no better way to feed a large group of people when you've got a bit of time and want to pull it off with minimal fuss. For the most part, roasting requires only a simple set-it-and-forget-it approach, or at least a set-it-and-check-with-a-digital-thermometer-occasionally approach, allowing you plenty of free time to throw back a cocktail with your guests, or, if you're like me, focus on side dishes while avoiding unnecessary social contact.

But roasts frighten many people, and, to a certain degree, rightfully so. We've all been exposed to far too many dry turkey breasts, overcooked standing rib roasts, and stringy chickens than we care to recall (yes, I'm looking at you, Mom). But it doesn't have to be that way! Roasting is, in fact, quite simple and nearly foolproof. All it takes is a bit of know-how.

## DIGITAL INSTANT-READ THERMOMETERS

Once you bite the bullet and spring for a good digital instant-read thermometer, it will quickly become your best friend in the kitchen. When buying one, look for a thermometer that gives accurate readings to within 1 degree or finer, has a nice large readout that you can quickly check (the less time spent with the oven door open, the better), is robust enough that it won't fail if you drop it on the floor or in the sink, and gives you a reading within a matter of seconds.

At around $96 (at the time of this writing), the Splash-Proof Thermapen from ThermoWorks is pricier than most, but it will last forever and it more than meets all the essential criteria (see page 61).

Folks often ask me whether it's worth purchasing a probe thermometer. You know, the kind that has a base that remains outside the oven and a probe on the end of a wire that you shove into your chicken or roast and leave in the oven as it cooks. Good models even have an alarm that will tell you when you've reached a particular target temp. Sounds great, right? No need to poke around in a hot oven.

The problem with these thermometers, however, is that it is nearly impossible to tell exactly where you should place the tip of the probe before you begin roasting. What may seem like it's going to be the thickest or coolest part of the chicken breast when you look at it in its raw state may not end up being the coolest—i.e., least-cooked—part of the chicken when it's done. What this means is that whether you use a leave-in probe or an instant-read, you're going to have to poke around in there when the chicken is close to done to find the coolest spot. For that reason, I use a leave-in probe as an early alarm system only, sticking with my instant-read to make the final call.

# { ROASTING POULTRY }

The problem with poultry, like many things in life, can be boiled down to two things: breasts and the government.

For some reason, years ago, poultry breeders got it into their heads that **most people like white meat**. As a result, birds have been bred for larger and larger breasts (that stick out farther and farther from their bodies). At the same time, the government got it into its head that people didn't want to kill themselves while cooking and started to recommend cooking poultry to the state beyond death known as "165°F." In this chapter, we'll find ways to circumvent both of these problems.

FACT: We love the taste of chicken. According to the USDA, about nine billion chickens are consumed in the United States each year. That's thirty chickens a year for every human being in the country, or approximately one breast, one leg, one wing, and a drumstick per person per week. That's a whole lotta bird (but we still complain every time some delicious creature like rabbit, snake, or alligator just "tastes like chicken"—that's hypocrisy for you).

For all that consumption, though, how many times a year do you sit back during a meal and say to yourself, "Mmm-mmm. This is a tasty chicken"? If the answer is fewer than thirty times per year (every single time you eat chicken, that is), then you could be doing at least a little bit better.

As with anything else, the key to perfect poultry is knowledge, care, and some practice. I can't help you with the last, but I just might be able to lend a hand in the first two categories. In researching the subject for this book, I roasted well over sixty birds, no two the same way—and that's not counting the hundreds, if not thousands, I've cooked in my lifetime.

Whether it's a young chicken with just enough meat on its bones to make a hearty meal for two or the Thanksgiving centerpiece for a family of twelve, the basics of selecting and cooking a chicken or a turkey are not all that different.

# The FOOD LAB's Complete Guide
## TO BUYING, STORING, AND ROASTING WHOLE POULTRY

A bird's a bird's a bird, right? Well, not necessarily—chickens and turkeys come in all shapes, sizes, and breeds, not to mention the various ways they are slaughtered, chilled, packaged, processed, and sold around the country. So you've got to make a lot of choices before you even begin to think about how to cook 'em. Here are the basics you need to know.

## SIZING

**Chickens seem to come in all sizes from peewee "game hens" all the way up to great big turkey-sized behemoths. Which one should I buy?**

Great question, and it depends what you want to do with it, as well as how many people you want to feed. For instance, broilers, fryers, and roasters all make great roasts, and they can be roasted pretty much the same way (timing will vary, but temperature will not). Pick the one you want depending on whether you're feeding just one other or a family of six. With chickens, when we're talking size, generally speaking, we're also talking age. Small, younger chickens have had less time to develop their muscles and connective tissue and, as a result, have more tender but less flavorful meat. Older birds tend to have more flavorful meat, but it's tougher and takes a longer time to break down properly.

In 2003, the USDA altered its classification system, shifting the entire thing back by a couple of weeks to make up for the fact that chickens these days are bred to reach maturity far faster than their predecessors did. The average chicken to reach the market these days is younger than three months of age. Oh, how fast they grow up!

The chart below lists the basic USDA categories and the best uses for each of them.

| CLASS | USDA DEFINITION | BEST USES |
|---|---|---|
| **Cornish Game Hen** | An immature chicken under 5 weeks of age and weighing no more than 2 pounds. (At least one parent of a Cornish game hen must be of the Rock Cornish breed.) | Single- or double-serving birds that should be stuffed and roasted, or grilled whole, or butterflied and grilled or pan-roasted. They have extremely tender meat and a very mild flavor. Plan on 1 hen per person. |
| **Broiler** | A chicken under 10 weeks of age that has yet to develop a hardened breastbone. Broilers weigh 1½ to 2 pounds. | With a very mild flavor and tender meat, these are best roasted, grilled, deep-fried, or pan-roasted. Small broilers can be treated like Cornish game hens, feeding 1 to 2 people. |
| **Fryer** | Similar to a broiler but larger, reaching up to 3½ pounds. This is the size I cook most often at home, and the size that most of the recipes in this book are designed for (for the soups and stocks, a stewing chicken can be used). | The perfect chicken for a family dinner, with enough meat to provide reasonable portions for 4 people, along with a carcass for making stock. The meat is tender but has a good deal of flavor. Best roasted, grilled, deep-fried, or pan-roasted. |
| **Roaster** | A chicken 3 to 5 months of age, weighing no more than 5 pounds, with a partially or fully hardened breastbone. | Great for serving a larger crowd; figure on about ¾ pound of carcass weight per person for reasonably sized portions. The meat is still tender but has a more robust flavor than that of younger chickens. Best roasted, grilled, deep-fried, pan-roasted, braised, or barbecued. |
| **Hen, Fowl, or Stewing Chicken** | A mature female chicken usually at least 10 months in age, generally weighing 6 pounds or more, with a fully hardened breastbone. | These chickens have noticeably tougher meat than younger ones, and the breast meat is particularly prone to drying out. They are best reserved for dishes like soups, and stews, stocks, and braises. The leg meat in particular works well in braises, where the large amount of connective tissue slowly converts to gelatin, adding richness and body. Some markets regularly carry stewing chickens (check out ethnic markets in particular), but you may have to ask your butcher to special-order one. |
| **Capon** | A male chicken under 8 months of age that has been castrated in order to promote more tender meat. | Their lack of hormone production makes capons the mildest tasting of all chickens, with very tender meat. They work in any recipe that calls for a broiler, fryer, or roaster, but they're uncommon in the United States. |
| **Rooster or Cock** | A mature male chicken with darkened meat and a fully hardened breastbone. | These guys make for pretty poor eating. They have smaller breasts than hens raised for the market, and what little meat they have tends to be dark, gamy, and stringy. Lucky then, that you'd have to go out of your way to find one. |

**What about turkeys? How big a bird should I get?**
Plan on about 1 pound of raw turkey per person, which translates to around half a pound of meat. In terms of flavor and ease of preparation, I find that the best birds are around 10 to 12 pounds, or even smaller. Much bigger, and they become very difficult to cook evenly. Large birds also take an inordinately long time, are difficult to transfer to and from the oven (not to mention trying to flip them), take up more oven space, and are more prone to drying out—all bad things when you've also got to deal with a houseful of family members, and Junior's just stolen Gramps's dentures and dropped them in Aunt Mabel's wine.

If you've got many mouths to feed, unless there's absolutely no way to get 'em in the oven, it's always a better idea to go with two smaller birds than one large one.

## READING THE LABELS

**There are so many labels and logos on the average supermarket chicken that it's hard to figure out what each means and which ones are important. What should I look for?**
Here's what you need to know.

- **"Hormone-Free"** means absolutely nothing. I repeat: absolutely nothing. By law, *no* chicken or turkey in the United States can be given any kind of hormones or steroids, so *every* chicken and turkey in the supermarket is completely free of added hormones. The labeling is a marketing gimmick to get you to think you are getting something special. It might as well read "deadly-cyanide free," because, yes, all poultry sold in this country is also free of deadly cyanide.
- **"Natural"** has very little meaning as well; it refers to birds that have no artificial colorings or additives and are minimally processed. Natural birds are routinely confined and raised in large batteries that offer no natural light or access to the outdoors. Unless you are buying rainbow-tinted birds, fresh meat—with no added ingredients—should be considered "natural." This is a self-enforced label and is not checked by third-party or government audits.
- **"No Antibiotics"** bears more weight than either of the two previous labels, indicating that the animals were raised without the use of antibiotics. There are arguments on both sides as to whether this is healthier for the consumer or for the birds.
- **"Fresh"** means that the meat has never been frozen (for poultry, freezing temperature is around 26°F, due to dissolved solids in its cells). Of course, some supermarkets do keep their stock cases colder than this minimum temperature, and you might find that the bird in the back of the case is frozen solid. A good way to determine whether or not the bird has been frozen is to check the packaging: Freezing damages cell structure and can cause interior liquids to leak out. If the packaging has lots of juices in it, chances are your bird was frozen. Move along.
- **"Cage-Free"** birds have been raised in large open barns rather than confined to small cages. However, this term does not guarantee they had any access to the outdoors, nor does it guarantee an improved stocking density (the number of chickens housed in a given space) or protection against debeaking, a painful procedure chickens are put through in order to prevent them from injuring each other when kept in close confinement. Chances are good that these birds were raised in cramped, crowded conditions.
- **Free-Range** or **Free-Roaming** birds have been raised in large open barns with limited access to

the outdoors via a door to an outdoor coop; often this is a single small door in a cavernous barn. While they are certainly better off than birds raised in cages, it's very likely that most "free-range" bids have not actually ever stepped foot outside. Even if they have, there's no guarantee that they've seen grass or pasture—that outside space can be dirt, gravel, or even concrete.

- **"Organic"** standards for birds are enforced by the government. By law, organic birds must be raised on a 100-percent-organic diet, must not be caged, and must have access to pasture and sunlight. Antibiotics are not allowed, and the animals must be "treated in a way that reduces stress," an ambiguous term that's generally agreed to mean a bit more space and an environment that promotes a few of their natural behaviors, like stretching their wings and enjoying dirt baths.

The table below summarizes all of this data:

| | CONVENTIONAL | NATURAL | CAGE-FREE | FREE-RANGE | ORGANIC |
|---|---|---|---|---|---|
| No hormones administered | ✓ | ✓ | ✓ | ✓ | ✓ |
| No animal by-products in bird feed | ✗ | ✗ | ✗ | ✗ | ✓ |
| No antibiotics administered | ✗ | ✗ | ✗ | ✗ | ✓ |
| No cages | ✗ | ✗ | ✓ | ✓ | ✓ |
| 100-percent-organic feed | ✗ | ✗ | ✗ | ✗ | ✓ |
| Access to outdoors | ✗ | ✗ | ✗ | ✓ | ✓ |
| Access to pasture and sunlight | ✗ | ✗ | ✗ | ✓ | ✓ |
| Stress-relieving environment | ✗ | ✗ | ✗ | ✗ | ✓ |
| Birds can be debeaked | ✓ | ✓ | ✓ | ✓ | ✓ |
| Audited by third party | ✗ | ✗ | ✗ | ✗ | ✓ |

The USDA makes no claims that organic foods are healthier than conventional foods, but it's pretty clear that organic production is healthier for the birds and the environment. If these matters concern you, choose Certified Organic birds at the market or, at the very least, birds that come from a reputable source that you trust. Many small and large farms that are both environmentally conscious and have a humane approach to animal welfare choose not to join the Organic program because of the fees involved or because they can't meet one of the standards (with small farms, this is often the "no antibiotics" standard, because the farmers will administer them to sick birds). Birds from such farms can still be a great choice.

**What the heck is an heirloom breed, and why would I ever want to buy a bird that's been passed down from generation to generation?**
Heirloom or heritage-breed birds come from pure genetic lines that can be traced back to a specific breed through several generations. What are the advantages of heirloom breeds? Well, the thing is this: we're a country obsessed with large breasts, and these days, most chickens and turkeys are bred for one thing—maximum white meat. If the breeders and poultry producers had their way, turkeys and chickens would resemble giant balloons of breast meat walking around on minuscule toothpick legs (or, better yet, sitting still and waiting for slaughter). Also, they'd have a dozen wings each. In the effort to grow an ever-increasing supply of lean breast meat, flavor has gone by the wayside. The modern chicken or turkey has extraordinarily large and extraordinarily bland breasts.

Heritage-breed birds represent an attempt to return to the old days of chicken and turkey farming, when the birds were a bit scrawnier but a whole

lot tastier. In blind tastings I've conducted, heritage-breed birds routinely beat out their modern counterparts in terms of flavor. The only downside? With thinner breasts and meat that's more prone to drying out, they're harder to cook—which means that a thermometer and very careful monitoring of temperature is more important than ever. With the recipes in this chapter, you'll get super tasty results even from supermarket birds, but I do recommend looking for heritage birds once you feel comfortable with roasting poultry.

## POST-SLAUGHTER PROCESSING

**What makes a "kosher" bird kosher?**
According to Jewish dietary laws, animals must be slaughtered in a particular way, which includes being washed, salted, and rinsed. The goal of this process is to remove excess blood from the meat.

In fact, any meat that you see in a supermarket was drained of blood immediately after slaughter (the red liquid in meat is not blood, by the way—see page 286). Kosher birds, however, get a few extra steps. After an initial wash in water, the birds are covered all over with coarse kosher salt. Through osmosis, additional blood and intracellular fluids are drawn out of the meat. Afterward, the birds are rinsed to remove excess salt and packaged to be sold.

As we'll soon find out, in the section on brining (see page 574), salt can have a powerful effect on meat, allowing it to retain a good 8 to 10 percent more moisture than an equivalent unsalted bird. If all you plan to do is cook your bird, kosher birds are a fantastic way of saving you the trouble of salting or brining them yourself. The act of koshering also means that your bird comes preseasoned with salt—a fact that leads to kosher birds often winning poorly controlled taste tests in which the competing non-kosher birds haven't been salted to an equivalent degree.

On the other hand, kosher birds give you less control over the finished product. What if you want to add other flavors with your brine? What if you want to make a salt-free stock from the leftover parts? I prefer to buy less-processed, nonkosher birds (which, of course, also give you the option of buying specific breeds and sizes of bird or air-chilled birds—see below), with the knowledge that if I want to salt or brine my bird, I can do it better myself.

More importantly: I have never seen a kosher bird that meets Certified Organic standards. Most are raised in high-density factory farms, just like other conventional birds.

**It's holiday time and I'm looking for a great turkey. I often see labels that say "self-basting" or "enhanced"— what do these mean?**

These labels are far more common with turkeys, but you may occasionally see them on chickens as well. The birds are injected with a flavored brine intended to help them retain more moisture as they cook—and it works. There's a reason why Butterballs stay so moist, even when they're drastically overcooked. My only problem with them is that the added liquid dilutes the natural flavor of the meat and often gives it a spongy quality. And many of the brands that offer "enhanced" birds, like Butterball or Jenny-O, use relatively flavorless factory-farmed birds, which doesn't help. Personally, I avoid these birds. Look for any fine print that says "enhanced," or check the ingredients list to make sure it doesn't include anything besides turkey or chicken.

**Sometimes I see chickens labeled "air-chilled" that sell for a premium at my market. Are they worth the extra cost?**

I believe they are. After slaughter, most conventional chickens are chilled by submerging them in ice water. It's an inexpensive, effective way to get them to a safe storage temperature rapidly, and for the chicken companies, it's got a bonus: the chickens absorb about 12 percent of their body weight in water and retain a full 4 percent at the time they are sold. But what does this mean for the consumer? Two things. First, when you buy a water-chilled chicken, part of what you're paying for is added water. This is especially the case if you buy your chicken packed in an airtight Cryovac-style bag (I'm sure you've noticed the copious liquid that leaks out when you open these bags). Second, your chicken will not cook as well. All that excess moisture it's holding on to is mostly close to the surface, particularly in the skin. As you cook the chicken, the moisture bleeds out, hindering browning and crisping. A water-chilled chicken will never get as crisp as an air-chilled chicken. For this reason, I absolutely avoid *any* chicken that's sold in a Cryovac bag and actively seek out air-chilled chickens.

Air-chilled chickens are cooled in blast coolers that rapidly circulate cold air around them. And most air-chilled chickens also come from reputable producers with humane standards. If you value crisp skin and good browning on your chicken (and why shouldn't you?), you'll probably find the extra cost of an air-chilled bird to be a reasonable exchange.

Turkeys are rarely air-chilled, so you'll have to take the extra time to dry them before roasting (see below).

**My supermarket carries only water-chilled chickens and turkeys. Is there anything I can do to get them to cook better?**

Certainly. The key is to remove as much excess moisture as possible before cooking them. As soon as you take the bird out of the package, rinse it and blot dry with paper towels inside and out, then place on a rack (set on a platter or baking sheet) in the fridge for several hours, or even overnight. The dry air of the fridge and the air circulation caused by its fan will help your chicken or turkey shed excess moisture—though don't expect it to ever get as crisp as a truly air-chilled bird.

**Once I get my bird home, is there any need to wash it or rinse it?**

If it's a nice air-chilled bird packaged on a tray or bought from the butcher and wrapped in paper, there's no need to wash it—in fact, all you're doing by washing it is soaking the skin, making it harder to cook and reducing its ability to crisp properly. Birds that come Cryovacked in tight plastic bags, on the other hand, are soaking in their own dripping juices. With these birds, I like to open the package in the sink (it'll save you a world of mess) and give the bird a quick once-over under cold running water to rinse away any of pinkish-red juices. Then it gets a thorough wipe down with paper towels, and into the fridge it goes (uncovered). Chicken that has been wrapped sometimes develops a very slight off odor even when it's perfectly safe to eat. In these cases, I also rinse it off before carefully drying and cooking it.

## CHICKEN PARTS

**Now that I've skipped ahead and read through the handy illustrated guide (page 583), I'm confident that I can break down my own chicken if I have to. But**

**what if I still want to buy chicken parts? What should I look for?**

Chicken parts can be a convenient way to get dinner on the table faster and more neatly, albeit a bit less frugally. But remember this: chicken producers *love* boneless, skinless breasts, and as a general rule, what's good for the producer is bad for the consumer. By selling boneless, skinless breasts, not only do they get to charge you a gigantic premium for what's essentially two minutes or less of work (pulling off the skin, cutting the breasts off the bone), but they can also transform those free rib cages into pricey packaged chicken stock.

Even if you're of the rare breed that doesn't like chicken skin, you should buy and cook chicken with the skin on. Why? It's a matter of moisture loss. Aside from cutlets and stuffed breasts, and, of course, chopped or ground meat for kebabs and stir-fries, pretty much any recipe that calls for boneless, skinless chicken breasts will work just as well and taste far better if you start with bone-in, skin-on chicken. Both the rib cage and the skin act as insulators, allowing the meat to cook more gently and evenly, as well as preventing it from losing too much moisture by keeping the surface covered. If you really don't want the skin, pull it off and throw it out *after* cooking. But save those rib cages for stock! I keep extra chicken parts in a gallon-sized zipper-lock bag in the freezer, then pull it out when it's full to make a big batch of stock (see page 187); even cooked rib cages will add flavor to the stock.

Legs are a slightly different story. I usually buy bone-in, skin-on legs because I just can't get enough of chicken skins, but if a recipe specifically calls for boneless thighs, I'll buy them boned. Boning thighs is not a fun task, no matter what anyone tells you.

## SAFETY AND STORAGE

**Is handling raw chicken and turkey really as dangerous as people say it is?**

Not nearly. There have been more reported cases of salmonella poisoning in the last decade from cantaloupe than from chicken. That said, it's always better to be safe than sorry. After working with raw poultry, *always* carefully scrub your cutting board, knife, hands, and any other surfaces the poultry may have come in contact with using warm soapy water, and dry them well. To avoid cross-contamination, never use the board or knife for other foods until after they've been cleaned.

**What is the best way to store poultry in the fridge?**

Poultry and other raw meats should always be stored on the bottom shelf of the fridge, toward the back. This is a safety measure intended to prevent any raw juices from dripping onto foods below the meat. Fresh poultry in a Cryovac package should remain fresh for a very long time—with turkey, as long as a couple of weeks. Once opened, however, it's best to use poultry as soon as possible. I try to use mine within 2 days of purchase.

**Is it a good idea to freeze poultry?**

Freezing poultry is perfectly fine, though you should expect it to lose some moisture and become tougher in the process—freezing causes large ice crystals to form within the meat that can pierce cells and cause fluid loss. If the meat isn't properly wrapped, you can also expect to see freezer burn over time—the sublimation of ice into water vapor that leaves the surface of the meat dry and stringy.

To get the best results when freezing poultry, you should first break down whole birds. Smaller parts freeze faster, minimizing ice-crystal formation. Airtight packaging is the only way to prevent freezer

burn, and though it may not seem like it, plastic wrap is actually "breathable" (not that you should try breathing through it!), allowing air to pass through it at a very slow rate—that's why that onion you wrapped in plastic wrap still manages to stink up the whole fridge. You're much better off using heavy-duty plastic bags specifically designed for the freezer or, better yet, a vacuum-sealing machine like the FoodSaver. If you don't have either of those options, wrap your meat tightly in a double or triple layer of plastic wrap, followed by a tight layer of aluminum foil (which *isn't* breathable). In any case, poultry can't be stored much longer than 2 months in the freezer before toughening becomes noticeable.

## WHAT ABOUT BRINING?

Let me start this off by saying I don't brine my turkey. Ever. Not for Thanksgiving, not for my Sunday supper, and certainly not for a quick weeknight meal. It's a personal decision, and you may well choose to do the opposite. Let me lay out both sides of the case.

It seems to me that as little as fifteen years ago, dry turkey was a given. The yearly Thanksgiving ritual at my family's table did not include any ill-mannered offspring crying out "DAAAAaaaad, Mom ruined the turkey again"—turkey wasn't something that could be ruined. It was always dry, tough, and stringy, and that was a fact of life. Then, about a decade ago, brining entered the scene.* Thanks to an overnight soak in a saltwater solution, gone were the days of dry breast meat, salvaged with extra servings of gravy. I, for one, welcomed our new moist-

---

*Or, more accurately, the technique that had been known to large swaths of the world for millennia, including China and Scandinavia, finally made inroads into North America's holiday menu.

breasted overlords. Even my mother could throw a turkey in the oven and pull out something edible a few hours later. It was positively magical!

These days, everybody and their grandmother (better known as the typical Thanksgiving gathering) has heard of brining, and more and more folks are doing it before Turkey Day. But it's not all pie and gravy. There are a few distinct and definite downsides to wet-brining, and many folks are now making the switch to dry-brining (aka extended salting). The question is, which method is best?

## How Brining Works

Before we get too far ahead of ourselves, let's do a quick recap of brining basics. The process involves soaking meat (usually lean meats like turkey, chicken, or pork chops) in a tub of heavily salted water overnight (most brines are in the 6- to 8-percent salt range by weight of the water). Over the course of the night, the meat absorbs some of that water. More important, that water stays put even when the meat is cooked. By brining meat, you can decrease the amount of total moisture loss by 30 to 40 percent.

To demonstrate, I cooked three identical turkey breasts in a 300°F oven to an internal temperature of 145°F. One was brined, one was soaked overnight in plain water, and the third was left as is. All three breasts came from nonkosher, nonenhanced birds (i.e., the birds were minimally processed). I charted their weights straight from the package, after brining, and then after cooking.

Both the bird soaked in brine and the one soaked in water gained a significant amount of weight prior to roasting, but the watered bird lost nearly all of that weight as it cooked and the brined bird retained significantly more. This cor-

responded to a juicier texture on eating. So, what's going on here?

Some sources attribute it all to osmosis—the tendency for water to move across a membrane from an area of low solute concentration to an area of high solute concentration. In this case, water moves from the brining vessel (low solute concentration) to the inside of the turkey's cells (where there are lots of proteins, minerals, and other fun biological goodies that dissolve in the water). But this theory is not quite accurate. If it were, then soaking a turkey in pure unsalted water should be more effective than soaking it in a brine, and we've seen that that is not the case. Moreover, if you soak a turkey in a ridiculously concentrated brine (I tested turkey in a fully saturated salt solution), according to the osmosis theory, it should dry out even more.

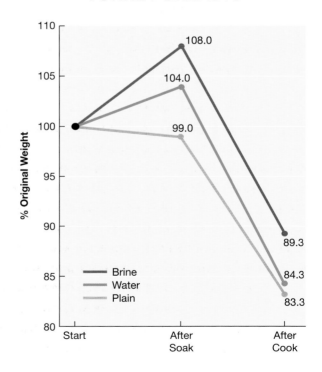

### WEIGHT LOSS IN COOKED TURKEY BREASTS

However, I found that despite turning the turkey inedibly salty, a highly concentrated 35-percent salt solution was just as effective as a more moderate 6-percent salt solution at helping the turkey retain moisture, indicating that there's something else going on here. To understand what's really happening, you have to look at the structure of turkey muscles. Muscles are made up of long bundled fibers, each one housed in a tough protein sheath. As the meat heats, the proteins that make up these sheaths will contract. Just like a squeezing a tube of toothpaste, this causes juices to be forced out of the bird. Heat the bird to much above 150°F or so, and you end up with dry, stringy meat.

Salt helps mitigate this shrinkage by dissolving some of the muscle proteins (mainly myosin). The muscle fibers loosen up, allowing them to absorb more moisture and, more important, they don't contract as much when heated, ensuring that more of that moisture stays in place as the turkey cooks. Sounds great, right? But there's a catch.

## The Problems with Brining

There are two major problems with brining. First off, it's a major pain in the butt. Not only does it require that you have a vessel big enough to submerge an entire turkey (common options are a cooler, a big bucket, or a couple of layered heavy-duty garbage bags tied together with hopes and prayers against breakage), but it also requires that you keep everything inside it—the turkey and the brine—cold for the entire process. With an extra-large bird, this can be for as long as a couple of days, meaning that you either give up using the main compartment of your fridge at the time of year that you most need it or you keep a constant supply of ice packs or ice rotating around the bird to keep it cold.

Second, as Harold McGee once pointed out to me, brining robs your bird of flavor. Think about it: The turkey is absorbing water and holding on to it. That 30 to 40 percent savings in moisture loss is not really turkey juices—it's plain old tap water. Many folks who eat brined birds have that very complaint: it's juicy, but the juice is watery.

I'd seen a few solutions (solutions, get it? ha-ha) offered for this problem, so I decided to test them all out side by side. I ran my tests on chicken breasts, which have essentially the same fat content and protein structure as turkey breasts but are smaller and easier to work with.

## Brining Solutions

By far the most common alternative to wet-brining is plain old salting (i.e., dry-brining). When you salt a turkey or chicken breast, meat juices are initially drawn out through the process of osmosis. As the salt dissolves in the juices, it forms what amounts to a very concentrated brine; see "How to Dry-Brine a Bird," page 579.

I've also heard people ask the very obvious question, "If brining introduces bland, boring tap water into the bird, why not brine in a more flavorful solution?"

Why not indeed? I decided to find out.

Here's what I tried:

- **Sample #1:** Plain (untreated)
- **Sample #2:** Brined overnight in a 6% salt solution
- **Sample #3:** Heavily salted overnight
- **Sample #4:** Brined overnight in chicken broth with a 6% salt content
- **Sample #5:** Brined overnight in cider with a 6% salt content
- **Sample #6:** Soaked overnight in plain water

(Samples #1 and #6 were included as a control to ensure that the brine and salt solutions were behaving as expected.)

As expected, the brined chicken samples held on to significantly more moisture than either the untreated breasts or the water-soaked breasts. (See also "Weight Loss in Cooked Chicken Breasts," page 359.) Indeed, in this test, the water-soaked breasts actually ended up drier on average than the plain breasts. Take a look at the carnage:

Water-soaked chicken breast.

Dry as the Gobi Desert (on an admittedly very-moist-for-a-desert day).

Then, take a look at the brined breast:

Brined chicken breast.

As plump and juicy as a benevolent aunt in a Disney film. Tasting it, there's a definite case of wet-sponge syndrome. Water comes out of it as you chew, giving you the illusion of juiciness, but the texture is a little too loose and the flavor a little bland.

Moving on to the salted breast, we find that it's still significantly moister than the nonsalted breast (though it was a couple of percentage points drier than the brined breast). Tasting it, it's undoubtedly more juicy and well seasoned, with a stronger chicken flavor. Texturewise, it's significantly different from both plain and brined turkey, with the smooth, dense-but-tender texture of lightly cured meat.

Visually, you can see clear signs of this curing with its decidedly pink hue:

Dry-brined chicken breast.

With a small chicken breast, the moist pink cured section extends nearly to the center of the breast. On a turkey, you'd only see it around the outer edges (which, serendipitously, happen to be the parts most prone to overcooking and drying out). While the brined breast was slightly juicier, flavorwise and texturewise, I'd take the salted chicken over the brined any day.

## What About Flavored Brines?

First off, don't try to brine your turkey or chicken in cider (or any other acidic marinade, for that matter). Just don't. The acid in the cider will begin the denaturization process of the meat, effectively "cooking" it without heat. The results? Ultradry meat with a wrinkled, completely desiccated exterior, like this:

Cider-brined chicken breast.

More interesting were the results of the broth brined chicken. It seems like the ultimate solution, right? If brining forces bland water into your meat, why not replace that water with flavorful broth?

Unfortunately, physics is a fickle mistress who refuses to be reined in. In tasting the broth-brined chicken next to the plain brined chicken, there was barely a noticeable difference in flavor: the broth-brined chicken still had the same hallmarks of a regular brined bird (juicy/wet texture, bland flavor). What the heck's going on?

There are two principles at work here. The first is that while to the naked eye broth is a pure liquid, in reality, a broth consists of water with a vast array of dissolved solids in it that contribute flavor. Most of these flavorful molecules are organic compounds that are relatively large in size—on a molecular scale, that is—while salt molecules are quite small. So, while salt can easily pass across the semipermeable membranes that make up the cells in animal tissue, larger molecules cannot.*

Additionally, there's an effect called "salting out," which occurs in water-based solutions containing both proteins and salt. Water molecules are attracted to salt ions and will selectively interact with them. The poor proteins, meanwhile, are left with only each other and end up forming large aggregate groups that make it even harder for them to get into the meat. When the salt breaks down muscle fibers sufficiently to allow the uptake of water, plenty of water and salt get into the meat, but very little protein does.

The result? Unless you are using an extraconcentrated homemade stock, the amount of flavorful compounds that make it inside your chicken or turkey will be very, very limited. Given the amount of stock you'd need to make this concentrated broth, it doesn't seem like a wise move.

---

*Good thing, too—otherwise you'd be leaking proteins and minerals out of your body every time you took a bath.

## What Does This All Mean?

Well, let me end the way I started: I don't brine my birds, because I like my birds to taste like birds, not like watered-down birds. Salting the meat is nearly as effective at preventing moisture loss, and the flavor gains are noticeable. Want to know the truth? Even advanced salting is not a necessary first step. I see it more as a safeguard against overcooking. It provides a little buffer in case you accidentally let that bird sit in the oven for an extra 15 minutes. As long as you are very careful about monitoring your bird, there's no reason to salt it in advance.

That said, it doesn't hurt to take precautions and let deliciousness, merriment, and family bonding ensue. You may not all be able to agree on whether the cranberries belong in the stuffing or on the side, but at least you can all agree that this is one darn tasty bird.

# HOW TO DRY-BRINE A BIRD

Salting poultry under its skin and letting it stand for a period of 24 to 48 hours in the refrigerator has much the same effect as brining. At first the salt draws liquid out of the meat (and this time it really is through osmosis), but then it dissolves in this extracted liquid, forming a concentrated bird-juice brine right on the surface of the bird that then goes to work at dissolving muscle fibers the same way as a regular brine. Eventually, as the muscle fibers get more and more relaxed, the liquid is reabsorbed. Over the course of a night or two, the salty solution can work its way several millimeters into the bird's flesh, helping it retain moisture and seasoning it more deeply. In some regards, it's more of a pain than regular brining (you have to loosen the skin from the meat), but it doesn't require the use of a massive cooler or ice-filled tub, and it doesn't dilute flavor in the way a regular brine does.

To dry-brine a bird, first carefully loosen the skin by running your hand or the handle of a wooden spoon between the skin and the breast meat, starting at the base of the breast. Then rub about 1 teaspoon of Diamond Crystal kosher salt per pound of meat all over its body, under its skin (or use one of the rubs in the chart on page 588). Place the bird on a rack set over a large plate or rimmed baking sheet and refrigerate uncovered, overnight (or for up to 48 hours if using a turkey). The next day, cook as directed, skipping or going light on the seasoning step.

# THERE'S THE RUB

When treating the skin of your turkey or chicken, there are a few options:

- **Going naked** is the easiest and will give you the crispest skin, particularly if you let the bird air-dry on a rack set in a rimmed baking sheet, uncovered, overnight in the fridge. Just don't let it dry for more than a day, or it'll turn papery and tough.

*continues*

- **Dry rubs made from salt mixed with spices and dried herbs** can add flavor to the skin. For best results, apply them the day before and let the bird air-dry overnight in the fridge. (For recipes, see pages 595–601).
- **Oil rubbed onto the skin** will get you a more even golden brown color, as it helps distribute heat from the hot oven air more evenly. It'll also help prevent the skin from drying out and turning leathery, though it will slightly decrease crispness.
- **Butter or an herb butter** will add lots of flavor to the skin (don't expect it to soak into the meat much, even if you spread it underneath instead the skin), but it'll also greatly reduce its crispness. Butter is about 18 percent water, which will cool down the skin as it evaporates. And the milk proteins present in butter will brown on their own, so poultry skin rubbed with butter will have a spottier appearance than skin rubbed with oil. Some people prefer this appearance (I do, on occasion).

## HOW TO ROAST A BIRD

Who doesn't love roast chicken? Crackly, crisp, salty skin. Moist, tender meat. Deep aromas filling the house. Little bits of fat and meat to tear off with your fingers or teeth as you linger over the last sips of your whiskey (whiskey goes with chicken, right?). It's about as classy and classic as food can get, and my go-to meal for company or the rare quiet night in with the wife and dog.

But, to be perfectly frank, most of the time, I *don't* like roast chicken, because most of the time, well, chicken, just isn't roasted very well. The problem is one I'm sure everyone of you has experienced: dry breast meat and it doesn't just apply to chicken—we've all also experienced dry turkey). I'm not talking about the kind that frays around the edges as soon as a carving knife comes close to it or that instantly turns to sawdust when it hits your tongue; I'm talking the kind that is just good enough that you can still smile and say nice things during dinner, but just bad enough that you wonder why the Pilgrims couldn't have eaten prime rib during that first fall.

**The problem, as we all know, is with overcooking.**

So first, let's take a quick look at what happens to chicken breast meat as it cooks:

- **Under 120°F:** The meat is still considered raw. Muscle cells are bundled up and aligned in long, straight cable-like fibrils wrapped in sheaths of elastic connective tissues, which are what gives meat its "grain."
- **At 120°F:** The protein myosin begins to coagulate, forcing some liquid out of the muscle cells, which then collects within the protein sheaths.
- **At 140°F:** The remaining proteins within the muscle cells coagulate, forcing all of the liquid out of the cells and into the protein sheaths. The coagulated proteins turn the meat firm and opaque. I like my chicken and turkey breasts cooked to 140°F.
- **At 150°F:** The proteins in the sheaths themselves (mainly collagen) rapidly coagulate and contract. All the water that was forced out of the cells and collected within the sheaths is now squeezed out of the meat completely. Despite government warnings to be sure to cook chicken to 165°F, in reality, above 150°F or so, muscle fibers

have become almost completely squeezed dry. **Congratulations, your dinner is now officially cardboard**.

On the other hand, leg meat must be cooked to at least 170°F. OK, that's a bit of an overstatement. It'll still be perfectly edible at around 160°F (any lower than that, and the abundant connective tissue will remain tough), but the juices will still be pink or red and the meat will not have yet reached optimal tenderness. Unlike breast meat, leg meat contains plenty of collagen. Given a high-enough tempera-

ture (say 160°F and above) and a long-enough time (say the 10 minutes it takes the legs to get from 160° to 170°F), the collagen will begin to convert into rich gelatin, keeping the meat moist and juicy even after the muscle fibers have shed most of their liquid.

So, the question is, how do you cook legs to 170°F without taking the breasts beyond 145°F? I use three different methods, depending on the situation and the bird at hand. Here they are, starting with the most effective, and most laborious.

## POULTRY COOKING METHOD 1
# (My Favorite): SPATCHCOCKING

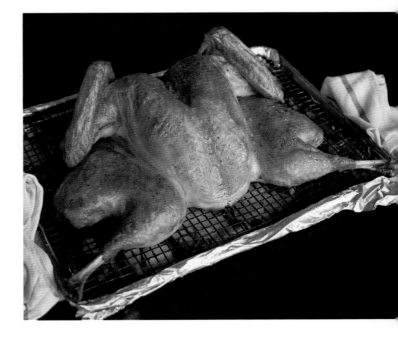

I'm gonna come right out and say it: this is the best way to cook a bird, hands down. It's now the *only* method I ever use. I understand that some folks like to see a whole bird arrive at the table looking like

a whole bird, but if I had my way, the gospel of spatchcocked birds would spread around the world.

To butterfly (a less fancy term for spatchcock) a bird, all you've got to do is use a pair of sharp poultry shears to cut out the spine, then flatten it, with its skin side up, by pressing down firmly on the breastbone. *Voilà*, that's it! It's a really simple operation that you'll get the hang of in no time, and it even works for turkeys.

Here's how cooking the bird works: Put the bird skin side up on a rack set in a rimmed baking sheet. Blast it in a hot oven (I'm talking 450°F), and you'll find that, miraculously, the breast will reach 150°F just as the legs reach 170°F and the skin reaches delicious. No brining, no salting, no flipping, no problems.

As I said, you do lose the prettiness of bringing a whole bird to the table for carving, but you gain the vastly preferable prettiness of perfectly cooked meat instead, and that's a trade-off I'll take any day. Its advantages are numerous.

## Advantage 1: Flat Shape = Even Cooking

Butterflying the bird and laying it out flat, with the legs spread out to the sides, means that what were once the most protected parts of the bird (the thighs and drumsticks) are now the most exposed. As a result, they cook faster—precisely what you want when your goal is cooking the dark meat to a higher temperature than the light meat.

As an added bonus, the bird doesn't take up nearly as much vertical space in your oven, which means that if you wanted to, you could even cook two birds at once. This is a much better strategy for moist meat than trying to cook one massive bird.

## Advantage 2: All the Skin on Top = Juicier Meat and Crisper Skin

A regular chicken (or turkey) can be approximated as a sphere, with the meat on the inside and the skin on the outside. Because it's resting on a roast-ing pan or baking sheet, one side of that sphere will always cook more than the other.

A spatchcocked chicken, on the other hand, resembles a cuboid, in which the top surface is skin and most of the volume is meat. This leads to three end results: First, all of the skin is exposed to the full heat of the oven the whole time. There is no skin hiding underneath, no underbelly to worry about. Second, there is ample room for the rendering fat to drip out from under the skin and into the pan below. This makes for skin that ends up thinner and crisper. Finally, all of that dripping fat distributes heat energy over the meat as it cooks, both helping it to cook more evenly and creating a temperature buffer, protecting the meat from drying out.

## Advantage 3: Thinner Profile = Faster Cooking

In terms of cooking, a sphere is the least efficient shape—that is, for a given mass, it's the shape that'll take the longest for heat to penetrate through to the center. Because of this, a regular roast chicken can take an hour or more to cook, a turkey several hours. But with a spatchcocked bird and its slim profile, you can blast it at 450°F and it'll cook through in about half the time. If I added up all the time I could have saved in Thanksgivings past using this method, I could perhaps—dare I say it—rule the world?

## Advantage 4: It's Easier to Carve

Carving a whole chicken can be a tricky affair. Its shape makes it tough to find an angle where you can get good leverage, and I usually resort to flip-flopping the bird around a few times as I carve it. A spatchcocked bird, on the other hand, is simple. The legs nearly fall off all on their own, requiring just a little tug and a single slice with the knife. Rather than having to flip or turn the bird to get at the wings, the laid-flat breasts expose them to you, making it easy to get them off without having to lift the chicken from the board. Even the breasts are easier to remove from the carcass, as it lies completely flat and still while you work.

### Advantage 5: Extra Bones = Better Gravy

It's always possible to make gravy or "jus" with nothing but canned chicken stock and drippings, but gravy is so much better when you have some real bones and meat to work with. Normally that means using the neck and giblets to flavor the broth while the bird roasts. You can still do that. But this way, you can add the entire bird's back to the mix, resulting in a far more flavorful broth.

## POULTRY COOKING METHOD 2
# (If You've Got Time to Spare): DIVIDE and CONQUER

You'll need to know how to break down a bird to use this method (or, to make it far easier, just buy parts individually at the store), as well as being completely willing to kiss any Norman Rockwell dreams of picture-perfect roast birds good-bye.

Separating the legs and the breasts from each other makes it a very simple matter to roast the parts together, removing the breasts from the oven once they reach their final temperature and cooking the legs until they reach theirs. After allowing everything to rest, all you need to do is bang the oven up to a raging 500°F, throw everything back inside for a few minutes to crisp up the skin, and dinner is served. To maximize even cooking, I make sure to cook the parts in a very gentle oven—as low a temperature as I have the time and patience for (see page 314 for an explanation on why a low oven promotes even cooking).

For the best results, it's always better to leave the breast meat attached to the bones and the skin on when you roast the breasts. This has nothing to do with any kind of mythical exchange of flavor between bones and meat (see "Bones," page 627) and everything to do with exposed surfaces. The more surface area of the meat is exposed, the more moisture it'll lose, but the bones and skin help insulate the fragile white meat, preventing it from drying out. The bones are much easier to remove after cooking, if you'd like to do so, and you can always pull off and throw out the skin if you don't want to eat it (or, better yet, pass it on to someone else in the family who'll be happy to take it off your hands).

This method also has a few advantages: namely, that you're not anatomically limited (you want 6 legs and 4 breasts?—no problem!), and the bird is less cumbersome to move around. Once it's butchered (or, even easier, just buy the parts you need prebutchered), all of the pieces are easy to pick up with tongs or your hands. If you want to really go the extra mile, this method also allows you to cook the breasts and legs by two completely different methods (see Thanksgiving Turkey Two Ways, page 617), giving you more options at the dinner table.

# (For the Traditional Look): The HOT STEEL

Say you don't want to break down your bird for whatever reason—you prefer the whole-bird presentation, or the thought of poultry shears makes you squeamish. I get it, sometimes you just need a different method. Well, here's a way in which you can start with a whole bird and end up with something that is evenly cooked with reasonably crisp skin, without the need to manhandle it before it enters the oven.

It does require one piece of specialized equipment: a pizza steel or stone. Place it on the lower rack of the oven and allow it to preheat for half an hour with your oven at full blast. Place the bird on a rack set on a rimmed baking sheet with the breasts facing up, then place the pan on top of the hot steel or stone. Immediately turn the oven down to 400°F. The retained heat in the steel or stone will give the legs a head start, and you'll find that your whole chicken will come to the right temperatures at pretty much the same time. Pretty easy, right? It won't crisp up quite as well as a spatchcocked bird, nor will it cook as fast, but if you want your bird coming to the table looking like the cover of the *Saturday Evening Post*, it's your best bet.

## SHOULD I USE THE CONVECTION SETTING IF MY OVEN HAS ONE?

A convection oven is nothing more than a regular oven with a fan mechanism inside it. The fan force-circulates air throughout the oven, overriding any natural convection currents that form through normal hot and cool zones in a regular oven. What this means is that the entire oven will be heated to a relatively even temperature. The fan also makes food cook faster and crisp better. How so?

In a regular oven, cool zones will naturally form around foods and in areas that are protected from radiation heat, such as below the rim of a roasting pan or inside the cavity of a turkey. A convection oven, on the other hand, forces hot air to circulate all around the food, constantly supplying it with plenty of heat energy. It also whisks away surface moisture, causing skin of crusts to dry out faster. This can significantly increase the rate of cooking, as well as make for better, more evenly browned poultry skin or meat crusts. If your oven has one, I recommend using the convection setting, particularly for roasting meats and poultry.

The drawback is that most recipes (including the ones in this book) are not specifically designed for convection ovens, so some adjustment is necessary to get them to work as advertised. The general rule of thumb is that if you are using a recipe developed for a conventional oven but cooking in a convection oven, reduce the oven temperature by around 25 degrees. The exact adjustment will vary depending on oven brand and model, so some testing may be required before you learn exactly how your oven works.

## WHAT MAKES CHICKEN SKIN CRISP AND GOLDEN?

Chicken skin is composed mainly of three elements: water, fat, and protein (mostly collagen). In order for it to become crisp, a few things must happen. First, the collagen must convert to gelatin. Next, the water must evaporate. Finally, the fat must render and run off. What you're left with is skin that's a crispy, golden brown shell of its former self.

To enhance the process, there are several things you can do. First, start with relatively dry chicken. Look for chickens that are labeled "air-chilled." Regular chickens are chilled in water, which adds extra moisture to the skin and can prevent good browning. Second, dry the skin well with paper towels. Better yet, if you have the time, let the chicken sit on a rack on a baking sheet, uncovered, in the refrigerator for a day. This will jump-start the drying process, allowing the skin to crisp faster in the oven. But letting it rest for longer than a day is not a good idea. Why is that?

Well, collagen breakdown is a time- and temperature-dependent process that requires the presence of moisture and a temperature of at least 160°F. For this reason, if you cook your chicken at too low a temperature or let the skin dry too much (by, say, allowing it to rest for several days uncovered in the fridge), you'll end up driving off the moisture required for the collagen to turn into gelatin. The skin becomes papery or leathery instead of crisp.

Aiding the rendering of the fat can give you a big boost in good crisping. To do this, you need to create channels for the rendering fat to escape your bird. The most effective means of doing this is to butterfly the bird. You end up with all the skin on top of the meat, exposing it all to the full blast of convective heat from the oven and leaving plenty of room underneath for the rendering fat to drip down and around the meat. If you are going to keep your bird whole, you should at the very least separate the skin on the breast from the meat to allow room for fat to drip out (see page 582).

To increase airflow and make heating more efficient, I also advise using a heavy-duty rimmed baking sheet with a wire rack set on it in stead of a roasting pan with a V-rack. The high sides of a roasting pan can interfere with airflow, leaving you with chicken or turkey that's still flabby and pale in its undercarriage. The only time I use a roasting pan instead of a baking sheet is for extremely large roasts, like a big turkey or a standing rib roast. If it fits on the baking sheet, I'm using the baking sheet.

If you want the skin of your roast chicken to stay crisp longer, remove it from the bird right after you take it out of the oven and serve it separately. This will prevent steam from the meat from softening it again.

# SEPARATING THE SKIN FOR CRISPER SKIN

Separating the skin from the meat of a roasting chicken allows the rendering fat to escape more easily, resulting in a crisper bird. It also allows you to season the bird underneath the skin. Here's how to do it.

### Step 1: Season the Exterior

Season the exterior of the bird well with salt and pepper.

### Step 2: Go in from the Bottom

Lift up the flap of skin at the bottom of the breasts, insert one or two fingers, and slowly work your way upward, separating the skin as you go and being careful not to tear it.

### Step 3: Meet in the Middle

Use the fingers of your other hand to go in through the neck end, separating the skin there. Your fingers should be able to meet in the middle once all the skin has been separated. Rub salt and pepper into the breast meat. Your bird is now ready to roast.

# BASTING AND TRUSSING: TWO METHODS THAT *DON'T* MAKE FOR JUICIER MEAT

I've seen it suggested that barding chicken by draping it with slices of a fatty meat, like bacon, or basting it by spooning melted butter or pan juices over the top as it cooks, will help it stay moist. There are two theories behind this. The first is that some of the fat will be absorbed into the breast meat. Poppycock. As our experiments have shown us, that breast meat is shrinking and actively forcing juices out—it certainly ain't absorbing anything! The second theory applies only to barding, and it's that the layer of fat will provide insulation to help the meat cook more gently and prevent it from drying out. This much is true, but here's the problem: as much as I love most things more when they're wrapped in bacon (yes, dear, I'm talking about you), bacon makes mild chicken and turkey taste like, well, bacon. And if I wanted bacon, I'd cook bacon. (Then again, if you are the type who *likes* that flavor in your turkey or chicken, go for it!) It also precludes the possibility of crisp skin.

In fact, not only does basting the breast with hot pan juices *increase* the rate at which it cooks, exacerbating the dryness, but the moisture in the juices or melted butter also keeps the skin from crisping properly. A far better solution is to brush your bird occasionally with room-temperature oil (or rendered duck or chicken fat, if you want to get fancy) as it cooks. This will help you achieve deeper, more even browning, but it won't affect moistness in any way.

Trussing—the act of tying up the bird's legs before cooking—is also an oft-recommended but totally pointless exercise. In fact, it has the *opposite* effect of what you want for a bird, effectively shielding the inner thighs and thereby making the slowest-to-cook part cook even more slowly. Chickens and turkeys should always be left as nature intended them: with their legs wide open, to allow for maximum heating via convection.

## LET IT REST!

Resting is key to making any roast poultry dinner, particularly a hectic one like Thanksgiving, both easier and tastier. Resting allows time for the meat to relax and the internal juices to redistribute themselves evenly throughout it. Additionally, the slightly cooler temperature will cause the meat's juices to thicken considerably, making them less likely to flow out of the bird when you carve it (for more on resting, see pages 306 and 870).

I let my birds rest until their internal temperature has dropped down to 143°F or less. For a chicken cooked to 150°F, this will take between 10 and 15 minutes; for a 10- to 12-pound turkey, it can take over 30 minutes. But consider that an added bonus: you've now got an extra half hour to do things like deglaze your pan drippings, heat up your casseroles, have a cocktail, and make whipped cream to cover up the fingerprints you left on the pumpkin pie.

## WHAT ABOUT FLAVOR?

To be perfectly frank, 90 percent of the time I roast a chicken, I rely on salt and pepper alone—if you've got yourself a really great chicken, its flavor should speak for itself. But what if you want to add a little extra something to it?

Here's the good news: once you have the basic roasting techniques down, adding flavor is as simple as a good rub or some herbs applied before cooking; the actual cooking method is exactly the same. Those dozens or hundreds of chicken recipes you can find online for herb-rubbed this or lemon that? They're really all the same old roast chicken with a few flavors added to it. What that means is that once you understand what makes chicken good (proper roasting technique, not overcooking the breasts, allowing the skin to render), how you flavor the bird is entirely up to you.

To get you started, I put together a little chart that shows you how to apply various flavors to your bird, along with a few specific recipe variations that I enjoy. Of course you can mix and match all of these flavoring agents as desired. In the chart, I haven't listed any salt, because in the two basic recipes that follow, the chicken is seasoned separately with salt. But you can just as easily add the salt directly to any herb rub or spice mixture to season and flavor the bird in one step.

| FLAVORING AGENT | HOW TO APPLY |
|---|---|
| **Tender, Leafy Herbs (parsley, basil, tarragon, cilantro, etc.)** | Finely chop ½ cup fresh herb leaves by hand or in a food processor. Combine with 1 to 2 tablespoons olive oil or melted butter to make a paste. Separate the skin covering the breasts from the meat. Rub the herb mixture all over the chicken, including between the breast skin and meat. Proceed as directed. |
| **Woody Herbs (thyme, rosemary, bay leaf, etc.)** | Separate the skin covering the breasts from the meat. Place whole herb sprigs inside the carcass (or under and on top of the chicken if butterflied). Proceed as directed. Discard the herb stems before serving. |
| **Alliums (garlic, shallots, scallions, chives)** | Finely mince enough alliums in a food processor or by hand to make 2 teaspoons to 2 tablespoons. Combine with 1 tablespoon olive oil or melted butter to make a paste. Separate the skin covering the breasts from the meat. Rub the herb mixture all over the chicken, including between the breast skin and meat. Proceed as directed. |
| **Spices** | Combine 1 to 4 teaspoons spices with 1 to 2 tablespoons olive oil or melted butter to make a paste, adding 1 to 2 tablespoons red or white wine vinegar if desired. Separate the skin covering the breasts from the meat. Rub the spice mixture all over the chicken, inside the cavity, and underneath the skin. Proceed as directed. |

| FLAVORING AGENT | HOW TO APPLY |
| --- | --- |
| Glazes and Marinades | Brush onto the skin during the last 10 to 15 minutes of roasting. Reserve the excess and serve with the chicken as a sauce. |
| Lemon | Separate the skin covering the breasts from the meat. Cut a lemon in half and rub the cut surfaces all over the chicken, including the cavity and under the skin. Then cut the lemon halves into slices and place underneath or inside the chicken. Proceed as directed. If desired, squeeze some lemon juice on top of the chicken halfway through roasting and just before serving (this will slightly soften the crisp skin but will add flavor).<br><br>Alternatively, add grated lemon zest to any chopped herb mix or spice blend. |

## SHOULD I MARINATE MY CHICKEN?

When I was in college, I remember receiving a mass e-mail from a housemate that said, "I'm leaving for the weekend, but there's a chicken in the fridge that's been marinating in Italian dressing for three days. Somebody should cook it—it'll be super-tender and juicy!" I snapped up the chicken, planning to use it as a surefire way to finally seduce the girl who lived down the hall. She'd be so impressed by its moistness and flavor that she'd instantly fall in love, we'd get married (and serve chicken at the reception, of course), have fourteen kids, and live happily ever after.

That I'm now happily married to an entirely different girl tells you something about how that worked out. Turns out that you *can* overmarinate chicken, particularly if using an acidic marinade (such as Italian dressing). The acid in the marinade will cause the proteins to denature, the same way that cooking them does. Given enough time, the denatured proteins will squeeze out moisture, and the chicken will take on a dry, chalky texture.

Just as with steak (see page 328), there's no need to marinate poultry for any more than half a day or so. Actually, far more effective use of a marinade for chicken and other poultry is to apply some of it to the roasting bird 10 minutes or so before you take it out of the oven and save most of it to apply *after* resting and carving, when it can really coat every surface and deliver flavor with each bite.

# HOW TO REMOVE THE WISHBONE
# FROM A CHICKEN OR A TURKEY

Removing the wishbone before roasting the bird makes carving it much easier. It's optional for a roast chicken but highly recommended when roasting a turkey.

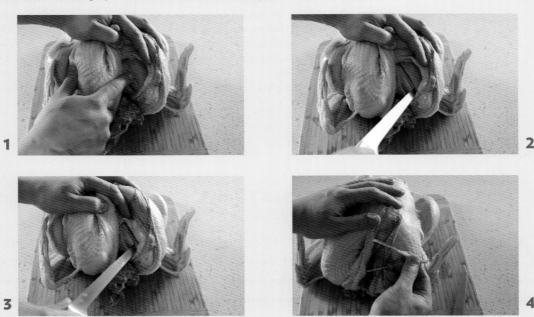

### Step 1: Locate the Wishbone

This step takes place *before* you start cooking the turkey. It's even more important with a turkey than with a chicken, making your carving job at the table far easier. Start by pulling back the flap of skin at the neck and finding the small Y-shaped bone that runs along the top of both breast halves.

### Step 2: First Incision

Make your first cut on one side of one branch in the wishbone with the tip of a sharp boning, chef's, or paring knife.

### Step 3: The Other Side

Repeat on the other side of the same branch, running the tip of the knife along it. Repeat on the opposite branch, making 4 incisions total.

### Step 4: Pry it Loose

Grab the top of the bone with your fingers or a dry kitchen towel and pry it toward you. It should come out with just a bit of a tug. If you are having trouble, locate the problem spots and use the tip of your knife to loosen it further. Once the bone is out, roast your turkey.

# ROASTED BUTTERFLIED CHICKEN

This is the easiest, most foolproof way to guarantee crisp skin and very moist meat from both the breast and the legs. As long as you don't mind that the bird won't arrive at the table looking like a whole chicken, you will not find a better way to roast a chicken.

NOTE: For the juiciest results and crispest skin, dry-brine the bird and air-dry it overnight (see page 579).

### SERVES 3 OR 4

1 whole chicken, 3½ to 4 pounds, butterflied according to the instructions on page 593

1 tablespoon vegetable, canola, or olive oil

Kosher salt and freshly ground black pepper

1. Adjust an oven rack to the upper-middle position and preheat the oven to 450°F.

2. Dry the chicken thoroughly with paper towels. Separate the skin from the breasts (see page 586), then rub the chicken evenly all over and under the skin with the oil. Season all sides with salt and pepper (go light on the salt if the bird has been dry-brined).

3. Set a wire rack on a rimmed baking sheet lined with aluminum foil. Position the chicken so that the breasts are in the center of the baking sheet and the legs are close to the edges. Roast until the thickest part of the breast close to the bone registers 145°F on an instant-read thermometer and the joint between thighs and body registers at least 160°F, 35 to 45 minutes.

4. Transfer the chicken to a cutting board, tent loosely with foil, and allow to rest for 10 minutes, then carve and serve.

# QUICK JUS
## FOR ROASTED BUTTERFLIED CHICKEN

**NOTE:** For this recipe, you'll need the neck and backbone from the butterflied chicken.

**MAKES ABOUT ½ CUP**

1 tablespoon vegetable or canola oil

Reserved chicken backbone and neck, roughly chopped with a cleaver

2 teaspoons chopped fresh thyme, rosemary, oregano, marjoram, or savory, or a mix (optional)

1 onion, roughly chopped

1 medium carrot, peeled and roughly chopped

1 stalk celery, roughly chopped

1 bay leaf

1 cup dry vermouth or sherry

1 cup water or homemade or low-sodium canned chicken stock

1 teaspoon soy sauce

3 tablespoons unsalted butter, cut into pieces

2 teaspoons lemon juice (from 1 lemon)

Kosher salt and freshly ground black pepper

1. While the chicken roasts, heat the oil in a small saucepan over high heat until shimmering. Add the chopped chicken bones and neck and cook, stirring frequently, until well browned, about 3 minutes. Add the herbs, if using, onion, carrot, and celery and cook, stirring frequently, until beginning to brown, about 3 minutes. Add the bay leaf and deglaze the pan with the vermouth and water, using a wooden spoon to scrape up any browned bits from the bottom of the pan. Reduce the heat to maintain a simmer and cook for 20 minutes.

2. Strain the sauce and return it to the pan. Boil over medium-high heat until reduced to approximately ⅓ cup, about 7 minutes. Off the heat, whisk in the soy sauce, butter, and lemon juice. Season to taste with salt and pepper. Keep warm until ready to serve.

# KNIFE SKILLS:
## How to Butterfly a Chicken or Turkey

**Butterflying any bird is pretty much the same, though you may have a tougher time getting through the bones of a large turkey than a chicken. If the idea of forcibly cutting through the bones with a pair of kitchen shears gives you the willies, ask your butcher to butterfly your turkey or chicken for you, making sure to save the backbone (and neck) to use in your sauce.**

**STEP 1: THE TOOLS** You'll need some good poultry shears. I like the Kitchen Shears from Kuhn-Rikon for their sharp blades, sturdy construction, and heavy spring, which helps them pop back out into place, ready for the next snip.

**STEP 2: FLIP AND SNIP** Position the bird so it's breast side down on the cutting board. Holding it firmly with one hand (a kitchen towel helps if the bird is slippery), make a cut down one side of the backbone, starting where the thigh meats the tail.

**STEP 3: CUT AROUND THE THIGH** If you are too far from the spine, you may hit the thigh bone: just shift the shears inward a bit toward the backbone to cut around the thigh bone instead of trying to crack through it. Then continue cutting through the ribs until you completely separate one side of the backbone.

**STEP 4: REPEAT ON THE SECOND SIDE** Make an identical cut down the other side of the backbone, being careful not to let your fingers get in the way of the shears as you grasp the bone.

**STEP 5: SURGERY COMPLETE** If all goes well, you will have completely removed the backbone. Use your fingers or the shears to remove any large pockets of fat and to clean out any exposed red marrow from the cut bones.

**STEP 6: FLIP AND TUCK** Flip the bird back over and tuck the wing tips underneath the breasts to help keep them in place.

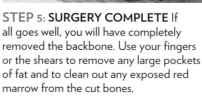

**STEP 7: PRESS DOWN** Splay the bird out in a manner that can only be described as inappropriate, then place your palms firmly upon the breasts and press down hard on the ridge of the breastbone until it lies flat. (With a turkey, you'll probably hear a few cracks.) The bird is now ready to roast.

# SIMPLE WHOLE ROAST CHICKEN

**If you value the presentation of a whole bird at the table, this is the recipe for you. You'll need a pizza steel or stone to help ensure that the legs cook as fast as the breasts.**

**NOTE**: For the juiciest results and crispest skin, dry-brine the bird and air-dry it overnight as described on page 579.

### SERVES 3 OR 4

1 whole chicken, 3½ to 4 pounds

1 tablespoon vegetable, canola, or olive oil

Kosher salt and freshly ground black pepper

1. Adjust an oven rack to the middle position and place a pizza steel or stone on it. Preheat the oven to 500°F for at least 30 minutes.

2. Place the chicken on a work surface and dry thoroughly with paper towels. Separate the skin from the breasts (see page 586), then rub the chicken evenly all over and under the skin with the oil. Season on all sides with salt and pepper (go light on the salt if the bird has been dry-brined). Set aside at room temperature while the oven preheats.

3. Line a heavy rimmed baking sheet with lightly crumpled aluminum foil. Place the chicken, breast side up, on a wire rack in the pan, put the baking sheet on the pizza steel or stone, and reduce the oven temperature to 350°F. Roast until the coolest part of the breast registers 145°F on an instant-read thermometer and the legs register at least 160°F, about 1 hour, brushing the chicken with the pan drippings halfway through cooking if desired. Remove from the oven and allow to rest, uncovered, for 15 minutes. Then carve, and serve.

# BUTTERY LEMON-HERB-RUBBED ROAST CHICKEN

**Butter, lemon zest, and herbs are a classic combination with roast chicken. True, adding butter will make the skin less crisp but the added flavor will compensate.**

**NOTE:** For the juiciest results and crispest skin, dry-brine the bird and air-dry it overnight as described on page 579.

**SERVES 3 OR 4**

¼ cup fresh parsley leaves

6 fresh sage leaves

1 tablespoon fresh rosemary leaves

1 scallion roughly chopped

1 medium clove garlic, minced or grated on a microplane (about 1 teaspoon)

2 teaspoons kosher salt

1 teaspoon freshly ground black pepper

1 tablespoon grated lemon zest (from 1 lemon)

2 tablespoons unsalted butter

1 whole chicken, 3½ to 4 pounds

Combine the parsley, sage, rosemary, scallion, garlic, salt, pepper, lemon zest, and butter in a food processor and process until a paste is formed, scraping down the sides as necessary. Separate the chicken skin from the breasts (see page 586). Rub the herb mixture all over the chicken and under the skin. Roast according to the Simple Whole Roast Chicken recipe (page 594), skipping step 2.

# JAMAICAN-JERK-RUBBED ROAST CHICKEN

**Jerk-style chicken is a specialty of Jamaica, where the chicken gets a wonderful smoky-sweet flavor from being slow-cooked over fresh pimento wood or laurel wood—the tree that bay leaves come from. We don't have access to those ingredients in our kitchens, but we do have what it takes to make the spicy allspice-scented marinade. Rather than roast the bird over laurel wood, I do the next best thing: roast it surrounded on all sides by thyme and bay leaves. Using the butterflied method gets you gorgeously crisp, charred skin.**

**NOTES:** For the juiciest results and best skin, dry-brine the bird and air-dry it overnight as described on page 579. Be extremely careful working with Scotch bonnets or habaneros: they are very hot and their oil can cause skin and eye irritations. Use a separate cutting board, wear latex gloves if you have them, and wash all surfaces and knives immediately after slicing them.

**SERVES 3 OR 4**

2 teaspoons ground allspice

1 teaspoon freshly ground black
  pepper

¼ teaspoon ground nutmeg

¼ teaspoon ground cinnamon

1 medium clove garlic, minced or
  grated on a microplane (about 1
  teaspoon)

1 scallion, roughly chopped

½ teaspoon minced fresh ginger

½ Scotch Bonnet or habanero
  pepper (see Note above)

1 teaspoon cider vinegar

1 teaspoon soy sauce

2 teaspoon kosher salt

1 tablespoon vegetable or canola oil

1 whole chicken, 3½ to 4 pounds,
  butterflied according to the
  directions on page 593

1 bunch thyme

6 bay leaves

1. Combine the allspice, pepper, nutmeg, cinnamon, garlic, scallion, ginger, chile, vinegar, soy sauce, salt, and oil in the bowl of a food processor or blender and blend until a rough wet paste is formed.

2. Separate the chicken skin from the breasts (see page 586). Spread the spice mixture evenly all over the chicken and under the skin. Roast according to the Roasted Butterflied Chicken recipe (page 591), skipping step 2 and placing the thyme sprigs and bay leaves inside the cavity, under the skin, and over the chicken before you put it in the oven. Discard the thyme stems and bay leaves before serving.

# SPICY LEMONGRASS-AND-TURMERIC-RUBBED ROAST CHICKEN

**Lemongrass and turmeric give this roast chicken a deep color and wonderful aroma. It's a little spicy, a little sweet, and very flavorful—even more so if you serve it with the chile sauce. You can use either roast chicken method for this recipe.**

NOTE: For the juiciest results and crispest skin, dry-brine the bird and air-dry it overnight as described on page 579.

## SERVES 3 OR 4

1 stalk lemongrass

2 teaspoons grated fresh ginger

2 medium cloves garlic minced or grated on a microplane (about 2 teaspoons)

1 tablespoon minced shallot (about ½ small shallot)

1 small fresh green Thai chile or ½ serrano chile

1 teaspoon ground turmeric

2 teaspoons kosher salt

1 teaspoon brown sugar

1 tablespoon vegetable or canola oil

1 whole chicken, 3½ to 4 pounds, butterflied according to the directions on page 593, if desired

1 recipe Thai-Style Sweet Chile Sauce (recipe follows; optional)

1. Trim off the bottom ½ inch of the lemongrass stalk and discard. Locate the place where the outer leaves begin to turn dry, about 4 inches up from the base, and cut them off. Discard any remaining dry outer leaves. Roughly chop the tender lemongrass core and add to the bowl of a food processor. Add the ginger, garlic, shallot, chile, turmeric, salt, sugar, and oil and process, scraping down the sides as necessary, until a paste is formed.

2. Separate the chicken skin from the breasts (see page 586). Spread the chile mixture evenly all over the chicken and under the skin. Roast according to either recipe (page 591 or 594), skipping step 2. Serve with the chile sauce if desired.

## Thai-Style Sweet Chile Sauce

**NOTE**: If you can't find fresh chiles, 2 teaspoons red pepper flakes can be used in their place.

**MAKES ABOUT ½ CUP**

2 medium cloves garlic, minced or grated on a microplane (about 2 teaspoons)

2 small fresh red Thai chiles, finely minced, or 1 red jalapeño or Serrano chile (see Note above)

½ cup palm or packed light brown sugar

¼ cup distilled white vinegar

¼ cup water

2 tablespoons Asian fish sauce

Combine all the ingredients in a small saucepan, bring to a simmer, and simmer gently until reduced by one-third, about 10 minutes. Once cooled, the sauce should have a syrupy consistency.

# PERUVIAN-STYLE ROAST CHICKEN

**This recipe re-creates the classic Peruvian-style roast chickens you find in those great rotisserie chains all over the country. While the rub works well with grilled chicken, it's also marvelous when the chicken is cooked in the oven. This is one case where I definitely prefer the deeper char and crisper skin delivered by the Roasted Butterflied Chicken method (see page 591) as opposed to the Simple Whole Roast Chicken (page 594).**

**NOTE**: For the juiciest results and crispest skin, dry-brine the bird and air-dry it overnight as described on page 579.

**SERVES 3 OR 4**

1 tablespoon ground cumin

1 tablespoon paprika

3 medium cloves garlic, minced or
grated on a microplane (about 1
tablespoon)

1 tablespoon distilled white vinegar

2 teaspoons kosher salt

1 teaspoon freshly ground black
pepper

1 tablespoon vegetable or canola oil

1 whole chicken, 3½ to 4 pounds,
butterflied according to the
directions on page 593

Peruvian-Style Spicy Jalapeño
Sauce (recipe follows; optional)

1. Combine the cumin, paprika, garlic, vinegar, salt, pepper, and oil in a small bowl and massage together with your fingertips.
2. Separate the chicken skin from the breasts (see page 586). Spread the spice mixture evenly all over the chicken and under the skin. Roast according to the Roasted Butterflied Chicken recipe (page 591), skipping step 2. Serve with the jalapeño sauce, if desired.

## Peruvian-Style Spicy Jalapeño Sauce

*A creamy sauce that's simultaneously spicy and cooling. It's excellent on roast or grilled chicken or meats of all kinds and great for dipping vegetables, or use it as a base for a salad dressing.*

**NOTE**: *Ají Amarillo* is a Peruvian yellow pepper with a mild heat. It can be found in paste or puree form in many Latin markets or online. If it is unavailable, simply omit it.

**MAKES ABOUT 1 CUP**

3 jalapeño chiles, roughly chopped

1 cup fresh cilantro leaves

2 medium cloves garlic, minced or
grated on a microplane (about 2
teaspoons)

½ cup mayonnaise

¼ cup sour cream

2 tablespoons *ají Amarillo* paste
(see Note above)

2 teaspoons fresh lime juice (1 lime)

1 teaspoons white vinegar

2 tablespoons extra-virgin olive oil

Kosher salt and freshly ground
black pepper

Combine the jalapeños, cilantro, garlic, mayonnaise, sour cream, chile paste, lime juice, and vinegar in the jar of a blender and blend on high speed until smooth. With the blender running, slowly drizzle in the olive oil. Season to taste with salt and pepper. The sauce can be stored in a sealed container in the refrigerator for up to 1 week.

# BARBECUE-GLAZED ROAST CHICKEN

**When I can't be out on my deck grilling but still crave the tangy, sweet, charred flavor of a good barbecue chicken, I'll whip out this version. It doesn't get the smoke from the grill, but it's an awesome alternative nonetheless. The key is to use a good spice rub, then glaze the chicken with a good barbecue sauce about 10 minutes before it's done. The sauce reduces to a caramelized, sticky glaze that coats every surface and makes for finger-licking flavor in every bite.**

NOTE: For the juiciest results and crispest skin, dry-brine the bird and air-dry it overnight as described on page 579.

### SERVES 3 OR 4

1 teaspoon paprika

½ teaspoon ground coriander

¼ teaspoon ground fennel

½ teaspoon ground cumin

½ teaspoon dried oregano

½ teaspoon freshly ground black
   pepper

1 medium clove garlic, minced or
   grated on a Microplane (about 1
   teaspoon)

2 teaspoons kosher salt

1 tablespoon vegetable or canola oil

1 whole chicken, 3½ to 4 pounds,
   butterflied according to the
   directions on page 593

About 1½ cups of your favorite
   barbecue sauce

1. Combine the paprika, coriander, fennel, cumin, oregano, pepper, garlic, salt, and oil in a small bowl and massage with your fingers to form a paste.

2. Separate the chicken skin from the breasts (see page 586). Spread the spice mixture evenly all over the chicken and under the skin. Roast according to the Roasted Butterflied Chicken recipe (page 591), skipping steps 2 and 4. About 15 minutes before the chicken has finished roasting, brush all over with 1 to 2 tablespoons of the barbecue sauce. Continue roasting until the sauce forms a sticky glaze, about 7 minutes. Brush more sauce and continue roasting until the chicken is cooked and the second layer of sauce has formed a sticky glaze, about 8 minutes longer. Remove the chicken from the oven and allow to rest for 10 minutes, then carve and serve, passing extra sauce at the table.

# TERIYAKI-GLAZED ROAST CHICKEN

**Some of my Japanese relatives may roll their eyes at me for this one—*teriyaki chicken? That's so inauthentic*. But you know what? It's also delicious, so who really cares? In this version, I combine a garlic-and-ginger-rubbed chicken with a sweet soy-and-sake glaze and serve the extra sauce at the table.**

**NOTE**: For the juiciest results and crispest skin, dry-brine the bird and air-dry it overnight as described on page 579.

Any sake will do for this recipe—no need to spring for the expensive stuff. Mirin is a sweet Japanese rice wine. It's available in most Asian markets but if you cannot find it, double the sugar and sake and proceed as directed.

**SERVES 3 OR 4**

½ cup Japanese soy sauce

½ cup sugar

½ cup sake (see Note above)

½ cup mirin (see Note above)

3 scallions, whites left whole, greens finely sliced

1 medium clove garlic, minced or grated on a microplane (about 1 teaspoon)

1 teaspoon minced fresh ginger

½ teaspoon kosher salt

1 tablespoon vegetable or canola oil

1 whole chicken, 3½ to 4 pounds, butterflied according to the directions on page 593

1. Combine the soy sauce, sugar, sake, mirin, and scallion whites in a small saucepan and heat over medium-high heat until barely simmering, then reduce the heat to maintain a gentle simmer and cook until reduced by half, about 30 minutes. Remove from the heat.

2. Meanwhile combine the garlic, ginger, salt, and oil in a small bowl and massage with your fingers to form a paste. Separate the chicken skin from the breasts (see page 586). Spread the garlic mixture evenly all over the chicken and under the skin. Roast according to the Roasted Butterflied Chicken recipe (page 591), skipping step 2. About 10 minutes before the chicken is done, brush it all over with 1 tablespoon of the sauce. Serve the chicken sprinkled with the scallion greens, with the remaining sauce on the side.

# KNIFE SKILLS:
## How to Carve a Chicken

Cooking a chicken is all well and good, and I suppose we *could* give way to our baser instincts and simply go at it with our bear claws and teeth once it hits the table, but carving is *so* much more civilized. (Carving a whole chicken is much easier if you remove the wishbone before roasting it; see page 590.) Here's how to carve a regular or butterflied chicken:

### To Carve a Whole Chicken

**STEP 1: SEPARATE THE LEG** Hold the chicken by one of the drumsticks using tongs, a towel, or your finger and separate the skin that holds it to the breast with the tip of a sharp knife, pulling it away as you go.

**STEP 2: FIND THE JOINT** Find the joint where the leg meets the hip by twisting the blade of your knife back and forth around the articulation until it slips in, then work the knife tip through to separate the leg entirely.

**STEP 3: REPEAT** Remove the second leg in the same manner.

**STEP 4: FLIP AND LOCATE** Flip the bird over and work the knife tip into the joint between one of the wings and the breast. Cut the wing away and repeat on the second side.

**STEP 5: SLICE ALONG THE BREAST-BONE** Flip the chicken over and, using the tip of your knife, follow the curve of the breastbone to release one breast half from the carcass, pulling it outward as you go and trying to remove as much meat as possible.

**STEP 6: KEEP PULLING** Continue to pull the breast outward as you separate the meat from the carcass until it is fully released. Repeat on the second side.

**STEP 7: READY TO SERVE** Your chicken is now ready to serve: 2 legs, 2 breast halves, and 2 wings. To break down the chicken further, you can split the thighs and drumsticks, as well as each breast half, making 10 serving pieces.

## To Carve a Butterflied Chicken

This is far easier than carving a whole chicken, as there is no awkward flipping, rotating, or funny angles involved. As with a whole chicken, the job is easier if you remove the wishbone before roasting the bird; see page 590.

**STEP 1: REMOVE THE LEGS** Pull on one of the drumsticks, using the side of your chef's knife to hold the rest of the chicken in place and slowly working the tip of the knife into the joint. The leg should come away almost on its own, requiring very little actual knife work. Repeat on the second side.

**STEP 2: SPLIT THE LEGS** Split the legs in half at the joint between the thigh and the drumstick if desired.

**STEP 3: REMOVE THE WINGS** With the breast nicely flattened and no cumbersome backbone to get in the way, removing the wings can be done without flipping the chicken over. Find the joint, work your knife into it by twisting the blade back and forth, and cut through it. Repeat on the other side.

**STEP 4: SLICE DOWN THE BREAST-BONE** Using the tip of your knife, follow the curve of the breastbone, releasing one breast half from the carcass, pulling it outward as you go and trying to remove as much meat as possible.

**STEP 5: KEEP PULLING** Continue to pull the breast outward as you separate the meat from the carcass until it is fully released. Repeat on the second side.

**STEP 6: SPLIT THE BREAST HALVES** Split each breast crosswise in half, to make 4 breast pieces.

**STEP 7: READY TO SERVE** Your chicken is now ready to serve: 2 legs, 2 breast halves, and 2 wings. To break down the chicken further, you can split each breast half.

# ROASTING TURKEY

**Turkey gets a bum rap, and I blame it all on Thanksgiving.**

Roast Turkey is one of the few foods that we often only prepare once a year, and, as a result, most people never really get too good at cooking it. Year after year, we gather around the holiday table, thinking about that turkey like we think about our family members: one of those things we just have to put up with before the wine and the pie kick in.

It's a crying shame, because turkey is one of my favorite birds to eat, and it's a great inexpensive option that should be on the roster year-round. (Turkey Association of America, you can mail that check to my home address.) It's got more flavor than chicken, and everyone knows that nothing beats leftover turkey for making soups, sandwiches, and all manner of treats for the rest of the week.

In theory, roasting a turkey is not all that different from roasting a chicken—the same basic problems and solutions apply—it's just a matter of adjusting scale and timing.

IF YOU follow food media, you may notice that every single year, every magazine, blog, and television show comes out with a brand-new recipe for roast turkey, claiming it as the be-all-end-all-world's-best-you'll-never-need-another recipe. Until the next year rolls around. Now, one *could* give them all the benefit of the doubt and assume that every year they're telling the truth. If so, what a happy world we live in, for, year after year, the quality of our roast turkeys is progressing on a never-ending, sure-and-steady upward path toward perfection.

Or you could go for the real answer: **we food writers are all liars.**

OK, so it's not so bad as all that. The truth is, there's no one best way to cook a turkey, and anybody who tells you different is selling something,

most likely a magazine or book (*wink wink*). There is a near-endless list of goals and restrictions, based on the tastes, skills, and time constraints of different home cooks, and thus a near-endless supply of recipes for turkey. Some people want that perfect golden brown centerpiece in the middle of the table. Some want their share of stuffing, moist with drippings. Others care only for the meat, pushing even the crispest, crackliest, saltiest bits of skin off to the side of their plates (we shall speak no more of these heathens).

Here are four different turkey recipes for four different scenarios.

## THE CLASSIC: STUFFED HERB-RUBBED ROAST TURKEY
### WITH GRAVY

**If you've got just one bird in your holiday roast arsenal, make it this one. It's got it all: moist breast and thigh meat; crisp, burnished skin; tasty gravy; and plenty of turkey-flavored stuffing, cooked (safely) inside the bird, to boot.**

**Most turkeys are too large and cumbersome to fit nicely on a rack in a rimmed baking sheet, like I'd use for a roast chicken. In this case, the handles of a roasting pan are worth the trade-off in poorer air circulation around the bird, because of the higher pan sides. And with the extra cooking time a turkey needs, the skin gets plenty crisp anyway.**

**NOTE:** For the juiciest results and crispest skin, dry-brine the bird and air-dry it overnight as described on page 579.

**SERVES 10 TO 12**

1 whole turkey, 10 to 12 pounds,
neck and giblets reserved for
gravy

Kosher salt and freshly ground
black pepper

12 tablespoons (1½ sticks) unsalted
butter

½ cup finely minced fresh parsley

1 tablespoon finely minced fresh
thyme (or 2 teaspoons dried
thyme)

1 tablespoon finely minced fresh
sage

1 tablespoon finely minced fresh
rosemary

2 medium cloves garlic, minced or
grated on a Microplane (about 2
teaspoons)

Kosher salt and freshly ground
black pepper

1 recipe Classic Sage and Sausage
Stuffing (page 622; optional)

1 tablespoon vegetable oil

1 large onion, roughly chopped

1 large carrot, peeled and roughly
chopped

3 stalks celery, roughly chopped

6 cups homemade or low-sodium
canned chicken or turkey stock; or
as needed

2 bay leaves

1 teaspoon soy sauce

¼ teaspoon Marmite

¼ cup all-purpose flour

1. Adjust an oven rack to the lowest position and place a pizza steel or stone on it. Place a heavy stainless steel roasting pan on the pizza steel or stone and preheat the oven to 500°F for at least 1 hour.

2. Meanwhile, when the oven is almost preheated, season the turkey on all sides with salt and pepper (go light on the salt if the bird has been dry-brined). Separate the skin from the breasts (see page 586).

3. Heat 8 tablespoons of the butter in a small skillet or microwave until just melted (it should bubble). Transfer to a small bowl. Whisk in the parsley, thyme, sage, rosemary, garlic, and a generous amount of salt and pepper. Rub this mixture evenly all over the bird and under the skin (it will harden and clump a bit as it hits the cold bird). Place the turkey on a V-rack.

4. If desired, stuff the turkey. Line the cavity with a double layer of cheesecloth. Fill with the stuffing, then use twine to tie the cheesecloth into a sack. Remove the sack and place on a plate. Microwave on high power until the center of the stuffing registers at least 180°F, about 10 minutes. Carefully return the stuffing to the turkey cavity. Stuff the cavity at the turkey neck with the reserved stuffing.

5. Remove the roasting pan from the oven, transfer the V-rack to the roasting pan, and immediately place it on the hot steel or stone, with the legs of the turkey facing the rear. Reduce the oven temperature to 300°F and roast until the turkey is golden brown, the deepest part of the breast registers 150°F on an instant-read thermometer, and the legs register at least 165°F, about 3 to 4 hours, spooning the browned butter from the roasting pan over the turkey every hour or so.

6. While the turkey is roasting, chop the neck into 1-inch chunks with a cleaver. Heat the oil in a medium saucepan over high heat until smoking. Add the turkey neck, onions, carrots, and celery and cook, stirring occasionally, until well browned, about 10 minutes. Add the stock, bay leaves, soy sauce, and Marmite and bring to a boil, then reduce to a simmer and simmer for 1 hour.

7. Strain the stock through a fine-mesh strainer into a large glass measure. You should have a little over 4 cups; if not, add more stock or water. Discard the solids and set the stock aside.

8. When the turkey is cooked, transfer the V-rack to a rimmed baking sheet. Pour the hot melted butter from the bottom of the roasting pan over the turkey. Tent with foil and allow it to rest for at least 30 minutes before carving. (If you stuffed the turkey, for presentation you can remove the stuffing from the turkey, discard the cheesecloth, and replace the stuffing in the turkey.)

9. Meanwhile, set the roasting pan over medium heat and add the reserved stock, scraping up the browned bits from the bottom of the pan with a wooden spoon. Pour the stock through a fine-mesh strainer into a 1-quart glass measure or a bowl.

10. Finely chop the turkey gizzard, heart, and liver if desired. Melt the remaining 4 tablespoons butter in a medium saucepan over medium heat. Add the chopped giblets, if using, and cook, stirring frequently, until just cooked through, about 1 minute. Add the flour and cook, stirring constantly, until golden brown, about 3 minutes. Whisking constantly, add the stock in a thin, steady stream. Bring to a boil, then reduce to a simmer and cook until the gravy is thickened and reduced to about 3 cups. Season to taste with salt and pepper and remove from the heat.

11. Carve the turkey and serve with the gravy and stuffing, if you made it.

*continues*

## STUFF IT!

Yⁿou may have heard from various reputable sources that stuffing a bird before roasting it is a bad idea. There's no problem with stuffing the neck cavity—it's in the interior of the bird that safety issues arise. Even though it might be safe to consume your stuffing itself at a lower cooking temperature than the turkey or chicken as the bird roasts, its raw juices can drip down into the stuffing, contaminating it. So, in order to be completely safe, your stuffing must come up to at least the same 145° to 150°F you're gonna cook your bird to. Unfortunately, because the stuffing is in the very center of the bird, by the time it is cooked through, your bird is overcooked.

There is, however, a solution, though it's a slightly tricky one: cook the bird from the outside *and* the inside. What you've got to do is stuff the bird with hot stuffing just before roasting. That's right: bring your stuffing all the way up to at least 180°F (to compensate for the heat it will lose while you're working with it) and, while it's still hot, jam it into the bird's cavity. The easiest way to do this is to form a cheesecloth pouch inside the turkey, stuff that pouch, tie it off, remove it, and microwave it on a plate, then put it back in the turkey before roasting.

Not only does the method give you stuffing that's perfectly safe to eat (so long as it never dips below 145°F while it is roasting, and it shouldn't), but it'll also help your turkey cook more evenly, insulating its breasts from the inside so that they cook a little more slowly and end up coming to temperature at the same time that the legs do. Of course, in my family, we still need an entire tray of stuffing on the side, because there can never be enough.

9

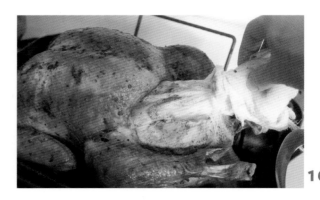

10

# KNIFE SKILLS:
## How to Carve a Turkey

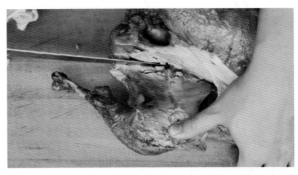

**STEP 1: REMOVE THE WISHBONE** (See page 590.)
**STEP 2: SLICE THE LEG** Once your turkey has rested and is ready to carve, use a sharp chef's knife or boning knife to slice through the skin between the leg and the breast. This is easier on a spatchcocked turkey than a regular roasted turkey. Use a kitchen towel to hold on to the turkey for leverage as needed, but make sure to always cut *away* from your free hand, or your guests may get more meat than they bargained for on their dinner plate.

**STEP 3: GET THE JOINT** Once the skin is cut, pull the entire leg away from the body. It should separate quite easily, displaying the socket joint where the thigh meets the hip. Cut through this joint with the tip of your knife, and the leg should be completely free. Just slice through the skin to release it.

**STEP 4: DIVIDE AND CONQUER** Locate the joint between the drumstick and the thigh by moving them back and forth and finding the articulation with your finger. Place the knife at the joint and cut through, jiggling the blade side to side a bit as you go until it slides through with relatively little pressure.

**STEP 5: REPEAT** Remove the other leg and separate it in the same manner.

**STEP 6: BONE THE THIGH**
Flip one thigh over so it's skin side down and cut one side of the bone with the tip of your knife to release a large chunk of meat.

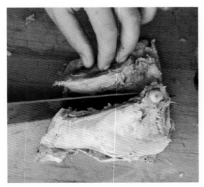

**STEP 7: REMOVE THE BONE** Repeat with the other side, taking as much meat off the bone as possible.

STEP 8: **SLICE THE THIGH MEAT** Slice the meat into serving pieces and transfer to a warmed serving platter, skin side up. Repeat with the other thigh pieces, then transfer the drumsticks to the platter as well.

STEP 9: **CLIP ITS WINGS** Locate the wing joint by articulating it, then slice through it with a sharp chef's knife. Repeat with the other wing, then separate the drumettes from the flats and transfer all 4 wing pieces to the platter.

STEP 10: **REMOVE THE BREAST** Start by running a sharp boning knife down one side of the breastbone, using a kitchen towel to hold the bird steady. Again, this is much easier with a spatchcocked bird than a whole turkey.

STEP 11: **FOLLOW THE BONE** Continue to work your knife into the breast, following the contour of the bones as closely as you can, until you reach a sharp curve in the breastbone.

STEP 12: **PULL AWAY** Once you've separated enough of the breast, you should be able to start prizing it loose from the carcass, using the side of your knife to push it outward as you continue to follow along the bones to remove as much meat as possible.

STEP 13: **SEPARATE THE BREAST** Eventually the breast should fall completely away, held only at the bottom edge. Slice through this edge to remove the breast entirely, then repeat steps 13 through 16 with the second side.

STEP 14: **SLICE THE BREAST** Slice the breast meat with a very sharp knife on a strong bias to create wide, even serving slices. Transfer the slices to the warmed platter.

STEP 15: **READY TO SERVE!** Arranged on a platter, ready to be presented at the table.

# THE EASIEST AND FASTEST: ROASTED BUTTERFLIED TURKEY
## WITH GRAVY

As with butterflied chicken, cooking a turkey like this solves pretty much every problem that a whole turkey has, making the actual roasting process completely idiot-proof. As long as you've got good kitchen shears, an instant-read thermometer, and a few brain cells to rub together, you should be able to put a perfect turkey on the table any day of the year.

With a quick-cooking chicken, you can go straight on a rack set in a foil-lined rimmed baking sheet. Do this with a turkey, and the drippings will start to burn before the turkey is done. To solve this problem, and add some flavor in the process, I spread a layer of chopped vegetables in my roasting pan. The vegetables release juices as they cook, preventing the drippings from burning and creating a flavorful base for you to add to your gravy at the end.

NOTE: For the juiciest results and crispest skin, dry-brine the bird and air-dry it overnight (see page 579).

**1** **2** **3** **4**

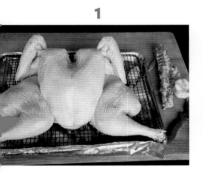   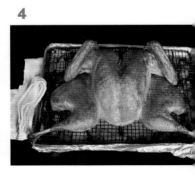

**SERVES 10 TO 12**

3 large onions, roughly chopped (about 6 cups)

3 large carrots, peeled and roughly chopped (about 4 cups)

4 stalks celery, roughly chopped (about 4 cups)

12 thyme sprigs

2 tablespoons vegetable oil

1 whole turkey, 12 to 14 pounds, butterflied according to the instructions on page 593, backbone, neck, and giblets reserved.

Kosher salt and freshly ground black pepper

6 cups homemade or low-sodium canned chicken or turkey stock

2 bay leaves

3 tablespoons unsalted butter

¼ cup all-purpose flour

1. Adjust an oven rack to the middle position and preheat the oven to 450°F. Line a rimmed baking sheet or broiler pan with aluminum foil. Scatter two-thirds of the onions, carrots, celery, and thyme sprigs across the bottom of the pan. Place a wire rack or slotted broiler rack on top of the vegetables.

2. Pat the turkey dry with paper towels. Loosen the turkey skin from the breasts (see page 586). Rub the turkey all over and under the skin with 1 tablespoon of the oil. Season liberally all over with salt and black pepper (go easy on the salt if the bird was dry-brined). Tuck the wing tips under the bird. Place the turkey on the rack, arranging so that it does not overhang the edges, and press down on the breastbone to flatten the breasts slightly.

3. Roast, rotating the pan occasionally, until an instant-read thermometer inserted into the deepest part of the breast registers 150°F and the thighs register at least 165°F, about 80 minutes. If the vegetables start to burn or smoke, add 1 cup water to the roasting pan.

4. While the turkey roasts, make the gravy: Roughly chop the neck, backbone, and giblets. Heat the remaining 1 tablespoon oil in a 3-quart saucepan over high heat until shimmering. Add the chopped turkey parts and cook, stirring occasionally, until lightly browned, about 5 minutes. Add the remaining onions, carrots, and celery and cook, stirring occasionally, until the vegetables start to soften and brown in spots, another 5 minutes or so. Add the stock, the remaining thyme, and the bay leaves and bring to a boil, then reduce to a bare simmer and cook for 45 minutes. Strain the stock through a fine mesh strainer into a 2-quart liquid measuring cup or a bowl; discard the solids. Skim off any fat from the surface of the stock.

5. Melt the butter in a 2-quart saucepan over medium-high heat. Add the flour and cook, stirring constantly, until golden brown, about 3 minutes. Whisking constantly, add the stock in a thin, steady stream until it is all incorporated. Bring to a boil, then reduce to a simmer and cook until reduced to about 4 cups, about 20 minutes. Season to taste with salt and pepper, cover, and keep warm.

6. When the turkey is cooked, remove from the oven and transfer the rack to a rimmed baking sheet. Tent the turkey with aluminum foil and allow to rest at room temperature for 20 minutes before carving.

7. Carefully pour any collected juices from the roasting pan through a fine-mesh strainer into a liquid measuring cup or a bowl. Skim off the fat and discard. Whisk the juices into the gravy.

8. Carve the turkey and serve with the gravy.

# THE SMALL-CROWD-PLEASER: EASY HERB-ROASTED TURKEY BREAST
## WITH STUFFING

Perhaps your family is small. Perhaps your friends all bailed on you the week before Thanksgiving. Perhaps only half your family eats meat. Perhaps you simply don't enjoy leftovers. (Weirdo.) Or perhaps you just had a craving for turkey in February and have no reason to cook a whole bird. Point is, there are any number of reasons you might have for not wanting to roast an entire turkey, but that doesn't mean you should be deprived of juicy meat, crispy skin, and turkey-saturated stuffing, right?

Cooking a turkey breast is far easier than cooking a whole turkey, since you only have a single final target temperature in mind, instead of having to worry about the legs and breasts cooking at different rates. As soon as that breast is at 150°F, you can pull it out of the oven and let it rest.

I'm the kind of guy who likes stuffing with my turkey, and this method makes it easy. Just plop the breast down on top of a pan of stuffing, throw it into a hot oven, and let it go. About halfway through cooking, the stuffing will threaten to burn. Do not panic—all you've got to do is remove the pan, transfer the turkey to a rack set on a rimmed baking sheet, and put it back in the oven on its own to finish roasting. Once the turkey is cooked, pour the pan juices over the stuffing and pop the stuffing back in the oven for one final crisp-up blast while the turkey rests.

The best part? The whole process goes from fridge to table in under two hours. This is the kind of stuff Sunday supper dreams are made of.

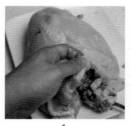

1　　　　2　　　　3　　　　4　　　　5

**NOTE:** For the juiciest results and crispest skin, dry-brine the bird and air-dry it overnight as described on page 579.

**SERVES 6 TO 8**

1 whole bone-in, skin-on turkey breast, 4 to 5 pounds, patted dry

1 recipe Classic Sage and Sausage Stuffing (page 622)

3 tablespoons unsalted butter, at room temperature

5 tablespoons minced fresh parsley

2 tablespoons minced fresh oregano

1 tablespoon kosher salt

½ teaspoon freshly ground black pepper

1. Adjust oven rack to middle position and preheat oven to 450°F. Using poultry shears, cut off any back portion still attached to the turkey breast. Fill the cavity under the turkey breast and under the flap of fat around the neck with stuffing. Transfer the remaining stuffing to a buttered 9- by 13-inch baking dish, then place the turkey on top.

2. Using your hands, carefully separate the turkey skin from the meat, starting at the bottom of the breast and being careful not to tear the skin (see page 586). Combine the butter with the parsley and oregano in a small bowl. Add the salt and pepper and stir with a fork until homogeneous. Rub the mixture evenly over and under the turkey skin.

3. Roast until the stuffing starts to brown, about 45 minutes. Remove from the oven and transfer the turkey to a wire rack set on a foil-lined rimmed baking sheet. Return the turkey to the oven and continue roasting until the skin is golden brown and crisp and the thickest part of the meat, near the bone, registers 145° to 150°F on an instant-read thermometer, about 30 minutes longer. Remove from the oven, transfer to a large plate, and let rest for 20 minutes.

4. Meanwhile, pour the pan juices over the stuffing. Return the stuffing to the oven and cook until it is golden brown on top and registers 160°F on the thermometer, about 15 minutes.

5. Carve the turkey, arrange over the stuffing, and serve.

**6**

**7**

**8**

**9**

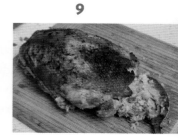

# THE PERFECTIONIST: THANKSGIVING TURKEY TWO WAYS

**We've already talked about the problems of cooking poultry, the main one being that legs and breasts need to cook to different temperatures.**

If you're not too worried about absolute perfection, the recipe on page 605 will do you just fine. The hot steel or stone will help the thighs cook fast enough to keep abreast of the, er, breast. If true perfection is what you're after though, you're better off breaking the turkey down into parts to be cooked individually. That way, the breast and legs can be cooked to precisely the right temperature. The divide and conquer method mostly works for turkey, save for one small problem: while it's far easier to cook the breast alone than a whole bird, the narrower, tapered end of the breast can still overcook and dry out, meaning that at least one family member is going to get stuck with subpar turkey (sorry, Granddad).

Lowering the oven temperature helps—roasting it at 250°F rather than cooking it at a normal 300° to 350°F range promotes more even cooking between the edges and center and between the thick and thin parts—but it's not quite enough. What I needed was a way to even out the shape of my turkey breast, and while I'm confident there are bioengineers hard at work developing turkeys with perfectly cylindrical breasts, for the time being, I had to resort to some kitchen surgery.

So, how do you take an unevenly shaped turkey breast and turn it into a perfect cylinder? Simple. Remove the breasts from the bones and put them together head-to-heels, then wrap the whole thing up with their breast skin and tie it up to roast slowly. Then, after the breast is cooked, brown it on the stove top to crisp up the skin.

This method offers a few distinct advantages:

- **Even cooking.** Because of its symmetrical shape, the turkey breast cooks through along its entire length at the same rate. Nobody gets stuck with a dry piece.
- **Better seasoning.** By removing the breasts from the carcass, you expose more surface area and you can then season the breasts on both sides before assembling the turkey roll.
- **Easier carving.** With no bones and an even shape, carving the turkey breast is as simple as slicing a tenderloin.
- **Better gravy.** With the bones from the breast at your disposal, it's easy to make a delicious very turkey-ey gravy.

It all sounds great, right? And it is, but after all the work I've done on improving those darn breasts, the legs are beginning to feel a little left out. Should they be content with plain old roasting? Well, sure—if you feel like it you can just add them to the baking sheet while the breast roll cooks and make sure they reach at least 160° to 170°F degrees before taking them out and following the same searing procedure. If you want to go the extra mile, though, my favorite way to serve Thanksgiving turkey legs is to braise them until falling-apart tender (for more on the science of braising, see page 239).

# THANKSGIVING TURKEY TWO WAYS

**NOTE:** You can start with a whole 12- to 15-pound turkey and butcher it yourself, or you can just buy turkey parts. If starting with a whole bird, save the back, neck, and giblets for the gravy if you are making the recipe on page 605.

**SERVES 10 TO 12**

1 whole bones-in, skin-on turkey
  breast, 3 to 4 pounds

Kosher salt and freshly ground
  black pepper

2 turkey legs

2 tablespoons oil

1 large onion, roughly chopped

1 large carrot, peeled and roughly
  chopped

3 stalks celery, roughly chopped

6 cups homemade or low-sodium
  canned chicken or turkey stock

2 bay leaves

1 teaspoon soy sauce

¼ teaspoon Marmite

1 recipe Dead-Simple Poultry
  Gravy (page 619) or gravy from
  Stuffed Herb-Rubbed Roast
  Turkey (page 605)

2 tablespoons unsalted butter

1. Using your hands, carefully remove the skin from the turkey breast, working very slowly and keeping it in one piece; use a knife as necessary. Using the back of the knife, scrape off any excess fat from the side of the skin.

2. Adjust the oven racks to the lower- and upper-middle positions and preheat the oven to 275°F. Remove the turkey breast halves from the bone using a sharp boning knife. Season liberally on all sides with salt and pepper. Place the breast halves against each other, tops to tails, to create a relatively even football-shaped roast. Spread the turkey skin out on the cutting board and place the breasts in the center. Wrap them up in the skin, overlapping it and tucking in the edges, and secure with 7 to 10 pieces of butcher's twine placed at 1-inch intervals, working from the ends in. Season on all sides with salt and pepper. Place the roll on a wire rack set on a rimmed baking sheet and set aside.

3. Season the turkey legs generously with salt and pepper. Heat the oil in a large Dutch oven over high heat until lightly smoking. Place the legs in the pot, skin side down, and cook, without moving them, until deep golden brown, about 8 minutes; lower the heat if the oil is smoking excessively.

4. Flip the legs over and add the onion, carrot, and celery. Pour the stock over the legs and add the bay, soy sauce, and Marmite. Bring to a boil, cover, and place on the lower oven rack. Cook until the meat is completely tender and starting to fall off the bone, about 3 hours.

5. Meanwhile, about 1 hour into the cooking, place the turkey breast on the upper oven rack and cook until the center of the breast registers 150°F on an instant-read thermometer, about 2 hours (it should finish around the same time as the legs).

6. When the turkey is done, remove the Dutch oven and baking sheet from the oven. Carefully transfer the legs to a large plate and tent loosely with foil. Strain the cooking liquid and add to the gravy; discard the solids.

7. Wipe the Dutch oven dry, add the butter, and heat over high heat until melted and browned. Add the turkey breast and cook, turning occasionally, until all sides are browned, about 8 minutes. Add to the plate with the legs, tent with foil, and allow to rest for 30 minutes before carving and serving with the gravy.

# GRAVY TIPS

The browned drippings form the base for a flavorful gravy.

The turkey recipes on pages 605 and 612 make their own gravy, but at *my* table, there's never enough gravy to go around. This gravy is just about the easiest way to get more without having to resort to that awful store-bought stuff. Here are a few tips:

- **You don't need to make your own stock.** Sure, in an ideal world, when you have the time and inclination, making your own stock by browning the chopped carcass and neck of your bird and simmering it with lots of vegetables is the best way to make gravy. But a good-quality low-sodium store-bought chicken stock makes a flavorful base for a homemade gravy that's far better than the jarred stuff. Even if you are planning on using your turkey neck and scraps (highly recommended!), use stock to simmer them instead of water, for an instant flavor boost.

- **Make your gravy in advance!** The gravy can be made at least a few days before Thanksgiving. Get your bird ahead of time, and you'll have the neck and giblets to work with. Make your gravy on Monday or Tuesday, then refrigerate it and don't even think about it until Turkey Day. It'll reheat well in a small saucepan or in the microwave (stir it every 30 seconds while microwaving to make sure it doesn't explode).

- **Reach for the umami bombs.** When used judiciously, Marmite and soy sauce can seriously increase flavor, adding depth and savoriness to your gravy. A quarter teaspoon of Marmite and a teaspoon of soy sauce for every quart of gravy is about the right amount (see page 245 for more on umami bombs).

- **Add aromatics.** If going the store-bought broth route, try first simmering it down with a couple of bay leaves, peppercorns, and some fresh herbs, like thyme or parsley stems. You'll be amazed at the depth of flavor it picks up with just a quick 30-minute simmer.

- **Deglaze Your Roasting Pan.** Remember that your turkey or chicken will give off plenty of flavorful liquids and solids while it's roasting. Look at the bottom of the pan when the bird is done—see the browned bits in there? That's called *fond*, and it is an instant gravy-booster. While your bird is resting, place the roasting pan over a burner and pour in some stock. Scrape up the browned bits with a wooden spoon, then strain and use this enhanced stock as the base for your gravy. If you make your gravy in advance, you can give it a last-minute boost by deglazing the pan with a little stock and whisking it into the gravy just before serving.

- **Thicken the Right Way.** To thicken 4 cups gravy, melt 4 tablespoons unsalted butter in a medium skillet over medium heat. Add ¼ cup all-purpose flour and cook, stirring constantly with a wooden spoon, until you have a nice golden blond color, which will add some nuttiness. Slowly add your stock, whisking constantly. The harder you whisk and the more slowly you add the stock, the smoother your gravy will be. Once you've added all the liquid, bring it up to a boil, then reduce it to a simmer and let it cook down, stirring occasionally, until it gets to the right consistency. Season it at the end with salt and pepper (seasoning too early can lead to the salt concentrating and becoming too strong).

## Dead-Simple Poultry Gravy

**MAKES ABOUT 3 CUPS**

4 tablespoons unsalted butter

¼ cup all-purpose

4 cups homemade or low-sodium
   canned chicken or turkey stock

1 teaspoon soy sauce

¼ teaspoon Marmite

Kosher salt and freshly ground
   black pepper

1. Melt the butter in a medium heavy-bottomed saucepan over medium high heat. Stir in the flour with a wooden spoon, then continue cooking, stirring constantly, until golden blond in color, about 2 minutes. Slowly add the stock in a thin stream, whisking vigorously, then whisk in the soy sauce and Marmite. Bring to a boil, reduce to a simmer, and cook, stirring occasionally, until reduced to 3 cups. Add any pan drippings from the turkey (or chicken) and simmer until reduced to the desired consistency. Season to taste with salt and pepper.

2. Serve immediately, or keep warm. The gravy can be made up to 1 week in advance and stored, covered, in the refrigerator. Reheat over medium-low heat, whisking occasionally, until fully hot.

# CLASSIC SAGE AND SAUSAGE STUFFING

**I'm a stuffing fiend.**

It's easily my favorite part of the holiday meal, and as far as I'm concerned, a side dish worth making any time of the year. For those of you oddballs out there who still refer to the stuff as "dressing," well, I'm not going to come down on either side of the whole nomenclature debate except to say that *three reputable sources give three different answers:*

- *The Oxford English Dictionary* says that "stuffing" is stuffed in a bird or joint (roast), while "dressing" is a more general term for seasoning that goes with food or sauce.
- *The Joy of Cooking* contends that they are one and the same, except one is in the bird and one is out.
- *The Food Lover's Companion* says the two terms can be used interchangeably.

With that out of the way, I expect to hear no more on that semantics discussion this holiday season.

So, moving on, stuffing. While it can be made with any number of bases, the most popular type (and my favorite) is made with bread, broth, eggs, and butter. Essentially it's best to think of stuffing as a savory bread pudding when constructing a recipe. The key to great bread pudding is to use the bread as a sponge to soak up as much flavorful liquid as possible. At the same time, through you don't want it to be spongy.

The cooked stuffing should have a moist, tender, custard-like texture. It should be firm enough to cut with a knife but soft enough to eat with a spoon, with a bit of space left over to soak up some gravy. Much of this has to do with how you pick and handle your bread, but before you get there, you've got to decide what kind of bread you are going to use. Whole-grain breads may have more flavor on their own, but they are rougher in texture than white-flour breads. Since in a stuffing the bread is more a vehicle for flavor than a flavor on its own, I prefer to use a white bread as it achieves a more custard-like texture. It's tempting to use a high-quality, crusty, chewy, large-holed fancy-pants artisanal bread, but the finer-hole structure of regular supermarket-style "Italian" or "French" bread (or just high-quality white sandwich bread) makes for better flavor absorption and retention, and that's what stuffing is all about.

After you've cubed your bread, the next stage is to dry it out. It may surprise you, but drying and staling are not the same thing (see "Drying Versus Staling," page 621). Though many recipes call for stale bread, what they're actually looking for is dry bread. Staling takes time. Luckily for us, drying is fast. I dry my bread by toasting it in a low (275°F) oven for about 45 minutes, tossing it a couple of times halfway through. By drying the bread like this, you make enough room in the cubes from two regular-

1

2

3

sized loaves (about 2½ pounds) to absorb a full 4 cups of chicken or turkey broth. It's so much broth that the stuffing will almost taste as if you baked it in the bird if you do it in a separate pan (for instructions on how to safely bake the stuffing in the bird, see page 607). I recommend starting it with foil on top to trap in some moisture, before removing the foil and crisping up the top.

The flavorings I go with are classic: butter (and plenty of it), sage sausage (you can get away with just sage for a nonmeaty version), onions, celery, and garlic. My sister likes to add dried cranberries and my mother likes to add chestnuts. They are, of course, both wrong.

## DRYING VERSUS STALING

Drying and staling are not the same thing. Here's the difference:

- **Drying** involves the evaporation of moisture from within a piece of bread. The structure of the bread remains more or less the same, though it become less pliable because of the moisture loss. Dry, not stale, bread will be crisp like a cracker and crumble into a fine powder. Bread that has dried out is very hard to refresh.
- **Staling** is the process by which moisture migrates out of swollen starch granules and into the spaces in the bread. The moisture-deprived starch molecules then recrystallize, forming tough structures within the bread. Stale, not dry, bread will taste leathery and chewy, not crackery or dry. Bread that has staled can be refreshed by heating it, causing the starch granules to reabsorb moisture.

It's quite possible for bread to stale without drying—just think about what happens to a loaf of preservative-free bread when you place it in the refrigerator overnight. Staling actually occurs much faster at cooler temperatures, which is why your bread will become leathery and chewy by the next morning no matter how tightly it was wrapped.

*continues*

In order to prevent both staling and drying, it's best to store bread tightly wrapped on the countertop or in a bread box if you're going to eat it within a day or two. For long-term storage, wrap your loaves in foil and pop 'em into the freezer. This will freeze the internal water molecules, preventing them from migrating out of the bread and thus stopping it from staling. Reheat frozen bread in a 300°F oven, wrapped in its foil, until warmed through.

# CLASSIC SAGE AND SAUSAGE STUFFING

**SERVES 10 TO 12**

2½ pounds (about 2 loaves) high-quality sandwich bread or soft Italian or French bread, cut into ¾-inch dice (about 5 quarts)

8 tablespoons (1 stick) unsalted butter

1½ pounds sage sausage, removed from casings

1 large onion, finely chopped (about 2 cups)

4 large stalks celery, finely chopped (about 2 cups)

2 medium cloves garlic, minced or grated on a Microplane (about 2 teaspoons)

¼ cup minced fresh sage (or 2 teaspoons dried sage)

4 cups homemade or low-sodium canned chicken or turkey stock

3 large eggs

¼ cup minced fresh parsley

Kosher salt and freshly ground black pepper

1. Adjust the oven racks to the lower- and upper-middle positions. Preheat the oven to 275°F. Spread the bread evenly on two rimmed baking sheets. Stagger the pans on the oven racks and bake, rotating the pans and stirring the bread cubes several times, until the bread is completely dried, about 50 minutes. Remove from the oven and allow to cool. Increase the oven temperature to 350°F.

2. Heat the butter in a large Dutch oven over medium-high heat until the foaming subsides (don't allow the butter to brown), about 2 minutes. Add the sausage and mash with a stiff whisk or potato masher to break it up into fine pieces (the largest pieces should be no greater than ¼ inch), then cook, stirring frequently, until only a few bits of pink remain, about 8 minutes. Add the onion, celery, garlic, and sage and cook, stirring frequently, until the vegetables are softened, about 10 minutes. Remove from the heat and add half of the stock.

3. Whisk the remaining stock, the eggs, and 3 tablespoons of the parsley in a medium bowl until homogeneous. Stirring constantly with a wooden spoon, slowly pour the egg mixture into the sausage mixture. Add the bread cubes and fold gently until evenly mixed.

4. Use part of the stuffing to stuff the bird if desired. Transfer the remaining stuffing to buttered 9- by 13-inch baking dish (or 10- by 14-inch oval dish). Cover tightly with aluminum foil and bake until an instant-read thermometer inserted into the center of the stuffing reads 150°F, about 45 minutes. Remove the foil and continue baking until golden brown and crisp on top. Sprinkle with the remaining tablespoon of parsley and serve.

# REALLY EASY CRANBERRY SAUCE

**I understand the appeal of canned jellied cranberry sauce.**

It plops out of the can, has those pretty ridges, and can be sliced up and placed right in the center of a plateful of curly parsley. It's got a kind of Betty Crocker appeal to it. But whole-berry sauce in a can or jar? Why, when homemade is so much better, and blindingly simple to do?

Here's why to make it yourself: first off, cranberries are extremely high in pectin. This is the cellular glue that holds plants together and is the primary jelling agent in jellies. Unlike most other berries, which require you to add powdered or liquid pectin to get the requisite jell level, cranberries already contain the perfect amount. That means that all you've got to do is cook them down with some sugar, and just a touch of water to get them started, and they basically do all the work themselves, setting into a jelly all on their own.

Cranberries and cranberry sauce also have an extremely long shelf life. In part due to their high acidity, in addition to naturally high levels of anti-microbial phenolic compounds, fresh cranberries can keep for weeks (if not months) in the refrigerator. I make my Thanksgiving cranberry sauce at least a week ahead of time. Then it sits in the fridge, no problem, and saves me from having to think about it on Turkey Day. Which is not to say you should restrict yourself to serving cranberry sauce only on Thanksgiving: it makes an awesome accompaniment for grilled or roasted pork and chicken, sausages, or meatballs.

Finally, making cranberry sauce yourself lets you adjust the flavorings any way you like 'em. I'm a purist at heart, so my sauce most often contains nothing but cranberries and sugar, with perhaps the occasional hint of cinnamon (cranberries contain spicy phenolic compounds similar to those in cinnamon, so the flavors go quite well together).

But here are a few more ideas:

- **Orange.** Replace the water in the recipe with orange juice. Add a couple teaspoons of grated orange zest along with the cranberries.
- **Ginger.** Add a teaspoon of grated fresh ginger along with the cranberries, then finish the sauce by stirring in a tablespoon of finely diced crystallized ginger.
- **Spices.** Cinnamon, as I mentioned, works well, as does grated nutmeg or ground allspice or cloves. Start with a pinch and work your way up until it

tastes the way you like it. A bit of vanilla or spiced rum added toward the end of cooking will also spice things up a bit.
- **Dried fruits.** A handful of raisins or currants can add texture and flavor. Add them right at the beginning to allow them to soften.
- **Nuts.** Toasted almonds, pecans, pistachios, or walnuts, roughly chopped and mixed into the sauce at the end, make a classic pairing.

# EASY CRANBERRY SAUCE

**MAKES ABOUT 2 CUPS**

4 cups fresh or frozen cranberries

½ cup water

1 cup sugar

¼ teaspoon ground cinnamon (optional)

¼ teaspoon kosher salt

1. Combine all the ingredients in a medium saucepan, bring to a boil over medium-high heat and cook, stirring occasionally with a wooden spoon, until the berries start to pop. Mash the berries against the side of the pan with the spoon, then continue to cook, stirring occasionally, until the berries are completely broken down and have achieved a jam-like consistency. Remove from the heat and allow to cool for about 30 minutes.
2. Stir in water in 1-tablespoon increments to adjust to the desired consistency.

# The FOOD LAB's Complete Guide
## TO BEEF ROASTS

You've already read the guide on steaks (see page 281), and most of that information applies here as well—grading, coloring, aging, and labeling are all identical whether you are talking about individual steaks and chops or entire roasts. Here's the additional information you need to know for the latter.

# THE FOUR BEEF ROASTS YOU SHOULD KNOW

**A**s with steaks, there are several cuts of beef for roasting that you'll find in the supermarket, but some don't have flavor or texture worth paying for. These are the ones that are worth their price tag.

| NAME | TENDER-NESS (on a scale of 1 to 10) | FLAVOR (on a scale of 1 to 10) | ALSO SOLD AS | WHERE IT'S CUT FROM | WHAT IT TASTES LIKE |
|---|---|---|---|---|---|
| Rib Roast | 7/10 | 9/10 | Prime rib | The rib primal. The main eye of meat is the longissimus dorsi, which runs down the back of the steer. The same muscle that rib-eye and strip steaks are cut from, it's often referred to as the loin. | This is the king of roasts. When it's full size, it comprises ribs 6 through 12 of the animal, counting back from the top of the neck, and can be large enough to feed 15 to 20 people. Usually it's sold as smaller 3- to 4-rib roast, and you can specify the "small end" (aka "loin end") or "large end." The small end tends to be a little bit leaner (and smaller), while the large end has more fat and a slightly chewier texture. I prefer the large end for its superior flavor. |
| Top Sirloin Roast | 8/10 | 6/10 | Top butt, center-cut roast; top round first cut, top round steak roast | The sirloin primal, which is behind the rib and short loin, near the cow's rear end. There are two major muscles in the top sirloin roast: the biceps femoris and the gluteus maximus. (Yep, that's what you think it is.) | Very tender and juicy, with moderately beefy flavor, though without the heavy marbling you'll find in a rib roast. Depending on how it's butchered, it can also house a pretty tough band of sinew running through its center. This is best discarded at the table—it can be overly chewy. |
| Tenderloin | 10/10 | 2/10 | Fillet, Châteaubriand (when sold as a center-cut roast) | The central section of the psoas major muscle in the short loin primal of the steer. Basically, the inside of the rib cage near the spine. | Tenderloin is prized for its buttery-tender texture, but what it gains in texture, it loses in flavor. Fact is, tenderloin is one of the least flavorful cuts on the steer, and requires a robust sauce or rub to enhance it. |
| Chuck Eye Roast | 5/10 | 6/10 | Boneless chuck roll | From the shoulder of the cow; it's the large eye of meat in the center of the chuck. | Quite flavorful and balanced but extremely fatty and full of connective tissue. Although it has a tendency to fall apart when you roast it, if you don't mind trimming it on your plate, this is an excellent inexpensive option for a roast. |

## BONES

**Here's one for you: bone-in or bone-out?**

I've always wondered about this one myself, so I ran a series of tests. Many chefs say that cooking meat on the bone is always a better idea because the bone contributes lots of flavor. I'm skeptical. First of all, most of the flavor in a bone is deep inside, in the marrow. If you ever make a stock out of just bones, you'll find that it's almost tasteless unless the bones are cracked first. And as far as flavor penetration goes, there's very little movement of molecules across a piece of meat. Even marinating overnight will only get you a couple millimeters of penetration (more on that in a moment). What chance does any flavor from the bone have of getting into the meat during the short roasting or pan-searing time?

To test this, I cooked four identical prime rib roasts. The first was cooked with the bone on. For the second, I removed the bone but then tied it back against the meat for cooking. For the third, I removed the bone and tied it back against the meat, but with an intervening piece of impermeable heavy-duty aluminum foil. The fourth was cooked without the bone.

Tasted side by side, the first three were indis-tinguishable from one another. The fourth, on the other hand, was a little tougher in the region next to where the bone used to be. What does this indicate? Well, first off, it means the flavor exchange theory is bunk—the completely intact piece of meat tasted exactly the same as the one with the intervening aluminum foil. But it also means that the bone serves at least one important function: it insulates the meat, slowing its cooking and providing less surface area to lose moisture.

Bottom line: the best way to cook your beef roast is to detach the bone and then tie it back on. You get the same cooking quality of a completely intact roast, with the advantage that once it's cooked, carving is as simple as cutting the string, removing the bone, and slicing. For pan-seared steaks, I opt for bone-in just because I like gnawing on the bones for dessert.

**What exactly is a prime rib roast?**

To locate the prime rib, start by cutting your favorite steer neatly down the center from head to tail. Set aside one half for another use. Place your hand on the back of the other half and feel your way back along its vertebrae until you start feeling ribs.

Count backwards to the sixth rib, and cut crosswise through the meat in front of it.

Then continue counting back until you get to rib 12 and cut behind it, again crosswise. Reserve the head and tail section for another use, saving the ribs you just cut out. Now saw the ribs off at about 13 to 16 inches down their length and set aside the belly section. Take off the hide, and what you're left with is the prime rib. It consists of seven full ribs with a large eye of meat running along their backside. This meat is part of the loin muscle of the cow, the same muscle that New York strip, rib-eye, and Delmonico steaks are cut from. It's also often referred to as a "standing rib roast," because, well, it comes from the ribs and it stands up as it roasts.

### Does "prime rib" have anything to do with Prime grade beef?

Glad you asked. The answer is no. The term "prime rib" has existed longer than the USDA's beef grading system, which classifies beef according to its potential tenderness and juiciness into various grades. The roast is called prime rib because it's what many butchers and consumers traditionally considered the best part of the cow. After the USDA began using its labeling system with the label "Prime" denoting the highest quality, things became a little confusing. It's possible to buy a prime rib that is also Prime grade, but it doesn't necessarily have to be so. My local Whole Foods sells Choice grade prime rib, for example, while the discount supermarket around the corner also carries Select grade prime rib.

### What's a good size roast to buy?

Generally, you want to aim for about a pound of bone-in prime rib per person, more if you have a hungry clan. A full seven-rib prime rib is a massive hunk of meat, between 20 to 30 pounds. That's too big to fit into my oven, which is why I, like most people, buy my prime rib in three- or four-rib sections. These sections have different names, depending on where they are cut from:

- **The Chuck End**: Ribs 6 through 9, from closer to the cow's shoulder (aka the chuck); referred to variously as the "chuck end," "blade end," or "second cut." It's got more separate musculature and more large hunks of fat than the loin end. Personally, I prefer this end, because I like to eat the fat in a well-roasted piece of beef.
- **The Loin End**: Ribs 10 through 12, taken from farther back and also known as the "small end" or "first cut." It's got a larger central eye of meat and less fat.

Depending on what part of the country you live in, your butcher may refer to these cuts as different things, but any butcher should know which ribs are which, so ask for a roast with "ribs 6 through 9" or "10 through 12," and you should be fine.

## COOKING TIME

### I've got my beef, so now can I just follow a timing chart to know when it's cooked?

No, no, no! Do yourself a favor and throw out every roasting timing chart you have. The only reliable way—I repeat, the *only* reliable way—to tell when your beef is done is to use an accurate thermometer, like the Thermapen from ThermoWorks. No matter what cooking method or oven temperature you use, as long as the center of your meat never goes above the right temperature, you'll be guaranteed the right results. Check out the sidebar on page 548 for a chart on what to aim for.

## SERVING A PICKY CROWD

**My dainty aunt likes her meat rare, but my tempestuous brother prefers his well-done. What's an understanding and generous host to do?**

First off, you should wonder how your brother was created out of the same genetic material as you. After that, you've got a couple options. Because most roast are not an even shape, you'll probably end up with a few pieces that are more cooked than others, even when you cook with very gentle heat. These slices generally come from near the ends of the roast, where energy is penetrating the meat through more surfaces. If those pieces are still not cooked enough for your bro, the best thing to do is take his slices, stick 'em on a rimmed baking sheet, and throw them back into the oven until they are the desired shade of dry.

## LEFTOVERS

**Gramps decided not to show up for dinner—what's the best way to store his leftovers?**

An all-too-common dilemma. For short-term storage, your best bet is to just tightly wrap the meat in plastic and keep it in the fridge. It'll last for around three days. Bear in mind that rare or medium-rare meat may turn brownish (remember myoglobin?) in the fridge but this isn't necessarily a sign of spoilage. For that, just follow your nose!

For longer-term storage of larger pieces of either cooked or uncooked meat, you'll want to freeze them. If you've got a vacuum-sealer, use it. Air is the enemy of frozen food, causing it to dry out in an irrevocable process known as freezer burn. If you don't have a vacuum-sealer, wrap the meat tightly, first in foil (plastic wrap is not airtight) and then a few layers of plastic wrap, and place it in the freezer. The plastic wrap helps keep the foil tightly against

the surface of the meat, while the foil prevents air from coming in contact with it.

Let your meat defrost in the refrigerator. This can take as long as a couple days for larger cuts of meat.

**Did I hear that right? Plastic wrap is not airproof?**

That's correct. Air can still travel through plastic wrap, though quite slowly.

**What about reheating?**

If you are wise, you will have been slicing the meat to order, leaving you with a large chunk of leftover roast rather than many thin slices. The best way to reheat a large chunk of meat (thicker than 1½ inches) is to treat it like you would a steak, which after all, it is: warm it in a low oven until it is about 10 degrees below the desired serving temperature (see the chart on page 548), then sear it in a hot skillet to crisp and brown the exterior. For thinner cuts, you can simply sear them in a skillet straight from the fridge, just like a steak. Alternatively, the microwave is surprisingly effective. Just remember this cardinal rule: no matter how you reheat it, do not let it rise above the initial temperature you cooked it to, or it will be overcooked.

If you've already sliced the whole roast extremely thin, your best option is sandwiches and salads. Still delicious!

**What about those chefs who cook their prime ribs sous-vide or sear them with various pieces of industrial machinery?**

Leave the fancy-pants methods to the fancy-pants. In my experience, cooking large roasts—whether prime rib or turkey—in a sous-vide water bath certainly guarantees perfectly, evenly cooked results, but the deep roasted flavor notes you get from meat roasted in the open air are completely absent. It's also a pain in the butt to vacuum-seal an entire prime rib. I much prefer mine done in a low-temperature oven.

As for torching, it looks really cool, but the results are not worth the trouble. Torching before roasting gives you a surface that's nearly burnt in spots and barely browned in others, while torching after roasting doesn't do nearly as good a job as a hot oven or roasting pan set over a couple burners.

## HOW TO ROAST BEEF (AKA PERFECT PRIME RIB)

A four-pound roast of well-marbled prime beef rib is not cheap. And while my friends provide me with as many mental and philosophical riches as a man could ask for, and my wife supplies an adequate amount of emotional wealth, dollars and cents are not something I, being a humble food writer, part with lightly. As such, when I buy a good quality piece of beef—and honestly, does beef get any better than prime rib?—I have a strong impetus not to mess it up, as do, I imagine, most of you. In writing this section, I decided to get through a lifetime's worth of messings-up so that I (and, I hope, you too!) will never again serve anything but a perfectly cooked roast.

First, a definition of perfection:

- **Commandment I:** The Perfect Prime Rib must have a deep brown, crisp, crackly, salty exterior crust.
- **Commandment II:** The gradient at the interface between the brown crust and the perfectly medium-rare interior of the Perfect Prime Rib must be absolutely minimized (as in, I don't want a layer of overcooked meat around the edges).

- **Commandment III:** The Perfect Prime Rib must retain as many juices as possible.
  - **Sub-Commandment i:** The Perfect Prime Rib must be cooked without the use of heavy or specialized equipment, including propane or oxyacetylene torches, sousvide machines, and C-vap ovens.

## Highs and Lows

Before I tried to start figuring out how to achieve all these goals simultaneously, it was helpful to note that when cooking beef to medium-rare, there are really only two temperatures that matter:

- 125°F is the temperature at which beef is medium-rare—that is, hot but still pink, cooked but still moist, and able to retain its juices. Any higher than that, and the muscle fibers start to rapidly shrink, forcing flavorful juices out of the meat and into the bottom of the roasting pan.
- 310°F is the temperature at which the Maillard reaction—that wonderfully complicated process by which amino acids and reducing sugars recombine to form enticing roasty aromas—really begins to take off. At this range, meat will quickly brown and crisp.

Ah—a dilemma revealed itself: In order to maximize browning, I had to cook the meat in a sufficiently hot oven—I tried 400°F. But the same time, I didn't want the interior to reach above 125°F. Since a big beef roast cooks from the outside in, by the time the center had reached 125°F (that is, 120°F in the oven, followed by a 5-degree rise in temperature after resting), there was a perfectly browned exterior, but the outermost layers had risen closer to around 165°F to 180°F, rendering them overcooked, gray, and dry, their juices having been squeezed out.

I was left with something that looked like this:

I know, I know—not pretty.

Score:
- **Commandment I**: Perfect Crust? Check.
- **Commandment II**: No Gray Zone? Negative.
- **Commandment III**: Full-on Juiciness? Negative.

OK, so what if I went to the opposite extreme, cooking the roast at a much lower temperature? I cooked another roast in a 200°F oven until the center reached 125°F. Well, just as with boiled eggs, the temperature at which you cook is directly related to the difference in temperatures between the center and the outside layers. In other words, by cooking it at a lower temperature, you minimize the proportion of beef that goes above the ideal final temperature. I was able to eliminate the gray band of overcooked meat almost completely. Of course, any browning was also right out the window, leaving me with a roast with pale, anemic exterior.

Again, not pretty.

Score:
- **Commandment I**: Perfect Crust? Negative
- **Commandment II**: No Gray Zone? Check.
- **Commandment III**: Full-on Juiciness? Unknown.

## The Myth of the Sear

Jump back a couple of decades and the solution to my dilemma would have been obvious. It was a commonly held belief (and still is by many home cooks and professional chefs alike) that in order to help a roast, steak, or chop retain moisture, you should first sear it, creating a crust that will "lock in the juices." Now, anyone who has read their Harold McGee or has ever seen juices squeeze out through the seared side of a steak after you flip it over on the grill knows that this can't possibly be completely true. But what about partially true? Could a sear actually help retain at least some of the juices? In order to test this, I cooked two roasts cut from the same rib sections, with comparable surface areas, weights, and fat contents according to the following processes:

- **Roast 1**: Seared in a pan in 3 tablespoons canola oil over high heat on the stovetop until a well-browned crust formed (about 15 minutes), then transferred to a 300°F oven and roasted to an internal temperature of 120°F, removed, and rested for 20 minutes (during which time the temperature in the center rose to 125°F and then dropped back down to 120°F).
- **Roast 2**: Roasted in a 300°F oven to an internal temperature of 120°F, removed, and seared in a pan in 3 tablespoons canola oil over high heat on the stovetop until a well-browned crust formed (about 8 minutes), then rested for 20 minutes (during which time the center temperature rose to 125°F and then dropped back down to 120°F).

## PERCENTAGE RETAINED WEIGHT VERSUS SEARING ORDER

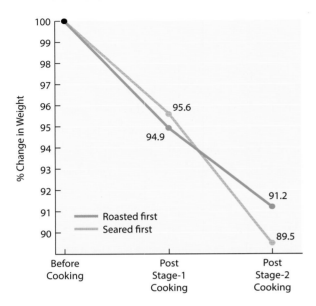

If searing does in fact "lock in juices," we would expect that the steak that was first seared and then roasted would retain more juices than the steak that was roasted and then seared. Unfortunately for old wives' tales, the exact opposite was the case. I carefully weighed each roast at each step of the process to gauge the amount of moisture and fat lost during cooking. These are the results:

The meat that was seared first and then roasted lost 1.68 percent more juices than the one that was roasted and then seared. It's not a huge difference, but the knowledge that searing conclusively does not lock in juices was liberating in the ways that it allowed me to think about the recipe.

Score:
- **Commandment I**: Perfect Crust? Check.
- **Commandment II**: No Gray Zone? Negative.
- **Commandment III**: Full-on Juiciness? Check.

### Inside and Out

So, great, you may be thinking—you can sear first or you can sear afterward, and it makes no difference. What's the big deal? Well, the big deal, as some of the more astute readers may have noticed in looking at the timing above, is that if you start with a raw roast, it takes around 15 minutes in the hot pan to get a well-browned crust, during which time the outer layers of the roast are busy heating up and overcooking, just like they did when roasted in a 400°F oven. But to get a well-browned crust after the prime rib has been roasted, you need only around 8 minutes in the pan. Why is this?

It all has to do with water.

In order for the surface of a roast to reach temperatures above the boiling point of water (212°F), it must first become completely desiccated. When you sear raw meat, about half of its stay in the skillet is spent just getting rid of excess moisture before browning can even begin to occur. You know that vigorous sizzling sound when a steak hits a pan? That's the sound of moisture evaporating and bubbling out from underneath the meat. A prime rib that has first been roasted, on the other hand, has had several hours in a hot oven, during which time the exterior has completely dried out, making searing much more efficient and thus giving all but the very exterior of the meat less chance of overcooking.

Taking what I had learned from both the oven-temperature testing and the searing testing into account, I knew what I had to do to fulfill all three commandments: my goal should be to cook the interior of the roast as slowly as possible (i.e., at as low a temperature as my oven could maintain), then sear it as fast as possible (i.e., at as high a heat as possible). But searing in a pan is not that practical for a roast bigger than a couple of ribs, so I needed a way to do this all in the oven.

While some recipes simply have you pump up the oven temperature toward the end of cooking, this is suboptimal. An oven can take 20 or 30 minutes to go from its lowest temperature to its highest temperature setting, during which time, once again, the outer layers of beef are busy overcooking. But then I thought, 20 to 30 minutes is exactly how long a rib roast needs to rest anyhow. What if I were to first cook it at a low temperature (200°F or lower), then take it out of the oven and allow it to rest while I heated the oven to its highest temperature (500° to 550°F), and pop it back in just long enough to achieve a crust?

What I achieved was nothing less than Prime Rib Perfection:

Score:

- **Commandment I**: Perfect Crust? Check.
- **Commandment II**: No Gray Zone? Check.
- **Commandment III**: Full-on Juiciness? Check.

As you can see above, a crisp brown crust, no gray overcooked meat, and a rosy pink from center to edges.

But wait—there's more!

The best part? I found that by using this two-stage method, I had a much larger window of time within which to serve the beef. Once I finished the initial low-temperature phase of cooking, as long as I kept the roast covered in foil, it would stay warm for over an hour. All I had to do was pop it into the 550°F oven for 8 minutes, and the roast would emerge hot, sizzling, and ready to carve—no need to rest it, since the only part that was affected here was the very exterior.

Family gatherings will never be the same. Now if only I could find a way to expose the rosy center under my sister's crusty exterior, we'd really have something to celebrate at the holidays!

## WHAT ABOUT JUS?

There's just one last question when it comes to prime rib: what about a great sauce to go with it?

Most recipes will call for a pan sauce of some sort, making use of the drippings that collect in the bottom of the roasting pan as the beef cooks. But here's the thing: my technique is specifically designed to produce *no* drippings whatsoever. That is, all of that moisture (and flavor) stays *inside* the beef, where it belongs. Truth be told, because of that fact, you really don't need any sauce at all, but some folks—traditionalists, let's call them—absolutely need a sauce with their meat. So, how do we get it?

The easiest solution I've found is to use some extra beef. By searing off a few hunks of oxtails in a Dutch oven, deglazing the drippings with wine and stock, adding some vegetables, and then roasting the whole lot in the oven with the prime rib, you can build a powerfully flavorful jus, with the added benefit of having a pile of fall-off-the-bone-tender braised oxtails to serve alongside that roast.

# KNIFE SKILLS:
## How to Carve a Bone-In Prime Rib

**If you were paying attention earlier, you know that the easiest way to carve a bone-in prime rib is to remove the bone and tie it back on before you begin roasting, so that it slips right off when ready to serve. But here's how you do it if you've roasted it bone-in.**

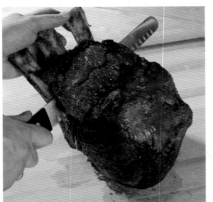

**STEP 1: LIFT AND SLICE**
Lift up the prime rib by the ends of the bones and place your carving knife in between the meat and the bones. Start to slice downward, holding the knife firmly against the bones.

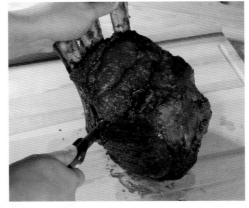

**STEP 2: CONTINUE SLICING**
Keep on slicing downward, hugging the bones the whole time.

**STEP 3: REMOVE THE BONES**
Cut through the base of the bones to completely separate the meat. You may have to work around a few ragged edges with the tip of a boning knife or paring knife.

**STEP 4: READY TO SLICE**
The meat is now bone free and ready to slice. Of course, those bones should be gnawed at, not thrown out!

**STEP 5: SLICE**
With the meat held upright, begin slicing across it in long, smooth strokes to cut off thin, even slices.

**STEP 6: READY TO SERVE**
Cut off only as many slices as you need—the beef will be easier to reheat and serve later on if you leave what you don't serve as one large piece.

1                2               3

# PERFECT ROAST PRIME RIB

**NOTES:** This recipe works for prime rib roasts of any size from 2 to 6 ribs. Plan on 1 pound of bone-in roast per guest (each rib adds 1½ to 2 pounds). For best results, use a dry-aged Prime grade or grass-fed roast.

To further improve the crust, season the roast with salt and pepper and allow it to air-dry in the refrigerator, uncovered, on a rack set on a rimmed baking sheet or in a roasting pan at least overnight, and up to 5 days.

If the timing goes off and your roast is ready long before your guests are, do not panic. Remove the roast from the oven and let it rest until you are about an hour away from serving, then reintroduce the roast to a 200°F oven for 30 minutes to rewarm it. Take it out of the oven, increase the heat to the highest possible setting, and continue with step 3.

**SERVES 3 TO 12,**
**DEPENDING ON SIZE**
**OF ROAST**

1 standing rib roast (prime rib), 3 to
   12 pounds (see Note above)
Kosher salt and freshly ground
   black pepper

1. Preheat the oven to the lowest possible temperature setting, 150°F or so (some ovens can't hold a temperature below 200°F). If desired, using a sharp chef's knife or a carving knife, cut off the ribs from the roast in a single slab (or have your butcher do this for you). Using butcher's twine, tie the bones securely to the roast.

2. Season the roast generously with salt and pepper (go lightly if you seasoned it ahead). Place the roast, fat cap up, on V-rack set in a large roasting pan. Place it in the oven and cook until the center registers 120°F on an instant-read thermometer for medium-rare, or 135°F for medium. In a 150°F oven, this will take 5½ to 6½ hours; in a 200°F oven, it will take 3½ to 4 hours. Remove the roast from the oven and tent tightly with aluminum foil. Place in a warm spot in the kitchen and allow to rest for at least 30 minutes, and up to 1½ hours. Meanwhile, heat the oven to highest possible temperature (500° to 550°F).

3. Ten minutes before your guests are ready to be served, remove the foil, place the roast in the hot oven, and cook until well browned and crisp on the exterior, 6 to 10 minutes.

4. Remove the roast from the oven and cut off and remove the butcher's twine if you used it. Remove the slab of bones and slice in between each rib. Set them on a serving platter. Carve the eye of meat into ¼-inch slices and arrange on the serving platter. Serve immediately.

## Oxtail Jus for Prime Rib

*This recipe, which can be made up to 5 days ahead, will produce a rich red wine jus that can be poured over your prime rib, along with a pile of tender pulled oxtail meat. That meat makes a perfect hors d'oeuvre for a fancy gathering. Serve it with toasted crusty bread a nd crunchy sea salt.*

**MAKES ABOUT 2 CUPS JUS AND ENOUGH PULLED OXTAIL MEAT TO SERVE 8 AS AN HORS D'OEUVRE**

1 tablespoon vegetable oil

3 pounds oxtails

1 large carrot, peeled and roughly chopped (about 1½ cups)

2 stalks celery, roughly chopped (about 1½ cups)

1 large onion, roughly chopped (about 1½ cups)

1 bottle (750 ml) dry red wine

2 bay leaves

4 sprigs fresh thyme

4 stems fresh parsley

4 cups homemade or low-sodium canned chicken stock

Kosher salt and freshly ground black pepper

1. Heat the oil in a large Dutch oven over high heat until lightly smoking. Add the oxtails and cook, flipping and stirring the pieces occasionally, until well browned on all surfaces, about 15 minutes. Using tongs, transfer to a large plate and set aside.

2. Add the carrot, celery, and onions to the pot and cook, stirring occasionally, until starting to lightly brown, about 8 minutes. Add the wine, bay leaves, thyme, and parsley and scrape up the browned bits from the bottom of the pot, then bring to a boil and cook until the liquid is reduced by half, about 10 minutes. Add the chicken stock and return the oxtails to the pot. Bring to a boil, reduce to a bare simmer, cover, and cook until the meat is starting to fall off the bones, 3 to 3½ hours.

3. Using tongs, transfer the oxtails to a large bowl. When they are cool enough to handle, shred the meat from the bones; discard the bones. Transfer the meat to a sealable storage container.

4. Strain the braising liquid through a fine-mesh strainer into a medium saucepan. Carefully skim the excess fat from the top with a ladle. Spoon a few tablespoons of the liquid over the shredded meat and season the meat to taste with salt and pepper; cover the meat and liquid and refrigerate until ready to serve. (Reheat the meat in the microwave or in a skillet before serving; see Note above.)

5. Shortly before serving the roast, return the strained braising liquid to a simmer and cook until reduced to 2 cups, about 15 minutes. Season to taste with salt and pepper. Serve the jus with the prime rib.

# SLOW-ROASTED BEEF TENDERLOIN

As with prime rib, the best method for roasting beef tenderloin is to start it out low and slow, then blast it with heat to brown its surface at the end. The problem is, with its relatively low fat content and small size, a beef tenderloin is far more prone to overcooking than a prime rib. Try and brown it in a hot oven after roasting it, and you'll end up cooking it to medium-well by the time a decently browned crust has developed.

The solution is a simple one: instead of finishing it in the oven, finish it quickly on the stovetop. The conductive heat of a hot skillet or Dutch oven is a far more effective means of energy transfer than the air inside an oven. By slow-roasting the tenderloin and then searing it on the stovetop, you get perfectly medium-rare meat from the edge, with a deep brown crust that adds some much-needed flavor to an otherwise relatively bland cut.

Tenderloin roasts can be served slathered with a compound butter just like a good steak (see page 325), or with horseradish cream sauce (page 640) alongside.

*Additional photograph on page 564.*

NOTE: A center-cut tenderloin roast is also referred to as Châteaubriand. Ask your butcher for a 2-pound center-cut tenderloin, or trim one yourself (see page 317). For the best results, tie and season the tenderloin, place it on a rack on a rimmed baking sheet and air-dry, uncovered, in the refrigerator at least overnight and up to 3 days.

### SERVES 4 TO 6

1 center-cut tenderloin roast (see Note above), about 2 pounds tied for roasting (see page 639)

Kosher salt and freshly ground black pepper

1 tablespoon oil

1 tablespoon unsalted butter

Horseradish cream sauce (page 640; optional)

1. Adjust an oven rack to the center position and preheat the oven to 275°F. Season the tenderloin liberally on all sides with salt and pepper (go lightly if you seasoned it ahead). Place on a wire rack set on a rimmed baking sheet and roast until the center of tenderloin reaches 120°F on an instant-read thermometer for medium-rare, or 130°F for medium, about 1 hour. Remove from the oven and transfer to a cutting board (the roast will appear gray and uncooked).

2. Heat the oil and butter in a 12-inch heavy-bottomed stainless steel or cast-iron skillet over high heat until the butter is browned and the fat is lightly smoking. Add the tenderloin and sear on all sides until well browned, about 5 minutes. If the oil and butter are beginning to burn or smoke too heavily, reduce the heat. Transfer the roast to a cutting board and allow to rest for 10 minutes.

3. Remove the twine, carve, and serve, with the horseradish sauce, if desired.

# HOW TO TIE A TENDERLOIN
# FOR ROASTING

Tying a beef tenderloin before roasting is not 100-percent essential, but it helps it keep a nice round shape, which, in turn, helps it to cook more evenly. Here's how it's done:

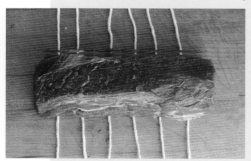

### Step 1: Lay Out the Twine

Lay out 12-inch-long lengths of kitchen twine at 1-inch intervals across your cutting board, using enough to span the length of the tenderloin. Place the tenderloin on top of the twine. Tie the first knot, starting at one end of the roast.

### Step 2: Keep It Tight

To ensure that the tie doesn't slip as you form the knot and tighten it, cross the ends over each other at least three times before tightening the loop and tying a simple granny knot.

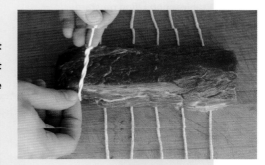

### Step 3: Work Toward the Center

Continue tying knots in the same manner, alternating ends and working from the outside to the center: this will help give the roast a more uniform shape.

### Step 4: Trim the Twine

Trim the ends of the twine with kitchen shears or a sharp knife.

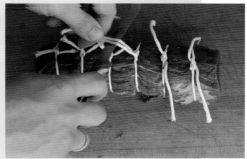

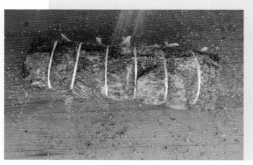

### Step 5: Season Liberally

Season the tenderloin liberally on all sides with salt and pepper. It is ready to roast.

# HORSERADISH CREAM SAUCE

**Just as with steaks, you can make easy pan sauces to go with roast beef by deglazing the juices left in the roasting pan. However, I prefer to serve mine with a cooling horseradish cream sauce. I love the richness it adds and the mustardy bite of the horseradish (who doesn't like mustard with beef?). Make it with homemade crème fraîche, and it'll really knock your socks off.**

## MAKES ABOUT 2 CUPS

2 cups crème fraîche, preferably
   homemade (page 123)

½ cup grated fresh horseradish
   (or substitute drained prepared
   horseradish)

1 tablespoon Dijon mustard

1 teaspoon white wine vinegar

Kosher salt and freshly ground
   black pepper

Combine the crème fraîche, horseradish, mustard, and vinegar in a medium bowl and whisk together. Season to taste with salt and lots of pepper. Refrigerate in a sealed container for at least 24 hours, and up to a week, for the flavor to develop.

1

2

3

# The FOOD LAB's Complete Guide to
## BUYING, STORING, AND COOKING LAMB

I think lamb is delicious, and so do many cultures. But in most of the United States it's a different story. Here's the average U.S. per capita consumption of a few different types of meat:

## MEAT CONSUMPTION BY TYPE

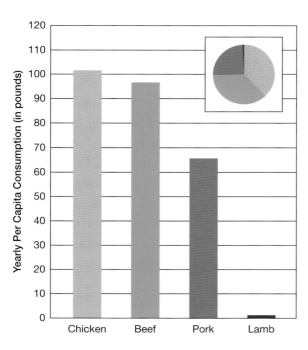

That's right. We eat more than 100 pounds of chicken per year each, but, we eat only 0.8 pound lamb per year, and the amount is getting smaller every year—in the 1970s, it was a larger, but still woefully tiny, 3 pounds per year. Not only that, but the vast majority of Americans don't eat a single bite of lamb all year. Most of this per capita consumption comes from minority communities—Greek, Muslim, Indian—who eat a ton of it, bringing up the overall average.

Even more distressing is the fact that lamb is what economists call an "inferior good," which means that its demand is inversely proportional to average consumer income: when people have money to spend, they'd rather spend it on beef and chicken breasts. A 2001 study from Kennesaw State University in Georgia found that for every 1 percent increase in consumer income, there was a corresponding 0.54 percent decrease in lamb consumption. It's an odd dichotomy, because while in many communities, lamb is seen as the "cheap" meat—the meat to buy when you can't afford beef—in high-end food and fancy supermarket circles, it's often far more expensive, and desirable, than beef.

I'm with the latter camp. I can think of precious few situations when I'd rather have a steak than a fatty, musky lamb chop. Or when I'd rather have a pot roast than a rich, slightly funky braised lamb shank. And when it comes to holiday roasts, prime rib may be the king of the table, but roasted leg of lamb is his wilder, more fun cousin.

Lamb marketers who have long known of the trouble with selling their product to consumers reluctant to leave the safety of their beloved chicken and beef, have responded by carefully breeding and raising lamb that is more suitable for the American palate, as well as selling it in forms that are increasingly easier to cook. Indeed, if you haven't attempted to cook lamb for yourself at home yet, you've really have no excuse.

What better time to start than now?

## DOMESTIC VERSUS IMPORTED

**I see lamb from Australia, New Zealand, and the United States at the butcher. What are the differences between these options, and is one better than another?**
There are major differences in terms of flavor, size, and price when it comes to American lamb versus lamb from Down Under. New Zealand/Australian lambs are quite small in size, with whole legs coming in at around 5 to 6 pounds. According to Mark Pastore, president of Pat LaFrieda, one of the most respected meat purveyors in the country, their size is a matter of both genetics and feed. The lambs are smaller to begin with, and they spend their lives grazing on grass. Grass gives them a more gamy flavor, which some people find off-putting, and they also tend to be lower in fat, making them a bit harder to cook properly—the legs in particular have a tendency to dry out. That said, if you're cooking for a smaller party—6 to 8 people or so—and you

value that gamy flavor over tenderness or richness, NZ or Aussie lamb is a good choice.

American lambs are larger, fattier, and sweeter in flavor. Most American lambs are fed on grass for most of their lives, supplemented with grain for the last 30 days before slaughter. The lamb at LaFrieda comes from Mennonite farms in Colorado that finish their lambs on a combination of grain, honey, alfalfa, wheat, and flaked corn. The results are a larger layer of protective fat around the legs and better marbling (the intramuscular fat that adds flavor and moistness to meat). Because of the grain supplements, American lamb tends to have a less funky but richer flavor, more similar to steak. A single leg of American lamb can weigh up to 15 pounds or so, with enough meat to feed over a dozen.

**I've read that grass-fed meat is always better—better tasting, better for the animal. Is there any truth in this?**
It depends. Some people prefer the gamier taste of 100-percent grass-fed lamb, others prefer the richer flavor and juicier meat of grain-finished lamb. As far as the health of the animal goes, while it's true that an animal that lived solely on grain would eventually develop health problems (much like a human who existed solely on hamburgers), grain finishing only occurs over the last 30 days of a lamb's life, after which it's going to be slaughtered anyway. That time period is not nearly long enough for the animal to develop any health problems that would cause it to suffer in any way. If you have no problem eating meat, you should have no problem eating grain-finished lamb or beef.

**What about the price differences?**
Unfortunately, American lamb tends to be more expensive than the imports, despite their long jour-

ney across the globe. It's a matter of scale. Australia and New Zealand's lamb output is several times greater than that of the United States. If you value tenderness and juiciness, the extra cost is probably worth it.

## BONES

**I'm confused by all of the butchering options I have when buying a lamb leg. What should I be looking for?**
Bone-in leg of lamb comes in two forms: shank end and sirloin end (occasionally you'll find a massive one for sale, with both the shank and sirloin). Shank-end legs start at just above the lamb's ankle and go to midway up the calf bone, while sirloin-end legs start at the hip and stop at around the knee.

I prefer the sirloin end because the meat is fattier and more tender, and the cut is more evenly shaped, making it easier to cook. On the other hand, the shank end tends to have slightly more flavorful meat. And its tapering shape is actually desirable for cooks who like being able to offer both medium-rare meat from the thick upper part and well-done meat from the thin lower part.

There are advantages to buying a bone-in lamb leg. It's generally cheaper per pound, even accounting for the weight of the bone. The bone can act as an insulator, making the whole thing cook more slowly and giving you a certain leeway in terms of hitting that medium-rare sweet spot. Contrary to what some believe, the bone does not actually add much flavor to the meat, though the meat directly around it will be a tad bit more tender, due to the fact that it will be less cooked.

But boned lamb leg also offers advantages. First of all, it's lighter, making the arduous task of lifting it in and out of the oven much easier on the back. It's also easier to calculate how much you need to feed your party. Finally—and this is probably the greatest advantage of all—it's far easier to carve: just cut straight through it into neat, even slices.

Butterflied leg of lamb is a boneless leg that has been split open and then opened out. This is how I prefer to purchase my lamb; it affords me the opportunity to season it both inside and out. Often this just means a quick rub with salt and pepper before rolling it up and tying it, but it can also mean more elaborate rubs or herb mixtures. If you choose to go with a butterflied leg, you'll need to know how to tie it up before roasting; see page 644.

**What about rack of lamb?**
Rack of lamb is the lamb equivalent of a prime rib of beef. It's that same muscle and same set of rib bones. The only difference is that a lamb is much smaller than a steer; hence the daintier proportions. And while beef rib bones are generally cut to within a few inches of the eye of meat, lamb rib bones are left longer, which makes for a stunning presentation—as well as giving you a convenient handle with which to hold your chops if you choose to eat caveman-style, gnawing the juiciest chunks of meat and fat off the bones at the end.

When it comes to buying racks, you've got two basic options: as is or frenched. "Frenched" is just a fancy way of saying "we stripped the meat from the ends of the ribs for you so they look all nice and pretty." Because it looks so nice, most lamb racks do come frenched, but to be honest, I *prefer* unfrenched racks because the juicy, fatty rib meat clinging to the bones is some of the tastiest stuff on the animal. Think of it as lamb bacon.

## SEASONING, ROLLING, AND TYING

**What are some good things to season my lamb with?**
Salt is a must, and just as with steak or a beef roast, the best time to salt your lamb is either the day

before roasting or immediately before cooking. If you've got the time, seasoning the lamb and letting it rest uncovered on a rack in a rimmed baking sheet in the fridge will season it more deeply, as well as dry out its exterior—allowing for superior browning.

With its robust flavor, lamb takes well to all kinds of spice mixes and aromatics. With a butterflied leg, you want to apply your seasoning to both the inner and outer surfaces before rolling it up. Here are a few of my favorite combinations:

- Lots of garlic, rosemary, and anchovies (see page 646)
- Olives and parsley (see page 646)
- Ground cumin and fennel (see page 647)
- Harrissa and garlic (see page 647)

**Why do I need to tie up my butterflied lamb leg?**
If you don't tie up a butterflied leg, it won't keep a regular shape during cooking. An irregular shape leads to uneven cooking. Uneven cooking leads to unhappy bellies. Unhappy bellies lead to lack of familial harmony. And lack of familial harmony leads to ruined holidays. Would you risk ruining a holiday for five minutes of work and the cost of a roll of butcher's twine?

**OK, I'm convinced. How do I do it?**
Simple. After laying your lamb out flat and seasoning it, roll it up again, with the fat on the exterior, then lay it seam side down over pieces of butcher's twine that you've already thoughtfully laid out in parallel lines on the cutting board at 1-inch intervals, each piece long enough to tie easily around the roast. Working from the ends toward the center, tie up the lamb. It's the same process as tying up a beef tenderloin (see page 639).

Your lamb is now ready to cook.

## COOKING IN THE OVEN

**How do I know when my lamb is done? Can I just follow one of those handy timetables, with X minutes per pound?**
Absolutely not! Ignore any and every timing chart you've ever seen—they don't work, because they don't take into account basic things like variances in shape and fat content, both of which can drastically affect how fast your meat cooks. Instead, get yourself a good digital instant-read thermometer. (Do I sound like a broken record here? Just do it!)

Doneness levels for lamb are pretty much the same as for beef, and they are the same whether you are talking leg or rack:

- **120°F (rare):** Bright red and slippery inside. The abundant intramuscular fat has yet to soften and render.
- **130°F (medium-rare):** The meat has begun to turn pink and is significantly firmer, juicier, moister, and more tender than either rare or medium meat.
- **140°F (medium):** Solid rosy pink and quite firm to the touch. Still moist but verging on dry. The fat is fully rendered at this stage, delivering plenty of flavor.
- **150°F (medium-well):** Pink but verging on gray. The moisture level has dropped precipitously, and the texture is chewy and fibrous. The fat has fully rendered and begun to collect outside the roast, carrying away flavor with it.
- **160°F (well-done):** Dry, gray, and lifeless. The moisture loss is up to 18 percent, and the fat is completely rendered.

Just like with beef, I personally recommend cooking lamb to at least medium-rare—it's hot enough that the abundant fat in the meat has begun to melt, lubricating and flavoring the meat. Rare lamb is tougher and less flavorful.

**So, if I'm cooking it in the oven, what temperature should I use?**

Just as with cooking any large piece of meat, you've got a decision to make right off the bat: do you want to cook hot or cool? Cooking in a high oven will obviously get dinner on the table much faster, but it'll also result in far more uneven cooking, with the outer layers of the meat overcooking and turning gray by the time the center is done. Now, I understand that some people don't mind this. I like having some juicy medium-rare meat and some tougher well-done meat on my plate, they say. Those of you who feel this way should be thankful—it makes cooking roasts very easy. Just bang it into a hot oven (around 400°F should do) and roast until the very center reaches the desired temperature.

But if you, like me, want your lamb evenly cooked from edges to center, the best thing to do is slow-roast it, just as when cooking prime rib (see Perfect Roast Prime Rib, page 635): Place it in a 200°F oven until it is within a few degrees of your desired serving temperature (use that thermometer!). Remove it, crank the oven up as far as it will go, and let it heat up, then throw the lamb back in for about 15 minutes to crisp up the well-rendered fat layer on the exterior.

**Do I have to let my lamb rest just like beef?**

Just as with a steak or a beef roast, lamb muscles tighten when they're hot. As they loosen up during resting, their ability to retain their juices increases. This means that more juice ends up in your meat and less on the cutting board. Allow lamb roasted entirely at a high temperature to rest for at least 20 minutes after removing it from the oven, and meat roasted low-and-slow to rest for at least 10 minutes.

## CARVING A BONE-IN LEG OF LAMB

When you've got a bone-in leg of lamb, you'll notice that the bone runs along one side of the bulk of the meat. You want to slice from the opposite side. Using a fork or tongs to hold the lamb steady, use a long, thin carving knife to cut the meat into thin slices: some of these slices may remain attached to the bone, but that's OK. Separate them by then making a single slice across the top and side of the bone. The slices should fall away neatly for you to serve.

# SLOW-ROASTED BONELESS LEG OF LAMB

**SERVES 10 TO 12**

1 butterflied boneless leg of lamb, 5
to 7 pounds

Kosher salt and freshly ground
black pepper

1. Adjust an oven rack to the center position and preheat the oven to 200°F. Open out the lamb leg on a cutting board. Season generously with salt and pepper inside and out (go light on the salt if you seasoned the meat ahead). Roll the lamb up and secure with kitchen twine.

2. Place the lamb on a wire rack set in a foil-lined rimmed baking sheet. Place it in the oven and cook until the center of the roast registers 125°F on an instant-read thermometer for medium-rare, or 135°F for medium, 2½ to 3 hours. Remove the roast from the oven and tent tightly with aluminum foil. Place in a warm spot in the kitchen and allow to rest for at least 30 minutes, and up to 1½ hours. Meanwhile, heat the oven to highest possible temperature (500° to 550°F).

3. Ten minutes before your guests are ready to be served, remove the foil, place the roast in the hot oven, and cook until well-browned and crisp on the exterior. 6 to 10 minutes. Remove from the oven and cut off and remove the butcher's twine. Slice the meat into ½-inch slices and serve immediately.

## SLOW-ROASTED BONELESS LEG OF LAMB WITH GARLIC, ROSEMARY, AND ANCHOVIES

*You may turn your nose up at the idea of combining anchovies with meat, but don't knock it until you've tried it! Anchovies are packed with glutamates and inosinates, both proteins that can greatly enhance the "meatiness" of meats. Your lamb ends up tasting more lamby than you'd ever expect it to, while the rosemary and garlic complement the natural funkiness of a good leg of lamb.*

Combine 12 cloves garlic, ¼ cup fresh rosemary leaves, 6 oil-packed anchovy fillets, a pinch of red pepper flakes, and ¼ cup extra-virgin olive oil in the bowl of a food processor and process until a paste is formed, scraping down the sides as necessary. Transfer the mixture to a small saucepan and cook over medium heat, stirring frequently, until just beginning to bubble, then cook, stirring constantly, until the garlic loses its raw flavor, about 1 minute. Transfer to a small bowl. Rub the leg of lamb thoroughly with the mixture before seasoning it in step 1; apply salt sparingly, as the anchovy mixture is salty.

## SLOW-ROASTED BONELESS LEG OF LAMB WITH OLIVES AND PARSLEY

Combine 1 cup pitted kalamata or Taggiasche olives, ½ cup fresh parsley leaves, 1 clove garlic, and ¼ cup extra-virgin olive oil in the bowl of a food processor and process until a paste is formed, scraping down the sides as necessary. Rub the leg of lamb thoroughly with the mixture before seasoning it in step 1; apply salt sparingly, as the olive mixture is salty.

### SLOW-ROASTED BONELESS LEG OF LAMB WITH CUMIN AND FENNEL

Whisk together 1 tablespoon ground toasted cumin seeds, 1 tablespoon ground toasted fennel seeds, 2 cloves of garlic minced, 2 teaspoons soy sauce, and ¼ cup olive oil in a small bowl. Rub the leg of lamb thoroughly with the mixture before seasoning it in step 1.

### SLOW-ROASTED BONELESS LEG OF LAMB WITH HARISSA AND GARLIC

Whisk together 3 tablespoons store-bought harissa paste, 1 clove garlic, minced, and ¼ cup olive oil in a small bowl. Rub the leg of lamb thoroughly with the mixture before seasoning it in step 1.

## COOKING RACK OF LAMB

Though a rack of lamb is technically a roast, it's actually much more useful to think of it as if it were a steak—which makes sense, because the proportions are about the same (a couple inches thick at most), as is the composition (tender meat with a good amount of intramuscular fat). So, when I cook a rack of lamb, unless I'm doing something fancy and involved like beer-cooler sous-vide (see page 385), I do it pretty much the same way I cook my butter-basted steak (see page 312): skip the oven completely and roast it in a moderately hot pan on the stovetop. The same basic rules apply.

1. Dry the rack and season it liberally either at least 45 minutes before cooking or immediately before cooking.
2. Don't bother letting it come to room temperature if cooking it immediately; it'll cook about the same either way.
3. Use the heaviest pan you've got, for more even heat distribution and a better sear.
4. Manage your temperature so that the lamb is perfectly browned just when the center hits medium-rare.
5. Don't crowd the pan—a couple of racks in a 12-inch skillet is about the most you can do at once.
6. Flip as often as you'd like—this leads to faster, more even cooking.
7. Don't add butter until close to the end, or it'll burn.
8. Make sure to brown the edges of the racks!
9. Let the meat rest before carving and serving.

The only other issue with a rack of lamb is its somewhat awkward shape. Because of the curvature of the bones, it's basically impossible to get good contact with the skillet on the inside curve of the bones. The solution is to simply not bother trying. Rather than letting the inside curve of the bones cook via the heat of the pan, it's better to cook them by basting them with hot fat.

Rack of lamb is expensive, which may have turned you away from cooking it in the past. But trust me, with this method and a good thermometer at your side, it's nearly impossible to mess it up. And you will enjoy the results.

# PAN-ROASTED RACK OF LAMB

**NOTES:** The sizing for and yield for this recipe are based on relatively small, more common Australian and New Zealand lamb racks. If you are using larger American lamb racks, you may only need 4 to 6 ribs total, depending on their size.

For best results, after seasoning the lamb in step 1, let it rest uncovered in the fridge overnight. Alternatively for faster results, season immediately before cooking, skipping the 45-minute post-salt resting phase. If resting it at all after salting, do not let it stand for any less than 45 minutes.

**SERVES 2 OR 3**

One 8- to 10-rib rack of lamb
(about 2 pounds), split into 2
half-racks
Kosher salt and freshly ground
black pepper
2 tablespoons vegetable oil
2 tablespoon unsalted butter
4 sprigs fresh thyme
1 medium shallot, roughly chopped
(about ¼ cup)

1. Season the lamb generously with salt and pepper on all sides. Set on a plate and let rest at room temperature for at least 45 minutes, and up to 2 hours (see Note above).

2. Heat the oil in a 12-inch stainless steel skillet over medium-high heat until shimmering. Lay the lamb racks in the pan, fat side down, and cook, turning occasionally with tongs, until starting to turn golden brown, about 4 minutes. Add the butter, thyme, and shallots, reduce the heat to medium, and continue to cook, turning and flipping the lamb occasionally, and basting regularly with the browned butter (to baste, tilt the pan toward you so that the butter pools at the bottom and spoon the butter over the lamb, taking care to hit any underbrowned spots repeatedly), until the lamb is deep golden brown and an instant-read thermometer inserted into the thickest part of each rack section registers 120°F for medium-rare, or 130°F for medium, 3 to 7 minutes longer. Transfer the lamb racks to a rack set on a rimmed baking sheet and let rest for 5 minutes.

3. To serve, reheat the fat in skillet until smoking and pour over the lamb to warm it. Carve and serve immediately.

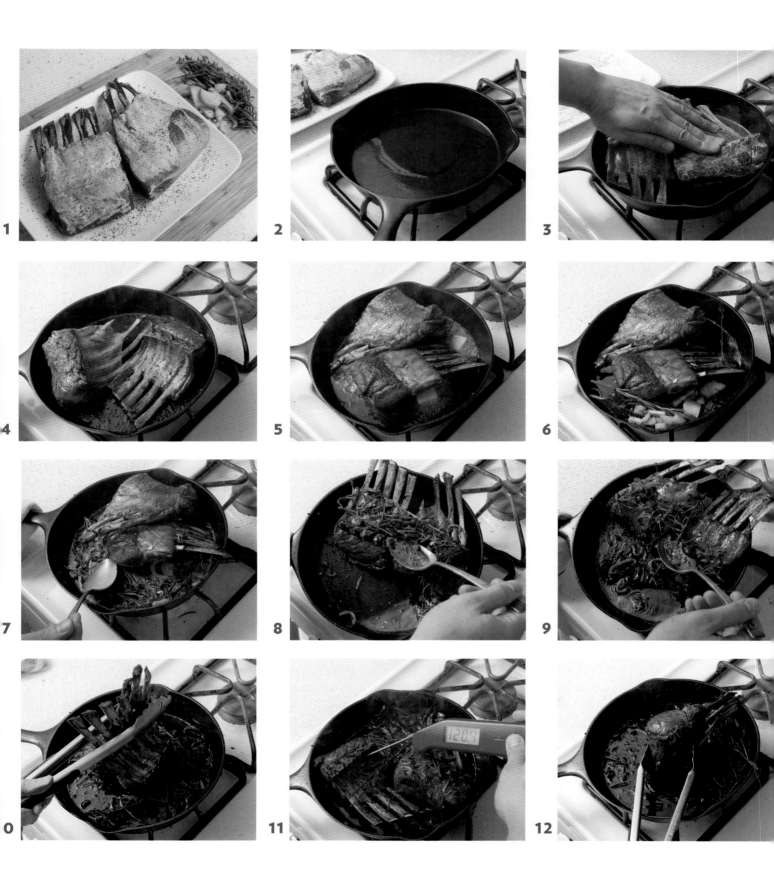

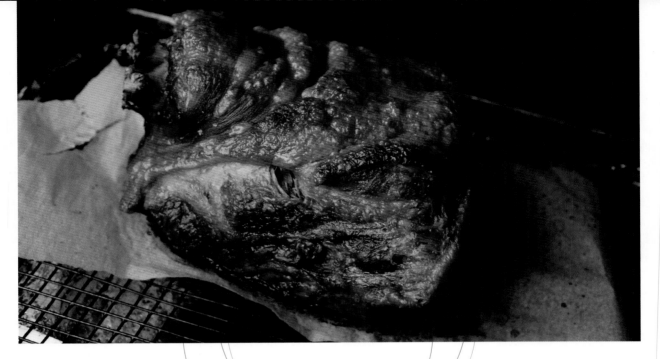

# ULTRA-CRISP ROAST PORK SHOULDER

**Pork shoulder is the lead of culinary alchemists,
just waiting, itching to be turned into gold.**

We're talking pork butt, in all of its juicy, porky, spoon-tender-in-the-middle, impossibly crisp-and-crusty-on-the-outside glory. The transformation of one of the cheapest cuts of meat in the butcher's display case into one of the most glorious, festive centerpieces imaginable is nothing short of a miracle. At least, it would be if we believed in such things. But we're here for the science of it, right? What makes pork get tender, how do we maximize that tenderness, and above all, how do we get the crisp, crackling skin that everyone fights over?

A full bone-in Boston butt is a formidable piece of meat, usually weighing in at around 8 to 12 pounds, riddled with a significant amount of connective tissue and inter-/intramuscular fat, all swathed in a thick, tough skin. Our goal is to make this tough piece of meat spoonably tender. How do you do that? Well, first we need to understand the difference between the two major types of muscles in an animal.

**Fast-twitch muscle** is the stuff that the animal rarely uses, except in short bursts: The breasts on a chicken that let it flap its wings rapidly when escaping danger. The loins on a cow that, well, barely get used at all. Fast-twitch muscle is characterized by tenderness (think chicken breast, pork chops, and New York strip steaks) and a fine-textured grain and is best prepared using fast-cooking methods like roasting, grilling, or sautéing (see Chapter 3). With fast-twitch muscle, optimal eating conditions are met pretty much as soon as you reach your final serving temperature (say, 145°F for a chicken breast or 125°F for a steak). As the meat is heated, it contracts, squeezing out moisture at a rate that's proportional to the temperature it's raised to. So, for example, you *know* that as soon as a steak hits 150°F, its muscle fibers have contracted enough to squeeze out about 12 percent of its moisture, and there's no turning back (see pages 295–96 for more details).

**Slow-Twitch muscle**, on the other hand, comprises the continually working muscles in an animal:

The shoulders and haunches that keep the animal upright and walking. The tail muscles that keep the flies off. The muscles around the flank that keep the animal breathing. Slow-twitch muscle is characterized by robust flavor and a very tough texture with lots of connective tissue that needs to be cooked for an extended period of time to break it down. With slow-twitch muscle, the tenderness of the finished product is dependent not only on the temperature at which it's cooked, but also on the length of time it is cooked. Beginning at around 160°F, the tough connective tissue collagen begins to break down into tender, juicy gelatin. The hotter the meat, the faster this breakdown occurs.

To put it simply: With fast-twitch muscle, **temperature** is the most important factor when cooking. With slow-twitch, both **time** and **temperature** affect the final product.

Whether we're talking pork shoulder, pot roast, braised turkey legs, or any other meat, the temperature ranges at which moisture expulsion and connective tissue breakdown take place are nearly identical. But exactly how long does it take collagen to break down into gelatin at a given temperature? I decided to find out.

Using a vacuum-sealer and a temperature-controlled water bath I cooked cubes of pork shoulder at 160°, 175°, 190°, and 205°F, keeping track of exactly how long it took to fully tenderize the meat at each temperature, and found that the amount of time needed to tenderize a piece of meat increases exponentially as the temperature is lowered. At 205°F (nearly as hot as a piece of meat can get), the cooking time was a mere 3 hours, while at 160°F, it took a full day and half!

If higher temperatures lead to faster breakdown of connective tissue, shouldn't you just blast your pork shoulder at the highest oven tempera-

ture it can take without burning the skin? Not so fast. Temperature has other effects too—namely, drying meat out. I roasted two identical pork shoulders until they were both equally tender, one at 375°F (which took about 3 hours), and the second at 250°F (which took about 8 hours). After roasting, I calculated the amount of total moisture lost by the meat by adding together the weight of the finished roast and the fatty drippings in the pan below and subtracting that from the initial weight of the roast.

## OVEN TEMPERATURE VERSUS WEIGHT LOSS IN PORK SHOULDER

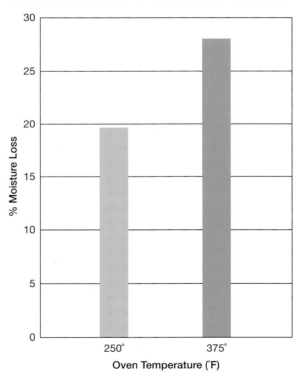

Turns out that at a higher temperature, a pork shoulder loses about 8 percent more juices than at a lower temperature, due to the muscle fibers contracting and squeezing out their contents. Eating the meat from the two roasts confirmed as much, though, to be honest, both were pretty crazy-juicy and moist. But, the high-temperature roast showed at least one definite advantage over the low-roast: the skin. In the hotter roast, the skin turned out crisp and crunchy (although not amazingly so), while in the low-temperature roast, the skin softened but was floppy and flaccid—a total bust.

So is there a way to get juicier meat *and* crisp skin? The problem is that cooking a great piece of pork skin requires two different approaches.

## SKIN JOB

There's a common misconception that animal skin—chicken skin, turkey skin, pork rinds—is made up entirely of fat. This is not true. There certainly is a lot of fat in the skin and directly underneath it (necessary to help warm-blooded animals maintain their body temperature), but skin also contains a great deal of water and connective proteins that, just like the connective tissues in slow-twitch muscles, must be broken down via long cooking.

On top of that, once the connective tissue has softened sufficiently, moisture must be forced out of it and the remaining proteins heated until they coagulate and stiffen up. It's the combination of these three things—connective tissue breakdown, moisture loss, and firming of proteins—that leads to crisp-but-not-tough skin.

When a pork shoulder is cooked at 375°F, all three of these things happen at about the same time. By the time the connective tissue has broken down, you've driven off enough moisture from the rind to render it hard and crunchy. In a 250°F oven, connective tissue breaks down for sure, but moisture loss and protein stiffening don't occur to a great enough degree to deliver a crisp finished product.

So clearly, once again, we should be cooking our

pork at a higher temperature, right? But hold on, dear, we got one more thing to consider. Patience.

## BUBBLE, BUBBLE

We all know what surface area is, right? Take a look at a close-up of a piece of the crisp skin from the pork cooked at 375°F:

Pork cooked at a steady temperature has skin that's crisp but smooth, with little surface area.

See how, despite a few wrinkles here and there, it's relatively smooth? Well, smooth objects have relatively low surface area given a particular volume, while wrinkled, bubbled, crinkly, curvy objects have a relatively high surface area given the same volume. And when it comes to texture, more surface area = more crunch. It's the same principle behind, say, scratching up the surface of potatoes before roasting them to get them extra-crisp (see page 474) or packing your burger extra-loose to give it a crisper exterior and more browning (see page 545).

When roasting at 375°F, because the dehydrating and protein setting is taking place at the same time that the connective tissue is breaking down, there's never really a stage when the skin is relatively soft.

It goes from being firm because of the connective tissue directly to being firm through dehydration.

On the other hand, after 8 hours in a 250°F oven, the skin has very little structural integrity—it's stretchy, soft, and easily bent. Indeed, if you looked at it under a microscope, you'd find that the structure of the skin very much resembled a whole bunch of interconnected balloons just waiting to be filled. How do you fill those balloons? Let heat do the work for you.

If you take that slow-cooked pork and bang it into a preheated 500°F oven, air and steam trapped within the skin will rapidly expand, causing millions of tiny bubbles to form. And here's the key: as the bubbles expand, their walls stretch out thinner and thinner; eventually they are so thin that the heat from the oven is able to quite rapidly set them into a permanent shape that won't collapse even when the pork is pulled out of the oven.

In this sense, pig skin is very much like raw pizza dough going into a hot oven: the high temperature causes gas expansion, which then gets trapped in a protein matrix that firms up in the heat of the oven to create a crunchy, crisp crust.

Have you ever seen anything so beautiful?

I generally prefer my meat relatively unadorned—good meat, salt, and pepper is all I need. Pork shoulder, on the other hand, is great at taking on flavor. Feel free to rub the meat and skin with your favorite spice blend or dry rub before roasting it, or—my method of choice, since it keeps your options open—keep it plain for roasting and then season the shredded tender meat before serving.

Actually I like to bring the thing whole to the table and allow dinners to pick and pull at it with their fingers, offering a few sauces to work with on the side. Try sweet-and-spicy nuoc cham, Chinese-style char siu, Cuban mojo, or a bright Argentinian chimichurri. Or, better yet, since this is a pork party, throw out a whole selection. Also see the suggestions in the note on page 655.

Shredded roasted pork shoulder is excellent on its own, even better in sandwiches with a bit of coleslaw, and it makes an excellent addition to soups, stews, taco fillings, Cuban sandwiches, empanada fillings, arepa stuffings, hash, omelets, etc.

It's nearly as difficult to mess up slow-cooked pork shoulder as it is to bring the sucker to the table without eating half the skin yourself before it arrives.

Slow-roasting followed by a blast at high heat creates skin that blisters and bubbles, adding surface area for extra crunch.

# NO BUTTS ABOUT IT

You may have seen large cuts of pork in the supermarket labeled "pork butt," but a few ninth-grade classes in physiology would have told you that what you were looking at was not a butt but a shoulder. What's up with the odd labeling?

Pork shoulder and pork butt are the same cut of meat, and it's an oddity of nomenclature, not anatomy, that makes them so. Turns out that in the early nineteenth century, New England was a pork-production powerhouse. The loins, bellies, and hams were eagerly snatched up by native New Englanders, but the far less desirable shoulder cut (obviously, the Yanks didn't know jack about BBQ; some argue they still don't) was packed into wooden barrels and shipped out across the country. The barrels came in different sizes, but the ones pork was packed into were of the size officially known as "butt" or "pipe." That'd be a 126-gallon barrel, half the size of a 252-gallon tun, larger than a 84-gallon firkin, and twice the size of a 63-gallon hogshead (which, incidentally, has nothing to do with actual hogs or heads).

The pork-filled butt-sized barrels shipped out across the country came to be known as Boston butts, a term that was soon applied to the meat inside, despite the fact that it actually came from the shoulder of the hog. To this day, the term is widely used. Depending on what part of the country you come from, you'll see pork shoulders labeled shoulder, butt, Boston butt, or blade roast, while the lower part of the front is sold as picnic shoulder. Ironically, you'll never see pork shoulder labeled "Boston butt" in Boston itself. Had our forebears deigned to ship pork in 10-gallon barrels, we might have found ourselves spooning slow-cooked pulled Boston firkin into our BBQ sandwiches, or perhaps making our Italian sausages out of Boston puncheon, from 84-gallon barrels. Or, if those shoulders were shipped to New Mexico in 18-gallon barrels, they'd be chowing down on chile verde made with Boston rundlet.

So then the question is, what do they call the anatomical butt of the pig? That'd be the ham.

# ULTRA-CRISP SLOW-ROASTED PORK SHOULDER

**This recipe is designed for bone-in, skin-on pork shoulder and produces very crisp skin, but it'll work just as well for a boneless skin-on shoulder, making your life a bit easier. Be careful when this thing comes out of the oven, because the sight of that glorious skin in all its crackling glory has been known to induce fainting spells.**

**NOTES**: If you want to serve your pork with a sauce, try Salsa Verde (page 396), Peruvian-Style Spicy Jalapeño Sauce (page 599), your favorite barbecue sauce, or Chimichurri Sauce (page 395), or offer your guests a choice of two or more of these.

Alternatively, do not use aluminum foil when roasting. After roasting, drain off the excess fat and deglaze the baking sheet by setting it over a single burner and adding 2 cups dry white wine or chicken stock, or a combination of both. Scrape up the browned bits, transfer the liquid to a small saucepan, season to taste, and, off the heat, whisk in 2 tablespoons butter. Serve this pan sauce with the pork.

### SERVES 8 TO 12

1 whole bone-in, skin-on pork
   shoulder, 8 to 12 pounds
Kosher salt and freshly ground
   black pepper

1. Adjust an oven rack to the middle position and preheat the oven to 250°F. Line a rimmed baking sheet with heavy-duty aluminum foil (see Note above) and set a wire rack in it. Lay a piece of parchment paper on top of the rack. Season the pork liberally on all sides with salt and pepper and place on the parchment paper. Roast until a knife or fork inserted into the side of the shoulder shows very little resistance when twisted, about 8 hours.

2. Remove the pork from the oven, tent with foil, and let rest at room temperature for at least 15 minutes, and up to 2 hours.

3. Meanwhile, increase the oven temperature to 500°F and allow to preheat. Return the pork to the oven and roast, rotating the pan every 5 minutes, until the skin is blistered and puffed, about 20 minutes. Remove from the oven, tent with foil, and allow to rest for an additional 15 minutes.

4. Serve by picking the meat in the kitchen or just bring the whole thing to the table and let guests pick the meat and crispy skin themselves, dipping it into the sauce of their choice (see Note above).

# CROWN ROAST OF PORK

**So it's holiday time again, and your little sister has adopted a pet calf,
so beef's off the menu; your mother can't stand the aroma of lamb;
and *everybody's* sick of turkey. What to do?**

Enter the crown roast of pork. Pretty, presentable, and delicious, it's the best pork option for those who prefer their pork a little leaner than, say, an all-belly porchetta (page 663), and who like it with the distinct chew and texture of a good meaty roast.

**What is a crown roast?**
A crown roast is nothing more than one or two regular bone-in pork loin roasts (that's the big muscle that runs down the back of the pig) formed into a circle with the ribs pointed skyward. Essentially, it's a long rack of pork chops joined together (or, more accurately, that have never been cut apart) and twisted into a crown shape.

**Ah, I got it. The same way that a prime rib of beef is like a bunch of rib-eye steaks left connected to each other, right?**
Exactly.

**And what's the point? Does it make cooking easier? Does it taste better in the end?**
The "crown" in a crown roast serves about as much purpose as the crown on a king: it's mostly aesthetic—a crown roast simply looks stunning when presented at the table. But it does aid in even cooking to a small degree. With the bones twisted so that they are all on the exterior of the roast, heat transfer to the meat is slowed, making for juicier, more evenly cooked meat in the end—though the trade-off is that the fatty cover around the meat will never get quite as crisp and browned as it would if you were to roast a whole rack of pork without forming it into a crown.

**So it's a bit of give-and-take. Say I want to go for it—how do I go about finding a crown?**
To form a crown with a single rack of ribs (about 10 ribs, enough to feed 6 to 8 normal-appetited people), you need to score the spaces in between

the ribs slightly so that they splay out. However, by doing this, you end up increasing the surface area of the pork, which can cause it to dry out more than it would if it were completely intact, and I don't recommend buying single-rack crown roasts for this reason. Better to buy a crown roast formed by two bone-in loins attached end to end, which are large enough to form a circle without unnecessary scoring.

When purchasing a crown roast, you will usually have to ask your butcher to form it for you—only very dedicated butchers are likely to have them formed and ready to go. You may have luck finding a ready-to-roast crown at a high-end supermarket, particularly around the holidays.

### How big a roast will I need?

Aim for about a rib and a half per person, or two per person if you're big eaters or looking for leftovers.

### I've got my crown roast home (and boy, was that heavy!). Now, how the heck do I cook this thing?

Well, remember—a crown roast is nothing more than a series of connected pork chops, fast-twitch muscle (see page 179). Like all fast-twitch muscles (say chicken breast, New York strip steak, or tuna loin), it has plenty of fine-textured muscle and not much connective tissue or fat. This means that internal temperature is the most important factor when it comes to cooking it. With little to no connective tissue to break down, as soon as it reaches its final temperature, it's done. Holding it at that temperature for an extended period of time will change it very little. The key is to get the entire roast, from edges to center, to around 140°F (medium, which is what I like my pork cooked to) while simultaneously crisping the exterior.

Luckily, we've already studied this very same engineering problem applied to prime rib (see "How to Roast Beef," page 630). Remember: the hotter your oven temperature, the more uneven your roasting will be. So, for example if you cook a crown roast in a 400°F oven, by the time the very center is at 140°F, the outer layers of the meat are well past the 165° to 180°F mark. Roast it in a 250°F oven, and you can get the entire thing pretty much exactly at 140°F from edges to center. Then all it takes after roasting is a rest and a quick bang into a 500°F oven to crisp up the fat on the exterior.

### Neat! And what about flavoring?

If you want to be all fancy-pants about it, you can add other seasonings to the exterior in addition to the kosher salt and black pepper I opt for. Herbs stuffed into the center would be nice, as would be garlic, shallots, citrus fruit—whatever tickles your fancy (pants). Some folks even like to fill the center with sausage or bread-based stuffing (like the Classic Sage and Sausage Stuffing on page 622. It's a fine thing to do if you have tons of guests to feed, and a solid stuffing like that will actually improve the cooking qualities of the pork, as it acts as an insulating barrier to heat. Do note, though, that it will dramatically increase cooking time—count on up to an hour more. Or, better yet, count on your thermometer.

### I'm the kind of person who likes to wear a hat just to go pick up the mail. What would you suggest for someone like me?

Go ahead and put cute little paper hats over the ends of the bones before serving to cover up the charring they will get (or, if you prefer, use foil hats while they cook to prevent them from charring). You can buy those paper hats online very cheaply. Personally, I like the primal nature of the way charred ribs look, enough so that I had my wife's engagement ring delivered to her on the bone of a wild boar chop. Isn't that romantic?

# CROWN ROAST OF PORK

**NOTE**: Ask the butcher for a crown roast at least a day or two in advance. Aim for around 1½ chops per person, or 2 chops per person if you want leftovers. Aromatics like minced garlic or chopped herbs can be added along with the salt and pepper if desired. To prevent the ends of the ribs from burning, you can cap each with a piece of aluminum foil.

**SERVES 10 TO 16**

1 crown roast of pork, 6 to 10 pounds (12 to 20 chops; see Note above)

Kosher salt and freshly ground black pepper

Caramalized Applesauce (recipe follows; optional)

1. Adjust an oven rack to the center position and preheat the oven to 250°F. Season the pork roast liberally with salt and pepper and place on a wire rack set on a rimmed baking sheet. Roast until the internal temperature reaches 140°F on an instant read thermometer, about 2 hours. Remove from the oven, tent with foil, and lets rest for at least 15 minutes, and up to 45 minutes.

2. Meanwhile, increase the oven temperature to 500°F and preheat it. Return the roast to the oven and cook until crisp and browned on the exterior, about 10 minutes. Remove from the oven, tent with foil, and allow to rest for 15 minutes.

3. Carve by slicing between the ribs and serve, with the applesauce if desired.

## Caramelized Apple Sauce

*Pork and apples go together like Winnie the Pooh and Piglet, and it's tough to imagine one without the other. Regular applesauce made by cooking down apples with a bit of butter and lemon juice can be great, but I prefer to add a bit more sweetness and acidic kick to mine. A touch of brown sugar tossed with the apples caramelizes as the apples cook, adding complexity and a hint of bitterness. The apple cider and cider vinegar cut through it all, making it almost reminiscent of an eastern North Carolina–style vinegar barbecue sauce—my favorite sauce for barbecued pork.*

**MAKES ABOUT 1½ CUPS**

4 Granny Smith apples, peeled, cored, and cut into ½-inch cubes

¼ cup packed brown sugar

2 tablespoons unsalted butter

2 tablespoons cider vinegar

½ cup apple cider

Kosher salt and freshly ground black pepper

1. Toss the apple slices with the brown sugar in a bowl until evenly coated. Set aside.

2. Heat the butter in a 12-inch nonstick or stainless steel skillet over medium-high heat until the foaming subsides. Add the apples and cook, tossing and stirring occasionally, until caramelized and softened, about 5 minutes. Add the vinegar and cider and cook, stirring occasionally, until the apples are broken down and the juices have thickened, another 5 minutes or so. Season to taste with salt and pepper.

# GLAZED PORK TENDERLOIN

**One-skillet meals are a real lifesaver on busy weekday nights,**

. . . or on weekends when you just want to spend that extra half hour playing with the dog instead of washing pots and pans. These pork tenderloin recipes feed a group of six, take about half an hour to prepare, and require nothing more than a skillet and a couple of mixing bowls.

It's tough to cook tenderloin medallions properly. Tenderloin is a lean cut, which means that heat travels through it rapidly, and it can go from being moist and juicy to dry and stringy with little warning. Because I'm the kind of guy who likes to hedge his bets and give himself the best odds possible, I *never* slice my pork tenderloin into medallions before cooking it. It's a surefire road to overcooked pork, which, if you've ever had it, tastes like broken dreams and unicorn tears. Not fun.

It's far more foolproof to cook your tenderloin whole, slicing it only when you serve it. With an ultratender and very thin cut like this, slow-roasting followed by a blast in the oven is not an option—even the quickest of blasts would be enough to overcook it internally. Instead, we have to opt for a more efficient browning method: searing in a skillet, then finishing it in a hot oven.

In order to maximize browning even more, I dredge the tenderloins in cornstarch. Cornstarch itself browns reasonably well, but, even more important, it absorbs excess moisture from the surface of the meat, allowing it to cook more efficiently. As it happens, cornstarch also creates the perfect surface for a tasty glaze to stick to. Think of it like coating your car with primer before you apply the paint.

I treat these glazes as something in between a glaze and a pan sauce: after browning the pork in the skillet, I pour in my glaze ingredients, using them to scrape up any browned bits from the bottom of the

pan. These browned bits bring a ton of flavor, and when I then return the tenderloins to the skillet and finish them off in the oven, spooning the glaze over them, it puts that flavor right back on the surface of the pork, where it belongs.

# APRICOT-GLAZED ROAST PORK TENDERLOIN WITH PRUNES AND FIGS

**SERVES 6**

¼ cup dried figs, quartered

¼ cup dried prunes, halved

¼ cup brandy

½ cup apricot preserves

½ teaspoon paprika

2 pork tenderloins, about 1½
   pounds each

Kosher salt and freshly ground
   black pepper

¼ cup cornstarch

2 tablespoons vegetable oil

1 teaspoon balsamic vinegar

2 tablespoons unsalted butter

1. Adjust an oven rack to the middle position and preheat the oven to 400°F. Combine the figs, prunes, brandy, preserves, and paprika in a small bowl and stir together. Set aside.

2. Pat the pork dry with a paper towel and season on all sides with salt and pepper. Place the cornstarch in a shallow dish and dredge the tenderloins in it, turning to coat. Set on a large plate.

3. Heat the oil in a 12-inch ovenproof nonstick or cast-iron skillet over high heat until shimmering. Add the pork tenderloins and cook, turning occasionally, until well browned on all sides, about 12 minutes (the pork may not fit initially, but just curve the tenderloins so they fit—they will shrink and straighten as they cook). Transfer to a large plate and set aside.

4. Pour the glaze mixture into the skillet and cook, scraping up any browned bits from the bottom of the pan, for about 2 minutes. Return the pork to the skillet and turn to coat. Transfer to the oven and cook, turning the pork to coat it in the glaze every 4 minutes, until the thickest part of pork registers 130°F on an instant-read thermometer, about 15 minutes. Transfer to a large plate, and allow to rest for 5 minutes.

5. Meanwhile, whisk the vinegar and butter into the glaze. Carve the pork and serve with the glaze.

## MAPLE-MUSTARD-GLAZED ROAST PORK TENDERLOIN

Omit the figs, prunes, apricot preserves, paprika, and balsamic vinegar. For the glaze, combine ⅓ cup maple syrup and 2 tablespoons whole-grain mustard with the brandy. Proceed as directed, finishing the glaze with the butter. Carve the pork and drizzle with the glaze.

# SLOW-ROASTED PORCHETTA

**Does anyone else feel like porchetta—the Italian slow-roasted, fennel-scented, juicy pork surrounded with crisp, crackling skin—is appearing everywhere these days?**

Not that I'm complaining. As far as I'm concerned, the more slow-cooked pork in my life, the better. Indeed, my goal is to get a porchetta on every table in America this year (and perhaps some beyond our borders as well). I'm counting on you all to help me achieve my vision of a United States of Porkdom.

Here are a few reasons why you should consider topping your dinner table with a porchetta roast:

- **It's delicious.** It's easily more delicious than turkey, pretty much definitely more delicious than prime rib, and arguably better than leg of lamb.
- **It looks awesome.** Other roasts can be imposing in the center of the table, but none are as geometrically perfect, as easy to carve, and as breathtakingly covered in crackly skin. Because I'm the husband of a well-proportioned mathematician, geometric symmetry is something I think about quite often and find aesthetically pleasing. In this case, it makes for easy, even cooking. No awkward thin regions that overcook or thick regions that stay raw in the center.
- **It helps avoid fights.** Holidays can be a bit trying for the whole family, especially when you're fighting over dark meat or light meat or who gets to gnaw on the rib bones. With porchetta, every single slice is exactly the same—by which I mean perfect.
- **It's forgiving.** Accidentally overcook red meat or poultry, and it'll be so dry you might as well serve the gravy-soaked contents of your paper recycling bin to your guests. Overcook porchetta and . . . wait, that's right, you pretty much *can't* over-

cook porchetta. Leave it in the oven for an extra hour or two? *No worries, it'll **still** taste fantastic.*
- **It's inexpensive.** Pork belly might cost you about $10 per pound—at a fancy butcher. More likely you'll find it for $4 to $5 a pound, at least a quarter the cost of a well-marbled prime rib. Want to serve aged prime rib? You must have some deep, deep pockets.
- **Leftover-porchetta sandwiches are freakin' awesome.** That's all there is to say about that one.

Not convinced yet? Read on, my friend.

## WHAT IS PORCHETTA?

Traditional porchetta is made by butchering a hog so that the boned-out loin is still attached to the boned-out belly. The meat is then carefully salted and rubbed with a garlic, herb, and spice mixture that features plenty of fennel and black pepper, along with ingredients like crushed red pepper, citrus zest, and rosemary, sage, and other piney-scented herbs (you can, of course, vary the mixture to suit your own tastes). By carefully rolling the two cuts up together, you end up with a single perfectly cylindrical roast with the fatty belly surrounding the lean loin, all covered in a layer of skin.

As the rolled porchetta rests, the salt slowly penetrates into the meat, dissolving the muscle protein myosin and altering its structure so that it's able to retain moisture more effectively, as well as giving it a slightly bouncier, more resilient texture (think sausage or ham, not rubber ball). When the pork is subsequently roasted, the fatty belly portion, rich in juices and connective tissues, ostensibly helps keep the relatively dry loin moist.

But we all know that this isn't really how cooking works. All the fat in the world surrounding a lean, tightly textured muscle like a pork loin will not help keep it moist if you cook it past 150°F or so, and, indeed, many porchettas I've had have had some unconscionably dry centers because of this. But belly, with its extensive network of connective tissue and abundant fat content, needs to be cooked to at least 160°F for a couple of hours in order for that tissue to slowly break down and for some of the fat to render.

Loin needs to stay below 150°F, belly needs to get above 160°F. You can see the problem here. So why do traditional porchetta recipes call for both belly and loin? My guess is that at the time porchetta was invented, hogs hadn't yet been bred to have large, lean loins, and thus there wasn't as big a distinction between the belly and the loin sections.

Both would have had plenty of fat and connective tissue, making both totally tasty even when cooked to a higher temperature. Today, we need a better solution, and here's one: discard the loin and go for an all-belly porchetta instead.

We all know that pork belly—the cut that the magnificence that is bacon comes from—is the king of pork cuts, and that pork is the king of meats, and that meats are the Masters of the Universe. This makes eating an all-belly porchetta somewhat akin to consuming an aromatic, crispy, salty slab of awesome seasoned with He-Man. Or something like that. You get the picture.

Tracking down an intact single belly shouldn't actually be too difficult. What you want is a whole boneless, rind-on belly with the rib meat still attached. It should weigh in at around 12 to 15 pounds or so. Your butcher should be able to order one for you, or if you live near a Chinatown, take a stroll into one of the butcher shops there—most likely they've got pork bellies in stock.

Once you've got your belly, everything else is a piece of cake; just give yourself enough time to execute. Assembling the porchetta itself should take no more than an hour, and once it's assembled, you can wrap it in plastic and store it in the fridge for up to three days (so long as the belly was quite fresh when you got it). It'll actually improve with age as the salt works its way through the meat.

# ALL-BELLY PORCHETTA
## WITH PORK-FAT-ROASTED POTATOES

**NOTES:** The herbs and aromatics can be varied according to taste. I find it easiest to work with a whole belly, but if a smaller roast is desired, split it in half and freeze one half while still raw. Wrapped tightly in foil and plastic wrap, it should keep for several months in the freezer. Thaw overnight in the refrigerator and proceed as instructed.

The porchetta can be cooked without the potatoes. But be sure to save the fat for roasting potatoes at another time.

**SERVES 12 TO 15**

1 whole boneless, rind-on pork belly, 12 to 15 pounds

2 tablespoons black peppercorns, toasted and ground (see page 666, steps 2 and 3)

3 tablespoons fennel seeds, toasted and ground

1 tablespoon peperoncini or red pepper flakes (optional)

3 tablespoons finely chopped fresh rosemary, sage, or thyme

12 medium cloves garlic, minced or grated on a Microplane (about ¼ cup)

Grated zest from 1 lemon or orange (optional)

Kosher salt

2 teaspoons baking powder

5 pounds russet (baking) or Yukon gold potatoes

Freshly ground black pepper

1. Following the step-by-step directions on page 666, season the pork belly with the black pepper, fennel, red pepper, rosemary, garlic, citrus zest, if using, and salt, shape into a rolled porchetta and rub with the baking powder and more salt. If the roast is too large and unwieldy, carefully slice it in half with a sharp chef's knife. Wrap the porchetta tightly in plastic and refrigerate for at least overnight, and up to 3 days. If desired, half the porchetta can be frozen at this point for future use (see Note above).

2. When ready to roast, adjust an oven rack to the lower-middle position and preheat the oven to 300°F. Place the pork on a V-rack set in a large roasting pan or, if cooking both halves at the same time, on a wire rack set on a rimmed baking sheet. Roast, basting occasionally with the pan drippings, until the center of the pork reaches 160° on an instant-read thermometer, about 2 hours.

3. Meanwhile, cut the potatoes into 2-inch chunks. Place in a large Dutch oven, cover with cold water, add 2 tablespoons salt, and bring to a boil over high heat. Reduce the heat and simmer until the potatoes are barely tender, about 10 minutes. Drain and transfer to a large bowl; set aside.

4. When the pork reaches 160°F, using pot holders, lift up the rack with the pork and set it aside. Pour the pan drippings over the potatoes. Season the potatoes with salt and pepper and toss to coat. If using a roasting pan, add the potatoes to the bottom of the pan, return the V-rack with the porchetta to the pan, and return it to the oven. If using a rimmed baking sheet, spread the potatoes on a second rimmed baking sheet. Return the porchetta to the first baking sheet and return it to the oven, placing the potatoes on a rack directly underneath.

5. Continue roasting, flipping the potatoes every 45 minutes or so, until a knife or skewer inserted into the pork shows very little resistance aside from the outer layer of skin, about 2 hours longer.

6. Increase the oven temperature to 500°F and continue roasting until the pork skin is completely crisp and blistered and the potatoes are crisp and golden, 20 to 30 minutes longer. Remove from the oven, tent with foil, and allow to rest for at least 15 minutes.

7. With a serrated knife, slice the pork into 1-inch-thick disks. Serve with the crisp potatoes. Porchetta can also be served at room temperature.

# STEP-BY-STEP:
## How to Form a Porchetta

**STEP 1: SPREAD OUT** Gather all your ingredients and give yourself plenty of space to work. There's nothing more frustrating than a cramped work space when you've got a big hunk of pork flopping around and a slippery knife on the edge of the cutting board. I clear out a big space in the middle of my coffee table or dining-room table.

**STEP 2: TOAST THE SPICES** Toasting the peppercorns and fennel seeds will add complexity to their flavor as the chemical compounds undergo a series of reactions, breaking down and recombining under the heat of a dry skillet. Toast them over a medium-high heat, stirring and tossing, until lightly browned and aromatic.

**STEP 3: GRIND THE SPICES** If you've got a kick-ass mortar and pestle like this one from Japan, use it. The ideal grind is coarse, not dusty, which is what a regular mortar and pestle will get you. You can also use a spice grinder and pulse the spices a few times; even a food processor will do the trick.

**STEP 4: SCORE THE MEAT** Pork bellies are thick. Given a couple weeks, salt and flavorings can penetrate deep into the meat (see: bacon and pancetta). We don't have that kind of time on our hands, so to hasten the flavoring process, turn the belly skin side down and score the meat deeply with a sharp, sharp knife.

**STEP 5: GO BOTH WAYS** Score it along both diagonals for optimal flavor absorption.

**STEP 6: MIX YOUR MOIST AROMATICS** Peperoncini are Italian hot pepper flakes. They are a little fancy and pricey, but it makes me feel good to use them. You can, of course, use regular old red pepper flakes, or if you don't want any heat at all, omit them completely. Finely chop the herbs. Grate the garlic on a Microplane grater (that's how I like to do it), or mince it by hand. If you'd like, you can add some grated lemon or orange zest to your blend.

**STEP 7: RUB IT IN** Salt your meat very generously (a light, even dusting), then add the rest of your aromatics and rub them deep into the grooves in the meat.

**STEP 8: TIE IT** Roll up your porchetta lengthwise (with a full pork belly, you should be able to just barely get the ends of the rind to touch; if you have trouble, don't worry—it's OK if they don't quite meet). If you're fancy, you can then tie up your whole porchetta with a single long piece of twine using butcher's knots, but for most of us, regular old double granny knots will do. The easiest way is to lay out foot-long lengths of twine at 1-inch intervals on your cutting board, then lay the roast on top. Working from the ends toward the center, tie your roast up as tightly as you can.

**STEP 9: TIED AND READY** You should have a nice, even log that's probably way too long to fit in your oven (unless you were wise and opted for a smaller piece of belly). We'll deal with that in a moment, but for now, notice how the pieces of string create indentations in the rind: this will be useful for portioning and slicing the roast later on.

**STEP 10: ADD THE BAKING POWDER** (not pictured) This is a trick I learned from my wife's aunt in Colombia, where pork belly is prepared rubbing the roast with a mixture of baking soda and salt, which raises the pH of the skin and causes some of its proteins to break down more readily, resulting in crisper skin. I find straight baking soda to have a slight soapy aftertaste, but using slightly alkaline baking powder instead works marvelously. I mix my baking powder at a ratio of 1 to 3 with kosher salt (by volume) before rubbing it all over the surface of the roast.

**STEP 11: SLICE IT IN HALF** If you want to roast your porchetta all in one go on a single baking sheet, you should slice it in half right now into more manageable pieces. Use a very sharp chef's knife or carving knife and try to do it in a single stroke instead of sawing, for the nicest presentation. Then wrap it very tightly in plastic and refrigerate at least overnight to give the salt and baking powder some time to work their magic. (You can, of course, skip this step if you're in a real rush to get pork in your mouth.)

**STEP 12: READY TO ROAST** When ready to go, preheat the oven to 300°F, with a rack in the lower-middle position, and remove the porchetta from the fridge. You can cook both pieces side by side on a single rimmed baking sheet with a wire rack set on top of it, or roast just one half in a roasting pan and freeze the other half. Because of its high fat content, porchetta freezes very well. Just be sure that it's wrapped airtight in foil and plastic or, better yet, vacuum-sealed in a FoodSaver-type bag, to prevent freezer burn.

TOMATO SAUCE,
MACARONI, AND

# THE
# SCIENCE
## *of*
# PASTA

**7**

"Everything in food is science. The only subjective part is when you eat it."—Alton Brown

# TOMATO SAUCE, MACARONI, AND THE
# SCIENCE *of* PASTA

## RECIPES IN THIS CHAPTER

MY WIFE AND **IRVING** THE DOORMAN HAVE GOT A PRETTY SWEET DEAL: ALL THEY HAVE TO DO IS NOTHING,

**a**nd they get hot, fresh food delivered to them several times a day. Of course, they do have to be content with eating, say, fried chicken and nothing else for a month as I test a recipe, but, all in all, they've pretty much got it made.

So you can imagine my surprise one day when I walked into the kitchen and saw my wife *cooking*, and my even greater surprise when I realized she was cooking pasta—*in our smallest pot, and at a simmer*. The water barely covered the noodles as she stirred them to keep them submerged.

"You can't do that!" I exclaimed in horror. "Obviously, my diminutive wife, you haven't cooked a lot of pasta in your time. Unless you use a giant pot of water at a rolling boil, your pasta will stick together. The starch will become too concentrated. It will cook unevenly. It will become mushy. It will be nine different sorts of horrible, each one worse than the one before. It is scientific fact that you will end up with an inedible starchy, sticky blob."

"Is that so?" was all she said as she turned back to the pot. Needless to say, my wife was right: the pasta was fine (though I declined to eat any more than a single tester piece, citing potential paradoxes in the space-time continuum as my reason). Indeed, she has precedence for her method. Even

Harold McGee, in a *New York Times* article in 2009, has mentioned the small-pot method of pasta cooking. So what gives? Exactly what *is* the best way to cook, sauce, and serve pasta? In this chapter, we'll get into all of that, along with discovering what I call the "mother sauces" of Italian-American cookery, the five simple sauces that form the base for countless variations: olive oil and garlic sauces, tomato sauces, pesto, cream sauces, and meat ragù.

But first, what exactly *is* pasta?

## { PASTA TRADITIONS }

At its simplest, pasta is nothing more than flour and water mixed together to form a dough, cut into shapes, and cooked in boiling water. As a food that has been made in China since the second century BC, in the Middle East since the ninth century, and Europe since at least the eleventh, it has one of the longest and most drawn out, and downright confusing, histories of any food around (though we can pretty safely say that Marco Polo did *not* play more than an apocryphal role in it). And that history is the realm of historians, not of cooks.

So why do I bring it up now? Only to illustrate my naiveté in telling my wife she was wrong for cooking it in a particular manner. In fact, you hear all sorts of things coming from folks who claim to be direct descendents of Signore Polo or perhaps acquaintances with the Pope's Personal Pasta Producer. "Always use fresh, not dried," or "Don't add too much sauce," or "Do not add oil to your water," or (my favorite), "Add salt only *after* it comes to a boil," often claiming tradition as the reason for doing so. Well, guess what? You don't need to listen to any of them. Indeed, cooking pasta in a large amount of boiling salted water for a matter of min-

utes is a relatively modern method. Prior to that, recipes called for cooking pasta for hours instead of minutes. Indeed, according to McGee, the term *al dente* didn't appear until after World War I. How's that for tradition?

Given its varied background, *I* say that you should cook pasta whichever way works best (just don't tell your *nonna*, if you have one).

MOST PASTA these days comes in two basic forms, dried and fresh.

**Fresh pasta**, made from wheat flour and eggs, is widely used in Northern Italy. The eggs add richness and color and improve the texture of the pasta, allowing it to become both tender and bouncy as it cooks. It's made by forming a stiff dough out of the eggs and flour, then rolling that dough repeatedly between two roller plates, getting progressively thinner with each roll, before finally cutting it into the desired shape. Since making fresh pasta requires time and specialized equipment, we won't be dealing with it much in this book. Instead, we're going to focus on the dried version.

**Dried pasta**, which originated in Southern Italy, is generally made from durum flour and water.

Durum is a high-protein flour that allows you to form a tough, malleable dough that holds its shape well. This is particularly important for the intricate folded or extruded shapes that dried pasta comes in. To form dried pasta into shapes, the stiff dough is pressed into a machine that forces it through metal dies that extrude it, then cut into smaller lengths.

The very best dried pastas have a distinct wheaty flavor and, more important, a rough texture that allows them to absorb sauce more easily once cooked. Of the major supermarket brands available in the United States, I prefer Barilla brand. When possible, though, I shop at Italian markets that import high-quality small-batch pasta from Italy. What's the difference? Many of the inexpensive mass-market brands use Teflon-coated dies in order to speed up production. These dies result in a smooth finish on the pasta. Traditional brass dies are slower to extrude but produce a rougher texture. When shopping for dried pasta, compare the texture of the noodles and pick the brand with the roughest-looking finish.

# The BEST Way
## TO COOK PASTA

If we go by traditional wisdom, pasta should be cooked in a large volume of salted water and added to the pot only after the water has reached a rolling boil. What's the reasoning here? There are four reasons generally cited:

- **Reason 1**. A large volume of water has a higher thermal mass than a smaller one. Thus, when you drop pasta into it, it cools less and returns to a boil much faster.

- **Reason 2**. A large volume of water at a rolling boil helps move the pasta pieces around so they don't stick to each other.
- **Reason 3**. A small volume of water will become too starchy as the pasta cooks. This will make the pasta stickier when you drain it.
- **Reason 4**. It's the way Grandma did it.

Let's break them down point by point and see if we can't make some sense of them.

## Reason 1

To test this, I brought three pots of water to a boil: one with 6 quarts of water, one with 3 quarts, and one with a mere quart and a half. After the water in each pot came to a boil, I added the pasta and waited for it to return to a boil. The three pots did so within seconds of each other. In fact, the pot with 3 quarts actually came back to a boil slightly faster than the one with 6 quarts—the exact opposite of what is supposed to happen. What gives?

To solve this mystery, we have to think about what's going on inside a pot of boiling water, what its energy inputs and outputs are. Imagine we have two pots of water on identical burners. One pot holds 2 gallons of boiling water and the other holds 2 quarts. The inputs are simple: the burner underneath each one is supplying a constant energy source. As long as the burners are set at high, the amount of energy they transfer to the pot-water system is consistent. What about energy loss? Well, that's going on too. First, there's energy being lost to the outside environment in the form of heat from the sides of the pot and the surface of the water. This amount of energy loss is proportional to the surface area of the pot-water system as well as its temperature. Since the temperature is staying at a constant 212°F, and the pots are (presumably) not changing size, that too is a constant. The other factor that contributes to energy loss is something called the *heat of vaporization*—that's the energy it takes to convert water into steam. Both pots of water are boiling, and the difference between energy-in and energy-out is compensated for by the energy used to boil that water.

So: energy-in, energy-out. With me so far?

Now, what happens when we add some room-temperature pasta to the pots? The temperature of the water immediately drops. How much it drops is inversely proportional to its total volume. The more water you have to begin with, the smaller the temperature change. A pound of pasta added to the 2-gallon pot will cause it to drop by only a degree or two, while a pound of pasta added to the 2-quart pot will cause a temperature drop four times as great (since a gallon is four times as big as a quart).

*Aha!* you must be thinking. *So the reasoning is correct. Lower volume means a bigger drop in temperature, which means a longer time to get back to a boil.*

On the face of it, it seems logical, but the problem is that it takes more energy to raise the temperature of 2 gallons of water than it does to raise the temperature of 2 quarts. How much more? Exactly four times more, in fact. And since our small pot dropped in temperature precisely four times as much as our large pot, it means that *both* pots will come back up to their boiling point in the same amount of time.

To make it simpler, think about this: if you heated your pasta to 212°F before dropping it into the pot of boiling water, the water would not cool at all and thus never lose a boil, no matter what volume of water you started with. So, the only energy we really need to add to the system—regardless of the size of the pot—is the energy it takes to bring the pound of pasta up to 212°F. And *that*, my friends, is a constant.

**So it's "*pasta la vista*" to Reason 1.**

**W**ant to *really* blow the mind of one of the large-pot-insisters? Lay this one on them: if you use too large a pot, the water will actually take *longer* to come back up to a boil than a small pot. Why is that? It's because the larger the pot, the more surface area it has, and the more energy is being constantly lost to the environment. Any amount of energy-out from the pot-water system has to be made up for with energy-in from the burner, which leaves less energy free to reheat that water. Indeed, use a pot large enough, and the energy-out will be so great that your burner will simply not be powerful enough to ever bring it to a full boil.

**Reason 2**

Drop the pasta into the water and just leave it there, and it will indeed stick to itself. But you know what? It'll do that *even in a really big pot with lots of boiling water.*

The problem is that excess starches on the pasta immediately start to hydrate and jell together. But if you rinse those starches off, dilute them, or allow them to cook enough that they begin to set, the problem completely disappears.

So the key is to stir the pasta a few times during the critical first minute or two in order to rinse off excess starch and make sure the pieces aren't sticking together, until the outer layers are fully cooked. After that, whether the pasta is swimming in a hot tub of water or just barely covered, absolutely no sticking occurs.

"Im*pasta*ble!" you cry? Try it out for yourself!

**That's goodbye to reason 2.**

**Reason 3**

I spent a few years working the pasta station at a restaurant known for its pasta., cooking dozens, if not hundreds, of portions on a given day. That's an awful lot of pasta. I cooked it all in one large six-slot pasta cooker that held about 15 gallons of water at a constant boil. At the beginning of the shift, the pasta water was clear. But as the night wore on, the water would get cloudier and cloudier, until by the end of the night, it was nearly opaque.

This cloudy, starchy pasta water is the line cook's secret weapon. You see, pasta water consists of starch granules and water—the exact same ingredients that go into a cornstarch slurry (you know, what you use to thicken your sauces?). Well, aside from just thickening a sauce, starch also acts as an emulsifier. It physically gets in the way of tiny fat molecules, preventing them from coalescing. This means that with a bit of pasta water, even an oil-based sauce like say, *aglio e olio*, or *cacio e pepe*, will emulsify into a light, creamy sauce that is much more efficient at coating the pasta, making the dish that much tastier. Think of pasta water as the diplomat of the pasta world—the guy who's there to help your sauce and your noodles get along. (Of course, this also means that go into any restaurant that serves a lot of pasta, and chances are, the later in the night it is, the better the consistency of your sauce will be!)

The water from pasta cooked in a smaller pot (*left*) is packed with more starch.

This starchy pasta cooking water improves the texture of your sauce, helping it cling to the cooked pasta.

Following that logic, our goal should be to get the water *as starchy as possible*, the more efficiently to bind the sauce with. I took a look at the water drained off the batch of pasta cooked in 1½ quarts against the one cooked in 3 quarts. Notice how much cloudier the one on the left is? "All the better to bind you with, my dear," I said out loud, just as my wife happened to walk into the kitchen. She declined to join me in my tasting, but side by side,

after saucing, the pasta cooked in less water resulted in a much better sauce consistency, and the sauce actually clung to the pasta better as well.

**Reason 3 debunked**.

## Reason 4

There are few times in life that I'm glad I don't have an Italian grandmother, but trying to explain low-water-pasta-cooking to her would be one of those times.

Now that I was completely satisfied that I could cook pasta in less water with no problems at all, I decided to try and take the method to the extreme. I knew that when you cook pasta, proteins denature and starches efficiently absorb water at temperatures as low as 180°F. So is it actually even necessary to boil the noodles the whole time? I covered a pot of penne with water by a couple inches (to account for the pasta expanding as it absorbed water), seasoned it with a bit of salt, and set it on a burner. After allowing it to come up to a simmer, I stirred it once to ensure that the pasta wasn't sticking to itself or the pot, immediately threw a lid on the thing, and shut off the burner.

I have to admit: even *I* was a little skeptical about this one. I mean, cook pasta without even boiling it? If this really worked, I'd never cook pasta the same way again. At the very least, I'd save a couple cents on my gas bill each month. I'd no longer have to be such a, ahem . . . *penne* pincher.

When the timer finally went off, I opened the lid and poked around a little. So far, so good. The pasta sure *looked* cooked, and tasting it revealed al dente perfection. Success! My wife won this round (but just let her *try* and change the way I cook burgers).

If you're really keen on saving time and energy,

you can do what I do: put half the water and the noodles in the pot and heat the rest of the water in an electric kettle as the first half heats up. Add the second half to the first, and you've got boiled water. All you've got left to do is stir, cover, and wait. Now *that's* using your *noodle*!

A few caveats:

- **Don't try this with fresh pasta.** Fresh egg pasta is simply too absorptive and lacks any structure until the egg proteins start to set.
- **For very long shapes, you'll need a tall pot.** There has to be enough water to completely submerge the pasta as it cooks, so for long shapes like spaghetti or fettuccine, you'll still need a big pot (or be willing to break the pasta in half).
- **Season the water.** Some people claim that adding salt helps raise the water's boiling point, thus cooking the pasta faster. Don't believe them. The difference you get is at most half a degree or so—nowhere near enough to make a difference, particularly because, as we now know, you don't even have to use boiling water. But salt is necessary for another reason: it makes the pasta taste good.

## SHOULD I ADD OIL TO MY PASTA WATER?

Some cookbooks advise you to add a glug or two of oil to the pasta water as the pasta cooks, claiming that it'll keep the pieces separated. Unfortunately, this is not true—how could it be when the oil just floats on top? Try it side by side: no matter how much oil you have your pasta behaves exactly the same way underneath.

What the oil on the surface *does* do is prevent the pasta water from boiling over. As the pasta cooks, more and more starch gets released into the pot, increasing the water's viscosity and allowing it to form ever more stable bubbles. Eventually these bubbles become so stable that they are able to push each other up, like a raft. The raft rises above the top of the water, and over it goes. Oil can help break the surface tension of the water, preventing these bubbles from forming in the first place. Of course, with our new no-boil pasta-cooking method, this is entirely moot: you never need to add oil.

What about oiling the pasta after it's been drained? *Bad idea.* Yes, it'll keep the pasta from sticking to itself as it sits around waiting to be sauced, but you know what? It'll also prevent the sauce from sticking to it.

When it comes to saucing pasta, it's *essential* that you have your sauce ready and waiting. As soon as you drain the pasta (reserve some of the pasta water), transfer it to the pan with the sauce and immediately start tossing it to coat, adding reserved pasta water as necessary to adjust the consistency.

# { ITALIAN-AMERICAN PASTA SAUCE }

Pasta without sauce is like the Lone Ranger without Tonto. Milli without Vanilli. Mario without Luigi. R2D2 without . . . You get the picture.

You've all heard of the French "mother sauces," right? Back in the early twentieth century. Auguste Escoffier, the granddaddy of French cuisine, classified all French sauces into five categories: béchamel (milk thickened with starch), espagnole (brown veal stock), velouté (a thickened white stock), hollandaise (an egg-yolk-and-butterfat emulsion), and tomate (tomato-based). His idea was that by learning how to properly make these five sauces, you would have the basics to make hundreds of derivative sauces. For instance, *sauce mornay* is a béchamel with the addition of cheese. *Sauce bordelaise* is espagnole sauce with reduced wine and bone marrow.

---

\* Even if you're the type who believes Carême to be the granddaddy, Escoffier is at the very least the strict uncle who lives in Canada to whom parents send their unruly kids to build their character over the summer.

Olive oil, garlic, and a splash of pasta cooking water makes one of the simplest pasta sauces around.

*Béarnaise* is Hollandaise made with reduced white wine, shallots, and tarragon (see page 321). And so on.

Well, over the years, I've noticed while making Italian-American (and many downright Italian) pasta dishes at home that there are also five basic sauces that form the root of just about every other sauce in the lexicon. Just as in French cuisine, by perfecting my techniques for these basic categories of sauces, I'd in effect be setting myself up to succeed in any number of variations based on them.

The five sauces are:

- **Olive oil and garlic**
- **Tomato (classic red sauce)**
- **Pesto**
- **Cream**
- **Meat-based (ragù bolognese)**

Let's go through all five of them here.

## Mother SAUCE #1:
### OLIVE OIL AND GARLIC

*Pasta aglio e olio* comes from the Abruzzo region of Italy, but this simple dish of pasta tossed with garlic sautéed in olive oil with a pinch of hot pepper flakes and a sprinkle of parsley can be found at pretty much any Italian-American restaurant in this country, and it is the base for dozens of common variations. Pasta with clam sauce, pasta primavera, and pasta with shrimp (or shrimp scampi, as some translationally challenged menus will call it) all stem from the same root.

Three flavors of garlic make their way into this bowl of pasta.

Often served with spaghetti, *aglio e olio* is about as simple as a pasta sauce gets. I prefer serving mine with shorter, stubbier pasta shapes like rotini, or with orecchiette, which will pick up plenty of sauce, but then, I'm a garlic-and-olive-oil fiend. Feel free to serve it with any shape you'd like. The key to the recipe is to start with great olive oil (see "All About Olive Oil," page 777) and to treat the garlic three different ways, first infusing the olive oil with toasted whole cloves to add sweet depth to the oil, then sautéing thin slices for small bursts of garlic flavor in the finished dish, and finally adding minced garlic at the very end for a spicy current to tie the whole thing together. The layering of flavors makes for deep, deep garlickiness. To this base, a pinch of red pepper flakes adds heat, and a handful of chopped parsley adds grassy freshness.

As for getting the sauce to coat the pasta effectively, that can be a little trickier. The problem with olive oil is that its emulsive properties are pretty low: it does not want to bind with water. And once you add your pasta to it, you end up with a thin sauce with a greasy oil slick on top.

*So what?* You might say. *Isn't all the flavor in there anyway?* Well, sort of. The flavor might all be there, but the real problem is that when the oil and water are separate like that, they tend to run off the pasta and collect at the bottom of the bowl. You end up with dry, bland pasta on top and a wet, soupy mess left over when you're finished with it. Indeed, it's exactly the same as the importance of creating a good emulsified vinaigrette when you're making a salad. Without one, you end up with underseasoned greens and broken oil and vinegar at the bottom of the salad bowl.

How do you fix it? Easy: a bit of butter. Butter has properties that allow it to emulsify quite well with water. What's more, it acts as a liaison, holding olive oil's hand and allowing it to come along for the ride, especially when it has a bit of help from the starch in the pasta cooking water.

Put it all together—the three different garlic treatments, the great olive oil, and the touch of butter to bind it—and you've got yourself a sauce, or sauce base, worth reckoning with.

# ALL ABOUT GARLIC

Along with its close cousin the onion, garlic is among the most widely used fresh aromatic ingredients in the world. There are only a few simple dos and don'ts you need to remember to get the most out of it.

## Shopping and Storage

Garlic comes in a few different forms at the supermarket:

- **Whole heads** will give you the best flavor and maximum lifespan. Look for heads that are hard, firm, and heavy. If there is any give at all, that's a good sign of spoilage underneath. Healthy garlic stored in a cool, dry place should last at least a month or two.

- **Pre-Peeled** cloves are another good option, particularly if you go through garlic quickly and appreciate the grab-and-go convenience, like I do. Peeled cloves must be refrigerated, which, according to some reports, can reduce garlic's aroma over time. I haven't noticed this, though do note that refrigerated peeled cloves have about half the lifespan of whole heads, so don't buy more than you need. I use peeled cloves because I'm as lazy as I can get away with without sacrificing quality (and you should be too!). Peeled garlic can be stored in a sealed container in the refrigerator for several weeks.

- **Prechopped garlic, garlic paste, garlic juice**, and other convenience products of their ilk should be roundly rejected by anyone with half a shred of good taste. Just as with onions, the aromatic compounds in garlic are formed through an enzymatic chemical reaction that occurs as soon as its cells are ruptured. So, to maximize garlic flavor, you need to cut it just before incorporating it into a dish. Precut garlic has none of the complexity and freshness of whole garlic cloves.

*continues*

- **Garlic powder** is garlic that has been dehydrated and granulated. It is not a viable substitute for fresh garlic, though sprinkled on pizza or incorporated into a barbecue-style rub, it adds a unique flavor all its own.

## Cutting Garlic

**Sliced garlic:** Garlic should be sliced with a sharp chef's or santoku knife. To slice it, cut off the root end, then thinly slice lengthwise into slivers. Sliced garlic is best when rapidly cooked until pale golden brown.

**Smashed garlic:** This is what I use when I want to slow-cook garlic and flavor an oil as the base for a sauce. You can eat the slow-cooked smashed cloves

or discard them after they're flavored the dish. To smash garlic cloves, place each garlic clove flat on your cutting board, hold the side of your chef's knife or a large cleaver against it, and press down on the side of the knife blade to smash the clove. Alternatively, you can use the bottom of a small skillet to smash garlic cloves.

**Minced garlic:** Minced garlic is the most common form you'll find in this book and most recipes. A garlic press will do the job reasonably well and may be a good investment if you often work with a ton of garlic. And you can use it with unpeeled garlic: simply throw the garlic clove, peel and all, into the hopper of the

press and squeeze down, and out comes garlic paste with the skin trapped nicely behind. The only problem is that this tool is essentially a unitasker: you can't really use it for anything else. So I skip the press and grate garlic on my Microplane zester—it works just as rapidly, is easier to clean, gives you a nice clean cut, and saves on drawer space.

---

\* A phrase first uttered by the inimitable Alton Brown.

# THREE FLAVORS OF GARLIC

**D**epending on how it's cooked, garlic can develop distinctly different aromas and flavors.

- **Raw garlic** has a pungent, powerful aroma and slightly spicy flavor. It should be used sparingly. It's best incorporated into powerful sauces like vinaigrettes or pesto, where it gets diluted with other flavors, or used in marinades, where it will then be briefly cooked when your proteins are seared or grilled.

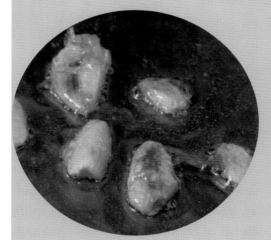

- **Slowly cooked garlic** develops an intense sweetness, much as caramelized onions do. It loses most of its pungency and trades it in for sweeter, roasted aromas. Whole heads can be roasted with a drizzle of oil in a low oven; individual cloves can be smashed and sizzled gently in oil until the oil picks up their flavor.

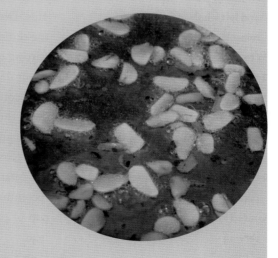

- **Rapidly cooked garlic** loses its hardest spicy edges but retains an oniony aroma. If you let it brown, it'll gain a few pleasing bitter notes—so long as you don't brown it *too* much.

# PASTA WITH OLIVE OIL AND THREE FLAVORS OF GARLIC

**SERVES 4 TO 6**

½ cup extra-virgin olive oil

12 medium cloves garlic, 4 smashed and left whole, 4 finely sliced, 4 minced or grated on a Microplane

¼ to ½ teaspoon red pepper flakes

2 tablespoons unsalted butter

1 pound short twisted or tubular pasta, such as gemelli, cavatappi, or rotini

Kosher salt

2 tablespoons chopped fresh parsley

Grated Parmigiano-Reggiano for serving

1. Heat the oil and 4 smashed garlic cloves in a 10-inch skillet over medium-high heat until the cloves are gently sizzling. Reduce the heat to medium-low and sizzle until the garlic is golden brown, 5 to 7 minutes. Discard the garlic cloves, leaving the oil in the skillet, and return the skillet to high heat. Add the sliced garlic and pepper flakes and cook, stirring constantly, until the garlic just begins to turn pale golden brown, about 45 seconds. Add the minced garlic and toss until fragrant, about 30 seconds. Remove from the heat and stir in the butter. Set aside.

2. Place the pasta in a large pot and cover with hot water. Add a large pinch of salt and bring to a boil over high heat stirring occasionally to keep the pasta from sticking. Continue to cook until the pasta is fully softened but retains a slight bite in the center (about 1 minute less than the box recommends after the pasta reaches a boil). Drain the pasta, reserving ½ cup of the cooking liquid, and return to the pot, set over medium-low heat.

3. Pour the sauce over the pasta, add half of the reserved pasta water, and stir until the sauce comes together and coats the pasta, about 2 minutes, adding more pasta water as necessary until the desired consistency is reached. Stir in the parsley and season to taste with salt. Serve immediately, passing Parmesan at table.

# PASTA WITH GARLICKY BROCCOLI, ANCHOVIES, AND BACON

**I love the way the garlicky sauce clings to and soaks into the crowns of broccoli—so much that I'm often tempted to leave out the pasta altogether and just double the (already generous) amount of broccoli. But the pasta offers nice textural contrast, as well as little nooks and crannies for bits of smoky bacon, intensely salty anchovy, and lemon zest to hide out in.**

## SERVES 4 TO 6

Kosher salt

1 pound broccoli, trimmed and cut into bite-sized florets

4 ounces bacon or pancetta, cut into ½-inch pieces

3 tablespoons extra-virgin olive oil

12 medium cloves garlic, 4 smashed and left whole, 4 finely sliced, 4 minced or grated on a Microplane

4 anchovy fillets, finely minced

Pinch of red pepper flakes

2 tablespoons unsalted butter

1 pound small cupped pasta such as orecchiette or shells

2 teaspoons grated lemon zest and 1 tablespoon juice (from 1 lemon)

1 ounce Parmigiano-Reggiano, grated, plus more for serving

Freshly ground black pepper

1. Bring a large pot of salted water to a boil. Add the broccoli and cook until bright green and tender-crisp, about 3 minutes. With a fine-mesh strainer, transfer to the insert of a salad spinner set in the sink and run under cold water to chill. Spin in the salad spinner in batches until thoroughly dried, then set aside.

2. Place the bacon in a 10-inch skillet, add ½ cup water, and bring to a boil over high heat, then reduce the heat to medium and cook, stirring occasionally, until the water has evaporated and the bacon is beginning to render its fat. Add the oil and 4 smashed garlic cloves and cook until the garlic is gently sizzling, about 1 minute. Reduce the heat to medium-low and cook until the garlic is golden brown and the bacon is crisp, 5 to 7 minutes. Discard the garlic cloves, leaving the oil and bacon in the skillet, and return the pan to high heat. Add the sliced garlic and cook, stirring constantly, until the garlic just begins to turn pale golden brown, about 45 seconds. Add the broccoli, immediately toss, and cook, stirring and tossing occasionally, until coated in the garlic-bacon oil and just starting to color in spots, about 1 minute. Add the minced garlic, anchovies, and pepper flakes and toss until fragrant, about 30 seconds. Remove from the heat and stir in the butter. Set aside.

3. Place the pasta in a large pot and cover with hot water. Add a large pinch of salt and bring to a boil over high heat, stirring occasionally to keep the pasta from sticking. Continue to cook until the pasta is fully softened but retains a slight bite in the center (about 1 minute less than the box recommends after the pasta reaches a boil). Drain the pasta, reserving ½ cup of the cooking liquid, and return to the pot, set over medium-low heat.

4. Add the broccoli, using a rubber spatula to scrape out any garlic and juices from the pan. Increase the heat to high, add half of the reserved pasta water, and stir until the sauce comes together and coats the pasta, about 2 minutes, adding more pasta water as necessary until the desired consistency is reached. Stir in the lemon zest and juice and cheese and season to taste with salt and pepper. Serve immediately, passing Parmesan at the table.

# PASTA WITH GARLIC AND LOTS OF VEGETABLES

This is a dish best made in the spring, when the bright green vegetables are at their crispest, sweetest, and brightest. Asparagus in particular is sweet and crisp when first picked. As it sits out of the soil, its sugars rapidly transform into starches. Even after a day, there is a striking difference in its sweetness, which is why winter asparagus—usually shipped long distances from warmer climates—is never as good as the stuff you get from the farmers' market in the spring.

Of course, once you know the basic process for this dish—blanching and chilling your vegetables, making your garlic oil, and then combining everything at the last minute—you can improvise, using whatever vegetables look best.

**SERVES 4 TO 6**

Kosher salt

8 ounces asparagus, ends trimmed and cut on a bias into 2-inch segments

1 medium zucchini, split lengthwise and cut into ¼-inch half-moons

1 medium summer squash, split lengthwise and cut into ¼-inch half-moons

1 cup shelled fava beans (optional)

1 cup broccoli florets cut into ½-inch pieces (optional)

1 cup frozen peas, thawed

1 cup grape or cherry tomatoes, split in half (optional)

¼ cup extra-virgin olive oil

12 medium cloves garlic, 4 smashed and left whole, 4 finely sliced, 4 minced or grated on a Microplane

Pinch of red pepper flakes

2 tablespoons unsalted butter

1. Bring a large pot of salted water to a boil. Add the asparagus and cook until bright green and tender-crisp, about 3 minutes. Using a fine-mesh strainer, transfer to a colander and run under cold water to chill. Repeat with the zucchini, summer squash, fava beans, and broccoli, one vegetable at a time. If using fava beans, peel them. Combine all the blanched vegetables in a large bowl, add the peas and cherry tomatoes, and set aside.

2. Heat the oil and 4 smashed garlic cloves in a 10-inch skillet over medium-high heat until the garlic is gently sizzling. Reduce the heat to medium-low and sizzle until the garlic is golden brown, 5 to 7 minutes. Discard the garlic cloves, leaving the oil in the skillet, and return the pan to high heat. Add the sliced garlic and cook, stirring, until the garlic just beings to turn pale golden brown, about 45 seconds. Add the minced garlic and pepper flakes and toss until fragrant, about 30 seconds. Remove from the heat and stir in the butter. Set aside.

3. Place the pasta in a large pot and cover with hot water. Add a large pinch of salt and bring to a boil over high heat, stirring occasionally to keep the pasta from sticking. Continue to cook until the pasta is fully softened but retains a slight bite in the center (about 1 minute less than the box recommends after the pasta reaches a boil). Drain the pasta, reserving ½ cup of the cooking liquid, and return to the pot, set over medium-low heat.

4. Add the vegetables, garlic sauce, and half of the reserved pasta water and stir until the sauce comes together and coats the pasta, about 2 minutes, adding more pasta water until as necessary until the desired

1 pound short twisted or tubular
  pasta such as gemelli, cavatappi,
  or rotini
¼ cup chopped fresh parsley
2 teaspoons grated lemon zest and
  1 tablespoon juice (from 1 lemon)
1 ounce Parmigiano-Reggiano,
  grated, plus more for serving
Freshly ground black pepper

consistency is reached. Stir in the parsley, lemon zest and juice, and cheese and season to taste with salt and pepper. Serve immediately, passing Parmesan at the table.

# PASTA WITH SHRIMP AND GARLIC

**Ask an Italian what "shrimp scampi" translates as,
and he may look at you a little funny.**

It's one of those quirks of translation like the "queso cheese" or "carne asada steak" you might find at a fast-food taco chain. Scampi *are* shrimp; a particularly large variety often cooked with white wine and garlic. But the false nomenclature for the dish is so ingrained at this point that there's no fighting it. I've even seen restaurant menus these days offering "Scampi Scampi," served with a wink and a nod.

As anybody who comes from a culture that eats shrimp with the shell on can tell you, the shell is where the shrimpiest, sweetest flavor is housed. So for my version of shrimp scampi, I extract that flavor by cooking their shells in olive oil along with the garlic. The shrimp-infused oil doubles up on the shrimp flavor, coating all the pasta with its fragrance. The shell-infusion technique is one I use every time I sauté shrimp.

# ALL ABOUT SHRIMP

First things first: if you've been buying precooked shrimp, or even peeled and deveined shrimp, stop right this instant! I mean it!

Precooked shrimp are unfailingly pre-*overcooked* shrimp and make it impossible to add flavor to a dish the way you can with raw shrimp. Shrimp that are raw but peeled and deveined are a small step up, but they often get mangled and beat up in the cleaning. And shrimp shells hold tons of flavor, so you are robbing yourself of some of the best part of the beast. You are much better off buying whole headless shrimp (or at the very least "EZ peel") and cleaning them yourself. It's a little more work, but worth the effort.

There are a number of choices to make when it comes to buying shrimp:

### Frozen Versus Fresh

The vast majority of shrimp are processed and frozen right at the farm or on the boat before they ever get anywhere near your fishmonger or supermarket. This means that the "fresh" shrimp you're seeing at the fish counter are simply frozen shrimp that have been defrosted and put on display. There's no way to know how long they've been there, so you're better off buying frozen shrimp and defrosting them at home. Placed in a bowl under cold running water, shrimp should take about 10 minutes to defrost. A small price to pay for freshness.

### Head-on Versus Headless

Normally I like to buy my products in an as-close-to-natural state as possible, but shrimp are a major exception. Shrimp heads contain enzymes that are held in check while the shrimp is alive. As soon as they die, however, those enzymes will slowly work their way into the shrimp's body, breaking down tissues and causing them to turn mushy. Even within a day or two the difference can be striking. Deheading the shrimp at sea before they are frozen prevents this mushiness. For this reason, unless I'm buying shrimp that I can verify were caught within half a day or so (and are preferably still alive and kicking), I'll opt for headless shrimp.

### IQF Versus Block-Frozen

IQF stands for Individually Quick Frozen, and the term means that each shrimp was frozen on its own before being bagged. Block shrimp come frozen together in a large block of ice. As a general rule, the faster you freeze something, the smaller the loss in textural quality, so go with the IQF. They also have the advantage that they are much quicker to defrost.

### Size

Forget labels like "medium," "large," or "jumbo": these are unregulated terms that are decided on by the packager or the supermarket. Instead, look for a set of two numbers, such as 26–30 or 16–20. These numbers indicate the number of individual shrimp that it takes up to make a pound. So a package labeled 16–20 will contain shrimp that weigh in at a little less than an ounce each. The smaller the number, the larger the shrimp. For superlarge shrimp, you may see a number like U–15, which means that it takes under 15 pieces to make up a pound. As far as flavor goes, there's not much difference among the sizes—look to individual recipes to specify the right size of shrimp for the job.

## Additional Ingredients

Shrimp, like scallops, are often treated with STP (sodium tripolyphosphate), a chemical intended to help them retain moisture. More than anything, this is a ploy to bulk up their weight and sell them at a higher profit. Check the ingredients list on packages of frozen shrimp. They should list shrimp, possibly salt, and nothing else.

# HOW TO CLEAN SHRIMP

Cleaning shrimp is a matter of peeling off the shells and removing the digestive tract that runs down their backs (often euphemistically referred to as the "vein"). If you want to save yourself some time, pick up EZ-peel shrimp, which have already had their shells cut and digestive tracts removed by machine, leaving you with the simple task of popping off their shells and feet. The shrimp will be a *little* more split open than they'd be if you did it by hand, but for some folks, that's a reasonable trade-off. If you want to go old school, here's how to do it.

## Step 1: Slit the Shell

If you are using fresh head-on shrimp, tear off the heads and reserve them for stock. Next, hold each shrimp flat against your cutting board and, using a very sharp paring knife, cut a shallow slit through its shell all the way down the center of its back.

## Step 2: Remove the Digestive Tract

Use the tip of your knife or a wooden skewer to carefully lift and pull out the digestive tract. The goal is to get it out in one piece so that it doesn't break and spill its contents onto the shrimp (if it does, simply rinse it off).

## Step 3: Remove the Shell and Legs

Pick up the shrimp and pull the shell halves out sideways. Once the sides are separated, grasp the shrimp at the segment just above the tail with one hand, then pull the rest of the shell away from that segment with the other. You should be left with a naked shrimp with the last tail segment still attached. The tail is customarily left on for aesthetic purposes in many preparations; I like to leave it on because I'm the kind of guy who picks up his shrimp with his fingers and pops them back before *chasing it down* with the tail itself. I love its sweet, flavorful crunch.

# PASTA WITH EXTRA-GARLICKY SHRIMP SCAMPI

**SERVES 4 TO 6**

1 pound large shrimp

12 medium cloves garlic, 4 minced
    or grated on a microplane, 4
    smashed and left whole, 4 finely
    sliced

½ cup extra virgin olive oil

Kosher salt

¼ to ½ teaspoon red pepper flakes

½ cup dry white wine

2 tablespoons unsalted butter

¼ cup chopped fresh parsley

2 teaspoons grated lemon zest and
    1 tablespoon juice (from 1 lemon)

1 pound short twisted or tubular
    pasta, such as gemelli, cavatappi,
    or rotini

Freshly ground black pepper

1. Peel the shrimp, leaving the last tail segments in place and reserving the shells. Place the shrimp in a large bowl, add the minced garlic, 2 tablespoons of the olive oil, and 1 teaspoon salt and toss to combine. Set aside.

2. Heat the remaining oil, the 4 smashed garlic cloves, and shrimp shells in a 12-inch skillet over medium-high heat until the garlic and shells are gently bubbling. Reduce the heat to medium-low and cook, tossing and stirring frequently, until fragrant, about 5 minutes. Strain the oil through fine-mesh strainer set over a bowl; discard the shells and garlic.

3. Return the infused oil to the skillet and heat over high heat until shimmering. Add the sliced garlic and pepper flakes and cook, stirring, until the garlic just begins to turn pale golden brown, about 45 seconds. Add the shrimp and cook, stirring and tossing until slightly pink, about 30 seconds. Add the wine and cook until the shrimp are nearly cooked through, about 1 minute longer. Remove from the heat, add the butter, parsley, and lemon zest and juice, and stir to combine. Set the sauce aside.

4. Place the pasta in a large pot and cover with hot water. Add a large pinch of salt and bring to a boil over high heat, stirring occasionally to keep the pasta from sticking. Continue to cook until the pasta is fully softened but retains a slight bite in the center (about 1 minute less than the box recommends after the pasta reaches a boil). Drain the pasta, reserving ½ cup of cooking liquid, and return to the pot, set over medium-low heat.

5. Pour the sauce and shrimp over the pasta, add half of the reserved pasta water, and stir until the sauce comes together and coats the pasta, about 2 minutes, adding more pasta water as necessary until the desired consistency is reached. Season to taste with salt and pepper and serve immediately.

# LINGUINE WITH FRESH CLAMS

My mother used to make spaghetti with clam sauce using clams from a can and plenty of bacon to liven things up. But clam spaghetti day was never a favorite among the Alt children, and I blame it on those canned clams. The problem with them is that during the canning process, they get cooked. Overcooked, that is. Open up a can of clams, and you're already dealing with clams that are rubbery and bland. The only way to truly appreciate clams is to start with fresh live ones (or, at the very least, freshly shucked or frozen clams) and cook them as briefly as possible.

The great thing about fresh, live, in-the-shell clams is that like they make their own sauce when you cook them. All you need is a little bit of stock or wine to get them going, and then the flavorful liquor inside bulks out the rest of your sauce. When you buy fresh clams, make sure that they are either tightly closed or that they clam up when you tap on them. Clams that are gaping open are dead and should be avoided.

**NOTE**: For best results, use fresh clams. If they are unavailable, you can substitute 12 ounces frozen or canned clams, thawed if necessary and drained. Add to the skillet in step 1, along with the wine and butter. Immediately remove from the heat and proceed as directed.

## SERVES 4 TO 6

Kosher salt

6 tablespoons extra-virgin olive oil

12 medium cloves garlic, 4 smashed and left whole, 4 finely sliced, 4 minced or grated on a Microplane

¼ to ½ teaspoon red pepper flakes

½ cup dry white wine

2 tablespoons butter

2 pounds littleneck clams

1 pound linguine

¼ cup chopped fresh parsley

2 teaspoons grated lemon zest and 1 tablespoon juice from 1 lemon

Freshly ground black pepper

1. Bring a large pot of well-salted water to a boil. Heat the oil and 4 smashed garlic cloves in a large saucepan over medium-high heat until the garlic cloves are gently sizzling. Reduce the heat to medium-low and sizzle until the garlic is golden brown, 5 to 7 minutes. Discard the garlic cloves, leaving the oil in the skillet, and return the pan to high heat. Add the sliced garlic and pepper flakes and cook, stirring, until the garlic just begins to turn pale golden brown, about 45 seconds. Add the minced garlic and cook, stirring, until fragrant, about 30 seconds. Add the wine, butter, and clams, cover, and cook, shaking the pan occasionally, until the clams open, about 6 minutes. Transfer the clams to a bowl, and set the sauce aside.

2. Cook the pasta in the boiling water until it fully softened but retains a slight bite in the center (about 1 minute less than the package recommends after the pasta reaches a boil). Drain the pasta, reserving ½ cup of cooking liquid, and return to the pot set over medium-low heat.

3. Pour the sauce over the pasta, add half of reserved pasta water, and stir until the sauce comes together and coats the pasta, about 2 minutes, adding more pasta water as necessary until the desired consistency is reached. Stir in the clams, parsley, and lemon zest and juice and season to taste with salt and pepper. Serve immediately.

# Mother SAUCE #2:
## CLASSIC RED SAUCE

A basic red sauce is an essential staple in any Western cook's pantry. Countless Italian-American restaurants are based on this sauce.

Marcella Hazan's recipe for tomato sauce may deliver the most culinary bang for your buck that you'll ever see. It's so simple it doesn't even need a full recipe—just simmer a 28-ounce can of whole tomatoes with 5 tablespoons unsalted butter and an onion split in half, crushing the tomatoes against the sides of the pot with a spoon—but the flavor you end up with is rich, fresh, and perfectly balanced. It's the butter that makes the difference. Unlike olive oil, butter contains natural emulsifiers that help keep the sauce nice and creamy. And the dairy sweetness works in tandem with the sweetness of the onions while rounding out the harsher acidic notes of the tomatoes.

Building from where Marcella leaves off, it's not a far jump to a classic Italian-American marinara sauce—tomato sauce flavored with garlic, oregano, and olive oil. Butter is still essential for smoothing out the rough edges of the acidic tomatoes, but here I like to substitute extra-virgin olive oil for half of it to bring some extra complexity into the mix. I make it in quadruple batches and store it in sealed Ball jars. Bottle while hot in sterile jars, seal them, and allow the sauce to cool to room temperature before refrigerating. It'll keep in the fridge for at least a month, ready to reheat and serve or incorporate into another recipe.

Here's the basic recipe, along with five variations it. The sauce also pops up in other places in this book, such as with my meatballs (page 540).

## DRIED VERSUS FRESH HERBS

Most recipes for marinara sauce call for either dried oregano or Italian seasoning, which is mostly dried oregano and basil. My immediate thought was replace the dried herbs with fresh. Imagine my surprise when I found after cooking two sauces side by side with dried oregano in one and fresh leaves in the other, there was barely any difference at all! Why was that?

Many chefs assert that fresh herbs are superior to dried herbs, and they're right—most of the time. Most herbs contain flavor compounds that are more volatile than water, which means that drying process that removes water also ends up removing flavor.

But it's not always the case, and here's why: savory herbs that tend to grow in hot, relatively dry climates—like oregano, for instance—have flavor compounds that are stable at high temperatures and are well

*continues*

contained within the leaf. They have to be, in order to withstand the high temperatures and lack of humidity in their natural environment. With these herbs, as long as you cook them for long enough to soften them, the flavor is just as good as with fresh—and a whole lot cheaper and easier to boot.

This chart shows you which herbs are best used fresh and which will fare just as well when used dried (in cooked applications).

| HERBS THAT ARE BEST USED FRESH | HERBS THAT CAN BE USED DRY (in cooked applications) |
| --- | --- |
| Parsley | Oregano |
| Basil | Rosemary |
| Mint | Marjoram |
| Cilantro | Bay leaf |
| Chervil | Thyme |
| Chives | Sage |
| Dill | Savory |
| Sorrel | |
| Tarragon | |

# CANNED TOMATOES

Do you shudder at the thought of making a fresh tomato sauce out of bland winter tomatoes? You should. Even at the absolute height of summer, it can be difficult to get a great tomato unless you grow it yourself, which leaves us with canned tomatoes. But what's the best type to use? You'll see five different versions at the supermarket:

- **Whole Peeled Tomatoes** are whole tomatoes that are peeled (either by steaming or by being treated with lye), then packed in tomato juice or tomato puree. Those packed in juice are less processed and therefore more versatile (tomatoes packed in puree will always have a "cooked" flavor, even if you use them straight out of the can). Sometimes calcium chloride, a firming agent, will be added to help prevent them from turning mushy. You'll also see them packed with basil leaves.
- **Diced Tomatoes** are whole peeled tomatoes that have been machine-diced, then packed in juice or puree. The main difference here is that with a greater exposed surface area, the calcium chloride can make the tomatoes too firm: they don't break down properly when cooking. I don't use them.
- **Crushed Tomatoes** can vary wildly from brand to brand. There are actually no controls on the labeling of crushed tomatoes, so one brand's "crushed" may be a chunky mash, while another's is a nearly smooth puree. Because of this, it's generally better to avoid crushed products, opting instead to crush your own whole tomatoes.

- **Tomato Puree** is a cooked and strained tomato product. It makes a good shortcut for quick-cooking sauces, but your sauce will lack the complexity you get from slowly reducing less-processed tomatoes. Leave the puree on the shelf.
- **Tomato Paste** is concentrated tomato juice. Fresh tomatoes are cooked, then the larger solids are strained out and the resulting juice is slowly cooked down to a moisture content of 76 percent or less. Tomato paste is great for adding a strong umami backbone to stews and braises, as well as for thickening them slightly.

So diced tomatoes are too firm, crushed tomatoes are too inconsistent, and tomato puree is too cooked—which is why in my pantry, you'll only see whole peeled tomatoes packed in juice (I prefer Muir Glen and Cento brands) and tomato paste.

# PERFECT EASY RED SAUCE

**SERVES 4**

2 tablespoons extra-virgin olive oil

2 tablespoons unsalted butter

1 medium onion, finely diced (about 1½ cups)

2 medium cloves garlic, minced or grated on a Microplane (about 2 teaspoons)

½ teaspoon dried oregano

Pinch of red pepper flakes

One 28-ounce can whole tomatoes packed in juice, crushed by hand, in a food processor, or with a potato masher into rough ½-inch chunks

1 stem fresh basil (optional)

Kosher salt

1. Heat the olive oil and butter in a medium saucepan over medium-high heat until the butter has melted and the foaming subsides. Add the onion and cook, stirring frequently, until softened but not browned, about 3 minutes. Add the garlic, oregano, and pepper flakes and cook, stirring, until fragrant, about 1 minute.

2. Add the tomatoes with their juice, and the basil, if using. Bring to a boil over high heat, reduce to a simmer, and cook, stirring occasionally, until the sauce is thickened and reduced to 4 cups, about 30 minutes. Season to taste with salt. The sauce will keep in a sealed container in the refrigerator for up to 1 week.

# PASTA WITH SAUSAGE AND RED-SAUCE-BRAISED BROCCOLI RABE

**Broccoli rabe and sausage is a classic Italian pairing. Putting it in a marinara sauce ain't, but I love the bitterness and complexity that broccoli rabe adds to the sauce as it slowly braises. This is not a place for al dente vegetables—cook down the rabe until it's ultratender!**

**SERVES 4 TO 6**

¼ cup extra-virgin olive oil

1 pound hot Italian sausage, preferably homemade (page 505), casings removed if necessary

1 pound broccoli rabe, trimmed and roughly chopped

2 tablespoons unsalted butter

1 medium onion, finely diced (about 1 cups)

2 medium cloves garlic, minced or grated on a Microplane (about 2 teaspoons)

½ teaspoon dried oregano

Pinch of red pepper flakes

One 28-ounce can whole tomatoes packed in juice, crushed by hand or with a potato masher into rough ½-inch chunks

1 stem fresh basil (optional)

Kosher salt

1 pound orecchiete or other small cupped pasta such as shells, or penne

Grated Parmigiano-Reggiano for serving

1. Heat 2 tablespoons of the olive oil in a large saucepan over high heat until lightly smoking. Add the sausage and cook, breaking it up with a wooden spoon or potato masher, until no longer pink, about 5 minutes. Add the broccoli rabe and cook, stirring frequently, until wilted, about 3 minutes. Transfer to a large bowl and set aside.

2. Add the remaining 2 tablespoons olive oil and the butter to the saucepan and heat over medium-high heat until the butter has melted and the foaming subsides. Add the onion and cook, stirring frequently, until softened but not browned, about 3 minutes. Add the garlic, oregano, and pepper flakes and cook, stirring, until fragrant, about 1 minute. Add the tomatoes, with their juice, and the basil, if using. Return the sausage and broccoli rabe to the pan and bring to a boil over high heat, then reduce to a simmer and cook, stirring occasionally, until the broccoli rabe is completely tender, about 30 minutes; top it up with a little water as necessary to keep the sauce from becoming too thick. Season to taste with salt and keep warm while you cook the pasta.

3. Place the pasta in a large pot and cover with hot water. Add a large pinch of salt and bring to a boil over high heat stirring occasionally to keep the pasta from sticking. Continue to cook until the pasta is fully softened but retains a slight bites in the center (about 1 minute less than the package recommends after the pasta reaches a boil). Drain the pasta, reserving 1½ cups of the cooking liquid, and return to the pot.

4. Add the sauce to the pasta and stir to combine, adding some of the reserved pasta water as necessary to thin the sauce to the desired consistency. Serve immediately, with grated Parmigiano-Reggiano.

# PUTTANESCA: THE BEST SPAGHETTI FOR A NIGHT IN
## (SPAGHETTI WITH GARLIC, ANCHOVIES, CAPERS, AND OLIVES)

Another classic based on a simple marinara sauce, *puttanesca* is a spicy, briny, salty dish named after the Italian prostitutes who ate it—or served it, or one of those things. It's perfect for those romantic winter nights spent in when there's no danger of anyone else besides your spouse, and perhaps the dog, smelling your breath. Add some good-quality tuna canned in olive oil (not the water-packed kind) for a complete meal.

**SERVES 4**

Kosher salt

¼ cup extra-virgin olive oil

2 tablespoons unsalted butter

1 medium onion, finely diced (about 1½ cups)

3 cloves garlic, thinly sliced

8 anchovy fillets, minced

½ teaspoon red pepper flakes

3 tablespoons capers, rinsed, patted dry, and roughly chopped

½ cup pitted olives, roughly chopped (any sharp, briny olive will do, such as Taggiasche, kalamata, or Manzanilla)

One 28-ounce can whole tomatoes packed in juice, crushed by hand or with a potato masher into rough ½-inch chunks

One 6-ounce can tuna packed in olive oil, drained (optional)

Freshly ground black pepper

1 pound spaghetti

2 tablespoons chopped fresh parsley

Grated Parmigiano-Reggiano or Pecorino Romano for serving

1. Bring a large pot of salted water to a boil over high heat. Heat the olive oil and butter in a large saucepan over medium-high heat until the butter has melted and the foaming subsides. Add the onion and cook, stirring frequently, until softened but not browned, about 3 minutes. Add the garlic, anchovies, and pepper flakes and cook, stirring, until the garlic is light golden, about 3 minutes. Add the capers, olives, and tomatoes, with their juice, and cook, stirring frequently, until the sauce has reduced and the oil starts separating, about 15 minutes.

2. Gently fold the tuna into the sauce, if using, and season to taste with salt and pepper. Set aside.

3. Cook the pasta in the boiling water until it is fully softened but retains a slight bite in the center (about 1 minute less than the package recommends after the pasta reaches a boil). Drain the pasta, reserving 1 cup of cooking liquid, and return to the pot.

4. Add the sauce to the pasta and stir to combine, adding some of the reserved pasta water as necessary to thin the sauce to the desired consistency. Serve immediately, topped with the parsley and grated cheese.

# PENNE ALLA VODKA
## WITH CHICKEN

**This is about as easy as marinara variations get. There are various stories for where the sauce got its origin—some say it was at New York restaurant Orsini in the 1970s, others claim it was a marketing trick devised by a vodka company in the 1980s. Either way, there are some very good culinary reasons why adding vodka to a sauce can make it more fragrant. As we learned while searching for the ultimate chili (see page 257), alcohol is more volatile than water, which means that at a given temperature, it'll produce more vapor, helping aromatic compounds from the food leap off the plate and into your nose, where you can smell them. A classic pink sauce, this also incorporates cream, its richness rounding out the acidity of the tomatoes. Adding slivered chicken to the mix turns the dish into a hearty meal.**

### SERVES 4

2 tablespoons extra-virgin olive oil

2 tablespoons unsalted butter

1 medium onion, finely diced (about 1½ cups)

2 medium cloves garlic, minced or grated on a Microplane (about 2 teaspoons)

½ teaspoon dried oregano

Pinch of red pepper flakes

One 28-ounce can whole tomatoes packed in juice, crushed by hand or with a potato masher into rough ½-inch chunks

1 stem fresh basil (optional)

½ cup heavy cream

¼ cup vodka

Kosher salt

1 pound penne, ziti, or other short tubular pasta

1 pound boneless, skinless chicken breasts, cut into ½-inch-wide slivers

2 tablespoons chopped fresh parsley

Grated Parmigiano-Reggiano for serving

1. Heat the olive oil and butter in a large saucepan over medium-high heat until the butter has melted and the foaming subsides. Add the onion and cook, stirring frequently, until softened but not browned, about 3 minutes. Add the garlic, oregano, and pepper flakes and cook, stirring, until fragrant, about 1 minute. Add the tomatoes, with their juice, and basil, if using. Bring to a boil over high heat, reduce to a simmer, and cook, stirring occasionally, until the sauce has thickened and reduced to 4 cups, about 30 minutes.

2. Transfer the sauce to the jar of a blender, discarding the basil, and add the cream and vodka. Starting on the lowest speed and gradually increasing to high, blend until completely smooth, about 30 seconds. Return to the saucepan and bring to a simmer over high heat, then reduce to a bare simmer and allow to reduce further while the pasta cooks.

3. Place the pasta in a large pot and cover with hot water. Add a large pinch of salt and bring to a boil over high heat, stirring occasionally to keep the pasta from sticking. Continue to cook until the pasta is fully softened but retains a slight bite in the center. Meanwhile about 2 minutes before the pasta is done, add the chicken to the sauce and stir to combine.

4. Drain the pasta, reserving 1½ cups of the cooking liquid, and return to the pot. Add the sauce, stir to combine, and cook over medium heat, stirring frequently, until the chicken is cooked through and the sauce has achieved the desired consistency, about 1 minute, adding some of the reserved pasta water as necessary. Serve immediately, topped with the parsley and grated cheese.

## DOES VODKA MAKE A DIFFERENCE?

**D**oes the vodka really add much to the sauce? Doesn't the alcohol all simmer off? Is it all just a ploy by the vodka manufacturers to get us to buy more of their hooch?

Harold McGee has a bit to say on the subject in his *On Food and Cooking* (get it *now* if you don't already own it). Check this out:

> *The alcohol molecule bears some resemblance to a sugar molecule, and indeed it has a slightly sweet taste. At high concentrations, those typical of distilled spirits and even some strong wines, alcohol is irritating and produces a pungent, "hot" sensation in the mouth, as well as in the nose. Its chemical compatibility with other aroma compounds means that concentrated alcohol tends to bind aromas in foods and drinks and inhibit their release into the air.*

Huh. I stopped reading when I got to that part and started scratching my head, because I know from past experience that adding alcohol to stews will increase their aroma. I tested it out in my Best Short-Rib Chili recipe (page 259). What's he on about, inhibiting aromas? But he quickly clears it all up:

> *But at very low concentrations, around 1 percent or less, alcohol actually enhances the release of fruity esters and other aroma molecules into the air.*

Aha! Now it made sense: concentration is an important factor when it comes to its effectiveness as a flavor enhancer. That jibes with my past experience. Adding a bit of alcohol at the end of cooking is a good idea for stews and chilis, but too much, and the booziness can become overpowering, leaving you smelling nothing but the alcohol instead of the aromas it is supposed to be carrying. Whiskey drinkers can tell you that diluting a dram from 40% ABV (Alcohol Percent by Volume) down to 30% or 20% ABV will also bring out aromatics that are otherwise hidden.

So does the same really happen to pasta with vodka sauce?

### The Testing

To test out the effects of alcohol concentration and cooking, I made a huge batch of "Sauced" columnist Josh Bousel's Vodka Cream Sauce, leaving out the vodka. I then divided it into smaller batches.

To one set of batches, I added varying concentrations of vodka, diluting the alcohol content to various levels starting at 4% ABV of the total sauce down to 1%, tasting the sauce immediately after adding the vodka. For the other set of batches, I did the same thing but allowed the sauce to simmer for 7 minutes after adding the vodka before tasting it.

Of the batches in which I tasted the sauce immediately after adding the vodka, none were great. The 4% was downright inedible, with a strong alcoholic aroma and bitter flavor. I'm not exactly sure where the bitterness came from. Perhaps masking the fruitier, sweeter aroma of the tomatoes caused their bitterness to come out more strongly? In any case, only when I got down to 2% ABV did the sauce become bearable. I very slightly preferred the 1% sauce over the completely alcohol-free one, but just barely.

Cooking the sauce made a huge difference. After the 7-minute simmer, even the 4% sauce was edible though the bright sweetness of the tomatoes didn't really start showing until I got to the 2% sauce (which, after simmering for a few minutes, must have settled down to closer to 1% in the end). The harsher flavors of the vodka had dissipated, the bitterness was gone, and I was left with a nicely balanced sauce that packed a little bit more heat and bright aroma than the vodka-free sauce.

So, to answer the question: Yes! Vodka does alter the flavor of the sauce in a pleasing way. It adds a touch of heat and a bit of a sharp bite that help balance the sweetness of the tomatoes and the cream. Is it absolutely necessary? No, but vodka sauce wouldn't be, well, vodka sauce without it.

# PASTA WITH CARAMELIZED EGGPLANT AND RICH TOMATO SAUCE
## (PASTA ALLA NORMA)

I'd had versions of *pasta alla Norma*, the classic Sicilian dish of pasta with tomatoes and eggplant, many times and always ended up scratching my head and thinking, "I just don't get it." I mean, there's nothing wrong with tomato sauce and caramelized eggplant, but what in the world is that bland, mild ricotta salata doing sitting on top? Surely you want a more flavorful aged cheese for grating, right?

It wasn't until I tasted it in Sicily while my wife and I were on our fourth honeymoon (or was it our fifth?—she drags me on these things every year, I lose track) that I got it: real ricotta salata is nothing like the stuff we generally find here. The Sicilian version is made with sheep's milk that is aged until it is intensely tangy, with a strong barnyard aroma that is really the backbone of the dish. I've found it stateside in specialty Italian markets, but if you can't find properly aged ricotta salata (smell it before you buy it—it should have a powerful aroma), you can replace it with an aged caciocavallo, a good sheep's-milk feta, or Pecorino Romano, though it will alter the profile of the dish a bit.

2 small Italian or Japanese
  eggplants, split lengthwise cut
  and into ½-inch half-moons

Kosher salt

6 tablespoons extra-virgin olive oil

2 tablespoons unsalted butter

1 medium onion, finely diced
  (about 1½ cups)

2 medium cloves garlic, minced or
  grated on a Microplane (about 2
  teaspoons)

½ teaspoon dried oregano

Pinch of red pepper flakes

2 tablespoons tomato paste

One 28-ounce can whole tomatoes
  packed in juice, crushed by hand
  or with a potato masher into
  rough ½-inch chunks

1 stem fresh basil, leaves removed,
  stem reserved (optional)

1 pound penne, ziti, or other short
  tubular pasta

Grated ricotta salata or feta cheese
  for serving

1. Toss the eggplant slices with 1 teaspoon salt in a large bowl, then transfer to the bowl of a salad spinner set in the sink and let stand for 30 minutes.

2. Meanwhile, heat 2 tablespoons of the olive oil and the butter in large saucepan over medium-high heat until the butter has melted and the foaming subsides. Add the onion and cook, stirring frequently, until softened but not browned, about 3 minutes. Add the garlic, oregano, and pepper flakes and cook, stirring, until fragrant, about 1 minute. Add the tomato paste and stir until homogeneous, about 30 seconds. Add the tomatoes, with their juice, and the basil stem, if using. Bring to a boil over high heat, reduce to a simmer, and cook, stirring occasionally, until the sauce has thickened and reduced to 4 cups, about 30 minutes. Set aside.

3. Spin the eggplant slices in the salad spinner to remove excess water, then place on a double layer of paper towels. Place another layer of paper towels on top and press down to extract as much moisture as possible.

4. Heat the remaining ¼ cup extra-virgin olive oil in a large nonstick or cast-iron skillet over medium-low heat until shimmering. Add the eggplant slices in a single layer (you may have to work in batches or in two pans) and cook, turning them and shaking the pan occasionally, until a deep caramelized brown on both sides and completely tender in the center, 7 to 10 minutes. Transfer to a paper-towel-lined plate and immediately season with salt.

5. Place the pasta in a large pot and cover with hot water. Add a large pinch of salt and bring to a boil over high heat, stirring occasionally to keep the pasta from sticking. Continue to cook until the pasta is fully softened but retains a slight bite in the center. Drain the pasta, reserving 1½ cups of cooking liquid, and return to the pot.

6. Add the sauce and stir to combine, adding some of the reserved pasta water as necessary to thin the sauce to the desired consistency. Fold in the caramelized eggplant. Serve immediately, topped with the basil leaves, if you have them, and grated cheese.

1

2

3

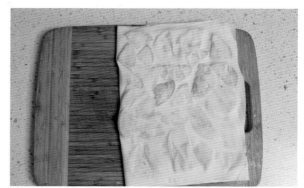

4

5

6

7

8

# EGGPLANT VARIETIES

I hated eggplant up until my early twenties. I think it's because I never had it cooked well. Unless treated right, eggplant is mushy, greasy, and insipid.* But when done right, it's meaty and substantial, with a subtle spicy bitterness and an unparalleled ability to absorb and complement other flavors. It also happens to be dirt cheap.

The best time of year to get it fresh is at the end of the summer, but unlike, say, inedibly bland winter tomatoes, even winter eggplants are perfectly serviceable. I cook with them pretty much year-round.

Eggplants come in a variety of shapes and sizes. Whatever variety you use, look for unblemished, smooth, firm skin and a hefty weight. When eggplants get too large, they become less dense, less flavorful, and more difficult to cook.

The most common varieties are:

- **Globe**: Large, deep purple, and relatively spongy, this is the most common variety and a great all-purpose choice. Excellent for dishes like eggplant Parmesan, where substantial wide slices are desired. These can also be roasted whole.
- **Italian**: Smaller, denser, and more flavorful than their larger cousins, firm Italian eggplants are the best choice for sautéing or grilling.
- **Japanese**: These are like the Italian variety, but with a longer, more slender shape. In Japan, the classic preparations include deep-frying, grilling, and broiling with a sweet miso-based glaze.
- **Chinese**: Long, skinny, and light purple in color, these dense eggplants are best after being parsteamed for braises and stews.
- **Thai**: Small, green, and crisp with an apple-like texture, Thai eggplants are one of the few varieties that are good eaten raw. In cooking, they're best added to dishes like curries and stir-fries right at the last minute and cooked just until heated through.

# DO EGGPLANTS HAVE SEX?

You've probably heard that you can tell the sex of an eggplant by looking at the dimple in its bottom, and that male eggplants have fewer seeds than females. While it's certainly possible to tell the sex of certain things by examining their bottoms (like mandrills or crabs), eggplants are *not* one of them—in reality, eggplants don't even have a sex.

So how *do* you find an eggplant with fewer seeds? The best way is to compare them by weight—the less

---

*My dislike could have also stemmed from the fact that the Eggplant Wizard destroyed me one too many times in Kid Icarus.

dense an eggplant is, the less seedy it will be. But eggplants that are less dense are harder to work with—they turn mushy more easily and absorb more oil while cooking—which makes this advice totally impractical.

Here's a better way to do it: just buy Italian eggplants. They are far less seedy than their larger American counterparts. They are also denser, less cumbersome to work with, and almost as widely available.

# WEEKNIGHT SPAGHETTI
## WITH MEAT SAUCE

In college, there was nothing easier than throwing a pound of ground beef into a pot, adding a jar of pasta sauce, and simmering them together, then tossing it all with pasta and calling it dinner. Tasty enough to be sure, but knowing what we know now, we can do a little better. For this sauce, I added carrot and celery to the onions to form the base of the marinara, along with the requisite garlic, oregano, and red pepper flakes. A couple of anchovy fillets add richness and meaty depth to the flavor. Anchovies contain both glutamates and inosinates—natural compounds that enhance the inherent meatiness of other ingredients, a factor further supported here by the use of glutamate-rich tomato paste. Chopping all the ingredients in the food processor makes short work of them.

As for the meat, I tried using straight-up ground beef but found that the texture was a little too tough without the benefit of a long, slow simmer (after all, I wanted this on the table within an hour or two). Instead, I decided to incorporate a technique frequently used for meat loaves and meatballs. By adding another element to the meat—mushrooms, in this case—and blending it right in, I was able to break up its texture, preventing it from getting tough while at the same time adding flavor.

1 small onion, quartered

1 small carrot, peeled and roughly
  chopped

1 stalk celery, roughly chopped

2 anchovy fillets (optional)

2 medium cloves garlic

½ teaspoon dried oregano

Pinch of red pepper flakes

8 ounces button mushrooms,
  stemmed and quartered

10 ounces ground chuck

2 tablespoons extra-virgin olive oil

2 tablespoons unsalted butter

2 tablespoons tomato paste

One 28-ounce can whole tomatoes
  packed in juice, crushed by hand,
  in a food processor, or with a
  potato masher into rough ½-inch
  chunks

1 tablespoon Asian fish sauce

¼ cup grated Parmigiano-
  Reggiano, plus more for serving

Kosher salt and freshly ground
  black pepper

1 pound spaghetti, linguine, penne,
  or other long skinny or short
  tubular pasta

Chopped fresh parsley or basil for
  serving

1. Place the onion, carrot, celery, anchovies, if using, garlic, oregano, and pepper flakes in the bowl of a food processor and pulse until finely chopped, 8 to 10 short pulses, scraping down the sides as necessary. Transfer to a bowl. Add the mushrooms to the empty food processor bowl of and pulse until finely chopped, 6 to 8 short pulses. Add the meat to the processor and pulse until the meat and mushrooms are evenly mixed, 6 to 8 short pulses. Set aside.

2. Heat the olive oil and butter in a Dutch oven over medium-high heat until the butter has melted and the foaming subsides. Add the chopped vegetables and cook, stirring frequently, until softened but not browned, about 5 minutes. Add the tomato paste and stir until homogeneous, about 1 minute. Add the meat/mushroom mixture and cook, stirring occasionally, until the moisture has completely evaporated and the mixture starts to sizzle, about 10 minutes. Add the tomatoes, with their juice, and bring to a boil over high heat, then reduce to a simmer and cook, stirring occasionally, until the sauce has reduced and thickened, about 30 minutes. Stir in the fish sauce and grated cheese and with salt and pepper to taste; keep warm.

3. Meanwhile, bring a large pot of well-salted water to a boil.

4. Cook the pasta according in the boiling water until it is fully softened but retains a slight bite in the center. Drain the pasta, reserving 1 cup of cooking liquid, and return to the pot. Add the sauce and stir to combine, adding some of the reserved pasta water as necessary to thin the sauce to the desired consistency. Serve immediately, topped with parsley and more Parmigiano-Reggiano.

# Mother SAUCE #3:

## PESTO

Most people know pesto in its classic Genovese version, made with basil, pine nuts, garlic, Parmesan, and plenty of olive oil. Traditionally pesto is made by grinding the ingredients to a paste in a mortar and pestle (*pesto* literally translates as "paste"), but these days the process is made much easier with the food processor. Making pesto can be as simple as dumping all the ingredients into the bowl and switching on the machine, but I've made a few upgrades to the classic recipe to address some problems I had with it.

First is color. When it is first made, pesto has a lovely emerald green color, but that color quickly fades to an unappetizing brown as it sits. This is due to oxidation reactions that occur when plant pigments are exposed to the air. To prevent this from happening, blanch your herbs in boiling water for about 30 seconds before dropping them into an ice bath, then drying and processing them. This blanching step will deactive the enzymes that are responsible for the oxidation reactions, and your pesto will stay bright green even after days of storage. I add the garlic directly to the blanching water along with

the basil leaves, which helps soften its harsh edges.

Check out these two pestos. The one on the left was made without blanching, while the leaves for the one on the right were blanched first. Over the course of a day or so, the difference in color becomes even more pronounced.

I like to mix in some baby spinach leaves with the basil when making pesto, which gives the sauce a slightly milder, more balanced flavor. At least in my opinion. Feel free to go 100-percent basil if you so desire. Aside from that, my only other addition is a grating of lemon zest, which adds a touch of brightness to what can be a very sharp sauce.

The great thing about pesto is that once you've got the basic process down (blanch, then blend), the recipe is almost infinitely adaptable. Want to make a walnut-parsley pesto? Easy, just replace the basil and spinach with parsley and the pine nuts with walnuts. Pistachio-arugula pesto would be fantastic too. You get the idea. I've included four variations here, but don't let that hold you back!

## STORING PESTO

Made with the blanch-then-process technique, pesto will remain bright and fresh in the refrigerator for several days. If you want to keep it for even longer, freeze it. The best way to do this is to fill an ice cube tray with the pesto and allow it to freeze overnight. The next day, pop out all the cubes and place them in a zipper-lock freezer bag. Pesto can be frozen for up to 6 months like this. To use it, cut a pesto cube into rough pieces on your cutting board, then thaw them in a skillet or directly in the saucepan with the cooked pasta.

# CLASSIC GENOVESE PESTO
## WITH BASIL AND PINE NUTS

**NOTE:** For a more intense flavor, omit the spinach and use additional basil in its place.

**MAKES ABOUT 1½ CUPS, SERVING 4**

2 ounces (about 3 loosely packed cups) fresh basil leaves

1 ounce (about 1½ loosely packed cups) baby spinach leaves

1 medium clove garlic

3 ounces (about ¾ cup) toasted pine nuts

2 ounces Parmigiano-Reggiano, grated (about 1 cup)

1 teaspoon grated lemon zest (from 1 lemon)

½ cup olive oil

Kosher salt and freshly ground black pepper

1. Bring a large pot of water to a boil. Prepare an ice bath. Add the basil, spinach, and garlic to the boiling water and blanch for 30 seconds. Drain, immediately transfer to the ice bath, and let cool completely.

2. Drain the basil, spinach, and garlic and transfer to a clean dish towel or triple layer of heavy-duty paper towels. Wrap tightly and squeeze out all the excess liquid.

3. Transfer to the bowl of a food processor, add the nuts, cheese, lemon zest, and olive oil, and process until a paste is formed, about 30 seconds, scraping down the sides as necessary. Season to taste with salt and pepper.

4. Serve immediately, or store in a sealed container in the refrigerator for up to 5 days.

*continues*

## ARUGULA AND WALNUT PESTO

**MAKES ABOUT 1½ CUPS, SERVING 4**

3 ounces (about 4½ packed cups) arugula leaves

1 medium clove garlic

3 ounces (about ¾ cup) toasted walnuts

2 ounces Parmigiano-Reggiano, grated (about 1 cup)

2 sun-dried tomatoes packed in olive oil

1 teaspoon grated lemon zest and 2 teaspoons juice (from 1 lemon)

½ cup olive oil

Kosher salt and freshly ground black pepper.

1. Blanch and chill the arugula and garlic as directed in step 1 of Classic Genovese Pesto (page 709); drain and squeeze out the excess liquid as in step 2.

2. Transfer to the bowl of a food processor, add the nuts, cheese, sun-dried tomatoes, lemon zest and juice, and oil and process until a paste is formed, about 30 seconds, scraping down the sides as necessary. Season to taste with salt and pepper. Serve or store as in step 4.

## ROASTED BELL PEPPER AND FETA PESTO WITH CHILES AND PEPITAS

**MAKES ABOUT 1½ CUPS, SERVING 4**

1 ounce (about 1½ loosely packed cups) fresh parsley leaves

2 tablespoons fresh oregano leaves

1 medium clove garlic

2 roasted bell peppers, peeled seeded, drained, and patted dry with paper towels

2 ounces feta cheese, roughly crumbled

1 small red serrano Thai bird chile

3 ounces (about ¾ cup) toasted pepitas

½ cup olive oil

Kosher salt and freshly ground black pepper

1. Blanch and chill the parsley, oregano, and garlic as directed in step 1 of Classic Genovese Pesto (page 709); drain and squeeze out the excess liquid as in step 2.

2. Transfer to the bowl of a food processor, add the bell peppers, cheese, chile, pepitas, and oil, and process until a paste is formed, about 30 seconds, scraping down the sides as necessary. Season to taste with salt and pepper. Serve or store as in step 4.

## TOMATO AND ALMOND PESTO WITH ANCHOVIES

**MAKES ABOUT 1½ CUPS, SERVING 4**

1 ounce (about 1½ loosely packed cups) fresh basil leaves

1 medium clove garlic

1 pint cherry tomatoes, halved

3 ounces roasted, skinned almonds (about ¾ cup)

2 ounces Parmigiano-Reggiano, grated (about 1 cup)

3 anchovy fillets

2 teaspoons balsamic vinegar

1 pickled peperoncini, stemmed

½ cup olive oil

Kosher salt and freshly ground black pepper

1. Blanch and chill the basil and garlic as directed in step 1 of Classic Genovese Pesto (page 709); drain, and squeeze out the excess liquid as in step 2.

2. Transfer to the bowl of a food processor, add the tomatoes, almonds, cheese, anchovies, vinegar, peperoncini, and oil and process until a paste is formed, about 30 seconds, scraping down the sides as necessary. Season to taste with salt and pepper. Serve or store as in step 4.

## SUN-DRIED TOMATO AND OLIVE PESTO WITH CAPERS

**MAKES ABOUT 1½ CUPS, SERVING 4**

2 ounces (about 3 loosely packed cups) fresh parsley leaves

1 medium clove garlic

3 ounces (about ½ cup) sun-dried tomatoes packed in oil

4 ounces (about 1 cup) pitted kalamata olives

2 tablespoons capers, rinsed and drained

2 ounces Pecorino Romano, grated (about 1 cup)

1 tablespoon red wine vinegar

Pinch of red pepper flakes

½ cup extra-virgin olive oil

Kosher salt and freshly ground black pepper

1. Blanch and chill the basil and garlic as directed in step 1 of Classic Genovese Pesto (page 709); drain and squeeze out the excess liquid as in step 2.

2. Transfer to the bowl of a food processor, add the tomatoes, olives, capers, cheese, vinegar, pepper flakes, and oil, and process until a paste is formed, about 30 seconds, scraping down the sides as necessary. Season to taste with salt and pepper. Serve or store as in step 4.

1

2

3

4

# Mother SAUCE #4:

## CREAM SAUCE

The origins of Alfredo sauce date back to early twentieth-century Rome, when the restaurateur Alfredo Di Lelio started serving it at his eponymous restaurant. The original version is quite similar to the Roman dish of *spaghetti cacio e pepe*—grated Pecorino Romano cheese and plenty of black pepper. It's made by tossing fresh semolina pasta with butter and cheese, along with some of the cooking water. The starch-laden cooking water helps the cheese emulsify and turn into a light, creamy sauce.

The American version we're more accustomed to these days usually includes eggs, heavy cream, and some sort of starchy thickener, making for a sauce that is far richer and creamier. I find it to be a little bit *too* rich for an everyday meal. My version bridges the gap between the two, using a small amount of heavy cream but omitting the egg (for my creamy mac and cheese recipe, on the other hand, extra

richness is not a problem; see page 714). As with *cacio e pepe*, I like to add a ton of black pepper to my sauce. The spiciness complements the cheese nicely.

The one issue that I've had with the sauce is the same one that always crops up when translating restaurant pasta dishes into recipes that work in the home kitchen: your home pasta water, used to cook just a single batch of pasta, is not as starchy as a restaurant's pasta water, which gets starchier and starchier as service progresses. This extra starch helps bind the emulsion. Without it, the cheese can quite easily break out, turning into a clumpy, gluey mess. To compensate for this, I add just the barest amount of cornstarch to the grated cheese before stirring it into the pasta. Wonderful on its own, this light and creamy sauce is also the perfect backdrop for any number of seasonal vegetables or aromatics; see the variations that follow.

# LIGHTER FETTUCINE ALFREDO

**SERVES 4**

2 ounces Pecorino Romano, grated
  (about 1 cup), plus extra for
  serving
½ teaspoon cornstarch
4 tablespoons unsalted butter, cut
  into chunks
Freshly ground black pepper
½ cup heavy cream
1 pound fresh fettucine
2 tablespoons chopped fresh chives
  or parsley
Kosher salt

1. Combine the cheese with the cornstarch in a small bowl, tossing to coat. Add the butter, ½ teaspoon pepper, and the cream. Bring a large pot of salted water to a boil.

2. Add the pasta to the boiling water and cook until al dente, about 1½ minutes. Drain, reserving 2 cups of the cooking liquid. Return the pasta to the pot and add the cheese mixture and 1 cup of the cooking liquid. Set over medium heat and cook, stirring constantly, until the sauce thickens and coats the pasta, about 2 minutes. Thin to the desired texture with more pasta cooking water. Stir in the herbs and season to taste with salt and more pepper. Serve immediately, passing extra cheese at the table.

## QUICK CREAMY PASTA WITH PROSCIUTTO, PEAS, AND ARUGULA

Omit the chives. Stir in 1 cup thawed frozen peas, 3 ounces prosciutto, cut into thin strips, and 3 cups loosely packed arugula leaves after adding the 1 cup cooking liquid and proceed as directed.

## QUICK CREAMY PASTA WITH LEMON ZEST AND ROSEMARY

Omit the chives. Stir in 1 teaspoon grated lemon zest and 1 tablespoon chopped fresh rosemary along with the cream and butter and proceed as directed.

# ULTRA-GOOEY STOVETOP MACARONI AND CHEESE

**No matter how much culinary training I've gone through,**

. . . no matter how many high-end ingredients I cook with or fancy restaurants I eat at, few things in the world can compete in terms of sheer deliciousness and childish pleasure with stovetop mac 'n' cheese. Who doesn't love gooey, cheesy, creamy, salty pasta, even when (or especially when?) it comes out of that blue box? For me—and I presume for many of you—it's a built-in taste memory, and a powerful one.

It's the texture that does it for me. No other mac 'n' cheese I've had has been quite so velvety smooth as the Kraft original. That said, in absolute terms, it does leave a bit to be desired in the flavor department. The ultimate goal? A cheese sauce with the creamy, gooey, oozy consistency of the blue box version but all the complex flavor of real cheese.

## Gimme a Break!

Cheese melts, right? So why not just throw some cheddar cheese into a pot with the pasta and heat it until it's at perfect sauce consistency? Anyone who's tried it can tell you: the cheese breaks, greasy slicks forming over a watery layer, with clumps of tough, rubbery cheese strands stuck together. It's not a pretty picture.

In order to understand why that happens, let's take a closer look at exactly what cheese is made of:

- **Water** is present to varying degrees. Young cheeses like Jack, young cheddars, and mozzarella have a relatively high water content—up to 80 percent. The longer a cheese is aged, the more moisture it loses, and the harder it becomes. Famous hard cheeses, like Parmigiano-Reggiano and Pecorino Romano, may contain as little as 30 percent water after several years of aging.
- **Milk fat** in solid cheese is dispersed in the form of microscopic globules kept suspended in a tight matrix of protein micelles (more on those in a second). Under around 90°F, this fat is solid. Because of this, and because of their suspension, these tiny globules don't come into contact with each other to form larger globules: cheeses stay creamy or crumbly, instead of greasy.
- **Protein micelles** are spherical bundles of milk proteins. Individual milk proteins (the main ones are four similar molecules called caseins) resemble little tadpoles with *hydrophobic* (water-avoiding) heads, and *hydrophillic* (water-seeking) tails. These proteins come together headfirst in bundles of several thousand, protecting their hydrophobic heads and exposing their hydrophillic tails to their watery surroundings. These micelles link together into long chains, forming a matrix that gives the cheese its structure.
- **Salt and other flavorings** make up the rest of the cheese. Salt can have a profound effect on the texture—saltier cheeses have had more moisture drawn out of the curd before being pressed, so they tend to be drier and firmer. Other flavorful compounds present in cheese are mostly intentional by-products of bacteria and aging.

In a well-aged cheese, all of these elements are in careful, stable balance. But heat throws the whole thing off. Everything may seem to be going all right at first—the cheese gradually softens, turning more and more liquid. Then, suddenly, at around 90°F, the liquefied fat comes together into greasy pools and separates from the water and proteins. As you continue to stir the melted cheese, the proteins—which are suspended in whatever water hasn't yet evaporated—glue themselves together with the help of calcium into long, tangled strands, forming the stretchy curds that you find in string cheese or stretched mozzarella. What was once whole and well has now completely separated into fat, protein, and water, and unless you've got a $5,000 homogenizer on hand, it *ain't* coming back together.

Cheese products like American and Velveeta have stabilizers added to them, along with extra liquid and protein, to keep them stable. I microwaved a small chunk of American cheese on a plate next to a block of extra-sharp cheddar. The American stayed smooth, while the cheddar broke. Perhaps we can learn some lessons from the former.

American cheese (*left*) has chemical salts that help it melt smoothly. Cheddar (*right*) breaks as it melts.

To get a cheesy sauce that's shiny and smooth, not greasy or stringy, requires three things:

- Keeping the fat globules from separating out and pooling
- Adding moisture to thin the texture

• Figuring out a way to keep the proteins from breaking apart and rejoining into long strands

Well, how the heck do you do all that? Luckily for us, **all of this has happened before, and it will all happen again**. In this case, I didn't want cheese that would go rapidly from solid to liquid I wanted cheese that softened linearly over time, which meant that a starch should be my thickener and stabilizer of choice.

Some cheese sauce recipes call for béchamel—a flour-thickened milk-based sauce—as the base. I don't like how it works out both in terms of texture (a cheesy béchamel is smooth and creamy but not gooey) and flavor (you can taste hints of the flour in the finished product). A purer starch like cornstarch is a definite step in the right direction, while replacing the regular milk (or heavy cream) with evaporated milk seals the deal.

Check out the difference between a béchamel-based sauce and one made with pure starch and evaporated milk:

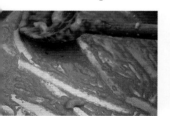 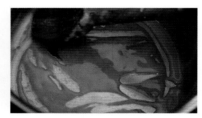

See, as the evaporated milk and starch mixture cooks, the starch molecules swell up, thickening the sauce, while the evaporated milk adds a concentrated source of milk proteins. This helps the entire mixture stay smooth and emulsified, resulting in a creamy sauce. The easiest way to incorporate the cornstarch is to toss it with the grated cheese. That way, when you add the cheese to the pot, the cornstarch is already dispersed enough that the cheese can't form annoying clumps. Want to get your sauce even shinier? Cutting your flavorful cheese with just a bit of American will introduce some full-strength emulsifying agents that'll get the sauce shiny enough to see your reflection in.

The sauce was great on its own, but when added to pasta, it didn't quite cling to the noodles the way I wanted it to. To fix this, I added a couple of eggs. Now, as the sauce cooks, the long, twisted proteins from the egg white begin to denature, unraveling and interconnecting with each other, thickening the sauce into what is essentially a very loose custard. The difference the eggs make in the sauce's coating ability is quite astonishing.

The best part? You don't even have to make a separate cheese sauce. Once the pasta is cooked, you can add all of your other ingredients directly to the pot and just stir over the burner until the sauce comes together on its own. What we've got here is a stovetop mac and cheese recipe that's only about 10 percent more cumbersome to make than the blue box (the only extra step is measuring a few ingredients) but tastes far, far better.

## WHY WON'T MY MAC 'N' CHEESE REHEAT?

Mac 'n' cheese is notoriously bad for reheating. Rather than a smooth, creamy sauce, you end up with a grainy, curdled, broken, unappetizing mess. It's all the pasta's fault. As we know, creating a stable cheese sauce requires the careful balance of fat to moisture, along with some emulsifying agents to help keep that fat and water getting along nicely together. Even though the pasta is completely cooked when it goes into the sauce, it has such a loose, sponge-like structure that it can continue to absorb water as it sits overnight in the refrigerator. This throws off the balance of the sauce, and the result is a sauce with too much fat that breaks out when you reheat it.

So is there a solution? Yes: just add back the water, duh. I've found that the best thing to do is add a few tablespoons of milk, which is essentially water with a bit of fat and a few proteins and sugars mixed in. The water content of the milk fixes the ratio, while the proteins help ensure that the sauce gets re-emulsified, as long as you stir while you reheat. Your pasta will *always* be mushier than it was in the first place, but sometimes mushy pasta can be a good thing.

# CHEESE CHART

The meltability of various cheeses can be affected by a number of factors, including their manufacture and their chemical makeup, but the most important thing is age. Young, moist cheeses tend to melt a whole lot better than older, drier ones. But what exactly happens when cheese melts? Most cheeses are made by adding bacteria and rennet\* to milk. The bacteria consume sugars, producing acidic by-products. Aside from lending tang and flavor, these acids, along with the rennet, cause the proteins in the milk (mainly casein) to denature. Imagine each protein as a tiny spool of wire that gets slowly unwound. The more it unwinds, the easier it is for it to get itself tangled up with other bits of wire. This is exactly what happens in cheese. The kinked wire–like proteins tangle up with each other, forming a stable matrix and giving the cheese structure. Trapped within this matrix are microscopic bits of solid fat and water.

As cheese is heated, the first part to go is the fat, which begins melting at around 90°F. Ever notice how a piece of cheese left out in the heat for too long forms tiny beads on its surface? Those are beads of milk fat. Continue to heat the cheese, and eventually enough of its protein bonds will break that it'll flow and spread like a liquid. Depending on the type of cheese, this takes place at anywhere from around 120°F, for super-melty high-moisture process cheeses like Velveeta, all the way up to 180°F and higher, for superdry cheeses like well-aged Parmigiano-Reggiano. Once the protein structure breaks down too much, individual microdroplets of fat and water coalesce, breaking out of the protein matrix and causing the cheese to completely break. Some cheeses, like feta or halloumi, have a protein structure so tight that no amount of heating will cause them to break or melt. Others have emulsifiers added to them to ensure that they melt smoothly at low temperatures without breaking (here's looking at you, American!). Still others need a bit of assistance from a recipe to remain stable.

Here's a chart of some of the more commonly available cheeses, along with their melting properties and best uses.

*continues*

---

\* Rennet is an enzyme derived from the lining of calves' stomachs or, increasingly common these days, from vegetarian sources (yes, most cheese is *not* vegetarian).

| CHEESE NAME | SLICING/ EATING PLAIN | MARINATING | FRYING | GRILLING | CRUMBLING | GRATING |
|---|---|---|---|---|---|---|
| American | | | | | | |
| American Munster (Muenster) | X | | | | | |
| Asiago (young) | X | | | | | |
| Asiago (aged over 1 year) | X | | | | X | X |
| Brie | X | | | | | |
| Cabrales (blue) | X | | | | X | |
| Camembert | X | | | | | |
| Cheddar (young) | X | | | | | X |
| Cheddar (aged 1 year or more) | X | | | | X | X |
| Colby | X | | | | X | X |
| Comté (Gruyère de Comté) | X | | | | | X |
| Cotija | | | | | X | X |
| Danish Blue | X | | | | X | |
| Domestic or Danish Fontina | X | | | | | X |
| Emmental (or Swiss cheese) | X | | | | | X |
| Feta | X | X | | | X | |
| Fourme D'Ambert (blue) | X | | | | X | |
| Gorgonzola | X | | | | X | X |

| ...ELTING | COUNTRY OF ORIGIN | TYPE OF MILK | FLAVOR |
|---|---|---|---|
| | United States | Cow | Very mild, salty, extremely meltable |
| | United States | Cow | Mild, creamy |
| | Italy | Cow | Tangy, milky |
| | Italy | Cow | Savory, nutty, and salty (like Parmesan) |
| | France | Cow | Strong aroma, creamy |
| (when young) | Spain | Cow, sheep, or goat | Salty, funky |
| | France | Cow | Strong aroma, creamy |
| | Great Britain/United States | Cow | Mildly nutty, creamy |
| | Great Britain/United States | Cow | Nutty, sharp |
| | United States | Cow | Mild, creamy, like a young cheddar |
| | France | Cow | Very nutty, savory |
| | Mexico | Cow | Mild, salty |
| | Denmark | Cow | Funky, tangy |
| | United States/Denmark | Cow | Mildly nutty |
| | Switzerland (or North America, New Zealand, and other countries) | Cow | Nutty, tangy |
| | Greece | Sheep, sheep/goat, or cow | Very salty |
| | France | Cow | Salty, funky |
| | Italy | Cow or goat | Tangy, salty |

*continues*

| CHEESE NAME | SLICING/ EATING PLAIN | MARINATING | FRYING | GRILLING | CRUMBLING | GRATING |
|---|---|---|---|---|---|---|
| Gouda (young) | X | | | | | |
| Gouda (aged over 1 year) | X | | | | X | X |
| Halloumi | X | | X | X | | |
| Havarti | X | | | | | |
| Italian Fontina | X | | | | | |
| Limburger | X | | | | | |
| Manchego Curado 3 (aged 3 to 6 months) | X | | | | | |
| Manchego Viejo (aged over 1 year) | X | X | | | X | |
| Maytag Blue | X | | | | X | X |
| Monterey Jack | X | | | | | X |
| Mozzarella | X | X | | | | X |
| Paneer | X | | X | X | X | |
| Parmigiano-Reggiano | X | | | | | X |
| Pecorino Romano | X | | | | | X |
| Provolone | X | | | | | X |
| Queso Oaxaca | X | X | | | | |
| Queso Panela (Queso Canasta, Queso de Frier) | | X | X | X | X | X |
| Roquefort (blue) | X | | | | X | X |
| Stilton | X | | | | X | |

| ELTING | COUNTRY OF ORIGIN | TYPE OF MILK | FLAVOR |
|---|---|---|---|
| | Netherlands | Cow | Creamy with a slight tang |
| | Netherlands | Cow | Savory, nutty, and salty (like Parmesan) |
| | Cyprus | Goat/sheep (or sometimes cow) | Mild, salty |
| | Denmark | Cow | Mildly nutty, creamy |
| | Italy | Cow | Pungent, salty, nutty |
| | Germany | Cow | Strong aroma, creamy |
| | Spain | Sheep | Mild, grassy, creamy |
| | Spain | Sheep | Tangy, grassy, salty |
| | United States | Cow | Salty and nutty with mild blue flavor |
| | United States | Cow | Mild, and slightly nutty, extremely meltable |
| | Italy | Cow or buffalo | Creamy, fresh |
| | India | Cow | Milky, bland |
| | Italy | Cow | Salty, savory, nutty |
| | Italy | Sheep | Salty, savory, nutty, grassy |
| (when young) | Italy | Cow | Mild to sharp |
| | Mexico | Cow | Milky, fresh |
| | Mexico | Cow | Mild, creamy |
| | France | Sheep | Very salty, strong blue-cheese flavor, grassy |
| | England | Cow | Very salty, strong blue-cheese flavor |

# ULTRA-GOOEY STOVETOP MAC 'N' CHEESE

**If desired, top the mac with toasted bread crumbs just before serving.**

**NOTES:** Use a good melting cheese or combination thereof, like American, cheddar, Jack, Fontina, young Swiss, Gruyere, Muenster, young provolone, and/or young Gouda, among others (see "Cheese Chart," pages 717–21). To reheat the pasta, add a few tablespoons of milk to the pan and cook, stirring gently, over medium-low heat until hot.

### SERVES 4 TO 6

1 pound elbow macaroni

Kosher salt

One 12-ounce can evaporated milk

2 large eggs

1 teaspoon Frank's RedHot or other
  hot sauce

1 teaspoon ground mustard

1 pound extra-sharp cheddar
  cheese, grated (see Note above)

8 ounces American cheese, cut into
  ½-inch cubes (see Note above)

1 tablespoon cornstarch

8 tablespoons (1 stick) unsalted
  butter, cut into 4 chunks

1. Place the macaroni in a large saucepan and cover it with salted water by 2 inches. Add a large pinch of salt and bring to a boil over high heat, stirring occasionally to keep the pasta from sticking. Cover the pan, remove from the heat, and let stand until the pasta is barely al dente, about 8 minutes.

2. Meanwhile, whisk together the evaporated milk, eggs, hot sauce, and mustard in a bowl until homogeneous. Toss the cheeses with the cornstarch in a large bowl until thoroughly combined.

3. When the pasta is cooked, drain it and return it to the saucepan. Place over low heat, add the butter, and stir until melted. Add the milk mixture and cheese mixture and cook, stirring constantly, until the cheese is completely melted and the mixture is hot and creamy. Season to taste with salt and more hot sauce. Serve immediately, topping with toasted bread crumbs if desired.

### STOVETOP MAC 'N' CHEESE WITH HAM AND PEAS

Stir in 1 cup cubed sautéed ham and 1 cup thawed frozen peas along with the milk and cheese mixtures in step 3.

### STOVETOP MAC 'N' CHEESE WITH BACON AND PICKLED JALAPEÑOS

Slice 6 strips bacon into ½-inch-wide pieces, place in a large skillet with ½ cup of water, and cook over medium heat, stirring occasionally, until crisp. Transfer the bacon and its rendered fat to a small bowl and set aside.

Follow the instructions for Stovetop Mac 'n' Cheese, reducing the butter to 6 tablespoons and stirring in the bacon, with its rendered fat, and ¼ cup sliced pickled jalapeños along with the milk and cheese mixtures in step 3.

### STOVETOP MAC 'N' CHEESE WITH BROCCOLI AND CAULIFLOWER

Stir in 1 cup blanched broccoli florets and 1 cup blanched cauliflower florets along with the milk and cheese mixtures in step 3.

*continues*

## STOVETOP MAC 'N' CHEESE SUPREME PIZZA-STYLE

Replace half of the cheddar cheese with mozzarella. Stir in 1 ounce of Parmesan cheese grated, 8 ounces Italian sausage, cooked and crumbled, ¼ cup pepperoni cut into ½-inch chunks, 4 ounces soppressata or salami, cut into ½-inch chunks, 1 cup roughly chopped drained canned tomatoes, ¼ cup sliced pitted black olives, and ¼ cup sliced jarred peperoncini, into the finished mac 'n' cheese. Top with chopped basil and drizzle of extra-virgin olive oil.

## STOVETOP MAC 'N' CHEESE WITH GREEN CHILE AND CHICKEN

Replace the cheddar cheese with pepper Jack. Stir in 2 cups shredded cooked chicken (leftover or rotisserie from the supermarket), one 3½-ounce can chopped green chiles (or ½ cup chopped roasted fresh green chiles), and 1 cup salsa verde into the finished mac 'n' cheese. Sprinkle with chopped fresh cilantro and scallions.

# CHEESY CHILI MAC

Here's a really important question: why doesn't chili mac always come with extra cheesy-goo? Now that we have good recipes for Easy Weeknight Ground Beef Chili (page 261) and Ultra-Gooey Stovetop Mac 'n' Cheese (page 722), it's easy to do. Just stir the two together, toss them into a casserole dish with some extra cheese, and bake off the whole thing.

**SERVES 4 TO 6**

8 tablespoons (1 stick) unsalted butter

1 medium onions, grated on the large holes of a box grater (about ¾ cup)

1 large clove garlic, minced or grated on a Microplane (about 2 teaspoons)

½ teaspoon dried oregano

Kosher salt

2 chipotle chiles in adobo sauce, finely chopped

1 anchovy fillet, mashed with the back of a fork

2 tablespoons chili powder (or ¼ cup Chile Paste, page 259)

1½ teaspoons ground cumin

¼ cup tomato paste

1 pound freshly ground chuck

One 14-ounce can whole tomatoes packed in juice, drained and chopped into ½-inch pieces

One 15-ounce can dark kidney beans, drained

1 cup homemade or low-sodium canned chicken stock (or water)

Freshly ground black pepper

1 pound elbow macaroni

One 12-ounce can evaporated milk

2 large eggs

1 teaspoon Frank's RedHot or other hot sauce

1 teaspoon ground mustard

8 ounces American cheese, cut into ½-inch cubes

1¼ pounds extra-sharp cheddar cheese, grated

1 tablespoon cornstarch

½ cup grated Parmigiano-Reggiano

2 tablespoons chopped fresh parsley or scallions

1. Melt 2 tablespoons of the butter in a large Dutch oven over medium-high heat. Add the onion, garlic, oregano, and a pinch of salt and cook, stirring frequently, until the onion is light golden brown, about 5 minutes. Add the chipotles, anchovy, chili powder, and cumin and cook, stirring, until aromatic, about 1 minute. Add the tomato paste and cook, stirring, until homogeneous, about 1 minute. Add the ground beef and cook, using a wooden spoon to break the meat into pieces and stirring frequently, until no longer pink (do not brown the beef), about 5 minutes.

2. Add the tomatoes, kidney beans, and stock season with salt and pepper, and stir to combine. Bring to a boil, reduce to a simmer, and cook, stirring occasionally, until the flavors have developed and the chili is lightly thickened, about 30 minutes. Remove from the heat.

3. While the chili simmers, place the macaroni in a large pot and cover it with salted water by 2 inches. Add a large pinch of salt and bring to a boil over high heat, stirring occasionally to keep the pasta from sticking. Cover the pot, remove from the heat, and let stand until the pasta is barely al dente, about 8 minutes.

4. Meanwhile, whisk together the evaporated milk, eggs, hot sauce, and mustard in a bowl until homogeneous. Toss the American cheese and 1 pound (four-fifths) of the cheddar with the cornstarch in a large bowl until thoroughly combined.

5. Adjust a broiler rack to 8 inches below the element and preheat the broiler to high. When the pasta is cooked, drain it and return it to the pot. Place it over low heat, add the remaining 6 tablespoons butter, and stir until melted. Add the milk and cheese mixtures and cook, stirring constantly, until the cheese is completely melted and the mixture is hot and creamy.

6. Stir the chili into the macaroni and cheese. Transfer to a large casserole dish (or two smaller ones). Top with the remaining cheddar and the Parmesan cheese. Broil until browned and bubbly, about 5 minutes. Let rest for 5 minutes, then top with the parsley and serve.

# Mother SAUCE #5:
## RAGÙ BOLOGNESE

The Bolognese sauce I knew growing up was not much more than a basic marinara tossed with a pound of ground beef. Tasty stuff, but decidedly not the real deal. (See page 705 for a better version of that quick weeknight meat sauce.) *Real* ragù Bolognese is the king of meat sauces. Deep, rich, rib-sticking, soul-satisfying, heart-warming, and yum-o are all words that have been used to describe it (I'd use five out of six of those descriptors). There are many myths and traditions when it comes to Bolognese. So many that to try and decipher them all and come up with a truly "authentic" version would invariably end in insulting at least half the population of Northern Italian grandmothers. Milk or no? What type of meat should I use? And what about wine—red or white?

I can't vouch for authenticity, but here's *my* take on all those questions.

### Meaty Matters

While the exact mix and cuts of meat can vary depending on who's making it, I like to use a mixture of three: ground lamb, for its intense flavor (ground beef does just fine); ground pork, for its mild fat; and ground veal, for the rich gelatin and tenderness it provides. (See more on this in the discussion of meat loaf, page 527.) Some folks like to brown their meat intensely before simmering, but I find that browning meat greatly diminishes the quality of its texture, turning it gritty and tough rather than silken smooth and rich. There's so much flavor concentration and boosting going on in this sauce that there's really no need to build flavor through browning.

In addition to the ground meat, I like to add a few chicken livers, which are traditionally called for in ragù Bolognese intended for special occasions. Frankly, if I'm putting in the time to make a sauce this complex, whatever occasion it is had best *make* itself special. Finely chopping the livers in the food processor before cooking helps them to melt into the sauce in a seamless way, adding their flavor without chunks of liver that can be off-putting to some.

As a meat sauce, Bolognese is virtually an exercise in umami, so I've added a few extra ingredients to my sauce—umami bombs—to up the savoriness. The usual suspects are here: anchovies, Marmite, and soy sauce are glutamate powerhouses. Then a few cubes of pancetta cooked down at the beginning and dash of Thai or Vietnamese fish sauce stirred in at the end of cooking add another element: *inosinate,* a naturally occurring chemical that works to increase the savory effect of glutamates (see "Glutamates, Inosinates, and the Umami Bombs," page 245). Don't worry, it won't taste fishy!

### Getting Saucy

A Bolognese should not be tomato sauce with meat in it. In fact, some recipes (like the extremely austere version in the classic *Silver Spoon* cookbook) don't call for tomatoes at all, or perhaps just a squeeze of paste. I prefer the sweetness and acidity tomatoes bring to the mix, though I can do without big chunks in the finished dish. I start by making a basic marinara sauce, which I puree until completely smooth so that it can fade into the meat as it simmers.

For the wine, either red or white works equally well—it's surprising how similar they taste after they've been simmered for a couple hours—as long as the wine is dry and unoaked. The only important thing is to make sure to reduce it *before* adding your other liquids (see "Do I Really Need to Reduce My Wine?" below). Low-sodium chicken stock makes up most of the rest of the liquid, along with some milk and cream—two more controversial ingredients. While very old recipes for Bolognese ragù seem to be dairy-free, nearly every modern version I've seen contains milk in some form, and I like what it does for the texture and richness of the final dish. Speaking of texture, here's another trick: a packet of gelatin dissolved in the chicken stock adds even more body to your sauce.

### Reduce, Reduce, Reduce!

Once all of the ingredients have made their way into the pot, the only thing left to do is to simmer it all down. This is a magical, wonderful, and occasionally harrowing process. The smells that fill your home will draw neighbors from miles around as the wine reduces, meat tenderizes, and vegetables melt into the background. But watching what the ragù does as it reduces might trouble you: it starts off wet and creamy and then, as it heat ups, the abundant fat from the meat, butter, and cream break out, form-

ing a crimson slick on the surface of the simmering liquid. The slick will grow and grow until it completely covers the sauce, and you'll probably think to yourself, "I should do something about that."

But don't reach for the skimmer just yet! As the sauce continues to cook down, the dissolved solids in the liquid will get more and more concentrated, until the liquid is finally thick enough to be able to re-absorb that oily layer and is transformed into a creamy, rich, emulsified sauce once again.

### Use Your Noodle

Bolognese sauce is fantastic with wide thick noodles like pappardelle or tagliatelle (see page 732), but my favorite way to eat it is in a traditional Lasagna Bolognese (page 735). If you've made yourself even one pot of Bolognese, then you understand the basic principles behind all manner of meaty ragùs (see, for example, the pork ragù on page 732). It boils down to a few key elements. First, don't brown your ground meat—it'll stay more tender. (Just as with chili, browning large chunks of meat is a better bet.) Second, up the savoriness by adding tons of umami bombs (soy sauce, Marmite, anchovies, cured meats, fish sauce). And third, reduce your sauce slowly to allow meat to tenderize and flavors to meld.

## DO I REALLY NEED TO REDUCE MY WINE?

Why is it necessary to reduce the wine before adding the stock? Doesn't all the alcohol burn off anyway during the long simmering period? Actually, no.

Despite the fact that alcohol has a lower boiling point than water (173°F versus 212°F), it's nearly impossible to burn off all the alcohol from a pot on a burner. This is because when alcohol and water are mixed, the alcohol actually *lowers* the boiling point of the water. Water molecules are like tiny magnets. Each one has two legs that are attracted to the heads of the other molecules. Stacked up like a human pyramid, they form a semi-rigid pattern that's fairly difficult to escape. Add some molecules of ethanol (alcohol) to the mix, and things become a little shakier (have you ever tried building a human pyramid while drunk?). The molecules of ethanol get in the way of the water molecules, so their bonds are not quite as secure. This makes it much easier for individual molecules of water to escape the surface of the liquid and evaporate.

Bring a pot of liquid with a 5-percent alcohol content to a boil, and the vapor being released is actually over more than 60 percent water. Reduce that alcohol content to 2 percent, and the vapor released becomes closer to 90 percent water. **The closer you get to removing all the alcohol from a solution, the harder and harder it becomes to remove it.** Which is to say, in order to remove *all* of the alcohol from a pot, you need to boil it down to nearly nothing.

So, to avoid an overly boozy sauce, the best thing to do is to add your wine or liquor first and allow it to reduce significantly before adding your remaining liquids to dilute the small amount of alcohol remaining.

By the way, for those of you who worry about the small amount of alcohol present in a finished dish: you needn't. As long as you properly reduce the wine before adding the remaining liquids, the amount of alcohol in the final product is less than you'll find in the average loaf of yeast-leavened bread!

# THE ULTIMATE BOLOGNESE SAUCE

**NOTES**: Pancetta is unsmoked Italian-style bacon. You can use unsmoked American bacon or prosciutto ends in its place. To make Bolognese without veal, increase the lamb or beef to 2 pounds and use an additional packet of gelatin.

4 ounces chicken livers

4 anchovy fillets

1 teaspoon Marmite

1 tablespoon soy sauce

2 cups whole milk

½ cup heavy cream

2 cups homemade or low-sodium
   canned chicken stock

¼ ounce (1 packet) powdered
   gelatin

¼ cup extra-virgin olive oil

4 medium cloves garlic minced or
   grated on a Microplane (about 4
   teaspoons)

2 teaspoons dried oregano

Large pinch of red pepper flakes
   (optional)

One 28-ounce can whole tomatoes
   packed in juice, crushed by hand
   or with a potato masher into
   rough ½-chunks

4 ounces pancetta, cut into ½-inch
   chunks (see Note above)

1 large onion, finely chopped
   (about 1½ cups)

2 carrots, peeled and cut into
   ¼-inch dice (about 1 cup)

3 stalks celery, and cut into ¼-inch
   dice (about 1 cup)

4 tablespoons unsalted butter

1 pound ground lamb (or 85/15
   ground beef)

1 pound ground pork

1 pound ground veal

½ cup fresh sage leaves, finely
   chopped

1. Combine the chicken livers and anchovies with the Marmite and soy sauce in the bowl of a food processor and pulse until finely ground, 8 to 10 short pulses. Transfer to a bowl and set aside. Combine the milk, heavy cream, and chicken stock in a bowl, sprinkle with the gelatin and set aside.

2. Heat 2 tablespoons of the oil in a medium saucepan over medium-high heat until shimmering. Add the garlic, oregano, and pepper flakes and cook, stirring, until fragrant, about 1 minute. Add the tomatoes, with their juice, and bring to a boil over high heat, then reduce to a simmer and cook, stirring occasionally, until the sauce has until thickened and reduced to 4 cups, about 30 minutes. Remove from the heat.

3. While the tomato sauce simmers, combine the remaining 2 tablespoons olive oil and the pancetta in a large Dutch oven or stock pot and cook over medium-high heat, stirring occasionally, until the pancetta is softened and the fat is translucent, about 6 minutes. Add the onion, carrots, and celery and cook, stirring frequently, until softened but not browned, about 10 minutes. Transfer to a large bowl.

4. Return the Dutch oven to medium-high heat, add the butter, and heat until the foaming subsides. Add the lamb, pork, veal, and sage and cook, stirring occasionally, until the meat is no longer pink (do not brown). Add the anchovy mixture, stir until homogeneous, and cook, stirring occasionally for 5 minutes. Add the pancetta/vegetable mixture and stir to combine. Add the wine and bring to a boil, then reduce to a simmer and cook until the wine is reduced by half, about 15 minutes.

5. While the wine simmers, blend the tomato sauce with an immersion blender until completely smooth. Alternatively, transfer to the jar of a standing blender and blend, starting on low speed and increasing to high, until smooth.

6. Add the tomato sauce, milk/cream mixture, bay leaves, half of the basil, and half of the parsley to the Dutch oven and stir to combine. Bring to a boil over high heat, then reduce to a bare simmer, cover with a lid set slightly ajar, and cook, stirring occasionally, until the sauce is thick and rich, about 2 hours. (It will start off creamy, then break, with a fat layer floating on top, and gradually re-emulsify as it reduces.)

7. Add the fish sauce and Parmigiano-Reggiano to the sauce and stir vigorously until completely emulsified. Season to taste with salt and pepper. Remove from the heat and allow to cool for 30 minutes.

1 bottle (750 ml) dry red or white
  wine
2 bay leaves
½ cup minced fresh basil
½ cup minced fresh parsley
1 tablespoon Asian fish sauce
2 ounces Parmigiano-Reggiano,
  finely grated (about 1 cup)
Kosher salt and freshly ground
  black pepper

8. Stir in the remaining parsley and basil. The Bolognese will keep for up to a week in a sealed container in the fridge and only improve with time.

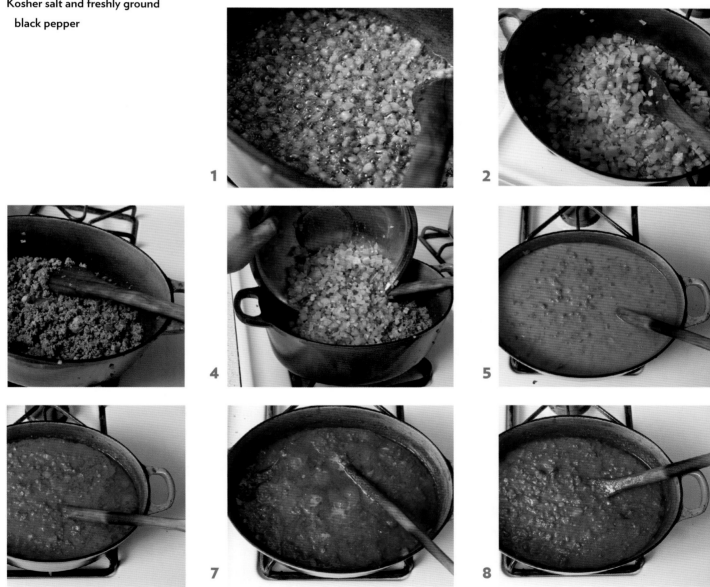

# FRESH PASTA WITH BOLOGNESE SAUCE

**SERVES 4**

5 cups **The Ultimate Bolognese Sauce** (page 729)

**Kosher salt**

1 pound fresh tagliatelle or pappardelle

**Grated Parmigiano-Reggiano for serving**

**Roughly torn or chopped fresh basil leaves for serving**

1. Heat the Bolognese sauce in a large saucepan until simmering. Keep warm.

2. Meanwhile, bring a large pot of salted water to a boil. Add the pasta and cook until al dente, about 1½ minutes. Drain, reserving 1 cup of the cooking water.

3. Add the pasta to the Bolognese sauce, along with ½ cup of the cooking water and bring to a simmer over high heat, stirring until the pasta is completely coated in sauce, add more cooking liquid as necessary to thin the sauce to the desired consistency. Serve immediately in warmed bowls, topped with grated cheese and basil.

# FRESH PASTA WITH PORK AND TOMATO RAGÙ

**NOTE:** Pancetta is unsmoked Italian-style bacon. You can use unsmoked American bacon or prosciutto ends in its place.

**SERVES 6 TO 8**

3 cups homemade or low-sodium canned chicken stock

¼ ounce (1 packet) powdered gelatin

2 tablespoons vegetable oil

2 pounds boneless pork shoulder, cut into 2-inch chunks

2 tablespoons extra-virgin olive oil

1. Place the chicken stock in a medium bowl, sprinkle with the gelatin, and set aside.

2. Heat the vegetable oil in a large Dutch oven over high heat until smoking. Add the pork and cook, without moving it until well browned on one side. Transfer to a large bowl and set aside to cool.

3. Return the Dutch oven to medium-high heat, add the olive oil, butter, and pancetta, and cook, stirring occasionally, until the pancetta is softened and the fat is translucent, about 6 minutes. Add the onions, carrots, and celery and cook, stirring frequently, until softened but not

2 tablespoons unsalted butter

4 ounces pancetta, cut into ½-inch chunks (see Note above)

1 large onion, finely chopped (about 1½ cups)

1 carrot, peeled and cut into ¼-inch dice (about ½ cup)

2 stalks celery, peeled and cut into ¼-inch dice (about ¾ cup)

4 medium cloves garlic, minced or grated on a Microplane (about 4 teaspoons)

Large pinch of red pepper flakes (optional)

3 tablespoons minced fresh rosemary, plus more for serving

2 anchovy fillets, finely chopped

1 teaspoon Marmite

1 tablespoon soy sauce

1 bottle (750 ml) dry white wine

One 28-ounce can whole tomatoes packed in juice, crushed by hand or with a potato masher into rough ½-inch chunks

2 bay leaves

1 tablespoon Asian fish sauce

2 ounces Parmigiano-Reggiano, finely grated (about 1 cup), plus more for serving

Kosher salt and freshly ground black pepper

1½ pounds fresh tagliatelle or pappardelle

browned, about 10 minutes. Add the garlic, red pepper, rosemary, anchovies, Marmite, and soy sauce and cook, stirring, until fragrant, about 30 seconds. Add the wine and bring to a boil, then reduce to a simmer and cook until the wine is reduced by half, about 15 minutes.

4. While the wine reduces, transfer one-quarter of the pork to the bowl of a food processor and pulse until roughly chopped, 6 to 8 short pulses. Transfer to a bowl and repeat until all the meat is chopped.

5. When the wine is reduced, add the chopped meat, tomatoes, with their juice, chicken stock, and bay leaves to the pot and bring to a boil over high heat, then reduce to a bare simmer, cover with a lid set slightly ajar, and cook, stirring occasionally, until the sauce is thick and rich and the meat is completely tender, about 2 hours.

6. Add the fish sauce and Parmigiano-Reggiano and stir vigorously until the sauce is completely emulsified. Season to taste with salt and pepper. Keep warm.

7. Bring a large pot of salted water to a boil. Add the pasta and cook until al dente, about 1½ minutes. Drain the pasta, reserving 1 cup of the cooking water. Add the pasta to the pork ragù, along with ½ cup of the cooking water, and bring to a simmer over high heat, stirring until the pasta is completely coated in sauce; add more cooking liquid as necessary to thin the sauce to the desired consistency. Serve immediately in warmed bowls, topped with more grated cheese and rosemary.

# { BAKED PASTA }

Here's something I've always wondered: when baking pasta, as in, say, lasagna or baked ziti, why do you always cook the pasta first? Aren't you inviting trouble by cooking it once, then proceeding to put it in a casserole and cooking it again? Well, there's the obvious first part of the answer to this question: pasta needs to absorb water as it cooks—a lot of water, around 80 percent of its own weight when perfectly al dente. So, add raw pasta directly to a baked pasta dish, and it will soften all right—it'll also suck up all of the moisture from the sauce, leaving it dry or broken.

**Here's the thing:** Dried pasta is made up of flour, water, and, on rare occasion, eggs. Essentially it's composed of starch and protein, and not much else. Starch molecules come aggregated into large granules that resemble little water balloons (see "How Starch Thickens," page 741). As they get heated in a moist environment, they continue to absorb more and more water, swelling up and becoming soft.

Meanwhile, the proteins in the pasta begin to denature, adding structure to the noodles (something that is much more obvious when cooking soft fresh egg-based pastas). When the stars are aligned, you'll manage to pull the pasta from the water just when the proteins have lent enough structure to keep the noodles strong and pliant and the starches have barely softened to the perfect stage—soft but with a bite—known as al dente.

But who's to say that these two phases, water absorption and protein denaturing, have to occur at the same time? H. Alexander Talbot and Aki Kamozawa of the fantastic blog Ideas in Food (blog.ideasinfood.com) asked themselves that very question, and what they found was this: You *don't* have to complete both processes simultaneously. In fact, if you leave uncooked pasta in lukewarm water for long enough, it'll absorb just as much as water as boiled pasta.

Here's what they had to say on the matter: "The drained [soaked] noodles held their shape, and since the starch had not been activated, they did not stick to one another and could be held without the addition of oil. Once we added the noodles to boiling salted water, we had perfectly cooked al dente pasta in just 60 seconds." Interesting indeed.

Unsoaked and soaked macaroni.

Macaroni soaked for varying degrees of time.

To try it out myself, I placed some macaroni in a bowl of warm tap water and allowed it to sit, pulling a piece out every 5 minutes to weigh how much water it had absorbed. After about 30 minutes, it had taken in just as much water as a piece of cooked boiled macaroni, all while remaining completely raw!

While the ability to cook presoaked pasta in just 60 seconds in itself is not all that exciting for a home cook (all it does is convert an 8-minute cooking process into a 30-minute soak plus 1 minute cooking process—hardly a time-saver), it's a very interesting application for restaurant cooks, who can have soaked pasta ready to be cooked in no time.

But what it *does* mean for a home cook is this: any time you are planning on baking pasta in a casserole, there is no need to precook it. All you have to do is soak it while you make your sauce, then combine the two and bake. Since the pasta's already hydrated, it won't rob your sauce of liquid, and the heat from the oven is more than enough to cook it while the casserole bakes. If you taste them side by side, you can't tell the difference between precooked pasta and simply soaked pasta. Think of what this means for lasagna! I know of at least six different common dental procedures that I'd rather have performed than to have to parcook lasagna noodles.

# TRADITIONAL LASAGNA BOLOGNESE

**The king of meat sauces deserves the mother of all pasta dishes, and this is it. You start with the Bolognese sauce, then layer it with creamy nutmeg-scented *besciamella* (that's Italian for *béchamel*, which is French for "white sauce") between layers of fresh pasta tinted green with spinach. Mine isn't *exactly* traditional, as I left out the spinach and also sneaked a bit of mozzarella into the *besciamella*.**

1

2

3

4

5

**SERVES 6 TO 10**

1 package (15 sheets) flat no-boil
   lasagna noodles

2 tablespoons unsalted butter

2 tablespoons all-purpose flour

2 medium cloves garlic, finely
   minced or grated on a Microplane
   (about 2 teaspoons)

2 cups whole milk

8 ounces whole-milk mozzarella
   cheese, grated

¼ teaspoon freshly grated nutmeg

Kosher salt and freshly ground
   black pepper

1 recipe The Ultimate Bolognese
   Sauce (page 729), warm

4 ounces Parmigiano-Reggiano,
   grated (about 2 cups)

2 tablespoons minced fresh basil or
   parsley (or a mix)

1. Place the lasagna noodles in a 9- by 13-inch baking dish and cover with warm water. Let soak, shaking the sheets gently every few minutes to prevent sticking, until lightly softened, about 15 minutes.

2. Meanwhile, heat the butter in a medium saucepan over medium-high heat, stirring occasionally until the foaming subsides; about 1 minute. Add the flour and whisk until the mixture is light blond in color and a slightly nutty aroma develops, about 1 minute. Add the garlic and stir to combine. Whisking constantly, add the milk in a steady stream until fully incorporated. Bring to a simmer (the mixture will thicken), then reduce the heat to low, add the cheese and nutmeg, and whisk until the cheese is fully melted. Whisking constantly, return to a simmer, then remove from the heat and season to taste with salt and pepper. Set aside.

3. Drain the soaked lasagna noodles and arrange in a single layer on paper towels or a clean kitchen towel to dry.

4. Adjust the oven racks to the lower-middle and lowest positions and preheat the oven to 375°F. Spread one-sixth (about 1⅓ cups) of the Bolognese over the bottom of 9- by 13-inch baking dish. Drizzle with one-sixth of the *besciamella* and sprinkle with ⅓-cup of the Parmigiano. Place 3 noodles in a single layer on top (the noodles will not quite touch each other; this is OK). Repeat with the remaining ingredients. The baking dish will be very full.

5. Place a foil-lined rimmed baking sheet on the lower oven rack to catch any drips, then place the lasagna on the rack above it and bake, rotating the pan half way through baking, until the edges are starting to crisp and the top is a bubbly, golden brown, about 45 minutes. Remove from the oven and allow to cool for 10 minutes. Sprinkle with the herbs and serve.

# CREAMY SPINACH AND MUSHROOM LASAGNA

**It was 2 a.m. on a chilly fall morning in our New York City apartment when my wife was suddenly awoken from her sleep by a loud clatter.**

She wearily dragged herself out of bed, narrowly missing stepping on the dog in her bleary-eyed walk out of the bedroom. I thought I'd gotten away scot-free, when she walked into the kitchen and caught me just as a drop of creamy sauce fell off my finger to the floor.

"What are you *doing* at two in the morning?!" she asked in her I'm-not-really-yelling-but-I-am-in-my-head voice.

"Um. . . ." I stammered. "Uh. . . ." I knew she'd never believe me if I told her the truth, so I decided to use my previously successful tactic of offering her an excuse before I offered an explanation. "I couldn't sleep!" That should satisfy her curiosity.

"Yes, but what are you doing, and why are you making so much noise, and why does it smell like béchamel and creamed spinach in here?"

"Well . . ., I couldn't sleep so I decided to make a lasagna."

She stared at me blankly for a moment, then turned around and shuffled back to the bedroom, muttering, "What did I marry?" under her breath.

**IF SHE** had given me a longer chance to explain, I would have been able to foist the blame squarely on the shoulders of Serious Eats community member KarmaFreeCooking. She'd started up a "Talk" thread titled "Vegetarian Lasagna Throwdown—Ideas to Win over Any Meat Eater," explaining that she'd been issued the challenge of bringing a vegetarian lasagna good enough to compete with a meaty one to a lasagna party.

I was not invited to this party, nor was I officially challenged, but, challenge accepted.

## The Layers

I started with a base of lightly creamed spinach. I considered going the easy route with frozen leaves, but figured that if the ultimate version is what we were after, and we were already putting in the not-insignificant amount of work required to construct a lasagna, using fresh spinach was not asking too much. Some spinach lasagnas have you blanch the leaves in boiling water, then wring out the excess. Far easier is to just wilt them in a pot along with some sautéed garlic and olive oil. From there, a hit of heavy cream and a grating of nutmeg is all they need.

Ricotta is a classic ingredient in an Italian-American lasagna, but I find the texture to be grainy and bland once cooked (mostly because store-bought ricotta just stinks). Instead, I use a trick I learned from *Cook's Illustrated*: replace the ricotta with some whole-fat cottage cheese pulsed in the food processor. It stays moist during baking and adds great tang to the finished dish. So I added the pulsed cottage cheese along with some chopped parsley and an egg to the spinach layer.

For the mushrooms, I made a classic *duxelles* by cooking chopped button mushrooms (you can also use cremini or shiitake) down with butter, shallots, thyme, and heavy cream. A dash of soy sauce added some meaty depth to them, while lemon juice brightened things up. Finally, a *besciamella* bound the whole thing together.

As with my lasagna Bolognese, using no-boil noodles presoaked in warm water save you the trouble of having to parcook the noodles. And, dear wife, I hope you like the results, because it's gonna be your lunch and dinner for the next four days.

# CREAMY SPINACH AND MUSHROOM LASAGNA

**NOTE:** The mushrooms can be chopped with a knife. Alternatively, break them up with your fingertips, or pulse them in a food processor. They should be chopped until no pieces larger than ¼ inch remain.

### SERVES 6 TO 10

8 tablespoons (1 stick) unsalted butter, plus more for greasing the baking dish

3 medium cloves garlic, minced or grated on a microplane (about 1 tablespoon)

2 pounds spinach, washed, tough stems removed, and roughly chopped

2 cups heavy cream

½ teaspoon freshly grated nutmeg

Kosher salt and freshly ground black pepper

1. Adjust the oven racks to the upper-middle and middle positions and preheat the oven to 400°F. Heat 3 tablespoons of the butter in a large saucepan over medium-high heat until the foaming subsides. Add the garlic and cook, stirring, until fragrant, about 30 seconds. Add the spinach in batches, allowing each batch to wilt before adding the next. Once all the spinach has been added, add 1 cup of the heavy cream and bring to a boil. Reduce to a strong simmer and cook, stirring frequently, until thickened and reduced, about 15 minutes. Add the nutmeg, season to taste with salt and pepper, and remove from the heat.

2. While the cream is reducing, combine the cottage cheese, egg, and 6 tablespoons of the parsley in the bowl of a food processor and process until combined and the cottage cheese is the texture of ricotta cheese, about 5 seconds. Transfer to a large bowl, add the cooked spinach, and mix well.

1 pound cottage cheese

1 large egg

½ cup chopped fresh parsley

1 package (15 sheets) flat no-boil lasagna noodles

1½ pounds button, cremini, or shiitake mushrooms, stems removed, and discarded if using shiitake, finely chopped (see Note above)

2 medium shallots, finely chopped (about ½ cup)

2 teaspoons minced fresh thyme

1 tablespoon soy sauce

2 teaspoons lemon juice (from 1 lemon)

2 tablespoons all-purpose flour

2 cups whole milk

12 ounces whole-milk mozzarella, grated

2 ounces Parmigiano-Reggiano, grated (about 1 cup)

3. Meanwhile, place the lasagna noodles in a 9- by 13-inch baking dish and cover with warm water. Allow to soak, agitating the noodles occasionally to prevent sticking, until lightly softened, about 15 minutes. Drain and arrange in a single layer on a paper towels or a clean kitchen towel to dry.

4. While the noodles soak, wash the spinach pan and return it to medium-high heat. Add 3 more tablespoons butter and heat until melted. Add the mushrooms and cook, stirring occasionally, until the liquid evaporates and the mushrooms start to sizzle, about 10 minutes. Add the shallots and thyme and cook, stirring frequently, until the shallots are softened, about 2 minutes. Add the soy sauce and lemon juice and stir to combine. Add the remaining 1 cup heavy cream, bring to a simmer, and cook until lightly thickened, about 3 minutes. Season to taste with salt and pepper and transfer to a bowl.

5. Wipe out the saucepan, return to medium-high heat, add the remaining 2 tablespoons butter, and heat until melted. Add the flour and cook, whisking constantly, until light golden blond. Slowly pour in the milk, whisking constantly. Bring to a simmer, then remove from the heat and stir in two-thirds of the mozzarella and the Parmesan, then season to taste with salt and pepper.

6. To assemble the lasagna, dry the baking dish and grease with butter. Spread 1 cup of the cheese sauce in the bottom of the dish. Lay 3 noodles on top of it, spacing them evenly (the noodles will not quite touch each other; this is fine). Top the noodles evenly with half of mushroom mixture and then with another 3 noodles. Top with half of spinach/cottage cheese mixture and another 3 noodles. Repeat the layers with the remaining mushroom mixture, spinach mixture, and noodles, ending with a layer of noodles. Pour the remaining cheese sauce over the top and spread it evenly. Sprinkle the remaining mozzarella evenly over the top.

7. Place a foil-lined baking sheet on the lower oven rack to catch any drips, then place the lasagna on the rack above it. Bake until the lasagna is bubbling around the edges, about 20 minutes. Turn on broiler on and broil until the top is lightly browned, about 5 minutes (if you have an under-oven broiler, transfer the lasagna to the broiler after baking). Let cool for 10 minutes, then slice and serve.

*continues*

**1**

**2**

**3**

**4**

**5**

**6**

# BAKED MAC 'N' CHEESE

**Unlike gooey stovetop mac 'n' cheese, baked mac 'n' cheese should have a tender, uniform, almost quiche-like texture—we're not looking for al dente noodles here.**

Of all the recipes in the book, this was perhaps the most vexing. I spent months—literally *months*—trying out various thickeners, emulsifiers, and techniques, using everything from mayonnaise to pure soy lecithin to tapioca to gelatin in an attempt to achieve the perfect tender texture, with an intense cheese flavor.

Nothing seemed to work until I realized that the key might not be in how the cheese is bound to the mac, but the *ratio*. So I decided to increase the amount of cheese. A lot. While a traditional mac

'n' cheese recipe might call for a pound of cheese per pound of pasta—at most a pound and a half—I decided to go with a full 2 pounds. After all, shouldn't great mac 'n' cheese really be all about the cheese? I stuck with the evaporated milk that had worked so well for my Ultra-Gooey Stovetop Mac 'n' Cheese (page 722), but for its creamier, softer baked cousin, a plain old white sauce (albeit one made with a *ton* of cheese along with the evaporated milk and eggs) was the way to go.

## HOW STARCH THICKENS

If you get right down to it, starch is a tiny molecule used by plants to store energy, but, more important, it's a vital tool in the culinary arsenal, used to thicken gravies, add body to stews, and prevent sauces from breaking and turning greasy. There are two basic varieties. *Amylose* resembles a long straight chain, made up of thousands of glucose molecules. *Amylopectin* resembles a small clump of weeds—bushy, with many entangled branches. When dispersed into a liquid, the long amylose molecules have a tendency to get

*continues*

tangled up with one another, sticking together and forming a loose matrix that adds viscosity. Amylopectin will do the same thing, but since it's so compact, it does so less efficiently. The amylose and amylopectin contents of various starches determine their thickening power.

In their raw state, starch molecules are bundled tightly together into granules. In order to activate their thickening power, these granules need to absorb water. As water gets heated, these tiny granules gradually swell up like miniature water balloons until finally, at around 130°F, they burst, spreading individual starch molecules throughout the liquid and thickening it. The starches continue to absorb water and swell further as the liquid is heated toward its boiling point. When adding starch to a liquid, it's essential to heat the sauce to serving temperature in order to gauge its thickening power.

There's one thing to remember: if you add dry starch quickly to hot liquids, it rapidly forms into clumps. The starches on the exterior of these clumps will swell and jell, preventing liquid from reaching the interior. For this reason, it's essential to add your starch in a form whereby the liquid can be absorbed evenly, either by combining it with a small amount of cold water to form a slurry or by combining it with a solid or liquid fat to form a beurre manié.

## Some Common Culinary Starches

There are a number of starches available in the supermarket. Flour and cornstarch are the most common, but potato starch and arrowroot can also be found. This chart shows you their thickening power, along with the best ways to incorporate them.

| STARCH | AMOUNT NEEDED TO THICKEN 1 CUP MILK TO THE CONSISTENCY OF HEAVY CREAM | BEST WAY TO INCORPORATE |
|---|---|---|
| Wheat Flour | 1 tablespoon | Cook in 1 tablespoon butter over medium heat, stirring constantly, until a light blond color is achieved (for certain applications, a darker color may be desired, but the darker flour is cooked, the less thickening power it has), then slowly whisk in the liquid. |
| Cornstarch | 1½ teaspoons | Combine with a small amount of cold liquid until a smooth slurry is formed, then whisk into hot liquid. |
| Arrowroot | 1 teaspoon | Combine with a small amount of cold liquid until a smooth slurry is formed, then whisk into hot liquid. |
| Potato Starch | ½ to ¾ teaspoon | Combine with a small amount of cold liquid until a smooth slurry is formed, then whisk into hot liquid. |

# EXPERIMENT:
## Adding Liquid to a Roux

The rate at which you add a liquid to a flour-and-butter roux can have a drastic effect on the smoothness of the final sauce, as well as its thickening power. To demonstrate this, try the following experiment.

## Materials

- **2 tablespoons butter**
- **2 tablespoons all-purpose flour**
- **2 cups milk**
- **1 pound cheese, grated**

## Procedure

Cook 1 tablespoon of the butter and 1 tablespoon of the flour in a small saucepan over medium heat, stirring constantly, until light golden blond, about 1 minute. Rapidly add 1 cup of the milk, whisking to combine. Allow to come to a boil, stir in half of the cheese, and set aside. Repeat with the remaining butter, flour, and milk, this time very slowly whisking in the milk over the course of 15 seconds. Allow to come to a boil, stir in the remaining cheese, and set aside.

## Results

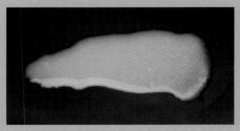

Rapid.

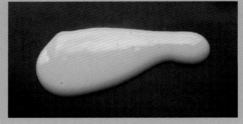

Gradual.

Taste the two sauces and examine the way they look when you put a spoonful on a plate. The sauce for which the milk was added slowly should be far smoother, with a glossy complexion, while the sauce for which the milk was added rapidly will be much thinner, with a grainy, broken appearance.

When adding liquid to a roux, the goal is to disperse the flour as evenly as possible within the liquid. By adding the liquid a bit at a time and incorporating it gradually, you ensure that the individual clumps of flour are broken up with the whisk. If you add the milk too fast, the flour clumps have plenty of space to flow around—many of them end up avoiding the wires of the whisk, resulting in a lumpier, thinner sauce.

So, what if you accidentally add your milk too fast? The simplest solution is to introduce even more vigorous mechanical stirring: a pass or two with an immersion blender or in a standing blender should do the trick, smoothing and thickening the sauce.

# CLASSIC BAKED MACARONI AND CHEESE

**NOTE:** Use a good melting cheese or combination thereof like American, cheddar, Jack, Fontina, young Swiss, Gruyère, Muenster, young provolone, and/or young Gouda, among others (see "Cheese Chart," pages 717–21).

**SERVES 6 TO 8**

1 pound elbow macaroni

Kosher salt

2 slices white bread, crusts removed, and torn into rough chunks

7 tablespoons unsalted butter

2 tablespoons all-purpose flour

One 12-ounce can evaporated whole milk

1½ cups whole milk

1 teaspoon Frank's RedHot or other hot sauce, or to taste

1 teaspoon ground mustard

1½ pounds extra-sharp cheddar cheese, grated

8 ounces American cheese, cut into ½-inch cubes

2 large eggs

1. Adjust an oven rack to the upper-middle position and preheat the oven to 375°F. Place the macaroni in a large bowl and cover with hot salted water by 3 or 4 inches. Let sit at room temperature until tender, about 30 minutes, stirring it after the first 5 minutes to prevent sticking. Drain.

2. While the pasta is soaking, combine the bread and 2 tablespoons of the butter in the bowl of a food processor and season with salt. Pulse until the bread is finely chopped, 10 to 12 short pulses. Set aside.

3. Melt the remaining 5 tablespoons of butter in a large saucepan over medium-high heat. Add the flour and cook, stirring constantly, until light golden blond, about 2 minutes. Whisking constantly, slowly add the evaporated milk, followed by the whole milk. Whisk in the hot sauce and mustard and bring to a simmer over medium-high heat, whisking occasionally to prevent scorching. Reduce heat to low, add the cheese all at once, and whisk until fully melted and smooth, about 3 minutes. Season to taste with salt and more hot sauce if desired.

4. Beat the eggs in a small bowl until homogeneous and frothy. Whisking constantly, pour in 1 cup of the cheese mixture, whisking until combined. Slowly pour the egg mixture into the cheese sauce, whisking constantly. Add the drained macaroni and stir to combine.

5. Transfer the mixture to a buttered 3-quart oval casserole dish or a 9-by 13-inch baking dish. Top with the bread crumbs and cover tightly with foil. Bake for 45 minutes. Remove the foil and bake until the bread crumbs are browned and the sauce is bubbling, about 15 minutes longer. Remove from the oven and allow to cool for 5 minutes, then serve.

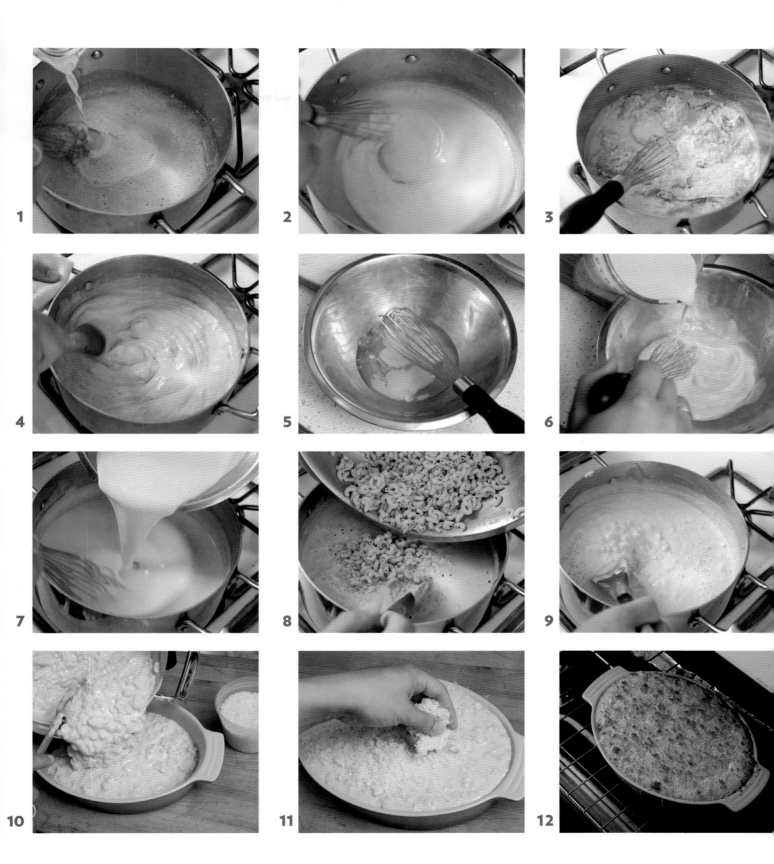

# CLASSIC BAKED ZITI

Now that we know how to make a basic marinara sauce and have learned how easy it is to soak, rather than boil, pasta for a baked casserole, it's just a short skip and a jump to classic baked ziti. The noodles get tossed with a pink mixture of tomato sauce, cream, and ricotta cheese, with a couple of eggs thrown in to lend structure to the casserole as it cooks. I also like to toss cubes of mozzarella cheese together with the pasta to form gooey, stretchy pockets. I top the whole thing with some more marinara, more cubes of mozzarella, and a grating of Parmesan.

This is the dish I make at the annual ski retreat that my friends and I take each year in New England. There are few pasta bakes that are easier to put together yet produce such ridiculously good results, particularly when it's snowing outside and you've got a whole cabinful of friends to feed.

### SERVES 6 TO 8

1 pound ziti, penne, or other thick
    tubular pasta
Double recipe Perfect Easy Red
    Sauce (page 695)
12 ounces whole-milk ricotta cheese
    (homemade, see page 154) or
    high-quality store-bought
3 ounces Parmigiano-Reggiano,
    finely grated (about 1½ cups)
2 large eggs, beaten
1 cup heavy cream
3 tablespoons minced fresh parsley
3 tablespoons minced fresh basil
Kosher salt and freshly ground
    black pepper
1 pound whole-milk mozzarella
    cheese, cut into rough ¼-inch
    cubes

1. Adjust an oven rack to the middle position and preheat the oven to 400°F. Place the ziti in a large bowl and cover with hot salted water by 3 or 4 inches. Let sit at room temperature for 30 minutes, stirring it after the first 5 minutes to prevent sticking. Drain.

2. Pour half of the marinara into a large pot, add the ricotta, half of the Parmigiano, the eggs, the cream, and half of the parsley and basil, and stir to combine. Season to taste with salt and pepper. Add the soaked ziti, along with half of the cheese cubes and stir until well combined. Transfer to a 13- by 9-inch baking dish and top with the remaining marinara sauce and mozzarella.

3. Cover tightly with aluminum foil and bake for 45 minutes. Remove the foil and bake until the cheese is beginning to brown, about 15 minutes longer. Remove from the oven and sprinkle with the remaining Parmigiano, then let cool for 10 minutes. Sprinkle with the remaining parsley and basil and serve.

# THE BEST GARLIC BREAD

There's not much to great garlic bread. As with marinara sauce, the key is to use a mixture of olive oil and butter. The milk solids in the butter help the bread brown evenly. Rather than spreading a butter mixture onto the bread, it's far easier to cook the garlic (along with some oregano and pepper flakes) in the butter and oil in a skillet large enough that you can dip the bread directly into the fragrant mix, giving it an even coating with no fuss.

Luckily for your garlic bread, it takes about 10 minutes to bake, which is exactly the same amount of time it takes a pasta bake to cool. Don't you just love it when things work out so nicely?

### SERVES 8 TO 10

2 loaves soft deli-style Italian bread, about 1 pound each

¼ cup extra-virgin olive oil

8 tablespoons (1 stick) unsalted butter

12 medium cloves garlic, minced or grated on a Microplane (about ¼ cup)

2 teaspoons dried oregano

½ teaspoon red pepper flakes

Kosher salt

2 ounces Parmigiano-Reggiano (optional)

1 tablespoon chopped fresh parsley leaves (optional)

1. Adjust the oven racks to the upper- and lower-middle positions and preheat the oven to 500°F. Split each loaf of bread lengthwise and then crosswise into 8 pieces.

2. Combine the olive oil and butter in a 12-inch skillet and heat over medium-high heat until the butter melts. Add the garlic, oregano, and pepper flakes and cook, stirring until fragrant, about 1 minute. Remove from the heat.

3. Dip the cut side of each piece of bread in the garlic butter and arrange them on two foil-lined rimmed baking sheets. Spoon any extra garlic butter evenly over the bread. Season with salt. If desired, use a Microplane to grate the Parmigiano-Reggiano evenly over the pieces of bread. Place in the oven and bake until toasted and bubbly, about 10 minutes. Sprinkle with fresh parsley, if desired. Serve immediately.

**1**

**2**

**3**

# A BETTER Way
## TO COOK RISOTTO

Being Colombian, my wife loves rice, and being of a diminutive frame prone to coldness, she also loves soup. So it's no wonder that risotto—which can be unpoetically described as soupy rice—lies somewhere between me and cheese sauce on her list of greatest loves.*

As such, I considered it my husbandly duty to discover not just how to make great risotto, but how to do it in the most efficient way possible. Everyone knows risotto as the self-saucing Italian rice dish with the notoriously tedious-to-prepare reputation. It's also often stodgy, thick, and heavy. What is perfect risotto? First off, it should be saucy in texture. A perfect plate of risotto should flow like lava if you tilt the plate. Spoon it onto a hot dish (and you must use a hot dish), and it should slowly spread out until it forms a perfectly level disk. Sticky, tacky, or, worse, gluey are words that should never enter your head when eating it.

If it can stack up into a clump like this . . .

---

\* Since we got married, I've given cheese sauce a run for its money but have yet to overtake it.

Risotto should never clump up like this.

...you've got problems.

Listen: I could give this discussion of risotto the typical sensationalist opening and craft some story about how everybody knows that to make great risotto, you've got to stir it gently and constantly, adding the hot broth to the rice one cup at a time and waiting until it's absorbed before you add the next. I could do that, but it'd be disingenuous. I mean, by this late stage in the game, is there anyone in the world beside hard-line Italians who doesn't know that you can make a bowl of luscious, al dente, perfectly *mantecato* (loose and creamy) risotto without preheating your broth or stirring constantly?

People have been saying and writing about it for years now. I'm strongly convinced that the myth only exists because of Italian grandmothers who used risotto as an excuse to keep an unnecessary kitchen helper occupied for half an hour or to escape from the rest of the family for a while. That said, I still had a ton of risotto questions left unanswered, so I decided to test just about every aspect of risotto I could think of to separate fact from fiction.

Which type of rice is best? How often do you really need to stir? Is toasting the rice necessary? And what about finishing the risotto with cream? So many questions, so many grains of rice, so little time. Let's get right to it, shall we?

## Rice Advice

First question: Which type of rice makes the best risotto?

Rice contains two molecules that make up its starch content, amylase and amylopectin. Generally speaking, rices with a higher proportion of amylopectin to amylase will tend to soften more completely and thicken their sauce more strongly. All risottos start with a short- to medium-grain form of rice high in amylopectin. It's the exact ratio of amylase to amylopectin that determines the final texture of your rice and sauce.

There are dozens of cultivars of short-grained rice used in Italy, but here in the United States, you're likely to see only four types that'll work for risotto:

*From left*: Bomba, Arborio, and Vailone Nano.

- **Bomba** is a Spanish rice used primarily for making creamy paellas. It's extremely short-grained, with a moderate level of amylopectin, and it makes a very fine risotto, despite the fact that it comes from the wrong country.

- **Arborio** is the most common risotto rice of choice. It's short-grained with almost zero amylose. It has a tendency to create a very thick sauce and can very easily be overcooked to the point of mush because of its lack of structure. Even perfectly cooked Arborio will tend to be relatively soft.
- **Carnaroli** and **Vialone Nano** are not quite as available as Arborio, but they are my favorite varieties of rice for risotto. They strike a good balance between creaminess and intact texture. If you can find one of these, use it.

You may see the words *fino* or *superfino* written on the packages of imported rices. While it'd be nice to imagine some Italian committee deciding exactly how fine a particular grain of rice is, it's not an indication of quality or attractiveness: it refers only to the width of the grain. You can mostly ignore these labels.

## The Basics: Adding Broth and Stirring

Basic instructions for old-school risotto: Heat up a large saucepan of stock and keep it at a bare simmer. Toast the rice briefly in butter and/or olive oil, then add a single ladleful of stock (or use wine for this first liquid addition) and stir slowly with a wooden spoon until the stock is absorbed. Add another ladleful and repeat. Continue doing this until the stock has all been absorbed, the rice is plump, and the broth is creamy. Remove from the heat and add cold butter and/or cream and/or Parmesan cheese while stirring vigorously to halt the cooking and add some extra richness and creaminess to the sauce.

This method works all right, but it's crazy-inefficient. First off, there's no need to heat up the broth in a separate pot. Sure, it'll shave a few minutes off the cooking time of the rice, but you add that time back and more in the amount of time it takes to heat up the stock, not to mention washing two pans instead of one. I've made risotto with stock straight from the fridge with no discernible difference in the final product.

What about adding the liquid all at once versus in batches, and stirring the whole time?

There are a couple of arguments in favor of adding in batches and stirring. First, when you add it in smaller batches, the grains of rice are kept in close contact with each other. More starch is rubbed off, and your risotto ends up creamier.

For now, though, we'll ignore this theory (and get back to it later). The second argument is that the technique helps the rice cook more evenly. This one happens to be true. Sort of.

Cook risotto in a standard risotto pot—that is, one that is relatively narrow across the bottom—and your rice and liquids stack. There's a big difference in height between the rice at the bottom of the pot and that at the top. The rice at the bottom, closer to the heat source, overcooks, while the rice at the top barely gets done.

Stirring helps prevent this, but there's an even easier way: use a wider, shallower pan. In a good wide skillet, the rice forms a fairly even thin layer over the bottom, which translates to much more even cooking. Using very low heat after initially bringing the liquid to a boil also helps. I found that by cooking the rice in a wide skillet like this, I could get perfect results by adding the rice and almost all of the liquid at once, covering the pan and cooking over very low heat until the rice was done, stirring just once during the process.

Cooking risotto in a wide skillet lets you get away with minimal stirring.

I could then thin out the sauce to the desired consistency with the remaining liquid, boiling it hard for just a moment to thicken it up properly.

Now, on to bigger, bolder questions.

## To Toast or Not to Toast?

First off: Butter, olive oil, or both? It's largely a matter of personal taste. I like the complexity that you get from using both fats as opposed to just one. There are some folks out there who claim that you add oil to the butter to prevent it from burning when you heat it, since butter starts burning in the low 300°F range, while most oils can get to 400°F or beyond before they start smoking. This is silliness and shouldn't be believed. A mixture of butter and oil will still burn at the same temperature as butter. I know, because I've tried it. It's the milk proteins in the butter that burn, and they don't care whether they're heated in oil or in pure butterfat.

The only reason to combine butter and oil is for flavor, and then you have to be careful not to burn the mixture when you heat it. Adding your rice or aromatics just as it stops foaming is key.

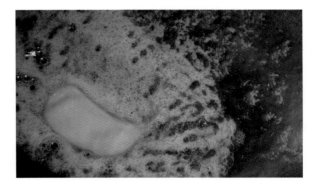

Don't let the butter burn before toasting your rice.

I've always understood that the point of toasting risotto is to help develop flavor. By adding the dried grains of rice to a pan of hot butter and olive oil, you develop some really nice nutty, toasty flavors. But what else is going on when you toast rice?

I cooked up two identical batches of risotto side by side. The first I made with absolutely no toasting. The liquid and rice went into the pan at the exact same time. For the second, I toasted the rice for 3 to 4 minutes before adding the liquid, during which time it acquired a faint golden hue and a nutty aroma.

Here's what I ended up with:

Untoasted rice (*left*) creates a creamier sauce than toasted rice (*right*).

Reserving the excess starch before toasting.

Obviously, there is something else going on while toasting: toasted rice produces a risotto that's noticeably less creamy than one made with untoasted rice. So, on the one hand, you've got great, super creamy rice but little toasted flavor. On the other, you've got rice with great nutty, toasty flavor but relatively little creaminess. The question is: how do you get your risotto both creamy and nutty?

## Isolating Starch

Here's my theory: I know that starch can break down under high heat Ever compare the thickening power of a very light roux to a dark-cooked roux? The blonder it is, the better is thickens. Perhaps a similar thing was happening to the starch in my rice as I toasted it, robbing it of it's thickening power.

To test this theory, I had to first isolate the starch used for thickening from the rest of the grain. Now, some folks claim that the starch that thickens the sauce in a risotto comes from within the rice grains themselves—indeed, they say, that's the very reason you have to stir the rice as it cooks. The jostling movement of the grains causes them to rub against each other, slowly scraping starch off and into the liquid. This could be true, but it doesn't exactly explain how many of the more modern no-stir risotto cooking methods work so well. Is it possible that this thickening starch is simply on the surface of the grains to begin with? There's a very easy way to test whether this is true or not: rinse the rice.

I ran my rice grains under a cold tap in a metal strainer, rubbing them and watching as a starchy, milky white liquid collected underneath. I then cooked them just as I had before (see page 752). What I ended up with was risotto with very little creaminess at all.

Cooking a second batch of rinsed rice according to the traditional method of stirring the entire time produced a risotto that was no more creamy. This confirmed the fact that, indeed, most of the starch that thickens a risotto resides in fine particles on the surface of the rice from the very beginning—stirring and jostling have little to do with its release. Even cooking is the only reason to stir risotto, and this fact handily provides us with the perfect solution for getting great toasted flavor in addition to perfect creaminess: simply remove the starch before toasting, then add it back before hydrating.

To test this, I cooked another batch of risotto, this time first dumping the raw rice into a bowl and pouring my cold broth on top of it. I agitated the rice to release all of the starch, then drained it in a fine-mesh strainer set over a bowl, reserving the starchy, cloudy broth on the side. (It was starchy enough that you could see white starch settling on the bottom.)

I toasted my rinsed rice in a mixture of butter and olive oil until it was just beginning to turn golden brown. Finally, I added the starchy liquid to the pan, brought it to a simmer, lidded it, and cooked it, stirring once in the middle. What I ended up with was pure win: risotto that was both perfectly creamy and nutty. All that was left was to finish it with some cream (I like to whip my cream first to introduce a bit of air into the mix for a lighter risotto) and cheese.

Of course, there are all kinds of flavor variants you can work in here. Vegetables, reconstituted dried mushrooms, fresh mushrooms, meats, saffron, other wines, miso paste—whatever. You've got the foundation, now go build your house. (Pro Tip: go high-low and stir in some nacho cheese sauce for an awesome treat.)

# BASIC ALMOST-NO-STIR RISOTTO

**NOTE:** I prefer Carnaroli rice for its slightly longer grains and firmer texture, but feel free to use any risotto-style rice, like Arborio, Vialone Nano, or even Bomba.

**SERVES 4 TO 6**

1½ cups (about 13½ ounces) risotto-style rice (see Note above)

4 cups homemade or low-sodium canned chicken stock

2 tablespoons unsalted butter, plus more for finishing if desired

2 tablespoons extra-virgin olive oil

2 medium cloves garlic, minced or grated on a Microplane (about 2 teaspoons)

2 small shallots, finely minced (about 2 tablespoons)

1 cup dry white wine (optional—can be replaced with extra broth)

¾ cup heavy cream, whipped to stiff peaks

3 ounces Parmigiano-Reggiano, finely grated (about 1½ cups)

Kosher salt and freshly ground black pepper

Chopped fresh herbs or other garnishes as desired (see the variations and recipes below)

1. Combine the rice and chicken stock in a large bowl. Agitate the rice with your fingers or a whisk to release the starch. Pour into a fine-mesh strainer set over a 2-quart liquid measure or large bowl and allow to drain for 5 minutes, stirring the rice occasionally. Reserve the stock.

2. Heat the butter and oil in a 12-inch heavy-bottomed skillet over medium-high heat until the foaming subsides. Add the rice and cook, stirring and tossing frequently, until all the liquid has evaporated, the fat is bubbling, and the rice has begun to take on a golden blond color and nutty aroma, about 5 minutes. Add the garlic and shallots and cook, stirring, until aromatic, about 1 minute. Add the wine, if using, stir once, and cook until reduced by half, about 5 minutes.

3. Give the reserved stock a good stir and pour all but 1 cup over the rice. Increase the heat to high and bring to a simmer. Stir the rice once, cover, and reduce the heat to the lowest setting. Cook the rice for 10 minutes, undisturbed. Stir once, shake the pan gently to redistribute the rice, cover, and continue cooking until the liquid is mostly absorbed and the rice is tender with just a faint bite, about 10 minutes longer.

4. Remove the lid, add the final cup of stock. Increase the heat to high, and cook, stirring and shaking the rice constantly, until thick and creamy. Off the heat, fold in the heavy cream and cheese. Season to taste with salt and pepper and stir in the herbs or other garnishes as desired. Serve immediately on hot plates.

### ALMOST-NO-STIR RISOTTO WITH CHERRY TOMATOES AND FETA

Stir 2 cups halved cherry tomatoes and 3 ounces feta crumbled, into the risotto just before serving.

### ALMOST-NO-STIR RISOTTO WITH CHORIZO AND BRUSSELS SPROUT LEAVES

Peel the leaves off 12 Brussels sprouts, trimming away and discarding the stem and core as you go (you should have about 2 cups leaves). Cut 3 ounces cured Spanish-style chorizo into ½-inch dice. Cook in a large skillet over medium heat until the fat is rendered and the chorizo is starting to crisp, about 4 minutes. Add the Brussels sprout leaves and cook, stirring, until wilted, about 2 minutes. Stir the chorizo and Brussels sprouts mixture into the risotto just before serving.

Spring Vegetable Risotto.

# SPRING VEGETABLE RISOTTO

As partners in crime, asparagus and risotto give Pinky and The Brain a run for their money in terms of sheer awesomeness. And while we're at it—oh, what the heck—let's grab a few more of my favorite spring vegetables as well.

Let's start with fava beans. The bane of every prep cook's existence, these mild-flavored, bright green beans need to be shucked not once, but twice. After popping them out of their big pods (that's the easy part), they then have to have their individual skins removed. It's not fun. Fortunately, there's an easy way to do it: blanch them first. After a brief boil in water, they not only slip out of their skins with an easy squeeze, but they actually achieve a brighter green color than they would if you blanched them postpeeling. That's a win-win. When shopping for favas, look for whole pods that are firm and snap if you start to bend them. Older fava pods will be spongy and bendy, and they contain older beans, which is not what you want.

Asparagus comes in a few different colors and sizes. You won't actually find much difference in flavor between the fat purple and the green varieties, but I like to mix them anyway because it makes the dish look prettier. White asparagus, on the other hand, does have a different flavor, both delicate and slightly bitter, with a deeper earthiness than its colored counterparts. I blanch my asparagus in the same water I blanch my fava beans in, which I eventually use to cook my risotto as well. That way any flavor that gets blanched out of the vegetables gets added right back to the rice as it cooks. Effectively, it's like making a quick vegetable stock.

Normally I wouldn't blanch zucchini—it's so bland and watery that boiling it renders it completely lifeless. But baby zucchini are more intense in flavor and take well to blanching.

Finally, snap peas are particularly bright and sweet in the spring. Just as with favas, look for whole pods that are stiff and snappy. They won't get any crunchier when they cook.

The only possibly tough part about this recipe is the fancy-pants morel or porcini mushrooms. Fresh morels and porcini are tough to find and, when you do, extraordinarily expensive. Luckily, this recipe is one of the rare cases where dried mushrooms are actually better.

The key to great-flavored risotto is to start with great-flavored liquid. Dried mushrooms offer you the perfect opportunity. Once you've blanched your vegetables, you can use that same flavorful water to rehydrate your mushrooms (the fastest way is to microwave the 'shrooms in the water; heat speeds up the hydration process). The water that comes off the mushrooms when you drain them should be deep, dark brown and intensely flavorful. This translates to deep, dark brown and intensely flavorful risotto.

Other than cooking the rice until it's nice and creamy, all that's left is to sauté your reconstituted mushrooms and stir them and the blanched vegetables into the pan. This risotto is bright and springy but still rib-sticking and filling—perfect for the occasional drizzly day in May.

**SERVES 4 TO 6**

8 ounces asparagus (white, green, or purple or a combination thereof), ends trimmed, stalks cut into 1-inch segments, tips reserved separately

8 ounces sugar snap peas, cut into ½-inch segments on the bias

8 ounces shelled fresh fava beans (still in their skins)

8 ounces baby zucchini, split lengthwise in half

2 ounces dried morel or porcini mushrooms

1½ cups (about 13½ ounces) risotto-style rice (see Note, page 754)

¼ cup extra-virgin olive oil

2 medium cloves garlic, minced or grated on a Microplane (about 2 teaspoons)

2 small shallots, finely minced (about 2 tablespoons)

1 cup dry white wine (optional—can be replaced with extra blanching water)

Kosher salt and freshly ground black pepper

¼ cup fresh parsley leaves, finely chopped

2 teaspoons grated zest and 1 tablespoon juice from 1 lemon

1. Bring 2 quarts lightly salted water to a boil in a medium pot. Prepare an ice bath. Working with one vegetable at a time, blanch the asparagus stalks, asparagus tips, snap peas, fava beans, and zucchini until just tender, 2 to 3 minutes (taste as they cook to confirm doneness). Transfer to the ice bath to stop the cooking, then drain them all and transfer to a bowl. Reserve the blanching water. Carefully peel the skins off the fava beans. Set the vegetables aside.

2. Add the mushrooms to a microwave-safe bowl and cover with 4 cups of the vegetable blanching water (reserve an additional 1 cup water if not using wine, and discard the rest). Microwave on high until just starting to simmer, about 5 minutes. Let steep for 10 minutes, then remove the mushrooms and carefully dry with paper towels. Reserve the mushroom liquid.

3. Combine the rice and mushroom liquid in a large bowl. Agitate the rice with your fingers or a whisk to release the starch. Pour into a fine-mesh strainer set over a 2-quart liquid measure or large bowl and allow to drain for 5 minutes, stirring the rice occasionally. Reserve the liquid.

4. Heat 3 tablespoons of the olive oil in 12-inch heavy-bottomed skillet over medium-high heat until shimmering. Add the rice and cook, stirring and tossing frequently, until all the liquid has evaporated, the oil is bubbling, and the rice has begun to take on a golden blond color and nutty aroma, about 5 minutes. Add the garlic and shallots and cook, stirring, until aromatic, about 1 minute. Add the wine, if using, stir once, and cook until reduced by half, about 5 minutes.

5. Give the reserved broth a good stir and pour all but 1 cup over the rice. Increase the heat to high and bring to a simmer. Stir the rice once, cover, and reduce the heat to the lowest setting. Cook the rice for 10 minutes, undisturbed. Stir once, shake the pan gently to redistribute the rice, cover, and continue cooking until the liquid is mostly absorbed and the rice is tender with just a faint bite, about 10 minutes longer.

6. While the rice is cooking, heat the remaining tablespoon of olive oil in a medium skillet over medium-high heat until shimmering. Add the dried mushrooms and cook, stirring occasionally, until faintly nutty and crisp in bits, about 2 minutes. Season to taste with salt and pepper and transfer to a plate.

7. Remove the lid from the rice, add the final cup of liquid, increase the heat to high, and cook, stirring and shaking the rice constantly, until thick and creamy. Fold in the vegetables, mushrooms, parsley, and lemon zest and juice. Season to taste with salt and pepper. Add water as necessary, stirring just until the risotto is creamy and loose. Serve immediately.

# GREEN RISOTTO
## WITH MUSHROOMS

Often I'll finish my risotto with a bit of whipped cream to lighten it and add some extra creaminess to the sauce, but trying to develop a good vegan version had me experimenting with alternative liquids. What I discovered was that cooked vegetable purees are an ideal liquid for loosening a risotto. Not only do they add sauciness to the dish, but they pack it with flavor as well.

One of my favorites? A bright green risotto finished with a puree of spinach and herbs. As with pesto, in order to get the greens to maintain their ultrabright color, I blanch them, here in vegetable stock, and shock them in ice water before pureeing them. This deactivates the enzymes that hasten the oxidation reactions that can turn chopped greens brown.

Doesn't bright green risotto just look awesome?

You can top the risotto with whatever you wish, but I like using mushrooms sautéed until deeply browned. A splash of lemon juice and soy sauce brightens their flavor.

**SERVES 4 TO 6**

½ cup loosely packed fresh parsley leaves

¼ cup loosely packed fresh tarragon leaves

6 cups homemade or canned vegetable stock or water

2 cups loosely packed spinach leaves

4 scallions, whites finely chopped, greens reserved separately

¼ cup extra-virgin olive oil

1½ cups (about 13½ ounces) risotto-style rice (see Note page 754)

2 medium cloves garlic, minced or grated on a Microplane (about 2 teaspoons)

2 tablespoons canola or vegetable oil

8 ounces mixed wild mushrooms, such as trumpet royale, chanterelle, morel, or oyster, cut in half if large

1 small shallot, minced (about 2 tablespoons)

1 teaspoon lemon juice and 1 teaspoon grated zest (from 1 lemon)

1 teaspoon soy sauce

Kosher salt and freshly ground black pepper

1. Finely chop 1 tablespoon each of the parsley and tarragon. Cover with a damp paper towel and refrigerate.

2. Bring the stock to a boil in a medium saucepan over high heat. Prepare a large ice bath. Add the spinach, the remaining parsley and tarragon leaves, and the scallion greens to the boiling water, pressing down with a wire-mesh spider to submerge them, and cook for 30 seconds, then lift out with the spider, transfer to the ice bath, and chill completely. Remove the stock from the heat.

3. Transfer the blanched greens to the jar of a blender and add ½ cup of the stock. Blend on high speed until completely smooth, about 30 seconds. Transfer to a small bowl and set aside.

4. Heat the olive oil in a 12-inch heavy-bottomed skillet over medium-high heat until shimmering. Add the rice and cook, stirring and tossing frequently, until all the liquid has evaporated, the oil is bubbling, and the rice has begun to take on a pale golden blond color and nutty aroma, about 3 minutes. Add the garlic and scallion whites and to cook, stirring, until aromatic, about 1 minute.

5. Add all but 1½ cups of the stock to the skillet. Stir the rice once, cover, and reduce the heat to the lowest setting. Cook the rice for 10 minutes, undisturbed. Stir once, shake the pan gently to redistribute the rice, cover, and continue cooking until the liquid is mostly absorbed and the rice is tender with just a faint bite, about 10 minutes longer.

6. Meanwhile, heat the canola oil in a 10-inch skillet over medium-high heat until shimmering. Add the mushrooms and cook, tossing occasionally, until well browned, about 5 minutes. Add the shallot and cook, stirring, until aromatic, about 30 seconds. Carefully add ¼ cup of the stock, the lemon juice, and soy sauce. Remove from the heat, toss to combine, and season to taste with salt and pepper. Set aside.

7. Remove the lid from the risotto, add the remaining broth, increase the heat to high, and cook, stirring and shaking the rice constantly, until thick and creamy. Off the heat, stir in the green puree, lemon zest, and chopped parsley and tarragon. Season to taste with salt and pepper. Serve immediately on hot plates, topping the risotto with the mushrooms and their pan juices.

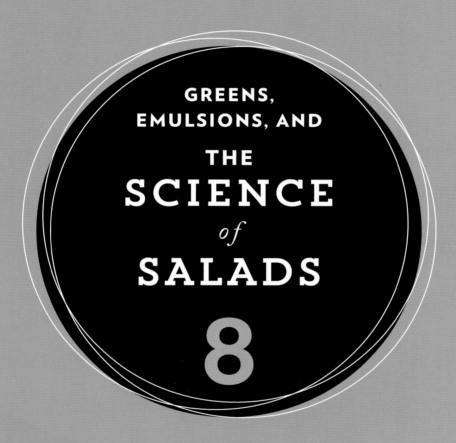

GREENS,
EMULSIONS, AND
THE
SCIENCE
*of*
SALADS
8

"A well-made salad must have a certain uniformity; it should make perfect sense for those ingredients to share a bowl."—Yotam Ottolenghi

# GREENS, EMULSIONS, AND THE
# SCIENCE *of* SALADS

## RECIPES IN THIS CHAPTER

I HAVE SOMETHING TO ADMIT: I'M **ADDICTED** TO SALAD.

I live in constant fear of the night that my wife wakes up in an empty bed and pads slowly out to the kitchen to catch me with a squeeze bottle of vinaigrette in one hand and a bowl of arugula in the other. I try to suppress my need for greens by forcing myself to cook more vegetables, but just as sometimes I don't feel like emptying the dishwasher and some days I just don't feel like talking to my wife during that long seventeen-floor elevator ride, there are days when laziness overwhelms me and I just can't get myself to cook *real* vegetables. Why should I, when that head of lettuce is just sitting there in the vegetable crisper, taunting me, whispering to me, "*I'm eeeeeeasy. Dress me, Kenji. Just reach out and dress me.*"

And then I give in. Who could resist salad's temptations? Who could deny that it's the unchallenged champion of easy, well-balanced meals, able to swoop in at moment's notice to add color, flavor, vibrancy, and all-important fiber to your dinner table? All it takes is some fresh greens and a good dressing (and no, it doesn't have to be store-bought).

But what exactly is a salad? It's these sorts of metaphysical questions that can really keep you preoccupied in the bath, so I'll make it easy. Whether they are mixed greens, vegetables, or meats, whether they are served cold, warm, or hot, there are two things that all salads have in common: they don't require any cutting or knife work at the table and they come with dressing—a sharply flavored mixture that is designed to coat the main ingredients, adding moisture and acidity. At its simplest, a salad can be tossed fresh greens, and from there, salad can go on to become as complex as you'd like, but don't worry—it's really not all that hard.

For those of you who are afraid of dipping your feet into the crazy world of salads, I've designed a six-step program that'll have you developing your *own* salad recipes in no time. The rules are basic and, as with all rules, are meant to be broken. And several of them are optional:

1. **Find the best, freshest greens you can get and treat them with care.** Nothing can ruin a salad like greens that are past their prime. Decide what type of greens you'd like (see "Picking Salad Greens," page 766), trim, wash, and store them carefully, and serve them before they even begin to expire.

2. **Pick a dressing style appropriate to your greens.** Salad dressings can be creamy or thin, mild and delicate, or sharp and pungent. Make sure that the dressing you're using enhances instead of competes with or overwhelms your greens.

3. **Add strongly flavored or aromatic garnishes (optional).** These are ingredients that give interest to the salad by releasing a burst of flavor in your mouth as you eat. My favorites are:

- *Thin shavings of pungent cheeses like Parmigiano-Reggiano, Pecorino Romano, or aged Gouda, or crumbled blue, feta, or goat cheese*
- *Tender herbs like parsley, basil, cilantro, dill, or chives*
- *Dried fruit such as raisins, currants, or cranberries*
- *Pungent vegetables such as raw onions or shallots*

- *Cured meats, like matchsticks of salami, Spanish chorizo, ham, or cooked bacon*
- *Pickled or cured things like olives, capers, or anchovies*

4. **Add "crunchies" for textural contrast (optional).** Well-seasoned croutons (see page 824) are great for this, as are toasted nuts or seeds, like almonds or sunflower seeds. To toast nuts or seeds, spread them on a rimmed baking sheet and pop them into the oven (or toaster oven) for about 10 minutes at 350°F, until they've taken on a bit of color and have an awesome, well, nutty aroma.

5. **Add supporting ingredients like raw or cooked fruits and vegetables or meat and seafood (optional).** Raw vegetables, such as thinly sliced peppers, grape tomatoes split in half, radish wedges, or grated carrots, make great accents for green salads, as do sliced chilled meats (like leftover steak or chicken) or bite-sized pieces of cold seafood (like shrimp, lobster, or squid). Roasted apples or pears are easy additions that can turn a simple salad into a full-on lunch entrée. Vegetables can, of course, completely supplant the greens in a salad, as in some chopped salads, salads of roasted vegetables (see the beet salads on

pages 794 and 796), salads of blanched and chilled vegetables (see Asparagus Salad, page 782), or salads made with cooked white or other dried beans (see page 839).

6. **Dress your salad properly and serve it immediately.** Greens begin to wilt the instant they are dressed. Wait until the last possible second to dress and season them, then toss them as gently as possible to coat. That means using a bowl and tossing with your hands (see "Dressing Salads," page 772).

As you go through these stages of salad development, it's important to keep in mind that more often than not, less is more. Does your salad really need cheese, anchovies, salami, onions, tomatoes, toasted nuts, *and* herbs? Probably not. I've provided a number of recipes for salads in this chapter, but I prefer to think of them more as blueprints—as a means of learning how to design your own salads to suit your own tastes.

# { PICKING SALAD GREENS }

I categorize salad greens into four different basic groups: crisp, peppery, mild, and bitter. In most cases, any member of one of these groups can be substituted for another member. So, for example, you can make a Caesar salad with iceberg lettuce without significantly altering the flavor profile, but you can't make it with hot and peppery arugula or bitter radicchio. Here are the most common lettuce varieties you'll find in these categories.

## Crisp Lettuces

Best served with creamy mayonnaise or dairy-based dressings.

- **Iceberg** got a bad rap in the 1990s when arugula came into vogue and it was suddenly seen as provincial or low-class. Not so. No other lettuce is as crisp or refreshing. While it may not deliver powerful flavor, it maintains its crunch even under duress. I can't think of anything else that can stand up so nicely to blue cheese dressing or a hot hamburger patty. The fact that it keeps for a couple of weeks in the fridge makes it a useful staple to have on hand.

- **Romaine**, also referred to as **Cos lettuce**, is the classic choice for Caesar salad. The pale yellow inner leaves are crisper and sweeter than the outer green leaves, and some people like to discard the darker leaves. It holds up nicely to creamy mayonnaise-based dressings. Closely related is **Little Gem** or **Sucrine** lettuce, a smaller, more tender variety.

- **Green leaf** and **red leaf**, along with other loose-leaf lettuces like **oak leaf, Lollo Rosso, Lollo Bionda,** and **Salad Bowl,** are far more delicate than Romaine or iceberg, with loosely packed leaves that are

Peppery arugula greens in a vinaigrette.

tender around the edges. Most varieties have a very mild flavor. Creamy dressings will work fine, but you'll want serve the salad as soon as possible after dressing, before the leaves turn limp. Mild vinaigrettes also work well.

- **Butter (Boston) lettuce** and its close cousin **Bibb lettuce** are the most tender of all, with large cup-shaped, mildly sweet leaves. As with green-leaf lettuces, you'll want to serve butter or Bibb lettuce as soon as possible after dressing.

### Peppery Greens

Best served with sharp or mild vinaigrettes.

- **Arugula**, sometimes called rocket, is the most widely available peppery green, and it ranges from relatively small, mild, and tender leaves to large, robustly peppery behemoths. **Sylvetta**, a wilder, spicier cousin, is more and more available these days. Arugula goes best with sharp vinaigrettes that won't get overwhelmed by its pepperiness. I buy my arugula prewashed in plastic clamshells so that I never have an excuse not to throw a quick side salad together for dinner.

- **Watercress** is a perennial weedy green well loved for its spicy bite. Its stems are quite hearty, but its leaves wilt relatively quickly after being picked—you should buy watercress no more than a day or two before you intend to use it. Other cress varieties like **garden cress** and **upland cress** can sometimes be found in high-end supermarkets in dirt-filled containers to be snipped and added to salad mixes as desired.
- **Mizuna,** also known as **Japanese mustard** or **spider mustard greens,** has texture similar to arugula but a much milder bite. When mature, it's best used for stir-fries, but the small greens are excellent in salads, dressed with a mild vinaigrette.

## Mild Greens

Best served with a mild vinaigrette.

- **Spinach** is one of my favorite greens to have on hand, as it's excellent either as a salad or quickly sautéed or steamed for a side dish. I prefer the milder, sweeter, more tender flat-leaf spinach (either the baby variety sold in plastic clamshells or the adult flat-leaf sold in bunches) to the tougher, more fibrous curly, which is better for cooking.
- **Tatsoi,** also called spinach mustard, has a mild, cabbage-like pungency faintly reminiscent of bok choy. It has small, round, tender leaves very similar in texture to spinach.
- **Mâche** is the French name for lamb's lettuce. It usually comes in tiny florets of 4 to 5 leaves attached at the roots. It's got a very mild flavor and is delicate, so it should be dressed lightly just prior to serving.

## Bitter Greens

Work well with any flavorful dressing, either creamy or vinaigrette-based.

- **Dandelion greens,** or the very similar Italian *puntarelle*, can range from mildly spicy to more-bitter-than-Mr.-Burns-on-tax-day. It's not always easy to tell, but, in general, paler, more tender leaves will have a milder flavor and larger, feathered, deep green leaves will be too bitter and tough to use in salads.
- **Belgian endive** is watery with a mild bitterness. It's great in chopped salads or served as individual leaves on a crudités platter alongside a bowl of creamy dip or dressing.
- **Curly endive,** also known as **frisée** or **chicory,** comes in small, feathery heads with deep- to pale-green fibrous outer leaves surrounding sweet, tender pale yellow center leaves. Obsessive-compulsive types or those with willing lackeys like to carefully pick away all but the most tender inner leaves. This is a great way to keep overzealous but undertalented helpers busy, though, really, a simple trim of the toughest green leaves will do.
- **Radicchio** resembles a small head of red cabbage. It's got an intensely bitter flavor that can be quite powerful in salads, though it also has an underlying sweetness that cuts through. Its sweetness can be amplified by grilling it or roasting it in a hot oven to caramelize it. One of my favorite salads is cold grilled radicchio with herbs and a simple vinaigrette.
- **Escarole,** or **broad-leaf endive,** vaguely resembles a larger version of curly endive and has a similarly faintly bitter flavor. As with curly endive, the tender pale green or yellow leaves are best. The heartier deep green leaves should be discarded.

# WASHING SALAD GREENS

Aside from careful selection and some basic trimming all you'll have to do with most salad greens is a quick wash to remove any dirt, sand, or bugs. By far the easiest way to do this is with a salad spinner. I like to use a large one—at least one gallon—so that you can prepare enough greens for four people at the same time.

To properly wash greens, remove the top of the salad spinner, leaving the basket in place. Fill it up with cold water, then submerge your greens and swish them around for 10 to 15 seconds. Carefully lift the basket out of the spinner. Any dirt and sand should be left behind in the bottom. Dump it out, and repeat the rinsing steps until the water is completely clear, then spin your greens until completely dry.

Whole head lettuces should be stored intact, but loose leaves should be washed immediately after purchase and stored in either their plastic clamshell container or rolled up in a paper towel and placed inside a plastic bag left slightly open.

## GREEN VEGETABLES IN SALADS

Green vegetables for salads should be blanched in boiling salted water, then shocked in ice water, so that they retain their bright color. Blanching them improves their texture and takes away their raw edge. Make sure to dry them carefully before adding them to a salad—excess water can ruin a balanced vinaigrette.

## FRUITS IN SALADS

Fruits offer many ways to add textural and flavor contrast to a regular green salad (as opposed to a fruit salad). Here are the categories I usually consider:

- **Raw fruits** are best when crisp and slightly acidic, like thinly sliced apples, pears, or young mangoes. Citrus suprêmes (citrus segments that have been cut away from the membranes; see "How to Cut Citrus Suprêmes," below) are delicious in salads.
- **Dried fruits** are quick and easy, providing concentrated bursts of sweetness and flavor in a green salad. I especially like the sweet-tart chew of dried cranberries, but don't overlook raisins, currants, dried apples, apricots, figs, and prunes.
- **Pan-roasted fruits** caramelized in a bit of butter and sugar (see page 789) add richness and complexity to lighter salads. I particularly like the combination of pan-roasted apples and pears with spicy greens. Use compact, crisp fruit, like apples, pears, quinces, or firm stone fruits.

## HOW TO CUT CITRUS SUPRÊMES

There are several reasons to cut your citrus fruits into pith-free segments, or suprêmes.

- The pith is bitter and can ruin the flavor of the fruit. I'm sure many a grapefruit hater would change his mind after tasting sweet pith-free segments the way they were intended.
- The membranes between the segments are papery, get stuck in your teeth, and add nothing to the flavor of the fruit.
- The slices can be incorporated much more attractively into a finished dish. Fruit salads will be tastier. Relishes and vinaigrettes can be eaten without having to pick out bits of membrane.
- Knowing how to do it makes you look way cool.

STEP 1: **REMOVE THE TOP AND BOTTOM OF THE FRUIT** Start by slicing the top and bottom off your fruit, exposing the flesh.

STEP 2: **START REMOVING THE SKIN** Stand the fruit on one of its cut surfaces and carefully cut away the skin with a sharp chef's or santoku knife, following the contours of the fruit to remove all the white pith but as little flesh as possible.

STEP 3: **REPEAT** Continue removing the skin and pith in this manner, working all the way around the fruit.

STEP 4: **TRIM AS NECESSARY** Once all the skin is removed, scan the fruit once more and remove any bits of pith left behind.

STEP 5: **SLICE ALONG THE MEMBRANES** Pick up the fruit, hold it over a bowl, and use your knife to make a single incision along the side of the membrane separating two segments.

STEP 6: **SLICE ALONG THE OTHER SIDE OF THE SEGMENT** Make another cut along the membrane on the other side of one segment. This should release the segment into the waiting bowl.

STEP 7: **REPEAT** Continue cutting along both sides of each segment until all of them have been released.

STEP 8: **JUICE** Squeeze the remaining membranes with your hands, or use a potato ricer, to extract any remaining juice. To store citrus segments, keep them in a sealed container in their juice.

STEP 9: **DRAIN AND USE** Before using the citrus segments, drain away the juice and reserve it for another use, such as mixing into the dressing or drinking alongside the salad. Cut the segments into smaller pieces, if desired

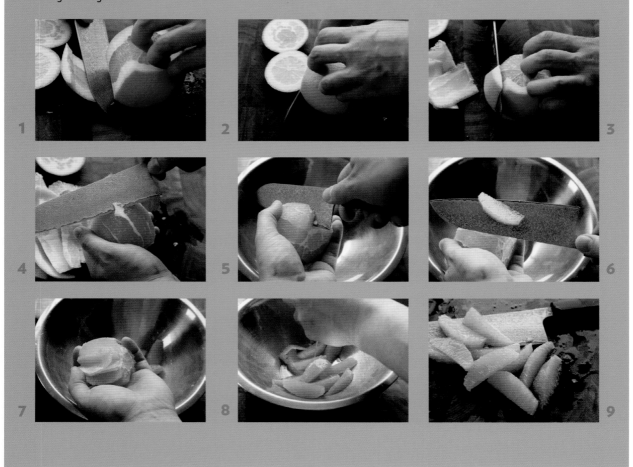

# DRESSING SALADS

A properly dressed salad is beautiful. There are few things that get my goat more than when a restaurant serves you undressed greens with a small dish of dressing on the side. For salads, "on the side" simply does not work, no matter how much of a control freak you are. Drizzling dressing over a salad delivers some leaves that are overloaded and others that have almost no dressing at all. What's the point of making a perfectly balanced, well-emulsified vinaigrette if its balance gets thrown off by poor distribution?

To properly dress a salad, start with a really large bowl—at least three times the volume of the amount of salad you are planning on dressing. Add the greens and less dressing than you think you need (you can always add more), along with a tiny pinch of salt and a few cracks of pepper (even salads should be seasoned properly). Gently toss the salad by scooping it up from underneath with your clean hands (never use harsh tongs on delicate greens), allowing the greens to both rub around the sides of the bowl and cascade down on top of one another. Once everything is thoroughly coated, taste the salad and add more dressing, salt, and/ or pepper if necessary. Now *that* is what a salad is supposed to taste like.

# SALAD DRESSINGS

While fresh leafy greens and vegetables may be the stars of a salad, it's the dressing that makes it. Think of dressing as the vermouth in lettuce's gin martini. It's not necessary, but it sure makes the whole thing go down much more smoothly.

Salad dressing comes in three basic categories:

- **Vinaigrettes** are emulsified mixtures of oil and acid—usually either vinegar or citrus juice—with other flavoring agents.
- **Mayonnaise-based dressings** begin with an emulsion as well, this time aided by egg yolks.

Because egg-yolk emulsions are extremely stable, mayonnaise-based dressings tend to be thicker and creamier.

- **Dairy-based dressings** start with a bacterially thickened dairy product like sour cream, crème fraîche, or buttermilk and add other flavoring agents to it.

When it comes to making any of these dressings, technique rules. Once you've got the basic methods and ratios down, they become infinitely adaptable.

# DRESSING Family #1:
## VINAIGRETTES

For me, the big question about vinaigrettes has never really been "how?" but "why?" Is emulsifying the oil and acid *really* necessary? Can't I just drizzle olive oil and vinegar over my greens, toss 'em in the bowl, and get the same result? *Why* must my vinaigrette be so carefully constructed? To get the answers to these questions, a bit of hard-core kitchen work was in order.

First things first: What exactly is an emulsion? At its most basic, it's what you get when you force two things that don't easily mix to form a homogeneous mixture. In cooking, this most

often occurs with oil and water (and, for all intents and purposes, vinegar or lemon juice can be considered water, as they behave in the same way). You can put them in a container together and stir them up, but eventually, like cats and dogs, they will separate and stick with their own kind. There are a couple of ways around this. The first is to disperse one of the two—the oil, say—into fine-enough droplets that water can completely surround them. Kind of like putting a single cat inside a ring of dogs—there's no way for it to escape and rejoin its feline friends. A common example of this kind of emulsion is homogenized milk, in which whole milk is forced at high pressure through a fine screen, breaking up its fat molecules into individual droplets that are suspended in the watery whey. This is called an oil-in-water emulsion, because the fat molecules are separated and completely surrounded by water molecules. Most familiar culinary emulsions are of this type, the most common exception being butter, which is a water-in-oil emulsion: tiny drops of water are completely suspended in butterfat (of course, once you incorporate that butter into a hollandaise sauce, you've converted it into an oil-in-water emulsion; see page 107 for more on hollandaise).

Simply mixing oil and vinegar forms an extremely unstable emulsion—no matter how thoroughly you mix them, no matter how much you separate the oil molecules, eventually they regroup and your emulsion will break. In order to form a stable emulsion, you need to add an emulsifying agent known as a surfactant.

Remember that cartoon CatDog? The one with the head of a cat on one end and the head of a dog on the other? Well, CatDog is kind of like a surfactant: he's got something that's attractive to both cats *and* dogs, which makes him a kind of feline-canine ambassador, allowing the two to mix together a little more easily. Culinary surfactants are molecules that have one end that is attractive to water (*hydrophilic*) and one that is attractive to oil (*hydrophobic*). Common kitchen surfactants include egg yolks, mustard, and honey, and it's easy to see the work of a surfactant in action.

The container on the left contains oil and balsamic vinegar mixed in a ratio of 3:1. The one on the right has the same ingredients, with the addition of a small amount of Dijon mustard. Both containers were sealed and shaken vigorously until the vinaigrette looked homogeneous. I then allowed them to rest at room temperature for 5 minutes. As you can see, the vinaigrette without the mustard separated much more rapidly than the one with mustard.

At this point, you're probably thinking what I'm thinking: this is all very neat, but what difference does it make to my salad? Good question.

I'd always been under the impression (and I'm not the only one) that a dressed salad eventually wilts because the acid in the vinegar attacks the leaves. To test this theory, I dressed ½ ounce of fresh salad greens with 1 teaspoon distilled white vinegar (5% acetic acid), another ½ ounce with plain water (as a control), and a third ½ ounce with olive oil, then let the leaves sit at room temperature for 10 minutes.

Oil-dressed greens.

Vinegar-dressed greens.

Surprise! Turns out that vinegar was not the culprit after all. The greens dressed with plain oil wilted significantly faster than those dressed with vinegar. In fact, the vinegar-coated greens fared pretty much just as well as those dressed with water!

The truth is that salad greens, like any leaf, spend their time exposed to the elements, and as such, need to be able to protect themselves from the rain. They do this via a thin, waxy cuticle: it's like a little built-in raincoat. But, this oily cuticle makes it very easy for the olive oil to penetrate the spaces between cells, causing damage to the leaf. It's the *oil*, not the vinegar, that causes greens to wilt (a fact that can actually be used to our advantages with certain tough greens like kale—see Marinated Kale Salads, page 825). So, to prevent your salad from turning soggy, you need to figure out a way to protect the leaves from the oil. An oil-in-water emulsion, where the oil is completely surrounded by vinegar molecules, should provide just that kind of protection.

I dressed another batch of salad greens with a shaken mixture of oil and vinegar and took an up-close-and-personal look at the results. Here's what I saw:

That's right. Drops of vinegar suspended above the surface of the leaves by larger drops of oil, like little blobs sitting in beanbag chairs. Lifting these leaves caused a cascade of vinegar to fall off the leaves, and examining the bottom of the bowl confirmed my fears: the oil stuck to the leaves and caused them to wilt, while the vinegar all sank to the bottom. Clearly, I needed a surfactant to keep my oil and vinegar emulsified.

I set up one last experiment, this time dressing two 1-ounce portions of salad greens side by side. The first was dressed with a homogenized mixture of 1 tablespoon olive oil, 1 teaspoon vinegar, and ½ teaspoon Dijon mustard. The second was dressed with just the oil and vinegar. After tossing the greens, I immediately placed each batch inside a funnel set over a small glass to catch any drippings.

Draining salad greens in a funnel allows us to see the difference between a properly emulsified vinaigrette and a poor one.

Almost immediately, the nonmustardy batch started dripping a steady trickle of vinegar into the glass, while the well-emulsified dressing stayed firmly in place. After only 10 minutes, the oil-and-vinegar glass had nearly a full teaspoon of vinegar in the bottom—almost the entire amount that I had put on the greens in the first place—and was starting to drip a few drops of oil as well. The other glass had shed at most a dozen drops.

The results were irrefutable: if you don't emulsify your vinaigrette, you end up with a pile of leaves wilting in oil and a pool of vinegar at the bottom of the salad bowl. An emulsified vinaigrette, on the other hand, uses the power of surfactants to help both the oil *and* the vinegar cling tightly to the leaves, giving you balanced flavor in every mouthful.

## OBSESSIVE-EMULSIVE

What about the ratio of oil to vinegar? I tried various ratios, everywhere from 1:4 to 4:1 oil to vinegar. In the end, the classic French recipe of three parts oil to one part vinegar proved to form the strongest, most stable emulsion with a nice, viscous, leaf-coating consistency. In some cases, I found the amount of vinegar a little too aggressive. But you can easily replace some of the vinegar with water to tone it down—or, if you want to bring a bit of a meaty bite to your salad (I often do), replace part of it with soy sauce.

As for the best emulsifier for the job, mustard is the most common surfactant, and it works best when you have at least 1 teaspoon per tablespoon of vinegar (you can add more if you'd like). Mayonnaise works even better, easily forming a creamy sauce, though it lacks the pleasant tang of mustard. For a sweeter dressing (say, on a beet salad or an asparagus salad), honey also works very well. Try adding honey and toasted crushed nuts to a basic vinaigrette. It rocks in more ways than one.

As for mixing, some advocate slowly whisking in the oil. Others shake it up in a jam jar. Still others insist on the blender. Well, after testing, I found that, not surprisingly, a blender will give you the tightest emulsion, though it can cause your olive oil to turn extremely bitter (see "The Bitter Blend," page 779) while the shake-it-in-a-jar version will be the weakest, lasting for only 30 minutes or so. But the truth of the matter is, your vinaigrette only needs to stay stable for the length of time it takes you to eat a salad.

I put the ingredients for my vinaigrette into a 1-pint squeeze bottle in the fridge and shake it up right before I use it. Or, as is more often the case, I take it out of the fridge and realize that, once again, my wife has finished off all but the last drop, forcing me to make more.

# ALL ABOUT OLIVE OIL

## How to Buy Olive Oil

Asking what olive oil you should buy is similar to asking what knife you should use, what car to drive, or what Beatles album to listen to: it's largely a matter of personal taste. Once you get past a certain base threshold of quality, whether to choose an oil that is buttery and rich or bright and grassy is largely up to you.

Olive oil comes in several different grades:

- **"Virgin" and "Extra-Virgin"** are standards set by the International Olive Oil Council, and they reflect a mark of quality. Virgin olive oils can contain up to 2 percent oleic acid. *Extra*-virgin olive oil contains no more than 0.8 percent oleic acid and it is subjectively deemed superior in flavor to standard virgin olive oil. Neither of these oils can be made from olives that have been heated to extract oil. The extra-virgin production of most countries accounts for between 5 and 10 percent of their total output, hence the relatively high price.
- **"First Cold Press"** indicates that the olive oil was pressed from unheated olives and that the oil came from their first pressing. To a large degree, this label overlaps the extra-virgin label.
- **"Pure" or "Light"** has nothing to do with olive oil's calorie content—it is only an indication of flavor. These olive oils are made from subsequent pressing of olives, or from olives that have been heated to extract more oil. The oil is then refined, so that none of the flavorful compounds found in virgin

*continues*

or extra-virgin oil are present—leaving you with a neutral oil with a high smoke point. Light olive oils tend to be far more expensive than vegetable or canola oils, which will work just as well or better for cooking. Leave these on the shelf.

Extra-virgin is pricey, so how do you find a good one? My advice is to locate a store that will let you try the oils before you buy, and taste at first without paying attention to sticker price. You may well find that the cheapest bottle in the shop suits you just fine. If you have the time, budget, and inclination, it can be fun to collect olive oils from various parts of the world. In the United States, it's now fairly easy to find olive oil imported from Italy, Spain, France, Morocco, and South America, along with our own domestic olive oils (mainly from California). I like to keep a few of my favorite bottles on hand: a sunny and grassy Spanish olive oil from Extremadura, like Merula or Oro San Carlos; a buttery, rich Italian oil, like Columela or Colavita; and a pungent and spicy one from California, like McEvoy Ranch, DaVero, or Séka Hills (you can order all of these online). One thing to note is that there have been reports that many olive oils claiming to be Italian are actually only *bottled* in Italy, the oils being sourced from other Mediterranean countries. I don't let these reports bother me: if I like the flavor of what's in the bottle, that's good enough for me.

If I were marooned in a strange city with only a supermarket in front of me and no chance of tasting before I buy, the brand I'd tend to gravitate toward is Colavita, which has a fine buttery, spicy nose and very little bitterness.

## How to Taste Olive Oil

In Deborah Krasner's fine book *The Flavors of Olive Oil*, she classifies olive oil into four distinct groups. Thinking of these flavors is an exercise that I find useful when shopping around. Her groups are: delicate and mild, fruity and fragrant, olivey and peppery, and leafy green and grassy. To these great descriptors, I'd also add buttery and rich.

When you taste olive oil, start by smelling it, noting its aroma, then place a bit on your tongue. Swirl it around your tongue to coat each part and try to pinpoint what it's doing to each section. Is that a hint of sweetness you detect? Are there bitter notes? How spicy is it? Finally, inhale a bit of oxygen through your mouth and draw it across your tongue to pull the aromas back to your soft palate and up into your nose. You should get an entirely new wave of flavors, which again multiply when you finally swallow the oil. Tasting good olive oil is not unlike tasting good wine.

## How to Store Olive Oil

You wouldn't believe the number of home kitchens I've walked into where the olive oil was stored next to or directly above the stove. In every single case, when I opened the bottle and smelled it, the olive oil was rancid.

As with all fats, the enemies of olive oil are heat, light, and air. When exposed to oxygen, long-chained fatty acids can break down into shorter pieces, lending the oil an off aroma. Heat and light both hasten the process. For the longest shelf life, olive oil should be stored in a dark container (preferably a metal can) in a cool, dark cabinet, as far away from the radiator or oven as possible. If you like to buy your olive oil in bulk, get it in gallon (or larger) cans and transfer some of it to a smaller container for daily use. I use carefully washed-and-dried dark-green wine bottles with small metal pizza-parlor-style olive oil pourers, so I can choose between different flavors depending on my mood.

If you plan on using a very special extra-virgin olive oil only on rare occasions, it's best to store it in the fridge. It may turn cloudy and solidify, but don't worry—it'll return to normal once it warms up to room temperature.

## Cooking with Olive Oil

You often hear that extra-virgin olive oil should only be used for flavoring and finishing, never for cooking. And this is true to a degree—you don't want to heat extra-virgin to the point where it starts to break down and develop bitter flavors. But cooking at relatively low temperatures with it—say, gently sweating onions or vegetables for the base of a sauce or soup—is totally fine, and, indeed, you can taste the difference in the finished product. That said, if you want to save a bit of cash, finishing a dish with extra-virgin oil is definitely the most efficient way to go.

So, **for best flavor**, cook with extra-virgin in gentle situations and finish the dish with some more drizzled on at the end. **For best value**, cook with a more neutral oil (such as canola, vegetable, or light olive oil), saving the extra-virgin until the very end. Never heat extra-virgin to shimmering or smoking temperatures.

## The Bitter Blend

The food processor may seem like the logical choice for making your vinaigrettes and mayonnaises more stable, and using high-quality extra-virgin olive oil seems like a no-brainer as well, but combine the two, and you've got a problem. You see, extra-virgin olive oil droplets are composed of many tiny fat fragments, many of which are bound tightly together, preventing our taste buds from picking them up. But whip the olive oil with enough vigor, by, using a food processor or blender, and you end up shearing those bitter-tasting fragments apart from each other. The result is a vinaigrette or mayonnaise with a markedly bitter taste. Not only that, but these tiny fragments actually *decrease* the efficacy of emulsifiers like mustard or lecithin, making your sauce more likely to break.

So what if you want to have an ultra stable mayonnaise that's strongly flavored with extra-virgin olive oil but has no bitterness? The key is to use a neutral-flavored oil like canola or vegetable to start your mayonnaise in the food processor. Once it's stable, transfer it to a bowl and whisk in some extra-virgin olive oil by hand. You'll get plenty of flavor but none of the bitterness.

# VINAIGRETTE RATIOS

Here's the really awesome thing: now that you know how a vinaigrette works, you'll never have to follow a recipe. As long as you get the ratio of your ingredients right and use the proper technique, you can flavor your vinaigrette any way that you'd like. Here's the most basic recipe for a cup of vinaigrette, in table form. One note: if you want to add herbs to your vinaigrette, it's best to add them just before use—they'll wilt and turn brown if you store the vinaigrette in the fridge.

| ACID (1 part) | EMULSIFIER (⅓ part) | OTHER FLAVORINGS | NEUTRAL OIL (2 to 3 parts) | FLAVORED OIL (up to 1 part; optional) |
|---|---|---|---|---|
| White or red wine vinegar<br>Balsamic vinegar<br>Sherry vinegar<br>Rice vinegar<br>Cider vinegar<br>Lemon juice (or a combination of lemon and another citrus)<br>Verjus<br>Soy sauce (in combination with an acid) | Mustard<br>Mayonnaise<br>Honey<br>Egg yolk | Minced shallots<br>Minced garlic<br>Minced fresh herbs (add just before serving)<br>Toasted and crushed nuts<br>Ground spices<br>Mashed anchovies | Canola<br>Vegetable<br>Grapeseed<br>Safflower | Extra-virgin olive oil<br>Nut oils (walnut, hazelnut, pecan, pistachio, almond)<br>Toasted sesame oil<br>Pumpkin seed oil |

To make any amount of the most basic vinaigrette, combine 1 part acid (or a combination of acid and water as desired), ⅓ part emulsifier, and any other flavorings you might like (the amount will vary by taste, but generally I go with around ¼ to ½ part), 3 parts neutral oil (or substitute up to 1 part of the neutral oil with a flavored oil. Neutral oil can be replaced 100% with extra-virgin olive oil, if desired). Shake everything in a sealed container, season with salt and pepper, and you're ready to get dressed.

# MILD LEMON- OR RED WINE–OLIVE OIL VINAIGRETTE

NOTE: Citrus-based vinaigrettes don't keep as long as vinegar-based ones do—the citrus juice will begin to ferment after about a week or so of refrigeration—so make them in smaller batches. Use this on mild or peppery greens or on simple blanched vegetables.

**MAKES ABOUT ½ CUP**

4 teaspoons lemon juice (from 1 lemon) or red wine vinegar

2 teaspoons water

1 teaspoon Dijon mustard

1 medium clove garlic, minced or grated on a Microplane (about 1 teaspoon)

1 small shallot, finely minced

6 tablespoons extra-virgin olive oil

¼ teaspoon kosher salt

¼ teaspoon freshly ground black pepper

Combine all the ingredients in a small jar or squeeze bottle. Seal it and shake vigorously until emulsified. The vinaigrette will keep in the refrigerator for up to 1 week if made with lemon juice, or up to 6 months if made with vinegar. Shake vigorously before each use.

# KNIFE SKILLS:
## How to Mince a Shallot

**Finely minced shallots are the secret ingredient of fancy restaurants. They go into everything from salad dressings to sautéed vegetables. Here's how you do it.**

**STEP 1: TRIM, SPLIT, AND PEEL**
Trim the non-root end off the shallot, then split it lengthwise. Remove the papery skin. Place the shallot half cut face down on the cutting board.

**STEP 2: MAKE THE VERTICAL CUTS** Holding a shallot half firmly in place and using your knuckles as a guide, make a series of very fine vertical cuts with the tip of a very sharp chef's, santoku, or paring knife, keeping the shallot intact at the root end.

**STEP 3: MAKE THE HORIZONTAL CUTS**
Hold the shallot from the top (never from the sides!) and make one or two horizontal incisions, keeping the shallot intact at the root end.

**STEP 4: MINCE** Make another series of vertical slices at a 90-degree angle to the original series. To chop it even finer, rock the knife back and forth across the minced shallot until reduced to the desired consistency.

# BASIC MIXED GREEN SALAD

**SERVES 4**

12 ounces (about 3 quarts) mixed salad greens, washed and dried

Kosher salt and freshly ground black pepper

½ cup Mild Lemon- or Red Wine– Olive Oil Vinaigrette (page 780), vigorously shaken

Combine the greens with a pinch of salt, a few cracks of pepper, and the vinaigrette in a large bowl and gently toss with clean hands until evenly coated. Serve immediately.

# ASPARAGUS SALAD
## WITH TOASTED ALMONDS AND GOAT CHEESE

**NOTE:** Asparagus skin can sometimes be tough or stringy. I like to peel my asparagus stalks, starting about 2 inches below the tips.

**SERVES 4**

Kosher salt

1½ pounds asparagus, ends trimmed, stalks peeled (see Note above), and cut into 2-inch pieces

½ cup toasted slivered almonds

1 medium shallot, finely sliced (about ¼ cup)

Freshly ground black pepper

½ cup Mild Lemon- or Red Wine– Olive Oil Vinaigrette (page 780), vigorously shaken

4 ounces goat cheese, crumbled

1. Bring a large pot of salted water to a rolling boil. Add the asparagus and cook until bright green and tender but still with a bit of bite, about 3 minutes. Drain in a colander and run under cold water until cool. Drain and dry in a salad spinner.

2. Transfer the asparagus to a large serving bowl and season with salt and pepper. Add the almonds, shallots, and half of the dressing and toss to coat. Sprinkle with the goat cheese and serve immediately, passing extra dressing at the table.

# HOW TO TOAST NUTS

Toasting nuts improves their flavor by adding a layer of complexity and their texture by making them crunchier. There are two ways to do it.

To toast nuts in a skillet, place the nuts in a dry skillet and cook over medium heat, tossing and stirring constantly, until the nuts darken a few shades. The more constantly you stir and flip, the more evenly they'll toast. Transfer to a bowl and allow to cool.

To toast nuts in the oven, spread the nuts on a rimmed baking sheet and bake in a preheated 350°F oven, giving them a stir every few minutes, until they darken a few shades about 10 minutes. Nuts toasted in the oven cook more evenly than those toasted in a skillet.

# SPRING VEGETABLE SALAD

The greatest part of a dish like this is that you can do pretty much everything ahead of time: blanch your vegetables, make the optional puree (see the Note), make the vinaigrette, even poach the eggs, and store them in the fridge. When you're ready to eat, just mix your vegetables (I add a few tender raw pea shoots to the salad as well) and toss them in the vinaigrette until coated. Lay them on top of your puree, add your eggs, and drizzle with a bit more vinaigrette (or just straight-up olive oil), and you're ready to dig in.

NOTES: Feel free to use whatever fresh green vegetables you can find. Young broccoli stalks, Brussels sprouts, fava beans, or fiddleheads would all work fine.

If desired, the asparagus peelings can be blanched until tender, then pureed in a blender with 2 tablespoons water and 1 tablespoon olive oil until smooth and used as an additional sauce for the dish.

½ cup Mild Lemon- or Red Wine–
    Olive Oil Vinaigrette (page 780)

1 teaspoon grated lemon zest plus
    a few dozen strips zest (from 1
    lemon)

1 tablespoon minced fresh parsley

Kosher salt

1 cup shelled fresh peas or
    defrosted frozen peas

2 cups sugar snap peas, strings
    removed, ends trimmed, and cut
    into ½-inch pieces on the bias

1 pound asparagus, ends trimmed,
    peeled, and cut into 2-inch pieces
    (see Note, page 782)

2 cups tender pea or snow pea
    shoots, thick stems removed

Freshly ground black pepper

4 poached eggs (see page 103)

¼ cup chopped mixed fresh herbs,
    such as parsley, tarragon, and
    chives

1. Combine the dressing with the lemon zest and parsley. Set aside.

2. Bring a large pot of salted water to a rolling boil. Prepare an ice bath. Blanch the peas in the boiling water until bright green and just tender, about 1 minute. With a wire-mesh strainer, transfer to the ice bath. Add the snap peas to the boiling water and cook until bright green and just tender, 1 to 1½ minutes. Transfer to the ice bath with the strainer. Add the asparagus to the boiling water and cook until bright green and just tender, about 1 minute. Drain and transfer to the ice bath.

3. Remove all the vegetables from the ice bath, drain, and transfer to a rimmed baking sheet lined with paper towels or a clean kitchen towel to dry.

4. Toss the peas, snow peas, asparagus, and pea shoots with three-quarters of the dressing in a large bowl. Season to taste with salt and pepper. Divide the salad evenly among four bowls. Top each with a poached egg. Spoon the remaining dressing over the eggs and season them with salt. Garnish the salads with the lemon zest strips and herbs, Serve immediately.

# THE RULES OF BLANCHING

There are no set rules for what vegetables to use in spring or summer salads, but there are some basics to bring them together perfectly. Here are the rules I go by when blanching those vegetables. This method will work for many of spring's finest green vegetables, including but not limited to peas, fava beans, asparagus, fiddlehead ferns, snow peas, and snap peas.

## Rule #1: Use a Big Pot with Lots of Water at a Rolling Boil

When you drop a green vegetable into a pot of boiling water, a number of changes occur. First, blanching destroys enough cellular structure to just barely tenderize the vegetable to the point that it has lost

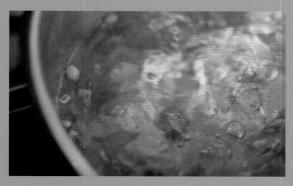

its raw, fibrous edge but still retains crunch. Second, intercellular gases expand and escape from the vegetable (you'll notice small bubbles coming out of, say, your asparagus stalks for a moment or two after dropping them into the hot water). This initial escape of gas is what causes the color of a vegetable to change from pale green to a vibrant bright green—the gas pockets that had been diffusing light suddenly disappear, allowing the full color of the chlorophyll pigment to stand out. At the same time, enzymes that would naturally break green pigments down into brown ones are destroyed.

This is why blanched vegetables appear brighter green and, more important, stay bright green much longer than raw vegetables. Of course, continue cooking too long, and the chlorophyll will eventually break down and your vegetables will go from bright green to a drab olive green or even brown. The goal is to effect those changes as quickly as possible, without allowing time for the chlorophyll to begin breaking down. That's why you want to use plenty of water—it retains its temperature better after you add the vegetables, which subsequently cook faster. ✳

## Rule #2: Blanch Each Vegetable Separately

Asparagus isn't exactly like snap peas. Snow peas are thinner than fiddleheads. Vegetables all take a slightly different amount of time to cook, depending on their size, density, etc. The only way to get all of your vegetables cooked perfectly is to cook them separately, though you can use the same pot and same water, of course. This takes us to . . .

## Rule #3: Cut All the Vegetables the Same Size

Each vegetable should be trimmed to pieces that are all the same basic size and shape so that they

*continues*

✳ Note that retaining temperature better is not the same as saying that it returns to a boil faster. As we found in Chapter 7 (page 676), a larger pot of water will actually return to a boil *more slowly* than a small pot, though it will not drop as much in temperature

will cook evenly. With snap peas, for instance, I remove the strings, cut off the tips, and then slice them on a bias into nice pea-sized pieces that cook quickly and evenly.

For asparagus, I'll often trim off the tips and cook them separately from the stalks, as the tips are so much narrower and more fragile. Fiddleheads can be cooked as is, as can shelled peas or fava beans. If you want to go real hard-core with your peas and favas, blanch them, then peel off the thin skin around each individual pea or fava. It's time-consuming, but you'll end up with pretty results.

### Rule #4: Trust Nothing Except Your Own Senses

When blanching vegetables, do not rely on a timer, do not rely on past experience—trust no one and nothing save your own eyes and mouth. Despite the best efforts of Big Ag, vegetables are still real, living organisms that are naturally diverse. The asparagus you're cooking today is different from the asparagus you cooked last week and will take a slightly different cooking time.

Watch carefully as the vegetables cook. Fish out pieces and taste them often, and as soon as they are ready, remove them with a wire mesh strainer and drop them into your ice bath.

### Rule #5: Shock the Vegetables in Ice Water and Dry Carefully

I've recently been reading conflicting reports on whether or not shocking in ice water is essential. It is, and it's easy enough to prove: Blanch a big ol' pile of peas, take them out, and put them into a bowl without shocking them in ice water. Let them cool. You'll find that the peas at the bottom and center of the pile will be overcooked by the time you dig 'em up.

This is because the reactions that cause a pea to lose its bright green color are not instantaneous. The peas have to be above a certain temperature for a certain amount of time to lose color. A single pea cooling at room temperature will rapidly cool to a safe zone. A pea in the middle of a pile of other really hot peas, however, may stay hot for a good fifteen minutes to half an hour, depending on the size of the pile. That's plenty of time for the pea to lose its color.

Moral of the story: if you are blanching more than one pea at a time, you should shock them in an ice bath, or at the very least spread them onto a large plate or rimmed baking sheet in a single layer to cool.

Then, as soon as the vegetables are chilled, remove them from the ice bath, let them drain, and lay them on paper towels or a clean kitchen towel to dry. The dressing you're going to apply to them sticks better to dry ingredients.

# FINGERLING POTATO SALAD
## WITH CREAMY VINAIGRETTE

I've made simple vinaigrette-dressed potato salads in the past—they tend to work pretty well with the firm low-starch fingerling-style potatoes you find in the spring. I actually like the cleaner, sharper flavors, which really allow the potatoes to shine. But they're never quite as satisfying as a real creamy, mayo-based potato salad texture-wise (like the Classic American Potato Salad on page 817). What to do? Why not use the power of the potatoes themselves to creamify my dressing?

I knew that the starch granules naturally present in a potato could be a powerful natural thickener, adding richness and creaminess to otherwise thin sauces. I initially tried mashing a few of my cooked fingerlings to see if I could get them to form a creamy coating, but it didn't work. Those small, young potatoes are so low in starch and so firm textured that they never really get smooth and creamy unless you go so far as to pass them through a tamis or fine-mesh strainer.

Much easier was to just add a single Yukon Gold potato to the mix and then fish out a few pieces of it after cooking all the potatoes, along with a bit of the starchy cooking liquid. I added the chunks and the liquid to the base for my dressing (a simple vinaigrette made with the vinegar, whole-grain mustard, shallots, and sweet pickle relish), then started smashing. Once the potatoes were relatively smooth, I slowly whisked in some extra-virgin olive oil. I ended up with a semi-loose yet creamy vinaigrette that had the advantage of bright, fresh acidity and a texture that didn't dilute any of the wonderful, subtle flavors of the fingerling potatoes.

Just a few more seasonings and textural elements—sugar, pepper, celery, parsley, capers, and sliced shallots—and my new light fingerling potato salad was born. And what a delicious baby it was.

---

\* Yes, this is really a word. Or it should be.

1½ pounds fingerling potatoes
(such as La Ratte or Russian
Banana), cut into ½-inch disks

1 large Yukon Gold potato (8 ounces),
peeled, quartered lengthwise, and
cut into ½-inch slices

Kosher salt

2½ tablespoons white wine vinegar

1 tablespoon whole-grain mustard

1 tablespoon pickle relish

1 tablespoon sugar, plus more if
desired

2 small shallots, 1 minced (about ¼
cup), 1 thinly sliced (about ¼ cup)

¼ cup extra-virgin olive oil

1 tablespoon capers, rinsed,
drained, and roughly chopped

2 stalks celery, finely diced

¼ cup roughly chopped fresh parsley

Freshly ground black pepper

1. Place the potatoes, 1 tablespoon salt, 1 tablespoon of the vinegar, and 3 cups tepid water in a large saucepan and bring to a boil over high heat, stirring occasionally, until the salt is dissolved. Reduce to a bare simmer and cook until the potatoes are completely tender and show no resistance when poked with a paring knife or cake tester, about 17 minutes. Drain, reserving ½ cup of the cooking liquid. Immediately toss the potatoes with ½ tablespoon of the vinegar in a bowl; set aside.

2. Combine the reserved cooking liquid with the remaining 1 tablespoon vinegar, the mustard, relish, sugar, and minced shallot in a large bowl. Add 5 to 6 pieces of cooked Yukon Gold potato and mash with a potato masher until smooth. Whisking constantly, add 3 tablespoons of the olive oil. Fold in the capers, celery, sliced shallot, parsley, and potatoes. Season to taste with salt, pepper, and more sugar if desired. Serve immediately, drizzling with the remaining tablespoon of olive oil. Or refrigerate, covered, for up to 3 days; allow to come to room temperature before serving.

## SQUEEZE BOTTLES WITH BUILT-IN RECIPES

My wife loves salad dressing, particularly the soy sauce–balsamic vinaigrette that I keep in constant supply in a squeeze bottle in the fridge. I know the recipe by heart, but problems arise when I'm out of town, my wife has a brand-new box of arugula, and the dressing's just run out.

Here's a little trick I devised to make sure that never happens again: I just write the recipe directly on the squeeze bottle. Since good vinaigrettes are all about the ratio of ingredients, it doesn't really matter if you measure them out precisely using measuring spoons and cups. Rather, I draw a line on the side of my squeeze bottle with a permanent marker indicating the proportions of ingredients. All my wife has to do is read the labels from the bottom of the bottle to the top, filling it as she goes along. Voilà! Perfect vinaigrette, no recipe to memorize, no measuring spoons or cups to clean.

I've started a collection of these built-in-recipe bottles, so that I'll always have an easy-to-refill supply of sauces and vinaigrettes on hand.

# ARUGULA AND PEAR SALAD
## WITH PARMIGIANO-REGGIANO AND SHARP BALSAMIC-SOY VINAIGRETTE

**This recipe adds two elements to a basic green salad. I always like to serve a sweet element and a salty element along with peppery greens for contrast. This is one case where a slightly underripe pear is preferable—it keeps its shape better while caramelizing in the butter and sugar mixture.**

**SERVES 4**

2 ripe but firm Bosc pears, halved, cored, and cut into ¼-inch slices

2 tablespoons sugar

1 tablespoon unsalted butter

2 quarts (about 8 ounces) baby arugula, mizuna, or watercress, washed and dried

2 ounces Parmigiano-Reggiano, shaved with a vegetable peeler into slivers

Kosher salt and freshly ground black pepper

½ cup Sharp Balsamic-Soy Vinaigrette; (recipe follows) vigorously shaken

1. Toss the pear slices with the sugar in a medium bowl until evenly coated. Heat the butter in a 12-inch nonstick skillet over medium-high heat until the foaming subsides. Add the pear slices in a single layer and cook, shaking the pan gently, until browned on the first side, about 1 minute. Carefully flip the slices with a thin flexible offset spatula and continue cooking until the second side is browned, about 1 minute longer. Slide the pears onto a large plate and allow to cool for 5 minutes.

2. Combine the pears, arugula, cheese, a pinch of salt, a few cracks of pepper, and the vinaigrette in a large bowl and gently toss with clean hands until evenly coated. Serve immediately.

## Sharp Balsamic-Soy Vinaigrette

**NOTE:** Use on simple salads made with spicy or bitter greens like arugula, watercress, or mizuna or a mesclun mix.

**MAKES ABOUT 1 CUP**

3 tablespoons balsamic vinegar

1 tablespoon soy sauce

4 teaspoons Dijon mustard

1 small shallot, minced or grated on a Microplane (about 1 tablespoon)

1 medium clove garlic, minced or grated on a Microplane (about 1 teaspoon)

½ cup canola oil

¼ cup extra-virgin olive oil

½ teaspoon salt

¼ teaspoon freshly ground black pepper

Combine all the ingredients in a small container or squeeze bottle. Seal the container and shake vigorously until emulsified. The vinaigrette will keep in the refrigerator for up to 3 months; shake vigorously before using.

# TOMATO AND MOZZARELLA SALAD
## WITH SHARP BALSAMIC-SOY VINAIGRETTE

**Salting the tomatoes before adding them to the salad draws out some of their juices, intensifying their meatiness. I like to then add this extracted juice to my vinaigrette, along with some extra-virgin olive oil, to make use of every last drop of flavor.**

**NOTE:** Use only the absolute ripest, peak-of-the-summer tomatoes and fresh mozzarella (preferably mozzarella di bufala, made from water buffalo milk) for this salad.

**SERVES 4**

1 small red onion, finely sliced
   (about ¾ cup; optional)
2 pounds very ripe tomatoes (about
   3 large), cut into 1½- to 2-inch
   chunks
2 teaspoons kosher salt
¼ cup Sharp Balsamic-Soy
   Vinaigrette (page 790),
   vigorously shaken
¼ cup extra-virgin olive oil
1 pound fresh mozzarella cheese,
   preferably mozzarella di bufala,
   cut or torn into 1-inch chunks
1 small bunch basil, leaves removed
   and roughly chopped or torn
   (about ½ chopped leaves)
Freshly ground black pepper

1. If using the onion, place in a medium bowl and cover with cold water. Set aside for 30 minutes.
2. Meanwhile, in a large bowl, toss the tomatoes with the salt. Transfer to a colander or strainer set over a large bowl and set aside for 30 minutes.
3. Combine the vinaigrette with 2 tablespoons of the juices from the tomatoes in a large bowl and whisk to combine (discard any remaining juices). Whisking constantly, slowly add the olive oil in a steady stream to emulsify. Drain the onions, if using, and add to the bowl. Add the tomatoes, cheese, and basil, season with plenty of pepper, and toss to combine. Serve.

**1**

**2**

**3**

**4**

**5**

# GREEN BEAN SALAD
## WITH RED ONION AND HAZELNUT VINAIGRETTE

**Green beans and nuts are a classic French combination. Here I use crisply blanched green beans with a honey-sweetened hazelnut vinaigrette. You can substitute almonds for the hazelnuts if you prefer. Red onions add pungency and freshness—soaking them in cold water removes a bit of their bite.**

### SERVES 4

1 medium red onion, finely sliced
   (about ¾ cup)

Kosher salt

1½ pounds green beans or haricots
   verts, end trimmed

¾ cup Hazelnut Vinaigrette (recipe
   follows)

Freshly ground black pepper

1. Place the onion in a medium bowl and cover with cold water. Set aside for 30 minutes, then drain.

2. Meanwhile, bring a large pot of salted water to a rolling boil. Prepare an ice bath. Add the green beans to the pot and cook until bright green and tender but still with a bit of bite, about 3 minutes. Drain and transfer to the ice bath to cool. Drain again and dry in a salad spinner.

3. Combine the drained onions, green beans, and vinaigrette in a bowl. Season to taste with salt and pepper and toss to combine. Serve immediately.

## Hazelnut Vinaigrette

### MAKES ABOUT 1½ CUPS

2 ounces (about ½ cup) hazelnuts,
   toasted and roughly chopped

3 tablespoons balsamic vinegar

1 tablespoon water

1 tablespoon Dijon mustard

1 tablespoon honey

1 small shallot, minced or grated
   on a Microplane (about 1
   tablespoon)

2 tablespoons minced fresh
   tarragon

½ cup extra-virgin olive oil

¼ cup canola oil

Kosher salt and freshly ground
   black pepper

Combine the hazelnuts, vinegar, water, mustard, honey, shallots, and tarragon in a medium bowl and whisk to combine. Set the bowl over a medium heavy saucepan lined with a dish towel to stabilize it and, whisking constantly, slowly drizzle in the olive and canola oil. The dressing should emulsify and thicken significantly. Season with the salt and pepper. The dressing will keep in a sealed container in the fridge for up to 2 weeks; shake vigorously before using.

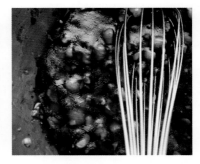

This recipe uses a trio of emulsifying agents—mustard, honey, and nuts—
and winds up extra-tight as a result.

# ROASTED PEAR SALAD
## WITH MIXED BITTER LETTUCES, BLUE CHEESE, POMEGRANATE, AND HAZELNUT VINAIGRETTE

**This is my mother's favorite salad. She asks for it at every holiday meal. I would be a bad son if I didn't include it in this book, since she's been begging for the recipe for years and I've never given it to her. This one's for you, Ma.**

**SERVES 4**

2 slightly underripe Bosc pears, halved, cored, and cut into ¼-inch slices

2 tablespoons sugar

1 tablespoon unsalted butter

2 Belgian endives, bottoms trimmed and separated into individual leaves

2 heads frisée, pale inner yellow leaves only, pulled apart by hand, rinsed and spun dry

3 cups (about 8 ounces) baby arugula leaves, rinsed and spun dry

3 to 4 tablespoons Hazelnut Vinaigrette (page 792)

Kosher salt and freshly ground black pepper

2 ounces Gorgonzola, Stilton, or Cabrales cheese, crumbled

About ½ cup pomegranate seeds (from 1 medium pomegranate)

1. Toss the pear slices with the sugar in a medium bowl until evenly coated. Heat the butter in a 12-inch nonstick skillet over medium-high heat until the foaming subsides. Add the pear slices in a single layer and cook, shaking the pan gently, until browned on the first side, about 1 minute. Carefully flip the slices with a thin, flexible offset spatula and continue cooking until the second side is browned, about 1 minute longer. Slide the pears onto a large plate and allow to cool for 5 minutes.

2. Combine the endive, frisée, arugula, and pears in a large bowl, drizzle with 3 tablespoons of the vinaigrette and season to taste with salt and pepper. Gently toss with clean hands until evenly coated with vinaigrette. Taste and add more vinaigrette, salt, and/or pepper as necessary. Add the cheese and pomegranate seeds and toss briefly. Serve immediately.

## TWO ROASTED BEET SALADS

Beets get their fair share of criticism from children and adults alike, and it's easy to understand why if you, like me, were exposed to the canned variety as a kid. Those are not easy to like. A freshly roasted beet, on the other hand, is something quite different. Sweet as candy, rich and earthy, with a great sorta-soft-sorta-crisp texture, they're one of my favorite vegetables. I make one or another form of beet salad a few times a year, and these two are among my wife's favorites. Just like her, they are pretty, colorful, and best at room temperature.

You can boil beets, but the process will rob them of flavor (notice how pink that water gets?—that's flavor going right down the drain). I've found that the best way to cook them is in the oven, in an airtight foil pouch. They steam as they cook, heating up the air in the pouch, which allows them to cook faster, with minimal moisture loss. Because you're using a dry cooking method, they barely lose any juices or flavor. And the foil pouch is a great way to add aromatics: a few sprigs of thyme or rosemary, some black pepper and olive oil, and perhaps some citrus zest. After roasting, they are extremely easy to peel—their skins slip right off under cool running water. To prevent staining your wooden cutting board, line it with a sheet of plastic wrap before working with beets.

## ROASTED BEET AND CITRUS SALAD
### WITH PINE NUT VINAIGRETTE

**Beets and citrus are a classic combination, and, luckily, they're in season together. This salad combines grapefruit, orange, roasted beets, rosemary, and bit of arugula for some peppery kick (you can use whatever herb or salad green you prefer). I like nuts with my beets, and pine nuts fit the bill just fine. A vinaigrette made with sherry vinegar, shallots, walnut oil, and a touch of agave nectar sweetens the whole thing.**

**SERVES 4**

2 pounds beets, greens and stems removed, scrubbed under cold running water

1 tablespoon extra-virgin olive oil

Kosher salt and freshly ground black pepper

4 sprigs fresh rosemary or thyme

1 recipe Pine Nut Vinaigrette (recipe follows)

1 grapefruit, peeled and cut into segments

1 orange, zest cut into thin strips, fruit cut into suprêmes (see page 770)

1 cup loosely packed arugula leaves, washed and spun dry

1. Adjust an oven rack to the middle position and preheat the oven to 375°F. Fold two 12- by 18-inch squares of heavy duty aluminum foil crosswise in half. Crimp the open left and right edges of each one together to form a tight seal; leave the top open. Toss the beets with the olive oil and season with salt and pepper. Divide evenly between the foil pouches. Add 2 herb sprigs to each pouch, then tightly crimp the tops of the pouches to seal together.

2. Place the pouches on a rimmed baking sheet and place in the oven. Cook until the beets are completely tender—a cake tester or toothpick inserted into a beet through the foil should show no resistance—about 1 hour. Carefully open the pouches and allow the beets to cool for 30 minutes.

3. Peel the beets under cold running water (the skin should slip right off) and pat dry. Cut into rough 1½-inch chunks.

4. Toss the beets with half of the dressing in a large bowl, then transfer to a serving plate. Add the grapefruit, orange, and arugula to the bowl, along with 1 more tablespoon dressing, toss, and season to taste with salt and pepper. Transfer to the serving plate. Drizzle the remaining dressing around the beets, top with the orange zest, and serve.

## Pine Nut Vinaigrette

**MAKES ABOUT ½ CUP**

2 tablespoons sherry vinegar

1 tablespoon agave nectar (or honey)

¼ cup toasted pine nuts

1 small shallot, finely minced (about 1 tablespoon)

¼ cup extra-virgin olive oil

1 tablespoon walnut oil

Kosher salt and freshly ground black pepper

Combine the vinegar, agave nectar, pine nuts, and shallots in a small bowl. Whisking constantly, slowly drizzle in the olive oil, followed by the walnut oil. The dressing should emulsify and thicken significantly. Season to taste with salt and pepper. The dressing will keep in a sealed container in the fridge for up to 2 weeks; shake vigorously before using.

# ROASTED BEET SALAD
## WITH GOAT CHEESE, EGGS, POMEGRANATE, AND MARCONA ALMOND VINAIGRETTE

Beets make me think of honey, and honey makes me think of Marcona almonds, so into the dressing they go, with a handful of pomegranate seeds to give you distinct bursts of sweet juiciness as you work your way through your bowl. Celery leaves are an underused part of this staple vegetable. Let's put 'em to use here. And for some sharp bite, slices of mild white onion. I love the way they turn pale pink when you toss them with the beets.

Just those five ingredients, perfectly dressed, would be enough for a nice balanced side dish, but the point here is a salad you can eat for lunch or dinner. Quarters of hard-boiled egg and a few chunks of creamy goat cheese round out the plate. Eat it fresh, or let it sit overnight and eat it the next day (make sure to add the eggs at the end, though, unless you don't mind pink beet-stained eggs)—either way, it'll be delicious.

**SERVES 4**

2 pounds beets, greens and stems removed, scrubbed under cold running water

1 tablespoon extra-virgin olive oil

Kosher salt and freshly ground black pepper

4 sprigs fresh rosemary or thyme

½ cup pomegranate seeds

2 small white onions, finely sliced (about ½ cup), rinsed in a sieve under warm water for 2 minutes

1 recipe Marcona Almond Vinaigrette (recipe follows)

4 ounces goat cheese, crumbled

2 to 3 hard-boiled eggs, (see page 102), quartered

½ cup leaves from the center of 1 bunch of celery

1. Adjust an oven rack to the middle position and preheat the oven to 375°F. Fold two 12- by 18-inch squares of heavy-duty aluminum foil crosswise in half. Crimp the open left and right edges of each one together to form a tight seal; leave the top open. Toss the beets with the olive oil and season with salt and pepper. Divide evenly between the foil pouches. Add 2 herb sprigs to each pouch, then tightly crimp the tops of the pouches to seal.

2. Place the pouches on a rimmed baking sheet and place in the oven. Cook until the beets are completely tender—a cake tester or toothpick inserted into a beet through the foil should show no resistance—about 1 hour. Carefully open the pouches and allow the beets to cool for 30 minutes.

3. Peel the beets under cold running water (the skin should slip right off). Cut into rough 1½-inch chunks.

4. Toss the beets, pomegranate seeds, onion, and dressing together in a large bowl. Transfer to a serving plate. Garnish with the goat cheese, hard-boiled eggs, and celery leaves. Serve immediately.

## Marcona Almond Vinaigrette

*I like to accentuate the natural dirt-candy sweetness of beets with a lightly sweetened dressing, and honey is the natural choice. It makes a great emulsifier, which means that your oil and vinegar should come together into a nice sauce-like consistency without you having to strain your wrist.*

**NOTE**: Marcona almonds can be found in many specialty food shops. Regular almonds can be used in their place.

**MAKES ABOUT ½ CUP**

2 tablespoons white wine vinegar

1 tablespoon honey

¼ cup toasted Marcona almonds

1 small shallot, finely minced (about 1 tablespoon)

5 tablespoons extra-virgin olive oil

Kosher salt and freshly ground black pepper

Combine the vinegar, honey, almonds, and shallots in a small bowl. Whisking constantly, slowly drizzle in the olive oil. The dressing should emulsify and thicken significantly. Season to taste with salt and pepper. The dressing will keep in a sealed container in the fridge for up to 2 weeks; shake vigorously before using.

# ENDIVE AND CHICORY SALAD
## WITH GRAPEFRUIT, CRANBERRIES, AND FIG AND PUMPKIN SEED VINAIGRETTE

**SERVES 4**

1 head chicory, dark green leaves
    removed and discarded, pale
    white and yellow sections washed,
    spun dry, and torn into 2-inch
    pieces

2 Belgian endives, bottoms
    trimmed, separated into leaves,
    and cut lengthwise into ½-inch-
    wide strips

1 ruby red grapefruit, cut into
    suprêmes (see page 770)

⅓ cup dried cranberries

½ cup Fig and Pumpkin Seed
    Vinaigrette (recipe follows)

Kosher salt and freshly ground
    black pepper

Combine the chicory, endive, grapefruit, cranberries, and dressing in a large bowl and toss to coat. Season to taste with salt and pepper. Serve immediately.

### Fig and Pumpkin Seed Vinaigrette

**NOTE:** Fig preserves can be found in most cheese shops and in the cheese section of many supermarkets. If unavailable, substitute any not-too-sweet fruit preserves, such as orange or grapefruit marmalade, apricot jam, or sour cherry jam.

**MAKES ABOUT 1 CUP**

3 tablespoons balsamic vinegar
1½ tablespoons fig preserves (see Note above)
⅓ cup toasted pumpkin seeds
1 medium shallot, finely minced about 2 tablespoons
½ cup extra-virgin olive oil
Kosher salt and freshly ground black pepper

Combine the vinegar, preserves, pumpkin seeds, and shallots in a small bowl. Whisking constantly, slowly drizzle in the olive oil. The dressing should emulsify and thicken significantly. Season to taste with salt and pepper. The dressing will keep in a sealed container in the fridge for up to 2 weeks; shake vigorously before using.

## KNIFE SKILLS:
## How to Prepare Chicory for Salads

**Bitter greens like chicory and frisée are sweetest and best tasting at their pale yellow core. The dark greens should be discarded or saved for braises and soups, as they can be tough and intensely bitter.**

STEP 1: **TRIM THE BOTTOM.** Trim just as much as you need to to release the leaves.

STEP 2: **SORT BY COLOR.** Find the yellowest bits and locate the point at which they begin to turn dark green.

STEP 3: **TEAR OFF THE GREENS.** Tear off the dark green bits and discard or save to add to soups.

STEP 4: **WASH AND USE.**

# KNIFE SKILLS:
## How to Prepare Endive for Salads

**Bitter endive leaves can be added to salads whole or sliced into slivers.**

**STEP 1: TRIM THE END**

**STEP 2: SEPARATE THE LEAVES**
Only remove the ones that come off easily.

**STEP 3: TRIM AGAIN AND REPEAT** Continue trimming the bottom a little at a time and removing the released leaves until you get to the very core.

**STEP 4: READY TO WASH AND ADD WHOLE** If you want slivers, go on.

**STEP 5: STACK THE LEAVES**
Stacking the leaves in neat piles makes it easier to cut them evenly.

**STEP 6: SLICE** Slice the leaves into slivers of the desired thickness.

**STEP 7: WASH** Place the endive in a salad spinner and wash under cold water.

**STEP 8: SPIN-DRY**

**STEP 9: READY TO USE**

# DRESSING Family #2:
## MAYONNAISE-BASED DRESSINGS

Pivotal, life-changing moments come in all forms. For some, it may be the day they slurped down their first littleneck clam at their father's side on Cape Cod, or perhaps the day they found out that Darth Vader was Luke's father. Maybe it was when they learned that playing outside can actually be *more* fun than watching *He-Man*, or when they discovered that light behaves as both a wave *and* a particle. For me, it was the first time I saw mayonnaise being made.*

When I was a kid, I never once thought about where mayonnaise came from. I mean, it's that kind of creamy, jiggly stuff that comes in a jar with a blue lid, right? I'd always just assumed it came from … some gigantic pump-action mayonnaise dispenser, perhaps in Wisconsin or Nebraska, one of those states that to my preadolescent mind seemed most likely to produce tons of mayo. I remember the very first time I saw mayonnaise being made. It was during a late-night infomercial for handheld immersion blenders (a new technology at the time, and the *It* kitchen gadget). The host put an

egg in the bottom of a cup, poured some oil on top, placed the immersion blender in there, pushed the button, and, within a matter of seconds, the egg and oil came together into creamy, opaque, white mayonnaise.

My wife and I have recently been discussing what we'd like to name our children. She, being South American, wants our firstborn daughter to have the beautiful Spanish name Salomé. I told her that she can name our first daughter Salami as long as I can name my first son Mayonnaise in honor of my favorite condiment. We'll see who gives in first.

As a sandwich spread or sauce, mayonnaise is a big divider. I used to be firmly on the "death before mayo" side of the divide—keep it away from the bread at all costs—but after having been slowly weaned onto it by means of excellent homemade versions, I've come to love it so much that I'll even abide the blue-topped jarred stuff from time to time. At its best, it is creamy, tangy, and light on the palate, with the ability to add richness to a dish without weighing it down. More often than not, though, it's either a poorly made, heavy, greasy, underseasoned goo or overly sweet, gloppy stuff from a jar. OK in a pinch, but hardly something

---

*OK, you win, I confess: these were all pivotal moments in my own life.

you'd want to, say, dip your asparagus into or use as the base of a Caesar salad dressing or tartar sauce.

So what is it that transforms two ingredients—egg yolks and oil—that are kind of icky (that's a technical term) to eat on their own into a luxuriously rich, tangy, creamy spread that's not greasy in the slightest, despite consisting of over 75 percent oil? It's called an *emulsion*, and it's one of the most important concepts to understand in the kitchen. An emulsion is what keeps your vinaigrettes clinging to your lettuce. It's what keeps your cheese smooth and stretchy when it melts (we touched a bit on cheesy emulsions in Chapter 7). An emulsion is, quite literally, what keeps your gravy boat flowing. Let's take a closer look at this fine stuff, shall we?

## Mayo Basics

In its loosest definition, mayonnaise is a flavored emulsion of minute particles of fat suspended in water. The tiny globules of suspended fat have a very difficult time flowing around once they are separated by a thin film of water, which is what gives mayonnaise its viscosity. For the record, small bits of fat will refract light to a much greater than a big pool of fat, which is what gives mayonnaise its opaque-white appearance. Think of it like the windshield of a car. When it's whole, light passes through it easily. But get a few cracks in it, and it becomes difficult to see through. Crack it enough times into small enough pieces, and it becomes opaque. Same thing with the fat in mayonnaise.

Normally when you mix fat molecules with water, no matter how thoroughly you combine them, like MIT nerds at an all-girls-college mixer, they eventually separate themselves and regroup. Because of their shape and electrical charges, fat molecules are mutually attracted to each other while simul-

taneously being repelled by water. This is where egg yolks come in. Egg yolks—which are complex fat and water emulsions in themselves—contain plenty of emulsifiers (agents that aid in getting fat and water to behave), the most important of which is lecithin, a phospholipid found in both the low-density lipoproteins (LDLs) and high density lipo-proteins (HDLs) abundant in eggs. Emulsifiers are long molecules that have a hydrophilic (water-loving, fat-hating) head, and a hydrophobic (water-hating, fat-loving) tail.

Setting your mixing bowl on top of a towel-lined pot will stabilize it during heavy whisking.

When egg yolks, water, and oil are whisked together, the fat-loving heads of the lecithin molecules bury themselves in the minute droplets of fat, leaving only their tails exposed. These tails repulse each other, preventing the fat droplets from coalescing and suddenly making the water seem much more attractive to them—a bit like adding a few kegs of beer to that nerd fest to mix things up a little. To make a traditional mayonnaise, egg yolks, water, salt, and a few flavorings—usually Dijon mustard and lemon juice or vinegar—are whisked together vigorously while the oil is simultaneously slowly drizzled into the mixture (a food processor makes this process nearly foolproof). As the oil falls into the bowl, the rapid action of the whisk quickly breaks it up into tiny droplets, which are kept suspended with the help of the emulsifiers in the egg yolk.

Here's what happens to that mayonnaise in the bowl as you add more and more oil to it:

- **When the oil and water is at a 1:1 ratio,** or one with less oil, there is no possibility of a stable emulsion forming. The fat won't break up and get coated by the water, nor will the water be able to suspend the fat within it. At this stage, your mayonnaise looks like a thin, cloudy liquid.
- **As the oil to water ratio approaches a 3:1,** your mixture begins to resemble a mayonnaise, albeit one that flows more like a vinaigrette. As more and more oil is incorporated into the emulsion, the mayonnaise starts to become opaque, because the tiny drops of oil refract light differently than a liquid mass of oil.
- **As the ratio passes a 5:1,** the mayonnaise begins to get much thicker—thick enough that the peaks will hold when you pull the whisk out of it. It seems counterintuitive: mayonnaise is thick, oil is thin, so adding oil to mayonnaise should make it thinner, right? Wrong. We know that oil droplets in a sea of oil can swim around and float past each other quite freely and that in an emulsion, they are trapped in a tight matrix of droplets separated by water. In order to flow, that water needs to be able to move freely around the system. As you add more and more oil to the mayonnaise, the water separating each droplet of oil gets stretched thinner and thinner, severely limiting its movement. Eventually, if you keep adding oil, the mayonnaise will start to turn from creamy and luscious to pasty and overly thick. Try it now, and it will coat your mouth like candle wax—there

isn't enough water in the emulsion to adequately coat each of the oil droplets, and they end up spilling out and breaking. Your mayonnaise turns greasy.

So, the key to a nice, creamy mayonnaise is to adjust the ratio of oil to water until you get the exact consistency you're looking for. Since I already know that, flavorwise, I like to use about 1 egg per cup (see "How Much Mayo Can I Make from One Egg?" page 806), the rest is just a matter of adding a little water a drop at a time to the finished product until it thins out to the consistency I'm after.

## Foolproofing

The rate at which the oil is added is a key factor in determining whether or not your mayonnaise will be successful. Looking back at the metaphor of a college mixer, imagine that only one or two MIT boys trickle into that girls-college mixer at a time. The emulsifiers have a fairly easier time of separating them from each other and getting them into the mix of things, completely surrounding them with girls. A steady stream of nerds is easy to mix, so long as they trickle into the party slowly. Now imagine the opposite: a whole group of them suddenly shows up at once, clinging tightly to each other. Suddenly it's much harder to get them to mix nicely. Not only that, but any nerds who have already been inserted into the fray will see this large group of nerds who just entered and have a strong desire to join them.

So it is with oil. Trickle it into the egg yolk base slowly, and you can form a strong, stable emulsion. Pour it in too rapidly, and you'll never be able to separate it into droplets small enough, and, what's worse, even if you've already formed a stable emulsion, you run the risk of breaking it. This is the great difficulty when it comes to mayonnaise, and it vexes even the best of cooks.

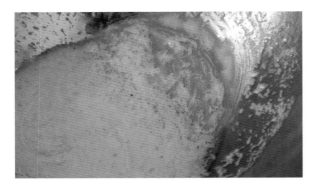

Mayo breaks if you try to add oil too fast.

Mayonnaise is one of my favorite foods. I'm not one of those folks who wakes up in the middle of the night and pulls a jar out of the fridge to eat with a spoon—at least not yet—but I've made a lot of it in my life. Some people swear by the food processor, but the easiest method by far is the very method that I witnessed on that infomercial all those years ago: using an immersion blender. By placing the egg yolks and other flavorings (usually mustard and lemon juice along with a splash of water to lighten the texture) in the bottom of a tall, narrow cup and carefully pouring oil on top, you create two distinct layers: water-based liquid with the fat floating on top of it. If you then slowly plunge the head of the blender to the bottom—into the water part—and flip the switch, you create a vortex that slowly but surely pulls the oil down into it, so the oil is fed into the blended egg yolks in a slow, steady stream. Before your very eyes, you see a creamy mayonnaise forming, starting from the bottom of the container and slowly working its way toward the top. If you don't own an immersion blender, get one, if only for this purpose!

## Foolproof Mayo Without a Hand Blender

OK, you're stubborn and you flat out refuse to buy an immersion blender. What then? Well, you could make your mayonnaise by hand (it's really, really tough), but if you've got a food processor, you're in luck. With enough practice, you can easily make mayonnaise in the food processor by dumping your eggs and flavorings in, then slowly trickling in the oil with the machine running. Problem is, it doesn't always work, particularly for small batches. The egg yolks ride up the sides of the processor bowl, which makes any attempt to form an emulsion with them an exercise in futility. But is there a foolproof way, one that ensures that the oil and the egg yolks all mix up nicely?

As I was scraping down egg yolks from the sides of my processor bowl and pondering this very question, a thought occurred to me: rather than stopping the processor every few seconds to scrape down egg yolks, why didn't I just introduce an element to the bowl that would scrape them down *for* me as the processor was running? And, on top of that, why didn't I make sure that the element I introduced *also* added oil to the mix at a very slow, steady rate? If I could do that, then the mayonnaise should basically make itself once I've added all the ingredients to the bowl, right?

Here's what I was thinking: frozen cubes of oil. By freezing the oil, I transformed it from a liquid into a solid that releases liquid at a slow and steady rate in the processor, all the while bouncing around against the walls to ensure that the egg yolks and flavorings don't stay perched up where the blades of the processor can't reach them.

I tested my theory by mixing up a batch of mayonnaise in the food processor, simply dumping all the other ingredients into it along with a few frozen cubes of oil. I switched on the processor and watched everything jump around. It was a little erratic at first as the oil cubes bounced around their prison, and then slowly but surely, things started to smooth out, and within moments, I had a full-on creamy, tangy, perfect mayonnaise.

The only other thing to think about is flavorings. A basic mayonnaise needs nothing more than some mustard and lemon juice, but more often than not I'll add garlic to the mix (a clove per egg is about right), as well as some extra-virgin olive oil. (Extra-virgin olive oil should *always* be whisked in by hand. Using an immersion blender or food processor will turn it bitter—see "The Bitter Blend," page 779.) Jump ahead to page 808 for some more flavored mayonnaise ideas.

1

2

3

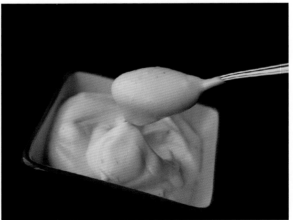

4

# HOW MUCH MAYO CAN I MAKE WITH ONE EGG?

Because lecithin is such a powerful emulsifier, you can create a very large amount of mayonnaise using a single egg yolk. The mayonnaise manufacturers gleaned this fact long ago, which is one of the reasons mayonnaise is so cheap: the most expensive component—the eggs—makes up only a tiny percentage of the finished product. In order to do this without the emulsion breaking, you need to be mindful of the ratio of oil to water. As the mayonnaise becomes thicker and thicker and is on the verge of breaking (just after the "pasty" stage), if you incorporate some water into the mix to reestablish the correct ratio, you can then continue to add more oil. Using this process, I've managed to make over a gallon of mayo with a single egg yolk.

That said, the ideal mayonnaise needs to have a certain amount of egg yolk in it for flavoring purposes—a mostly oil mayo just doesn't taste right. I find that the ideal ratio is a single large egg yolk for each cup of mayonnaise.

# MAYONNAISE VERSUS AIOLI

Any time I dine out at a fancy restaurant and see the chef using the word "aioli" when he or she really means "mayonnaise," I make it a point to inform the waiter, my wife, and perhaps a few of the surrounding tables of the chef's loose lexical morals and the liberties he or she is taking by obfuscating two of the world's great sauces. The word "aioli" comes from the Occitan and is a contraction of *ai* (garlic) and *oli* (oil). A true aioli is made by smashing garlic cloves in a mortar with a pestle, then slowly drizzling in olive oil a drop at a time until a smooth emulsion is formed. It's an intensely spicy, pungent sauce often served with seafood and croutons or boiled potatoes. The Spanish version, *allioli*, is commonly served with olives, grilled meats, or grilled vegetables.

These days, it's perfectly acceptable to call a garlic-flavored mayonnaise made with egg yolks and mustard "aioli," but it must contain *some* garlic. So why do restaurant menus refer to a creamy, emulsified, egg-based sauce as aioli when there's not a hint of garlic in it? It's a matter of public perception. Despite the fact that mayonnaise at one point was considered a staple of *haute cuisine*, its use on menus fell out of fashion because it became associated too strongly with cheap everyday food. I'm a fancy restaurant chef—I can't very well serve the same stuff people are slathering on their sandwiches!

Luckily, this silliness seems to be disappearing, with more and more chefs unafraid to love mayonnaise for what it is: creamy, rich, and delicious. I am certain that my midmeal lectures to waitstaff—despite the dirty looks it earns me from my lovely wife—have played no small role in effecting this change, and I intend to soon move on to ensuring that bruschetta is never again pronounced with a soft "sh" sound.

# FOOLPROOF HOMEMADE MAYONNAISE

**NOTES:** You can whisk in additional lemon juice to taste after the mayonnaise is finished if desired. Make sure to season it pretty aggressively: mayonnaise tastes very flat and greasy without enough salt. This mayonnaise can also be made in a regular blender or in a standing mixer fitted with a whisk attachment.

## MAKES 2 CUPS

2 large egg yolks

2 teaspoons Dijon mustard

1 tablespoon lemon juice (from 1 lemon) or more to taste

1 medium clove garlic, minced or grated on a Microplane (about 1 teaspoon; optional)

About 2 tablespoons water

1 cup canola oil

1 cup extra-virgin olive oil

Kosher salt and freshly ground black pepper

### TO MAKE THE MAYONNAISE WITH AN IMMERSION BLENDER

1. Combine the egg yolks, mustard, lemon juice, garlic (if using), and 1 tablespoon water in a tall, narrow cup just wide enough that the head of the blender fits in the bottom. Carefully pour the canola oil on top. Slowly submerge the head of the blender, reaching the bottom of the cup. Holding the cup flat and steady, turn on the blender. It should create a vortex, slowly pulling the oil down and creating a smooth, creamy mayonnaise.

2. Slowly lift the head until all the oil is incorporated. Scrape the mixture out into a medium bowl set in a heavy saucepan lined with a towel to stabilize it. Whisking constantly, slowly drizzle in the olive oil. Add salt and pepper to taste and whisk to combine. Whisk in up to 1 tablespoon more water, until the desired consistency is reached. The mayonnaise can be stored in a sealed container in the refrigerator for up to 2 weeks.

### TO MAKE THE MAYONNAISE WITH A FOOD PROCESSOR

1. Pour the canola oil into 4 to 6 compartments of an ice cube tray and place in the freezer until fully frozen.

2. Combine the egg yolks, mustard, lemon juice, garlic if using, and 1 tablespoon water in the bowl of a food processor. Add 2 of the frozen oil cubes and run the machine until the large chunks are broken down, about 5 seconds. Remove the lid and scrape down the lid and sides with a rubber spatula. Add the remaining frozen oil cubes and run the machine again until the mayonnaise is smooth, about 5 seconds longer.

3. Transfer the contents to a medium bowl set in a heavy saucepan lined with a towel to stabilize it. Whisking constantly, slowly drizzle in the olive oil. Add salt and pepper to taste and whisk to combine. Whisk in up to 1 tablespoon more water, until the desired consistency is reached. The mayonnaise can be stored in a sealed container in the refrigerator for up to 2 weeks.

# FLAVORED MAYONNAISE

Once you've got the basic process for constructing a mayonnaise down, it opens up the possibilities for a world of flavor variations. Here are a few of my favorites:

| TYPE OF MAYONNAISE | PROCEDURES | BEST USES |
| --- | --- | --- |
| **Garlic Mayonnaise** (see "Mayonnaise Versus Aioli" page 806 | Combine 1 cup mayonnaise, and 2 to 4 cloves garlic minced or grated on a Microplane, in a food processor and pulse to combine. | With simple green vegetables like boiled green beans or asparagus<br>On burgers<br>With roasted or boiled potatoes<br>Stirred into soups<br>With grilled or seared chicken or fish<br>In sandwiches |
| **Roasted Red Pepper Mayonnaise** | Thoroughly drain and dry ½ cup roughly chopped jarred roasted red peppers. Add to a food processor, along with 1 cup mayonnaise and 2 cloves minced or grated garlic, and process until smooth. | With grilled or seared chicken or fish<br>In sandwiches |
| **Caesar Salad Dressing** | When making the mayonnaise in a food processor, add 4 anchovy fillets, 2 cloves minced or grated garlic, 2 teaspoons Worcestershire sauce, and 2 ounces Parmigiano-Reggiano finely grated. | As a dip for raw vegetables<br>With cold chicken or other meats<br>In sandwiches and wraps |
| **Garlic-Herb Mayonnaise** | Combine 1 cup mayonnaise, 2 cloves minced or grated garlic, and ¼ cup mixed fresh tender herb leaves, such as parsley, tarragon, chervil, dill, and/or basil in a food processor and process until the herbs are finely chopped, then add 2 tablespoons thinly sliced fresh chives and pulse briefly to combine. | As a dip for raw vegetables<br>In sandwiches and wraps<br>With sausages and lamb<br>In cold seafood salads, like shrimp, lobster, or crab |
| **Horseradish Mayonnaise** | Add ¼ cup drained prepared horseradish and 1 tablespoon Dijon mustard to 1 cup mayonnaise and whisk to combine. | With cold leftover roasted meat<br>On burgers<br>In roast beef or roast lamb sandwiches<br>In potato salad |

| TYPE OF MAYONNAISE | PROCEDURES | BEST USES |
|---|---|---|
| Chipotle-Lime Mayonnaise | Combine 1 cup mayonnaise, 1 tablespoon fresh lime juice, and 2 chipotle chiles packed in adobo sauce, along with 2 tablespoons of the adobo sauce to a food processor and process to combine. Mix in 2 tablespoons minced fresh cilantro if desired. | On burgers<br>In tacos<br>In roasted meat sandwiches<br>As a dip for fried foods, like fish, French fries, or onion rings |
| Tartar Sauce | Combine 1 cup mayonnaise, 3 tablespoons chopped rinsed capers, 1 medium shallot, finely diced, 2 tablespoons minced cornichon pickles (or 2 tablespoons sweet pickle relish), 1 teaspoon sugar, and a couple of tablespoons of chopped fresh parsley in a medium bowl. Stir to combine and season to taste with pepper. | As a dip for fried fish and other seafood |
| Bacon Mayonnaise | Replace ¼ cup of the canola oil in the basic mayonnaise recipe with ¼ cup rendered bacon fat. Once the mayonnaise is formed, add 4 strips cooked bacon, crumbled, and blend or process to combine. Stir in 2 scallions, finely sliced. | On burgers<br>In sandwiches |
| Sun-Dried Tomato Mayonnaise | Combine 1 cup mayonnaise, 2 cloves minced or grated garlic, and ½ cup drained sun-dried tomatoes in a food processor and process until nearly smooth. Blend in 2 tablespoons minced fresh parsley if desired. | As a dip for raw vegetables<br>In sandwiches and wraps |
| Spicy Garlic-Chili Mayonnaise | Whisk together 1 cup mayonnaise, 2 cloves minced or grated garlic, and 3 tablespoons of your favorite Asian chile sauce, such as gochujang, Chinese chile-garlic sauce, sambal oelek, or Sriracha. | On burgers<br>As a dip for fried fish and other seafood<br>With grilled meats or seafood |
| Honey-Miso Mayonnaise | Whisk together ¾ cup mayonnaise, ¼ cup white miso paste, 2 teaspoons rice wine vinegar, and 2 tablespoons honey in a bowl until smooth. | As a dip for fried fish and other seafood |

# WINTER GREENS SALAD
## WITH WALNUTS, APPLES, AND PARMESAN-ANCHOVY DRESSING

Ever since I tasted April Bloomfield's awesome fall green salad at **The John Dory**, that combination of crisp, bitter greens and savory anchovy dressing has been one of my favorites.

Winter greens by their very nature are hearty in both texture and flavor. Radicchio, with its dark purple, frilly, cabbage-like leaves, is the bitterest of the lot, so I like to cut it with sweeter greens like Belgian endive. The latter are grown completely underground to induce a process called *etiolation*, a natural occurrence in plants that grow in low-light conditions. In their effort to reach the light, rapid growth takes place, resulting in weaker cell structure and no chlorophyll formation. This is good for us when more tender leaves and a less bitter flavor are what we're after. Tight pale yellow or pure white endives are what to look for.

Similarly, frisée (or curly endive) is sweetest at its core, which was protected from sunlight, so the small leafy stems are still pale yellow and tender. For the best-tasting frisée salads, discard the tougher dark green outer leaves (or save them for soup) and use just the pale green and yellow centers.

The dressing is a heavy-on-the-anchovy variation of a classic Caesar dressing: mayo-based, with lemon juice and Worcestershire sauce adding acidity and bright depth. With a few slivers of apples for sweetness and some crunchy toasted walnuts, it is a simple salad that manages to hit you with enough levels of flavor and texture to serve as a full-on meal.

## SERVES 4

½ cup mayonnaise, preferably homemade (page 807)

1 ounce Parmigiano-Reggiano, finely grated (about ½ cup)

6 anchovy fillets, mashed into a paste with the back of a fork

2 teaspoons lemon juice (from 1 lemon)

1 teaspoon Worcestershire sauce

Kosher salt and freshly ground black pepper

2 Belgian endives, core removed and leaves cut into ⅛-inch slivers

1 head radicchio, cored and finely sliced

4 cups loosely packed pale green and yellow frisée (curly endive) fronds (from about 2 heads)

1 large tart apple, such as Fuji or Granny Smith, cored and cut into ⅛-inch-thick matchsticks

¼ cup chopped fresh parsley

2 cups toasted walnuts

1. Combine the mayonnaise, Parmesan, mashed anchovies, lemon juice, and Worcestershire sauce in a small bowl and whisk together. Season to taste with salt and pepper.

2. Toss the endive, radicchio, frisée, apples, and parsley in a large bowl with dressing to taste. Season to taste with salt and pepper. And the nuts and toss briefly. Serve immediately.

# KNIFE SKILLS:
## How to Prepare Radicchio for Salads

**STEP 1: CUT IT IN HALF** Split the radicchio along its vertical axis.

**STEP 2: START CUTTING OUT THE CORE** Cut around the core in one half, angling the knife blade in toward the center of the head.

**STEP 3: REMOVE THE CORE** Cut around the other side at a 90-degree angle to the first to remove the core in a single wedge-shaped piece. Repeat with the other half.

**STEP 4: SLICE THE RADICCHIO**

**STEP 5: READY TO DRESS**

1  2  3  4  5

# CLASSIC AMERICAN POTATO SALAD

**Potato salad—big deal, right?**

It's kinda like background music in a restaurant—something to keep you and your fellow diners distracted and occupied during the awkward silences before the main course arrives. You put an obligatory spoonful on your paper plate and poke at it with a plastic fork until the burgers are done. At least that's what **most** potato salad is. The problem is, it's such a simple dish that it's often made without thought. Boil the potatoes, toss them with some mayo, add a few dollops of whatever condiment catches your fancy, and throw it into a bowl.

But a well-constructed potato salad can be as interesting as the burger it precedes (and you know by now how I feel about burgers). Tangy, salty, and sweet, with a texture that's simultaneously creamy, crunchy, and fluffy, a perfect potato salad should taste featherlight, despite being made with potatoes and mayo, two of the heaviest ingredients around.

The way I see it, there are three things that can go wrong with a potato salad. Screw up any one of these, and you're quickly going south:

- **The potatoes are not flavored all the way to the center.** In a good potato salad, the pieces of potato should be seasoned all the way through. Their hearty, earthy flavor does fine on its own or with a bit of salt when hot—but when cold, it comes across as heavy and bland. Without plenty of acid and a bit of sweetness to brighten it up, potato salad is dead in the water.

- **The potatoes are under- or overcooked.** If there's one thing I can't stand, it's al dente potatoes. Potatoes should *not* be crunchy or firm. Nor do you want your potato salad to be cold mashed potatoes. The perfect piece of potato should be tender and fluffy all the way through, with the edges just barely beginning to break down, adding a bit of potato flavor and body to the dressing.

- **The salad is underseasoned.** Foods that are served cold need to be seasoned more aggressively than foods that are served hot—our taste buds are less receptive at colder temperatures. Combine this with the heaviness of potatoes, and it makes sense that a potato salad needs to have more vinegar, sugar, spice, and salt than most other dishes. But balance is key. All the elements need to come together, instead of competing.

## Hot and Cold

The first step is getting the texture just right. Potatoes are made up of a series of cells that contain starch granules. These cells are glued together with pectin. As the potatoes cook, the pectin slowly breaks down and the starch granules start absorbing water. When you overcook them, the first thing that happens is the pectin breaks down too much. The potato cells start falling away, and the whole thing turns mushy. Welcome to cold-mashed-potato city, population: you. Overcook them even more, and the starch granules will swell so much that they'll begin to burst, turning a mildly offensive bowl of cold mashed potato salad into an outright disrespectful bowl of gluey, inedible goo. Undercook potatoes, on the other hand, and they remain crunchy, and crunchy potatoes are grounds for immediate ejection from the backyard.

It gets even more complex: since potatoes heat up from the exterior toward the center, it's possible to have a potato that's simultaneously overcooked *and* undercooked. The best way to accomplish this feat of culinary indecency is to drop your cut potatoes into a pot of already-boiling water. When potatoes start in hot water, the outside will rapidly overcook before the center has even got the chill off it. Make a salad with these, and you'll have crunchy nuggets of uncooked potato swimming in a sea of gluey mash. No thanks. If you start the potatoes in cold water, the potatoes heat up evenly right along with the water, so that by the time they are perfectly cooked in the center, the edges have just barely startied to break apart—which is not a bad thing. I like a little

bit of broken-up potato to thicken and flavor the dressing. This, by the way, is the best method *any* time you boil potatoes.

Of course, even with a cold start, one problem remains: potatoes require constant vigilance—they can go from undercooked to overcooked in an instant. Get distracted for just a minute (say, to chase after your puppy who's just peed on your backpack, then hidden your notebook under the couch), and you've got gluey potatoes on your hands. There is a way to solve this problem, which we'll get back to, but first . . .

## 'Tis the Season

Using russet potatoes is key for this style of potato salad. Not only do they absorb dressing more easily, but their starches actually help to bind the salad better. To prove that they do absorb seasoning better, I boiled russet and red potatoes, cut them into cubes, and tossed them with a dressing tinted with a bit of green food coloring. You can see for yourself which absorbed it better:

For some time, I'd thought that it's better to season your potatoes when they are hot, but I'd never actually figured out *why*. Do they actually absorb more seasoning, or could it just be a psychosomatic effect? To find out, I cooked three separate batches of potatoes, this time using green food coloring as a stand-in for the salt and vinegar. The first batch I cooked in green-colored water. The second batch I cooked in plain water, then "seasoned" with green-colored water after draining them, while they were still hot. I allowed the last batch to cool completely before tossing with green water. After all the batches cooled completely, I cut a cube from each one in half to see how far the food coloring had penetrated.

The potato that was cooked in seasoned water and the one that was "seasoned" while still hot were a light shade of green all the way to their centers, while the potato that was seasoned after cooling was mostly pale in the center, with a single green streak where a natural fault in the potato existed.

The reason for this is twofold. First, the cooked starch on the surface of the potatoes hardens and gelatinizes as it cools, making it harder for anything to penetrate. Second, as the potatoes cool, they contract and tighten up slightly, making it harder for any seasoning to work its way into the center, even if it manages to make it past the gelatinized starch sheath on the exterior.

As you will have noticed, whether the potatoes were cooked in seasoned water or seasoned imme-diately after coming out of plain water made very little difference in terms of flavor penetration. So, you might as well just add the vinegar to the potatoes after they're cooked, right? But there's actually a very good reason to add a little vinegar to the potatoes cooking water: it helps prevent them from overcooking, something we'll see again when exploring French fries in Chapter 9 (page 904). Pectin breaks down much more slowly in acidic environments. I found that adding a tablespoon of vinegar per quart of water to my potato pot prevented the potatoes from becoming mushy, even when slightly overcooked.

Potatoes boiled in plain water break down.

Potatoes boiled in water with vinegar retain their shape.

I make my dressing while my boiled potatoes are cooling.

## Balancing Act

Once the potatoes were perfectly cooked, light, and bright, the rest was simple: balancing flavors. Nothing too hard-core nerdy here. Rice wine vinegar is my favorite all-purpose vinegar, and it works well in potato salad. Two tablespoons in the cooking water, another to dress the hot potatoes, and a final two in the mayonnaise mixture added plenty of layered brightness. Mayonnaise—be it store-bought or homemade (page 807)—is a must. A cup and and a quarter is less than average for 4 pounds of potatoes, but I like to keep the mayo a little light. By stirring the salad vigorously, you can bash off the corners of the potatoes, which get mashed up and increase the ratio of creamy dressing to tender potato chunks. For heat, I added a few tablespoons of whole-grain mustard (Dijon or even yellow would work fine if you prefer).

Pickles are a point of contention in potato salad. I like to use chopped cornichons in mine, mostly because that's the type of pickle I usually have in my fridge. Chopped dills, bread-and-butters, or even a couple scoops of pickle relish also work just fine. Chopped celery and red onions add necessary crunch to the mix. To be honest, though, once the potatoes are properly cooked and seasoned, the dressing itself is very much a matter of personal taste. I like sugar in mine, others don't. Then again, whether or not it needs black pepper is not a matter of taste—put in the pepper.

There are few dishes much humbler than potato salad, but if you want to gussy it up a bit, you could do worse than to add a handful of chopped fresh herbs. Parsley and chives work great. I sometimes add scallion greens, because they seem to spontaneously generate in my vegetable drawer. If you've saved your celery leaves, you can go fancy by using them. Now I know that there are those who like to add pickle juice. Those who like to add garlic. Those who add sour cream. Really, all those things could be great, and as far as flavorings go, there's no right way to make a potato salad. The keys to remember are:

- Use russet potatoes.
- Cut them evenly and start them in cold water, seasoned with salt, sugar, and vinegar (1 tablespoon of each per quart of water).
- Season the potatoes again with vinegar as soon as they come out of the water.
- Use bold flavors, because cold food tastes bland without them (see "Cold Confusion," page 816).

With these four simple tips in mind, you're free to do whatever the heck you like with your potato salads. Well, whatever you want within the boundaries of the law.

# EXPERIMENT:
## Cold Confusion

Our taste buds are extremely sensitive to the temperature of foods being served. How many times have you eaten cold pizza out of the fridge the next morning and thought to yourself, "How could this cold, clammy, bland thing be the remains of the well-seasoned flavor bomb I was eating last night?" OK, admittedly, on the mornings I'm eating cold pizza, it's usually because I'm too hungover to care. But to prove to yourself that seasoning *is* temperature-dependent, try this little experiment yourself.

### Materials

- **3 pounds carrots, peeled and cut into 1-inch chunks**
- **3 tablespoons butter**
- **Kosher salt**
- **Blender**

### Procedure

1. Put the carrots in a large saucepan and cover them in cold water (do not add salt). Bring to a boil over high heat and cook until the carrots are very tender.

2. Drain the carrots, reserving about 2 cups of the liquid. Place the carrots and butter in a blender, add the liquid, and puree (make sure to start out slow and gradually increase the speed to avoid a blowout). You should have about 4 cups carrot puree.

3. Divide the puree into 4 even parts. Mix ½ teaspoon salt into the first batch, 1 teaspoon into the second, 1½ teaspoons into the third, and 2 teaspoons salt into the final batch. Have a panel of tasters taste each one and write down which they feel is the best seasoned (not the saltiest, but the best according to their own palate).

4. Refrigerate the puree overnight and repeat the tasting when the puree is cold.

### Results

If your friends are anything like mine, then they will have on average chosen a saltier batch when the puree was served cold than they did when it was served hot. This is because hot foods stimulate your taste buds (not to mention create aromatic vapors) more easily than cold foods do. Cold foods need to be seasoned more aggressively than hot foods. So when seasoning your food, always make sure to taste it *at the temperature it is going to be served.*

# CLASSIC AMERICAN POTATO SALAD
## DONE RIGHT

### SERVES 8 TO 10

4 pounds russet (baking) potatoes, peeled and cut into ¾-inch cubes

Kosher salt

¼ cup sugar

6 tablespoons rice wine vinegar

3 stalks celery, finely diced (about 1 cup)

1 medium red onion, finely diced (about ½ cup)

4 scallions, green parts only, thinly sliced (about ½ cup; optional)

¼ cup fresh parsley leaves, minced (optional)

¼ cup chopped cornichons

2 tablespoons whole-grain mustard, or to taste

1¼ cups mayonnaise, preferably homemade (page 807)

Freshly ground black pepper

1. Put 2 quarts water in a large saucepan, add the potatoes, 2 tablespoons salt, 2 tablespoons of the sugar, and 2 tablespoons vinegar, and bring to a boil over high heat. Reduce to a simmer and cook, stirring occasionally, until the potatoes are completely tender, about 10 minutes. Drain the potatoes and transfer them to a rimmed baking sheet. Spread into an even layer, then sprinkle with 2 more tablespoons vinegar. Allow to cool to room temperature, about 30 minutes.

2. Combine the remaining 2 tablespoons sugar, the remaining 2 tablespoons vinegar, the celery, onion, scallions, if using, parsley, if using, pickles, mustard, and mayonnaise in a large bowl. Stir with a rubber spatula to combine. Fold in the potatoes. Season to taste with salt and pepper. Cover and let rest in the refrigerator for at least 1 hour, and up to 3 days, before serving.

# KNIFE SKILLS:
## How to Cut Celery

**I had a friend in college who believed that celery was the worst vegetable, and I admit it—as sticks, it isn't all that exciting.**

Add some blue cheese or Green Goddess dressing, and I'll happily eat them, but it's really only when you start adding celery to other dishes that it reveals its true purpose: Best Supporting Role.

My celery-hating friend enjoyed eating at restaurants, and many of them were undoubtedly flavoring any number of sauces, stews, salads, soups, and braises with the vegetable. Combined with pungent onions and sweet, earthy carrots, celery, with its slight bitter edge, forms the backbone of half the dishes in the Western repertoire. Potato salad or a lobster roll wouldn't be the same without its distinctive crunch and fresh flavor, and the Chinese learned long ago that celery is particularly good in a spicy stir-fry. Even the leaves can be used as a flavorful garnish.

This guide will help teach you to cut celery into all of the major shapes and sizes.

### Shopping and Storage

When buying celery, look for heads with tightly bundled stalks still attached at the root and a bright green to yellowish-green color. Skip any that have bruised brown spots or look overly fibrous. Avoid those in sealed packs, which can often hide blemishes. A good grocery store will keep its celery stalks lightly misted with water to keep them fresh and crisp.

Once at home, celery can wilt in a matter of days. It's best to store it in a slightly open plastic bag or a perforated plastic bag to help it retain moisture but still give it room to breathe. Use the vegetable crisper drawer if you've got one. Properly stored, celery should last up to a week and a half. Stalks that have started to go limp can be revived by cutting them off and standing them cut end down in a cup of water in the fridge.

If you want to use the leaves as garnish, pick the pale yellow ones closest to the center of the bunch (darker green leaves can be tough or fibrous) and store them in a container of water with a few ice cubes in the fridge until ready to use. They make a great addition to mixed green salads.

STEP 1: **EQUIPMENT (See photo on page 818.)** You'll need a sharp chef's knife or santoku knife and, if you want to get extra fancy, a vegetable peeler.

**2**

STEP 2: **SEPARATE THE STALKS** Separate the individual stalks by gently pulling them out from the bottom until they snap off.

STEP 3: **CLEAN AND TRIM** Wash the stalks under cold running water to remove any dirt and debris, then trim the large white section off the bottom of each stalk (reserve for stock, compost, or discard as you wish).

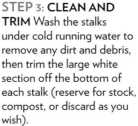

**3**

**4**

STEP 4: **PEEL THE STALKS (optional)** When using celery for gently cooked preparations or in larger batons or chunks, the fibrous skin on the outer surface can be distracting. Using a vegetable peeler, peel each stalk by holding the base against the cutting board and pulling the peeler down the length with a smooth, even motion. Repeat until the entire outer surface is clean.

**5**

STEP 5: **TO CUT LARGE CHUNKS** Large chunks are primarily used for stocks and sauces that will eventually be strained, or for rustic stews. Using your chef's knife, cut the stalks into 1- to 1½-inch pieces.

**6**

STEP 6: **TO CUT THIN SLICES** For slices to use in salads or sautés, cut the stalks into ⅛- to ¼-inch half-moons.

STEP 7: **TO CUT BIAS SLICES** Slicing the half-moons on a bias (at an angle) will yield slightly larger pieces perfect for things like stir-fries or hearty sautés.

**7**

*continues*

**STEP 8: TO CUT FINE DICE** Use the tip of your chef's knife to split the stalk lengthwise, keeping it attached at the leaf end (cutting it crosswise into 2 or 3 shorter sections can help if this step is difficult for you at first).

**STEP 9: CONTINUE TO SPLIT** Split the stalk a few more times: the total number of splits will determine the size of your dice. For medium dice, just split the stalk in half. For finer dice, split into quarters. For brunoise, split into ⅛- to ¹⁄₁₆-inch-wide strips.

**STEP 10: TO CUT DICE** Rotate the stalk and cut crosswise in to dice. The split section of the stalk should hold together as you cut, keeping the strips together.

**STEP 11: TO CUT BATONS** To make batons for soups or salads, split the stalk lengthwise as you would for fine dice, then cut crosswise into 1½- to 2-inch pieces.

**STEP 12: DONE** Batons, large dice, slices, and fine dice, ready to cook or incorporate into salads.

# THE BEST EGG SALAD

The best egg salad starts with perfect hard boiled eggs, with yolks that are just set, but not chalky or dry. Luckily, we've got those eggs covered. The key is to start them in boiling water, then quickly drop the temperature down so that they cook gently and evenly through to the core (see page 102). Once you've got those eggs, all it takes is some mayo to bind them, along with celery, red onion, parsley, and a hint of lemon juice to flavor them.

To chop the eggs, I tried several different methods, including pressing them through a grater to get very fine bits, pulsing them in the food processor, and mashing them with a potato masher or a whisk. In the end, the most primal method produced the best end results: wash your hands well, then get right into that bowl, squishing the eggs between your fingers to get a roughly textured salad.

**SERVES 4**

6 hard boiled eggs, chilled and peeled

¼ cup mayonnaise, homemade (page 807) or store-bought

½ teaspoon zest and 1½ teaspoons fresh juice from 1 lemon

⅓ cup finely diced celery (about 1 small stalk)

¼ cup finely diced red onion

1 tablespoon minced fresh parsley leaves or chives

Kosher salt and freshly ground black pepper

Combine the eggs, mayonnaise, lemon zest and juice, celery, red onion, and parsley or chives in a medium bowl. Using your hands, squeeze the eggs through your fingers, mixing the contents of the bowl until reduced to the desired consistency; alternatively, smash and mix with a firm whisk or potato masher. Season generously with salt and pepper. Serve immediately or store in a sealed container in the refrigerator for up to 3 days.

 1

 2

 3

# CREAMY COLESLAW

Most of the coleslaw I ate growing up was of the wet variety. Soggy and dripping, it left a pool in the bottom of the serving bowl and a runny puddle on your plate, inevitably contaminating your fried chicken or macaroni and cheese. Now, where I come from, "wet" is not an adjective that any self-respecting man would like applied to his food. So what's the key to great, flavorful, nonwet coleslaw? Yep, you guessed it, *osmosis*.

Osmosis is the transfer of liquids across a permeable membrane. It occurs when the concentration of solutes (that's science-speak for "stuff dissolved in liquid") on one side of the membrane is higher than on the other. Water will shift across the membrane to try and balance out this difference. Despite its firm appearance, cabbage is actually one of the wettest vegetables around—a whopping 93 percent of its weight is made up of water. Compare that to say, 79 percent in peas or potatoes, and you begin to get an idea of why coleslaw is always so wet. Getting rid of this excess water is a simple procedure: just salt the cabbage and let it rest for an hour or so, then squeeze it dry.

The remaining ingredients in my coleslaw are pretty standard. Carrots and onions add a bit of sweetness and pungency to the base, while the dressing is a balanced sweet, creamy, tangy blend of mayonnaise (preferably homemade), sugar, cider vinegar, and Dijon mustard.

## MAKES ABOUT 4 CUPS, SERVING 6 TO 8

1 medium head green or white cabbage, cored and shredded (about 8 cups)

1 large carrot, peeled and shredded on the large holes of a box grater

Kosher salt

1 medium red onion, halved and finely sliced

1 cup mayonnaise, preferably homemade (page 807)

2 tablespoons sugar, or to taste

¼ cup cider vinegar

2 tablespoons Dijon mustard

Freshly ground black pepper

1. Toss the cabbage and carrots with 2 tablespoons salt in a large bowl. Transfer to a colander, set it in the sink, and allow to drain for at least 1 hour, and up to 3 hours.

2. Rinse the cabbage and carrots thoroughly and place in the center of a clean kitchen towel. Gather the corners of the towel into a bundle and twist over the sink (or a large measuring cup or bowl) to wring the cabbage and carrots completely dry. Transfer to a large bowl.

3. Add the red onion, mayonnaise, sugar, vinegar, mustard, and a generous sprinkling of pepper to the cabbage and toss thoroughly. Season to taste with salt and more pepper and or sugar if desired. The coleslaw can be served immediately, but for better flavor, refrigerate for at least 2 hours, and up to overnight, to let the flavors mingle and the cabbage to wilt. Retoss just before serving.

Salting the shredded vegetables and squeezing them in a clean kitchen towel removes excess water for better texture and more intense flavor.

# CAESAR SALAD

It's Fourth of July, 1924, in Tijuana. Caesar Cardini, an Italian-Mexican restaurateur who recently left San Diego to run his eponymous restaurant south of the border, where Prohibition laws haven't stemmed the flow of booze-related revenue, is under pressure from a crazy holiday-related rush on the restaurant. The hungry patrons have wiped out his larder, so, the story goes, Caesar is forced to invent a dish on the spot based on the ingredients he has on hand. He decides to serve simple leaves of Romaine lettuce tossed with croutons in a dressing made tableside with egg yolks, Worcestershire sauce, olive oil, garlic, lemon juice, and Parmesan cheese. The dish is a hit, and history is made.

While this account may contain much that is apocryphal, or at least wildly inaccurate, my question is this: why is it that all of these semimythical food-origin stories—burgers, Caesar salad, Buffalo wings—have to involve sort some shot-in-the-dark form of recipe development akin to winning the lottery? For once, couldn't we have a great dish that was created through years of hard research and perfection? Whatever happened to the American dream, the hard work and the payoff at the end?

Suffice it to say, the dish was in fact invented in Mexico, and it was not named after a Roman emperor as I'd always thought growing up. And there's no denying that it's an awesome salad, packed with savory umami notes from the Worcestershire, Parmesan, and anchovies (not one of the original ingredients, but widely accepted these days as necessary—hey, maybe there *is* something to perfecting recipes, even those that come like a bolt of lightning, after all!), with a satisfying crunch from the Romaine and crisp croutons, it's the type of salad that even a hard-core meatatarian could enjoy. While the original dressing may have been a loosely-whisked-together vinaigrette made tableside, I like to make my Caesar dressing with a base of a strongly emulsified mayonnaise. It coats the leaves much better and doesn't end up in the bottom of the bowl. Caesar dressing, by the way, makes a great dip too.

**SERVES 4**

2 to 3 romaine lettuce hearts, separated into individual leaves, larger leaves cut in half crosswise

½ recipe (about ¾ cup) Caesar Salad Dressing (recipe follows)

1 recipe Garlic Parmesan Croutons (recipe follows)

1 ounce Parmigiano-Reggiano, grated (about ½ cup)

1. Wash the lettuce, and dry carefully by laying out on a paper towel–lined baking sheet.
2. With clean hands, gently toss the lettuce and dressing together in a large serving bowl. Add the croutons and toss gently to combine. Sprinkle with the cheese and serve immediately.

**1**

**2**

**3**

## Caesar Salad Dressing

### MAKES ABOUT 1½ CUPS

1 cup mayonnaise, preferably
   homemade (page 807)

2 ounces Parmigiano-Reggiano,
   finely grated (about 1 cup)

2 teaspoons Worcestershire sauce

4 anchovy fillets

2 medium cloves garlic, minced or
   grated on a Microplane (about 2
   teaspoons)

½ cup extra-virgin olive oil

Kosher salt and freshly ground
   black pepper

Combine the mayonnaise, cheese, Worcestershire, anchovies, and garlic in the bowl of a food processor and process until homogeneous, about 15 seconds, scraping down the sides once or twice during processing as necessary. Using a rubber spatula, transfer the mixture to a medium bowl. Whisking constantly, slowly drizzle in the olive oil. Whisk in water a teaspoon at a time until the dressing is just thin enough to flow slowly off a spoon. Season to taste with salt and pepper. The dressing will keep in a sealed container in the refrigerator for up to 1 week.

## Garlic Parmesan Croutons

*These croutons make a great addition to chopped salads, as well as to soups. The croutons can be stored at room temperature in a zipper-lock bag for up to 2 weeks; wait until they are completely cool before bagging them.*

### MAKES ABOUT 4 CUPS

3 tablespoons extra-virgin olive oil

1 medium clove garlic, minced or
   grated on a Microplane (about 1
   teaspoon)

½ loaf ciabatta or hearty Italian
   bread, cut into ½-inch cubes
   (about 4 cups)

Kosher salt and freshly ground
   black pepper

1 ounce Parmigiano-Reggiano,
   grated (about ½ cup)

1. Adjust an oven rack to the middle position and preheat the oven to 350°F. Whisk the olive oil and garlic together in a large bowl until thoroughly mixed. Toss the croutons in the oil until evenly coated. Season with salt and pepper and spread out on a rimmed baking sheet. Bake, flipping halfway through cooking, until dry and lightly browned, about 20 minutes.

2. While the croutons are hot, transfer to a large bowl and toss with the cheese to coat.

# MARINATED KALE SALADS

**I'm not sure exactly when or where marinated
kale salads became a thing, but if forced
to guess, I'd put my wager on 2009, in Brooklyn.**

That's certainly where I first started seeing it on menus. These days, it's common enough that even friends who don't cook and don't believe in Brooklyn have heard about it and probably tried it.

Kale leaves are roughly chopped, massaged with dressing and salt, and allowed to sit. The beautiful thing about these salads is that kale is robust enough that the salad stays crisp and crunchy even after sitting in the fridge for days. You can make it and eat it over the course of a few days with no loss in quality.

## Try a Little Tenderness

To make marinated kale salads, I start by removing the major stems from a bunch of kale, then shredding the leaves into bite-sized strips.

Some folks make the false assumption that it's the acid in the dressing or marinade that causes the tough leaves to tenderize. Actually, it's the oil that does the job. (See "Vinaigrettes," page 773, for more on this.) Plant leaves naturally have a waxy cuticle on them in order to protect them from rain. Haven't you seen rainwater falling on a leaf? It rolls straight off, like water off a duck's back. But this cuticle is oil soluble, so when you massage oil into a pile of kale leaves, it removes the coating, allowing the cells underneath to acquire some controlled damage, thereby softening them. The question then is this: is it necessary to pretenderize the greens with straight oil before dressing them, or can the dressing alone do the job?

I tried it both ways, making a couple big batches at the office for folks to try (any day when there are extra greens in the office is a happy one). The first I tossed with olive oil, salt, and pepper, massaging the oil into the leaves and letting the kale rest for half an hour before tossing it with the dressing. The second one I tossed with the dressing alone (adding

As kale sits dressed in oil it slowly softens, turning from tough to tender, until it is salad-ready.

extra olive oil, salt, and pepper to compensate) and served it up immediately.

The results were pretty conclusive: presoftened greens have a superior texture, coming out tender and crisp, as opposed to fibrous and chewy. The difference was not so great that anyone rejected the unsoftened batch, though, so if I'm in a real rush, I'll go with the direct-dressing route.

Presoftened kale on the left, fresh, fibrous kale on the right.

## HOW TO PREPARE KALE FOR SALADS

To prepare kale for salad, start by cutting or tearing out the large central stems and discarding them. Next, hold a bunch of leaves in one hand and slice through them at the desired thickness. Repeat with the remaining leaves. Wash the greens carefully and spin dry.

# MARINATED KALE SALAD
## WITH CHICKPEAS AND SUMAC ONIONS

**NOTES:** Like all kale salads, this one keeps very well, retaining its crunch and developing flavor in the fridge over a couple days.

Sumac (see page 506) can be found in spice stores or Middle Eastern grocers. Simply omit it if unavailable.

**SERVES 4**

1 pound (about 2 bunches) Tuscan or curly kale, tough stems removed and leaves roughly chopped (about 4 quarts loosely packed)

3 tablespoons olive oil

Kosher salt

1 small red onion, thinly sliced (about ½ cup) and rinsed in a sieve under warm water for 2 minutes

1 teaspoon ground sumac (see Note above)

½ teaspoon toasted sesame seeds

1 tablespoon lemon juice (from 1 lemon)

1 medium clove garlic, minced or grated on a Microplane (about 1 teaspoon)

2 teaspoons Dijon mustard

One 14-ounce can chickpeas, drained and rinsed

Freshly ground black pepper

1. Massage the kale with the olive oil and 1 teaspoon salt in a large bowl, making sure to coat all the surfaces and kneading to help break down the tougher pieces, about 2 minutes. Set aside at room temperature until the kale is softened, at least 15 minutes, and up to 1 hour.

2. Combine the onions with the sumac and sesame seeds. Season to taste with salt. Combine the lemon juice, garlic, and mustard in a small bowl.

3. Once the kale is wilted, add the lemon juice mixture and chickpeas and toss to combine. Season to taste with salt and pepper. Serve topped with the sumac onions. Leftovers will keep in a sealed container in the refrigerator for up to 5 days; retoss before serving.

*continues*

## MARINATED KALE SALAD WITH SHALLOTS AND KIDNEY BEANS

Massage the kale with the olive oil and 1 teaspoon salt as directed above and let stand until softened. Combine 1 tablespoon red wine vinegar, 1 medium clove garlic, minced or grated on a Microplane, and 2 teaspoons Dijon mustard in a small bowl. Toss the wilted kale with the vinegar mixture, 1 large shallot, thinly sliced (about 1 cup) and rinsed in a sieve under warm water for 2 minutes, and one 14-ounce can kidney beans, drained and rinsed. Season to taste with salt and pepper and serve. Leftovers will keep in a sealed container, refrigerated, for up to 5 days. Retoss before serving.

# KALE CAESAR SALAD

The idea of a kale Caesar salad is a natural extension of the marinated kale salad (see page 825). Caesar dressing, which naturally pairs with slightly bitter, very crunchy lettuces, seems like a perfect partner in crime. And it is.

A typical Caesar salad comes with large, crunchy croutons. In this version, rather than large chunks, I break up the bread into very small pieces using a food processor. Once tossed with a bit of olive oil and baked until crisp, the croutons become ultra crunchy because of their increased surface area. When you toss them with the salad, they adhere to the greens. Every bite you take includes these little bits of sweet, toasty, olive oil–coated crunch.

And the greatest part of the recipe? Store the dressed kale in the fridge and the croutons in a sealed container on the countertop (they get soggy if you store them with the salad). The dressed kale will stay crisp for at least three days, meaning whenever you want a perfectly dressed, crisp and crunchy salad, it's as easy as opening the container, sprinkling on the croutons, and serving.

It's dangerously simple, but overindulging in kale has never been a great fear of mine.

**SERVES 4**

1 pound (about 2 bunches) Tuscan or curly kale, tough stems removed and leaves roughly chopped (about 4 quarts loosely packed)

5 tablespoons extra-virgin olive oil

Kosher salt

5 ounces hearty bread, roughly torn into 1-inch pieces (about 3 cups)

Freshly ground black pepper

⅔ cup mayonnaise, preferably homemade (see page 807)

6 anchovy fillets

1 medium clove garlic, minced or grated on a Microplane (about 1 teaspoon)

1½ ounces Parmigiano-Reggiano, finely grated (about ¾ cup)

2 teaspoons Worcestershire sauce

2 tablespoons lemon juice (from 1 lemon)

1 small white onion or 2 shallots, finely sliced

1. Adjust an oven rack to the middle position and preheat the oven to 350°F. Massage the kale with 3 tablespoons of the olive oil and 1 teaspoon salt in a large bowl, making sure to coat all surfaces and kneading to help break down the tougher pieces, about 2 minutes. Set aside while you prepare the croutons and dressing.

2. Combine the bread pieces with the remaining 2 tablespoons olive oil in the bowl of a food processor and pulse until broken down into pea-sized pieces. Season to taste with salt and pepper and pulse once or twice to combine. Spread out on a rimmed baking sheet and bake until the croutons are pale golden brown and crisp, about 20 minutes. Set aside.

3. Meanwhile, wipe out the food processor bowl. Add the mayonnaise, anchovies, garlic, cheese, Worcestershire sauce, and lemon juice and process until smooth. Season to taste with salt and pepper if necessary.

4. Add the onions, dressing, and half of the croutons to the bowl of wilted kale. Toss with your hands until thoroughly coated. Serve sprinkled with the remaining croutons.

# TAMING ONION'S BITE

Soaking onions doesn't remove much pungency.

Rinsing under hot water is the best way to remove an onion's bite, leaving behind just its sweetness.

Let's say you happen to have an extra-pungent onion—it happens to the best of us—is there a way to tame it?

I tried out a few different methods, from submerging onions in cold water for times ranging from 10 minutes to 2 hours, to chilling them, to letting them air out on the counter.

Soaking the onions in a container just led to onion-scented liquid in the container, without much of a decrease in the aroma of the onions themselves—perhaps if I'd used an unreasonably small amount of onion in an unreasonably large container it would have diluted it more efficiently. Air-drying led to milder flavor, but also to dried out onions and a papery texture.

The best method turned out to be the fastest and easiest: just rinse away all those extra-pungent compounds under running water—warm water. The speed of chemical and physical reactions increases with temperature. Using warm water causes onions to release their volatile compounds faster—about 45 seconds is enough to rid even the most pungent onions of their kick.

The next question on your mind might be, but doesn't hot water make the onion go all limp?

Nope. Even if you use the hottest tap water, it generally comes out at around 140° to 150°F or so. Pectin, the main carbohydrate "glue" that holds plant cells together, doesn't break down until around 183°F. There are other bits of the onion that, given enough time, will begin to soften at hot tap water temperatures, but it takes a long, long time.

Don't worry, your onions are safe here.

# DRESSING Family #3:
## DAIRY-BASED DRESSINGS

These dressings are by far the easiest to make, because dairy products, as a general rule, come pre-emulsified. That's right—the creamy milk you're drinking is creamy precisely because it has fat distributed through it in tiny, tiny droplets. With a dairy-based dressing, you never have to worry about it breaking. On the flip side, dairy-based dressings don't have nearly as long a shelf life as constructed emulsions like a vinaigrette or a mayonnaise.

Cultured milk products, like sour cream, yogurt, or crème fraîche, are the best bases—their thickness makes them ideal for coating leaves and vegetables evenly, while their tanginess is a natural in salads.

## ICEBERG WEDGE SALAD

**The iceberg salad is the exception that proves the rule. According to my definition of a salad, an iceberg wedge is technically *not* one, since it needs to be cut with a knife and fork at the table. But it's close enough, and I'll take delicious over pedantic any day of the week.**

### SERVES 4

1 head iceberg lettuce, cut into quarters, core removed

1 recipe Three-Ingredient Blue Cheese Dressing (recipe follows)

4 ounces grape tomatoes, cut in half (about 1 cup)

8 slices bacon, cooked and crumbled

Place the iceberg wedges on individual serving plates. Drizzle each wedge with one-quarter of the dressing. Divide the grape tomatoes and bacon evenly among the plates. Serve immediately.

## Three-Ingredient Blue Cheese Dressing

**This is about the simplest blue cheese dressing you can make, but it relies on high-quality, really sharp and flavorful blue cheese. Do *not* try this with the cheapest Danish blue you can find—you will be disappointed. To maximize texture and chunkiness in the dressing, I like to form a base with the tangy buttermilk, creamy mayonnaise, and half of the blue cheese, reserving the other half to crumble and mix in.**

**NOTE:** Use a sharp blue cheese like Gorgonzola, Roquefort, Fourme d'Ambert, or Stilton.

**MAKES ABOUT 1½ CUPS**

½ cup buttermilk

½ cup mayonnaise, preferably homemade (page 807)

8 ounces sharp blue cheese (see Note above), finely crumbled (about 2 cups)

Kosher salt and freshly ground black pepper

Combine the buttermilk, mayonnaise, and half the Gorgonzola in the bowl of a food processor and process until smooth, about 15 seconds. Transfer to a medium bowl and fold in the remaining Gorgonzola with a rubber spatula. Season to taste with salt and pepper. The dressing will keep in a sealed container in the refrigerator for up to 1 week.

# CHOPPED SALADS

**Chopped salads are a lot like the A-Team:**

. . . a ragtag group of individuals with totally conflicting personalities that somehow manage to come together in a beautiful way to make the world a better place.

And there're no buts about it—a chopped salad takes more time to put together than a simple green salad. First you've got an ingredients list that's a good three to five times longer, and then you've got to chop, drain, and dress all the vegetables. But what it lacks in brevity, it makes up for in heartiness. Nothing beats a chopped salad for lunch on a picnic (bring all the prepared ingredients with you, with the dressing in a separate container, and toss just before serving), or as a simple but tasty meal on a warm summer night.

Good chopped salads are all about balancing flavors and textures. Obviously you want plenty of crunchy elements—crisp lettuces and vegetables—along with intensely flavored bites like cheese, nuts, and cured meats. I've included some complete recipes for a few of the most classic chopped salads, but I sincerely hope it won't end there! With their minimal fuss and inexpensive ingredients, chopped salads are one of the best forums for beginning cooks to experiment in terms of combining flavors and textures to discover a style that suits them best. I hope you take the chance. The following chart will show you the best way to treat chopped salad ingredients so you can put together your own combinations. When constructing a chopped salad, I always try to mix together two to three base ingredients and a secondary ingredient or two that contrasts the base ingredients texturally and flavorwise, along with a couple of flavorful accents.

The real key to a successful chopped salad is controlling moisture. Here's what happens if you don't:

Undrained vegetables leave a pool of watered-down dressing in the salad bowl.

Very moist ingredients like tomatoes and cucumbers should be salted in advance and left to drain in a strainer for at least half an hour before being blotted dry.

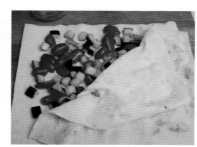

# CHOPPED SALAD INGREDIENTS

| INGREDIENT | HOW TO PREPARE | ROLE IN SALAD |
|---|---|---|
| Romaine or iceberg lettuce, radicchio, escarole, Belgian endive, or curly endive | Wash, dry, and chop into 1-inch pieces. | Base ingredient |
| Cucumbers | Peel, split lengthwise in half, remove the seeds with a sharp spoon, and chop into ½-inch pieces. Toss with ½ teaspoon salt per pound and let drain in a colander for 30 minutes. | Base or secondary ingredient |
| Tomatoes, cherry tomatoes, and grape tomatoes | Split small tomatoes in half or into quarters. Remove the seeds from larger tomatoes and chop into ½-inch dice. Toss tomatoes with ½ teaspoon salt per pound and let drain in a colander for 30 minutes. | Base or secondary ingredient |
| Canned beans (black beans, cannellini beans, chickpeas, kidney beans, etc.) | Drain, rinse, and carefully dry. | Base or secondary ingredient |
| Radishes | Scrub and quarter. | Base or secondary ingredient |
| Crisp vegetables like celery, zucchini, squash, fennel, jicama, or hearts of palm | Peel if necessary and cut into ½-inch dice. | Base or secondary ingredient |
| Nuts and seeds (walnuts, almonds, hazelnuts, peanuts, macadamia nuts, sunflower seeds, etc.) | Toast, then roughly chop if larger than ½ inch: leave whole if small. | Base or secondary ingredient |
| Avocado | Halve, pit, peel, and cut into ½-inch cubes. | Base or secondary ingredient |
| Cabbage | Shred or cut into ½-inch dice. Toss with ½ teaspoon salt per pound and let drain in a colander for 30 minutes. | Base or secondary ingredient |
| Corn | Cut from the ears, blanch in boiling salted water for 1 minute, drain, and cool. | Base or secondary ingredient |

| INGREDIENT | HOW TO PREPARE | ROLE IN SALAD |
|---|---|---|
| **Pasta** | Use bite-sized shapes. Cook, drain, and cool. | Base or secondary ingredient |
| **Green vegetables like broccoli, asparagus, snap peas, and green beans** | Blanch in 1 gallon boiling water with ½ cup kosher salt until just tender-crisp; cool under cold running water, then drain and carefully dry. | Base or secondary ingredient |
| **Red onions, sweet onions, or scallions** | Thinly slice and soak in cold water for 30 minutes; drain. | Secondary ingredient |
| **Bell peppers (green, red, yellow, orange)** | Cut into ½-inch cubes. | Secondary ingredient |
| **Carrots** | Peel and cut into ½-inch cubes or ¼-inch, thick matchsticks, or shred on the large holes of a box grater. | Secondary ingredient |
| **Citrus fruit** | Cut into suprêmes (see page 770). | Secondary ingredient |
| **Crisp acidic fruits like apples, pears, and young mangoes** | Cut into ½-inch cubes. | Secondary ingredient |
| **Eggs** | Hard-boil and roughly chop. | Secondary ingredient |
| **Poached or roasted chicken, turkey, or ham** | Cut into ½-inch cubes. | Secondary ingredient |
| **Canned tuna** | Drain and roughly shred. | Secondary ingredient |
| **Salami, pepperoni, soppressata, chorizo, ham, or other dry-cured meats** | Cut into ¼-inch cubes. | Flavorful accent |
| **Olives** | Buy prepitted (or pit them yourself) and split in half or into quarters. | Flavorful accent |

*continues*

| INGREDIENT | HOW TO PREPARE | ROLE IN SALAD |
|---|---|---|
| Capers | Rinse and dry. | Flavorful accent |
| Tender herbs like parsley, chives, basil, mint, chervil, tarragon, cilantro, or dill | Wash, dry, and roughly chop. | Flavorful accent |
| Jarred hot peppers (peperoncini), roasted red peppers, sun-dried tomatoes, or other pickled or preserved jarred vegetables | Roughly chop. | Flavorful accent |
| Semi-firm cheese like feta, provolone, Manchego, or cheddar—see "Cheese Chart," pages 717–21 | Cut into ½-inch cubes. | Flavorful accent |
| Bacon | Cook until crisp (see page 131) and crumble. | Flavorful accent |
| Cured fish like anchovies or sardines | Roughly chop. | Flavorful accent |
| Dried fruits | If larger than ½ inch, roughly chop. | Flavorful accent |

Depending on the ingredients, a chopped salad can be dressed with any of the three basic salad dressing styles. A lemony vinaigrette best brings out the fresh flavors of a light cucumber-and-tomato-based Greek salad, while a mayo-based creamy Italian dressing is de rigueur for a classic Italian-American antipasti salad. Tangy buttermilk ranch dressing is great as a dipping sauce on a vegetable platter, but it's also the dressing of choice for a ranch-style Cobb salad.

# CHOPPED GREEK SALAD

**While your corner pizza shop or deli might sell an iceberg lettuce salad topped with olives, tomatoes, cucumber, and feta cheese as "Greek," a true Greek salad is a lettuce-free chopped salad lightly dressed in a lemony vinaigrette. This is one of my favorite late-summer side dishes, when tomatoes are at the height of awesomeness.**

**SERVES 4**

8 ounces grape tomatoes, cut in
half (about 2 cups)

1 large cucumber, peeled, halved
lengthwise-seeded, and cut into
½-inch dice (about 2 cups)

Kosher salt and freshly ground
black pepper

1 medium red onion, finely sliced
(about ¾ cup)

1 large green or red bell pepper, cut
into ½-inch cubes

½ cup pitted kalamata olives, split
in half

3 ounces feta cheese, crumbled

½ cup loosely packed fresh parsley
leaves, roughly chopped

2 teaspoons chopped fresh oregano

⅓ cup Mild Lemon- or Red Wine–
Olive Oil Vinaigrette (page 780)

1. Toss the tomatoes and cucumbers with ½ teaspoon salt and a few grinds of black pepper. Transfer to a colander set in the sink and let drain for 30 minutes. Meanwhile, place the red onion in a small bowl cover with cold water, and let stand for 30 minutes. Then rinse and drain.

2. Carefully dry all the vegetables with paper towels. Combine the tomatoes and cucumbers, red onions, bell pepper, olives, feta, parsley, and oregano in a large bowl. Drizzle with the dressing and season to taste with salt and pepper. Toss thoroughly and serve.

## HOW TO BUY AND PREPARE CUCUMBERS

Cucumbers are one of the oldest cultivated vegetables, and one of my favorites. Peeled, cut, and sprinkled with a little salt, they are simultaneously savory and refreshing. They're fantastic marinated overnight in a little soy sauce, sesame oil, and red pepper flakes; the salt in the soy sauce will draw out some of their liquid, so they get a superconcentrated flavor.

As a stir-fry ingredient, cucumbers are one of the most underused.

Oddly enough, cucumbers are one of the two foods my wife can't stand (the other is tomatoes), so chicken stir-fried with cucumbers, fermented bean paste, and Sichuan peppercorns is one of my quick, easy, and delicious go-to staples when she's out on a girls' night.

While you can certainly simply slice whole cukes into salads, peeling and seeding them helps you make the most of their flavor and texture.

*continues*

## Shopping and Storage

In the supermarket, you're usually faced with three choices:

- **American cucumbers** are dense fleshed and flavorful. They have thicker skins than most other cucumbers, so I recommend removing it. In any case, definitely give them a good scrub under cool running water to remove some of the food-grade wax that they usually come coated in. They also have lots of watery seeds that should be scraped out before use.

- **English cucumbers** usually come individually shrink-wrapped, which means that there's no need to scrub them before eating (they are not waxed). The skins are thinner than those of American cucumbers and so can be consumed with no problem. Although English cucumbers are usually relatively seedless, they are also much more watery than their American counterparts. They are more convenient to prepare but less flavorful.

- **Kirby cucumbers** are like small versions of American cucumbers. Thick-skinned and relatively seedless, they have the strongest flavor of the three types and a texture that can sometimes border on tough. They are best pickled.

Fresh whole cucumbers can be stored in the vegetable crisper drawer for at least a week and often much longer. Cut cucumber pieces should be stored wrapped in a moist paper towel inside an airtight plastic bag or container to prevent moisture loss. Eat cut cucumbers within 3 days.

# KNIFE SKILLS:
## How to Cut Cucumbers

**STEP 1: SPLIT THE CUCUMBER IN HALF** Split the cucumber lengthwise in half.

**STEP 2: REMOVE THE SEEDS AND SPLIT FURTHER** Cucumber seeds can be watery and bland, so they should be scraped out with a spoon; then cut the halves lengthwise into spears of the desired width.

**STEP 3: CUT CROSSWISE** Rotate the cucumber 90 degrees and cut the strips into cubes.

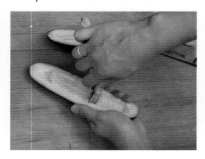

# WHITE BEAN AND MANCHEGO CHEESE SALAD

**Crunchy celery and red onion contrast beautifully with the creamy white beans in this simple salad. Chopped parsley and cubes of salty Manchego cheese (you can substitute another semi-firm cheese if you'd prefer; see "Cheese Chart," pages 717–21) round out the flavors. Of course, you can use whatever canned beans you'd like (see page 195), or cook them yourself (see page 255). It's essential that your beans are fully dried before you put the salad together, to avoid a watery salad.**

### SERVES 4 TO 6

Two 14½-ounce cans white beans (or 4 cups cooked white beans), drained, rinsed, and dried with paper towels or a clean kitchen towel

2 large stalks celery, peeled and cut into ½-inch cubes (about 1½ cups)

1 medium red onion, thinly sliced (about ¾ cup)

½ cup roughly chopped fresh parsley

8 ounces Manchego cheese, cut into ½-inch cubes

½ cup Mild Lemon- or Red Wine–Olive Oil Vinaigrette (page 780)

Kosher salt and freshly ground black pepper

Combine the beans, celery, red onion, parsley, and cheese in a large bowl. Add the vinaigrette, season to taste with salt and pepper, and toss to combine. Serve immediately.

# RESTAURANT-STYLE CHOPPED ANTIPASTI SALAD

**SERVES 4 AS A LIGHT MEAL**

8 ounces grape tomatoes, cut in half (about 2 cups)

Kosher salt and freshly ground black pepper

1 medium red onion, finely sliced (about ¾ cup)

One 14-ounce can chickpeas, drained and rinsed

One 6-ounce jar peperoncini, drained and coarsely chopped

One 6-ounce jar roasted red peppers, drained, rinsed, and chopped into ¼-inch pieces

3 stalks celery, peeled and cut into ½-inch dice (about 1½ cups)

8 ounces Genoa salami, cut into ¼-inch cubes (about 2 cups)

6 ounces sharp provolone cheese, cut into ¼-inch cubes (about 1½ cups)

1 head Romaine lettuce, trimmed and chopped into ½-inch pieces (about 3 cups)

1 recipe Creamy Italian Dressing (recipe follows)

1. Toss the tomatoes with ¼ teaspoon salt and a few grinds of pepper. Transfer to a colander set in the sink and let drain for 30 minutes. Meanwhile, place the red onion in a small bowl, cover with cold water, and let stand for 30 minutes, then rinse and drain.

2. Carefully dry the drained tomatoes, onions, chickpeas, peperoncini, and red peppers with paper towels. Transfer to a large bowl, add the celery, salami, provolone, Romaine, and dressing, season to taste with salt and pepper, and toss thoroughly to combine. Serve immediately.

**1**

**2**

**3**

## Creamy Italian Dressing

*If you've only ever used store-bought Italian dressing, this one's for you. It straddles the line between a vinaigrette and a mayonnaise. Mix it up properly, and it'll stay nice and creamy for several hours, though, unlike a mayonnaise, it'll eventually break and separate again.*

**NOTE:** Use on crisp, watery greens like iceberg or Romaine.

### MAKES ABOUT 1 CUP

½ cup mayonnaise, preferably homemade (page 807)

2 tablespoons lemon juice (from 1 lemon)

1 small shallot, finely minced (about 1 tablespoon)

1 medium clove garlic, minced or grated on a Microplane (about 1 teaspoon)

2 tablespoons minced fresh basil

2 teaspoons minced fresh oregano

½ teaspoon red pepper flakes

6 tablespoons extra-virgin olive oil

Combine all the ingredients in a squeeze bottle or container with a tight-fitting lid. Shake vigorously to emulsify. The dressing will keep in a sealed container in the refrigerator for up to 1 week; shake vigorously before using.

# CHOPPED RANCH COBB SALAD

The classic Cobb salad is one of those more-calories-than-any-entrée-on-the-menu dishes that lets you feel somewhat virtuous all the while knowing that really you just want to stuff your face with bacon, avocado, and blue cheese. There ain't nothing wrong with that. If a Cobb salad feels a little uncomposed, that's because it is: it's nothing more than a handful of ingredients all thrown together on the same plate. The key to a great Cobb salad is to make sure that each one of those components is perfect.

The chicken in a Cobb should be supremely moist and tender. You can accomplish this by poaching the chicken in not-too-hot water until it registers exactly 150°F on an instant-read thermometer. At this stage, it's cooked through but still retains most of its moisture, without any stringiness to speak of.

The remainder of the ingredients—the bacon, the eggs, the dressing—are pretty straightforward, and luckily we already have great techniques that'll take care of them. If you are planning this salad for a party or a picnic, all the ingredients can be prepped in advance to be plated at the last moment, though for best results, you should cook the bacon and dice the avocado just before serving.

**NOTE:** Leftover roast chicken can be used in place of the poached chicken.

## SERVES 4 AS A MAIN COURSE

2 boneless, skinless chicken breasts, about 8 ounces each

Kosher salt

2 heads Romaine lettuce, roughly chopped into 1-inch pieces (about 3 quarts)

1 recipe Buttermilk Ranch Dressing (recipe follows)

Freshly ground black pepper

8 strips bacon, cooked and crumbled

4 hard-boiled eggs (see page 102), roughly chopped

1 avocado, halved, pitted, peeled, and diced into ½-inch cubes

1 large tomato, diced into ½-inch cubes

6 ounces Roquefort cheese, crumbled

2 tablespoons finely minced fresh chives

1. Place the chicken breasts in a large saucepan and cover with 2 quarts cold water and season with 2 tablespoons salt. Bring to a boil over high heat, reduce to a bare simmer, and cook until an instant-read thermometer inserted into the thickest part of a breast registers 150°F, about 10 minutes. Drain the chicken and run under cold water until cool enough to handle. Pat dry and dice into ½-inch chunks.

2. Toss the lettuce with half of the dressing in a bowl. Season to taste with salt and pepper. Divide the lettuce among four serving plates and top with the chicken, bacon, eggs, avocado, tomato, and cheese, keeping each ingredient separate. Season to taste with salt and pepper and sprinkle with the chives. Serve immediately, passing the remaining dressing at the table.

## Buttermilk Ranch Dressing

**MAKES ABOUT 1 CUP**

½ cup low-fat or skim cultured
buttermilk

½ cup sour cream

2 teaspoons lemon juice (from 1
lemon)

1 medium clove garlic, minced or
grated on a Microplane (about 1
teaspoon)

1 teaspoon Dijon mustard

2 tablespoons minced fresh chives

2 tablespoons minced fresh cilantro

1 teaspoon freshly ground black
pepper

Pinch of cayenne pepper

Kosher salt

Whisk together the buttermilk, sour cream, lemon juice, garlic, mustard, chives, cilantro, black pepper, and cayenne, in a small bowl. Season to taste with salt. The dressing will keep in a sealed container in the refrigerator for up to 1 week; whisk or shake before using.

BATTER,
BREADINGS, AND
THE
SCIENCE
*of*
FRYING

9

"I don't care if you're frying dog s*&$. If it comes out of the fryer, put some salt on it."—Ken Oringer

# BATTER, BREADINGS, AND THE
## SCIENCE *of* FRYING

### RECIPES IN THIS CHAPTER

I've been hiding the fried chicken from you.

Now I know what you're thinking: *How could you do such a thing to me? My love for juicy chicken coated in a crisp crust knows no bounds. Whether it's a thick, crunchy coating with eleven secret herbs and spices or paper-thin crackling skin that unites with the flesh underneath to achieve that cosmic oneness so coveted by fried chicken aficionados like myself, there is nothing—I mean nothing—I'd rather be doing in this sweet, fair world right now than sinking my teeth into a golden brown thigh, feeling the snap of the skin against my lips, the salty golden juices dribbling down my chin. If you'd only let me, I'd eat fried chicken for breakfast, lunch, and dinner, and several meals in between.*

And therein lies the problem. As passionate as I am, I'm a man of science, and in order for me to perform truly scientific tests on fried chicken, the stuff has to stick around at least long enough for me to document and measure it. This simply doesn't happen when you're in the vicinity. For the past year, I've resorted to waiting until you're out of town before cranking up the deep fryer to do my testing. You come back from trips with the smell of chicken fat still lingering in the air, the bony remains of my experiments sitting at the bottom of the trash can.

It's cruel, I know, and I promise, dear, that I gain only a small bit of sadistic pleasure out of doing it. Do I not make it up to you by supplying you with increasingly better versions of fried chicken on special occasions like Christmas or Colonel Sanders's birthday? Fact is, you're not alone here. I love fried chicken. You love fried chicken. Irving our doorman loves fried chicken. Our dog loves fried chicken almost as much as he loves chasing his own tail. Is there anyone in this fair country who *doesn't* love fried chicken? As a food writer, recipe developer, kitchen scientist, and lover of American food, I see it as my responsibility—nay, my *duty*—to give fried chicken its due. To see exactly what it is that makes it tick. To decipher what's going on underneath that crisp golden brown shell, and to deliver to my wife the culmination of those studies.

But first, a few words on deep-frying.

## { WHAT IS DEEP-FRYING? }

If there's one cooking technique that home cooks fear more than any other, it's deep-frying. I get it. Bringing a pot full of oil up to 375°F on your stovetop is a frightening prospect. But what if I told you that despite the violent bubbling, alchemic transformation, and incredible deliciousness that results from a fry-job-done-well, frying is actually a remarkably simple process and one of the easiest techniques to perform in the kitchen, given a bit of know-how?

Crispy Beer-Battered Fried Cod (page 892).

I mean, think about it: who does most of the deep-frying in this country? The least-trained kitchen workers, that's who. Why do you think the fry station is the very first place that most beginning restaurant cooks are assigned to? All those faultlessly fried clams you get from roadside shacks in New England in the summer—guess who's cooking them? Hint: it's *not* a four-star chef. Most likely it a high school kid trying to make a few bucks over summer vacation. And if *they* can do it, then so can you.

Here's what happens when you drop a piece of food into a deep fryer:

- **Dehydration.** Free water inside foods and in batters or breadings will evaporate at 212°F. As soon as your food hits the oil in a deep fryer, which for most recipes ranges from 300° to 400°F, moisture will rapidly convert into steam, releasing itself in a violent cascade of bubbles. This escape of mois-

ture is what you see when you first lower food into a fryer. Within a few minutes (depending on the thickness of your food and temperature of your fryer), most of the free moisture in your food will have completely evaporated and the bubbling will slow down. After this, bound water from inside the food—that is, water that takes more energy to escape from its cellular prisons—will continue to be released in small streams of bubbles. Eventually, after all free and bound moisture has been expelled, you will no longer see bubbles coming from your food. At this stage, your potato chip is about as fried as it's gonna get.

- **Expansion.** This phenomenon occurs in foods that are coated in a batter or dredged in a mixture leavened with baking powder, whipped egg whites, or other ingredients that cause the formation of air bubbles. Hot air takes up more space than cold air, so the rapid change in temperature that occurs when you drop food into a fryer causes the air bubbles inside the batter around a piece of fried food to expand. In that way, it's very similar to the way that a ball of dough will puff when put into a hot oven. This expansion brings lightness and crispness to fried foods.

- **Protein coagulation.** Cooking in hot oil precipitates the rapid coagulation of proteins. Just as proteins set in a loaf of bread or a pancake, giving it more structure and rigidity, so do they set in the batter or breading coating a piece of fried food. It's this protein matrix—usually comprised of gluten in a flour-based batter, or egg proteins in a basic breading—that gives rigid structure to fried foods, transforming the batter or coating into a firm solid.

- **Browning and caramelization.** The Maillard reaction—the complex string of chemical reactions that gives flavor and color to well-browned foods—as well as caramelization—the similar reaction that occurs when sugars are heated—take place rapidly at normal frying temperatures. This is what gives fried foods their enticing golden brown color and delicious flavor.

- **Oil absorption.** As water is forced out of food through evaporation, it leaves spaces behind. What moves in to take the place of that water? The only thing that can: oil from the fryer. It's an inevitable part of frying, essential to the flavor of the finished food. And, despite what many books may tell you, frying at higher temperatures will *not* reduce the amount of oil your food absorbs (quite the opposite, in fact, see page 861).

Seem complicated? It's not. The beautiful thing about deep-frying is that once you've got the right amount of oil heated to the right temperature, all of these things happen on their own, with very little direction needed from you, the cook.

## THE WOK: THE BEST VESSEL FOR DEEP-FRYING

Frying in a Dutch oven works reasonably well, but it's got its problems: its straight sides make maneuvering food inside it difficult. You could go out and get a dedicated deep fryer, but do you really have room on your counter for that?＊ Here's a better alternative: I'd be willing to wager that anyone who complains about how difficult and messy it is to deep-fry at home has never tried deep-frying in a wok. Why don't people fry at home? The most common answers are: it's messy, it's expensive ("What do I do with all the leftover oil?"), and it's unhealthy. Well, a wok can help solve your first two problems; you're on your own for the third. Frying adds fat to your food, period—try eating a few fewer French fries or only one piece of fried chicken if you don't want the extra calories.

*continues*

＊And do you really want to have to consider the havoc that could be wreaked on your waistline if you had easy access at all times to deep-fried foods?

The flared sides of a wok offer several advantages over a straight-sided saucepan or Dutch oven:

- **There is less mess.** If you've ever tried deep-frying in a Dutch oven, you know that your range gets splattered with little droplets of oil splashing out from the sizzling food inside the pot. The sloping sides of a wok, on the other hand, extend out a good three inches or so from where the edges of the oil are, catching those droplets and keeping your counter neat and clean.
- **It's easier to maneuver.** To get the crispest food possible, it's important to keep the food moving (more on that below). Many times, you also need to flip foods while frying. The flared shape of a wok makes it easy to reach in with a spider or chopsticks, and it gives you plenty of room to work in.
- **There is less chance of a spill-over.** Having a pot of hot oil bubble over the rim of a Dutch oven ain't fun. It's dangerous, the oil will probably catch fire, and, at the very least, it'll make a huge mess. It ranks up there with the old hand-in-the-blender or dog-in-the-dishwasher as worst kitchen nightmare ever. It happens when you add too much moist or cold food to a too-full pot of oil. The food rapidly releases bubbles of water vapor, those bubbles pile up on top of each other, and over the edges they go. Since a wok widens out at the top, there is much more volume for those bubbles to expand into, so their surface area increases, weakening their structure, and they pop before they get a chance to go up and over.
- **It's easier to keep the oil clean, making the technique more economical.** The edges of a Dutch oven can harbor burnt bread crumbs, little bits of French fries, and other such baddies. In a wok, there's no place to hide. Food particles left in hot oil are the main reason the oil breaks down and becomes unusable. Oil that's carefully cleaned as you cook should last for at least a dozen frying sessions, if not more.

## THE KEYS TO PERFECT DEEP-FRYING

No matter what cooking vessel you choose, here are ten tips to ensure that your frying will be successful:

### Use a Thermometer

There is no other way to ensure that your oil is at the right temperature. Depending on what you're cooking, you'll want to use different oil temperatures. For example, French fries fried at 300°F will never crisp up, and chicken fried at 425°F will burn on the exterior before it's cooked through. A thermometer is the only way to guarantee that you're cooking things right. You can get a dedicated deep-frying thermometer, but if you already own a Thermapen (and you should!), it'll do the job even better.

## Don't Fear the Fat!

A hot wok of oil is nothing to be trifled with, but just like a pit bull, it can sense fear. Timid novice fryers often decide to keep their hands a safe distance from the oil by dropping the food into it from a height. The food ends up splashing hot oil out of the wok, and onto their skin and clothes, making them even more scared the next time. The goal when adding food to a fryer should be to minimize splashing. You do this by bringing your hand (or tongs, if you're using them) as close to the surface of the oil as possible before dropping in the food. For small pieces of food, a couple inches or less, this means getting your fingers to about an inch of above the oil as you add each one. For larger pieces—a whole fish fillet, say—it means dipping in one end and then gently lowering the fillet into the oil until only the last inch sticks out before dropping it. Lowering foods into the oil a piece at a time will also keep battered items from sticking together in a large mass.

## Avoid Crowds

Adding too much food to the hot oil in a single batch will cause the oil temperature to drop rapidly, reducing the effectiveness of your fry. Your food won't crisp properly, and batter-coated foods may lose some of their coating. A good rule of thumb is to never add more than a half pound of refrigerator-temperature food per quart of hot oil. So, if you want to cook that 1-pound batch of fries, you've gotta use a full half-gallon of oil, or go in batches (I suggest batches). Of course, frozen food should be fried in even smaller batches.

## Dry = Good

As we've discussed, deep-frying is essentially a process of dehydration. The hot oil causes water to rapidly transform to steam, which escapes and allows the crust to form. So it stands to reason that the drier your food is to begin with, the more effectively it'll fry. Surface moisture can also cause undue bubbling and more rapid breakdown of the oil. For best results, all solid foods to be fried should first be patted dry or coated with a batter or breading. And make sure to allow excess batter to drip off foods before adding them to the oil.

## Keep It Clean

The more you use a batch of oil, the less effective its frying ability will become. The main factors that affect oil breakdown are small particles of food and moisture. To extend the life of your fry oil, you should constantly clean it. Whenever I fry, I keep a wire-mesh spider nearby to fish out any bread crumbs, bits of tempura batter, or other food particles between (or even during!) batches. To remove this debris, I start by swirling the oil in a clockwise direction with the spider, then flip the spider around and give it a pass in the counterclockwise direction: the flow of the oil should force the majority of food particles into its mesh. Deposit the debris in a metal bowl kept handy (do NOT throw into a garbage can with a plastic liner!), and repeat until the oil is clean. After each frying session, pour the oil through a fine-mesh strainer lined with paper towels or cheesecloth set over a bowl to get rid of any leftover detritus.

## Keep It Moving

Ever notice how when you're in a cool swimming pool, if you stand still, you'll feel a little warmer until someone swims by and creates a current around you that cools you down again? Well, the opposite thing happens with cold food in hot oil. If you allow it to sit still, a pocket of cooler oil will develop around pieces of food, reducing the effectiveness of the fry. By constantly agitating the food and moving it around, you're continuously exposing it to fresh hot oil. Your food will fry more evenly, and come out crisper than the simpler dunk-and-sit method. A wire-mesh spider or a pair of long chopsticks is the best tool for this job.

## Choose Your Oil Wisely

The best oils for deep-frying are relatively cheap, flavorless oils with high smoke points. Flavorful oils like sesame oil or extra-virgin olive oil contain compounds that cause them to smoke at far below the effective frying temperature for most foods. Other oils have their adherents, but peanut oil or peanut oil cut with a bit of lard, bacon fat, and shortening is my frying medium of choice. See "All About Oil," below for more details.

## Drain Quickly, and Use Paper Towels!

While it may seem logical to drain fried foods on a metal rack, it's actually far more effective to drain them on a paper-towel-lined plate or bowl. When they are set on a rack, very little oil actually drips out of the food—the oil's surface tension keeps it in place. A paper towel, on the other hand, wicks oil away through capillary action, effectively drawing more fat out of the food and helping it stay crisp

longer. In fact, in a side-by-side test, I found that paper towels drew out nearly *four times* as much oil from the food as a simple rest on the rack. To get the least greasy food possible, drop your food directly from the fryer onto a paper-towel-lined plate, tray, or bowl, flip it to blot oil on both sides, and then quickly transfer it to a rack, to allow for circulation (steam building up under the food can soften its crust).

Rest your fried food on paper towels to wick away grease.

## Season Immediately

I worked for a chef who was fond of saying, "I don't care if you've just fried dog sh*t. If it comes out of the fryer, you season it the second it does!" And he's right. Salt sticks to and dissolves more rapidly on hot surfaces, so the sooner you season your fried food, the better it'll taste down the line.

## Reuse Your Oil

To save your oil, skim it and let it cool in the wok, then pour it through a fine-mesh strainer lined with cheesecloth or a paper towel. From there, pour it right back into its original bottle (or an empty soda bottle) through a funnel. Seal the cap, and store it in a cool, dark cabinet until the next time you need it. If it starts getting very dark or produces foamy bubbles on its surface when you start heating it, it's past its prime and should be disposed of.

## ALL ABOUT OIL

**What's the best oil for deep-frying?**

There's a baffling array of fats on supermarket shelves these days, from supposedly heart-healthy olive and canola oils high in Omega-3 fatty acids to expensive designer oils, like avocado or grapeseed, to solid-at-room-temperature fats like vegetable shortening and lard. Which fat is best for frying? Which produces the crispest crust and best flavor?

I decided to find out the only way I know how: try 'em all.

I fried a dozen batches of chicken using the following fats: shortening, lard, canola oil, olive oil, peanut oil, sunflower oil, corn oil, palm oil, avocado oil, generic "vegetable" oil (usually a mix of soybean and corn oil), grapeseed oil, and bacon fat.

I immediately noticed a direct correlation between the level of saturated fat in a given cooking medium and how crisp the chicken got. Chicken cooked in highly saturated lard (40 percent satu-rated fat), shortening (31 percent), bacon fat (40 percent), or palm oil (81 percent) was by far the crunchiest. This seems like a good thing—until you actually let it cool a bit and eat it. Because those fats are all close to solid at body temperature, they leave your mouth with an unappetizing waxy coating. With lighter foods like tempura-style vegetables or fish, this coating is especially noticeable.

On the opposite end of the spectrum, chicken fried in highly unsaturated fats like grapeseed (10–12 percent saturated fat), olive (13 percent), corn (13 percent), sunflower (10 percent), avocado (12 percent), or vegetable (around 13 percent) oil suffered from the opposite problem: the chicken simply didn't crisp up as well. The winner? Peanut oil, with its moderately high level of saturated fat (17 percent) and clean, neutral flavor. The chicken fried up clean and crisp, without any of the mouth-coating waxi-ness of the highly saturated fats. It's my fat of choice for almost all frying projects, not just chicken.

## SATURATED VERSUS UNSATURATED FAT

**W**e hear the terms "saturated fat" and "unsaturated fat" thrown about often, but what do they really mean, and how do they affect your cooking?

Like most organic compounds, fats are pretty complicated molecules. They're naturally kinked and wound up, but if you were to straighten one out, it would resemble the letter E, with a molecule of glycerol forming the spine and long chains of carbon atoms called "fatty acids" forming the three arms. It's the exact makeup of these arms that determines whether or not a fat is saturated.

A carbon atom can form four bonds with other atoms. In saturated fats, every carbon atom in the chain is bonded to two hydrogen atoms, along with the carbon atom preceding it and following it.* In unsaturated fats, one or more of the carbon atoms is bound only to a single hydrogen. In place of the missing hydrogen,

*continues*

---

*With the exception of the final carbon atom, which is bound to three hydrogen atoms, and the first, which is bound to two oxygen atoms (one with a double bond).

Highly saturated fats like lard or shortening are solid at room temperature and melt into a liquid when heated.

it forms a double bond with a neighboring carbon atom. Monounsaturated fats contain a single double-carbon bond, while polyunsaturated fats contain two or more double-carbon bonds.

Because saturated-fat molecules are straight, they can stack together more tightly and efficiently, rendering most of them solid at room temperature. That's why fats with a high proportion of saturated fat, such as butter, shortening, animal fats, and palm oil, will be solid and opaque at room temperature, turning clear and liquid only when they are heated. Highly unsaturated fats like canola or olive oil, on the other hand, remain clear and liquid at room temperature because their molecules have a tough time packing together in an organized manner.

It gets even more complicated when you take into account trans and cis fats, terms used to describe the geometry of unsaturated fats. **Trans fats** are unsaturated fats in which the double-carbon bond is formed in such a way that the resulting fatty acid is straight. A **cis fat** has its double-carbon bond formed in a way that makes the fatty acid kink, like a boomerang. While both formations occur in nature (trans fats are commonly found in animal fats), trans fats are far more abundant in hydrogenated fats—manmade fats in which the hydrogen atoms have been forced to bond with polyunsaturated fats to increase their saturation. Margarine and some types of shortening are examples of hydrogenated fats.

Current research indicates that trans fats have been positively linked to coronary artery disease. As a result, artificial trans fats are banned in some areas. Modern shortening products are manufactured in ways that minimize the amount of trans fat that makes it into the final tub.

**My mother used to keep a jar of bacon grease in her fridge that she swore was the key to the best fried chicken. Anything to that?**

Indeed! Bacon fat is not only highly saturated (making for that extra-crunchy coating), but it also adds flavor of its own. This can be a great thing for dishes like fried chicken or chicken-fried steak, as long as you don't use so much that it becomes overpowering. A ratio of 1:7 is ideal (that is, for every 4 cups peanut oil you use, replace ½ cup of it with rendered bacon fat). Avoid using bacon fat for more delicate things like vegetables or fish.

**I've heard that if you fry fish in oil, it can make the oil smell fishy. Any truth in that?**

At one point or another, you've probably walked into a restaurant or perhaps a neighbor's home and immediately caught a whiff of rancid, fishy-smelling oil. You may even have said to your host (or if you are more restrained than I have been in the past, to yourself), "Oof, somebody frying fish in here?"

Well, here's the thing: I've got both a fried fish shop and a fried chicken shop right on my block in Harlem. Oddly enough, it's the fried chicken joint that smells like rancid, fishy oil, while the fried fish shop smells only of fresh seafood. What gives? Turns out that the "fishy oil" smell you get from fried foods has nothing to do with the fish itself; it's caused by the inexorable breakdown of fat molecules.

**Hang on a minute. Breakdown of fat molecules? Sounds to me like you're talking about oxidation and hydrolysis. Can you explain yourself?**

No problem. Remember high school biology, where we learned that a fat molecule is made up of three fatty acids attached to a glycerol backbone, all arranged in a large upper-case-E shape? Well, the problem is that these fat molecules are not exactly stable. Given exposure to oxygen and enough time, they break down. And this gradual breakdown is sped up by exposure to heat, light, and air. Unfortunately, when you fry, all three of these types of exposure occur in abundance. On its own, exposure to oxygen causes *oxidation*, a process that causes the large fat molecules to break down, resulting in many smaller molecules, among them ketones and short-chain fatty acids. These are the foul-smelling molecules that are the true cause of that fishy smell in fry shops, and this type of reaction occurs even when you don't add any food to the hot oil. In extreme cases, it can even occur in poorly stored bottles of oil (for this reason, you should *never* store your oil near the stove—you're simply inviting rancidification).

Once you actually start frying things, it gets worse. *Hydrolysis*, a reaction that occurs when you combine water, oil, and heat (i.e., fry something), compounds and speeds up the effects of oxidation. That's why oil used for frying eventually breaks down and becomes stinky and unusable. Depending on how hot you fry and how much food you fry at a time, a container of oil will get anywhere from a half dozen to a few score of uses before it becomes unusable.

Finally, the last way in which oil breaks down is a process called *saponification*—literally, the conversion of oil to soap, and when we say soap, we're not talking Ivory or Dove bars, we're talking about the chemical definition: a chemical salt of a fatty acid.* Soaps are a surfactant, which means they have a hydrophobic (oil-loving/water-hating) end and a

---

*Although, yep, bath soaps do contain chemical soaps in addition to lathering agents, fragrances, exfoliants, and their ilk.

hydrophilic (water-loving/oil-hating) end. They are the peacemakers of the oil and water world, allowing the two to coexist without separating, as they are wont to do.

In this case, though, peaceful coexistence is a bad thing: the more surfactants your oil contains, the more water it can hold, the faster hydrolysis will occur, the lower the smoke point of your oil will be, and the less efficient it will be as a frying medium.

As it turns out, the fried chicken joint in my neighborhood smells fishy *not* because it is storing old fish under the counter—but because it doesn't filter or change the oil frequently enough. The fried fish shop, on the other hand, changes its oil regularly, leaving only the aroma of its fresh fish to linger in the air (guess which restaurant has the longer line).

**What about at home? How many times can I reuse my oil?**
When frying at home, you can expect to use the same batch of oil for six to eight frying sessions before it begins to break down. Certain foods will cause oil to break down faster than others. In general, the smaller the particles in your breading or batter, the faster your oil will break down. Thus, fried chicken that's dredged in flour will ruin your oil faster than eggplant slices breaded in chunkier bread crumbs—which in turn will break down your oil faster than onion rings dipped into a batter that turns solid when fried.

**How come a restaurant is able to reuse its oil so many more times than I can at home?**
This is because of one of the major advantages commercial deep fryers have over home setups: they are not heated from the bottom. Restaurant deep fryers

have electric or gas-powered heating elements that are several inches above the base of the fryer. What does this mean? One of the biggest problems with deep-frying is the buildup of detritus in the fat. Bits of batter, flour, and bread crumbs all fall off your food when you drop it in the fryer and stay there after you take your cooked food out. What happens to this debris? Eventually it completely dehydrates and subsequently sinks to the very bottom of the oil. In a restaurant fryer, this isn't much of a problem: the debris sits in the relatively cool section of oil, underneath the heating element. At home, however, your wok or pot of oil is heated directly from underneath. Fallen particles burn, wreaking havoc on the quality of your oil and sticking to subsequent batches of food you drop into it.

So, what can you do at home? The key to preventing the oil from ruining your meal is to be very meticulous about cleaning it between batches of food and between frying. There are also some countertop small-scale electric deep fryers that have heating elements that work like the restaurant fryers, allowing you to get more out of your oil in the long run. However, they also take up counter space and most take a long time to heat up. The trade-off is largely a matter of personal preference.

**Any other advantages to a restaurant deep fryer?**
Restaurant deep-fat fryers are designed for volume: most have a capacity of at least 10 gallons. At home, you're more likely to be cooking with a couple of quarts of oil at most—about twenty times less. The advantage of using a ton of oil is easier temperature management. Drop a handful of room-temperature French fries into 10 gallons of 375°F oil, and it'll lose *at most* a degree or two. Do the same in two quarts of oil, and you're looking at a drop closer to

50 degrees. So at home, you have to heat your oil hotter in order to compensate for this loss.

**OK, I think I've got it: New oil = good, old oil = bad, right?**
Not necessarily! You may think that using fresh new oil is the best way to fry foods, and you would be forgiven for thinking that. Forgiven, but wrong. Here's why:

Completely fresh oil is highly hydrophobic: it doesn't want to get anywhere *near* water. Any food that you drop into a deep fryer is bound to have a very large percentage of water in it (after all, the whole point of frying is to drive off water), which means that the oil is *not* going to like it. It hates it so much, in fact, that it has trouble getting close to its surface. Have you ever noticed that when you drop battered food into fresh oil, there's a shiny bubble that forms around the food? That's a layer of water vapor rapidly escaping from its surface and preventing the fat from getting too close. Because the fat can't come in contact with the food, heat transfer is inefficient with fresh oil. This means longer cooking times, less crispness, and less "fried" flavor (remember, fried flavor comes from a combination of browning, dehydration, and fat absorption—see "What Is Deep-Frying?," page 849).

Slightly older oil, on the other hand, has got a few surfactants in the mix—those molecules that allow fat and water to come close to each other. Because of that, older oil is better able to penetrate foods, cooking them far faster and giving you crisper, better—flavored crusts.

As any longtime fry cook will tell you, you should always save a bit of the old fry oil to add to the new batch if you want to make sure your foods come out at optimum crispness right from the first batch of fresh oil. For home cooks, this amount can be as little as a tablespoon of old oil per quart of new oil.

**What should I do with the oil in between batches of frying?**
Just as with fresh oil, used oil should be stored in a cool, dark, relatively airtight environment. If you are planning to do a bunch of frying over the course of a few days, this can be as easy as straining it through a fine-mesh strainer lined with cheesecloth or a paper towel into a pot with a metal lid (not glass—glass lets in light) and keeping it in a cool corner of the kitchen. For longer-term storage, strain the used oil through a fine-mesh strainer, then funnel it back into its original packaging. Seal tightly and store it in a cool, dark cabinet.

**And what should I do once the oil finally *does* reach the point where it can't be reused any more? How do I get rid of it?**
Discarding used oil can be a real pain in the butt. Small amounts, say, less than a half cup or so, can be poured down the drain *with plenty of soap and warm water* (the soap helps the oil emulsify with the water, preventing it from sticking to and coating the insides of your pipes), but larger amounts require a bit more care.

The absolute best way to get rid of used oil is to donate it to an organization that collects spent oil to be used as fuel for specially adapted cars (in Boston, we used to call them McNugget mobiles, because the exhaust smells like a fast food kitchen). Unfortunately, these aren't that easy to come by. The easiest way for the home cook to discard used oil is to save its original container, funnel the cooled used oil back into it, screw on the cap, and dispose of it with the solid garbage.

# KEYS TO MAXIMIZING OIL USE

Here's a quick-and-dirty guide to maximizing the lifespan of your oil.

- **Watch the temperature.** Don't let oil get past its smoke point, where rapid breakdown will occur.
- **Remove excess batter and breading meticulously during and after frying.** Small particles of batter, bread crumbs, and, especially, flour can collect in the bottom of your cooking vessel, causing the oil to break down.
- **Carefully remove debris after frying.** Use a fine-mesh strainer to fish out any debris from your oil while it is still hot. For maximum effectiveness, strain the oil through a fine-mesh strainer lined with cheesecloth or a paper towel between each use to completely clean it.
- **Store the oil in a cool, dark, dry place.** For short-term storage (up to a few days), a pot with a metal lid in a cool corner of the kitchen is fine. For long-term storage, return it to its original container, seal tightly, and store it in a cool, dark cabinet.

# THE SMOKE POINTS OF COMMON OILS

Every oil has a smoke point, the temperature at which wisps of smoke will appear on its surface, and a flash point, the temperature at which actual flames will start dancing across the top.

Oils used for deep-frying should never be heated to either of these temperatures, for both safety and flavor reasons. Here are the smoke points of most common oils, along with the percentage of saturated fat they contain. There are many reasons to pick various oils for frying jobs. Some folks choose oils with lower saturated fat contents (like olive, canola, or rapeseed oil) for health reasons. But who are we kidding? We don't eat fried foods for their health benefits. Others pick oils high in saturated fats with relatively high smoke points for their superior frying ability. It's up to you, but I know which way I lean. My oil of choice for deep-frying is almost always peanut oil, which has a high smoke point and enough saturated fat to give a perfectly crisp crust without it tasting waxy or heavy, as shortening, lard, or other animal fats can.

The use of the term "vegetable oil" is not strictly regulated—it can contain any combination of certain oils. Practically, however, it's almost always a combination of corn, canola, and or sunflower oil, and its smoke point will fall somewhere in the 400° to 450°F range.

| FAT | SMOKE POINT | PERCENTAGE OF SATURATED FAT |
|-----|-------------|----------------------------|
| Butter | 300° to 350°F | 62% |
| Coconut Oil | 350°F | 86% |
| Vegetable Shortening | 360°F | 31% |
| Lard | 370°F | 40% |
| Extra-Virgin Olive Oil | 375° to 410°F | 13% |
| Canola Oil | 400° to 425°F | 7% |
| Sesame Oil | 410°F | 14% |
| Light Olive Oil | 425°F | 13% |
| Peanut Oil | 440°F | 17% |
| Sunflower Oil | 440°F | 10% |
| Corn Oil | 450°F | 13% |
| Palm Oil | 450°F | 81% |
| Soybean Oil | 495°F | 14% |
| Safflower Oil | 510°F | 9% |
| Avocado Oil | 520°F | 12% |

## THE MYTH OF THE FRY: USING HOTTER OIL DOES *NOT* LEAD TO LESS OIL IN YOUR FOOD

Crack open nearly any book about deep-frying and you'll find this advice: "Make sure that your oil is hot enough before adding your food, or it will absorb fat and become greasy." The theory is that as long as your oil as hot enough, as soon as you lower your food into it, the outward pressure of water vapor bubbles escaping the food will prevent oil from rushing into the food, and therefore your food will remain grease free. At first glance, it makes sense, right? I mean, we've all eaten bad fried food that's come out of a too-cool fryer, and indeed it *does* taste heavy and greasy. But is it actually because it contains more grease? A study reported in the *Journal of Food Process Engineering* says differently.

Turns out that the truth is quite the opposite: the hotter you fry your food, the *more* oil it will absorb. See, most foods that you throw in the fryer—whether batter-coated food, potatoes, or a hunk of chicken—are filled with water. They are literally saturated with the stuff. Imagine that French fry, for example, as a hotel with no vacancies—every single room is filled up with a water molecule. In order for any oil at all to penetrate the potato and take up residence, some of that water must first check out. If you think about this, you already know it: drop a piece of cold potato into a cold pot of oil. Does it absorb any of that oil? *Nope.* Wash the potato off, and it's as if the oil was never there. Now, here's the thing: water is pretty happy with its cellular accommodations. The only way to get it to leave is through forceful eviction, namely by adding some energy to it in the form of heat. When you drop a piece of potato into hot oil, energy from the oil is transferred to the water inside it, which will eventually absorb so much energy that it leaps from within the potato's cells and escapes in a bubble of vapor—thus freeing up a room for the oil to check into.

The water in a piece of food being fried exists in two forms: Free water will easily escape, jumping out of the food at relatively low temperature. Bound water, on the other hand, requires significantly more energy and a higher temperature before it escapes. Heat a potato slice up to 275°F, and despite the fact that the temperature is well above the boiling point of water, some of the bound water will remain inside until you get it even hotter. So, the hotter you fry a food, the more water will escape, and the more room is left for oil to be absorbed.

This was all pretty shocking news to me, so I did what any good skeptic would do: I tested it. I filled my wok with two quarts of oil and weighed it on a precise scale. Next, I heated the oil to 275°F and maintained that temp while I fried chicken for a fixed period of time. After removing the chicken, I weighed the oil remaining in the wok. I then repeated the test, this time maintaining the oil at 325°F while cooking the chicken. After repeating the test a few times, the results confirmed what I had read: the hotter the temperature the more oil the chicken absorbed.

There's an explanation commonly given for why foods cooked at a higher temperature supposedly absorb less fat: the outward pressure of water vapor rapidly escaping from the food prevents the influx of oil. This may be true while the food is actually in the hot oil, but as soon as it's removed, its temperature drops rapidly. What once was positive pressure being exerted from inside the food reverses itself and results in a partial vacuum within a matter of milliseconds. Rather than pushing water vapor out, the food rapidly sucks the oil clinging to its surface into its interior. Even the fastest fry cook in the world can't drain his onion rings fast enough to stop this influx of oil. Up to 70 percent of the oil absorbed by fried foods is absorbed within the first few seconds *after* it comes out of the fryer.

But despite having absorbed less oil, the chicken cooked at 275°F, which was limp and oily, *tasted* much greasier than the crisp chicken cooked at 325°F. Turns out that what we describe as "greasy" actually has nothing to do with the total amount of grease in a food—it's just an illusion. Rather, it's the combination of surface oil and a mushy, moist breading or batter in our mouths that gives us the sensation of greasiness or heaviness. Crisp, well-fried batter, breading, or chicken skin may contain more fat, but it sure doesn't taste that way.

Lesson learned: When frying, cooking at a higher temperature is absolutely essential if you want your food to be crisp and to not *taste* greasy, but don't fool yourself into thinking there's less fat in it!

## BATTERS AND BREADINGS

Have you ever dropped a naked skinless chicken breast into the deep fryer? I strongly advise against it. The moment it enters a vat full of 400°F oil, a couple of things start happening. First, the water content will rapidly convert to steam, bubbling out like a geyser, and the chicken's outer tissues become drier and drier. At the same time, the soft network of folded proteins in its musculature will begin to denature and tighten, firming its flesh and squeezing out juices. Pull it out a minute or two later, and you'll discover that it's become quite stiff, with a layer of desiccated meat a good ¼ inch thick surrounding it. This is when you'll quite rightfully say to yourself, "Ah, I wish I had battered that first."

Batters are made by combining some sort of flour—usually wheat flour, though cornstarch and rice flour are not uncommon—with a liquid and optional leavening or binding ingredients like eggs and baking powder. They coat foods in a thick, goopy layer. Breadings consist of multiple layers. Generally a single layer of flour is applied directly to the food to ensure that its surface is dry and rough so that the second layer—the liquid binder—will adhere properly. That layer generally consists of beaten eggs or a dairy product of some kind. The last layer gives the food texture. It can consist of a plain ground grain (like the flour or cornmeal in a traditional fried chicken breading), ground nuts, or any number of dry ground bread or bread-like products such as bread crumbs, crackers, or breakfast cereals.

No matter how your breading or batter is constructed, it serves the same function: adding a layer of "stuff" around the item being fried means the oil has a tough time coming in direct contact with it, and thus has a hard time transferring energy to it. All the energy being transferred to the food has to go through the medium of a thick air-pocket-filled coating. Just as the air-filled insulation in your house helps mitigate the effects of harsh external conditions on the air temperature inside, so do batters and breadings help the food underneath cook more gently and evenly, rather than burning or becoming desiccated by the fiercely energetic oil.

Of course, while the food inside is gently cooking, the precise opposite is happening to the batter or breading: it's drying out and its structure is getting firmer and firmer. Frying is essentially a drying process. Batters and breadings are formulated to dry out in a particularly graceful way. Rather than burning or turning leathery, a nice airy batter forms a delicately crisp, air-filled web of teeny-tiny bubbles—a solid foam that provides substance and crunch. Breadings work similarly, though rather than foamy in structure, they're craggy. The nooks and crannies in a good bread-crumb coating vastly increase the surface area of the food being fried, giving you more crunch in each bite. In the ideal world, a batter or breading becomes perfectly crisp just as the food inside—say, a slice of onion or a delicate piece of fish—approaches the ideal level of doneness. Achieving this balance is the mark of a good fry cook.

The recipes in this chapter will cover all of the basic types of breadings and batters, as well some other forms of breading and batter-free frying.

# THE PROS AND CONS OF FIVE COMMON BREADINGS AND BATTERS

| COATING | HOW IT'S DONE | PROS | CONS | CLASSIC USES | CRISPNESS LEVEL (1 to 10, low to high) |
|---|---|---|---|---|---|
| **Breading: Flour Dredge** | Brined or soaked (often in buttermilk) pieces of food are tossed in seasoned flour and fried. | When done well, produces plenty of crunchy, dark brown crust. | A little messy (you often end up breading your hands). Causes extremely rapid breakdown of oil. | Southern-style fried chicken, chicken-fried steak | 8 |
| **Breading: Standard Bread Crumb** | Food is dredged in flour, followed by beaten eggs, followed by dried bread-crumbs. | Very easy, though it requires a few pans for dredging. Achieves a very crisp, solid, airtight crust that absorbs sauces well. | Bread crumbs can sometimes be too flavorful, obscuring the food they coat. Standard bread crumbs can get soft fairly rapidly. Causes fairly rapid breakdown of oil. | Chicken Parmesan, schnitzel | 5 |
| **Breading: Panko Bread Crumb** | Same as for standard breading. | Panko crumbs have tons of surface, area, leading to exceptionally crisp coatings. | Panko can occasionally be hard to find. Very thick coating—food underneath must be quite robust. | Traditionally, Japanese-style tonkatsu (fried chicken or pork cutlets) | 9 |
| **Batter: Beer** | Seasoned (sometimes leavened) flour is mixed with beer (and sometimes eggs) to create a thick pancake-like batter (the beer promotes browning, while its bubbles help keep the batter light). Beer-battered items can be redredged in flour for increased crispness. | Great flavor. Thick, thus good at protecting delicate foods (like fish). Easy to make and relatively stable after mixing. Very slow oil breakdown if plain (no second flour dredge). | Doesn't achieve same crispness as some other batters. Quite a few ingredients required. Batter must be used quickly after it's made. Coating can turn soft fairly rapidly if plain (no second flour dredge). Rapid oil breakdown if second flour dredge is applied. | Fried fish, onion rings | 5 |
| **Batter: Cornstarch/ Thin Tempura-Style** | High-starch/low protein flour (such as a wheat flour/cornstarch mix) is combined with ice-cold water (sometimes soda water, or sometimes egg) and rapidly mixed, leaving the batter still lumpy. Foods are immediately dipped and fried briefly. | Extremely crisp. High surface area means lots of crunchy bits. Low protein batter means less browning, allowing flavor of delicate foods like vegetables or shrimp to come through. Moderately slow oil breakdown. | Difficult to mix batter correctly (it's very easy to over-or undermix). Batter must be used immediately. | Vegetables, shrimp, Korean-style fried chicken | 8 |

# COATING Style 1:
## FLOUR DREDGING

### Southern-Style Fried Chicken

I know how passionate people can get about fried chicken, and I'm not one to tell you who makes the best, but if you were to ask Ed Levine, the Serious Eats overlord, he'd tell you that it's Gus's, a sixty-seven-year-old institution in Mason, Tennessee. They serve fried chicken that he describes as incredibly crunchy, with a crisp, craggy crust, juicy meat, and a "cosmic oneness" between the breading and the skin. We're talking fried chicken so good that you have to resort to metaphysics to make sense of it.

For me, as a kid growing up in New York, fried chicken came from one place, and one place only: those grease-stained cardboard buckets peddled by

the Colonel himself. To my young mind, KFC's extra-crispy was about as good as it got. I distinctly remember eating it: picking the coating off in big, fat chunks; tasting the spicy, salty grease; and shredding the meat underneath with my fingers and delivering it to my waiting mouth. It was heavenly.

But times have changed, and as is often the case, revisiting those fond childhood memories results only in disappointment and disillusionment. All over the country, there's a fried chicken and soul food renaissance going on. Even the fanciest restaurants in New York are adding it to their menus. My eyes and my taste buds have been opened to what fried chicken truly *can* be. I may still dig

the ultracrunchy, well-spiced crust that KFC puts on its birds, but that's about the only thing it has going for it. Flaccid skin, dry and stringy breast meat, and chicken that tastes like, well, it's hard to tell if it really tastes like *anything* once you get rid of the crust.

That said, stylistically, it can't be faulted. So I figured that I could somehow manage to take what the Colonel started and bring it to its ultimate conclusion—that is, deep chicken flavor; a flab-free skin; juicy, tender meat; and crisp, spicy coating—I might just be able to recapture those first fleeting childhood tastes of fried chicken as I remembered them.

## Inside Out

I started with a working recipe of chicken pieces simply dipped in buttermilk and tossed in flour seasoned with salt and black pepper, then fried in peanut oil at 325°F until cooked through. A few problems immediately became clear. First off, timing: By the time my chicken was cooked through (that's 150°F in the breasts and 165°F in the legs), the outer crust was a dark brown, bordering on black in spots. Not only that, but it didn't have nearly as much crunch as I wanted. Finally, the meat underneath the crust wasn't completely desiccated, but I wouldn't exactly describe it as moist, not to mention its rather bland flavor. I decided to fix my chicken from the inside out.

The problem is that with fried chicken, the crisp well-seasoned coating is merely a surface treatment. None of that flavor penetrates very deeply. Surely

---

*For those of you squeamish about "undercooked" chicken or who insist that breast meat *must* be cooked to 165°F to be safe and tasty, please see page 388, for a discussion on *real world* food safety, which is quite different from what the U.S. government would have you believe.

Looks crisp outside, but inside this chicken is dry.

brining and/or marinating should help with that problem? Brining is the process by which a lean meat (most often chicken, turkey, or pork) is submerged in a saltwater solution. As the meat sits, the saltwater will slowly dissolve key muscle proteins—most notably myosin, a protein that acts as a sort of glue, holding muscle fibers together). As the myosin dissolves, three things take place:

- **First, the ability of the meat to hold onto moisture increases.** You can imagine meat as a series of long, skinny toothpaste tubes tied together. As you cook the meat, the tubes of toothpaste get squeezed, pushing out valuable juices. Breading will help mitigate this effect to a degree by slowing down the transfer of energy to the meat, but a significant amount of squeezing is still going to occur regardless of how well breaded the chicken is. Myosin is one of the key proteins responsible for this squeezing action, so by dissolving it, you prevent a lot of moisture loss from taking place.
- **Second, brining alters the texture of the meat by allowing dissolved proteins to cross-link with each other.** This is the main principle behind sausage making—dissolved proteins can bond with each other, creating a pleasantly bouncy, tender texture. By brining a chicken breast or a pork chop, you're in effect giving it a very light cure—the

same process that converts a raw ham into a supple prosciutto.

- **Third, as the brine slowly works its way into the meat, it seasons it beyond just the very surface.** An overnight brine will penetrate a few millimeters into the meat, giving you built-in seasoning before you ever get to the breading. Brines also improve juiciness by increasing the muscles' ability to retain moisture. My normal brining for chicken breast is anywhere from 30 minutes to 2 hours. In this case, however, a much, much longer brining time was necessary in order to completely mitigate the effects of high-temperature frying, delivering a uniquely smooth, juicy texture to the meat.

A full 6 hours submerged in salt/sugar water produced the beauty below. Weighing the meat confirmed that an overnight-brined-then-fried bird loses about 9 percent less moisture than an unbrined bird does and is significantly tastier.

Unbrined chicken on the left versus
brined chicken on the right.

I've experimented with tossing certain animal preparations with a mixture of baking powder and salt a day in advance in order to improve their crispness (see All-Belly Porchetta, page 664). The salt acts as a brine, while the baking powder raises the pH of the skin, causing it to brown more efficiently and the thin film of protein-rich liquid around it to form microbubbles that can add crispness. I tried this method on my fried chicken, but it ended up drying the skin out too much, making it tough to get the breading to remain attached down the line.

Knowing that I'd be soaking my chicken in buttermilk the next day anyway, I wondered if I'd be able to kill two birds with one stone by replacing the water in the brine with buttermilk. Not only did the chicken come out just as moist as with water brine, it was actually significantly more tender as well, due to the tenderizing effects of buttermilk on food (soaking it for more than one night led to chicken that was so tender that it bordered on mush). Finally, hitting the buttermilk with spices helped build flavor right into the surface of the bird. I played around a bit with the mix before arriving at a blend of cayenne pepper and paprika (for their heat and peppery flavor), garlic powder,✳ a bit of dried oregano, and a healthy slug of freshly ground black pepper. The Colonel may use eleven secret herbs and spices in his chicken recipe, but five was quite enough for me (and both my wife and my doorman heartily concurred).

### Crust Lust

Next up: add some extra crunch to that crust. I reasoned that there were a few ways to do this. First off, I wanted to increase the crust's thickness. I tried double-dipping my chicken—that is, dredging the brined chicken in flour (seasoned with the same spice blend as my brine), dipping it back into the

---

✳ Some folks shun garlic powder, saying that it's nothing like real garlic. I agree: garlic powder is nothing like real garlic. But that doesn't mean it doesn't have its culinary uses. It's particularly effective in spice rubs and breadings, where fresh garlic would be difficult to incorporate, due to its texture.

buttermilk, and then dredging it once more in flour before frying, a method chef Thomas Keller uses for his justifiably famous fried chicken at Ad Hoc. This worked marginally better—that second coat definitely developed more crags than the first coat did. But it also made for an extremely thick breading that had a tendency to fall off the breast because of its heft.✳

A double coating of flour creates a thick crust that falls off the chicken.

Much better was to simply add a bit of extra structure to the breading in the form of an egg mixed into the buttermilk.

My crust was certainly thick enough now, but I ran into another problem: rather than crisp and crunchy, it was bordering on tough, almost rock-like in its density. Knowing that gluten—the network of proteins formed when flour meets water—was the most likely culprit, I sought out ways to minimize its formation. First and foremost: cut the protein-rich wheat flour with cornstarch, a pure starch that adds moisture-absorbing capabilities to the breading without adding excess protein. Replacing a quarter

of the flour worked well. Adding a couple teaspoons of baking powder to the mix helped bring a bit of air to the mix, forming a crust that was lighter and crisper, with increased surface area (and we all know that more surface area = more crispness, right?).

Finally, I used a trick that a friend, a former employee of the Chick-fil-A Southern fast-food fried-chicken chain had told me about. He'd mentioned that once the chicken was breaded, the later batches always come out better than the earlier ones as bits of the flour mixture clumped together, making for an extra-craggy coat. Adding a couple tablespoons of buttermilk to the breading mix and working it in with my fingertips before dredging the chicken simulated this effect nicely.✳

Adding buttermilk to the dredge creates the extra-craggy surface on the left.

The last problem—the coating overcooking long before the chicken is cooked through to the center—was simple to solve. Just fry the chicken until golden brown, then transfer it to a hot oven to finish cooking at a gentler pace. The result is chicken with a deep brown, craggy crust that's shatteringly crisp but not tough and that breaks away to meat that bursts with intensely seasoned juices underneath.

---

✳ You may notice the redness of the center of the chicken. This is not because it is undercooked, but because I cracked the bone when cutting it open, revealing some of the chicken's red marrow. Occasionally bones may snap or crack on their own, or while you are breaking down the chicken, leaving a few red spots inside the chicken even when it is fully cooked. This should not alarm you.

---

✳ This method is also employed in *Cook's Country* magazine's fried chicken recipe.

# EVEN-CRUNCHIER FRIED CHICKEN?

I was chatting with my friend the San Francisco chef Anthony Myint a while back, when he mentioned that whenever he has leftover fried chicken, he'll fry it in hot oil again the second day, resulting in chicken that's even better than it was the first day. Normally there's a limit to how much moisture you can drive out the coating on a piece of fried chicken, defined by how long you leave it in the oil. Let it get too hot, and the exterior will start to burn. But if you let it cool overnight, some of the moisture from the inner layers of coating will work their way toward the exterior. When you fry it again the second day, this moisture is driven off, leaving you with an extra-thick layer of dehydrated coating around your chicken.

Not only is it a great way to treat leftovers, but if you've got the time, allowing your regular fried chicken to rest at room temperature or in the refrigerator for an hour or two before *re*frying it will give you the same results, all in a single day's work.

## THE REJECTS

I wasn't kidding when I said that I tested this recipe thoroughly. I went through more than fifty whole chickens and over a hundred separate iterations and tests to get there, at times filling up all of my cooling racks with samples so that I was then forced to drain some of the chicken on a dish rack. I made every conceivable bad version of fried chicken along the way.

# EXPERIMENT:
## Is It Best to Rest?

Here's a question that has always vexed me: many existing recipes for fried chicken advise you to let the chicken rest for a half hour or so after breading it before frying it. But very few offer an explanation as to why, which led me to question the very practice. Is fried chicken really better when you let it rest? I cooked a few batches of chicken side by side to find out.

### Materials

- **See the recipe for Extra-Crunchy Southern Fried Chicken (page 871).**

### Procedure

Follow the recipe as directed through step 3, but do not heat the oil yet. In step 4, dredge half the chicken pieces in the seasoned flour, place them on a rack set on a rimmed baking sheet and let sit undisturbed for at least 30 minutes, and up to 1 hour. Heat up the oil as directed in step 3. Dredge the remaining chicken pieces and immediately fry and finish baking all the chicken as directed, being careful to keep track of which batch is which.

Taste both batches of chicken and note their texture.

### Results and Analysis

While neither batch will be *bad*, the chicken that was allowed to rest before frying should end up significantly tougher with a hard, brittle crust instead of the crisp, crunchy crust of the chicken dropped into the fryer immediately after dredging. Why is this?

Once again, it's our friend gluten.* As the chicken rests, the flour in the dredging gradually absorbs moisture from the buttermilk and the surface of the chicken. As it absorbs moisture, its proteins begin to unfold and link up with each other, forming a sheath that grows tougher and tougher as it sits. Let it sit for too long, and you'll get a crust hard enough to crack your teeth. Fried chicken should be crisp, never tough, and to achieve this goal, you want to get your chicken from flour to oil as rapidly as possible. Of course, this introduces a problem: freshly dredged chicken will release a ton of dry flour into the oil, which will cause the oil to go bad very fast. My advice is to shake the bejeezus out of those chicken pieces in a metal strainer as they come out of the flour mix so that there are as few excess particles of flour as possible.

---

*You'll find that like a stomped Koopa Troopa or that ring of scum around the bathtub, he just keeps popping up, over and over.

# EXTRA-CRUNCHY SOUTHERN FRIED CHICKEN

**SERVES 4**

2 tablespoons paprika

2 tablespoons freshly ground black pepper

2 teaspoons garlic powder

2 teaspoons dried oregano

½ teaspoon cayenne pepper

1 cup buttermilk

1 large egg

Kosher salt

One whole chicken, about 4 pounds, cut into 10 pieces (see "How to Break Down a Chicken," page 363) or 3½ pounds bone-in, skin-on breasts, legs, drumsticks, and/or wings

1½ cups all-purpose flour

½ cup cornstarch

1 teaspoon baking powder

4 cups vegetable shortening or peanut oil

1. Combine the paprika, black pepper, garlic powder, oregano, and cayenne in a small bowl and mix thoroughly with a fork.

2. Whisk the buttermilk, egg, 1 tablespoon salt, and 2 tablespoons of the spice mixture in a large bowl. Add the chicken pieces and toss and turn to coat. Transfer the contents of the bowl to a gallon-sized zipper-lock freezer bag and refrigerate for at least 4 hours, and up to overnight, flipping the bag occasionally to redistribute the contents and coat the chicken evenly.

3. Whisk together the flour, cornstarch, baking powder, 2 teaspoons salt, and the remaining spice mixture in a large bowl. Add 3 tablespoons of the marinade from the zipper-lock bag and work it into the flour with your fingertips. Remove one piece of chicken from the bag, allowing excess buttermilk to drip off, drop the chicken into the flour mixture, and toss to coat. Continue adding chicken pieces to the flour mixture one at a time until they are all in the bowl. Toss the chicken until every piece is thoroughly coated, pressing with your hands to get the flour to adhere in a thick layer.

4. Adjust an oven rack to the middle position and preheat the oven to 350°F. Heat the shortening or oil to 425°F in a 12-inch straight-sided cast-iron chicken fryer or a large wok over medium-high heat. Adjust the heat as necessary to maintain the temperature, being careful not to let the fat get any hotter.

5. One piece at a time, transfer the coated chicken to a fine-mesh strainer and shake to remove excess flour. Transfer to a wire rack set on a rimmed baking sheet. Once all the chicken pieces are coated, place skin side down in the pan. The temperature should drop to 300°F; adjust the heat to maintain the temperature at 300°F for the duration of the cooking. Fry the chicken until it's a deep golden brown on the first side, about 6 minutes; do not move the chicken or start checking for doneness until it has fried for at least 3 minutes, or you may knock off the coating. Carefully flip the chicken pieces with tongs and cook until the second side is golden brown, about 4 minutes longer.

*continues*

**1**

**2**

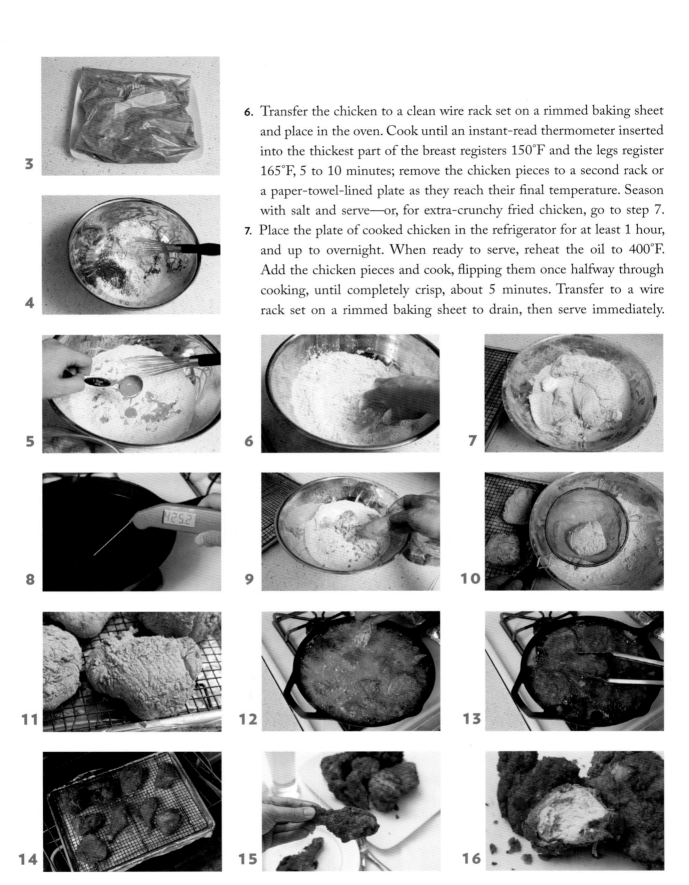

6. Transfer the chicken to a clean wire rack set on a rimmed baking sheet and place in the oven. Cook until an instant-read thermometer inserted into the thickest part of the breast registers 150°F and the legs register 165°F, 5 to 10 minutes; remove the chicken pieces to a second rack or a paper-towel-lined plate as they reach their final temperature. Season with salt and serve—or, for extra-crunchy fried chicken, go to step 7.

7. Place the plate of cooked chicken in the refrigerator for at least 1 hour, and up to overnight. When ready to serve, reheat the oil to 400°F. Add the chicken pieces and cook, flipping them once halfway through cooking, until completely crisp, about 5 minutes. Transfer to a wire rack set on a rimmed baking sheet to drain, then serve immediately.

# EXTRA-CRUNCHY CHICKEN-FRIED STEAK
## WITH CREAM GRAVY

Once you've perfected crispy fried chicken, it's only a hop, skip, and a jump to Texas-style chicken-fried steak—that is, beef steak that's fried in the style of fried chicken and served with a creamy, peppery white gravy. The key to great chicken-fried steak is starting with the right cut. We went into cuts of beef for steaks in detail in Chapter 3 (see page 328), but for now, we're just looking for an inexpensive cut with plenty of beefy flavor. Since chicken-fried steak both gets pounded thin (a technique that tenderizes the meat) and goes for a dunk in my buttermilk brine, tenderness won't be an issue—we can pick a cut based on its flavor, confident that it'll soften up before we fry it. Similarly, pounding and brining improve juiciness, so even when the steak comes out medium or well-done, as it does when it's fried, it'll still be plenty juicy.

My go-to cut for this is flap meat (also sold as sirloin tips), though bottom round or top sirloin work will also work well if you can't find it. Chicken is naturally tender, steak needs a bit more help. Lightly scoring the meat against the grain with a sharp knife will shorten long muscle fibers, resulting in a more tender finished product. The cream gravy served with the steak is nearly identical to the Creamy Sausage Gravy on page 164; just leave out the sausage. If you want to go extra-decadent, leave it in.

This steak is brined and dredged in the same manner as the fried chicken. Refer to photos there for a closer look at the process.

**SERVES 4**

**For the Steak**

2 tablespoons paprika

2 tablespoons freshly ground black pepper

2 teaspoons garlic powder

2 teaspoons dried oregano

½ teaspoons cayenne pepper

1 pound flap meat (also sold as sirloin tips; see Note above), cut into four 4-ounce steaks

1 cup buttermilk

1 large egg

Kosher salt

1½ cups all-purpose flour

½ cup cornstarch

1 teaspoon baking powder

4 cups vegetable shortening or peanut oil

**For the Cream Gravy**

2 tablespoons unsalted butter

1 small onion, diced

2 medium cloves garlic, minced or grated on a Microplane (about 2 teaspoons)

2 tablespoons all-purpose flour

1 cup whole milk

¾ cup heavy cream

Kosher salt and freshly ground black pepper

1. Combine the paprika, black pepper, garlic powder, oregano, and cayenne in a small bowl and mix thoroughly with a fork.

2. Place each steak on a cutting board and score at 1-inch intervals against the grain, cutting ¼ inch deep into the meat; flip and repeat. One at a time, sandwich each steak between two pieces of plastic wrap and pound with a meat pounder or the bottom of a heavy skillet until approximately ¼ inch thick.

3. Whisk the buttermilk, egg, 1 tablespoon salt, and 2 tablespoons of the spice mixture in a large bowl. Add the steaks and turn to coat. Transfer the contents of the bowl to a gallon-sized zipper-lock freezer bag and refrigerate for at least 4 hours, and up to overnight, flipping the bag occasionally to redistribute the contents and coat the steak evenly.

4. When ready to fry the steaks, make the gravy: Heat the butter in a 10-inch heavy-bottomed nonstick skillet over medium-high until foamy. Add the onion and cook until softened, about 4 minutes (lower the heat if the butter starts to brown). Add the garlic and cook, stirring, until fragrant, about 30 seconds. Add the flour and cook, stirring constantly, until fully absorbed, about 1 minute. Add the milk in a thin stream, whisking constantly. Add the cream, whisking, and bring to a simmer, whisking constantly. Simmer, whisking, until thickened, about 3 minutes. Season to taste with salt and plenty of pepper. Keep warm while you fry the steaks.

5. Whisk together the flour, cornstarch, baking powder, 2 teaspoons salt, and the remaining spice mixture in a large bowl. Add 3 tablespoons of the marinade from the zipper-lock bag and work it into the flour with your fingertips. Remove the steaks from the bag, allowing the excess buttermilk to drip off, drop the steaks into the flour mixture, and toss and flip until thoroughly coated, pressing with your hand to get the flour to adhere in a thick layer. Shake the steaks over the bowl to remove excess flour, then transfer to a large plate.

6. Adjust an oven rack to the middle position and preheat the oven to 175°F. Heat the shortening or oil to 425°F in a wok or 12-inch cast-iron skillet over medium-high heat, about 6 minutes. Adjust the heat as necessary to maintain the temperature, being careful not to let the fat get any hotter.

7. Carefully lower 2 steaks into the pan. Adjust the heat to maintain the temperature at 325°F for the duration of cooking. Fry the steaks without moving them for 2 minutes, then carefully agitate the steaks with a wire-

mesh spider, being careful not to knock off any breading, and cook until the bottom is a deep golden brown, about 4 minutes. Carefully flip the steaks and continue to cook until the second side is golden brown, about 3 minutes longer.

8. Transfer the steaks to a paper-towel-lined plate to drain for 30 seconds, flipping them once, then transfer to a wire rack set on a rimmed baking sheet and transfer to the oven to keep warm. Repeat with the remaining 2 steaks. Serve with the cream gravy.

## POUNDING MEAT

The easiest way to pound meat into a thin, even layer without making a mess is to place it between two sheets of plastic wrap or inside a heavy-duty zipper-lock bag with the sides split open with a knife. This will ensure that the steak doesn't stick to your cutting board or your pounding device, allowing it to thin easily and evenly. A meat pounder is nice to have if you've got the space, but a heavy 8-inch skillet will do the job perfectly adequately.

# EXTRA-CRUNCHY FRIED CHICKEN SANDWICHES

Chick-fil-A, the fast food chain started in Atlanta, GA, has a well-deserved near-cult-like following. The classic Chick-fil-A sandwich is a thing of simple beauty: A juicy, salty, crisply fried chicken breast. A buttered and toasted soft, sweet bun. Two dill pickle chips. That's all there is to it.

What makes it great is the perfection of each of the elements: That crisp golden brown crust, spiced just right with a perfect sweet-salty-savory-hot balance. The way it coats that breast underneath, a chicken breast that defies all we know about chicken. This is no dry, stringy, bland chicken bosom; it is a breast of unparalleled juiciness, with a dense, meaty texture and deeply seasoned flavor. Bring all of the elements together, and you've got a sandwich that is nearly impossible to improve upon. Of course, nearly impossible means slightly possible. We'll take that chance.

How are we going to improve it? By using our own fried chicken recipe, of course. Luckily, modifying our existing recipe into one that works well in a sandwich is a relatively painless procedure. All we need to do is start with smaller pieces of chicken—a single breast half, split horizontally to yield two 3- to 4-ounce portions, is the ideal size for a sandwich—and pair it with just the right type of butter-toasted soft bun and pickle.

The bun is your typical hamburger bun. Soft and slightly sweet, with a fluffy Wonder Bread–like texture. It measures in at around 4½ inches in diameter, which puts it right in the range of Arnold Hamburger Rolls (sold under the name Oroweat west of the Rockies). Toasted in a skillet in just a bit of melted butter, it's a perfect taste-alike to the Chick-fil-A buns.

As for the pickles, I tried a few different brands of crinkle-cut dill chips. Heinz had the right flavor, but the chips were too small—I could've added a few extras, I suppose, but I feel like the two-pickle-per-sandwich rule that Chik-fil-A has laid out is a wise and unbreakable law. Instead, I turned to Vlasic Ovals Hamburger Dill Chips, which have a larger surface area and the same salty-vinegary-garlicky flavor.

2 tablespoons paprika

2 tablespoons freshly ground black pepper

2 teaspoons garlic powder

2 teaspoons dried oregano

½ teaspoon cayenne pepper

1 cup buttermilk

1 large egg

Kosher salt

3 boneless, skinless chicken breast halves, 6 to 8 ounces each, horizontally split in half, to make 6 cutlets (see page 367)

6 cups vegetable shortening or peanut oil

1½ cups all-purpose flour

½ cup cornstarch

1 teaspoon baking powder

6 soft hamburger buns, toasted in butter

12 dill pickle chips

1. Combine the paprika, black pepper, garlic powder, oregano, and cayenne, in a small bowl and mix thoroughly with a fork.

2. Whisk the buttermilk, egg, 1 tablespoon salt, and 2 tablespoons of the spice mixture in a medium bowl. Add the chicken pieces and toss and turn to coat. Transfer the contents of the bowl to a gallon-sized zipper-lock freezer bag and refrigerate for at least 4 hours, and up to overnight, flipping the bag occasionally to redistribute the contents and coat the chicken evenly.

3. Heat the shortening or oil to 375°F in a large wok, deep fryer, or Dutch oven.

4. Meanwhile, whisk together the flour, cornstarch, baking powder, 2 teaspoons salt, and the remaining spice mixture in a large bowl. Add 3 tablespoons of the marinade from the zipper-lock-bag and work into the flour with your fingertips.

5. Remove one piece of chicken from the bag, allowing excess buttermilk to drip off, drop the chicken into the flour mixture, and toss to coat. Continue adding chicken pieces to the flour mixture one at a time until they are all in the bowl. Toss the chicken until every piece is thoroughly coated, pressing with your hands to get the flour to adhere in a thick layer. Transfer one of the coated cutlets to a fine mesh strainer and shake to remove excess flour and then, with your hands or a pair of tongs, slowly lower it into the hot oil. Repeat with remaining breasts. Cook, turning the cutlets occasionally, until golden brown and crisp on both sides and cooked through, about 4 minutes. Transfer to a paper-towel-lined plate.

6. Place 2 pickles on each bottom bun and a fried chicken cutlet on top. Close the sandwiches, then cover with an overturned bowl or aluminum foil and allow to rest for 2 minutes to steam the buns. Serve immediately.

1          2          3          4

# COATING Style 2:
## BREAD-CRUMB COATING

# CHICKEN PARMESAN

**Fried breaded chicken cutlets are good enough on their own, but add some great marinara sauce and a layer of gooey melted cheese? It's like extra-crispy meat-based pizza. Ah, only in America.**

**And the best part? Chicken Parm is *easy*. The only slightly irritating part is the breading. Standard breading consists of three distinct layers: flour, egg, and bread crumbs. Here's what they're for.**

- **Bread crumbs** make up the outermost layer, and they perform two functions. First, the many nooks and crannies formed by the crumbs increases the overall surface area of the chicken (see "Fractals, Panko, and Bread-Crumb Coatings," page 882). It also serves as an insulator, preventing the chicken from overcooking and drying out. Of course, bread crumbs won't stick without . . .

- **Eggs**. They form the adhesive layer, and they're perfect for the job. They start out as a viscous liquid, but as they fry, they form a solid gel, ensuring that the crumbs stay put. The eggs, however, would have a hard time sticking to the food without . . .
- **Flour**. Like a coating of primer before you add paint, the flour coats the food being breaded and begins to absorb some of its moisture, hydrating and forming a thin layer of sticky, irregularly lumpy gel. It's this gel that the eggs latch on to.

**Chicken Parmesan is one of the easiest and tastiest ways to get the hang of the breading process. For me, the most annoying part of breading foods is accidentally breading your fingers as you do it. All you've got to remember in order to prevent this is to use one hand for dry and one hand for wet.
Like this:**

1. Using your right hand (or your left if you are left-handed), pick up a piece of food and transfer it to the bowl of flour. Scoop up some flour and toss it on top of the food, then toss the food around until it's nicely coated.
2. Still using your right hand, pick up the food from the flour, shake it a bit to get rid of the excess, and drop it into the egg bowl. This time, use your left hand to move the food around until it's well coated.
3. Still using your left hand, pick up the eggy food, let it drip a bit, and drop it into the bowl with the bread crumbs. Here's the tricky part: If you were to now use your right hand to pick up the food and flip it, you'd get egg on it. Use your left hand, and you end up coating it in bread crumbs. Here's what to do: Use your right hand to pick up some extra bread crumbs from around the food and drop them on top, then carefully spread them around until you can pick up the food without egging your hand. Flip the food and repeat, pressing it into the crumbs to coat it thoroughly. It pays to have more bread crumbs than you need in the bowl.
4. Pick up the food with your right hand and transfer it to a plate or a rack, ready to be fried.

**As far as cooking the chicken goes, I found that shallow-frying it in a wide skillet was less messy and easier to clean up than deep-frying it. I also used panko-style crumbs, which I seasoned myself—far easier than making your bread crumbs, and much better than the sandy "Italian-style" crumbs from the supermarket (see "Fractals, Panko, and Bread-Crumb Coatings," page 882).**

Next up: the sauce and cheese. I opted to use my basic marinara sauce for this dish; its richness and deep flavor stand up nicely to the crunchy chicken. For cheese, a grating of fresh mozzarella applied before baking is traditional. Despite the nomenclature, Parmesan cheese does not always make an appearance in this dish. But that's not gonna stop us. I like applying good-quality Parmesan at three different stages. First, I incorporate some into the breading. It takes on a sweet, nutty flavor as it fries—much better than bread crumbs alone. Next, I mix some in with the mozzarella before topping the chicken with it before baking. Finally, I add a handful of grated Parm to the chicken after it comes out of the oven. The residual heat of the chicken softens the cheese slightly, but you still get plenty of intense, salty hits as you eat it.

While many restaurants opt to blanket the entire piece of chicken in a thick layer of mozzarella, that ends up softening the coating too much. Instead, I found that laying the chicken cutlets out in a casserole dish, then spooning on sauce in a line down the center and adding a layer of cheese left the ends of the cutlets protruding so that they maintained at least *some* of that crispness you worked hard to create.

I'd put this version of chicken Parmesan up against any Little Italy version in the country.

**SERVES 4**

2 boneless, skinless chicken breast
   halves, about 8 ounces each, split
   horizontally in half, to make 4
   cutlets (see page 367)

Kosher salt and freshly ground
   black pepper

1½ cups panko-style bread crumbs

2 teaspoons dried oregano

2 ounces Parmigiano-Reggiano,
   finely grated (about 1 cup)

½ cup all-purpose flour

2 large eggs, beaten

1 cup vegetable oil

1 recipe Perfect Easy Red Sauce
   (page 695)

8 ounces mozzarella cheese, grated

2 tablespoons minced fresh basil

2 tablespoons minced fresh parsley

1. Adjust an oven rack to the center position and preheat the oven to 375°F. One at a time, place each cutlet between two sheets of plastic wrap and gently pound with a meat pounder or the bottom of a heavy skillet to an even ¼- to ⅛-inch thick. Season with salt and pepper and set aside.

2. Combine the bread crumbs, oregano, and ¼ cup of the Parmigiano-Reggiano in a shallow bowl or pie plate. Place the flour and the eggs in separate shallow bowls or pie plates. Use your right hand to pick up one chicken cutlet and add it to the bowl of flour. Use your left hand to coat it evenly with flour, then use your right hand to pick up the chicken, shake off excess flour, and add to the eggs. Turn the chicken with your left hand until evenly coated, then use your left hand to transfer it to the bread crumbs. Lift some crumbs with your right hand and press them onto the top of the chicken, then use your right hand to turn the chicken several times, pressing it into the crumbs until evenly coated. Transfer to a wire rack set on a rimmed baking sheet. Repeat with the remaining cutlets.

3. Heat the oil in a 12-inch nonstick or cast-iron skillet over high heat until it reaches 350°F on an instant-read thermometer (the corner of a chicken cutlet dipped into it should sizzle vigorously). Carefully add the chicken cutlets (you may need to work in two batches) and cook until the first side is golden brown, about 3 minutes, shaking the pan gently and adjusting the heat as necessary to maintain a constant temperature. Carefully flip the chicken with tongs and cook until the second side is golden brown, about 2 minutes longer. Transfer the cutlets to a paper-towel-lined tray to absorb excess oil.

4. Spread half the sauce over the bottom of a large oven-safe serving platter or baking dish. Add the chicken cutlets, shingling them slightly as necessary. Spread the remaining sauce evenly down the center of the cutlets, leaving the edges exposed. Scatter the mozzarella and half of the remaining Parmigiano-Reggiano evenly over the sauce. Bake until the cheese has melted and just started to brown, about 15 minutes. Remove from the oven and allow to rest for 5 minutes.

5. Sprinkle the chicken with the remaining Parmigiano-Reggiano, the basil, and parsley and serve.

## FRACTALS, PANKO, AND BREAD-CRUMB COATINGS

Have you heard of Mandelbrot fractals? They're computer-generated images that appear on a small scale very much as they do on a large scale. Fractals are something that occur in nature quite often—the outlines of a cloud, for instance, or the leaves on a fern. One well-known fractal effect pertains to coastlines. When you look at a coastline from far away and measure it, you'll measure a certain perimeter. As you zoom closer and closer, you realize there are tiny inlets or curves in the beach that weren't visible from far away. When measured again, these bumps add length to the total perimeter. This is a phenomenon known as the Richardson Effect, which basically says that the more precisely you measure a coastline, the longer the measurement gets. And the more bumpy or irregular the surface is, the more pronounced the effect.

Well, the same applies to breaded foods. Though a nonbreaded chicken cutlet and a breaded chicken cutlet have essentially the same mass and volume, because of the irregular edges of the breaded cutlet, it actually has far greater surface area than the nude chicken.

Panko, Japanese-style bread crumbs, can further enhance this effect. In contrast to sandy or coarse crumbs, panko crumbs are wide, craggly flakes that jut out wildly when applied, providing over twice as much surface area as regular bread crumbs. Now that's adding some *major* crunch!

# EGGPLANT PARMESAN CASSEROLE

If you had asked the fourteen-year-old me what my three least favorite things in the world were, I would have answered, "My sister, She-Ra, and eggplant." But I would have rather sat through a She-Ra marathon *with* my sister than be forced to down a bite of the slimy, bitter vegetable. Later on down the line, though, I realized that it's not eggplant itself that's horrible, it was more just that my mom didn't really know how to cook it (sorry, Mom)—and she's not the only one.

Indeed, most of the ingredients in eggplant Parmesan are tough to mess up. And I'm not talking the traditional Sicilian style where unbreaded slices of eggplant are deep-fried in olive oil and delicately layered with tomatoes and mozzarella—I'm talking the all-American version. The kind where meaty slabs of eggplant are breaded and fried before being layered in a casserole dish with a cooked tomato sauce loaded with mozzarella and Parmesan cheese. The best bites of the dish are the parts where the fried breading soaks up the sweet tomato sauce and swells in between the layers of meaty eggplant and gooey cheese.

Tomato sauce? No problem (see page 693). Gooey mozzarella? I can get that. Breading and deep-frying? A bit messy, but nothing overwhelming. But properly cooking eggplant? Not so simple. Even if you manage to purge the slices properly of their bitter liquid (not an easy task), they're still so airy and spongy

---

\* I won't say which one, but you know who you are, Aya.

that they instantly absorb any and all oil. Cook them too much, and they turn to mush. Don't cook them enough, and they are tough, with a tannic, astringent bite.

What happens if you try to fry plain raw eggplant? To find out, I weighed out a 24-gram slice of eggplant and placed it in a bowl of oil. Twenty minutes later, I weighed it again.

As you can see, it absorbed a full 92 percent of its weight in oil! If you've ever tried sautéing raw eggplant, you know that it almost instantly absorbs all the oil in the pan, sticks to the bottom, and burns. Like a sponge, the cells of the slices are held together in a very loose network with plenty of air in between. Before you can even think about cooking them, you need to figure out a way to remove that air.

To find the best way, I tried five different methods:

- **Salting, resting, and pressing the slices first** gets rid of moisture through osmosis. Like a leaky water balloon, as the moisture leaves the eggplant, its structure weakens, allowing you eventually to press out the excess air. This works fairly well, but it requires quite a bit of pressure, it's easy to over- or undersalt the slices, and sometimes you're still left with uncompressed sections in the center of the slices, which means undercooked, astringent finished results. The method works for recipes with caramelized eggplant and rich tomato sauce like Pasta Alla Norma (page 701) where you're going to be cooking the eggplant slowly and thoroughly, but in this case, it's too unpredictable.
- **Steaming the eggplant slices** in a bamboo steamer will rapidly soften them to the point that you can easily compress them. It also makes them soggy and mushy. It's a technique better suited for eggplant that's going to be braised or mashed.
- **Roasting eggplant slices uncovered** is tough to do. Cook them dry, and they turn leathery and tough. Oil them before roasting, and the usual problem occurs: the oil is instantly sucked in, and the eggplant slices still end up leathery and tough and greasy.
- **Roasting the eggplant slices covered by a layer of paper towels** in between two baking sheets pans is by far the most successful in-the-oven way to do it. Lining a baking sheet with paper towels (or a clean kitchen towel), placing the slices on top, covering them with another layer of paper towels, and then with another baking sheet ensures that the slices cook evenly, and the paper towels absorb excess moisture while at the same time keeping them just moist enough to prevent them from turning leathery.

- **Microwaving** is my go-to method. It's fast and consistent. (See "How Microwave Ovens Work" on page 887.) Just lay the slices (or cubes) of eggplant on a microwave-safe plate lined with a couple paper towels. Lay some more paper towels on top, followed by a heavy plate, and microwave on high for 5 to 10 minutes, until the eggplant has given up excess moisture through steam and completely collapsed. You can stack multiple plates with paper towels between them to cook more eggplant at the same time.

So, microwaving it is, though you can prep yours sandwiched between baking sheets oven if you wish. In under 10 minutes, I had all the eggplant I needed fully precooked. It was just a matter of pressing out excess moisture, and the slices were ready to be breaded and fried.

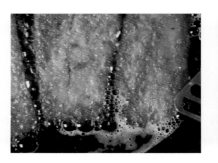

As with my chicken Parmesan, seasoned panko bread crumbs are the way to go here. When all is said and done, the cross section of your eggplant slices should look something like this: crisp and golden brown, with a dense, meaty, fully cooked interior.

What's that? You mean after you've done all that, you *still* have to make sauce, grate cheese, assemble, and bake this sucker? That's right, young grasshopper: you gotta pay your dues if you wanna sing the blues.

The sauce and cheese mixture are identical to the one used in Chicken Parmesan (page 878)—the only difference is in the layering. American eggplant Parmesan is generally baked casserole style, like a lasagna, with the fried eggplant taking the place of noodles. I see no reason to veer from tradition in that regard.

FUN FACT: Despite the fact that the dish is called eggplant Parmigiana, it actually has nothing to do with Parma, the city in Emilia-Romagna that produces both prosciutto di Parma and Parmigiano-Reggiano. Depending on who you ask, the name comes from either the use of Parmesan cheese or from the Sicilian word *parmiciana*, a reference to window shutters and the way the eggplant slices overlap each other like slats of wood. I'll leave it to the Italians to fight over the etymology.

### SERVES 4

1 large eggplant (about 1 pound), sliced lengthwise into ½-inch slices

Kosher salt and freshly ground black pepper

1½ cups panko-style bread crumbs

2 teaspoons dried oregano

4 ounces Parmigiano-Reggiano, finely grated (about 2 cups)

½ cup all-purpose flour

2 large eggs, beaten

1 cup vegetable oil

Double recipe Perfect Easy Red Sauce (page 695), warm

1 pound mozzarella cheese, grated

2 tablespoons minced fresh basil

2 tablespoons minced fresh parsley

1. Season the eggplant slices lightly on both sides with salt and pepper. Place a double layer of paper towels or a clean kitchen towel on a large microwave-safe plate and lay a single layer of eggplant slices on top. Top with two more layers of paper towels or another kitchen towel. Top with a second large plate. Microwave on high power until the eggplant is easily compressed, about 3 minutes (be careful, the plates will be hot).

2. Working with one piece of eggplant at a time, press the slices firmly between paper towels until compressed. Set aside on a large tray. Repeat the microwaving and pressing steps until all the slices are compressed.

3. Combine the bread crumbs, oregano, and ¼ cup of the Parmigiano-Reggiano in a shallow bowl or pie plate. Place the flour and eggs in separate shallow bowls or pie plates. Use your right hand to pick up one piece of eggplant and add it to the bowl of flour. Use your left hand to coat it evenly with flour, then use your right hand to pick up the eggplant, shake off excess flour, and add to the eggs. Turn the eggplant with your left hand until evenly coated, then use your left hand to transfer it to the bread crumbs. Lift some crumbs with your right hand and press them onto the top of the eggplant, then use your right hand to turn the eggplant several times, pressing it into the crumbs until evenly coated.

Transfer to a wire rack set on a rimmed baking sheet. Repeat with the remaining slices.

4. Adjust an oven rack to the center position and preheat the oven to 375°F. Line a rimmed baking sheet with a double layer of paper towels. Heat the oil in a 12-inch nonstick or cast-iron skillet over heat until it reaches 375°F on an instant-read thermometer. Carefully slide 3 or 4 eggplant slices into the hot oil in a single layer. Cook, shaking the pan occasionally, until the first side is golden brown and crisp, about 2½ minutes. Using tongs, carefully flip the eggplant and continue cooking, shaking the pan occasionally, until the second side is crisp, about 1½ minutes longer. Transfer the slices to the paper-towel-lined baking sheet and immediately season with salt. Repeat with the remaining eggplant slices.

5. Spread one-quarter of the tomato sauce evenly over the bottom of a 9- by 13-inch glass baking dish. Add one-third of the eggplant slices in a single layer (they can overlap a little bit). Press down to form an even layer. Add another one-quarter of the sauce and spread it evenly. Scatter one-third of the mozzarella and one-third of the remaining Parmesan evenly over the sauce. Repeat with two more layers each of eggplant, sauce, and cheese, reserving the remaining ¼ cup Parmesan.

6. Cover with foil and bake for 20 minutes. Remove the foil and bake until light golden brown and bubbling on the surface, about 20 minutes longer. Sprinkle the reserved parmesan over the top and allow to rest for 15 minutes.

7. Sprinkle with the basil and parsley and serve.

Gooey, cheesy Italian American–style Eggplant Parmesan Casserole (page 882)

# HOW MICROWAVE OVENS WORK

A microwave oven works by bombarding food with electromagnetic radiation in the microwave-frequency spectrum. That may sound scary, but bear in mind that not all electromagnetic radiation is bad. Both heat and visible light, for instance, consist of electromagnetic radiation in a frequency spectrum that our eyes or heat-sensitive nerves are able to detect.

Charged molecules—such as water molecules—will have a tendency to try to align themselves with the electric field created by a microwave, so as the long waves of a microwave ✳ pass by them, they will rapidly flip back and forth as they try to stay aligned. The resultant friction cooks your food. Microwaves can penetrate deeply into solid matter—up to several centimeters—though the denser and thicker the food, the less penetration a microwave will get. Dense, relatively dry items, like, say, your sister's My Little Pony dolls, can take a long time to heat up (not that I'd know through personal experience). Porous, moist slices of eggplant, on the other hand, are microwave gold, cooking evenly and rapidly.

Because microwave ovens allow so little energy to be lost to the outside environment (unlike the way, for example, a gas burner will heat up the room), they are extremely efficient at heating water. But there's one thing to be aware of: it's called superheating, and it is as cool as it sounds. Heat up water in a blemish-free container with minimal disturbance, and because of a lack of nucleation points (see page 99), it's possible to heat it well beyond its boiling point without it ever boiling. As soon as some turbulence is introduced, though—a little wobble from the turntable, for example—bubbles burst forth, sending hot water all over the inside of your microwave. This doesn't happen on the stovetop, since heating from the bottom of the pot creates lots of convection currents (the movement that occurs between relatively hot and cool regions of liquid or gas). You can avoid this by sticking a wooden spoon into your cup of water in the microwave to provide nucleation sites.

It's a lot like my lovely wife, who will quietly suppress tiny annoyances until suddenly the slightest disturbance sends her into an all-out rage. Unfortunately, the wooden spoon method does not work on her.

---

✳ Microwaves have a wavelength of from several inches up to a foot.

# COATING Style 3:

## BEER BATTER

### Beer-Battered Fried Fish

Here's the thing: Fish is extremely delicate, while frying is horribly violent. Dropping a hunk of fish into a pot of hot oil is like throwing an Ewok into a cage match with a Terminator: it doesn't stand a rat's chance in hell.

This is particularly true of white-fleshed, slow-moving, bottom-dwelling fish like cod or halibut. These large ichthyoids spend the vast majority of their time slowly grazing along the ocean floor, like gigantic cows of the sea. As such, their muscles don't get much of a workout. Just as with land animals, the less they use their muscles, the more delicate they'll

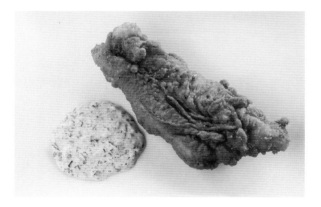

be in both flavor and texture. It's this mildness that is prized above all in these white-fleshed fish, and it's a characteristic that we should attempt to maximize during cooking. Just like steak and chicken, fish flesh will dry out and toughen if cooked to too high a temperature. Our goal with a fish like cod is to bring it up to around 150° to 160°F—just hot enough to break down the thin, film-like membranes of connective tissue between the layers of muscle but not hot enough to dry it out.

This is why you want to batter fish before dropping it into the oil. Battering is all about mitigating the energy transfer from the hot oil in order to gently steam the food on the interior while simultaneously causing the proteins in the batter to coagulate and eventually dehydrate, forming a crisp crust. Even though the fish is completely submerged in a pot of 350°F oil, it cooks relatively gently and evenly from all sides because of its coat of batter. This gives you, the intrepid cook, quite a bit of leeway, making perfectly tender flesh not just a distinct possibility, but in fact quite easy (and for those of you who worry about the smell of fried fish, don't. It's not as bad as you think—see page 857).

## What's in a Batter?

A batter consists primarily of two ingredients: flour and water. The loose proteins in the flour gradually link up with each other in the presence of water, creating gluten, which is what causes your batter to stick to the food being fried, as well as to itself. Use too much flour or stir the batter too vigorously, and you get too much gluten, which retains liquid and fat, weighing your batter down and turning it chewy or greasy (see "Experiment: Gluten Development in Batter," page 899). Similarly, the temperature of your ingredients can have a profound effect: keep your liquids ice-cold until you mix them into the flour, in order to minimize gluten development.

It's the proportion and manner in which these ingredients are mixed, as well as what other ingredients are added, that determine how crisp and light your final fried product will be.

Here are some common ingredient substitutions and additions:

- **Using beer or soda water** in place of water adds carbonation, which helps to leaven the batter. As the tiny bubbles of carbon dioxide are heated, they expand, creating a more open texture. Beer also adds flavorful compounds and carbohydrates that aid in browning.
- **Eggs** add a concentrated source of protein, allowing you to create a firm structure using less flour, resulting in the characteristic thin, wispy, and crisp coating in traditional Japanese tempura.
- **Baking powder and baking soda** form carbon dioxide when dissolved and heated (for baking soda, you also need another acid source). This puffs and lightens the batter.
- **Other grains, like rice flour or cornstarch or corn flour,** can have varying effects. Rice flour and cornstarch can be used to dilute the protein concentration of pure wheat flour, giving the batter a lighter structure (you still need at least bit of protein, or it won't have any structure at all). Corn flour has larger grains than wheat flour, which add the crunch you get in a hush puppy or good corn dog.

The first order of business with any batter is to get the consistency right. Too thick, and it comes out bready; too thin, and it doesn't offer enough protection. It's also essential to balance leavening power and gluten development. Too little leavening, and you get a hard, tough shell; too much, and your batter will overinflate and strip itself off your food.

Rather than using straight-up flour for my batter, I use a combo of flour and cornstarch, which reduces the amount of gluten formed—the protein network that can cause a batter to become leathery and tough. Gluten formation is also increased with excessive stirring, so mixing the batter with a whisk or a pair of chopsticks just until it barely comes together is the way to go. A few spots of raw flour are perfectly fine.

There are a couple reasons to use beer. First off, sugars present in the beer will increase the brownability of the batter. The bubbles are also essential— they create the tiny, tiny pockets inside a good batter that add to our perception of crunchiness; it's really just a little boost for the baking powder, which performs a similar function.

There's another element in there that's helping to keep my batter nice and light and crisp: the alcohol.

Vodka (even the cheap stuff!) is the secret to extra-crispy batters.

Regulars of the now-closed Lenox Liquors on Lenox at 133rd will recognize Georgi as the cheapest vodka they offer, while fans of Heston Blumenthal will recognize vodka as one of the ingredients in his Perfect Fish & Chips recipe. If you are in the small group of people who recognize both this bottle and its context, then we are kindred spirits and I welcome you to my home for fried fish any day of the year. When Heston presented the idea, initially the thought was that the volatility of the vodka (that is, its propensity to evaporate quickly) would cause it to jump out of the batter faster as it fried, allowing the batter to dehydrate more quickly and thus brown faster and also crisp up better. At that task, it serves admirably. If you add alcohol to your batter, it dries out faster than if you just use water. Indeed, increasing the alcohol content by, say, adding a shot or two of 80-proof vodka in place of some of the beer can accelerate this process significantly, resulting in a lighter, crisper coating.

There's an even more important factor it brings to the table: limiting gluten development. Gluten will develop in the presence of water, but not alcohol. Replacing part of the liquid in a batter with alcohol will allow you to achieve a batter with the exact same texture when raw but with significantly less gluten development, leading to crisper structure when fried.

I experimented using a few different coating methods—flouring before battering, battering the fish straight up, etc. I found that the most effective method, the one that resulted in the best balance between crispness and lightness, was to give the fish

---

*It's a brilliant recipe that you should check out, by the way. You can find it in his book *In Search of Perfection*, from the BBC series of the same name. It's also easily found online.

a quick coat in the flour mixture, followed by a dip in the batter, and then a second dip into the flour before lowering it into the fryer.

I admit, the method is not the neatest. You're going to end up breading your hands, and once the fish has come out of the drippy batter and back into the flour, it's important to work fast before the coating all starts to drip off. I find the easiest method is to drop the battered fish into the flour, throw some more flour on top to coat, and then pick it up by scooping under it and tossing it back and forth between your hands to get rid of excess flour. From there, it goes straight into a wok (or Dutch oven) full of hot oil.

Finally, it's important to make sure your beer is ice-cold, for three reasons:

1. Cold liquids hold their carbonation better.
2. Cold liquids inhibit the formation of gluten.
3. The recipe only calls for 1 cup of beer, so you're gonna have to drink the leftovers.

## CAN OIL BOIL?

We've all heard the phrase "boiled in oil," and we've certainly seen a pot of oil bubbling vigorously when you add food to it. But can oil truly boil? Technically, yes. Practically, no. Boiling, as we discuss on page 96, is the conversion of a liquid to a gas. Depending on how tightly their molecules are stuck together, various liquids boil at various temperatures. Water boils at 212°F, while extremely volatile liquid nitrogen boils at negative 320°F! The boiling point of oil is far greater. In reality, oil begins to smoke and will eventually catch on fire long before it can possibly reach its boiling temperature. The **smoke point** (the temperature at which wisps of smoke begin to appear above the oil; see page 860) and the **flash point** (the temperature at which the oil actually catches on fire) can vary from oil to oil but are generally in the 375° to 550°F range—well below the actual boiling temperature of oil.

What you are seeing in a pot of "boiling oil" is *not* the oil boiling. You're seeing the water content of the food you are frying in it boiling and bubbling its way up through the oil, giving you the illusion that the oil is bubbling. As soon as you take the food (and, therefore, the water) out, the boiling will stop.*

---

*So the next time someone threatens to throw you in boiling oil, just tell them, in your haughtiest tone, "Actually, it'll be the water content in my body doing the boiling, *not* the oil." Then run.

# BEER-BATTERED FRIED COD

**Serve with Thin and Crispy French Fries (page 910).**

**NOTE:** This recipe will work for any flaky white fish, such as, haddock, pollock, or even halibut or striped bass.

**SERVES 4**

1½ cups all-purpose flour

½ cup cornstarch

1 teaspoon baking powder

¼ teaspoon baking soda

Kosher salt

¼ teaspoon paprika

¾ cup light-flavored beer (such as
PBR or Budweiser), ice-cold

¼ cup 80-proof vodka

4 cups peanut oil

1 recipe Extra-Tangy Tartar Sauce
(recipe follows)

1 pound cod fillet, cut into four
4-ounce pieces

1. Heat the oil to 350°F in a large wok or cast-iron skillet over medium-high heat. Whisk together 1 cup of the flour, the cornstarch, baking powder, baking soda, 2 teaspoons salt, and paprika in a large bowl. Combine the beer and vodka in a small bowl.

2. Add the remaining ½ cup flour to a large bowl. Toss the fish pieces in the flour until evenly coated. Transfer to a wire rack set on a rimmed baking sheet.

3. Slowly add the beer mixture to the flour mixture, whisking just until the batter has texture of thick paint (you may not need all of the beer). The batter should leave a trail if you drip it back into the bowl off the whisk. Do not overmix; a few small lumps are OK.

4. Transfer the fish to the batter and turn to coat. Pick up one piece of cod from one edge, allowing excess batter to drip back into the bowl, quickly dip it into the bowl of flour and turn to coat both sides, and then carefully transfer it to the hot oil, lowering it in slowly to prevent splashes. Repeat until all 4 pieces are in the oil. Cook, shaking the pan gently and and agitating the oil with a wire-mesh spider constantly, flipping the fish halfway through cooking, until the cod is golden brown and crisp on all sides, about 8 minutes.

5. Transfer the fish to a paper-towel-lined plate and season immediately with salt. Serve with the tartar sauce.

## Extra-Tangy Tartar Sauce

**NOTE:** Cornichons are small, vinegary French pickles. They can usually be found in the olive section of a supermarket, or near the mustards and pickles. For a slightly sweeter sauce, substitute 2 tablespoons prepared sweet pickle relish for the cornichons.

**MAKES ABOUT 1 CUP**

¾ cup mayonnaise, preferably homemade (page 807)

1 medium shallot, finely minced (about 2 tablespoons)

3 tablespoons capers, drained, patted dry, and finely minced

6 to 8 cornichons, finely minced (about 2 tablespoons; see Note above)

1 teaspoon sugar

2 tablespoons finely minced fresh parsley

½ teaspoon freshly ground black pepper

Kosher salt

Combine the mayonnaise, shallot, capers, cornichons, sugar, parsley, and pepper in a small bowl. Season to taste with salt. Transfer to an airtight container and refrigerate for at least 1 hour before using. Tartar sauce will keep for up to 1 week in the refrigerator.

Fried Fish Sandwich with Creamy Slaw and Tartar Sauce (page 896).

# FRIED FISH SANDWICHES
## WITH CREAMY SLAW AND TARTAR SAUCE

**NOTE**: This recipe will also work for any flaky white fish, such as haddock, pollock, or even halibut or striped bass.

**SERVES 6**

**For the Slaw**

1 small head cabbage, cored and finely shredded (about 6 cups)

½ small red onion, thinly sliced (about ½ cup)

Kosher salt and freshly ground black pepper

2 teaspoons cider vinegar

1 teaspoon Dijon mustard

3 tablespoons mayonnaise, preferably homemade (page 807)

1 tablespoon sugar

**For the Sandwiches**

4 cups peanut oil

1½ cups all-purpose flour

½ cup cornstarch

1 teaspoon baking powder

¼ teaspoon baking soda

Kosher salt

¼ teaspoon paprika

¾ cup light-flavored beer (such as PBR or Budweiser), ice-cold

¼ cup 80-proof vodka

18 ounces cod fillet, cut into six 3-ounce pieces

6 soft hamburger buns, toasted in butter

6 tablespoons Extra-Tangy Tartar Sauce (page 894)

1. **To make the slaw:** Toss the cabbage and onion with 1 teaspoon salt and lots of pepper in a bowl and set aside. Combine the vinegar, mustard, mayonnaise, and sugar in a medium bowl and set aside for at least 15 minutes.

2. To finish the slaw, pick up the salted cabbage and onions in batches, squeeze out the excess moisture, transfer to the bowl with the dressing. Toss to combine and season to taste with more salt and pepper if desired. Set aside.

3. **To make the sandwiches:** Heat the oil to 350°F in a large wok or cast-iron skillet over medium-high heat. Whisk together 1 cup of the flour, the cornstarch, baking powder, baking soda, 2 teaspoons salt, and paprika in a large bowl. Combine the beer and vodka in a small bowl.

4. Add the remaining ½ cup flour to a large bowl. Toss the fish pieces in the flour until evenly coated. Transfer to a wire rack set over a rimmed baking sheet.

5. Slowly add the beer mixture to the flour mixture, whisking constantly just until the batter has texture of thick paint (you may not need all of the beer). The batter should leave a trail if you drip it back into the bowl off the whisk. Do not overmix; a few small lumps are OK.

6. Transfer the fish to the batter and turn to coat. Pick up one piece of cod from one edge, allowing excess batter to drip back into the bowl, quickly dip it into the bowl of flour and turn to coat both sides, and then carefully transfer it to the hot oil, lowering it in slowly to prevent splashes. Repeat until all 4 pieces are in the oil. Cook, shaking the pan gently and and agitating the oil with a wire-mesh spider constantly, flipping the fish halfway through cooking, until cod is golden brown and crisp on all sides, about 8 minutes. Transfer the fish to a paper-towel-lined plate and season immediately with salt.

7. Place a small pile of slaw on the bottom half of each bun. Top with a piece of fish and a dollop of tartar sauce. Close the buns. Serve with extra slaw and sauce on the side.

- **Not enough batter.** When there's too little batter, the onion is exposed to the full ravaging power of the oil. Its sugars rapidly caramelize and then burn, while tissues dry out, turning papery and tough.

- **Too much batter.** This is almost worse than having batter that's too thin. Instead of staying light and crisp, an onion ring with too much batter will retain too much internal moisture, and as soon as it comes out of the oil, the batter starts getting soggy.

## Onion Rings

At their physical core, onion rings couldn't be more different from fried fish. But at their philosophical core, they are one and the same. In each case, the goal is to prevent the browning and toughening of the main ingredient being fried (that would be the onions or the cod) while simultaneously adding textural contrast and flavor to the exterior.

It's always difficult to decide whether to get onion rings or fries (get a combo if they'll let you!). Proper beer-battered onion rings, with a substantial crisp crust covering a sweet, tender, thick ring of onion, are one of life's three greatest pleasures (and the only one that can be enjoyed legally, incidentally), but how often do you get perfect rings? These are the four most common ways that a good beer-battered onion ring turns into a bad one:

- **The "split shell."** This occurs when everything appears to be going fine until all of a sudden, through some as-yet-undiscovered mechanism, the batter crust spontaneously splits in half. Oil rushes into the gap, rendering the onion leathery and burnt.

- **The dreaded worm.** This is the most heinous of onion ring crimes. It occurs when the onions aren't cooked thoroughly, so that rather than breaking off cleanly with each bite, you're left with a long worm of onion in your mouth and the hollow shell left behind in your hand.

Dealing with the batter problems is a snap—we've already got an awesome recipe for light, crisp, lacy, just-thick-enough batter for our fried cod. But what about splitting and worming? Splitting was a tough case to crack. What could cause the batter shell to break open like that? To figure it out, I carefully dis-

sected an afflicted ring with a set of tweezers and discovered that it's not the batter that's the problem, it's the onion. Every layer inside an onion is separated from the next by a thin, papery membrane—you can quite easily see it if you rub the inside of a raw onion ring—the membrane will slip off.

Onions have a thin membrane between each layer.

Because of their thinness and lack of structure, these membranes shrink much more than the ring itself during cooking and it's this shrinkage that tears a hole in the partially set batter, allowing oil to rush inside. Removing the membranes before battering solved the problem, but it was a tedious process—about as much fun as trying to brush my dog's teeth, and much less cute. Soaking the rings in water for half an hour before attempting the separation helped, but I found it was far better to place the onion rings in the freezer. When vegetables are frozen, their water content crystallizes into large, jagged shards of ice, puncturing cells, which results in limp vegetables. In most cases, this is a bad thing—that's why frozen vegetables are almost never as good as fresh. With onions destined for

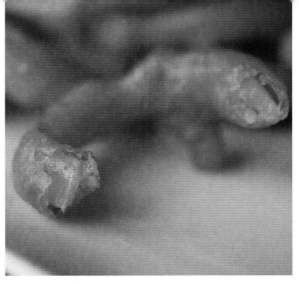

Onion rings should have a crisp coating and break cleanly when you bite into them.

the batter, however, this is not a defect—indeed, aside from making the inner membranes easier to remove, freezing tenderized the rings to the point that they could be broken quite easily when bitten; I'd inadvertently ended up solving my worming problem as well!

I was so ecstatic at the breakthrough that the only logical course of action was to commemorate the discovery with a celebratory batch of perfectly crisp, perfectly tender, worm- and crack-free, golden brown, beer-scented, sweet-and-salty onion rings.

# EXPERIMENT:
## Gluten Development in Batter

Just as in a kneaded bread dough, gluten—the network of interconnected flour proteins—can form in a heavily mixed batter. Need proof? Try this little test.

### Materials

- **See ingredients list for Foolproof Onion Rings, page 900.**

### Procedure

Follow the recipe through step 3. Divide the batter in half and whisk one half of it for an extra minute. Proceed with the recipe as directed, using regular batter and the overmixed batter, and making sure to keep the rings separate from each when you fry them.

### Results

Taste the rings side by side. You'll find that the rings with the regular batter are light and crisp, while the rings with the overwhisked batter are chewier, denser, and doughier.

As you continue to whisk a batter, protein molecules in the flour (*gliadin* and *glutenin*) form tighter and tighter bonds with each other. Eventually those bonds are so tight that even the leavening power of baking powder is not enough to lighten and leaven the batter—it stays dense. Interconnected proteins also turn the texture leathery instead of crisp and tender. Lesson learned: *do not overmix batter.*

# FOOLPROOF ONION RINGS

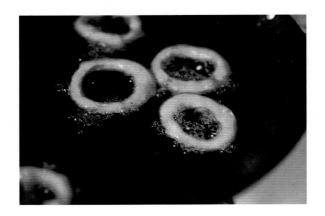

**SERVES 4**

2 large onions, cut into ½-inch
  rounds

2 quarts peanut oil

1 cup all-purpose flour

½ cup cornstarch

1 teaspoon baking powder

¼ teaspoon baking soda

½ teaspoon paprika

¾ cup light-flavored beer (such as
  PBR or Budweiser), ice-cold

¼ cup 80-proof vodka

Kosher salt

1. Separate the onion rounds into individual rings. Place in a gallon-sized zipper-lock freezer bag and put them in the freezer until completely frozen, at least 1 hour (they can stay in the freezer for up to 1 month).

2. When ready to fry, remove the onion rings from the freezer bag, transfer to a bowl, and thaw under tepid running water. Transfer to a rimmed baking sheet lined with a clean kitchen towel or several layers of paper towels and dry the rings thoroughly. Carefully peel off the inner papery membrane from each ring and discard (the rings will be very floppy). Set aside.

3. Preheat the oil to 375°F in a large wok or a Dutch oven over medium-high heat. Combine the flour, cornstarch, baking powder, baking soda, and paprika in a medium bowl and whisk together. Combine the beer and vodka in a small bowl.

4. Slowly add the beer mixture to the flour mixture, whisking constantly until the batter has texture of thick paint (you may not need all of the beer). The batter should leave a trail if you drip it back into the bowl off the whisk. Do not overmix; a few small lumps are OK. Dip one onion ring in the batter, making sure that all surfaces are coated, lift it out, letting the excess batter drip off, and add it to the hot oil by slowly lowering it in with your fingers until just one side is sticking out, then dropping it in. Repeat until half of the rings are in the oil. Fry, flipping the rings halfway through cooking, until they are deep golden brown, about 4 minutes. Transfer the rings to a large mixing bowl lined with paper towels and toss while sprinkling salt over them. The fried rings can be placed on a rack on a rimmed baking sheet and kept hot in a 200°F oven while you fry the remaining rings. Serve the rings immediately.

# COATING Style 4:
## THIN BATTERS

### Japanese-Style Tempura

Tempura-style batters were originally brought to Japan by Portuguese missionaries in the sixteenth century.* Since then, tempura has been perfected to a near art form by Japanese chefs. At the best tempura houses in Japan, all of your courses will be cooked by a single tempura chef who spent years in apprenticeship before ever being allowed to touch the batter or fry oil.

Tempura chefs are sort of like the Jedi of the cooking world: they must deftly perform with the utmost skill and precision, using extremely dangerous tools, all while maintaining a calm, serene demeanor. It is an elegant technique, from a more civilized time. The bad news is that you, I, and the vast majority of people in the world are never going to become as great as the masters who spend their entire lives training. But the good news is that we can get about 90 percent of the way there right off the bat.

The key characteristics of a tempura-style batter are extreme lightness of color and texture: good tempura should be pale blond with an extraordinarily lacy, light, and crisp coating. To achieve this takes just a little more care than other types of batter. Traditional tempura batter is made by combining flour

The chef at Tsunahachi, one of Tokyo's finest tempura restaurants, serves a tempura-fried shrimp.

(usually a mix of wheat flour and lower-protein-rice flour—I use wheat flour and cornstarch instead) with eggs and ice-cold water. The batter is mixed until *just barely combined* so that plenty of pockets of dry flour remain and virtually no gluten development occurs. A tempura batter has a lifespan of only moments before the flour becomes too saturated with water and a fresh batter must be made. But there are ways we can improve on this fickleness, so long as we aren't married to tradition.

First off, using the old vodka-in-the-batter trick (which by now you may be sick of) works very well, limiting the rate of gluten formation so that the batter can sit a bit longer before it goes bad. So does replacing the ice water with club soda, a trick I learned from my old chef Ken Oringer, at Clio Restaurant in Boston. But the real key is in the process: rather than simply dumping the dry and wet ingredients into a bowl and whisking them together, I found that by adding the wet ingredients to the dry then immediately lifting up the bowl and shaking it with one hand while simultaneously rapidly stirring with a pair of chopsticks, I could get all of the ingredients incorporated while minimizing the amount of flour that is completely moistened by the liquid.

---

*The word "tempura" itself comes from the Portuguese, as do many other Japanese words. According to Harold McGee's *On Food and Cooking*, "*tempora*" means "period of time" and refers to the fasting seasons during which fried fish was consumed in place of meat. These days, the word refers to any battered and fried item cooked in the manner of tempura fish, much like Americans have their "chicken-fried steak"— steak cooked in the manner of fried chicken.

**NOTE:** For instructions on how to prepare ingredients for frying, see page 863.

**SERVES 4**

2 quarts peanut oil or vegetable
 shortening

½ cup cornstarch

½ cup all-purpose flour

Kosher salt

1 large egg

¼ cup 80-proof vodka

½ cup ice-cold club soda

4 cups thinly sliced vegetables or 1
 pound shrimp (see Note above)

Lemon wedges or 1 recipe Honey-
 Miso Mayonnaise (recipe follows)

1. Heat the oil to 375°F in a large wok over high heat, then adjust the heat as necessary to maintain the temperature. Line a large plate or baking sheet with a double layer of paper towels.

2. Combine the cornstarch, flour, and 1 teaspoon salt in a large bowl and stir with chopsticks to blend. Combine the egg and vodka in a small bowl and whisk until completely homogeneous. Add the club soda and stir with chopsticks until barely combined. Immediately add to the bowl with the flour and, holding the bowl with one hand and the chopsticks in the other, shake the bowl back and forth while vigorously stirring with the chopsticks until the liquid and dry ingredients are just barely combined. There should still be many bubbles and pockets of dry flour.

3. Add the vegetables (and/or shrimp) to the batter and fold with your hand to coat. Pick up the vegetables a few pieces at a time, allowing excess batter to drip off, and transfer to the hot oil, getting your hand as close as possible to the surface before letting go in order to minimize splashing. Increase the heat to high to maintain the temperature as close to 350°F as possible, and add the remaining vegetables (and/or shrimp) a few pieces at a time. Immediately start agitating them with chopsticks or a wine-mesh spider, separating the vegetables, flipping them, and constantly exposing them to fresh oil. Continue frying until the batter is completely crisp and pale blond, about 1 minute.

4. Transfer the tempura to a paper-towel-lined plate or baking sheet and immediately sprinkle with salt. Serve with lemon wedges or honey-miso mayonnaise.

## Honey-Miso Mayonnaise

*At Clio, Ken Oringer used to serve the house vegetable tempura with a choice of two dipping sauces: a traditional tentsuyu (made with a Japanese bonito-and-kelp broth, soy sauce, and mirin) and a honey-miso aioli. This recipe is based on the latter, though I've stripped it down a bit, into a simple five-ingredient mayonnaise. The result is a balanced sweet-and-savory sauce that is light enough to go perfectly with fried shrimp and vegetables but tasty enough that, well, you'll want to eat it with a spoon in the middle of the night by the pale glow of the refrigerator light.*

**NOTE:** For this recipe, it's important that you use white miso paste (preferably Kyoto-style *saikyo* miso). Darker miso paste is too strong in flavor and will throw the sauce out of balance.

## MAKES ABOUT 1 CUP

1 large egg yolk

¼ cup white miso paste (see Note above)

2 teaspoons rice wine vinegar, plus more to taste

4 teaspoons honey, plus more to taste

¾ cup vegetable oil

Up to 1 tablespoon water

Combine the egg yolk, miso paste, vinegar, and honey in a tall narrow cup that will just fit the head of your immersion blender. Carefully pour in the oil, so that it floats on top of the other ingredients. Insert the blender into the bottom of the cup, turn on the blender, and slowly draw the head up through the oil: a thick emulsion should form as you do so. Transfer the mayonnaise to a bowl and whisk in more vinegar and honey to taste. Add up to 1 tablespoon water to thin the sauce to the desired consistency. It should be thick and cling to your finger or a spoon but not feel pasty or waxy on your tongue.

## HOW TO PREPARE COMMON TEMPURA INGREDIENTS

| INGREDIENT | PREPARATION |
| --- | --- |
| Green Beans | Trim the ends |
| Mushrooms | Clean and thinly slice, or leave thin mushrooms like shiitake or oyster whole |
| Bell Peppers | Cut into ½-inch-wide rings or strips |
| Zucchini and Summer Squash | Cut into ½-inch rounds or sticks |
| Onions | Cut into ½-inch rings |
| Eggplant | Cut into ½-inch rounds |
| Sweet Potatoes | Peel and cut into ¼-inch slices |
| Butternut Squash | Peel, seed, and cut into ¼-inch slices |
| Okra | Trim the stem ends |
| Broccoli and Cauliflower | Cut into 1-inch florets |
| Carrots | Peel and cut into ¼-inch slices or planks |
| Shrimp | Peel, leaving the final tail section intact if desired, and remove the legs, flatten each shrimp, and insert a wooden skewer lengthwise to keep it straight while it fries; remove the skewers after cooking. |

# FRENCH-FRIED POTATOES

**There's a reason why nearly a third of all potatoes grown in the United States make their way into a fry basket:** *fried potatoes are spectacular.*

No other food achieves quite the same balance of crisp exterior and fluffy interior without the need for any sort of external breading or batter. It all has to do with the natural balances of starches and moisture in the spuds. But making a perfect French fry is not as simple as dunking a potato in hot oil for a few minutes.✳ For the rest of this chapter, we're going to talk about how to achieve crisp, golden nirvana every time.

## French Fries

The intricacies involved in taking two simple ingredients—potatoes and oil—and applying science, heat, and a bit of blind faith are so complex that it boggles the mind. It took me a good decade to finally decode their secrets, to achieve that holy grail of burger joint cookery: the perfect French

✳The popular In-N-Out chain of fast-food hamburgers prepares their French fries in this manner—they are cut, rinsed, fried, and served. As anyone can tell you, despite how great their burgers are, their limp, pale fries leave much to be desired. A second trip to the fryer would do wonders for them.

fry. A substantial, crisp, grease-free crust that cracks open with a puff of steam revealing a tender, almost fluffy center.

There are four basic criteria that define a perfect French fry:

A perfect French fry should have a crisp crust that breaks instead of bending.

- **Perfect Fry Factor #1: The exterior must be very crisp but not tough.** In order to achieve such crispness, the surface structure of a fry must be riddled with microbubbles. It's these tiny crisp bubbles that increase the surface area of the fry, making it extra crunchy. Ideally, this layer should only be as thick as it needs to be to add crispness. Any thicker, and you start running into leathery or tough territory.
- **Perfect Fry Factor #2: The interior must be intact and fluffy and have a strong potato flavor.** Fries

with a pasty, mealy, or gummy interior or, even worse, the dreaded state known as "hollow-fry" (when the interior is missing entirely) are an automatic fail.

- **Perfect Fry Factor #3: The fry must be an even light golden blond.** Fries that are too dark or are spotty have an off-putting burnt flavor. Light golden but perfectly crisp is how I want my fries to be.
- **Perfect Fry Factor #4: The fry must stay crisp and tasty for at least as long as it takes you to eat a full serving.** Fries that comes straight out of the fryer are almost always perfectly crisp. The true test of a great fry is whether or not it is still crisp and edible a few minutes later, after it's been sitting on your plate. The bendy fry pictured above fails that test.

First, a few decisions. For potato variety, russet is what you want. Its high starch content means that it'll fry up crisper than waxier varieties like Yukon Gold or red skins. It'll also maintain a fluffier interior once cooked. For size, ¼ to ⅜ inch thick is good, optimizing the ratio of crisp crust while maintaining enough soft center to provide good potato flavor.

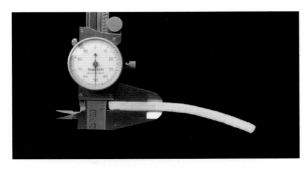

¼-inch is the ideal thickness for a French fry.

On to the cooking. Classic French technique will have you believe that the road to perfect fries involves frying once at a relatively low temperature (between 275° and 325°F), followed by a resting period and then a second fry at a higher temperature (between 350° and 400°F). The most common explanation I've heard for this is that the first low-temperature fry allows the fries to soften through to the center, while the secondary fry crisps up their exterior. I decided to put this theory to the test by cooking three identical batches of fries:

- The first I cooked per the French technique (a two-stage fry, the first at 275°F and the second at 375°F).
- For the second, I replaced the low-temperature fry with a trip to a pot of boiling water, then followed up by frying at 375°F as usual.
- For the third, I skipped the primary step altogether, simply dropping the potatoes into 375°F oil.

If the only purpose of the first fry were to cook the potatoes through to the center, then potatoes parcooked via another method should work just as well. Conversely, a potato that is *not* parcooked should not be evenly cooked to the center.

The results? The boiled-then-fried potatoes were crisp, but the layer of crispness was paper-thin and quickly softened. The single-fry potatoes were quite similar, though slightly less fluffy inside. Still, they were cooked through, no problem. The double-fried fries had a substantial, thick crust that stayed crisp for a while, proving that there's something more going on during that initial fry than simple softening. Indeed, using the set of calipers that my mother had so thoughtfully given me several years ago to try and draw me out of restaurant kitchens and into a much more sensible career, like mechanical engineering or gunsmithing, I was able to determine that the crisp layer on a double-fried fry was more than twice as thick as the one on a boiled-then-fried fry, though still not quite as thick as I would have liked it to be.

A double-cooked French fry has an exterior crust that is at least twice as thick as a single-fried fry.

To crack the case, I had to take a closer look at what I was dealing with, starting with putting a potato under a microscope.

## The Anatomy of a Potato

Like all plants and animals, potatoes are composed of cells. The cells are held together by pectin, a form of sugar that acts as a glue. Within the cells are starch molecules—large sponge-like molecules composed of many simple sugars bundled together. Starch molecules, in turn, stick together in starch granules. When starch granules are exposed to water and heat, they begin to swell, eventually bursting and releasing a shower of swollen starch molecules. This water can come from the outside (in the case of a boiled potato) or from inside the potato itself (in the case of a double-fried potato), and that bursting of starch granules is essential to forming a thick crust: it's the sticky, gelatinized starches that form the framework for the bubbly crust.

So the path to perfect fries seems easy—just burst a ton of starch granules, and you're home free, right? Not that simple. If your potato contains too many simple sugars, it'll brown long before it crisps. Starches and simple sugars will naturally convert their forms back and forth, depending on storage conditions. You can see this effect most dramatically with spring vegetables like peas and asparagus, which come off the vine packed with sugar but become noticeably less sweet and more starchy even twenty-four hours after they've been picked.

If potatoes have too much sugar, they won't crisp properly, and they'll become an unattractive dark brown as the sugars overcaramelize in the fryer, developing acrid, bitter flavors.

The effect is even more dramatic if you try it with potato chips. Unsoaked, unblanched chips end up dark, dark brown, while blanched chips come out nearly transparent.

The other difficulty in bursting starch granules is that if the pectin glue holding the cells together has broken down too much before the starch granules have had a chance to burst and release their sticky innards, they will fall apart and crumble before they get a chance to crisp.

It's the breakdown of pectin that in some cases, nearly too horrible to mention, causes the dreaded condition known as hollow-fry.

The dreaded "hollow fry."

There are a lot of things that the McDonald's corporation does wrong, but also some it does right. French fries are one of them. Through millions of dollars' worth of research and a partnership with the J. R. Simplot company—the inventor of the modern frozen French fry—they long ago discovered the key to both washing away excess simple sugars and ensuring that the pectin doesn't break down during frying: parcook the potatoes in 170°F water for precisely 15 minutes. This accomplishes two things: First, it washes away excess simple sugars. Second, and more important, it strengthens the pectin with the aid of a natural enzyme called *pectin methylesterase* (PME). According to an article in the *Journal of Agricultural and Food Chemistry*, PME induces calcium and magnesium to act as a sort of buttress for the pectin. They strengthen the pectin's hold on the potato cell's walls, which helps the potatoes stay firmer and more intact even as their starch granules swell and burst. Like most enzymes, PME is only active within a certain temperature range, acting faster and faster as the temperature gets higher and higher until, like a switch, it shuts off completely once it reaches a certain level.

Think of PME as little factory workers hard at work building cars. As their floor manager, if you apply a bit of pressure to them (in the form of heat), at first that will get them to work faster. Cars will come off the production line at a faster clip. But apply too much pressure (by overheating), and the little enzymes just won't be able to take it any more, throwing down their tools and walking out. Production slows to a halt. For PME, that shut-off point is just slightly above 170°F.

Unfortunately, most home cooks don't have an easy way to maintain a water bath at exactly 170°F for the requisite 15 minutes. I needed to find an alternate way to maintain the pectin structure of the potato while still releasing the starch molecules, and it struck me: it's as easy as apple pie.

What's apple pie got to do with French fries? Well, anyone who's ever baked an apple pie knows that different apples cook differently. Some retain their shape, others turn to mush. The difference largely has to do with their acidity. Thus, supertart apples like Granny Smith will stay fully intact, while sweeter apples like Macoun will almost completely dissolve. Just like in a potato, apple cells are held together by pectin, and, as it turns out, acidic environments can reduce or even prevent the breakdown of pectin.

So what if rather than trying to fiddle with temperature, I relied on the use of acid to help the potatoes keep their structure? I brought two pots of cut potatoes to a boil side by side, the first in plain water and the second in water spiked with vinegar at a ratio of 1 tablespoon per quart. Here's what I saw:

While the plain-water-cooked fries had broken down by the time they were cooked through, the fries cooked in the vinegar-spiked water stayed perfectly intact, even after boiling them for 50 percent longer than the other fries. Despite their smooth-looking exteriors, I knew that by boiling them for so long, I'd burst plenty of starch gran-

ules. With the excess sugars washed away and the pectin strengthened and ready to buttress the thick, crisp walls my fries would develop in the deep fryer, all that remained was to give them a first fry at 325°F to burst any remaining starch granules and begin crust formation, followed by a second fry at 375°F to bring them up to a perfectly crisp, golden blond.

Frying them up proved it: they came out positively *riddled* with tiny, crisp microbubbles:

Moreover, they were crisp enough that they *stayed* crisp for a full 10 minutes after frying.

# WHY FREEZE FRENCH FRIES?

The best way to preserve fries is to freeze them after the first frying stage. You can then fry them the second time straight from the freezer. But do they lose quality? To test this, I tried freezing half a batch of fries and letting them sit in the freezer overnight before frying them up and tasting them against their unfrozen counterparts. The results were surprising: the frozen fries were actually better, with a distinctly fluffier interior. Why?

Freezing potatoes causes their moisture to convert to ice, forming sharp, jagged crystals. These crystals damage the cell structure of the potatoes, making it easier for water to be released and convert to steam when they are heated, and this results in a drier, fluffier interior. The best part? Because freezing actually improved them, I could do the initial blanching and frying steps in large batches, freeze them, and have a constant supply of ready-to-fry potatoes in my freezer, just like Ronald himself!

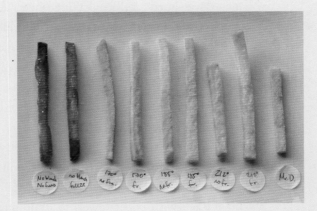

In this image: left to right, potatoes fried without blanching, along with potatoes blanched to various temperatures and a real McDonald's French fry at the end. Notice the dramatic difference the simple blanching step can make in the final product.

# THIN AND CRISPY FRENCH FRIES

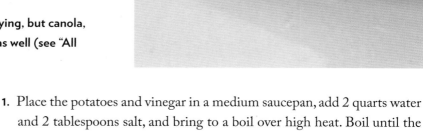

**NOTES:** For best results, it's imperative that you use an accurate instant-read thermometer and a timer during frying. After the optional freezing in step 2, the fries can be stored in a zipper-lock freezer bag in the freezer for up to 2 months; cook them directly from frozen, as in step 3.

Peanut oil is the best oil for deep-frying, but canola, vegetable oil, or shortening will work as well (see "All About Oil," page 855).

**SERVES 4**

2 pounds russet (baking) potatoes (about 4 large), peeled and cut into ¼-inch-thick fries (keep in a bowl of water until ready to cook)

2 tablespoons distilled white vinegar

Kosher salt

2 quarts peanut oil

1. Place the potatoes and vinegar in a medium saucepan, add 2 quarts water and 2 tablespoons salt, and bring to a boil over high heat. Boil until the potatoes are fully tender but not falling apart, about 10 minutes. Drain and spread them on a paper-towel-lined rimmed baking sheet. Allow to dry for at least 5 minutes.

2. Meanwhile, heat the oil to 400°F in a Dutch oven or large wok over high heat. Add one-third of the fries to the hot oil (the oil temperature should drop to around 360°F) and cook for exactly 50 seconds, agitating the potatoes occasionally with a wire-mesh spider, then transfer to a second paper-towel-lined rimmed baking sheet. Repeat with the remaining potatoes (working in two more batches), allowing the oil to return to 400°F before each addition. Allow the potatoes to cool to room temperature, at least 30 minutes; set the pot of oil aside. The potatoes can be kept at room temperature for up to 4 hours or, for the best results, frozen in a single layer at least overnight. Then, for longer storage, transfer to a zipper-lock freezer bag.

3. Return the oil to 400°F over high heat. Fry half of the potatoes until crisp and light golden brown, about 3½ minutes, adjusting the heat as necessary to maintain the oil at around 360°F. Transfer the potatoes to a bowl lined with paper towels to drain and immediately season with salt. The cooked fries can be kept hot and crisp on a wire rack set on a baking sheet in a 200°F oven while you cook the second batch. Serve immediately.

# THE ULTIMATE QUINTUPLE-COOKED THICK AND CRISP STEAK FRIES

I've never been a fan of steak fries. The ratio of crisp crust to fluffy interior is all off for me. I *like* that crunchy, slightly greasy crust, and with a thick steak fry, you get so little of it compared to the vast expanse of relatively bland interior. But what if there were a way to increase the crispness of the exterior? To build up a crust even more substantial than with my regular thin and crispy fries?

Here's a thought: if double-frying fries gives them a nice thick crust, would triple-frying or even quadruple-frying improve them even more? Only one way to find out. I made several batches of thick-cut fries (I'm talking ½ inch thick), using my thin-and-crispy fry technique as the baseline. The first batch I made exactly according to the directions. For the second, I fried them once at 360°F for 50 seconds and allowed them to cool, then fried them again for another 50 seconds and allowed them to cool, and finally fried them a third time until completely crisp and golden brown. For the third and fourth batches, I increased the total number of frying stages to four and five respectively.* Turns out that indeed you *can* increase the crispness of a fry with repeated fryings.

See, with each stage of the fry, you burst more and more starch granules. The starch molecules fly out and gelatinize when they come in contact with water from the potato. Subsequent cooling allows those gelatinized starches to recrystallize, in effect staling like old bread (see "Drying Versus Staling," page 621). With repeated fryings, these layers of crystallized starches build up into a substantial layer. Cooling the potatoes between fries also prevents them from overbrowning with each subsequent fry. Only during the very last fry do you leave them in the oil long enough that the gelatinized/crystallized starch layers are fully dehydrated, rendering them crisp and golden brown.

I'm not going to lie: these fries are a pain in the butt to make. They are a project, and you've got to devote a significant amount of time to them. But man, are they killer. This is a path you don't want to start down unless you are prepared to be eternally spoiled for regular French fries. You've been warned.

NOTES: For best results, it's imperative that you use an accurate instant-read thermometer and timer during frying. After the optional freezing in step 3, the fries can be stored in a zipper-lock freezer bag in the freezer for up to 2 months; cook them directly from frozen, as in step 4.

Peanut oil is the best oil for deep-frying, but canola, vegetable oil, or shortening will work as well (see "All About Oil," page 855). For these fries, I like to leave the skin on the ends of the potatoes to give you a bit of skin on each fry.

---

*Does this remind anyone else of the razor-blade battles when Schick came out with its four-bladed Quattro to compete with Gilette's Mach 3, and Gilette responded with a five-bladed razor? When will it all end?!?

**SERVES 4**

2 pounds russet (baking) potatoes (about 4 large), peeled (see Note above) and cut into ½-inch-thick fries (keep in a bowl of water until ready to cook)

2 tablespoons distilled white vinegar

Kosher salt

2 quarts peanut oil

1. Place the potatoes and vinegar in a medium saucepan, add 2 quarts water and 2 tablespoons salt, and bring to a boil over high heat. Boil until the potatoes are fully tender but not falling apart, about 10 minutes. Drain and spread them on a paper-towel-lined rimmed baking sheet. Allow to dry for at least 5 minutes.

2. Meanwhile, heat the oil to 400°F in a Dutch oven or large wok over high heat. Add one-third of the fries to the hot oil (the oil temperature should drop to around 360°F) and cook for exactly 50 seconds, agitating the potatoes occasionally with a wire-mesh spider, then transfer to a second paper-towel-lined rimmed baking sheet. Repeat with the remaining potatoes (working in two more batches), allowing the oil to return to 400°F before each addition. Allow the potatoes to cool to room temperature, at least 30 minutes.

3. Repeat step 2 twice more, allowing the fries to cool for 30 minutes after each fry. Set the pot of oil aside. At the end of this stage, when the potatoes have been boiled once and fried three times, they can be stored at room temperature for up to 4 hours or, for best results, frozen in a single layer at least overnight. (Then, for longer storage, transfer to a zipper-lock freezer bag.)

4. Return the oil to 400°F over high heat. Fry half of the potatoes until crisp and light golden brown, about 3½ minutes, adjusting the heat as necessary to maintain the oil at around 360°F. Transfer the potatoes to a bowl lined with paper towels to drain and immediately season with salt. The cooked fries can be kept hot and crisp on a wire rack set on a baking sheet in a 200°F oven while you cook the second batch. Serve immediately.

 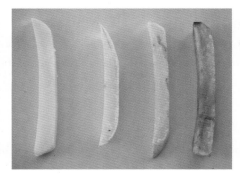 

Each successive frying session delivers crisper French fries.

# JOËL ROBUCHON'S SLOW-COOKED FRENCH FRIES

If you're a member of the online community (and who isn't these days?), you've probably read about French chef Joël Robuchon's magic French fry recipe. The idea is simple: rather than double-frying French fries (once cool, once hot), just do the whole process in a single go by putting the potatoes in cold oil, putting the pot on a burner, and letting it go. The potatoes slowly cook through from the outside and, over the course of an hour or so, end up golden brown and crisp.

The problem, however, is that with regular fries, the gelatinized starch has a chance to recrystallize in between frying sessions, allowing the fries to become much more structurally sound and crisp. The Robuchon recipe, easy as it is, doesn't produce fries that are anywhere near as crisp as blanched-then-double-fried potatoes. Like all things in life, it's a trade-off. Robuchon's method gets a 2 in effort and an 8 in flavor, while the blanch-double-fry method is more like a 7 in effort and a 9½ in flavor. Truth be told, I often opt for the lazy approach.

To do it, cut your potatoes and rinse them in water, then carefully dry. Put them in your wok or Dutch oven and cover with oil by an inch or two. Cook the fries over medium heat until they are very soft, about 35 minutes, stirring once or twice in the first 5 to 10 minutes. Increase the heat to high and let them continue to cook for 5 minutes undisturbed, then give them a few gentle stirs and keep cooking until golden brown and crisp, about 10 minutes longer. Drain on a paper-towel-lined plate and immediately season with salt.

# CRUNCHY OVEN FRIES

In need of a quick fix? Crunchy oven fries don't live up to the simple-yet-lofty culinary standards set by a French fry, but they're equally tasty in their own way. Since they're cooked in the relatively-low-energy environment of the oven (remember, 375°F oil is far more efficient at transferring heat than even a 500°F oven), you need to take a few extra steps to ensure proper crust formation.

First you've gotta help them thicken their skins. I tried simply battering and baking some, but it's a no-go. The watery potatoes sog out their skins far too quickly. Starting them out by boiling them in vinegar-spiked water is a much better first step, allowing the crust to stay intact after frying.

Potatoes baked from raw (*left*) end up with thin, soggy crusts. Parcooked potatoes (*right*) stay crisp after baking.

Then I tried dozens of variations of coating, from thick egg-based batters to simple dustings, and in the end, I found the best method is to toss the boiled potatoes with a bit of buttermilk, which is just thick enough to coat them in a layer of liquid that then forms a thin sheath when combined with the flour and cornstarch I dredge them in. Of the dozen kinds of flour I tried, flour, cornstarch, and potato starch were the favorites. Flour on its own, with its relatively high protein content (around 12 percent for all-purpose), forms a crust that's a little too leathery. Cornstarch and potato starch on their own are powdery and pale. Combining flour with cornstarch dilutes the flour's protein content, creating a thinner, more delicate crunch that's further enhanced by a bit of baking powder.

Finally, to best emulate frying in the oven, you need to add the potatoes to a fully preheated pan with a layer of oil in it, so that they sizzle upon contact.

**NOTE:** These seasoned planks are also great deep-fried. Follow the recipe as directed through step 3 (skip preheating the oven and the rimmed baking sheet), then deep-fry the potatoes in two batches in 2 quarts peanut oil heated to 400°F in a wok or Dutch oven until golden brown and crisp, about 2½ minutes.

**SERVES 4**

1½ pounds russet (baking) potatoes (about 3 large), scrubbed

1½ tablespoons distilled white vinegar

Kosher salt

½ cup vegetable oil

1½ cups buttermilk

1 teaspoon garlic powder

1 teaspoon paprika

1 teaspoon freshly ground black pepper

¼ teaspoon cayenne pepper

¾ cup all-purpose flour

½ cup cornstarch

1 teaspoon baking powder

1. Adjust an oven rack to the lower-middle position and preheat the oven to 400°F. Split the potatoes in half lengthwise. Place one half cut side down on the cutting board and slice into planks ⅓ to ½ inch thick. Repeat with the remaining potato halves.

2. Place the potatoes and vinegar in a medium saucepan, add 2 quarts water and 2 tablespoons salt, and bring to a boil over high heat. Boil until the potatoes are fully tender but not falling apart, about 10 minutes. Drain, then transfer them to a medium bowl, add the buttermilk, and toss gently to combine (it's OK if some potatoes break). Allow to sit for 5 minutes.

3. Meanwhile, add the oil to a rimmed baking sheet and place it in the oven to preheat. Combine the garlic powder, paprika, black pepper, cayenne, flour, cornstarch, baking powder, and 1 tablespoon salt in a large bowl and whisk to combine.

4. Drain the potatoes and return them to the bowl. Sprinkle half the flour mixture over them and turn a few times. Sprinkle the remaining flour mixture over them and fold gently until all the potatoes are coated. Let them sit in the flour mixture for at least 5 minutes, tossing occasionally, until a thick layer of coating has built up around each plank.

5. Working in batches, transfer the potatoes to a fine-mesh strainer and shake gently over the sink to remove the excess flour, then transfer to a large bowl.

6. Carefully remove the baking sheet from the oven (the oil should be lightly smoking) and add the potato planks in a single layer. Return the baking sheet to the oven and bake until the bottom side of the potatoes is light golden brown, about 10 minutes. Remove from the oven and flip the potatoes using a thin flexible spatula. Return to the oven and continue to bake until both sides are deep golden brown and crisp, 10 to 15 minutes longer. Drain on paper towels, season with salt to taste, and serve immediately.

*continues*

# CRUNCHY OVEN FRIES (page 914)

# ACKNOWLEDGMENTS

Believe it or not, this unreasonably large book started as a very manageably sized book nearly five years ago. Actually, I take that back. This unreasonably large book started out as a teeny-tiny blog post about boiling eggs that ended up transforming my life in the way that a meat grinder can transform pork shoulder into a sausage. With the help of dozens and dozens of people, I was systematically broken down, re-formed, stuffed, extruded, stretched, kneaded, seasoned, and perhaps even cooked a little, and ultimately came out the other end as a better writer, better cook, better photographer, and better person.

I'd like to thank my wife, Adriana, who had to put up with an apartment that smelled eternally of hamburgers and roasted chickens and Brussels sprouts and steak and everything else that ended up going into the bellies of our friends and neighbors. She's humored me when I dragged her along on 18-meal-per-day "research" trips. She's been OK with the fact that I've cheated on her—several times, in fact—with a hamburger-shaped mistress. She's spent years going to sleep in an empty bed while I clack away on my computer only to be woken up deep in the middle of the night when I decided that I just needed to fry another batch of chicken wings to answer a burning (literally) question. All this and she's still urging me to start writing another book.

I don't know how my family puts up with my near-fascist control over our holiday menus, but I'd like to apologize for monopolizing the kitchen year after year in pursuit of ever-better roast turkey and stuffing. Next year, you all get a one-time pass. Aya, you can put as many cranberries as you'd like in the stuffing. Pico, you can make your mashed potatoes lumpy. Fred, you can poke and pick at any pot or bowl in the kitchen you'd like (and I'll even let you interrupt my carefully regimented cooking schedule to mix yourself a martini or two, so long as you make one for me as well). Koji, you can continue to entertain us all with your magic tricks and Keiko, well . . . I'm sorry, but the turkey is still mine. I think we can both agree it'll all be better that way.

It's true that this book started out as a blog post about eggs, but none of this—the blog post, the book, the online column—would have existed without Ed Levine, who has been by far the most generous and supportive boss I've ever had. It was at his suggestion that I started writing a food science column. He was the one who came up with "The Food Lab" as the title. He gave me the boost I needed to get off my butt and start working on this book. Yes, he offered me a job, but what I've got at Serious Eats is more than just a job. It was an open platform, a playground for food nerds, and Ed and I are still figuring out what it's transforming into as we enter our fifth year of what has been an extraordinarily and mutually fruitful relationship. Heck, even his wife, Vicky, is my agent.

Speaking of which, you couldn't ask for a better agent or advocate than Vicky Bijur. She's not only represented me and my interests with great vigor and zeal, she's also been the first to edit my words, the first to offer opinions on layout and design, and the first to tell me when I'm about to make yet another bad business decision.

With the exception of someone tricking Adri into loving me, taking a job at Serious Eats was the best decision I've ever made in my life and in no small part due to the people. Robyn Lee, Carey Jones, Erin Zimmer, Adam Kuban, and the rest of the original crew made work into an amusement park every day.

I couldn't produce the work I do without the generous criticism (constructive or otherwise) of everyone who's worked there, past and present.

Thanks to Christine Kim and to Carly Gilfoil, for all the help they lent in researching, prepping, shopping, and cleaning during research and photo shoots, and thanks to Conor Murray for pointing out exactly how poorly I planned my charts and graphs before his editing made them legitimately useful.

I've had many mentors in the restaurant world over the years and all of them have helped me grow my interest in food. Barbara Lynch, Jason Bond, and Dave Bazirgan were my first real chefs. The ones who took me in and gave me a job despite the fact that I was an overeducated and underskilled smartass who barely knew how to hold a knife. They'd be my first picks for Drill Sergeants at culinary boot camp. It was under Ken Oringer's mentorship that I honed my knife skills, became a perfectionist at the stove, and developed my palate to the point where I could finally begin to think about calling myself a cook.

I'd like to thank Chris Kimball, Jack Bishop, Keith Dresser, Erin McMurrer, and the other editors and cooks at *Cook's Illustrated* and *America's Test Kitchen* for giving me my first schooling in the world of recipe development, food writing, video, and television. Sheryl Julian had faith in my abilities as a freelance journalist when I gave her absolutely no reason to, and Jolyon Helterman is the one who actually taught me those skills.

Any writer who tells you that their voice and style are all their own are lying to you. We're all influenced by who we read. My writing is a stew of mannerisms, jokes, and styles that I've mashed together from all over the place. Jeffrey Steingarten and Michael Ruhlman taught me that gonzo journalism is alive and well in the food world. Hugh Fearnley Whittingstall and Anthony Bourdain reminded me that good food writing, like all good writing, has to be personal. Jacques Pepin's mastery of technique and his ability to express not just how to do something but why you should care is breathtaking (his "Complete Techniques" makes for some riveting beach reading. Trust me).

Whenever I'm in a rut where my words just aren't light or humorous enough I reread a Douglas Adams or Kurt Vonnegut, Jr., novel, or perhaps watch a few Monty Python sketches. Even more than my own efforts, these are the folks that I really owe my writing and recipe testing style to, and I hope my words do them proud.

I owe a big debt of gratitude to Russ Parsons (*How to Read a French Fry*), Aki Kamozawa and Alex Talbot (Ideas in Food), Robert Wolke (*What Einstein Told His Cook*), Dave Arnold (*Liquid Intelligence*) and Nathan Myhrvold and the entire Modernist Cuisine team for leading the charge when it comes to good food science education and advocacy for the public.

I'd like to thank my editor, Maria Guarnaschelli, who, despite her fearsome reputation, turned out to be my biggest supporter. I trembled as I turned in my original 800-page manuscript for what was supposed to be a 300-page product. She turned around and said "I like it. Can you write me 600 more?" (Are you kidding? I had to work to get it down to 800 in the first place!) I've been with her for five years and three assistants now, and she has never once said "no" to me, even when she probably should have.

Finally, I wouldn't be the person I am today if I hadn't snuck into the living room every morning at 6 a.m. in my Winnie-the-Pooh pajamas to catch an episode of *Mr. Wizard's World* on Nickelodeon. So here's to you Don Herbert, for inspiring generations of young nerdlings to go out and become full-fledged nerds.

# INDEX

Note: Page references in *italics* indicate photographs.

marinating, 327–28, *328*

medium, temperature for, *295*, 296

medium-rare, temperature for, *295*, 296

medium-well, temperature for, *295*, 296

pan-searing, rules for, 309

poking with fork, experiments on, *296*, 296–97

Porterhouse (*See* Porterhouse Steaks)

prime-grade, about, 283

rare, temperature for, 295, *295*

rested, carved, appearance of, *308*

resting, before serving, 306–8

resting times, 308

rib-eye (*See* Rib-Eye Steaks)

rib section numerical designations, 300

room temperature, before cooking, 292–93

salting, 289–91, *290*

searing, best cooking fats for, 293–94

searing, flavor gained from, 292

searing, heat through conduction, *31*

searing, myths about, 291–92

select-grade, about, 283

short ribs (*See* Short Rib(s))

skirt (*See* Skirt Steak(s))

slicing against the grain, 340–41, *341*

standard- and commercial-grade, about, 283

strip (*See* Strip Steaks)

T-bone (*See* T-Bone Steaks)

tenderloin (*See* Tenderloin Steaks)

testing doneness, with cut-and-peek method, 297

thick-cut, best pan-searing methods, 297

thick-cut, buying, advice about, 289

thick-cut, grilling directions, 323–24, *324*

tri-tip (*See* Tri-Tip)

unrested, carved, appearance of, *306*

USDA grades, 282–83

utility, cutter, and canner grade, about, 283

well-done, temperature for, *295*, 296

wet-aging, about, 304

Beer

batter, deep-frying with, 864, 888–91

-Battered Fried Cod, *850, 892*, 892–94

-battered onion rings, best cooking methods, 897–99

-battered onion rings, common mistakes, *897*, 897–98, *898*

and Brats, Cooler-Cooked, 398

Foolproof Onion Rings, 900, *900*

Mustard, and Sauerkraut, Grilled or Pan-Roasted Bratwurst with, 520

Beet(s)

best cooking methods, 794

cooking techniques, 407

Roasted, and Citrus Salad with Pine Nut Vinaigrette, *794*, 794–95

Roasted, Salad with Goat Cheese, Eggs, Pomegranate, and Marcona Almond Vinaigrette, *796*, 796–97

Belgian endive. *See* Endive (Belgian)

Bench scraper, 65

Bibb lettuce, about, 767

Biscuits

Bacon Parmesan, 163

buttermilk, best methods for, 158–62, *160–62*

Buttermilk, Super-Flaky, 163

Cheddar Cheese and Scallion, *158, 160–62*, 163

Cream, Easy, 165

Blender, immersion

about, 62

pureeing soups with, 203

Blender, stand

about, 64

pureeing soups with, 203, 205

Blueberry Pancakes, 150

Blue Cheese

Butter Seasoning, 326

Dressing, Three-Ingredient, 832, *832*

Mixed Bitter Lettuces, Pomegranate, and Hazelnut Vinaigrette, Roasted Pear Salad with, *793*, *793*

Boiling water

under cover, experiment on, 242

at high altitude, 98–99

myths about, 99

science of, 96–97, 675

Bok Choy

cooking techniques, 407

micro-steaming, 423

Bolognese

Sauce, Fresh Pasta with, 732, *732*

Sauce, The Ultimate, *729*, 729–31, *731*

Traditional Lasagna, *735*, 735–36, *736*

Bottles, squeeze, 72

Bowls, prep, 66

Chicken (*continued*)

breasts, dry-brined, appearance of, *578*

breasts, resting times, 308

brining, before deep-frying, 866–67, *867*

"cage-free" label, 569, 570

"certified organic" label, 571

connective tissue in, 180

conventional, meaning of, 570

cutlets, how to prepare, *367*, 367–68

cutlets, pan-roasted, sauces for, 368–70, *369*

cutlets, quick cooking method for, 366, *366*

deep-frying, twice, 869

destroying salmonella in, 361

dry-brining, 579–80

and dumplings, best cooking methods, 235, *236*

fat on, 180

"free-range" or "free-roaming" label, 569–70

freezing, 574

"fresh" label, 569

fried, Southern-style, best ways to prepare, 865–68

heirloom breeds, 571

"hormone-free" label, 569

internal temperatures, quick guide to, 362

kosher, about, 571–72

labeling terms, 569–71

legs, cooking temperatures, 581

marinating, 589

muscles in, 179–80

"natural" label, 569, 570

"no antibiotics" label, 569

"organic" label, 570–71

parts, buying, 573

parts, pan-roasted, sauces for, 368–70, *369*

parts, pan-searing, best methods for, 357–58

parts, roasting methods, 583

parts, saving for stock, 573

preparing, for chopped salads, 835

raw, handling, 574

raw, rinsing, 573

raw, storing, 574

resting, after breading and before frying, 870

resting, before serving, 587

roasted, flavoring agents for, 588–89

roasting methods, 580–84

roasting on hot steel, 584

salting versus brining, 358–60

"self-basting," or "enhanced" label, 572

skillet-braised, best methods for, 247, *247*

skin, collagen in, 585

skin, crispy, obtaining, 585–86, *586*

skin, fat in, 180

skin, preparing for cooking, 579–80

skin, seasoning under, 586, *586*

skin, separating from meat, 586, *586*

skin-on, buying and cooking, 573

soup, best cooking methods, 188–89

stock, best cooking methods, 180–83, *181, 182, 183*

stock, clear versus cloudy, 183

stock, for vegetable soup recipes, 202–3

stock, ideal cooking time, 182

stock, saving chicken parts for, 186

temperature, time, and bacterial reduction, 362

thighs, boning, note about, 573

"water-chilled" label, 573

Chicken (recipes)

Barbecue-Glazed Roast Chicken, 600

Buttery Lemon-Herb-Rubbed Roast Chicken, 595

Chicken and Dumplings, *235, 236*, 237

Chicken Parmesan, *878*, 878–81, *880*

Chopped Ranch Cobb Salad, 841–42

Cooler-Cooked Chicken with Lemon or Sun-Dried Tomato Vinaigrette, 396–97

Easy Skillet-Braised Chicken with Mushrooms and Bacon, 252

Easy Skillet-Braised Chicken with Peppers and Onions, 251

Easy Skillet Braised Chicken with Tomatoes, Olives, and Capers, *247,* 248

Easy Skillet Braised Chicken with White Wine, Fennel, and Pancetta, 249

Extra-Crunchy Fried Chicken Sandwiches, *876,* 876–77, *877*

Extra-Crunchy Southern Fried Chicken, *871,* 871–72, *872*

Jamaican-Jerk-Rubbed Roast Chicken, 595–96, *596*

Pan-Roasted Chicken Parts, 365, *365*

Penne alla Vodka with Chicken, 699, *699*

Peruvian-Style Roast Chicken, 598–99

Quick Chicken Stock, 187

Quick Jus for Roasted Butterflied Chicken, 592

Roasted Butterflied Chicken, 591, *591*

Dressings. *See also* Vinaigrettes
    Blue Cheese, Three-Ingredient, 832, *832*
    Buttermilk Ranch, 843
    Caesar Salad, 808, 824, *824*
    choosing, for salads, 765
    dairy-based, about, 773, 831
    Italian, Creamy, 841
    mayonnaise-based, about, 773
    mayonnaise-based, best ways to prepare, 801–5, *805*
    three basic categories, 773
    tossing with salad greens, 766, 772
Drinks
    Homemade Hot Chocolate Mix, *172,* 173
    hot chocolate, best ways to prepare, 172–73
    storing in refrigerator, 77
Dry-brining poultry, 579–80
Duck fat, about, 475
Dumplings
    boiling in stock, *31*
    Chicken and, *235, 236, 237*
    for soups, best cooking methods, 235

Eggplant
    Caramelized, and Rich Tomato Sauce, Pasta with (Pasta
        alla Norma), 701–2, *703*
    Chinese, about, 704
    cooking techniques, 407
    with fewer seeds, choosing, 704–5
    globe, about, 704
    Italian, about, 704
    Japanese, about, 704
    Parmesan Casserole, 882–85, *884, 885, 886*
    preparing, for tempura, 903
    raw, removing excess air from, 883
    Thai, about, 704
Egg(s). *See also* Egg(s) (recipes)
    blind taste tests on, 92–94
    boiled, overcooked, *97*
    boiled, science of, 95–100
    boiling, experiments on, *96, 97, 97*
    brown versus white, 91
    Certified Humane label, 92
    Certified Organic label, 92
    culinary definition, 87
    Free-Range, Free-Roaming, and Cage-Free labels, 92

    fresh, advantages of, 94, *94*
    fresh versus aged, 88
    fried, best cooking methods, *114,* 114–15
    fried, loose egg whites in, *89*
    gauging freshness of, 90–91
    hard-boiled, best cooking methods, 102
    hard-boiled, peeling shell from, 90, *90*
    hard-boiling, tip for, 89–90
    labeling of, 91–92
    for meat loaf recipes, 528, 529
    Natural label, 92
    old, water test for, *91*
    older, myths about, 89
    Omega-3 Enriched label, 92
    omelets, best cooking methods, 124
    pack dates and sell-by dates, 90
    pasteurized, about, 91
    poached, best cooking methods, 103–5, *104–5*
    poached, common questions about, 105–6
    poached, cooking in advance, 106
    poached, strainer-to-pan technique, 104, *105,* 106
    preparing, for chopped salads, 835
    salting, effect of, *119,* 119–20, *120*
    scrambled, best cooking methods, 117–19
    scrambled, creamy versus fluffy, 117–19
    shelf life, 90
    sizes and weights, 9, 88
    soft-boiled, best cooking methods, 100
    storing in refrigerator, 76
    unrefrigerated, shelf life of, 91
    USDA grades, 88
    USDA weight standards, 88
    whites, about, 87
    whites, cooking temperatures, 97
    whites, whipping, science of, 145–46
    yolks, about, 87
    yolks, cooking temperatures, 100
    yolks, emulsifiers in, 802
Egg(s) (recipes)
    The Best Egg Salad, 821, *821*
    Chopped Ranch Cobb Salad, 841–42
    Creamy Scrambled Eggs, 122, *122*
    Diner-Style Asparagus, Shallot, and Goat Cheese
        Omelet, 126–27
    Diner-Style Ham and Cheese Omelets, *125,* 125–26

Ginger
    adding to cranberry sauce, 624
    Quick Chickpea and Spinach Stew with, *449*, 449–50
Glace (reduced stock), 186
Glazes
    flavoring roast chicken with, 589
    Orange–Cream Cheese, 170–71
Gliadin, 899
Glucosinolates, 430
Glutamates, 246
Gluten
    development, in batter, 899
    formation of, 142
Glutenin, 899
Goat Cheese
    Asparagus, and Shallot Omelet, Diner-Style, 126–27
    Eggs, Pomegranate, and Marcona Almond Vinaigrette,
        Roasted Beet Salad with, *796*, 796–97
    and Toasted Almonds, Asparagus Salad with, 782, *782*
Gorgonzola cheese, flavor of, and best uses for, 718–19
Gouda cheese, flavor of, and best uses for, 720–21
Grains. *See also specific grains*
    storing, 79
Grapefruit
    Cranberries, and Fig and Pumpkin Seed Vinaigrette,
        Endive and Chicory Salad with, *798*, 798–99
    Roasted Beet and Citrus Salad with Pine Nut Vinai-
        grette, *794*, 794–95
    Vinaigrette, 400
Gravy
    The Classic: Stuffed Herb-Rubbed Roast Turkey with,
        *604, 605*, 605–7, *608–9*
    Cream, *873*, 873–75
    The Easiest and Fastest: Roasted Butterflied Turkey
        with, *612*, 612–13
    Poultry, Dead-Simple, 619, *619*
    preparing, tips for, 618–19
    Sausage, Creamy, 164
Grease, bacon, deep-frying with, 857
Greek Salad, Chopped, 836–37, *837*
Green and red leaf lettuce, about, 766–67
Green Bean(s)
    Braised String Beans with Bacon, 446, *446*
    Casserole, Upgraded, *416*, 416–17, *417*
    cooking techniques, 407

    Cryo-Blanched, with Fried Garlic, 443, *443*
    Micro-Steamed, with Olives and Almonds, 429
    micro-steaming, 423
    preparing, for chopped salads, 835
    preparing, for soups, 190
    preparing, for tempura, 903
    Salad with Red Onion and Hazelnut Vinaigrette, 792
Greens. *See also specific greens*
    Basic Mixed Green Salad, 782
    bitter, cooking techniques, 407
    bitter, for salad, 768
    choosing, for salads, 765
    crisp lettuces, for salad, 766–67
    hearty, cooking techniques, 407
    mild, for salad, 768
    peppery, for salad, 767–68
    preparing, for soups, 190
    salad, preventing wilting of, 775
    salad, storing, 769
    salad, washing, 769
    tender, preparing, for omelets, 130
Gruyère cheese
    Fast French Onion Soup, *223*, 226, *226*
    Traditional French Onion Soup, 234

Halloumi cheese, flavor of, and best uses for, 720–21
Ham
    Bean, and Kale Stew, Easy, 275, *275*
    and Cheese Omelets, Diner-Style, *125*, 125–26
    Eggs Benedict, *112*, 112–13
    Hot Buttered Peas with Prosciutto, Pine Nuts, and Gar-
        lic, 414
    and Peas, Stovetop Mac 'n' Cheese with, 723
    preparing, for chopped salads, 835
    preparing, for omelets, 130
    Quick Creamy Pasta with Prosciutto, Peas, and Aru-
        gula, 713
    and Scallions, Hot Buttered Snap Peas with, 412
Hanger Steaks
    about, 332
    best way to cook, 330
    cooked, appearance of, *340*
    cooking methods, 332–33
    Cooler-Cooked, with Chimichurri, 394
    flavor of, 330

Mangoes, preparing, for chopped salads, 835

Maple

    Bacon Waffles, 157

    -Mustard-Glazed Pan-Seared Pork Chops, 356

    -Mustard-Glazed Roast Pork Tenderloin, 662

    -Sage Breakfast Sausage, 507, *507*

Marinades

    acid for, 327–28

    aromatics for, 328

    flavoring roast chicken with, 589

    Herb and Garlic, 342

    Honey-Mustard, 345

    oil for, 327

    for poultry, working with, 589

    salt and proteases for, 328

    Spicy Thai-Style, 344

    Steak House–Style, 343

    working with, 328

Marmite

    flavoring Bolognese with, 727

    flavoring chili with, 257, 264

    flavoring meat loaf with, 530

    glutamates in, 246

Marsala-Mushroom Pan Sauce, 369, *369*

Mayonnaise

    adding oil to, 803–5

    Bacon, 809

    broken, appearance of, *804*

    Caesar Salad, 808

    Chipotle-Lime, 809

    compared with aioli, 806

    Extra-Tangy Tartar Sauce, 894, *895*

    flavorings for, 805

    flavor variations, 808–9

    Foolproof Homemade, *801, 805,* 807

    Fry Sauce, 553

    Garlic, 808

    Garlic-Chili, Spicy, 809

    Garlic-Herb, 808

    Honey-Miso, 809, 902–3

    Horseradish, 808

    made with a food processor, 804–5

    made with an immersion blender, 804

    for potato salad, 815

    quantity made with one egg yolk, 806

    Roasted Red Pepper, 808

    Sun-Dried Tomato, 809

    Tartar Sauce, 809

Mayonnaise-based dressings

    about, 773

    best ways to prepare, 801–5, *805*

    Caesar Salad Dressing, 824, *824*

    Creamy Italian Dressing, 841

    Three-Ingredient Blue Cheese Dressing, 832, *832*

Maytag blue cheese, flavor of, and best uses for, 720–21

Measurements

    metric, 501

    weight and volume conversions, 9

    weight versus volume, 73

Meat. *See also* Beef; Lamb; Pork; Veal

    adding to salads, 765

    for Bolognese sauce, 727

    braising sous-vide, effect on juiciness, 387

    cathespin enzymes in, 385

    chemical cures for, 494

    chopping, by hand, *492,* 492–93

    cooked, pouring hot pan drippings over, 391

    curing, methods for, 494–95

    defrosting, 77

    dehydration/fermentation process, 495

    dry-cured, preparing, for chopped salads, 835

    fast-twitch muscles in, 650

    grinding, basic tips for, 488

    grinding, in a food processor, 491–92, *492*

    grinding, in a meat grinder, 491

    grinding methods, 486–87

    grinding your own, benefits of, 485–86

    ground, anatomy of, 497

    ground, effect of salt and time on, 502

    pounding, 875

    preground, notes about, 486

    raw, packaging for freezer, 77

    raw, storing in refrigerator, 76

    raw, testing for seasoning, 535

    resting, after sous-vide cooking, 391

    salting, science of, 497–98

    salting versus brining, 358–60

    for sausages, 495–96

Salads (*continued*)

Roasted Beet, with Goat Cheese, Eggs, Pomegranate, and Marcona Almond Vinaigrette, *796,* 796–97

Roasted Beet and Citrus, with Pine Nut Vinaigrette, *794,* 794–95

Roasted Pear, with Mixed Bitter Lettuces, Blue Cheese, Pomegranate, and Hazelnut Vinaigrette, 793, *793*

Spring Vegetable, *783,* 783–84

Tomato and Mozzarella, with Sharp Balsamic-Soy Vinaigrette, *790,* 790–91, *791*

White Bean and Manchego Cheese, 839

Winter Greens, with Walnuts, Apples, and Parmesan-Anchovy Dressing, *810,* 810–11

Salad spinner, 68

Salami

preparing, for chopped salads, 835

Restaurant-Style Chopped Antipasti Salad, *840,* 840–41

Salmon

coho, about, 379

common cooking mistakes, 371–72, *371–72*

cooking temperatures and texture, 372–73

cooking with skin on, 373

Cooler-Cooked Olive Oil–Poached, 399

fillet, cross section, appearance of, 372, *372*

fillets, boning, 378, *378*

keeping skin intact while cooking, 373–76, *374–76*

king, about, 379

obtaining perfectly crisp skin, 375–76

pale-orange-red flesh, about, 372

skin, about, 372

sockeye, about, 379

subcutaneous fat, about, 372

Ultra-Crisp-Skinned Pan-Roasted Fish Fillets, 380, *381*

Salsa

Classic Pico de Gallo, 351

Santa Maria–Style, 348

Verde, 396

Salsify, cooking techniques, 408

Salt

added to eggs, effect of, *119,* 119–20, *120*

adding to burger meat, *546,* 546–47, *547*

adding to lamb, 643–44

adding to pasta water, 678

adding to simmering water, 99

applying to meat, science of, 497–98

dry-brining poultry with, 579–80

for marinades, 328

seasoning beef steaks with, 289–91, *290*

seasoning meat loaf mix with, 529

seasoning sausage meat with, 496–97, 502

seasoning soups with, 204

solutions, versus brining, 358–60

storing, 80

types of, 81

volume and weight conversions, 9

Saltcellar, 65–66

"Salting out," 578

Sandwiches. *See also* Burgers

Extra-Crunchy Fried Chicken, *876,* 876–77, *877*

Fried Fish, with Creamy Slaw and Tartar Sauce, *895,* 896

grilled cheese, best cooking methods, 207, *207*

Grilled Cheese, Extra-Cheesy, *207,* 208–9

Leftover Meat Loaf, *534,* 534–35

Santa Maria–Style Salsa, 348

Saponification, 857–58

Sardines, preparing, for chopped salads, 836

Saturated fat

about, 855–56

in common oils, 861

Sauces. *See also* Pan Sauces; Pasta Sauces; Salsa

Béarnaise, about, 321

Béarnaise, Foolproof, 322

Caramelized Apple, 659

cheese, smooth, rules for, 715–16

Chimichurri, 395

Cranberry, Easy, *623,* 624

cranberry, flavoring ideas, 623–24

cranberry, homemade, benefits of, 623

Cream Gravy, *873,* 873–75

Creamy Sausage Gravy, 164

Dead-Simple Poultry Gravy, 619, *619*

Dill-Lemon Crème Fraîche, 382

Fry, 553

gravy, tips for preparing, 618–19

hollandaise, best preparation methods, 107–10, *108*

Hollandaise, Foolproof, *107, 108,* 111

Honey-Miso Mayonnaise, 902–3

Horseradish Cream, 640, *640*

Oxtail Jus for Prime Rib, 636–37, *637*

preparing, for prime rib, 633

# CONVERSIONS

## COMMON INGREDIENTS BY VOLUME AND MASS*

| INGREDIENT | TYPE | AMOUNT | WEIGHT |
|---|---|---|---|
| **Water-Based Liquids** (including water, wine, milk, buttermilk, yogurt, etc.) | | 1 cup = 16 tablespoons | 8 ounces (227 grams) |
| **Eggs** | Jumbo<br>Extra Large<br>Large<br>Medium<br>Small<br>Peewee | | 2.5 ounces (71 grams)<br>2.25 ounces (64 grams)<br>2 ounces (57 grams)<br>1.75 ounces (50 grams)<br>1.5 ounces (43 grams)<br>1.25 ounces (35 grams) |
| **Flour** | All-purpose<br>Cake/pastry<br>Bread | 1 cup | 5 ounces (142 grams)<br>4.5 ounces (128 grams)<br>5.5 ounces (156 grams) |
| **Sugar** | Granulated<br>Brown (light or dark)<br>Confectioners' | 1 cup | 6.5 ounces (184 grams)<br>7 ounces (198 grams)<br>4.5 ounces (128 grams) |
| **Salt** | Table<br>Diamond Crystal kosher<br>Morton's kosher | 1 teaspoon | 0.25 ounce (7 grams)<br>0.125 ounce (3.5 grams)<br>0.175 ounce (5 grams) |
| **Instant Yeast** | | 1 teaspoon | 0.125 ounce (3.5 grams) |
| **Butter** | | 1 tablespoon = ⅛ stick | 0.5 ounce (14 grams) |

*Note: In standard U.S. recipes, liquids are measured in fluid ounces (volume), while dry ingredients are measured in regular ounces (weight).

## VOLUME EQUIVALENCIES

3 teaspoons = 1 tablespoon

2 tablespoons = 1 fluid ounce

16 tablespoons = 1 cup (8 fluid ounces)

2 cups = 1 pint (16 fluid ounces)

4 cups = 1 quart (32 fluid ounces)

1 quart = 0.95 liters

4 quarts = 1 gallon